Handbook of Vegetable Science and Technology

FOOD SCIENCE AND TECHNOLOGY

A Series of Monographs, Textbooks, and Reference Books

1. Flavor Research: Principles and Techniques, *R. Teranishi, I. Hornstein, P. Issenberg, and E. L. Wick*
2. Principles of Enzymology for the Food Sciences, *John R. Whitaker*
3. Low-Temperature Preservation of Foods and Living Matter, *Owen R. Fennema, William D. Powrie, and Elmer H. Marth*
4. Principles of Food Science
   Part I: Food Chemistry, *edited by Owen R. Fennema*
   Part II: Physical Methods of Food Preservation, *Marcus Karel, Owen R. Fennema, and Daryl B. Lund*
5. Food Emulsions, *edited by Stig E. Friberg*
6. Nutritional and Safety Aspects of Food Processing, *edited by Steven R. Tannenbaum*
7. Flavor Research: Recent Advances, *edited by R. Teranishi, Robert A. Flath, and Hiroshi Sugisawa*
8. Computer-Aided Techniques in Food Technology, *edited by Israel Saguy*
9. Handbook of Tropical Foods, *edited by Harvey T. Chan*
10. Antimicrobials in Foods, *edited by Alfred Larry Branen and P. Michael Davidson*
11. Food Constituents and Food Residues: Their Chromatographic Determination, *edited by James F. Lawrence*
12. Aspartame: Physiology and Biochemistry, *edited by Lewis D. Stegink and L. J. Filer, Jr.*
13. Handbook of Vitamins: Nutritional, Biochemical, and Clinical Aspects, *edited by Lawrence J. Machlin*
14. Starch Conversion Technology, *edited by G. M. A. van Beynum and J. A. Roels*
15. Food Chemistry: Second Edition, Revised and Expanded, *edited by Owen R. Fennema*
16. Sensory Evaluation of Food: Statistical Methods and Procedures, *Michael O'Mahony*
17. Alternative Sweetners, *edited by Lyn O'Brien Nabors and Robert C. Gelardi*
18. Citrus Fruits and Their Products: Analysis and Technology, *S. V. Ting and Russell L. Rouseff*

19. Engineering Properties of Foods, *edited by M. A. Rao and S. S. H. Rizvi*
20. Umami: A Basic Taste, *edited by Yojiro Kawamura and Morley R. Kare*
21. Food Biotechnology, *edited by Dietrich Knorr*
22. Food Texture: Instrumental and Sensory Measurement, *edited by Howard R. Moskowitz*
23. Seafoods and Fish Oils in Human Health and Disease, *John E. Kinsella*
24. Postharvest Physiology of Vegetables, *edited by J. Weichmann*
25. Handbook of Dietary Fiber: An Applied Approach, *Mark L. Dreher*
26. Food Toxicology, Parts A and B, *Jose M. Concon*
27. Modern Carbohydrate Chemistry, *Roger W. Binkley*
28. Trace Minerals in Foods, *edited by Kenneth T. Smith*
29. Protein Quality and the Effects of Processing, *edited by R. Dixon Phillips and John W. Finley*
30. Adulteration of Fruit Juice Beverages, *edited by Steven Nagy, John A. Attaway, and Martha E. Rhodes*
31. Foodborne Bacterial Pathogens, *edited by Michael P. Doyle*
32. Legumes: Chemistry, Technology, and Human Nutrition, *edited by Ruth H. Matthews*
33. Industrialization of Indigenous Fermented Foods, *edited by Keith H. Steinkraus*
34. International Food Regulation Handbook: Policy • Science • Law, *edited by Roger D. Middlekauff and Philippe Shubik*
35. Food Additives, *edited by A. Larry Branen, P. Michael Davidson, and Seppo Salminen*
36. Safety of Irradiated Foods, *J. F. Diehl*
37. Omega-3 Fatty Acids in Health and Disease, *edited by Robert S. Lees and Marcus Karel*
38. Food Emulsions: Second Edition, Revised and Expanded, *edited by Kåre Larsson and Stig E. Friberg*
39. Seafood: Effects of Technology on Nutrition, *George M. Pigott and Barbee W. Tucker*
40. Handbook of Vitamins: Second Edition, Revised and Expanded, *edited by Lawrence J. Machlin*
41. Handbook of Cereal Science and Technology, *Klaus J. Lorenz and Karel Kulp*
42. Food Processing Operations and Scale-Up, *Kenneth J. Valentas, Leon Levine, and J. Peter Clark*
43. Fish Quality Control by Computer Vision, *edited by L. F. Pau and R. Olafsson*
44. Volatile Compounds in Foods and Beverages, *edited by Henk Maarse*
45. Instrumental Methods for Quality Assurance in Foods, *edited by Daniel Y. C. Fung and Richard F. Matthews*
46. Listeria, Listeriosis, and Food Safety, *Elliot T. Ryser and Elmer H. Marth*
47. Acesulfame-K, *edited by D. G. Mayer and F. H. Kemper*
48. Alternative Sweeteners: Second Edition, Revised and Expanded, *edited by Lyn O'Brien Nabors and Robert C. Gelardi*
49. Food Extrusion Science and Technology, *edited by Jozef L. Kokini, Chi-Tang Ho, and Mukund V. Karwe*
50. Surimi Technology, *edited by Tyre C. Lanier and Chong M. Lee*
51. Handbook of Food Engineering, *edited by Dennis R. Heldman and Daryl B. Lund*
52. Food Analysis by HPLC, *edited by Leo M. L. Nollet*
53. Fatty Acids in Foods and Their Health Implications, *edited by Ching Kuang Chow*
54. *Clostridium botulinum*: Ecology and Control in Foods, *edited by Andreas H. W. Hauschild and Karen L. Dodds*

55. Cereals in Breadmaking: A Molecular Colloidal Approach, *Ann- Charlotte Eliasson and Kåre Larsson*
56. Low-Calorie Foods Handbook, *edited by Aaron M. Altschul*
57. Antimicrobials in Foods: Second Edition, Revised and Expanded, *edited by P. Michael Davidson and Alfred Larry Branen*
58. Lactic Acid Bacteria, *edited by Seppo Salminen and Atte von Wright*
59. Rice Science and Technology, *edited by Wayne E. Marshall and James I. Wadsworth*
60. Food Biosensor Analysis, *edited by Gabriele Wagner and George G. Guilbault*
61. Principles of Enzymology for the Food Sciences: Second Edition, *John R. Whitaker*
62. Carbohydrate Polyesters as Fat Substitutes, *edited by Casimir C. Akoh and Barry G. Swanson*
63. Engineering Properties of Foods: Second Edition, Revised and Expanded, *edited by M. A. Rao and S. S. H. Rizvi*
64. Handbook of Brewing, *edited by William A. Hardwick*
65. Analyzing Food for Nutrition Labeling and Hazardous Contaminants, *edited by Ike J. Jeon and William G. Ikins*
66. Ingredient Interactions: Effects on Food Quality, *edited by Anilkumar G. Gaonkar*
67. Food Polysaccharides and Their Applications, *edited by Alistair M. Stephen*
68. Safety of Irradiated Foods: Second Edition, Revised and Expanded, *J. F. Diehl*
69. Nutrition Labeling Handbook, *edited by Ralph Shapiro*
70. Handbook of Fruit Science and Technology: Production, Composition, Storage, and Processing, *edited by D. K. Salunkhe and S. S. Kadam*
71. Food Antioxidants: Technological, Toxicological, and Health Perspectives, *edited by D. L. Madhavi, S. S. Deshpande, and D. K. Salunkhe*
72. Freezing Effects on Food Quality, *edited by Lester E. Jeremiah*
73. Handbook of Indigenous Fermented Foods: Second Edition, Revised and Expanded, *edited by Keith H. Steinkraus*
74. Carbohydrates in Food, *edited by Ann-Charlotte Eliasson*
75. Baked Goods Freshness: Technology, Evaluation, and Inhibition of Staling, *edited by Ronald E. Hebeda and Henry F. Zobel*
76. Food Chemistry: Third Edition, *edited by Owen R. Fennema*
77. Handbook of Food Analysis: Volumes 1 and 2, *edited by Leo M. L. Nollet*
78. Computerized Control Systems in the Food Industry, *edited by Gauri S. Mittal*
79. Techniques for Analyzing Food Aroma, *edited by Ray Marsili*
80. Food Proteins and Their Applications, *edited by Srinivasan Damodaran and Alain Paraf*
81. Food Emulsions: Third Edition, Revised and Expanded, *edited by Stig E. Friberg and Kåre Larsson*
82. Nonthermal Preservation of Foods, *Gustavo V. Barbosa-Cánovas, Usha R. Pothakamury, Enrique Palou, and Barry G. Swanson*
83. Applied Dairy Microbiology, *edited by Elmer H. Marth and James L. Steele*
84. Milk and Dairy Product Technology, *Edgar Spreer*
85. Lactic Acid Bacteria: Microbiology and Functional Aspects, Second Edition, Revised and Expanded, *edited by Seppo Salminen and Atte von Wright*
86. Handbook of Vegetable Science and Technology: Production, Composition, Storage, and Processing, *edited by D. K. Salunkhe and S. S. Kadam*

Additional Volumes in Preparation

Spice Science and Technology, *Kenji Hirasa and Mitsuo Takemasa*

Food Lipids: Chemistry, Nutrition, and Biotechnology, *edited by Casimir C. Akoh and David B. Min*

Polysaccharide Association Structure, *edited by Reginald Walter*

Handbook of Vegetable Science and Technology

Production, Composition, Storage, and Processing

edited by

D.K. SALUNKHE

Utah State University
Logan, Utah

S.S. KADAM

Mahatma Phule Agricultural University
Rahuri, India

CRC Press
Taylor & Francis Group
Boca Raton London New York

CRC Press is an imprint of the
Taylor & Francis Group, an **informa** business

CRC Press
Taylor & Francis Group
6000 Broken Sound Parkway NW, Suite 300
Boca Raton, FL 33487-2742

First issued in paperback 2019

ISBN-13: 978-0-8247-0105-5 (hbk)
ISBN-13: 978-0-367-40056-9 (pbk)

Library of Congress Cataloging-in-Publication Data

Handbook of vegetable science and technology : production,
composition, storage, and processing / edited by D. K. Salunkhe, S. S. Kadam.
p. cm. -- (Food science and technology ; v. 86)
Includes bibliographical references and index.
ISBN 0-8247-0105-4 (alk. paper)
1. Vegetables. 2. Truck farming. 3. Vegetables--Postharvest technology.
4. Vegetables--Processing. I. Salunkhe, D. K. II. Kadam, S. S. III. Series:
Food science and technology (Marcel Dekker, Inc.) ; 86.
SB320.9.H35 1998
635--dc21 97-46799
CIP

Visit the Taylor & Francis Web site at
http://www.taylorandfrancis.com

and the CRC Press Web site at
http://www.crcpress.com

To Dr. L. H. Pollard,
Professor of Horticulture, Utah State University,
Logan, Utah

and

Dr. S. H. Wittwer,
Professor of Horticulture, Michigan State University,
East Lansing, Michigan

Preface

The world's most urgent need is to increase the production of nutritious food so that we may adequately feed the hungry people of the planet. A major and often neglected step toward offering a greater volume of nutritious foods is to prevent loss of food between the time of harvesting and consumption. According to a report published by the National Research Council of the National Academy of Sciences (Washington, D.C., 1978), postharvest losses may be as high as 30–40% in both developed and developing nations. With application of adequate technology to prevent their deterioration after harvest, supplies of fresh fruits and vegetables can be increased to the extent of their existing postharvest losses.

Fresh vegetables and fruits are vital sources of minerals, vitamins, and dietary fibers. Both contain nutritionally important compounds, such as vitamins, that cannot be synthesized. They supply certain constituents that other foods do not. Vegetables and fruits contribute over 90% of dietary vitamin C. Green vegetables are a rich source of vitamin A. Similarly, niacin and folic acid (which are required for normal body functions) are present in significant quantity. Because vegetables and fruits are perishable products with high metabolic activity during the postharvest period, proper postharvest handling plays an important role in increasing their availability.

Recent developments in agriculture have contributed significantly to the improvement of vegetable production throughout the world. Similarly, remarkable improvements have been made in the postharvest handling of various vegetables and control of their market diseases. Storage practices have been developed to protect the vegetables and add to consumer appeal. The development of sizing equipment, conveyors, and package fillers all contribute to the success of vegetable handling. New chemicals more effective in decay control have been developed by the chemical industry to serve the fruit and vegetable industry. Improvements in refrigerated rail wagons, trucks, and trailers have helped to reduce losses during transport. All of the information above is scattered in many recent research papers, reviews, bulletins, and books. There was a need to have information on production and postharvest technology of vegetable crops compiled in one volume. This book will be useful to students of horticulture, marketing, food processing and engineering, food science, and nutrition as well as growers, processors, and shippers of vegetables in both developed and developing countries.

D. K. Salunkhe
S. S. Kadam

Contents

Contributors

R. N. Adsule Department of Agricultural Chemistry and Soil Science, Mahatma Phule Agricultural University, Rahuri, India

J. K. Chavan Department of Biochemistry, Mahatma Phule Agricultural University, Rahuri, India

B. B. Desai Department of Biochemistry, Mahatma Phule Agricultural University, Rahuri, India

U. T. Desai Department of Horticulture, Mahatma Phule Agricultural University, Rahuri, India

S. S. Deshpande IDEXX Laboratories, Inc., Sunnyvale, California

V. M. Dhamane Department of Plant Pathology, College of Agriculture, Kolhapur, India

V. M. Ghorpade Industrial Agricultural Products Center, University of Nebraska, Lincoln, Nebraska

S. P. Ghosh Indian Council of Agricultural Research, New Delhi, India

M. A. Hanna Department of Biological Systems Engineering, University of Nebraska, Lincoln, Nebraska

S. J. Jadhav Food Processing Development Center, Processing Services Division, Food and Rural Development Department, Alberta Agriculture, Leduc, Alberta, Canada

N. D. Jambhale Department of Botany, Mahatma Phule Agricultural University, Rahuri, India

S. S. Kadam Department of Food Science and Technology, Mahatma Phule Agricultural University, Rahuri, India

K. M. Kate Department of Plant Pathology, Mahatma Phule Agricultural University, Rahuri, India

B. G. Keskar Department of Horticulture, Mahatma Phule Agricultural University, Rahuri, India

P. M. Kotecha Department of Food Science and Technology, Mahatma Phule Agricultural University, Rahuri, India

Pushpa R. Kulkarni Food and Fermentation Technology Division, Department of Chemical Technology, University of Bombay, Bombay, India

K. E. Lawande Department of Horticulture, Mahatma Phule Agricultural University, Rahuri, India

D. L. Madhavi Department of Natural Resources and Environmental Sciences, University of Illinois, Urbana, Illinois

S. D. Masalkar Department of Horticulture, Mahatma Phule Agricultural University, Rahuri, India

A. M. Musmade Department of Horticulture, Mahatma Phule Agricultural University, Rahuri, India

Y. S. Nerkar Mahatma Phule Agricultural University, Rahuri, India

Y. R. Parulekar Department of Horticulture, Konkan Agricultural University, Dapoli, India

J. C. Rajput Department of Horticulture, Konkan Agricultural University, Dapoli, India

N. Rangavajhyala Department of Food Science and Technology, University of Nebraska, Lincoln, Nebraska

D. K. Salunkhe Department of Nutrition and Food Science, Utah State University, Logan, Utah

S. K. Sathe Department of Nutrition, Food and Movement Sciences, Florida State University, Tallahassee, Florida

P. N. Satwadhar College of Agricultural Technology, Marathwada Agricultural University, Parbhani, India

D. M. Sawant Department of Plant Pathology, Mahatma Phule Agricultural University, Rahuri, India

A. R. Sawate College of Agricultural Technology, Marathwada Agricultural University, Parbhani, India

K. G. Shinde Department of Horticulture, Mahatma Phule Agricultural University, Rahuri, India

Rekha S. Singhal Food and Fermentation Technology Division, Department of Chemical Technology, University of Bombay, Bombay, India

D. B. Wankhede College of Agricultural Technology, Marathwada Agricultural University, Parbhani, India

S. D. Warade Department of Horticulture, Mahatma Phule Agricultural University, Rahuri, India

Handbook of Vegetable Science and Technology

1

Introduction

D. K. SALUNKHE
Utah State University, Logan, Utah

S. S. KADAM
Mahatma Phule Agricultural University, Rahuri, India

Hunger is a widespread and growing phenomenon, and billions of people suffer from severe malnutrition in Asian, African, and Latin American countries (1). There is a nearly 20-year difference in the life expectancies of rich and poor nations. Apart from caloric needs, there is a severe shortage of food materials, such as fruits and vegetables, tubers, root crops, and fruit nuts, which are the most important plant foods to supply humans with many of their nutritional requirements, including minerals, vitamins, proteins, starches, fats, and sugars. They provide crude fiber and bulk as well as a variety of flavors and odors. Vegetables, with the increasing recognition of their value in the human diet, are gaining commercial importance (1).

Many vegetables are grown on different continents. These include roots, tubers, bulbs, leafy vegetables, beans, melons, squashes, corn, mushrooms and many other vegetables (Table 1). Yamaguchi (2) classified vegetables based on botany, end uses, or a combination of both (Tables 2 and 3) as well as growing requirements such as salt tolerance and tolerance to soil acidity. Weichmann (3) classified vegetables according to respiratory behavior (Table 4). Vegetables have also been classified according to metabolic characteristics (2,3) and sensitivity to chilling temperature during postharvest storage (3).

The worldwide production of roots and tubers and other vegetables, including carrots and melons (4), is presented in Table 5. China, India, the United States, Turkey, and Italy are the major vegetable-producing countries in the world (Table 6). Among the vegetables, including roots and tubers, potato ranks first in production, followed by cassava, sweet potato, cabbage, onion, and melons (Table 5). Even though potato is included as a vegetable, it is used as a staple food in many countries of Europe and Latin America. Similarly, cassava stands second in production, but its production is concentrated mainly in African, Asian, and South and Central American countries, where it is utilized as a staple food.

The nutritional value of vegetables as a vital source of essential minerals, vitamins, and

TABLE 1 Important Vegetables Grown Worldwide

Common name	Scientific name
Amaranth	*Amaranthus tricolor* L.
Artichoke	
Globe	*Cynara scolymus* L.
Jerusalem	*Helianthus tuberosus* L.
Asparagus	*Asparagus officinalis* L.
Basella	*Basella alba* L.
Bitter gourd	*Momordica charantia* L.
Bottle gourd	*Lagenaria siceraria* (Mol.) Standl
Breadfruit	*Artocarpus altilis* L.
Brinjal (Eggplant)	*Solanum melongena* L.
Broad Bean	*Vicia faba* L.
Brussels sprout	*Brassica oleracea* var. *gemmifera*
Cabbage	*Brassica oleracea* var. *capitata*
Capsicum	*Capsium annuum* L.
Cardoon	*Cynara cardunculus* L.
Carrot	*Daucus carota* var. *sativus* (Hoffm.) Arcong.
Cassava	*Manihot esculenta* Crantz
Cauliflower	*Brassica oleracea* var. botrytis
Celery	*Apium graveolens* L. var. *dulce* (Mill.) Pers.
Chicory	*Chichorium intybus* L.
Chinese cabbage	*Brassica chinensis* L.
Cluster bean	*Cyamopsis tetragonoloba* (L.) Taub.
Cowpea	*Vigna unguiculata* (L.) Walp.
Cucumber	*Cucumis sativus* L.
Drumstick	*Moringa oleifera* L.
Elephantfoot yam	*Amorphophallus campanulatus* L.
Endive	*Cichorium endivia* L.
Fenugreek	*Trigonella* sp.
French bean	*Phaseolus vulgaris* L.
Garden beet	*Beta vulgaris* var. *rubra*
Garlic	*Allium sativum* L.
Hyacinth bean	*Lablab purpureus* (L.) Sweet
Indian squash	*Praecitrullus fistulosus*
Ivy gourd	*Coccinia indica*
Jackfruit	*Artocarpus heterophyllus* L.
Kale	*Brassica oleracea* var. *doephala*
Kakrol	*Momordica dioica* L. *M. cochin-chinesis* L.
Knolkhol (Kohlrabi)	*Brassica oleracea* var. *gongylodes*
Leek	*Allium ampeloprasum* L. var. *porrum*
Lettuce	*Lactuca sativa* L.
Lima bean	*Phaseolus lunatus* L.
Muskmelon	*Cucumis melo* L.
New Zealand spinach	*Tetragonia tetragoniodes* L.
Okra	*Abelmoschus esculentus* (L.) Moench
Onion	*Allium cepa* L.
Parsley	*Potroselinum crispum* (Mill.) Nym.

TABLE 1 Continued

Common name	Scientific name
Parsnip	*Pastinaca sativa* L.
Pea	*Pisum sativum*
Plantain	*Musa paradisiaca*
Pointed gourd	*Trichosanthes dioica* Roxb.
Potato	*Solanum tuberosum* L.
Pumpkin	*Cucurbita moschata* (Duch. Poir)
Radish	*Raphanus sativus* L.
Ridge gourd	*Lufa acutangula* (L.)
Rutabaga	*Brassica napus* var. *napobrassica*
Snake gourd	*Trichosanthes cucumerina* (L.)
Spinach	*Spinacea oleracea* L.
Spinach beet	*Beta vulgaris* var. *bengalensis*
Sponge gourd	*Luffa cylindrica* L.
Sprouting broccoli	*Brassica oleracea* var. *italica*
Summer squash	*Cucurbita pepo* L.
Sweet potato	*Ipomoea batatas* (L.) Lam.
Swiss chard	*Beta vulgaris* var. *cicla*
Taro	*Colocasia esculenta* (L.) Schott
Tomato	*Lycopersicon escutentum* Mill.
Turnip	*Brassica rapa* var. *rapifera*
Watermelon	*Citrullus lunatus* (Thunb.)
Water spinach	*Ipomoea aquatica* L.
Wax gourd	*Benincasa hispida* (Thunb.)
Winged bean	*Psophocarpus tetragonolobus* L. DC
Yam	*Dioscorea* sp.

dietary fiber has been well recognized (5). In addition to these constituents, they also supply fair amounts of carbohydrates, proteins, and energy (Table 7).

Vegetables play a particularly important role in human nutrition in supplying certain constituents in which other food materials are deficient. They neutralize the acid substances produced in the course of digestion of meat, cheese, and other high-energy foods.

The nutritional composition of vegetables depends upon several factors. It is influenced by genetic as well as environmental factors, such as temperature, light, moisture, and the nutritional status of the soil in which the vegetable is grown (6,7). It is also influenced by the cultural practices, stage of maturity, postharvest handling, and storage conditions. Moreover, many vegetables are processed before they are used for consumption. Processing methods such as cooking and canning also influence the nutrient content of vegetables (8–10).

Ascorbic acid (vitamin C) is the principal vitamin supplied by fruits and vegetables in the diet. About 90% of the human dietary vitamin C requirement is obtained from fruits and vegetables (Fig.1). Capsicum, cauliflower, and bitter gourd are rich sources of vitamin C (Table 8) (5). An average adult requires about 50 mg of vitamin C per day, and less than 100 g of many vegetables contains this amount.

Vegetables are rich sources of vitamin A, thiamin, niacin, folic acid, and β-carotene. Vitamin A (β-carotene) is essential for the normal functioning of the visual process and the struc-

TABLE 2 Vegetable Classification According to Plant Organ Used

Root vegetables	Carrot, celeriac, garlic, horseradish, parsnip, cassava, radish, rutabaga, salsify, sweet potato, table beet, turnip
Stem vegetables	Asparagus, Irish potato, Jerusalem artichoke, kohlrabi, taro, yam
Leafy vegetables	Amaranth, cabbage, cardoon, celery, Chinese cabbage, chard, chicory, chive, dandelion, endive, onion, leek, lettuce, kale, mustard, New Zealand spinach, shallot
Flower vegetables	Cauliflower, broccoli, globe artichoke
Immature fruit vegetables	Beans, cucumber, eggplant, gherkin, okra, pea, pepper, squash, sweet corn
Mature fruit vegetables	Pepper, pumpkin, melon, tomato

Source: Refs. 2 and 3.

ture of the eye, and a prolonged deficiency of the vitamin can lead to blindness. Vegetables do not contain an active vitamin A compound, retinol, but only carotenoids, which are converted into active retinol in the body. Carrot, sweet potato, amaranth, basella, cowpea leaves, and spinach are important sources of vitamin A (Table 8). Some vegetables are also rich in calcium, phosphorus, and iron (Table 9).

A substantial proportion of the carbohydrates found in vegetables is present as dietary fiber in the form of cellulose, hemicellulose, pectic substances, and lignin (11). This is not digested but passes through the intestinal system of humans, as they are not capable of secreting the digestive enzymes necessary to break down the polymer complexes into simpler units in the forms absorbed by the intestinal tract (1). Humans lack enzymes like cellulase, hemicellulase, and pectinase. Fiber was once considered to be an unnecessary component of the human diet, although believed to relieve constipation. Epidemiological evidence shows that dietary fiber can be important in preventing several diseases, especially the fiber found in leafy vegetables such as celery, cabbage, spinach, and lettuce, characterized by high water content and a high percentage

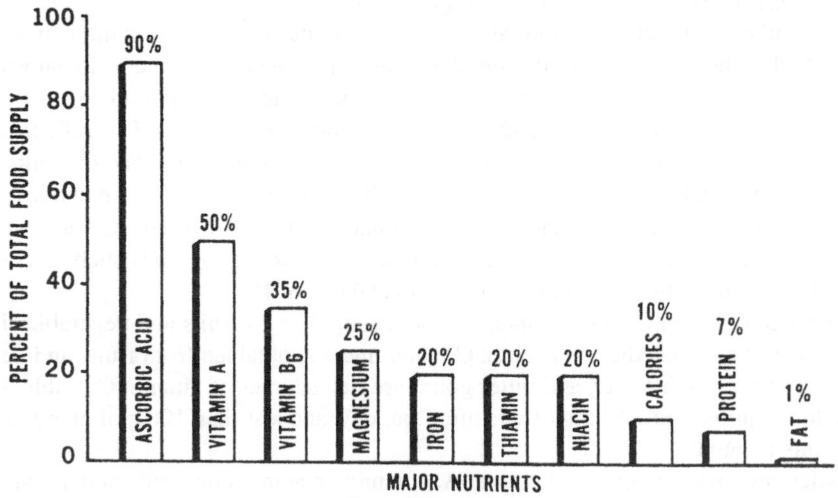

FIGURE 1 Nutritional contribution of fruits and vegetables compared to percentage of total food supply. (From Ref. 5.)

TABLE 3 Classification of Vegetables Based on End Use, Botany, or a Combination of Both

1.	Leafy vegetables	Chard, collard, dandelion, kale, mustard, New Zealand spinach, spinach
2.	Salad crops	Celery, chicory, endive, lettuce, watercress
3.	Cole crops	Brussels sprout, cabbage, Chinese cabbage, cauliflower, kohlrabi, sprouting broccoli
4.	Root crops	Beet, carrot, celeriac, parsnip, radish, rutabaga, salsify, sweet potato
5.	Bulb crops	Garlic, leek, onion, shallot
6.	Peas and beans	Chickpea, cluster bean, French bean, hyacianth bean, lima bean, peas, pigeon pea, winged bean
7.	Cucurbits	Cucumber, muskmelon, pumpkin, squash, watermelon
8.	Solanaceous fruits	Brinjal (eggplant), pepper, and tomato
9.	White or Irish potato	
10.	Corn	

Source: Ref. 2.

of fiber. Most developed countries encourage increased fiber consumption, and in the United States, the United Kingdom, and Germany increased consumption of fruits and vegetables is specifically advised.

Many therapeutic drugs in use in modern medicine originated as plant extracts. Certain vegetable components exert pharmacological or therapeutic effects. Flavonoids present in vegetables have been shown to have anticarcinogenic activity (12) and the capacity to reduce hypercholesterolemia (13). These also provide biochemical protection to cells against damage from carcinogenic substances (12). Some research indicated that the antioxidant properties of β-carotene may also play a role in the prevention of some forms of cancer (6). Experimental evidence indicates that vegetables such as onions and garlic contain allylic compounds—allyl methyl disulfide, allyl methyl trisulfide, and diallyl sulfides—which inhibit stomach and colon cancers. Cruciferous vegetables such as cabbage, cauliflower, and broccoli are rich in indoles and isothiocyanates, which inhibit carcinogenesis in the lungs and esophagus. Green leafy vege-

TABLE 4 Classification of Vegetables According to Respiration

Class	Respiration intensity at 10°C (mg CO_2/kg/hr)	Vegetables
Very low	<10	Onion
Low	10–20	Cabbage, cucumber, melon, table beet, turnip
Moderate	20–40	Carrot, celeriac, celery, Chinese cabbage, gherkin, kohlrabi, leek, pepper, rhubarb
High	40–70	Asparagus, chicory, eggplant, endive, fennel, lettuce, radish
Very high	70–100	Bean, Brussels sprout, mushroom, spinach, savoy cabbage
Extremely high	>100	Broccoli, chicory, pea, parsley, sweet corn

Source: Ref. 3.

TABLE 5 Worldwide Production of Vegetables

Vegetable	Production (1000 MT)
Roots and tubers	603,195
Potato	288,183
Sweet potato	123,750
Cassava	153,628
Yams	28,126
Carrot	13,977
Artichoke	1,137
Vegetables, including melons	465,457
Tomato	70,623
Cabbage	40,414
Cucumber and gherkins	18,326
Cauliflower	6,754
Pumpkin, squash, gourds	8,019
Eggplant	8,682
Green peppers	10,630
Onions (dry)	29,961
Garlic	7,624
Green beans	3,087
Green peas	4,602
Watermelons	27,063
Cantaloupes and other melons	12,976

Source: Ref. 4.

TABLE 6 Major Vegetable[a] - Producing Countries

Country	Production (1000 MT)
China	125,509
India	60,010
United States	32,660
Turkey	18,468
Italy	13,035
CIS	10,450
Spain	9,945
Egypt	7,474
Mexico	5,651
Nigeria	5,495
Total worldwide	465,457

[a]Vegetables (except roots and tubers) including carrot and melons.
Source: Ref. 4.

tables, which are rich in chlorophyll and chlorophyllides, are antimutagenic and antitumorigenic in humans. Psoralins and coumarins found in umbelliferous vegetables such as celery, parsley, and parsnip bind to hepatic enzymes and inhibit mutagenicity in the experimental animals. Carotenoids—α-carotene, β-carotene, lycopene, lutein, and zeaxanthin present in carrots, tomatoes, and corn—have been shown to be anticarcinogenic in nature. Micronutrients such as selenium and zinc repair the damage done by mutations. Likewise, thiamine and vitamin B_6 inhibit human platelet aggregation. Tannins and phenolic compounds found in vegetables are antimutagenic in humans (14).

There is a growing health consciousness with recognition that a diet based primarily on animal products is not healthful. Hence, there is a significant increase in the consumption of vegetables. This has resulted in an increase in international trade. Many agro-based industries use vegetables as their raw material. Tomatoes and potatoes are two major vegetables processed into a wide range of products. Frozen vegetables are also of great commercial importance, as they retain most of the qualities of fresh vegetables. In addition, vegetables are utilized by many food-processing units for the production of vegetable sauces and fermented and pickled vegetables (6).

Many tropical vegetables, such as sweet potato, taro, yam, chilies, and pumpkins, have good keeping quality. These vegetables are cultivated in tropical countries of Asia, Africa, and South America and are new to many people in Europe and America. Export of these vegetables is growing steadily (7).

TABLE 7 Proximate Composition (per 100 g edible portion) of Some Important Vegetables

Common name	Energy (kcal)	Moisture (g)	Protein (g)	Fat (g)	Carbohydrates (g)
Bitter gourd	25	92.4	1.6	0.2	4.2
Brinjal (Eggplant)	24	92.7	1.4	0.3	4.0
Cabbage	24	92.4	1.3	0.2	5.4
Capsicum	22	93.4	1.2	0.2	4.0
Carrot	42	82.2	1.1	0.2	9.7
Cassava	157	59.4	0.7	0.2	38.1
Cauliflower	27	91.0	2.7	0.2	5.2
Celery	17	94.1	0.9	0.1	3.9
Cucumber	18	96.3	0.4	0.1	2.5
French bean	32	90.1	1.9	0.2	7.1
Garlic	30	62.0	6.3	0.1	29.8
Lettuce	14	95.1	1.2	0.2	2.5
Muskmelon	17	95.2	0.3	0.2	3.5
Okra	35	89.6	1.9	0.2	6.4
Onion	50	86.8	1.2	0.1	11.1
Peas	84	78.0	6.3	0.4	14.4
Potato	97	74.7	1.6	0.1	22.6
Spinach	26	90.7	3.2	0.3	4.3
Sweet potato	114	59.4	0.7	0.2	38.1
Tomato	22	93.5	1.1	0.2	4.7
Watermelon	26	92.6	0.5	0.2	6.4
Yam	102	74.0	1.5	0.2	24.0

Source: Refs. 5–7.

Table 8 Vitamin Content (per 100 g edible portion) of Some Important Vegetables

Common name	Vitamin A (IU)	Thiamine (mg)	Riboflavin (mg)	Niacin (mg)	Ascorbic acid (mg)
Bitter gourd	416	0.07	0.09	0.5	88
Brinjal (Eggplant)	244	0.04	0.11	0.9	12
Cabbage	130	0.05	0.35	0.3	47
Capsicum	900	0.06	0.06	0.5	128
Carrot	11000	0.06	0.05	0.6	8
Cassava	0	0.05	0.10	0.3	25
Cauliflower	60	0.11	0.10	0.7	78
Celery	240	0.03	0.03	0.3	9
Cucumber	—	0.03	—	0.2	7
French bean	600	0.08	0.11	0.5	19
Garlic	Trace	0.06	0.23	0.4	13
Lettuce	900	0.06	0.06	0.3	8
Muskmelon	558	0.11	0.08	0.3	26
Okra	172	0.07	0.10	0.6	13
Onion	Trace	0.08	0.01	0.4	11
Peas	640	0.35	0.14	2.9	27
Potato	24	0.10	0.01	1.2	17
Spinach	8100	0.10	0.20	0.6	51
Sweet potato	8800	0.10	0.06	0.6	21
Tomato	900	0.06	0.04	0.7	23
Watermelon	590	0.03	0.03	0.2	7
Yam	—	0.1	0.01	0.8	15

Source: Refs. 5–7—Not reported.

Some vegetables like tomato, capsicum, cucumber, and French beans are in demand in the West during winter. The tropical and subtropical countries producing these crops have been able to meet the demand to a great extent by exporting these commodities. With the improvement in postharvest handling and storage of vegetables using modern techniques like cold chain, refrigerated and controlled atmosphere storage, and rapid transportation by land, sea, and air, international trade in such perishable commodities is increasing steadily (7).

The United States imports tropical vegetables from Mexico and Latin American countries. The European countries import these vegetables from the Middle East and some countries in North Africa. The Persian Gulf countries import vegetables mostly from the Indian subcontinent and some Southeast Asian countries. Apart from fresh vegetables, processed products also enjoy good international trade. Taiwan is a major source of processed vegetables for several countries (6).

Fruits and vegetables are highly perishable food products (15,16). Water loss and postharvest decay account for most losses, estimated to be more than 20–50% in the tropics and subtropics (8,9). Postharvest technology of fruits and vegetables is critical in the battle to minimize crop wastage. Wastage of fruits and vegetables is so great in some instances that between the field and the consumer (Fig. 2), bountiful amounts of highly nutritious crops are reduced to heaps of refuse. Lack of understanding of the postharvest technology of such crops affects both supplies and profits, even in many developed countries. Reduction of high postharvest losses of fruits and vegetables, therefore, entails the integration of a variety of aspects, such as the botany

TABLE 9 Mineral Content (per 100 g edible portion) of Some Important Vegetables

Common name	Calcium (mg)	Phospohorus (mg)	Iron (mg)
Bitter gourd	20	70	1.8
Brinjal (Eggplant)	18	47	0.9
Cabbage	49	29	0.4
Capsicum	9	22	0.7
Carrot	37	36	0.7
Cassava	50	40	0.9
Cauliflower	25	56	1.1
Celery	39	28	0.3
Cucumber	10	25	1.5
French bean	56	44	0.8
Garlic	30	310	1.3
Lettuce	35	26	2.0
Muskmelon	32	14	1.4
Okra	66	56	1.5
Onion	47	50	0.7
Peas	26	116	1.9
Potato	10	40	0.7
Spinach	93	51	3.1
Sweet potato	32	47	0.7
Tomato	13	27	0.5
Watermelon	7	10	0.5
Yam	12	35	0.8

Source: Refs. 5–7.

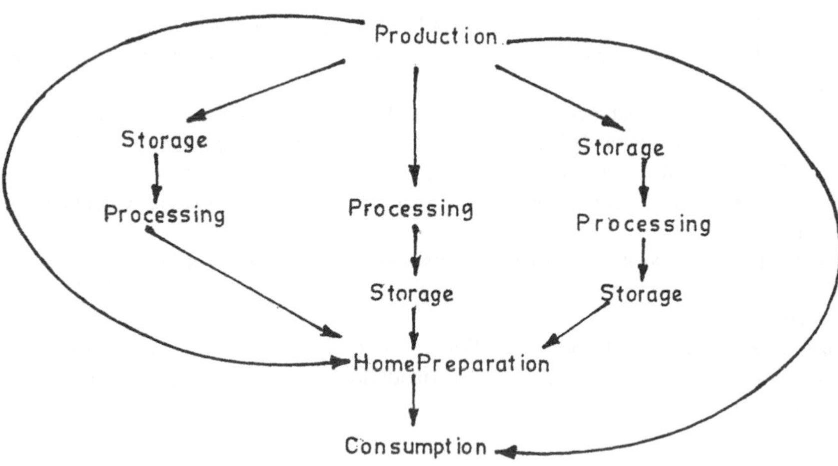

FIGURE 2 Possible routes from production to consumption of vegetable. (From Ref. 5.)

of an individual crop and cultivar, its physiology and biochemistry, the types of infecting pathogens and pests, and the various feasible technological measures to be adopted to reduce these losses. These include harvesting, handling, and maturity, low-temperature storage and environmental control (controlled/modifed and hypobaric storage), irradiation, use of chemicals and fungicides, packaging techniques, and processing (canning, freezing, drying, etc.) of fresh vegetables into suitable products such as juices and baby foods with improved storage characteristics. Information about all these areas is scattered among many research papers, reviews, and books of recent origin. In this book, an attempt has been made to compile information on production, postharvest handling, chemical composition, storage, and processing aspects of vegetables grown throughout the world. Such technological developments can eventually be utilized to increase production and to reduce postharvest losses in both developed and developing countries.

REFERENCES

1. Salunkhe, D. K., and S. S. Kadam, Fruits in human nutrition, in *Handbook of Fruits Science and Technology* (D. K. Salunkhe and S. S. Kadam, eds.), Marcel Dekker Inc., New York, 1995, p. 614.
2. Yamaguchi, M., *World Vegetables: Principles, Production, and Nutritive Values*, AVI Pub. Co. Inc., Westport, CT, 1983, p. 24.
3. Weichmann, J., *Postharvest Physiology of Vegetables*, Marcel Dekker Inc., New York, 1987, p. 5.
4. FAO, *Production Yearbook* (Statistics Series), Food and Agriculture Organization, Rome, Italy, 1993.
5. Salunkhe, D. K., H. R. Bolin, and N. R. Reddy, *Storage, Processing, and Nutritional Quality of Fruits and Vegetables, Vol. 1, Fresh Fruits and Vegetables, 2nd ed., CRC Press, Boca Raton, FL, 1991, p. 1.*
6. Roy, S. K., and A. K. Chakraborti, Vegetables of temperate climate: Commercial and dietary importance, in *Encyclopaedia of Food Science, Food Technology, and Nutrition* (R. Macrae, R. K. Robinson, and M. J. Sandler, eds.), Academic Press, London, 1993, p. 715.
7. Roy, S. K., and A. K. Chakraborti, Vegetables of tropical climates—commercial and dietary importance, in *Encyclopaedia of Food Science, Food Technology, and Nutrition* (R. M. Macrae, R. K. Robinson, and M. J. Sandler, eds.), Academic Press, London, 1993, p. 4743.
8. Salunkhe, D. K., H. R. Bolin, and N. R. Reddy, *Storage, Processing, and Nutritional Quality of Fruits and Vegetables*, Vol. II, *Processed Products*, CRC Press, Boca Raton, Fl, 1991, p. 27.
9. Salunkhe, D. K., and B. B. Desai, *Postharvest Biotechnology of Vegetables*, Vol. I, CRC Press, Boca Raton, FL, 1984, p. 168.
10. Salunkhe, D. K., and B. B. Desai, *Postharvest Biotechnology of Vegetables*, Vol. II, CRC Press, Boca Raton, FL, 1984, p. 147.
11. Dreher, M. L., *Handbook of Dietary Fiber*, Marcel Dekker, Inc., New York, 1987, p. 28.
12. Kuhnau, J., The flavonoids: A class of semi-essential food components: Their role in human nutrition, *World Rev. Nutr. Diet 24*:117 (1976).
13. Mumma, R. O., and A. J. Verlangieri, *In vivo* sulfatation of cholesterol by ascorbic acid 3-sulfate as possible explanation for the hypocholesteremic effects of ascorbic acid, *Fed. Proc. 30*:370 (1971).
14. Huang, M. T., O. Toshihiko, C. T. Ho, and R. T. Rosen, *Food Phytochemicals for Cancer Prevention I, Fruits and Vegetables*, ACS Symposium Series, 546 American Chemical Society, Washington, DC, 1994.
15. Sparks, W. C., Losses in potatoes and lesser fruits and vegetables, in *Proceedings of the National Food Loss Conference,* September 12–15, 1976, Boise, ID (M. V. Zachringer ed.) University of Idaho, Moscow, 1976.
16. Hardenburg, R. E., A. E. Watada, and C. Y. Wang, *The Commercial Storage of Fruits, Vegetables and Florist and Nursery Stocks*, Agricultural Handbook 66, U.S. Department of Agriculture, Washington, DC, 1986, p. 6.

2

Potato

S. J. Jadhav
Alberta Agriculture, Leduc, Alberta, Canada

S. S. Kadam
Mahatma Phule Agricultural University, Rahuri, India

I. INTRODUCTION

Potato (*Solanum tuberosum* L.), with an annual production of nearly 300 million metric tons, is one of the major food crops grown in a wide variety of soils and climatic conditions (1). It is the most important dicotyledonous source of human food. It ranks as the fourth major food crop of the world, exceeded only by wheat, rice, and maize (2). The dry matter production of potatoes per unit area exceeds that of wheat, barley, and maize (1). Because of increasing yield per unit area of land, total potato production has increased in both developed and developing countries in the past 40 years. In addition, the rate of production in developing countries has increased significantly more than that of developed countries (3,4). The production figures of potatoes in major producing countries are presented in Table 1. Per capita availability of potato production is highest in Europe, and especially in Eastern Europe, where a large share of total production is fed to livestock (1). The countries of Western Europe, the United States, and Japan have the highest potato yields in the world (3).

II. BOTANY

A. Morphology and Anatomy

Potato (*Solanum tuberosum* L.) belongs to the family Solanaceae (5), which also includes tomato, tobacco, pepper, eggplant, petunia, black nightshade, belladonna, and others. Out of several hundred species of *Solanum*, only potato (*S. tuberosum*) and a few others bear tubers. The potato is a herbaceous dicotyledon (Fig. 1), sometimes regarded as a perennial because of its ability to reproduce vegetatively, although it is cultivated as an annual crop. The tubers are modified, thickened, underground stems, their size, shape, and color varying according to the cultivar (6–9). On the surface of the tubers are the eyes, from which arise the growing buds.

TABLE 1 Major Potato-Producing Countries of the World

	Production × 1000 metric tons			
Country	1979–81	1985	1989	1993
USSR	76706	73000	72000	64838
Poland	39508	36546	34390	36271
China	25415	45528	30045	35050
United States	14923	18331	16659	19024
India	9377	12642	14500	15718
Germany	19465	20204	16978	12074
United Kingdom	6601	6850	6369	7069
France	6735	7814	5750	5801
Turkey	2957	4110	3900	4650
Japan	3299	3735	3700	3800
Canada	2626	2953	2754	3333
Colombia	1931	2017	2737	2860
Brazil	2002	1989	2104	2365
Argentina	1836	2000	2600	2000

Source: Refs. 1, 3.

The potato tuber is formed at the tip of the stolon (rhizome) as a lateral proliferation of storage tissue resulting from rapid cell division and enlargement. Early in its development, the integument (periderm) is only a few cells thick, but later it becomes massive with a long micropyle. The mature integument consists of three layers: (a) an external region of a single layer of epidermal cells, (b) an intermediate area consisting of an external and an internal region, and (c) an internal layer of cells adjoining the endosperm. The principal areas in the mature tuber from the exterior inwards are the periderm, the cortex, the vascular cylinder perimedullary zone, and the central pith (Fig. 2). The periderm is 6–10 cell layers thick, acting as a protective area over the surface of the tuber. Small, lenticel-like structures occur over the surface of the tuber. These develop in the tissue under the stomatas and are initiated in the young tuber when it still has an epidermis. Periderm thickness varies considerably between different varieties. Cultural conditions also, however, influence the thickness of the periderm, rendering this characteristic too variable to be used for variety identification. Underlying the periderm, a narrow layer of parenchyma tissue is present. Vascular storage parenchyma high in starch content lies within a shell of cortex. The size of parenchyma cells changes with advancement in maturity of the tubers. The pith or water core is located at the center of the tuber, which consists of cells containing less starch than is found in the vascular area and the innermost part of the cortex. The external features of the stem show eyes, which are rudimentary scale leaves, or leaf scar with axillary buds. Wound healing develops under cut, bruised, or torn surfaces. Suberin forms within 3–5 days in walls of living cells under the wound. A cambium layer developing under the suberized cells gives rise to a wound periderm. The tuber surface permits or excludes the entrance of pathogen, regulates the rate of gas exchange or water loss, and protects against mechanical damage (4).

B. Cultivars

Potato varieties differ greatly in time of maturity, appearance, yield, quality, and resistance to diseases and pests. Old potato varieties, grown for more than 50 years, originated as

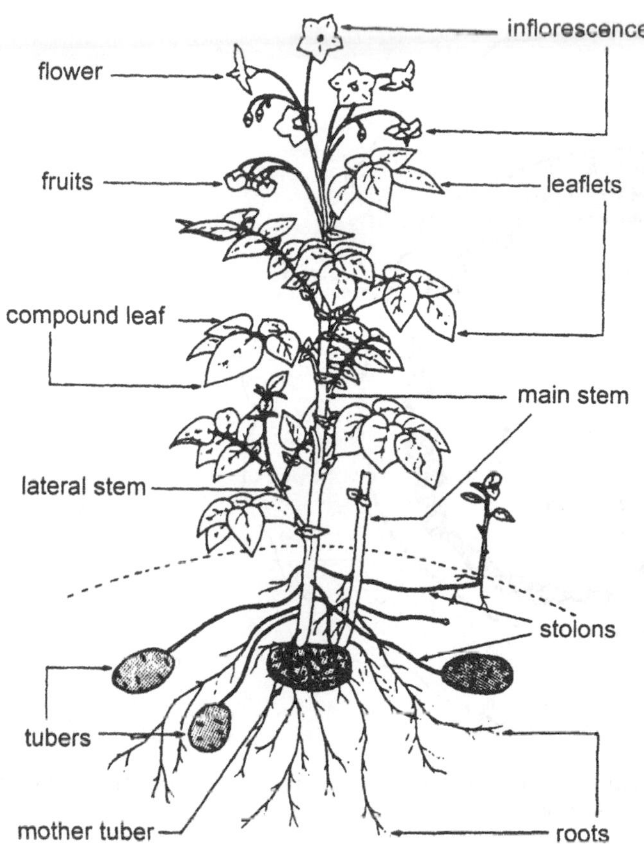

FIGURE 1 The potato plant leaves, stems and tubers. (Courtesy of International Potato Center, Lima, Peru.)

chance seedlings. They have more deep eyes, tend to be rough and irregular in shape, and have little resistance to diseases or insect pests. The U.S. Department of Agriculture and various state experiment stations introduced "new" varieties after 1925. These have few eyes, a smooth, uniform shape, and higher yields than old varieties, as well as resistance to some diseases and pests. Thompson and Kelly (10) described the following commercial varieties of potato grown in the United States. These include Katahdin, Red Pontiac, Russet Burbank or Netted Gem, Irish Cobbler, White Rose, Bliss Triumph, Kennebec, Cherokee, and Chippewa (10).

Russet Rural, Red McClure, Green Mountain, Superior, Norchip Norgold Russett, and Norland are some other popular old varieties grown in the United States and Canada, but they have limited adaptability. Tigchelaar (11) described Belleisle, a new variety originated by the Canada Department of Agriculture Research Station. It has high–specific gravity, oval tubers that are bruise resistant and resistant to common scab, black wart, and *Fusarium sambucinum.*

Both old local potato varieties, such as Phulwa, Darjeeling Red Round, Craig's Defiance, Great Scot, and Satha, as well as the improved varieties, mainly developed at the Central Potato Research Institute, Simla, India, such as Hybrid 9, 45, 208, 209, 2236, Kufri Red, Kufri Safed, Kufri Kuber, Kufri Kisan, Kufri Kundan, Kufri Kumar, Kufri Jeewan, and Kufri Chandramukhi, are grown in different parts of India (12).

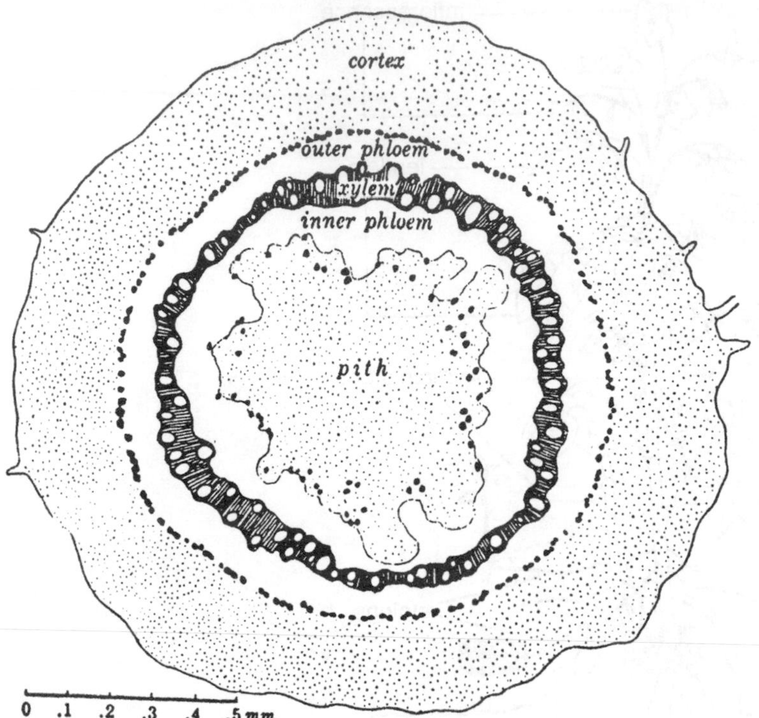

FIGURE 2 Diagrammatic transection of a potato tuber showing proportional amounts of tissue in the main areas. (From Ref. 4.)

III. PRODUCTION

A. Soil and Climate

Climate plays an important role in influencing potato yields. The climatic factors that influence potato yields are temperature, light intensity and photoperiod, rainfall, and length of growing season.

The potato is classified as a cool-season crop and thrives best with moderate temperatures during the growing season. The potato production zones by climate in developing countries are shown in Table 2. The best growth is achieved when the mean July temperature is about 20°C. Temperature influences the rate of absorption of plant nutrients and their translocation within the plant, the rate of photosynthesis, and the rate of respiration, and it also hastens early growth. Temperatures during growing determine the length of the growing season. At low temperatures, the rate of respiration is less than the rate of photosynthesis, resulting in more accumulation of carbohydrates in the tubers and an increase in the specific gravity of tubers.

The influence of light on growth and yield of potatoes is dependent upon intensity, quality, and day length. Effects of the photoperiod on the growth and development of potatoes have been studied extensively (12–14). During short days, the light-demanding growth stage of the potato is completed faster than on a long day. During the stage of tuber formation, potato plants prefer long days. At this stage, the plant devotes a large portion of its carbohydrates resulting from photosynthesis to the growth of tubers and less to vegetative growth. A long photoperiod

TABLE 2 Potato Production Zones by Climate in Developing Countries

Climate	Estimate of production (%)	Growing season	Observation
Hot summer; cool to cold winter, unreliable	5	Cooler months	Irrigation
Hot, dry year round	6	Cooler months	Irrigation
Hot, wet summer; hot, dry winter	2	Dry season	Irrigation
Tropical rain forest	1–2	Rainy season	Humid
Warm to hot, wet summer	35	Winter	Lowland Asia
Hot, dry summer; cool, wet winter	2	Various	Dry winter irrigation
Cool to warm, wet summer; cool, wet winter	10	Summer	Temperate
Cool to warm, wet summer; cool to cold, dry winter	12	Summer	Temperate
Highlands (<1500 m)	21	Various	Complex production

Source: Annual Report of International Potato Center, Lima, Peru, 1988.

increases the maturity period of potato plants, whereas with a short period they tend to mature earlier.

Adequate moisture is required for steady growth and maximum yield of potatoes. In areas that receive little rainfall, additional irrigation water should be distributed so as to provide adequate moisture for the growth of the plants.

The soil factors that influence tuber growth are structure, water-holding capacity, aeration, temperature, drainage, and the nutrient-supplying capacity of the soil. Potatoes can be grown in all types of soil except saline and alkaline soils. Sandy loam and loamy soils rich in organic matter are most suitable for potato cultivation. In general, loam soil produces high–specific gravity tubers, probably because they have optimum soil moisture and uniform temperature and structure relationships for potato production than heavy-textured soils. Sandy soils are conducive to rapid growth, provided sufficient nutrients and moisture are present, resulting in early maturity of the crop. Soils that are too sandy and especially those low in organic matter usually are not desirable for high yields, as a result of their low water-holding capacity. To obtain high yields, soil should be loose and friable with good drainage and aeration. Such soils are easy to till and are not favorable for the development of blight rot of tubers. Potatoes do not grow well on heavy textured or undrained soils. Potatoes grown on some fertile clay soils may produce good yields, but not on sticky soils that hinder digging and marketing of potatoes due to the soil adhering to them.

The sensitivity of potato plants to soil aeration has been well demonstrated (15). Potato plants respond well in soils that have 50% or more pore or air space.

The optimum soil pH for potatoes is about 5.0–5.5. Acidic soil conditions tend to limit potato scab disease, which is favored by alkalinity. Successful production of potatoes can be achieved on soils with neutral or alkaline pH. The effects of soil pH on potato yield are probably associated with the availability and uptake of plant nutrients.

Potatoes should be cultivated in well-prepared, pulverized, moist soil. Soils that lack certain physical and chemical characteristics are not suitable for potato cultivation. The moisture content of the soil is an important factor that determines the yield of potatoes. Potatoes require a uniform supply of moisture throughout the growing season for maximum yields. The most critical period is at the beginning of tuberization (10). Excess soil moisture may, however, produce low dry matter yields. Fluctuation in soil moisture tends to promote unequal growth in vines and tubers, resulting in lower yields. The soil moisture should be maintained at about 60% of field capacity for optimum yields. To obtain potatoes high in dry matter, water is withheld late in the season to help them become mature at harvest. The specific gravity of the tubers is reduced when they are supplied with excessive irrigation late in the season.

B. Fertilization

Potatoes require large quantities of mineral nutrients, particularly nitrogen and potassium, for maximum growth. The kind and amount of fertilizers required for maximum yield vary depending upon soil type, soil fertility, climate, crop rotation, variety, length of growing season, and moisture supply. The ratio of nitrogen, phosphorus, and potash is important. The most widely and recommended ratios are 1-2-1, 1-2-2-, 2-3-3, and 1-1-1 (17). The results of studies on the effect of different levels of fertilizer on yield of potatoes are presented (17) in Table 3. In many areas, fertilizers are applied before or at the time of planting. In other areas, in addition to applied fertilizers at the time of planting, a side-dressing application is made during the growing season.

Phosphorus application to potatoes has some effect on dry matter. At the early stage of growth, the need for phosphorus is critical. High-yielding potato cultivars require large amounts of potassium for growth and development. Potassium removal by the potato crop is high, but the response to potassium application is less than that of nitrogen and phosphorus. Potassium sulfate as a potassium fertilizer is often recommended for use on potatoes, because it is thought that it causes less of a problem with tuber quality than with potassium chloride in terms of specific gravity. In addition to the reduction in specific gravity with high rates of potassium application, further reduction has been shown with the use of potassium chloride over the use of K_2SO_4 (18). Micronutrients do affect potato quality and yield when present in deficient quantities.

C. Weed Control

In many potato-growing areas, it is common practice to control weeds early in the season with an application of certain chemicals to the soil. Such chemicals are usually applied as sprays to the soil before the potatoes emerge, but after weeds are in the seedling stage. Intercultural

TABLE 3 Influence of Fertilizer Application (4-14-8) on Yields of Three Cultivars in a New Red Latosol of Brasilla D.F., 1981

	Aracy		Achat		Bintje	
Fertilizer (t/ha)	1980	1981	1980	1981	1980	1981
0	12.1	12.2	9.9	3.4	11.7	4.8
2	48.5	24.5	32.7	14.6	25.8	9.6
4	46.9	30.0	41.2	18.9	31.4	15.6
6	50.1	29.9	42.4	23.8	33.6	14.6
8	46.8	32.8	45.4	21.9	29.6	14.6
Tukey 5%	14.24	5.71	16.96	3.66	7.21	4.79

Source: Ref. 17.

operation like hoeing, weeding, and earthing-up help in checking the weeds. Linuron, simazine, and patoran are more effective for weed control in the spring crop because of their persistent effects (19). Dinobo compounds such as dinitro-*o*-secondary butyl phenol and dinitro-*o*-secondary amyl phenols are available under various trade names. Premerge and sinox-PE are effective for preemergence weed control (20,21).

D. Diseases

Potato disease consists of an interaction between a host (the potato) and a pathogen (bacterium, fungus, virus, mycoplasma, nematode, or adverse environment) that impairs the productivity or usefulness of the crop. The host-pathogen interaction is influenced by the environment acting on either the potato or the pathogen or both and is determined by the genetic capabilities of (a) the potato, being either susceptible or resistant, and (b) the pathogen, being pathogenic or nonpathogenic. Disease or adverse environment may severely affect production or quality of potatoes. Various diseases caused by bacteria, fungi, viruses, mycoplasma, and nematodes have been compiled by the American Phytopathological Society and control measures suggested (22).

Three important bacterial diseases of the potato are black leg (*Erwinia phytophthora* Apple or *E. caratovora* Jones), ring rot (*Corynebacterium sepedonicium* (Spiedk and Kotth) (Skapt and Burkh), and brown rot or bacterial wilt (*Pseudomonas solanacearum* E.K. Smith). Small differences in the susceptibility of potato tubers of different varieties to black leg and ring rot have excluded these diseases from the breeding programs, because there are no known good sources of resistance to these diseases. Black leg is controlled by producing disease-free stocks from stem cutting, and ring rot is controlled by isolating disease-free tubers to produce new "seed" stocks. *Solanum phureja*, a diploid clone of the cultivated species, has shown significant resistance to a serious pathogen of warm soil, namely, *P. solanacearum*. Common scab, caused by *Streptomyces scabies* (Thaxt) Waksm and Henrici, reduces tuber quality and consumer acceptance. Significant varietal difference in the susceptibility to common scab has made it possible to breed a potato variety for resistance to scab.

The potato is affected by several fungi, by rust (*Puccinia pitteriana*), and by smut (*Thecaphora solani*)—the latter two in Columbia and Peru, respectively. Among fungal diseases, the most important are wart disease (*Synchytrium endobioticum*), "late blight" (*Phytophthora infestans*), and storage rots, namely, dry rot (*Fusarium solani* var. *coeruleum*), gangrene (*Phoma exigua* var. *foveata*), pink rot (*Phytophthora erythroseptica*), black scurf (*Corticum solani*), skin spot (*Oospora pustulans*), and Verticillium wilt (*Verticillium alboatrum*) (23,24).

E. Insect Pests and Nematodes

1. Insect Pests

Many insect pests damage the potato. It is obviously difficult to estimate yield losses caused by pests. However, pest damage to potatoes in the field and in storage is a major constraint for potato production. Potato pests cause damage in a number of ways. Some, such as the Colorado beetle and the lady beetle, defoliate the plant, whereas aphids remove sap, may inject toxic saliva, act as virus vectors (25–27), and reduce tuber yields. Others, such as wireworms, eat tubers, considerably decreasing the value of the crop. Several insecticides used to control potato pests include chlorinated hydrocarbon, organo-phosphorus, carbamates, agaricides, and soil fumigants. The application of these insecticides to control various pests depends upon the availability and relative susceptibility of pests to different insecticides (28,29).

2. Nematodes

Nematodes pathogenic to potato occur in all climates and cause serious crop losses, but much of this damage is unrecognized or attributed to other causes (30). The most common potato pests include cyst nematodes, *Globodera* (Heterodera) spp. (namely, *G. rostochionsis* and *G. pallida*), root-knot nematodes (Meloidogyne sp.), false root-knot nematodes (*Nacobbus* sp.) and migratory nematodes (*Trichodorus* sp.), and aphids (*Myzus persicae* and *Macrosiphum euphorbiae*) (31–37). Resistance to all these different disease and pests is being sought under various climatic conditions and regions of the countries. Howard (24) reported a method for screening potatoes for resistance to *Globodera* sp. under laboratory conditions. These parasites have infected potatoes since 1880 and cause an average 11% loss of potatoes worldwide (30).

F. Maturity

Storage and conditioning of tubers are influenced by tuber maturity. Completely mature potatoes are more desirable for processing than less mature potatoes. Potatoes increase in their specific gravity and yield on complete maturation. The way to obtain complete maturity in potatoes is by early planting, late harvesting, and slow killing of potato vines. Mosher (38) suggested that mature tubers can be obtained by using early-maturing varieties, covering seed pieces low, planting early, using suggested rates of nitrogen, and using low rates of potash fertilizer when possible. Delayed harvesting of potatoes tends to increase the specific gravity and yield of tubers. Mature tubers usually result in higher-quality processed product as well as in higher yield of product per acre. Immature tubers are low in specific gravity and yields. They are subject to skinning and bruising. Potato tubers are harvested before maturity due to favorable market prices at that time and also to avoid the danger of freezing temperatures.

G. Harvesting

The time of harvest can be adjusted to suit market prices. This is a very important part of potato production. In developed countries, harvesting is done by two-row mechanical diggers or mechanical harvesters. In developing countries, manual harvesting is a common practice. In two-row digger harvesting, potatoes are usually dropped to the ground behind the digger and picked up later by hand. They are placed into containers, sacks, or directly into crates and loaded into trucks and transported to a packing shed or storage. The mechanical harvester digs the potatoes, separates them from soil, vines, and stones, and deposits the tubers into containers or trucks to be taken to the storage or packaging shed. Losses of potatoes depend upon the maturity stage at harvesting and the method of harvesting.

H. Yield

Potato yield is related to several other characteristics, such as the duration of crop growth, time of maturity and harvest, plant density, the number of plants, and the weight of individual tubers. Investigations of the frequency of early-maturing seedlings under long-day conditions suggested that the time of potato maturity depends upon a number of genes, and varieties are heterozygous for most of these (39). According to Moorby (40), a quantitative understanding of crop behavior can be used to predict and assess the yield potentialities of unusual genotypes. A parallel approach to conventional growth analysis offers advantages. Crop yield of potatoes was higher in plots planted in mid-June than those planted at the end of May—110 days after planting, the yield was higher with the earlier planting dates. The yield at 130 days after planting was not significantly different from the yield at 110 days after planting. Burton (41) surveyed the history and various factors in influencing the yield of potatoes.

I. Tuber Quality

The quality of potato tubers is related to numerous factors including morphology, structure, and chemical composition, which in turn influence the nutritional, sensory, and processing quality of potatoes (40–44). Of all the factors affecting tuber quality, the most important are environment during crop growth, variety, the cultural practices employed, irrigation, fertilization, and the use of other agricultural chemicals, such as pesticides and growth regulators. According to Gray and Hughes (39), the quality and nutritional value of a tuber at harvest is a result of the effects of varietal, cultural, and environmental factors on the growth of the potato crop. The quality of potato tubers associated with morphology and external appearance include tuber size, shape, depth and appearance of the skin, depth of the eyes, flesh and skin color, and greening. Tuber shape, skin, and flesh color and depth of the eyes are largely influenced by varietal and environmental factors.

IV. GRADING OF POTATOES

The desirable quality characteristics of potatoes are dictated by the intended use, whereas the acceptability of raw potatoes is determined primarily by size, shape, color, and attractiveness of the tubers; the quality of processed products is evaluated in terms of color, flavor, and texture. High-quality processed products can be made only from good-quality raw potatoes. Uniformity of size, shape, and composition is essential, and in many advanced countries rigid specifications for raw potatoes used for processing have been set.

Many types of standards apply to horticultural products in the United States. Some of these have been drawn up by trade associations and are voluntary, while others are government standards issued by state, country, or municipal authorities, which are often mandatory. The most important groups of standards are those developed by the U.S. Department of Agriculture (45). According to Schoenemann (46), the use of grades in marketing of potatoes is not mandatory unless a state or region is operating under a set marketing order or compulsory grading regulations, specifying various grades (47). In the United States, fresh potatoes are marketed under five different federal grade standards (47,48): U.S. Fancy, U.S. No. 1, U.S. Commercial, U.S. No. 2, and unclassified.

Grade standards of potatoes now in use are mostly based on factors affecting the external or internal quality of potatoes. It is well established that the dry matter content or specific gravity of a tuber significantly influences its cooking and processing qualities. The possibility of utilizing specific gravity and its application for grading in the potato chip industry has been indicated (48–53). Schoenemann (46) suggested that potatoes segregated for high dry matter content could be utilized more effectively for processing purposes.

V. NUTRITIONAL COMPOSITION

Although the potato is rich in carbohydrates, it also provides significant quantities of other nutrients, such as proteins, minerals (iron), and vitamins (B complex and vitamin C (53–61). Burton (58) reported that 100 g fresh weight of potato provided 2.1 g protein (N × 6.25), 0.3 MJ energy, 25 mg vitamin C, 0.1 mg thiamine, 0.02 mg riboflavin, 0.5 mg nicotinic acid, and 1.0 mg iron. A report of the National Food Survey Committee (59), London, stated that on an average in the United Kingdom, potato consumption accounted for 4–4.5% of daily energy and protein intake and over 25% of the daily requirement of ascorbic acid.

The nutritional or chemical composition of potato tubers varies with variety, storage, growing season, soil type, preharvest nutrition, and method of analysis used. According to Mondy (60), the average nutritional composition of the potato is as follows: water, 80%; carbohydrate, 18%; protein, 2%; lipid, 0.1%; minerals, vitamins, etc., < 0.1%. A study carried out at the Animal Nutrition Laboratory at Cornell (60) in which rats were fed a lifetime diet comparable to that eaten by one third of the population of the United States indicated that rats fed a diet rich in potatoes had the best survival in old age and the greatest mean lifespan.

A. Carbohydrates

Carbohydrates constitute about 80% (range, 63–86%) of the total solids found in potatoes (13, 62, 63). They are the constituent of highest concentration other than water and consist largely of starch. Although the potato is an important source of energy, its major disadvantage is how much of it (approximately 3.5 kg) would have to be consumed to meet daily requirements. One medium-sized potato yields about 100 kcal. The contribution of the potato to energy intake, however, varies markedly, depending upon whether it is cooked in water or fat (oil); boiled potatoes provide about 0.3 MJ per 100 g fresh weight, whereas chips or French fries provide up to 0.6 MJ, providing 5 and 15%, respectively, of the daily requirements at U.K. levels of consumption.

Like other tuber crops, starch is the major component of potato tubers (Table 4). Potato starch contains amylose and amylopectin, with 0.093% phosphorus (62). Much of the starch is present in starch granules. The amylose content increases with maturity (63). Fertilization of crops influences the starch content of tubers (64). The starch content in potatoes varies according to storage temperature (65), diseases and application of chemicals (66). The amylose content in starch ranges from 18.5 to 32.0%. Sucrose, glucose, and fructose comprise the major sugars of potato. Traces of ketoheptose, melibiose, melezitose, and raffinose have been detected in potato tubers (63,67). Large amounts of sugars accumulate during low-temperature storage. Sprouting increases the sugar content in the potato (68). The sugar concentration is higher at the center of the tuber than in the outer region. Potatoes stored under a nitrogen atmosphere do not accumulate sugars, but instead lose starch content.

Potatoes contain cellulose, pectic substances, hemicellulose, and nonstarch polysaccharides to the extent of about 0.2–3.0%. Cellulose constitutes 10–12% of the nonstarch polysac-

TABLE 4 Chemical Composition of Potato Tubers (Dry Matter Basis)

Constituent	Reported range (%)	Average range (%)
Starch	60–80	70
Sucrose	0.25–15	0.5–1.0
Reducing sugars	0.25–3.0	0.5–2.0
Total N	1.0–2.0	1.0–2.0
Protein N	0.1–1.0	0.5–1.0
Fat	0.1–1.0	0.3–0.5
Dietary fiber	3–8	6–8
Minerals	4–6	4–6

Source: Ref. 58.

charides. Pectic substances range from 0.7 to 1.5%. The skin contains about 10 times more pectin than does the flesh. The application of auxin stimulates synthesis of pectin at the expense of cellulose (69). Pectic substances include protopectin, soluble pectin, and pectic acid. Protopectin constitutes about 70% of the total pectic substance. Storage of the potato increases soluble pectin and decreases protopectin. Soluble pectin accounts for about 10% of total pectic substances. The pectic acid fraction constitutes 13.25% of total pectic substances. Hemicellulose contains glucuronic acid, xylose, galacturonic acid, and arabinose. It has been shown that 1% of the total nonstarch polysaccharide of potato is present as hemicellulose. The dietary fiber content ranges between 1 and 2% in fresh potatoes (70–79). Often dietary fiber includes starch that is resistant to hydrolysis by enzymes used to remove starch prior to dietary fiber determination. Jones et al. (74) reported that there was little resistant starch in raw potato, but that it formed 20–50% by weight of total dietary fiber of cooked potato. However, it is not known whether or not this resistant starch is digested in the human digestive tract (75–79). Fresh potato has a dietary fiber content similar to that of the sweet potato, but somewhat lower than that of other roots and tubers, and much lower than that of cereals and legumes. Unpeeled potatoes contain higher amounts of dietary fiber than peeled raw or boiled potatoes (74).

Potato has a lower average energy content than other tubers and raw cereals. However, significant variation exists in the energy values of commercial potato varieties (80,81). When calculated on the basis of a moisture content equivalent to that of dry staples, the energy content of potatoes is similar to that of cereals or legumes (82). Because the potato is a low–energy density food, it would be advantageous to include it in the diet of the population of the developed world where obesity is prevalent (83). Starch provides most of the energy supplied by the potato. The digestibility of potato starch is low in raw potatoes, but is markedly improved during cooking or processing (74).

B. Proteins and Amino Acids

The nitrogen content of potatoes ranges from 1 to 2% of dry weight. There is an inverse relationship between the distribution of starch and nitrogen. The nitrogen content in tubers is influenced by the cultivar and environmental conditions under which the potato crop is grown. The protein nitrogen in potato tuber is mainly contributed by the salt-soluble globulin fraction (84). Gel electrophoretic studies of potato proteins have revealed several protein components in potato tubers (85). The protein fraction, which constitutes from one half to two thirds of total nitrogen, is present as free amino acid. Asparagine and glutamine are present in approximately equal amounts and together constitute about one half of the total amino acids. The amount of lysine in potatoes is similar to that in typical animal protein (86) (Table 5). Chang and Avery (87) found that the nutritive value of potato protein was superior to that of rice. Weight gains and protein efficiency ratios were higher in those rats fed with the potato diet.

Protein content in the potato is comparable to that of root and tuber samples (88,89). Potatoes contain a higher concentration of lysine than cereals (90,91). Thus, potatoes can supplement a cereal-based diet. It has been shown that 100 g of potato can supply 7, 6, and 5% of daily energy and 12, 11, and 10% of the daily protein needs of children aged 1–2, 2–3, and 3–5 years, respectively (89). The potato is a well-balanced food in terms of protein and energy. Potatoes baked in their skins or roasted or fried in fat contain more calories and protein than boiled potatoes due to the loss of nutrients during boiling. The nutritive value of potato protein varies considerably between lots of the same variety, but little is known about the influence of variety, cultural practices, or climatic and environmental factors on potato quality.

The presence of several enzymes such as polyphenol oxidase, peroxidase, catalase, esterase, proteolytic enzymes, invertase, phosphorylase, and ascorbic acid oxidases has been

TABLE 5 Essential Amino Acid Composition of Potato Tubers

Amino Acid	Reported range (mg/g)
Histidine	1–4
Isoleucine	2.7–4.2
Leucine	3.9–6.1
Lysine	4.2–5.7
Met + Cys	1.0–2.9
Phenylalanine + tyrosine	5.8–8.2
Threonine	2.5–3.8
Tryptophan	1.2–1.4
Valine	5.1–6.4

Source: Ref. 86.

documented in potato tubers. These enzymes influence the processing properties of tubers and are involved in sprouting.

C. Lipids

The amount of lipid present in the potato is small—approximately 0.1% on a fresh weight basis. Mondy (60) stated that the nutritional importance of lipid in the potato could not be judged solely by its quantity, especially considering its role in membrane structures.

D. Vitamins

Potatoes are an excellent source of ascorbic acid, thiamin, niacin, and pyridoxine and its derivatives (72). Fresh potatoes may contain 30 mg or more ascorbic acid per 100 g when newly harvested, although values decline when potatoes are stored (Fig. 3), cooked, or processed (93). Potatoes contain higher quantities of ascorbic acid than do carrots, onions, or pumpkins. The vitamin content of potatoes depends upon variety (92), soil type (93), nitrogen fertilization (94), date of harvesting (95), and phosphorus application (96).

According to Mondy (60), of the vitamins included in the recommended daily dietary allowances of the National Research Council, potatoes offer substantial amounts of ascorbic acid, niacin, thiamine, and riboflavin. Potatoes contribute more vitamin C to our food supply than any other major food. The vitamin C content of the potato varies markedly with variety, maturity, preharvest mineral nutrition of the crop, soil type, and storage conditions employed (93).

According to Rosenberg (97) vitamin C in the potato tuber is present in both the reduced (ascorbic acid) and the oxidized state (dehydroascorbic acid), but the content of the latter is usually low. On cooking, dehydroascorbic acid is readily converted irreversibly to diketogulonic acid, resulting in a loss of vitamin C. Both temperature and length of storage significantly influence the vitamin C content. Carter and Carpenter (98) reported that only 23% of the total niacin in cooked potato was found to be in an available form.

E. Minerals

The potato is a good source of iron and magnesium and contributes some trace mineral elements lacking in milk. The iron content of potato is comparable to that found in other roots and tubers.

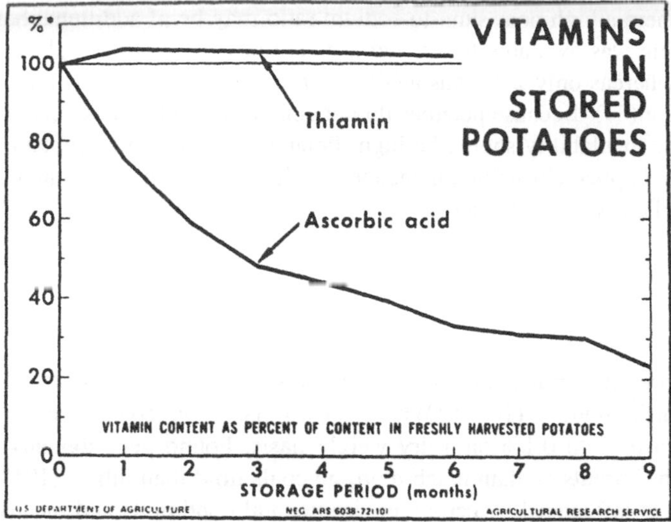

FIGURE 3 Vitamin C losses in potatoes during storage. (From Ref. 58.)

A positive correlation was found between ascorbic acid content of potatoes and the amount of iron solubilized from potatoes by gastric juice in vitro (97). The potato appears to have a moderate iron availability superior to that of other vegetable foods. Potatoes are a good source of phosphorus (99) (Table 6). A relatively small percentage of the total phosphorus in potatoes occurs in the form of phytic acid. Phytic acid is known to interact with calcuim, iron, and zinc in the form of phytate, thus rendering them unavailable for absorption into the body. Approximately 25% of the total phosphorus was found in the phytic acid (99). A mean of 8.3% of the total phosphorus was found in phytic acid among 23 samples of potato grown in India. The lower phytic acid content in the potato would result in a higher availability of the minerals found in a meal including these potatoes.

Potatoes are a poor source of calcium, but a rich source of potassium (97). Sodium content is also low. Potatoes can therefore be used in diets designed to restrict sodium intake in patients

TABLE 6 Mineral and Vitamin Contents in Potato Tubers (100 g edible portion)

	Content (mg)	
Constituent	(Ref. 60)	(Ref. 58)
Minerals		
Calcium	9	7
Phosphorus	50	53
Iron	0.8	0.6
Vitamins		
Thiamine	0.10	0.09
Riboflavin	0.04	0.03
Niacin	1.5	1.5
Ascorbic acid	20	16

with high blood pressure, where a high potassium-to-sodium ratio may be of additional benefit. It was found that 97% of zinc was available to rats fed a potato-based diet with a phytic acid content of 0.23 mg/100 g, whereas only 23% was available in a corn-based diet with a phytic acid content of 9.93 mg/100 g (87). Because potatoes do not contain excessive fiber and have a low phytic acid content, zinc availability should be high. Potatoes are known to contain other minerals such as magnesium, copper, chromium, manganese, selenium, and molybdenum. They are considered to be an excellent source of fluoride.

F. Antinutritional/Toxic Compounds

1. Glycoalkaloids

The steroidal glycoalkaloid present in potato tubers has been shown to be a mixture of α-solanine and α-chaconine (Fig. 4). Earlier reviews (100–103) have shown that the glycoalkaloids present in normal tubers range from 0.01 to 0.1% on a dry weight basis. Potato peel contains more glycoalkaloids than the flesh. Sprouts contain much more glycoalkaloid than tubers (101). The formation of glycoalkaloids is influenced by certain environmental conditions and mechanical injury. Glycoalkaloids of potato are not destroyed by cooking, baking, or frying. Jadhav and Salunkhe (102) have reviewed the literature on the formation, distribution, and control of glycoalkaloids in potato tubers and have evaluated their toxicity. Potatoes containing more than 0.1% glycoalkaloid (dry weight basis) are considered to be unfit for human consumption. The alkaloids are regarded as normal constituents in all solanaceous plants, but in high concentration they are recognized as toxins and teratogens. Glycoalkaloids are cholinesterase inhibitors (102), which can cause headache, nausea, diarrhea, and serious illness if consumed in concentrations greated than 2.5 mg/kg body weight. Concentrations greater than 3 mg/kg body weight can be lethal to humans (103).

2. Protease Inhibitors

Potato tubers contain protease inhibitors, including trypsin inhibitor, chymotrypsin inhibitor (104), hemagglutinin activity (105), and kallikrein inhibitor (106). The tubers are known to contain high concentrations of protease inhibitors (107–110). Lau et al. (111) reported enterokinase-inhibiting activity in potato tubers. Since enterokinase initiates the cascade reaction that activates digestive proteinases in animals, its inhibitor would also probably be a potent inhibitor of digestive proteolysis. The amino acid compositions of different protease inhibitors found in potato are very similar (112). These inhibitors are implicated in the defense mechanisms against pest attack (107). Chymotrypsin inhibitor functions as a storage protein during potato plant development (112). The trypsin and chymotrypsin inhibitors are implicated in field resistance to late blight (113).

Ryan and Hass (108) reported that boiling, microwave cooking, or baking potatoes destroyed most of the protease inhibitor activity, but the carboxypeptidase inhibitor was extremely stable in all three methods of cooking. Huang et al. (114) reported that significant chymotrypsin inhibitor activity also survived baking and boiling, although trypsin inhibitor activity was completely destroyed.

Livingston et al. (115) found that potato chymotrypsin inhibitor was responsible for poor nitrogen utilization in pigs. Partial cooking reduced inhibition by one third, but steaming at 100°C for 20 minutes completely destroyed inhibitor activity. Digestibility of nitrogen in raw potato was 32.8%, and that in cooked potatoes was 89.8% (115). The digestibility of nitrogen in partially cooked potatoes was 48% of the completely cooked sample. These studies indicated that the low availability of nitrogen from raw potatoes is due to antinutritional factors.

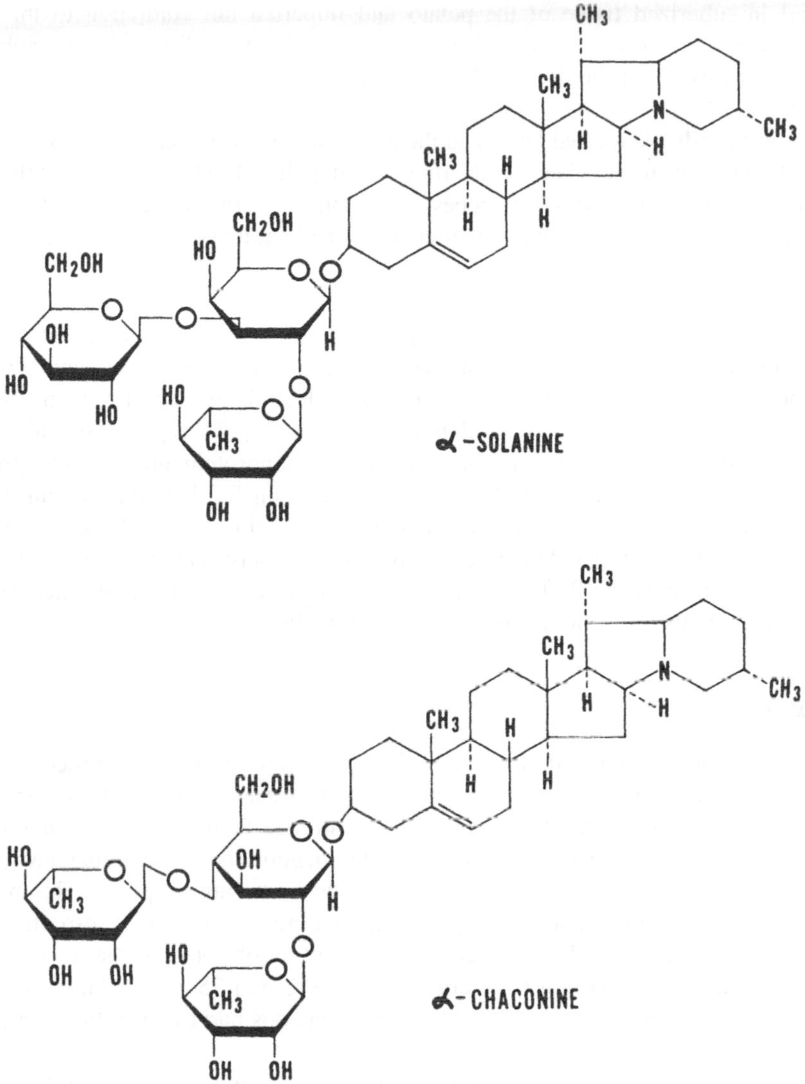

FIGURE 4 Structures of α–solanine and α–chaconine. (From Ref. 4.)

3. Lectins

Lectins from potato have been shown to agglutinate erythrocytes of several animals as well as humans (116,117). The potato tuber lectin has a saccharide specificity similar to that of wheat germ agglutinins (117). However, very few studies are available on the nutritional significance of potato lectin. It is anticipated that under certain conditions, potato lectin could also exert a toxic effect.

4. Phenolic Compounds

Many reports (118–121) indicate the presence of phenolic compounds in potato tubers, including monohydric phenols, coumarins, anthocyanins and flavones, and polyphenols. Tannins are mostly localized in suberized tissue of the potato and impart a tan coloration to the skin. Coumarins have been implicated in the discoloration of cooked potato. Tyrosine, a monohydric phenol, constitutes 0.1–0.3% of the dry weight of the potato, whereas chlorogenic acid constitutes 0.025–0.15% (118).

Phenolic compounds are located largely in the cortex and peel tissues of the potato. High levels of phenols have been associated with after-cooking discoloration. This discoloration occurs mostly in the stem end of cooked potatoes, where phenols are more concentrated (121). Mondy et al. (122) found a high positive correlation between phenolic content and bitterness and astringency.

Considering its nutritional quality, the potato is an ideal food for the year 2000 (60). Zgorska (123) described the factors affecting the quality of table potatoes, giving data on the contents of dry matter, starch, reducing sugars, sucrose, phenolics, citric acid, and tyrosine, together with their relationship to quality characteristics such as darkening of raw and cooked flesh, chip color, and discoloration after tuber damage. Klein et al. (124) reported the nutritional composition and quality of potatoes in terms of total nitrogen, nonprotein nitrogen, total protein, amino acids, minerals, and firmness as influenced by magnesium fertilization. The nutritional composition of the tuber was largely related to the cultivar, the climatic conditions during the growing season, and the storage conditions. Chip color was closely related to the contents of reducing sugar and sucrose, while flesh darkening of the tuber and discoloration after damage were related to the contents of phenolic compounds and tyrosine.

VI. STORAGE

Potatoes are generally stored before they are used for either table purposes or processing (4). After harvest, they are transported to storage facilities by carts, trucks, trains, or ships, depending upon the quantity being shipped and the distance to the storage facilities. A large number of tubers are bruised, skinned, or injured during harvest and transportation (125). Immediately after harvest, potatoes are stored at 10–16°C and high humidity for curing purposes. This process stimulates the growth of periderm, which helps in the healing of wounds. It also stimulates thickening of the periderm. This reduces weight loss and rotting of potatoes due to storage rot organisms. It has been shown that at a termperature of 12.8°C and 75–85% relative humidity (RH), suberization occurs within a week (126). The harvest and postharvest operations for potato tubers are shown in Fig. 5.

The production of potatoes throughout the year is virtually impossible in most developing countries. Potatoes are frequently stored to prevent seasonal gluts and to increase the availability of potatoes to consumers throughout the year. Potato storage facilities maintain tubers in their most edible and marketable conditions by preventing large moisture losses, spoilage by pathogens, attacks by insects and animals, and sprout growth. Several changes in nutrients occur during storage (127–131). The changes vary according to storage conditions and potato variety. Potatoes are normally stored at temperatures of 4–20°C. Lower temperatures inhibit sprout growth. However, lower-temperature storage results in an accumulation of sugars. Before storage, tubers are generally allowed to undergo wound-healing processes (126). Potatoes used for domestic consumption should be stored at about 5°C to avoid sprouting and accumulation of sugars. Tubers for later use in the food-processing industry should be maintained at about 10°C, which avoids disease and excessive sprouting and prevents a high accumulation of reducing sugars.

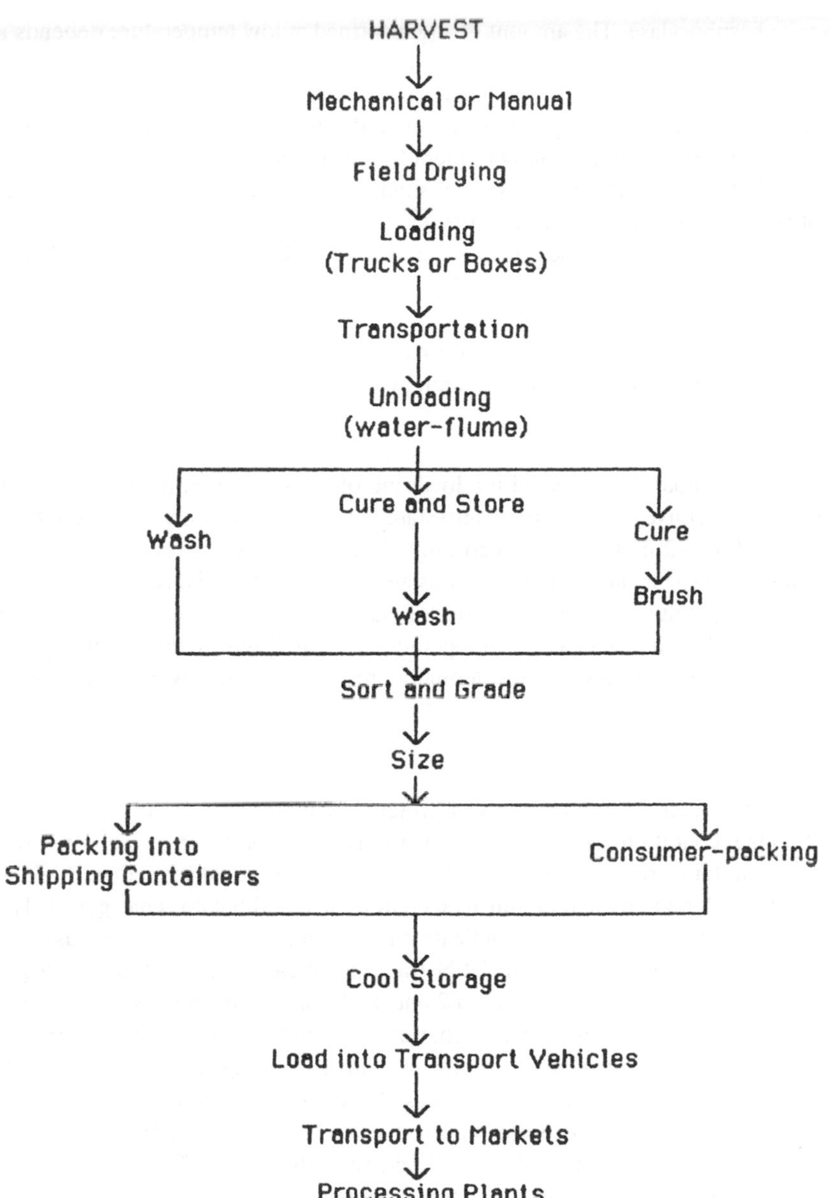

FIGURE 5 Harvest and postharvest operation for potato tubers. (From Ref. 4.)

A. Changes in Nutrient Composition During Storage

1. Sugars

It is established that the sugar content of potatoes increases when they are stored at comparatively low temperatures (132–134). Both sucrose and reducing sugars accumulate in different proportions at different temperatures. This has been attributed to the conversion of starch into sugars by enzyme phosphorylase. The amount of sugar formed at low temperature depends upon cultivar, maturity of tubers, prestorage treatments, and storage temperature (135–143). A large increase in sugars, primarily reducing sugars, occurs in potatoes stored at 1.1–2.2°C. The increase in CO_2 content in the storage atmosphere reduces the accumulation of sugars at low-temperature storage. Samatous and Schwimmer (144) reported that when potatoes are stored in nitrogen at 0°C, there is complete suppression of sugar accumulation. Irradiation increases the sugar content of tubers stored at low temperatures.

A large portion of stored potatoes goes to industries making chips and French fries in developed countries. Hence, potatoes stored at low temperatures and having high amounts of sugars are reconditioned at temperatures of 15.0–26.0°C. The relative humidity (RH) is maintained between 75 and 90%. The conditioning of potatoes results in a decrease in phosphorylase activity and sugars are converted to starch by starch synthetase (143).

2. Starch

The starch content of potatoes decreases with a lowering of the storage temperature. The starch content may increase via conversion of sugars into starch at high temperatures (>10°C). Potatoes stored at 1.1–13.3°C for 2–3 months contained only about 70% of their original starch (143). The change of starch into sugars at low temperatures and subsequent partial resynthesis of starch from sugars at high temperatures affect the quality of starch and the texture of cooked potatoes. The degradation of starch into sugars at low temperatures is catalyzed by the enzyme phosphorylase, whereas the reverse process at high temperatures is catalyzed by the enzyme starch synthetase.

3. Proteins

It is known that potatoes contain low amounts of proteins, ranging from 1.5 to 2.5% on a fresh weight basis. Storage conditions, especially temperature, may have an effect on proteins. Potatoes stored at room temperature contain a higher amount of amide nitrogen than those stored at 0, 4.4, or 10°C (143). Although the total nitrogen content of stored potatoes has generally been reported to change very little, many reports indicate changes in individual nitrogenous constituents, but these have been studied mainly in NPN fraction. Mica (145,146) found that protein nitrogen decreased with length of storage at both 2 and 10°C, although changes in total nitrogen were small. The free amino acid content was higher at the end of storage. However, Habib and Brown (147) reported little or no change in free amino acid composition of four cultivars stored at about 4°C, but reconditioning at 23°C caused a marked decrease in total free amino acids and complete loss of arginine, histidine, and lysine. Fitzpatric et al. (148) found a general increase in all free amino acids after cold storage and reconditioning. They attributed this to metabolic degradation of protein occurring as tubers sprouted during later stages of reconditioning. Many reports (143,145,146) present data on decreases in NPN or free amino acid content during cold storage.

4. Vitamins

Loss in ascorbic acid during storage has been reported (147–152). Thomas et al. (150) reported ascorbic acid to be stable during and after irradiation. A loss of ascorbic acid during cold storage has been reported (151–152). Losses have been found to take place most rapidly during the early period of storage. Linneman et al. (133) studied the effects of high storage temperatures on the ascorbic acid content of potatoes. Losses from potatoes in traditional stores in developing countries are likely to be lower than those during low-temperature storage. Losses of B vitamins, such as folic acid, have been reported during storage (151). However, an increase in pyridoxine in potatoes during storage has been demonstrated (154). Augustin et al. (151) pointed out that it is not known whether this increase is due to synthesis of the vitamin or its release from a bound form during the early stages of storage. Addo and Augustin (155) provided further evidence to support the synthesis of vitamin B_6 during storage. Barker and Mapson (156) stored potatoes in nitrogen and found that the content of ascorbic acid was almost stabilized by the exclusion of oxygen. Storage of potatoes at low temperature has little effect on thiamine and riboflavin contents.

5. Others

Storage of potatoes at room temperature increases fatty acid content (157), followed by a marked decrease when storage is extended. Low-temperature storage results in an increase in malic acid and citric acid in tubers. No significant changes are reported in the mineral content of potatoes during storage. Yamaguchi et al. (158) observed no significant changes in the contents of calcium, iron, or phosphorus in white potatoes held at 5 or 10°C for 30 weeks.

B. Effects of Storage on Processing Quality of Potatoes

Mazza et al. (159) attempted to relate changes in sucrose, reducing sugars, ascorbic acid, protein, and nonprotein nitrogen contents with the processing quality (chip and French fry color) of Alberta (commercially grown and stored), Kennebec and Norchip (both potato chippers), and Russet Burbank (a French fry cultivar) potatoes during growth and long-term storage. Correlation analysis of chip color, dry matter, sucrose, reducing sugars, ascorbic acid, protein, and storage temperature data showed that while dry matter, reducing sugars, and sucrose were significant in determining chip color of freshly harvested potatoes, reducing sugars, tuber temperature, and sucrose were important in determining the chip color of stored tubers. The relative importance of each parameter varied with the cultivar and the age of the potato tubers.

The reducing sugar content of potatoes, in particular, and processing quality, in general, are markedly affected by cultivar and environmental conditions during both the growing season and storage. Mazza (160) described the effect of several cultural and environmental factors on potato processing quality, including date of planting, soil type, soil reaction (pH), soil moisture, season, location, mineral nutrition, cultivation and weed control, control of disease and insect pests, temperature during growing season, time and method of killing vines, time of harvest, degree of bruising and other mechanical injuries, temperature of tubers going into storage, storage temperature, relative humidity and ventilation, length of storage, and cultivar.

Optimum conditions for controlling the color of chips or French fries also require stringent control of weight loss of potatoes during storage. Cargill et al. (161) reported on the influence of RH on weight loss during 300 days of storage of Kennebec potatoes held at 4.4°C and 95% RH or 4.4°C and 80% RH. The weight loss at 80% RH was more than double that at 95% RH. A difference in weight loss during a 200-day storage period between 80 and 95% RH represented a 10% loss, corresponding to a monetary loss of about $16,000 to the grower. A significant correlation with chip color was obtained for dry matter, reducing sugars, and sucrose levels for all potatoes tested (160).

C. Greening of Potatoes During Storage

Synthesis of chlorophyll in the peridermal layers of tubers exposed to light leads to "greening," which markedly reduces the product's acceptability. Sometimes the greened tubers taste bitter when cooked owing to the parallel synthesis of glycoalkaloids. Greening can occur in the retail outlets (162). Gull and Isenberg (163) demonstrated that light intensities as low as 3–11 W m^{-2}

for as short a period as 24 hours could induce greening. The development of the green color is influenced by variety (164), stage of maturity (165), and temperature (163). No greening was found at 5°C, and it was extensive at 20°C, the greater effect being observed in immature tubers (166). Lewis and Rowberry (167) noted that the major cause of greening in the field was insufficient cover over the tubers at planting. The effect of greening was aggravated in potato varieties whose tubers were formed near the surface. Clumping of stems resulted from the use of large seed planted widely spaced (168). Less severe competition due to wide spacing probably allowed more tubers to set, some of which were forced to the surface, exposing them to light and thus turning them green.

Greening in potatoes is often associated with formation of steroid alkloids, which can cause off-flavors on cooking at concentrations of 15–20 mg/100 g (169). This concentration is 5–10 times higher than that occurring in normal potatoes. Most of the glycoalkaloids are concentrated in the skin, and in prepared potatoes this is usually too low to cause any nutritional hazards or poisoning. Synthesis of glycoalkaloids is markedly influenced by variety (170), temperature of storage (171–175), and intensity of light (171,176–178). Zitnak and Johnston (169) recorded concentrations of glycoalkaloids as high as 35 mg/100 g in the Lenape cultivar, which was withdrawn from commercial cultivation for this reason.

Jadhav and Salunkhe (179) reviewed research done on the control of chlorophyll and glycoalkaloid formation, describing several physicochemical methods. Various measures that can be used include cultivar selection; packaging; treatment with wax, oil, soap and surfactants, and various chemicals; controlled and hypobaric storage; and ionizing radiation. Wu and Salunkhe (180) reported that hot paraffin wax effectively controlled the formation of chlorophyll and glycoalkaloids in potato (Fig. 6). The potatoes were treated with paraffin wax at 60, 80, 100, 120, 140, and 160°C for half a second and exposed to fluorescent light (200 fc) for 10 days at 16°C and 60% RH. The synthesis of chlorophyll and glycoalkaloids was not inhibited at 60 and 80°C; it was significantly inhibited at 100 and 120°C and was almost completely inhibited at 140 and 160°C. Heating of tubers at 160°C in air for 3–5 minutes and subsequent

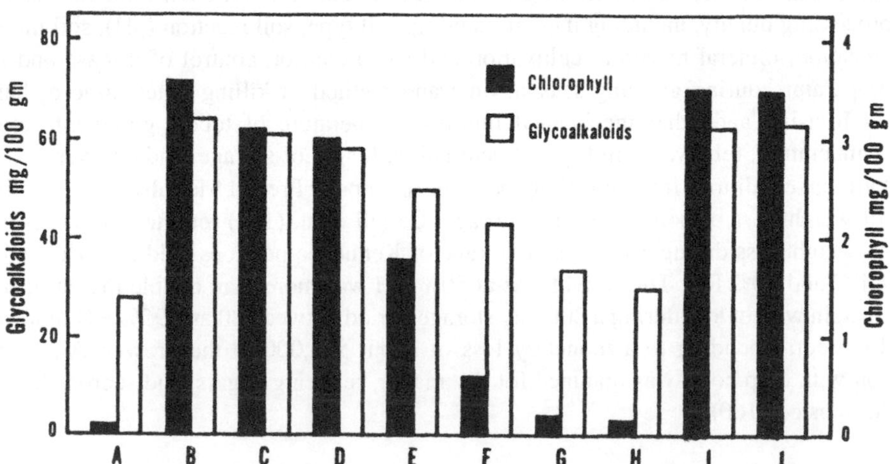

FIGURE 6 Effects of waxing and heating at different temperatures on chlorophyll and glycoalkaloid formation in peels (fresh) of Russet Burbank tubers after exposure to 2152 lx light intensity for 10 days at 16°C and 60% RH. (A) Original (zero time) sample; (B) control (nonwaxed); (C)–(H) waxing at 60, 100, 120, 140, and 160°C; (I) heating with air at 160°C for 3 min.; (J) heating with air at 160°C for 5 min. (From Ref. 180.)

exposure to light did not prevent formation of chlorophyll and glycoalkaloids. The combined treatment of waxing and heating effectively retarded the development of chlorophyll and alkaloids. Wu and Salunkhe (181) dipped potato tubers in corn oil at 20, 60, 100, and 160°C for one half second and removed excess oil with tissue paper. Oiling at 22°C reduced chlorophyll by 93–100% and reduced glycoalkaloid formation by 92–97%. At elevated temperatures (60, 100, 160°C), the synthesis of chlorophyll and alkaloid was completely inhibited (Fig. 7). These authors further noted that treatment of potatoes with corn oil, peanut oil, olive oil, vegetable oil, or mineral oil at 22°C was equally effective (181), but the tubers appeared oily and the possibility of the development of oxidative rancidity of oil was indicated. To decrease the amount of oil used, the corn oil was diluted with acetone. Treatment with one half, one fourth, and one eighth oil significantly and effectively inhibited the formation of chlorophyll and glycoalkaloids.

Jadhav and Salunkhe (182) reported that postharvest applications of Phosfon, Phosfon-S, Amchem 72-A42, Amchem 70-334, and Telone (250, 500, and 1000 ppm) significantly inhibited glycoalkaloid and chlorophyll formation. Amchem 72-A42 was the most effective chemical in preventing the synthesis of both glycoalkaloids and chlorophyll.

Forsyth and Eaves (183) evaluated the effect of CO_2 on the greening of Sebago potatoes in controlled atmosphere storage (Table 7). They noticed no immediate greening in an atmosphere of 100% N_2 or 75% CO_2, and the storage time was extended up to 1 week without obvious development of greening. Potatoes stored under 100% N_2, but not those stored under 75% CO_2, developed off-flavor due to formation of glycoalkaloids. It was concluded that 15% or higher concentrations of CO_2 prevented greening without affecting palatability. Patil (184) observed no significant effect of CO_2 on the formation of potato glycoalkaloids, while chlorophyll content was decreased by 33% of values for control tubers. Jadhav and Salunkhe (179) reported that potato tubers stored at 126 mmHg did not turn green. Treatments at 253, 380, 507, and 633

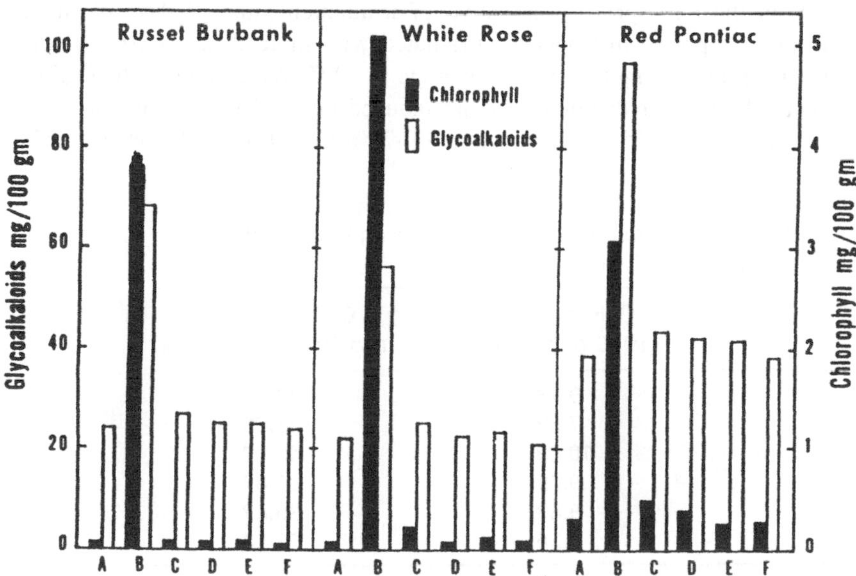

Figure 7 Effects of oil dipping on chlorophyll and glycoalkaloid formation in fresh peels of potato tubers after exposure to 2152 lx light intensity for 10 days at 16°C and 60% RH. (A) Original (zero time) sample; (B) control (nontreated); (C) to (F) oiling at 22, 60, 100, and 160°C. (From Ref. 180.)

TABLE 7 Effects of CO_2 on Greening in Controlled Atmosphere Storage

Present CO_2			
Mean	Range	Greening	Market grade
0.03	—	Severe	Below grade
7.5	4–10	Medium	Below grade
12.0	7–15	Slight	Below grade
14.9	9–22	Very slight	33.3% below[a]
20.3	12–26	None	Canada grade[a]

Note: Potatoes were of cultivar Sebago. Greening scores were based on eye observations after 48 hours of illumination (2260 lx).
[a]Graded by Canada Department of Agriculture Inspector 168 hours after commencement of experiment. Rejections are done to an excess of green coloration.
Source: Ref. 183.

mmHg pressure were ineffective in controlling formation of chlorophyll. Patil (184) found no differrences in glycoalkaloid levels among tubers subjected to any of the storage pressure treatments and the control (Table 8). Ziegler et al. (185) exposed tubers to 3012 lx of continuous illumination after irradiation for 12 days and found increasing inhibition of greening with increasing doses of irradiation from 0 to 400 krad (Table 9). The irradiation decreased greening irrespective of CO_2 treatment after 4 days of illumination, and the greening decreased with increasing CO_2 in the atmosphere irrespective of irradiation after 12 days of illumination. Madsen et al. (186) reported that potatoes irradiated with more than 1×10 krad of gamma radiation failed to sprout during 6 months of storage at 10°C. Wu and Salunkhe (187) reported inhibitory effects of gamma radiation on wound-induced glycoalkaloid formation in 13 potato cultivars: whereas a dose of 25 krad resulted in 11–79% inhibition in Russet Burbank, at 200

TABLE 8 Effects of Subatmospheric Pressure Storage and Light Intensity on Chlorophyll and Glycoalkaloids

Treatment pressure (mmHg)	Chlorophyll[a] (mg/100 g fresh peel)	Glycoalkaloid[b] (mg/100 g fresh peel)
Control	9.43	46.34
633	7.52	42.66
507	10.89	40.82
380	9.79	51.87
253	9.83	35.82
126	0.26	52.94

Note: Potato tubers (cultivar Russet Burbank) were exposed to 2260 lx light intensity for 15 days.
[a]Determined by AOAC method.
[b]Determined by the sulfuric acid-formaldehyde reagent method.
Source: Ref. 184.

TABLE 9 Effects of Gamma Irradiation on Chlorophyll Formation

Irradiation dose (krad)	Chlorophyll content[a] (% of control)		
	4 days	8 days	12 days
0	100	100	100
50	50	59	80
100	29	70	65
200	24	24	35
400	14	24	26

Note: Chlorophyll formation of potatoes (cultivar Kennebec) during 12-day illumination (3012 lx).
[a]Determined by Ziegler and Egle method.
Source: Ref. 185.

krad inhibition was 81–92%. Gamma radiation had no effect on existing alkaloids or on light-induced glycoalkaloid formation in tubers.

D. Sprouting of Potatoes During Storage

Potatoes exhibit sprouting when stored at temperatures of 10–20°C (188). Sprout growth is slow at temperatures of 5°C and less. Above 5°C, increasing temperature causes increased sprout growth up to about 20°C; at higher temperatures, sprout growth rate decreases. However, storage of potatoes at temperatures <10°C causes an increase in sugar content, resulting in increased browning of heat-processed products.

It has been shown that ascorbic acid content changes in tubers during sprouting (189,190). There is a decrease in ascorbic acid content during the early stages of sprout growth, followed by a temperature increase and by another decrease (189). Bantan et al. (190) found a higher ascorbic acid content in sprouts than in the rest of the tuber after 8 months of storage. However, conflicting reports are available on contents of other vitamins in tubers during sprouting (191). Yamaguchi et al. (174) found no changes in thiamine in White Rose potatoes even after 30 weeks of storage. Leichsenring (93) noted a slight increase in thiamine over 24 weeks of storage even on a dry weight basis, although the overall trend for one of the four varieties (Chippewa) studied showed a slight decrease.

Several chemical inhibitors have been found to inhibit sprouting in potatoes (192–194). The use of maleic hydrazide in the control of potato sprouting is well documented (195–202). Other chemicals used to control sprouting include chloroisopropyl carbamate (CIPC), isopropyl *N*-chlorophenyl carbamate (Chloroprophan, IPC) (203–207), tetrachloronitrobenzene (tecnazene) (208), alcohols (209), methyl ester of naphthalene acetic acid (MENA), and other inhibitors such as abscisic acid (212).

Irradiation is a very potent sprout inhibitor (213–220). Sparrow and Christenses (213) observed that certain irradiation doses gave excellent sprout control for as long as 15 months at 4.4°C. Dallyn and Sawyer (207) found that a 10 krad dose inhibited sprouting at storage temperatures up to 21.1°C. At a storage temperature of 4.4°C, sprouting was inhibited by a 5 krad dose. It has been suggested that irradiation might be economically advantageous in tropical countries, where it could be used in combination with cold storage at 10–15°C rather than conventional cold storage at 2–4°C.

E. Controlling Storage Losses

Potatoes, like other tuber crops, are subjected to postharvest losses owing to their continuing metabolism, damage sustained during harvesting and handling, rotting, shriveling, and sprouting (221–241). In the Dominican Republic, Mansfield (225) estimated that 7.5% of weight loss due to dehydration and pathogenic infection over a 1- to 2-day period with the total losses reaching about 30% in less than 15 days. In contrast, in the United States, with improved storage systems involving temperature and humidity control, the total loss (including weight loss) was less than 13% after 11 months of storage (226). Weight loss during storage due to dehydration of immature tubers is comparatively higher than for mature tubers (226). Postharvest losses of potatoes have been estimated to vary from 5 to 40%. According to an FAO report (227), losses of potatoes in cold storage amounted to about 8%, while on the farm losses occurred to the extent of 20–40%. Such losses can be due to internal blackening (230), chilling injury (229), heat injury (228), mechanical injury (232), greening (233,234), or postharvest diseases (235). Various methods have been suggested to control these losses (236–241).

1. Curing

One of the most simple and effective ways of reducing water loss and decay during postharvest storage of potatoes is curing. Injured or bruised surfaces are allowed to heal and form periderm. During the process of curing, wounds are healed by producing a new cork cambium, thus preventing infections by pathogenic organisms.

Some water loss takes place during curing. The removal of decaying products prior to curing and storage ensures a greater percentage of usable product after storage. Curing of potatoes effectively reduces postharvest losses during storage by providing protection against infections by pathogenic organisms. Successful curing of potatoes can be achieved by subjecting potatoes at 8–20°C and at 85% RH. Care should be taken to avoid the condensation of water on the tubers during curing. The optimum conditions for curing root tubers and bulb vegetables are given in Table 10.

2. Low-Temperature Storage

To prevent excessive shrinkage and rotting, potatoes are generally stored for the first week at a temperature of 10.0–15.6°C and RH of 85–95% to permit suberization and formation of wound periderm (240). Most of the shrinkage takes place during the first month of storage. It is therefore important to allow cuts and bruises to heal rapidly and then to lower the storage temperature. Following the curing period, the temperature is lowered to 3.3–4.4°C to prevent sprouting after

TABLE 10 Optimum Conditions for Curing Root, Tuber, and Bulb Vegetables

Commodity	Temperature (degrees)		Relative Humidity (%)	Duration (days)
	Celsius	Fahrenheit		
Potato	15–20	59–68	85–90	5–10
Sweet potato	30–32	85–90	85–90	4–7
Yam	32–40	90–104	90–100	1–4
Cassava	30–40	86–104	90–95	2–5
Onion and garlic	35–45	95–113	60–75	0.5–1

Source: Ref. 4.

the rest period. Most potato cultivars can be stored for 6 months or longer without sprouting if held at 4.4°C. Although potatoes freeze at about -1.7°C, temperature injuries may result from prolonged storage below -1.7°C. Storage at 4.4°C is not desirable for tubers to be used for chips or frozen French fries, which develop an undesirable dark brown color due to accumulated sugars. Temperatures of about 10.0–12.8°C and 90% RH are suitable conditions for storing potatoes to be used for manufacturing chips or French fries. The internal discoloration ("Mahogany browning") develops in all potato cultivars when stored for 20 weeks or longer at 0–1.1°C. Pantastico et al. (241) recommended storing potatoes at temperatures of 3.3–4.4°C at 85% RH for about 34 weeks.

3. Controlled Atmospheric Storage

According to Ryall and Lipton (229), controlled atmosphere (CA) storage is not advantageous for potatoes destined for table use. The concentrations of ≤ 5% oxygen inhibit periderm formation and wound healing, and oxygen levels of ≤ 1% cause off-flavors and greatly increase decay and surface mold and blackheart during 1 week at moderate temperatures of 15–20°C. At 5°C, the deleterious effects of low temperature were less pronounced or absent (242). High CO_2 (10% or more) enhanced decay at 4.4°C and aggravated the effects of low oxygen at higher temperatures (243,244). Oxygen at ≤10% increased sprouting of potatoes in storage but had no discernible effect on the subsequent field performance of the seed potatoes (245). Use of 12% CO_2 (8% or more) and low O_2 for 6 months resulted in the complete failure of seed potatoes. The combination of high CO_2 (8% or more), low O_2 (5% or less), and low storage temperature (O°C) had the most serious adverse effects (246). Although high CO_2 (15–20%) prevented greening of prepackaged potatoes at 10.0–15.6°C, the incidence of bacterial soft rot increased even at lower CO_2 (10%) and lower temperature (4.4°C). Yamaguchi et al. (247) reported that susceptibility of potatoes to black spot increased with even lower CO_2 concentrations (5%) and a very short exposure period of just once a day. Thus, CA storage of potatoes hardly appears to be beneficial.

4. Chemical Control of Losses

Several chemicals have been found to suppress sprouting effectively in potato and onion during storage at higher temperatures. Van Niekerk (248) demonstrated that effective suppression of sprouting and consequent reduction in weight loss could be obtained over an 8-month storage period in a thatched-roof structure with a temperature range of 10–20°C and 34–70% RH using a mixture of propham [isopropyl-*N*-phenylcarbamate (IPPC)] and chloropropham [3-chloroisopropyl-*N*-phenylcarbamate (CIPC)]. The most important sprout suppressant is CIPC. Others in commercial use are maleic hydrazide, nonylalcohol, MENA, and tecnazene [2,3,4,6-tetrachloronitrobenzene (TCNB)], although the weakest inhibitor has the advantage that it does not inhibit suberization in curing and may be used for seed potatoes (225). CIPC is probably the most widely used chemical sprout inhibitor for the storage of potatoes. It may be applied as a dust, water dip, vapor, or aerosol. Since CIPC interferes with periderm formation, it should be applied only after curing. In the United States the tolerance of CIPC in raw and processed potatoes is 50 μg/g. Kennedy and Smith (196) found that 3 lbs. of maleic hydrazide (MH) per acre used as a foliage spray prevented sprouting of potatoes stored at 10.0°C. MH is applied when the lowest leaves begin to turn yellow and die (238). The application of CIPC to potato tubers in storage prevented sprouting for several months (205). Salunkhe and Wu (249) reviewed the literature on chemical modifications in several fruits and vegetables, including the effects of MH and other growth inhibitors on the potato. Arteca (250) found that vacuum infiltration of 1 and 2% $CaCl_2$ reduced chlorophyll accumulation in Katahdin potatoes 50–60% and 70–80%, respectively, related to controls, and $CaCl_2$ at 3 or 4% induced internal breakdown. Maximum uptake of $CaCl_2$ was obtained by placing the tubers under a vacuum (-0.9 atm) for 30 minutes with a soak time of 15 minutes. The infiltration of calcium ions helped to maintain tuber qualtiy in light, allowing consumers maximum visibility of the product.

The effects of GA, IAA, and kinetin on ascorbic acid and the total sugar content of potatoes, associated with sprouting during storage, were investigated by Kumar et al. (251). The ascorbic acid content decreased and the total sugar content increased progressively during storage at 5–10°C, but the increase in ascorbic acid and total sugar content did not change when the tubers were transferred to room temperature storage (30 ± 2°C). Both ascorbic acid and sugar contents were higher in potatoes treated with GA, corresponding to earlier visible growth of sprouts, than in the IAA- and kinetin-treated tubers, which corresponded to delayed visible growth of sprouts.

5. Irradiation

Low-dose irradiation of potatoes produced no detrimental effects on potato flavor (252). Panalaskas and Pellefier (253) found that specified levels of gamma radiation did not cause consistent variations in the ascorbic acid content of potatoes. Mikaelsen and Roer (254) reported that the vitamin C content decreased in both the irradiated and the nonirradiated potatoes during the first 7 months of storage at 5°C but was restored after this period, and that the ascorbic acid levels of the irradiated potato samples were higher than those under control. Ogata et al. (255) and Tatsumi et al. (256) reported the results of their experiments on the mechanism of browning. These investigators (256) showed that the O-diphenol content increased in irradiated potato tubers (Table 11) and the rate of increase was greater in the cortex and vascular bundles than in the pith. The ascorbic acid content decreased with increasing levels of irradiation dose, the rate of decrease being greater in the cortex and vascular bundles than in the pith. Sparrow and Christensen (213) also demonstrated that the storage life of potatoes could be extended by the

TABLE 11 Ascorbic Acid and *O*-Diphenol Contents after Irradiation

Doses (krad)	Irradiated 1 day after harvest[a]			Irradiated 3 months after harvest[a]		
	A	B	C	A	B	C
			Ascorbic acid			
0	18.2	18.5	19.3	9.3	9.8	7.6
10	11.3	12.5	15.5	8.9	8.3	6.7
20	15.4	11.0	11.7	7.2	7.1	7.3
40	9.9	10.4	10.1	7.6	7.6	7.3
			O-Diphenol			
0	3.2	3.4	2.0	5.7	3.6	1.2
10	8.4	7.2	4.4	6.0	5.7	0.6
20	10.8	7.6	4.8	5.4	6.6	0.9
40	11.7	8.4	4.6	3.9	6.0	0.0

A, cortex; B, vascular bundle; C, pith.
[a]mg/100 g fresh weight of potato tubers.
Source: Ref. 256.

use of ionizing radiation. Their observation on the inhibition of sprouting of potatoes by gamma-irradiation has been confirmed by other investigators (257–260).

Palmer et al. (261,262) studied the wholesomeness of irradiated potatoes. They investigated the effects of an irradiated potato diet on reproduction and longevity in rats. No significant effects attributed to the feeding of irradiated potatoes were observed on growth, food consumption, fertility, or reproductive performance. The values for litter data, fecal losses, and the incidence of malformations were comparable in all groups. No evidence of adverse effect on spermatogenesis (dominant lethal assay) or on the incidence of chromosomal aberrations was noted. According to Wills et al. (263), irradiation of potato and onion is more expensive than treatment with chemical sprout inhibitors like CIPC and MH.

F. Systems of Potato Storage

Ideal potato storage will ensure minimum losses in quantity and preservation of quality for use, processing, or propagation. Storage involves considerations of climate and weather, design of the storage equipment, control of the environment in the store, economics, and related factors (259). Research carried out at the International Potato Center (CIP) has shown that diffused-light storage (Fig. 8), as compared to conventional seed tuber storage in the dark, improves seed quality so that potato yields are increased by about 18%.

Factors that affect the inherent storage and market life of potatoes at the time of harvest include variations in potential length of dormancy, cultural practices, selection of harvest maturity in relation to dry matter content, presence or absence of disease, and method of harvesting (255). Booth and Proctor (264) reviewed storage problems and handling practices of potatoes in the tropics. Various methods were suggested for potato storage (265–267). Von Glerke (268) outlined different types of storage and summarized conditions for satisfactory storage of potato tubers, including adequate maturity of crop; healthy, dry, and undamaged tubers; and a slow lowering of temperature and careful control of the storage environment. The improved postharvest handling and storage of potatoes increased final tuber yields by about 20–30% (269). Researchers at the International Potato Center, Lima, Peru, are investigating simple stores with ambient air ventilation rather than complex storage with refrigerated air. A successful use of outside air requires that it be cooler than the temperature required for 6 hours daily. An adequate and timely curing is important to heal wounds caused during harvest and handling and can be achieved with controlled ventilation at about 13–15°C and high RH (85%). An ideal storage structure invented at the International Potato Center, Lima, Peru, is shown in Figure 9.

VII. PROCESSING

Potatoes undergoing various types of processing must have certain characteristics. This has resulted in the development of potato varieties grown solely for the purpose of processing, e.g., Dutch variety Record, having yellow flesh, which is used to manufacture crisps (chips) in the United Kingdom. The biochemical composition of the potato tuber, especially the dry matter (DM) content, carbohydrate (reducing and nonreducing sugars), discoloration of raw flesh, textural characteristics of the tuber, browning, and after-cooking blackening, constitute the important quality parameters for processing of potatoes (270–272).

The quality of processed product is significantly influenced by texture, color, and taste (270). The texture depends upon the size of the starch granules (273–283), the starch and pectin contents (277–280, 284–292), environmental conditions (293–302), and storage conditions (303–305). It is also influenced by starch size (308–310) and cell wall composition (313–318).

Figure 8 Diffused light storage of potatoes. (Courtesy of CIP, Lima, Peru.)

Figure 9 Improved Potato Storage in Developed Countries (Courtesy of CIP, Lima, Peru, From CIP Annual Report, 1981, IPC, Lima, Peru, 1981.)

The color of processed products depends upon enzymatic browning during the cutting and peeling of potatoes (319–324) and nonenzymatic browning (328–336) and blackening (337–339) during processing.

A. Commercial Processing

During the past three decades, the proportion of potato crop that is processed for domestic consumption has increased considerably (4). Processed potatoes were used for the following products: chips (crisps), 42%; frozen French fries, 36%; dehydrated products, 4%; miscellaneous uses, 4%; canned new potatoes, 2% (273).

1. Prepeeled Potato

Prepeeled potatoes are preserved from discoloration and are cold-stored (340). In developing countries, the production of prepeeled potatoes has increased during the last decade. They are extremely perishable, have a relatively short shelf life, and are used in restaurants, canteens, and retail establishments. These potatoes may be left whole or cut into strips for fresh frying.

Potatoes used in this way are washed to remove dirt and they are peeled by various methods. These include abrasion peeling, lye peeling, steam peeling, or a combination of lye and steam peeling and infrared peeling (341–344). The potatoes are washed to remove any peel left and are trimmed to remove eyes, residual peel, and damaged, diseased, or green areas. The prepeeled potatoes are immersed in sulfite solution for a few minutes to prevent enzymatic browning reactions. They are then drained, packaged, and refrigerated. Losses in nutrients depend upon the method of peeling (345–357). The losses also vary according to the size and shape of the tuber, the depth of the eyes, and the length of storage. Weight loss varies from 5 to 24%, with an average weight loss of 8–12%. Damage in the form of penetrating cracks followed by superficial crushing and bruising results in significant weight loss.

2. Potato Chips

Several factors influence the yield and quality of chips prepared from potatoes (353–357). The major problem associated with the chip industry is the maintenance of the desired color of the product (358,359). Chip color is the result of the browning reaction between sugars and other constituents such as amino acids. Potatoes stored at 20°C for 3 weeks yielded darker chips than those held at room temperature due to the conversion of starch to sugars by phosphorylase enzymes. The process for making potato chips is shown in Figure 10. Several treatments such as (a) a hot solution of sodium bisulfite or (b) a combination of sodium bisulfite, citric acid, and phosphoric acid, dilute HCl, hot water, HCl and phosphoric acid solution (360–367) and glucose oxidase (363–365) have been recommended to improve the quality of chips. Potatoes with high reducing sugars produce darker chips when fried in oil. These potato slices are partially fried and then subjected to infrared heat or microwave heat to remove excess moisture. Temperatures above 121°C induce browning, and the degree of darkening increases with time. Hence the time and temperature of frying are critical factors influencing the color of the finished product (366,367).

Several oils can be used for frying sliced potatoes, including soybean oil, palm oil, and safflower or groundnut oil. The amount of oil absorbed by the chips is influenced by the amylose-amylopectin ratio. The amount and quality of oil present in the chip influence the shelf life of chips. Other factors such as exposure to air or light, high-temperature contamination with metals, and poor packaging also influence the keeping quality of chips (368–372). Antioxidants can be added to oil or to the salt applied to chips to delay rancidity and extend shelf life. A number of antioxidants such as BHA and BHT have been found useful in the chip industry. The

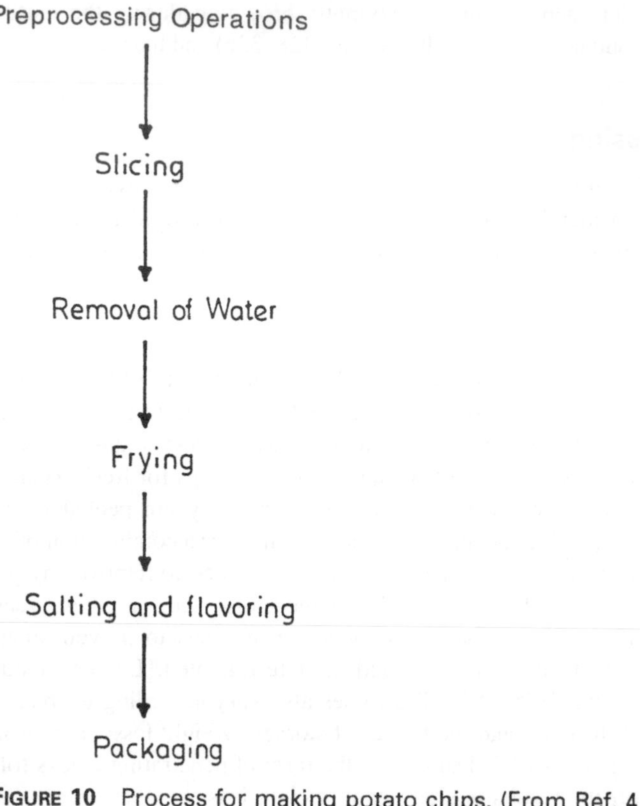

FIGURE 10 Process for making potato chips. (From Ref. 4.)

texture of chips can be improved by adding sodium acid pyrophosphate to the frying oil, which increases crispness. However, it produces an acidic flavor. Details of the various steps involved in chip making from potatoes are outlined by Smith (367). Several reports (373–380) are available on changes in nutrient composition when chips are prepared from raw potatoes. Chip making results in losses of amino acids and ascorbic acid.

3. French Fries

French fries are important products of potato-processing industries. A large percentage of frozen fries are served in restaurants and institutions (381,382). These are prepared for serving by finish frying in deep fat. As for potato chips, the reducing sugar content should be low to avoid dark fried pieces. Sodium acid pyrophosphate has been reported to prevent darkening in French fries. Weaver et al. (382) reported that a low concentration of ammonia can cause frozen potato products to turn from a light color to gray or black. Products that are discolored due to ammonia may be decolorized by using a citric acid dip while they are frozen. The addition of sucrose prevents the acid taste in treated potato products.

French fries are prepared from good-quality potatoes. The process of preparation of French fries includes washing and peeling, trimming, sorting and cutting, blanching, frying, defatting, cooling, freezing, and packaging (Fig. 11). Peeling can be done with lye or steam or infrared treatment (383–389). Blanching prior to frying results in a more uniform color of the fried product, reduced fat absorption by gelatinizing the surface layer of starch, reduced frying time, and improved texture of the fried product (381). Sodium acid pyrophosphate or calcium lactate

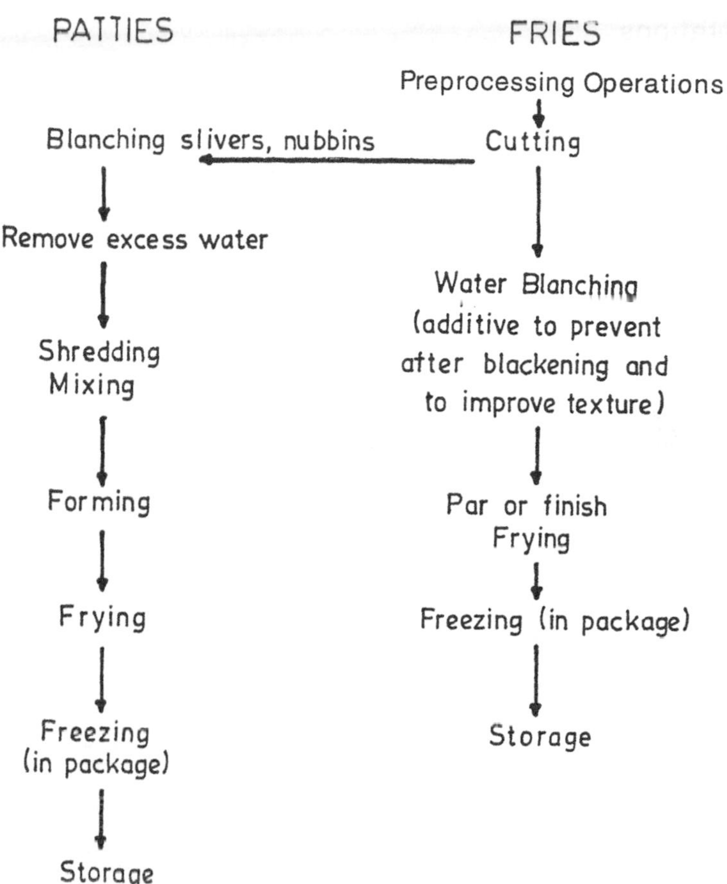

FIGURE 11 Process for making frozen patties and French fries. (From Ref. 4.)

can be used during blanching to improve texture. The blanched produce is fried in oil at 177–187°C. Excess fat is removed from the fried product and the product is air cooled. Difluorodichloromethane (CF_2Cl_2) can be used to remove excess oil on the surface of the product. Freezing of the product can be performed before or after packaging.

4. Other Frozen Products

Products other than French fries prepared from potatoes include puffed, hashed-brown, scalloped, mashed, and sliced potatoes. These are made from the slivers and nubbins left from the French fry line and from chopped or sliced small potatoes (340).

5. Canned Products

Potatoes are canned in different forms including whole, diced, sliced, strips, and julienne. A good canning potato should not slough or disintegrate during processing. Potatoes of low specific gravity (<1.075) are preferred for canning. Sloughing of the canned product can be prevented by adding calcium chloride, which tends to firm the tissue. The canning process outlined by Smith and Davis (340) is given in Figure 12. They (340) also described the preparation of the following canned potato products: potato pancakes, potato soup, canned French fried potatoes,

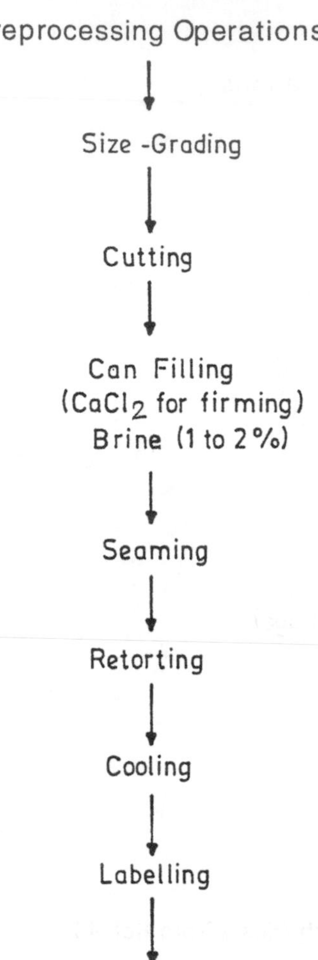

FIGURE 12 Process involved in canning of potatoes. (From Ref. 4.)

and shoestring potatoes. For canned corned beef hash and beef stew, raw chopped or diced potatoes or reconstituted dehydrated diced potatoes are used (390,391).

6. Fabricated French Fries and Chips

Preformed French fries (Fig. 13) are a relatively new product, sold in the United States and The Netherlands, which consists of a mixture of dehydrated potatoes and other ingredients that can be reconstituted rapidly in cold water to form a doughlike material. The dough is extruded in square cross sections and cut at the desired length while being extruded. The simulated French fry strips are deep-fat fried, resulting in a product very uniform in color, shape, form, and texture (340).

Fabricated chips (Fig. 14) are also available. Markakis et al. (391) described the process of making potato chip–like products. Dough prepared from dried potatoes, gelatinized maize, gluten, and oil is shaped into discs, dried to 12% moisture, and deep-fat fried. Various products of this type have been commercially prepared (392–394). Such products exhibit homogeneity or uniformity. However, the flavor of the fried product does not resemble that prepared from raw

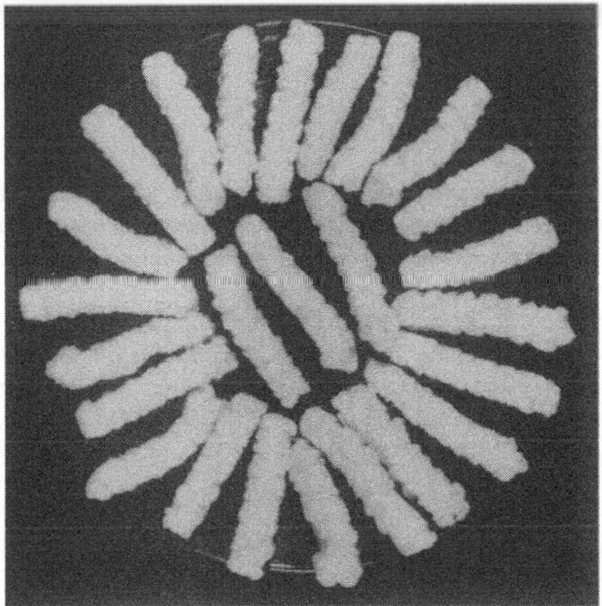

FIGURE 13 Preformed French fries. (Courtesy of Dr. B. N. Wankier, Ore-Ida Potato Processor, Ontario, OR.)

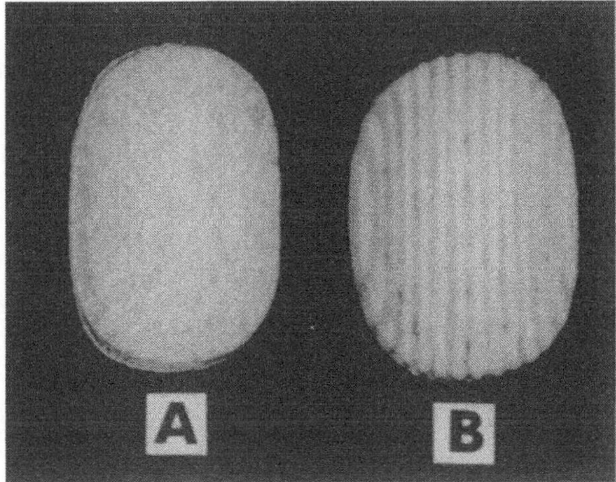

FIGURE 14 Preformed chips (A) Regular plain potato chips; (B) preformed (wrinkled potato chips). (Courtesy of Dr. B. N. Wankier, Ore-Ida Potato Processor, Ontario, OR.)

potatoes. Several methods have been suggested to improve the flavor (395), texture (396), crispness (397), and shelf life of such fabricated products.

7. Dehydrated Products

Potato Granules

Potato granules, one of the important products prepared from dehydrated mashed potatoes, contain 6–7% moisture. They are reconstituted to a texture that is either dry and mealy or moist and creamy, according to individual preference. Many reviews (399–401) outline the process of production of granules from potatoes.

Even though several processes have been developed for the production of potato granules, the add-back process is the one commercially employed in the United States (Fig. 15). In this process, cooked potatoes are partially dried by adding back previously dried granules to give a moist mix, which, after holding, can be satisfactorily granulated to a fine powder (402). A method

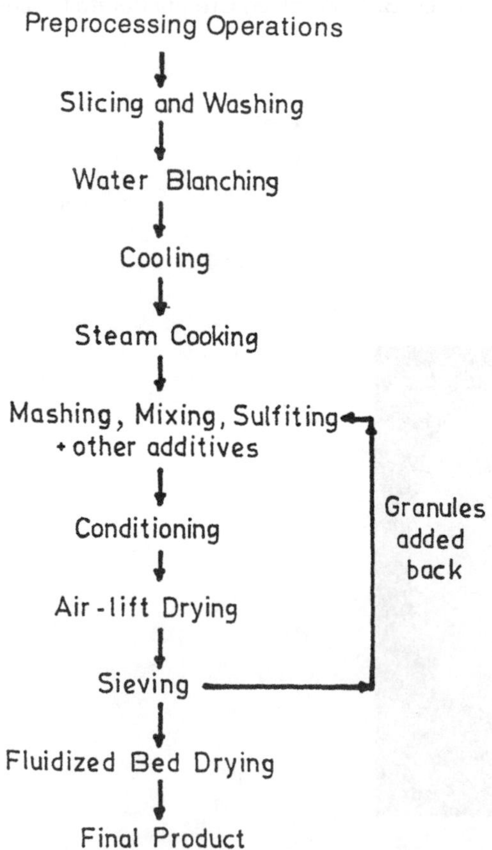

FIGURE 15 Process for producing dehydrated granules. (From Ref. 4.)

for the production of dehydrated potato granules by a straight-through freeze-thaw process without add-back of a large proportion of the product has been described (403). These granules have been found superior in nutritional quality (404) as well as in flavor, color, and texture (403). The freeze-thaw granules exhibited higher water-binding capacity, lower bulk density, and larger particle size than those from the add-back process (405).

When preparing granules, it is necessary to minimize rupturing of the cells. Excess rupture of cells releases starch and causes products to become unduly sticky or pasty. Granulation is significantly improved by conditioning of the moist mix. This includes adjustment of temperature and moisture. Potter (406) reported a decrease in soluble starch during conditioning. Changes in the physical properties of starch during conditioning play an important role in the preparation of potato granules.

Nonenzymatic browning is a common problem with dehydrated products of potatoes. Hence, using potatoes with low sugar content, sulfiting, drying the product to low moisture content, and avoiding high storage temperature are common methods for controlling nonenzymatic browning. Oxidative deterioration is markedly influenced by storage temperature. Fats present in potatoes contribute to oxidative deterioration. Packing in nitrogen greatly retards oxidative deterioration but has little effect on the rate of browning. Although lowering the moisture content retards nonenzymatic browning, it accelerates oxidative deterioration (407). The use of antioxidants such as BHA and BHQ has been found to be helpful in retarding oxidation in potato granules.

Potato Flakes

Dehydrated potato flakes are prepared by applying cooked mashed potatoes to the surface of a drum drier fitted with applicator rolls, drying the deposited layer of potato solids rapidly to the desired final moisture content, and breaking the sheet of dehydrated potato solids into a suitable size for packing (408). During this process (Fig. 16), cells are ruptured to a considerable extent. However, the reconstituted product is acceptably mealy because of the precooking and cooling treatments to which the potatoes are subjected during processing and the addition of an emulsifier. Since flakes are dried quickly, the potato cells are easily rehydrated and the potato starch retains its high absorption power. The flakes can be reconstituted with cold water.

Potato flakes rehydrate rapidly with boiling water, resulting in excessive cell rupture and a pasty texture. When potato flakes are broken into smaller particles for economical packaging, a certain amount of cell rupture occurs along the edges of the flakes. The gelatinized starch released from these cells would result in pasty or rubbery texture if it had not been subjected to retrogradation in the precooking and cooling steps. An emulsifier added to the potatoes reacts with the released amylose molecules, forming a starch emulsifier complex with less solubility and reduced stickiness. Potato flakes made using the precooking/cooling process and containing added monoglyceride emulsifier are more mealy because the water used for reconstitution is not strongly held by the intercellular material. As a result, more water penetrates the cell wall of intact cells, expanding them and thus creating more mealy, less cohesive mashed potatoes. If more of the reconstitution liquid is held between intact cells, the texture of the mash will be pasty or rubbery or gummy.

Many additives are employed in making potato flakes in order to improve texture and extend the shelf life of the product. These include sodium bisulfite (to retard nonenzymatic browning), monoglyceride emulsifier, antioxidants, and chelating agents (408). Dehydrated potato flakes can be fortified at the mash stage prior to dehydration (409). However, ascorbate reacts with potato protein, causing a pink Schiff base compound. Pinking occurs erratically and takes time to appear after dehydration. Flake fortification can be accomplished by mixing potato

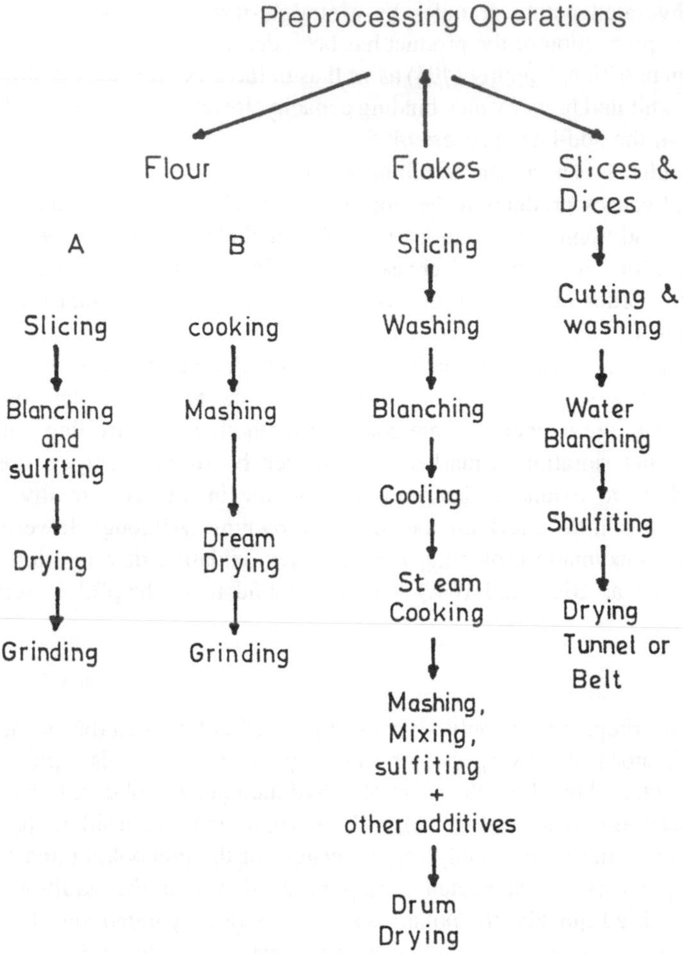

FIGURE 16 Processes for producing dehydrated and flour products. (From Ref. 4.)

flakes with vitamined flakes of approximately the same size and shape. The vitamin flakes are actually 50–75% fat, containing water-soluble vitamins and minerals.

The shelf life of potato flakes depends upon the chemical composition, the variety, the extent of cooking, the drying conditions, the amount of water used in processing, and the residual level of antioxidants (especially SO_2) used to prevent off-flavor development. Bitterness is related to phenolic compounds. The off-flavors in stored flakes are associated with hexanal and other aldehydes produced by oxidation of the natural fat of potatoes and branched aldehydes such as 2- and 3-methyl butanal, resulting from reactions of amino acids. Sapers (410) reported that the haylike off-flavor found in potato flakes results from oxidation rather than from nonenzymatic browning. It is reported that flakes treated with emulsifier (0.66%), BHA (150 ppm), BHT (150 ppm), and SO_2 (40 ppm) had the best after-storage quality.

Augustin et al. (411) reported that the retention of thiamine in flake processing is higher than that in granule processing even though sulfites are used. Ascorbic acid content decreases gradually when flakes are placed in storage. They further reported nutrient losses during preparation of potato flakes and holding of the reconstituted product (412). Retention of vitamin C and folic acid was 63 and 56%, respectively, after 1 hour on the steam table (412).

Attempts have been made to fortify potato flakes with iron (413) and proteins. However, fortification with seven iron compounds resulted in after-cooking darkening and objectionable off-flavor during storage. Potato flakes have a problem of reconstitution with boiling water. Hence, these cannot be served as a hot product, and they cannot be mixed with milk. Potato flakes can be used to form French fries for restaurant use using a dry mix combined with other dry binding agents. Flakes can also be ground to produce flour used as an ingredient in soups, baby foods, and baked goods.

Diced Potato

Diced potatoes prepared from whole potatoes by slicing, followed by blanching and dehydration, are used in canned meat products. One of the most important problems associated with the preparation of diced potatoes is graying or darkening. This occurs in dry or reconstituted products. The presence of tyrosine has been implicated in discoloration. The enzyme tyrosinase catalyzes the conversion of tyrosine into quinones, resulting in the development of brown pigment. Iron salt is known to increase discoloration. Sodium pyrophosphate treatment can prevent after-cooking darkening in dehydrated potatoes. Certain commercial cultivars exhibit this phenomenon, whereas others do not. The color of the product either in dry form or reconstituted after processing or storage is influenced by a nonenzymatic browning reaction. During dehydration and/or storage, amino acids react with sugars. In storage, dehydrated diced potatoes tend to develop a reddish-brown discoloration, accompanied by the production of CO_2 and bitter off-flavors.

Potatoes with high total solid content tend to slough excessively on reconstitution unless processing procedures are adjusted to suit these conditions. Potatoes with low solid content tend to be firmer and do not lend themselves well to end-use requirements where mealy characteristics are desirable. Other factors such as insect damage and pathological or physiological disorder influence the quality of diced potatoes.

Potato Flour

Potato flour, commonly used in the baking industry, is prepared (Fig. 16) by dehydration of peeled, cooked potatoes on a single drum drier equipped with applicator rolls. The thin, dried sheet of potato solids is then ground to the desired fineness. The single drum drier is one of the most efficient means of dehydrating potatoes. By spreading the mash into a thin sheet, extremely rapid evaporation of water can be achieved. Potato flours are available in two forms: granular and fine flours. Granular potato flour is normally used in operations where freedom from lumping is desired.

In general, flour is used in bread to the extent of 6%. At 2–3%, it helps materially to preserve freshness due to increased water absorption afforded by potato flour. Potato flour is used in cookies, doughnuts, cakes and cake mixes, and snack foods (415–421). The use of potato flour in such products results in a substantial increase in quality when added up to 5%. Potato flour is used as an ingredient in many dehydrated soup mixes. Bushway et al. (422) showed that 1.5% potato flour and 1.5% potato starch can be used in the manufacture of frankfurters to increase the tenderness and juiciness of the frankfurters.

8. Miscellaneous Products

Potatoes serve as a raw material for several other food and industrial products. The industrial waste from potato processing may be utilized for cattle feed or for isolation of some chemicals.

Potatoes have been used for the production of several miscellaneous food products such as canned potato salad, canned beef hash, canned beef stew, canned French fries, pancakes, soups, chip bars, chip confections, potato nuts, potato puffs, sponge dehydrated potatoes, and potato snack items (423–427).

B. Traditional/Domestic Processing

Potatoes are a staple food in many countries, including Rwanda, Nepal, Tibet, China, Peru, Bolivia, Ecuador, and many countries in Eastern Europe. In addition, potatoes serve as a major side dish to principal staples. This pattern is evident in Central America, where corn, tortillas, and beans form the mainstay of the diet. Potatoes are also consumed as a vegetable side dish on a regular basis, but in comparatively smaller quantities and not every day (e.g., in India, Pakistan, and Bangladesh). In some countries, like Indonesia, Malaysia, and the Philippines, potatoes are consumed as a small item among many other dishes prepared to complement the basic staple foods.

Potatoes are generally cooked before consumption. There are various methods of cooking potatoes. These include boiling with or without peels, baking, frying, and microwave cooking. Similarly, potatoes are boiled with or without the peel and then macerated into potato mash, which is used for the preparation of various traditional products. These products are fried before consumption. This section summarizes in brief various traditional methods of processing potatoes.

1. Peeling

Potato peeling is one of the important traditional processes commonly employed before direct consumption of boiled potatoes or processing of peeled potatoes into various products. The method of peeling significantly influences the quantity of loss through peeling. The average weight loss from peeling has been reported to range from 5 to 24% (428,429). Augustin et al. (430) found that peel fractions of whole tubers when the peel was removed with a domestic peeling knife either before or after cooking were 6, 2, and 10% for peeled raw potatoes, potatoes peeled after boiling, and potatoes peeled after baking, respectively. It was suggested that potatoes should be peeled after and not before boiling when cooked domestically to reduce losses.

Peeling results in the loss of some minerals and trace elements (431–436), crude fibers (430), riboflavin (430), and amino acids (435). The loss of nutrients through peeling removed from boiled potatoes, expressed as a percentage of total tuber nutrients, was lower in all cases than that from raw potatoes. The weight loss varies from 5 to 24%, with an average weight loss of 8–12%. Damage in the form of penetrating cracks followed by superficial crushing and bruising results in a significant loss of weight.

2. Methods of Cooking

Boiling

Boiling is the most common method of preparing potatoes domestically worldwide. In developing countries, potatoes are usually boiled intact by the consumer, whereas in developed countries potatoes are peeled and cooked. Peeling of potatoes before boiling results in a higher loss of nutrients (437,438). It has been shown that cooking potatoes by boiling increases the digestibility of the starch (439). However, the presence of starch that is resistant to enzymatic hydrolysis in vitro has been reported (440,441). Cooking results in the loss of nitrogen. Peeled potatoes when cooked exhibited more nitrogen loss than unpeeled potatoes (442). Jaswal (443) observed negligible losses of free amino acids. These losses were attributed to carbohydrate–amino acid

TABLE 12 Nutrient Losses (%) in Boiled and Steamed Potatoes

	Cooking Method	
Constituent	Boiled	Steamed
Dry matter	9.4	4.0
Protein	—	—
Calcium	16.8	9.6
Magnesium	18.8	14.0
Phosphorus	18.3	14.0
Iron	18.3	11.7

Source: Refs. 448, 451.

interactions. Changes in dietary fiber content due to boiling of potatoes have been reported by several workers (444, 445). Significant losses in ascorbic acid have been reported due to boiling of peeled or unpeeled potatoes (446). Changes in B group vitamin contents vary according to boiling conditions (447). Losses of minerals during boiling, except in the case of iron, are probably smaller than losses due to careless peeling before cooking (448). Nutrient losses in boiled potatoes were less than in steamed potatoes (Table 12).

Frying

Potatoes are generally boiled or blanched before frying. The boiled potatoes are peeled (or peeled potatoes are boiled) and then fried in oil or margarine. The fried products are used for consumption. As frying results in a decrease in moisture content, nutrients are concentrated. Hence, changes occurring due to frying are difficult to assess when nutrient contents are assessed on a fresh weight basis. Fenton (449) and Richardson et al. (450) observed ascorbic acid retentions of 67% in fried potatoes. Domah et al. (451) studied the effects of frying potatoes at 140°C for 10, 20, and 30 minutes and at 180°C for 5 minutes on the ascorbic acid content in potatoes (Table 13). The retention of total vitamin C was good and in fact better than that in boiled potatoes. However, ascorbic acid is oxidized to dehydroascorbic acid (DAA) more rapidly

TABLE 13 Effects of Frying on Stability of Ascorbic Acid (AA) and Dehydroascorbic Acid (DAA) in Potatoes

Sample	Dry matter	DAA (mg/100 g dry matter)	AA (mg/100 g dry matter)	Total content of vitamin C (mg/100 g dry matter)
Raw, peeled potatoes (before frying)	26.20	7.4	44.6	52.0
Fried potatoes				
140°C/10 min	83.01	29.7	20.6	50.3
140°C/20 min	84.00	33.7	7.3	41.0
140°C/30 min	88.00	42.7	0.0	42.7
180°C/5 min	89.10	42.8	0.0	42.8

Source: Ref. 451.

TABLE 14 Percentages of Weight Changes and Vitamin C Retained in Canadian Potatoes After Different Types of Cooking

Preparation	% Weight changes	% Vitamin C retained
Boiled in skin	94 ± 2	81 ± 10
Boiled pared before cooking	99 ± 2	73 ± 8
Boiled, mashed + milk	120 ± 2	77 ± 10
Boiled, mashed + margarine + milk	123 ± 2	72 ± 12
Boiled, browned	72 ± 2	28 ± 11
Fried	57 ± 5	72 ± 14
Baked	85 ± 2	78 ± 9
Raw, browned	68 ± 4	60 ± 13
Scalloped + cheese	123 ± 6	80 ± 15
Scalloped	113 ± 6	68 ± 10

Source: Ref. 454.

TABLE 15 Changes in Nutrient Composition of Potatoes Baked by Microwave and Conventional Means

		Retention (%)	
Nutrient	Tissue	Conventional (60 min/400° F)	Microwave (3 min)
Protein	Cortex	87	94
	Pith	87	84
Total amino acids	Cortex	97	102
	Pith	109	96
Free amino acids	Cortex	85	108
	Pith	108	83
Potassium	Cortex	85	101
	Pith	122	113
Phosphorus	Cortex	88	98
	Pith	110	100
Calcium	Cortex	98	102
	Pith	104	117
Magnesium	Cortex	90	104
	Pith	104	99
Manganese	Cortex	89	91
	Pith	104	109
Iron	Cortex	88	98
	Pith	122	90
Copper	Cortex	93	96
	Pith	93	105
Zinc	Cortex	91	100
	Pith	136	100

Source: Ref. 456.

by boiling than by frying, but hydrolysis of DAA is slowed by the dehydration of the product during frying, and therefore DAA accumulated in fried potatoes. During boiling, DAA is converted to 2,3-diketogulonic acid. Frying decreases mineral content significantly in both the cortical and the pith areas, with most of the loss occurring in the cortical areas (10–45%).

Baking

Baking of potatoes is one of the common methods of potato processing. Potatoes are generally baked in their skin. Several studies have been conducted to study the effects of baking on nutrient retention in potatoes (452). Kahn and Halliday (453) compared baked (in skin), parabaked, and French fried potatoes and observed ascorbic acid retention of 80, 80, and 77%, respectively. After standing approximately 1 hour, the retention rates were 41, 52, and 71%, respectively. Pelletier et al. (454) found that the cooking method used has a significant effect on the retention of vitamin C, with boiled potatoes retaining about 80% compared to only about 30% for hashed browns (Table 14). During the baking of potatoes, movement of minerals such as potassium, phosphorus, and iron toward the interior tissues has been demonstrated by Mondy and Ponnampalam (434), Baking increased the contents of potassium (14–23%), phosphorus (2–9%), and iron (2–8%) in the interior pith tissue.

Microwave Cooking

The use of microwave cooking has been on the rise, and it is estimated that by the year 2000, microwave ranges will be common in food-service units (455). In conventional cooking, heat is applied to the outside of food by convection, radiation, or conduction and then conducted to the interior of the food. Heat generated from within the food by a series of molecular vibrations is the basis for microwave cooking. The advantages of microwave cooking include higher energy efficiency, greater time savings, convenience, and easy clean-up. Disadvantages include greater cooking losses and less palatability (455).

A comparison between conventional and microwave baking of potatoes was made by Klein and Mondy (456). Conventional cooking primarily resulted in the migration of some nutrients from the cortex to the pith, whereas microwave cooking resulted in the loss of volatiles from the interior pith tissue (Table 15). It appears that baking potatoes conventionally is less nutritious with respect to nitrogenous and mineral constituents, especially when the skin and adhering cortical tissue are not consumed.

REFERENCES

1. Salunkhe, D. K., and B. B. Desai, *Postharvest Biotechnology of Vegetables,* FL, 1985, p. 238.
2. Hooker, W. J., Ed., *Compendium of Potato Diseases,* American Phytopathological Society, St. Paul, MN, 1986.
3. F.A.O. Production Yearbooks, Food and Agriculture Organization, Rome, 1993.
4. Salunkhe, D. K., S. J. Jadhav, and S. S. Kadam, *Potato: Production, Processing and Products,* CRC Press, Boca Raton, FL, 1991.
5. Kay, D. E., Root crops, in *Tropical Product Institute—Crop and Product Digest No. 2,* Tropical Products Institute, London, 1973, p. 100.
6. Urgent, D., The potato, *Science 170*:1161 (1970).
7. Pushkarnath, *Potato in India,* Indian Council of Agricultural Research, New Delhi, 1969.
8. Choudhary, B., *Vegetables,* National Book Trust, New Delhi, p. 214.
9. Yawalkar, K. S., Vegetable Crops of India, 2nd ed., Agri-Horticultural Publishing House, Nagpur, India.
10. Thompson, H. C., and W. C. Kelly, *The Potato in Vegetable Crops,* 5th ed., McGraw-Hill, New York, 1957, p. 372.

11. Tigchelaar, E. C., New vegetable varieties List xxi. The Garden Seed Research Committee. American Seed Trade Association. *HortScience 15*(5):565 (1980).
12. Pushkarnath, Potato varieties in India, *Exp. Agric. 6*:181 (1970).
13. Driver, C. M., and J. C. Hawkes, *Photoperiodism in the Potato,* Imperial Bureau Plant Breeding and Plant Genetics School of Agriculture, Cambridge, England, 1943.
14. Krug, H., On the photoperiodic response of potato varieties, *Eur. Potato J. 3*(47):107 (1960).
15. Bodlaender, K. B. A., Influence of temperature, radiation and photoperiod on development and yield. The growth of potato, Proc. 10th Easter School Agric. Sci., University of Nortingham, 1963, p. 190.
16. Bushnell, J., Sensitivity of the potato plant to soil aeration, *J. Agron.* 27:4 (1935).
17. Lindbergue, C., and T. Campos, Internal report, Centro National de Pesequisqde Hortalicas EMBRAPA, Brasilia, D.P., Brazil, 1981.
18. Smith, O., and L. B. Nash, Potato quality. I. Relation of fertilizers and rotational systems to specific gravity and cooking quality, *Am. Potato J. 17*:163 (1940).
19. Kishore, H., and M. D. Upadhya, Potato in India: present status and problems, in *Recent Technology in Potato Improvement and Production* (Nagaich, B. B., ed.), Central Potato Research Institute, Simla, 1977.
20. Bodganova, L. S., Effectiveness of using herbicides during the cultivation of potatoes, *Nauchn Tr. Sev-Zapadn. Nauchno-Issled. Inst. Sel'sk. Khoz. 20*:141 (1971).
21. Leszozynski, W., G. Lisinsk, G. Sobkowicz, and J. Rola, Effect of herbicides used in potato farming for the control of dicotyledon weeds on the chemical composition of potato tubers, *Zest. Nauk. Wyzszq. Szk. Roln. Wroclawiu, Roln. 29*:163 (1972).
22. Hooker, W. J., *Compendium of Potato Diseases,* American Phytopathological Society, St. Paul, MN, 1986, p. 125.
23. Howard, H. W. The production of new varieties, in *The Potato Crop: The Scientific Basis for Improvement* (P. M. Harris, ed.), Chapman and Hall, London, 1978, p. 607.
24. Howard, H. W., *Genetics of the Potato, Solanum tuberosum,* Logos Press Ltd., London, 1970.
25. Kennedy, J. S., M. F. Day, and V. F. Eastop, *A Consensus of Aphids as Vectors of Plant Viruses,* Commonwealth Institute on Entomology, London, 1962, p. 114.
26. Bottrell, D. G., *Integrated Pest Management,* Council on Environmental Quality, Washington, DC, 1979, p. 120.
27. Eastop, V. F., Worldwide importance of aphids as virus vector, in *Aphids as Virus Vectors* (F. Harris and K. Maramorosch, eds.), Academic Press, New York, 1974, p. 3.
28. Cisneros, F. H., Integrated pest control: New approaches to the priority components, in *Proc. Inst. Congress Research for the Potato in the year 2000,* International Potato Centre, Lima, Peru, 1983.
29. Shelton, A. M., J. A. Wyman, and A. J. Mayor, Effects of commonly used insecticide on potato tuberworm and its associate parasites and predators in potato, *J. Econ. Entomol. 74*:303 (1981).
30. Krishna Prasad, K. S., Nematode problems of potato, in *Plant Parasitic Nematodes of India—Problems and Progress* (G. Swarup and D. R. Dasgupta, eds.) IARI, New Delhi, 1986.
31. Jatala, P., and P. R. Rowe, Reaction of 62 tuber-bearing *Solanum* species to root-knot nematodes, Meloidogyne incognila acrita, *J. Nematol. 8*:290 (1983).
32. Spears, J. F., *The Golden Nematode Handbook: Survey Laboratory Control and Quarantine Procedures,* U.S. Department of Agriculture, Agricultural Handbook, 1981, p. 353.
33. Fawkner, L. R., and H. M. Darling, Pathological histology, hosts and culture of the potato rot nematode, *Phytopathology 51*:778 (1967).
34. Koen, H., Notes on the host range ecology and population dynamics of *Pratylenchus brachyrus, Nematologica 13*:118 (1967).
35. Robde, R. A., and W. R. Jenkins, Host range of a species of Trichodorus and its host-parasite relationships on tomato, *Phytopathology 47*:295 (1957).
36. Winslow, R. D., and R. J. Willis, Nematode diseases of potato, in *Economic Nematology* (J. H. Webster, ed.), Academic Press, New York, 1972, p. 17.
37. Miller, P. M., and A. Hawkins, Long term effects of pre-plant fumigation on potato field, *Am. Potato J. 46*:387 (1969).

38. Mosher, P. N., For better potato quality, *Maine Ext. Ser. Bull.*, 1959, p. 474.
39. Gray, D., and J. C. Hughes, Tuber qualtiy, in *The Potato Crop—The Scientific Basis for Improvement* (P. M. Harris, ed.), Chapman and Hall, London, 1978, p. 504.
40. Moorby, J., The physiology of growth and tuber yield, in *Potato Crop—The Scientific Basis for Improvement* (P. M. Harris, ed.), Chapman and Hall, London, 1978, p.153.
41. Burton, W. G., The potato, in *A Survey of Its History and of the Factors Influencing Its Yield, Nutritive Value, Quality and Storage* (H. Veenman and N. V. Zonen, eds.), European Association of Potato Research, Wageningen, The Netherlands, 1966.
42. Sanderson, J. B., and P. R. White, Effect of in-row spacing on potato tuber yield, sizes and numbers measured at 10 weekly sampling dates, *Am. Potato J.* 59:484 (1982).
43. Rioux, R., J. Gosselin, and H. Genereux, Effect of planting date on potatoes grown in short seasons, *Can. J. Plant Sci. 61*:417 (1981).
44. Howard, H. W., Factors influencing the quality of warehouse potatoes I. The genotype, *Potato Res. 17*:490 (1974).
45. Athey, V. D., *Quality of Horticultural Products,* Butterworths, London, 1975, p. 22.
46. Schoenemann, J. A., Grading, packing and marketing of potatoes in *Potatoes: Production, Storage, Processing* (O. Smith, ed.), AVI Pub. Co., Westport, CT, 1968, p. 359.
47. Self-Help Stabilization Programs with Use of Marketing Agreements and Orders, U.S. Department of Agriculture, Agric. Sta. Bull. No. 479, 1961.
48. United States Standards for Potatoes. U.S. Department of Agriculture, Agric. Market Serv. U.S. Govt. Printing Office, Washington, DC, 1958.
49. Miller, C. J., Price and quality of table stock potato, *Am. Potato J. 43*:22 (1968).
50. Perry, A. L., Potential profits for packaging Maine potatoes to the U.S. Fancy grade, *Maine Agric. Exp. Stn. Bull.* No. 545, 1956.
51. Kunkel, R., J. Gregory, and A. M. Binkley, Mechanical separation of potatoes into specific gravity groups show promise for potato chip industry, *Am. Potato J. 28*:690 (1951).
52. Potato Preferences Among Household Consumers, U.S. Dept. of Agriculture, 667, 1948.
53. Augustin, J., Variation in the nutritional composition of fresh potatoes, *J. Food Sci. 40*:1295 (1975).
54. McCay, C. M., J. B. McCay, and O. Smith, The nutritive value of potatoes, in *Potato Processing* (W. F. Talburt and O. Smith, eds.), AVI Pub., Westport, CT, 1975.
55. Mondy, N. I., R. L. Koch, and S. Chandra, Influence of nitrogen fertilization on potato discoloration in relation to chemical composition. I. Phenols and ascorbic acid, *J. Agric. Food Chem. 27*:418 (1979).
56. Watt, B. K., and A. L. Merill, *Composition of Foods: Raw Processed and Prepared,* Agriculture Handbook No. 8, U.S. Department of Agriculture, Washington, DC, 1964.
57. Howard, H. W., *Genetics of the Potato* (*Solanum tuberosum*), Logos Press Ltd., London, 1970.
58. Burton, W. G., The potato, in *A Survey of its History and of the Factors Influencing its Yield, Nutritive Value, Quality and Storage* (H. Veenman and N. V. Zonen, eds). European Association of Potato Research, Wageningen, 1966.
59. *Household Food Consumption and Expenditure,* Annu. Rep. National Food Survey Committee, Her Majesty's Stationery Office, London, 1972.
60. Mondy, N. I., Factors affecting the nutritional quality of potatoes in *Proc. Int. Congress—Research for the Potato in the year 2000* (W. J. Hooker, ed.), International Potato Center, 1983, p. 136.
61. Woodward, C. F., and E. A. Talley, Review of nitrogenous constituents of the potato: Nutritive value of the essential amino acids, *Am. Potato J. 30*:205 (1953).
62. Banks, W., and C. Greenwood, The starch of the tuber and shoots of the sprouting potato, *Biochem. J. 73*:237 (1959).
63. Schwimmer, S., A. Vevenue, W. Western, and A. Potter, Survey of major and minor sugar and starch components of white potato, *J. Agric. Food Chem.* 2:1284 (1954).
64. Petrov, A. A., Effects of forms of nitrogen fertilizers on the level of starch and ascorbic acid on potato tubers, *Nauch Rab. Molodykn. Veh. Chuvash Sel. Skokhoz. Inst. 1*:65 (1970).

65. Samotus, B., and M. Palasinski, Transformation of carbohydrate in potato tubers transferred from low to high temperature during storage, *Zesz. Nauc. Wyzsz. Szk. Roln Krakowie Roln. 10:*81 (1964).
66. Fisher, A. M., and O. T. Physhtaleva, Effect of chlorocholine chloride on the yield and food qualities of potato, *Tr. Nauchno-Issled Inst. Kracv. Patol. Alma-Ala, 26*:183 (1974).
67. Verma, S. C., K. C. Joshi, and T. R. Sharma, Some observations on the quality of potato varieties grown in India, *Abstr. Conf. Pap. Trienee Conf. Eur. Assoc. Potato Res. 6*:162 (1975).
68. Mondy, N. I., and R. Ponnampalam, Effect of sprout inhibitor isopropyl N-(3-chlorophenyl carbamate) on total glycoalkaloid content of potatoes, *J. Food Sci., 50*:258 (1985).
69. Buffel, K., and A. Carlier, The cell wall composition of hydrated potato tissue and action of auxins, *Agriculture* (Louvain) *4*:157 (1956).
70. Trowell, H., D. A. T. Southgate, T. M. S. Wolever, A. R. Leeds. M. A. Gassull, and D. J. A. Jenkins, Dietary fiber redefined, *Lancet 1*:967 (1976).
71. Paul, A. A., and D. A. T. Southgate, *McCance and Widdowson's The Composition of Foods,* 4th ed., MRC Spec. Rep. 297 Her Majesty's Stationery Office, London, 1988.
72. Finglas, P. M., and R. M. Faulks, A new look at potatoes, *Nutr. Food Sci. 92*:12 (1985).
73. Willis, R. B. H., J. S. K. Lim, and H. Greenfield, Variation in nutrient composition of Australian retail potatoes over a 12 month period, *J. Sci. Food Agric. 35*:1012 (1984).
74. Jones, G. D., D. R. Briggs, M. L. Wahlquist, and L. M. Flentje, Dietary fiber content of Australian foods. I. Potato. *Food Technol. Austr. 37*:81 (1985).
75. Dreher, M. L., C. Breedon, and P. H. Orr, Percent starch hydolysis and dietary fiber content of chipped and baked potatoes, *Nutr. Rep. Intl. 28:*687 (1985).
76. Sandberg, A. S., *Dietary Fiber: Determination and Physiological Effects,* Dept. of Clinical Nutrition and Department of Surgery II, University of Goteborg, Sweden, 1982.
77. Reistad, R., Content and composition of non-starch polysaccharides in some Norwegian plant foods, *Food Chem. 12*:45 (1983).
78. Theander, O., Advances in the chemical characterization and analytical determination of dietary fiber components in *Dietary Fiber* (G. G. Birch and K. J. Parkar, eds.), Applied Science Pub., New York, 1983.
79. Englyst, H., H. S. Wiggins, and J. H. Cummings, Determination of non-starch polysaccharides in plant foods by gas chromatography of constituent sugars as alditol acetates, *Analyst 107*:307 (1982).
80. Toma, R. B., P. H. Orr, B. D'Appolonia, R. Dintzis, and M. M. Takekhia, Physical and chemical properties of potato peel as a source of dietary fiber in bread, *J. Food Sci. 44*:1403 (1979).
81. Toma, R. B., J. Augustin, R. L. Shaw, R. H. True, and J. M. Hogan, Proximate composition of freshly harvested and stored potatoes (*Solanum tuberosum* L.), *J. Food Sci. 43*:1702 (1978).
82. Doughty, J., Water: the hidden ingredient, *Appropr. Technol. 8:*11 (1982).
83. WHO, Energy and Protein Requirements, Rep. Joint FAO/WHO UNV. Expert Consultation, WHO Tech. Rep. Ser. No. 724, World Health Organization, Geneva, 1985.
84. Groot, E., L. Jansen, A. Kentie, H. Oosterhuis, and H. A. Trap, A new protein in potato, *Biochim. Biophys. Acta. 1*:410 (1947).
85. Loeschcke, V., and H. Stegeman, Potato varieties and their electrophorograms characteristics, *Eur. Potato J. 9*:111 (1966).
86. Hughes, J. C., Chemistry and after cooking discoloration in potato, *Natl. Inst. Agric. Bot. J. 9*:235 (1962).
87. Chang, Y. O., and E. E. Avery, Nutritive value of potato proteins, *J. Am. Diet. Assoc. 55*:565 (1969).
88. Mondy, N. I. and C. B. Munshi, Chemical composition of potato as affected by herbicide: enzymatic discoloration, phenols and ascorbic acid content, *J. Food Sci. 53*:475 (1988).
89. Cameron, M., and Y. Hofvander, *Manual on Feeding Infants and Young Children,* 2nd ed., Protein Calorie Advisory Group, United Nations, 1976.
90. Kaldy, M. S., Protein yield of various crops as related to protein value, *Econ. Bot. 26*:142 (1972).
91. Chick, H., and E. B. Slack, Distribution and nutritive value of nitrogenous substances in potato, *Biochem. J. 45*:211 (1951).

92. Swaminathan, K., and Pushkarnath, Nutritive value of Indian potato varieites, *Indian Potato J. 4*:76 (1962).
93. Leichsenring, J. M., *Factors Influencing the Nutritive Value of Potatoes,* Minnesota Tech. Bull No. 96, University of Minnesota, Agricultural Experiment Station, Minneapolis, MN, 1951.
94. Augustin, J., R. B. Toma, R. H. True, R. L. Shaw, C. Teitzel, S. R. Johnson, and P. Orr, Composition of raw and cooked potato peel and flesh: proximate and vitamin composition, *J. Food Sci. 44*:805 (1979).
95. Augustin, J., B. G. Swanson, C. Teitzel, S. R. Johnson, S. F. Pometto, W. E. Artz, C. P. Huang, and C. Schomaker, Changes in nutrient compostion during commercial processing of frozen potato products, *J. Food Sci. 44*:807 (1979).
96. Klein, L. B., S. Chandra, and N. I. Mondy, The effect of phosphorus fertilization on the chemical quality of Katandin potatoes, *Am. Potato J. 57*:259 (1980).
97. Rosenberg, H. R., Potato, in *Chemistry and Physiology of Vitamins,* Interscience, New York, 1942.
98. Carter, E. G. A., and K. J. Carpenter, The availability of niacin in foods, *Fed. Proc. Abstr. Part I,* CA 1536, 1980, p. 557.
99. Quick, W. A., and P. H. Li, Phosphorus balance in potato tubers, *Potato Res. 19*:305 (1976).
100. Maga, J. A., Potato alkaloids, *Crit. Rev. Food Sci. Nutr. 12*:371 (1980).
101. Gull, D. D., and F. M. Isenberg, Chlorophyll and solanine content and distribution in four varieties of potato tubers, *Proc. Am. Soc. Hort. Sci., 75*:545 (1960).
102. Jadhav, S. J., and D. K. Salunkhe, Formation and control of chlorophyll and glycoalkaloids in tubers of *Solanum tuberosum* L. and evaluation of glycoalkaloid toxicity, *Adv. Food Res. 21*:307 (1975).
103. Morris, S. C., and T. H. Lee, The toxicity and teratogenicity of Solanaceae glycoalkaloids particularly those of the potato (*Solanum tuberosum*): A review, *Food Technol. Aust. 36*:118 (1984).
104. Balls, A., and C. Ryan, Concerning a crystalline chemotrypsin inhibitor from potatoes and its binding capacity for the enzyme, *J. Biol. Chem. 238*:2976 (1963).
105. Marinkovich, V., Purification and characterization of the hemagglutinins present in potatoes, *J. Immunol. 93*:732 (1964).
106. Werle, E., W. Appel, and W. Happ, The Kallikrein inactivation of potatoes and its differentiation from proteinase inhibitor, *Ferment. Forsch. 10*:127 (1959).
107. Santarius, K., and H. D. Belitz, Protease activity in potato plants, *Planta 141*:145 (1978).
108. Ryan, C. A., and G. M. Hass, Structural, evolutionary and nutritional properties of proteinase inhibitors from potatoes, in *Antinutrients and Natural Toxicants in Foods* (R. L. Ory, ed.), Food and Nutrition Press, Westport, CT 1981.
109. Hass, G. M., J. E. Derr, D. J. Makus, and C. A. Ryan, Purification and characterization of the carboxypeptidase inhibitors from potatoes, *Plant Physiol. 64*:1022 (1979).
110. Richardson, M., The protease inhibitors of plants and macro-organisms, *Phytochemistry 16*:159 (1977).
111. Lau, A., H. Ako, and M. Werner-Washburne, Survey of plants for enterokinase inhibitors, *Biochem. Biophys. Res. Commun. 92*:1243 (1989).
112. Ryan, C. A., T. Kuo, G. Pearce, and R. Kunkel, Variability in the concentration of 3 heat stable protease inhibitor proteins in potato tubers, *Am. Potato J. 53*:443 (1976).
113. Peng, J. H., Increase activity of proteinase inhibitors in disease resistance to *Phytophthora infestans, Diss. Abstr. Intl.* B. *26*:24 (1975).
114. Huang, D. Y., B. G. Swanson, and C. A. Ryan. Stability of proteinase inhibitors in potato tubers during cooking. *J. Food Sci. 46*:287 (1981).
115. Livingston, R. M., B. A. Baird, T. Atkinson, and R. M. J. Crafts, The effect of either raw or boiled liquid extract from potato (*Solanum tuberosum*) on digestibility of diet on barley in pigs, *J. Sci. Food Agric. 31*:695 (1980).
116. Goldstein, J. J., and C. E. Hynes, The lectins: Carbohydrate binding proteins of plants and animals, in *Advances in Carbohydrate Chemistry and Biochemistry* (R. S. Tipson and D. Horton, eds.), Academic Press, New York, 1978.

117. Kilpatric, D. C., Isolation of lectin from pericarp of potato (*Solanum tuberosum*), *Biochem. J. 191*:273 (1980).
118. Clark, R., J. Kuc, R. Henze, and F. Quackenbush, The nature of fungitoxicity of an amino addition product of chlorogenic acid. *Phytochemistry 49:*594 (1959).
119. Zucker, M., Influence of light on synthesis of protein and of chlorogenic acid in potato tuber tissue, *Plant Physiol. 38*:575 (1963).
120. Craft, C., and W. Audia, Phenolic substances and barrier formation in potato, *Bot. Gaz.* (Chicago) *123*:211 (1962).
121. Reeve, R. M., E. Hautala, and M. L. Weaver, Anatomy and compositional variation with potatoes II. Phenolics enzymes and other minor components, *Am. Potato J. 46*:347 (1969).
122. Mondy, N. I., C. Metcalf, and R. L. Plaisted, Potato flavour as related to chemical composition I. polyphenols and ascorbic acid, *J. Food Sci. 41*:459 (1971).
123. Zgorska, K., Factors affecting the quality of table potatoes *Ziemniak* 183-206, 1979.
124. Klein, L. B., S. Chandra, and N. I. Mondy, Effect of magnesium fertilization on the quality of potatoes: Total N, non-protein, N, protein amino acids minerals and firmness, *J. Agric. Food Chem. 30*:754 (1982).
125. Wolfe, J. A., *The Potato in the Human Nutrition,* Cambridge University Press, London, 1987, p. 83.
126. Smith, O., *Studies of Potato Storage,* Cornell University Agricultural Exp. Station Bulletin, 553, 1933.
127. Thomas, P., Wound induced suberization and periderin development in potato tubers as affected by temperature and gamma irradiation, *Potato Res. 15*:155 (1982).
128. Burton, W. G., Postharvest behavior and storage of potatoes in *Applied Biology,* Vol. 2 (T. H. Coaker, ed.), Academic Press, New York, 1966.
129. Booth, R. H., and R. L. Shaw, *Principles of Potato Storage,* International Potato Centre, Lima, 1981.
130. Faulks, R. M., N. M. Griffith, M. A. White, and K. I. Tomlins, Influence of site, variety and storage on nutritional composition and cooked texture of potato, *J. Sci. Food Agric. 33*:589 (1982).
131. Fitzpatric, T. J., and W. L. Porter, Changes in sugars and amino acids in chips made from fresh stored and reconditioned potatoes, *Am. Potato J. 43*:238 (1966).
132. Smith, O., Studies of potato storage, Cornell University, Agricultural Experiment Station Bulletin, 553, 1933.
133. Linnemann, A. R., A. Van Es, and K. J. Hartmans, Changes in the content of L-ascorbic acid, glucose, fructose, sucrose and total glycoalkaloids in potatoes stored at 7, 16 and 28°C, *Potato Res. 23*:271 1985).
134. Picha, D. H., Influence of storage duration and temperature on sweet potato sugar content and chip colour, *J. Food Sci. 51*:239 (1986).
135. Plaza, S. G., R. J. Sueldo, M. Crupkin, and C. A. Barassi, Changes in composition of potatoes stored in clamps, *J. Food Sci. 50:*1254 (1985).
136. Wills, R. B. H., J. S. K. Lim, and H. Greenfield, Variation in nutrient composition of Australian raw potato over a 12 month period, *J. Sci. Food. Agric. 35*:1012 (1984).
137. Augustin, J., S. R. Johnson, C. Teitzel, R. M. Toma, R. L. Shaw, R. H. True, J. M. Hogan, and R. M. Deutsch, Vitamin composition of freshly harvested and stored potatoes, *J. Food Sci. 43*:1566 (1978).
138. Shekhar, V. C., W. M. Iritani, and R. Arteca, Changes in ascorbic acid content during growth and short term storage of potato tubers (*Solanum tuberosum*), *Am. Potato J.* 56:663 (1978).
139. Singh, M., and S. C. Verma, Postharvest technology and utilization of potato, in *Postharvest Technology and Utilization of Potato* (H. Kishore, ed.), International Potato Center, Region VI, New Delhi, 1979.
140. Sweeney, J. P., P. A. Hepner, and S. Y. Libeck, Organic acid, amino acid and ascorbic acid content of potatoes as affected by storage conditions, *Am. Potato J. 46*:463 (1969).
141. Talley, E. A., R. B. Toma, and P. H. Orr, Amino acid composition of freshly harvested and stored potato, *Am. Potato J. 61*:267 (1984).
142. Toma, R. B., J. Augustin, R. L. Shaw, R. H. True, and J. M. Hogan, Proximate compostion of freshly harvested and stored potatoes, *J. Food Sci. 43*:1702 (1978).

143. Talburt, W. F., and O. Smith, *Potato Processing,* 4th ed., Van Nostrand Reinhold, New York, 1987.
144. Samatous, B., and S. Schwimmer, Changes in carbohydrate and phosphorus content of potato tuber during storage in nitrogen, *J. Food Sci. 28*:163 (1963).
145. Mica, B., Changes in content of glucose, fructose, sucrose and lysine in potatoes during storage and boiling, *Rostl. Vyroba 24*:35 (1978).
146. Mica, B., Effect of storage and boiling on the content and free amino acids in potatoes, *Rostl. Vyroba 24*:731 (1978).
147. Habib, A. T., and H. D. Brown, Factors influencing the colour of potato chips, *Food Technol. 10*:332 (1956).
148. Fitzpatric, T. J., E. A. Talley, W. L. Porter, and H. J. Murphy, Chemical composition of potatoes. III Relationship between specific gravity and nitrogenous constituents, *Am. Potato J. 41*:75 (1964).
149. Burton, W. G., in *The Potato* (Veenman, H., and N. V. Zonin, eds.), Wageningen, The Netherlands, 1978.
150. Thomas, P., A. N. Srirangarajan, M. R. Joshi, and M. T. Janave, Storage deterioration in gamma irradiated and unirradiated Indian potato cultivars under refrigeration and tropical temperature, *Potato Res. 22*:261 (1979).
151. Augustin, J., S. R. Johnson, C. Teitzel, R. H. True, J. M. Hogan, R. B. Toma, R. L. Shaw, and R. M. Deutsch, Changes in nutrient composition of potatoes during home preparation II. Vitamins, *Am. Potato J. 55*:653 (1978).
152. Roine, P., K. Wichmann, and Z. Vihavainen, The content and stability of ascorbic acid in different potato varieties in Finland, *Suom. Maataloustiet Seuran Julk. 83*:71 (1955).
153. Mareschi, J. P., J. P. Belliot, C. Fourton, and K. E. Gey, Decrease in vitamin C content in Bintje potatoes during storage and conventional cooking procedures, *Int. J. Vitamin Nutr. Res. 55*:402 (1983).
154. Page, E., and F. M. Hanning, Vitamin B_6 and niacin in potatoes, *J. Am. Diet. Assoc.* 42:42 (1963).
155. Addo, C., and J. Augustin, Changes in the vitamin B_6 content in potato during storage, *J. Food Sci. 53*:749 (1988).
156. Baker, J., and L. W. Mapson, The ascorbic acid content of potato tubers III. The influence of storage in nitrogen, air and pure oxygen, *New Phytol. 51*:90 (1952).
157. Cheng, F. C. and D. Muneta, Lipid composition of potato as affected by storage and potassium fertilization, *Am. Potato J. 55*:441 (1978).
158. Yamaguchi, M., J. W. Perdue, and J. H. MacGillivray, Nutrient composition of "White Rose" potatoes during growth and after storage. *Am. Potato J. 37*:73 (1960).
159. Mazza, G., J. Hung, and M. J. Dench, Processing/nutritional quality changes in potato tubers during growth and long term storage, *Can. Inst. Food Sci. Technol. 16*(1):39 (1983).
160. Mazza, G., Correlations between quality parameters of potato during growth and long term storage, *Am. Potato J. 60:*145 (1983).
161. Cargill, B. F., D. R. Heldman, and C. L. Bedford, Influence of environmental storage conditions on potato chip quality, *Potato Chipper 30*:10 (1971).
162. Smith, O, Culinary quality and nutritive value of potatoes, in *Potatoes: Production, Storage, Processing* (O. Smith, ed.), AVI Pub., Westport, CT, 1968, p. 498.
163. Gull, D. D., and F. M. Isenberg, Light and off-flavour development in potato tubers exposed to fluorescent light, *Proc. Am. Soc. Hortic. Sci. 71*:446 (1958).
164. Akeley, R. V., G. V. C. Houghland, and A. E. Schark, Genetic differences in potato tuber greening, *Am. Potato J. 39*:409 (1962).
165. Yamaguchi, M., D. L. Hughes, and F. D. Howard, Effect of season, storage temperature during light exposure on chlorophyll accumulation of "White Rose" potatoes, *Proc. Am. Soc. Hortic. Sci. 75*:529 (1960).
166. Larsen, E. C., Investigation on cause and prevention of greening in potato tubers, *Idaho Agric. Exp. Stn. Res. Bull.* No. 16, 1949.
167. Lewis, W. C., and R. G. Rowbery, Some effects of depth of planting and time and height of hilling on 'Kennebec' and "Sebago" potatoes, *Am. Potato J. 50*:301 (1973).

168. Bleasdale, J. K. A., and R. Thompson, Potatoes, *Annu. Rep. Natl. Veg. Res. Stn. Welesbourne, England,* 37, 1964.
169. Zitnak, A., and G. R. Johnston, Glycolalkaloid content of B 5141-6 potatoes, *Am. Potato J.* 47:256 (1970).
170. Patil, B. C., D. K. Salunkhe, and B. Singh, Metabolism of solanine and chlorophyll in potato tubers as affected by light and specific chemicals, *J. Food Sci. 36*:474 (1971).
171. Salunkhe, D. K., M. T. Wu, and S. J. Jadhav, Effect of light and temperature on the formation of solanine in potato slices, *J. Food Sci. 37*:969 (1972).
172. Hilton, R. J., Factors in relation to tuber quality in potato: Preliminary trials on bitterness in Netted Gem. Potato, *Sci. Agric. 31*:61 (1951).
173. Zitnak, A., *The Influence of Certain Treatments upon Solanin Synthesis in Potatoes,* M.S. Thesis, University of Alberta, Edmonton, 1953.
174. Yamaguchi, M., D. L. Hughes, and F. D. Howard. Effect of season, storage temperature during light exposure on chlorophyll accumulation of white Rose potatoes. *Proc. Am. Soc, Hortic. Sci.* 75:529 (1966).
175. Buck, R. W. Jr., and V. Akeley, Effect of maturity, storage temperature and storage time on greening of potato tubers, *Am. Potato J. 44*:56 (1957).
176. Lilijemark, A., and E. Widoff, Greening and solanine development of white potato in fluorescent light, *Am. Potato J. 37*:379 (1960).
177. Patil, B. C., B. Singh, and D. K. Salunkhe, Formation of chlorophyll and solanine in Irish potato (*Solanum tuberosum* L.) tubers and their control by gamma radiation and CO_2 enriched packaging, *Lebensm. Wiss Technol. 4*:123 (1971).
178. Jeppsen, R. B., D. K. Salunkhe, and S. J. Jadhav, Formation and anatomical distribution of chlorophyll and solanine in potato tubers and their control by chemical and physical treatments, The 33rd Annu. Inst. Food Tech. Meet. 153, (1973).
179. Jadhav, S. J., and D. K. Salunkhe, Formation and control of chlorophyll and glycoalkaloids in tubers of *Solanum tuberosum* L. and evaluation of glycoalkaloid toxicity, *Adv. Food Res. 21*:307 (1975).
180. Wu, M. T., and D. K. Salunkhe, Control of chlorophyll and solanine syntheses and sprouting of potato tubers by hot paraffin wax, *J. Food Sci. 37*:629 (1972).
181. Wu, M. T., and D. K. Salunkhe, Control of chlorophyll and solanine formation in potato tubers by oil and diluted oil treatments, *HortScience* 7:466 (1972).
182. Jadhav, S. J., and D. K. Salunkhe, Effects of certain chemicals on photo-induction of chlorophyll and glycoalkaloid synthesis and on sprouting of potato tubers, *Can. Inst. Food Sci. Techno. J.* 7:178 (1974).
183. Forsyth, F. R., and C. A. Haves, Greening of potatoes, CA cure, *Food Technol. 22*:48 (1968).
184. Patil, B. C., *Formation and Control of Chlorophyll and Solanine in Tubers of Solanum tuberosum L. and evaluation of Solanine Toxicity,* Ph.D. thesis, Utah State University, Logan, 1972.
185. Ziegter, R., S. H. Schanderi, and P. Markakis, Gamma irradiation and enriched CO_2 atmosphere storage effect on light induced greening of potatoes, *J. Food Sci. 33*:533 (1968).
186. Madsen, K. A., D. K. Salunkhe, and M. Simon, Morphological and biochemical changes in gamma irradiated carrots and potatoes, *Radiation Res. 10*:48 (1959).
187. Wu, M. T., and D. K. Salunkhe, Effect of gamma irradiation on wound induced glycoalkaloid formation in potato tubers, *Lebensm. Wiss Technol. 10*:141 (1977).
188. Boguchi, S., and D. C. Nelson, Length of dormancy and sprouting characteristics of ten potato cultivars, *Am. Potato J. 57*:151 (1980).
189. Burton, W. G., and A. R. Wilson, the sugar content of sprout growth of tubers of potato cultivar, 'Record' grown in different localities when stored at 10, 15 and 20°C, *Potato Res. 21*:145 (1978).
190. Bantan, S., M. Krapez, and M. Vardjan, Variation in ascorbic acid during development and storage of potato tubers of potato Cv. Vesna and Bintije, *Biol. Vestn. 25*:1 (1977).
191. Meiklejohn, J. The vitamin B_1 content of potatoes. *Biochem. J. 37*:349 (1943).
192. Boyd, I. G. M., J. Dalziel, and H. J. Duncan, Studies on potato sprout suppresants 5. The effect of chlorpropham contamination on the performance of seed potatoes, *Potato Res. 25*:51 (1982).

193. Struckmeyer, B. E., G. G. Weis, and J. A. Schoenemann, Effect of two forms of maleic hydrazide on cell wall structure at the midsection stem, and bud ends of the cortical and perimedullary regions of Russet Burbank tubers, *Am. Potato J. 58*:611 (1981).
194. Sawyer, R. L., and J. P. Malagamba, Sprouting inhibition, in *Potato Processing* (Talburt, W. F., and O. Smith, eds.), Van Nostrand Reinhold, New York, 1987, p. 183.
195. Kennedy, E. J., and O. Smith, Response of the potato to field application of maleic hydrazide, *Am. Potato J. 28*:701 (1951).
196. Kennedy, E. J., and O. Smith, Response of seven varieties of potato to foliar application of maleic hydrazide, *Proc. Am. Soc. Hortic. Sci. 61*:395 (1953).
197. Franklin, E. W., and N. R. Thompson, Some effects of maleic hydrazide on stored potatoes, *Am. Potato J. 30:*289 (1953).
198. Salunkhe, D. K., and S. H. Wittwer, The influence of a postharvest foliar spray of maleic hydrazide on the specific gravity of Irish cobbler potatoes and quality of their chips, Summary U.S. Rubber Co., MHIS 613:21 (1952).
199. Weis, G. G., J. A. Schoenemann, and M. D. Groskopp, Influence of time of application of maleic hydrazide on the yield and qulaity of Russet Burbank potatoes, *Am. Potato J. 57:*197 (1980).
200. Kunkel, R., N. M. Holsted, D. C. Mitchell, T. S. Russell, and R. E. Thornton, Maleic hydrazide studies on potato in Washington's Columbia basin, *Am. Potato J. 56*:470 (1979).
201. Davis, J. R., and M. D. Groskopp, Yield and quality of Russet Burbank potato as influenced by interactions of rhizoctonia, maleic hydrazide and PCNB, *Am. Potato J. 58*:227 (1981).
202. Mondy, N. I., A. Tymiak, and S. Chandra, Inhibition of glycoalkaloid formation in potato tubers by sprout inhibitor maleic hydrazide, *J. Food Sci. 43*:1033 (1978).
203. Covsini, D., G. I. Stallknecht, and W. C. Sparks, A simplified method for determining sprout inhibiting levels of chloroprophan (CIPC) in potato, *J. Agric. Food Chem. 26*:990 (1978).
204. Van Vliet, W. F., and H. Sparenberg, The treatment of potato tubers with sprout inhibitors, *Potato Res. 13*:223 (1970).
205. Marth, P. C., and E. S. Schultz, A new sprout inhibitor for potato tubers, *Am. Potato J. 29*:263 (1952).
206. Kim, M. S., E. E. Swing, and J. B. Siezcka, Effects of chloroprophan (CIPC) on sprouting of individual potato eyes on plant emergence, *Am. Potato J. 49*:420 (1972).
207. Dallyn, S., and Sawyer, R. L., Physiological effects of ionizing radiation on onions and potatoes, *Cornell University Contract No. DA* 19-129-QM-755, *Final Report,* January 1959.
208. Ellison, J. H., Inhibition of potato sprouting by 2,3,5,6 tetrachloronitrobenzene and methylester of naphtha leneacetic acid, *Am. Potato J. 29*:176 (1952).
209. Sawyer, R. L., and S. Dallyn, Vaporized chemicals control sprouting in stored potatoes, *Farm Res. 23*(3):6 (1957).
210. Ellison, J. H., and H. S. Cunningham, Effect of sprout inhibitors on the incidence of fusarium dry rot and sprouting of potato tubers, *Am. Potato J. 30*:10 (1953).
211. Thompson, N. R., and D. R. Isleib, Sprout inhibition of bulk stored potatoes, *Am. Potato J. 36*:32 (1959).
212. Sukumaran, N. P., and S. S. Grewal, and M. S. Virk, Inhibition of mustard seed germination as bioassay for growth inhibitory substances in potato tubers, *J. Indian Potato Assoc. 5*:13 (1978).
213. Sparrow, A. H., and E. Christensen, Improved storage quality of potato tubers following exposure to gamma irradiation of Cobalt 60, *Nucleonics 12*(3):16 (1954).
214. Brownell, L. E., C. H. Burns, F. G. Gustafson, D. Isleib, and W. J. Hooker, Storage properties of gamma irradiated potatoes, *Proc. 17th Annu. Potato Utilization Conf.,* 1956, p. 2.
215. Sparks, W. C., and W. M. Iritani, The effects of gamma rays from fission product waste on storage losses of Russet Burbank potatoes, *Idaho Agric. Exp. Stn. Res. Bull. 60* (1964).
216. Bellomonte, G., A. Guadiano, E. Sanzini, L. Boniforti, S. Civalleri, S. Giammarioli, G. Gilardi, L. Leeli, A. Massa, and M. Mosca, Treatment of food with gamma radiation I. Biochemical studies on potato, *Riv. Soc. Ital. Sci. Aliment. 7*:157 (1978).
217. Bergers, W. W. A. Investigation of the content of phenolic and alkaloidal compounds of gamma irradiated potatoes during storage, *Food Chem.* 7:47 (1981).

218. Guo, A. X., G. Z. Wang, and Y. Wang, Biochemical effect of irradiation of potato, onion and garlic in storage I. Changes of major nutrients during storage, *Yuang TZu Neng Nung Yeh Ying Yana,* 1 1/6 (1981).
219. Janave, M. T., and P. Thomas, Influence of postharvest storage, temperature and gamma irradiation on potato carolenoids, *Potato Res.* 22:365 (1979).
220. Mazon-Matanzo, M. P., and G. J. Fernandez, Effect of gamma radiation on potato (Solanum *tuberosum* L.) tuber preservation during storage period, *Junta Energ. Nucl.* 354 (1976).
221. Proctor, F. J., J. P. Goodliffe, and D. G. Coursey, Postharvest losses of vegetables and their control in the tropics, in *Vegetable Productivity* (C. R. W. Spalding, ed.), MacMillan, London, 1981, p. 139.
222. Booth, R. H., Postharvest deterioration of tropical root crops: losses and their control, *Trop. Sci. 16*:49 (1974).
223. Pantastico, E. B., and O. K. Bautista, Postharvest handling of tropical vegetable crops, *Hort. Science 11*:122 (1976).
224. Harvey, J. M., Reduction of losses in fresh market fruits and vegetables, *Ann. Rev. Phytopathol. 16*:321 (1978).
225. Mansfield, J., Postharvest losses of potatoes quantified during marketing process, in Report of the Seminar on the Reduction of Postharvest Food Losses in Caribbean and Central America, August 8-11, 1977.
226. NRC Report on Steering Committee for Study on Postharvest Food Losses in Developing Countries, National Research Council, National Academy of Sciences, Washington DC, 1978.
227. FAO, *Analysis of an FAO Survey on Postharvest Crops Losses in Developing Countries,* Food and Agriculture Organization, Rome, 1977.
228. Blackbeard, J., Do it yourself damage testing, *Arable Farming 8*:16 (1981).
229. Ryall, A. L., and W. J. Lipton, *Handling, Transportation and Storage of Fruits and Vegetables,* Vol. I. Vegetables and Melons, AVI Pub. Co., Westport, CT, 1972.
230. Howard, F. D., M. Yamaguchi, and J. E. Knott, Carbon dioxide as a factor in susceptibility of potatoes to black spot from bruising, *Proc. 16th Intl. Hortic. Congr.,* 582, 1962.
231. Lorenz, O. A., F. H. Takatori, H. Timm, J. W. Oswald, T. Bowman, F. S. Fullmer, M. Snyder, and H. Hall, *Potato Fertilization and Black Spot Studies,* Vegetable Series No. 88, Santa Maria Valley, Department of Vegetable Crops, University of California, Davis, 1957.
232. Wu, M. T., and D. K. Salunkhe, Changes in glycoalkaloid content following mechanical injury in potato tubers, *J. Am. Soc. Hort. Sci. 101*(3):329 (1976).
233. Isenberg, F. M., and D. D. Gull, Potato greening under articficial light, *Cornell Univ. Exp. Bull.* 1033 (1959).
234. Hardenburg, R. E., Greening of potatoes during marketing—a review, *Am. Potato J. 41*:215 (1964).
235. Eckert, J. W., P. P. Rubio, A. K. Mattoo, and A. K. Thompson, Diseases of troical crops and their control, in *Postharvest Physiology, Handling and Utilization of Tropical and Subtropical Fruits and Vegetables* (Er. B. Pantastico, ed.), AVI Pub. Co., Westport, CT, 1975, p. 415.
236. Thompson, A. K., M. B. Bhatti, and P. P. Rubio, *Harvesting, Handling and Utilization of Tropical and Subtropical Fruits and Vegetables* (Er. B. Pantastico, ed.), AVI. Pub. Co., Westport, CT, 1975, p. 236.
237. Ware, G. W., and J. P. McCollum, *Raising Vegetables,* Interstate, Danville, IL, 1959.
238. Thompson, H. C., and W. C. Kelly, *Vegetable Crops,* 15th ed. McGraw-Hill, New York, 1957.
239. Smith, O., Studies of potato storage, *Cornell Agric. Exp. Stn. Bull.* 553 (1933).
240. Smith, O., Transport and storage of potatoes, in *Potato Processing* (W. F. Talburt and O. Smith, eds.), Van Nostrand Reinhold, New York, 1987, p. 7.
241. Pantastico, Er. B., T. K. Chattopadhyay, and H. Subramanyam, Storage and commercial storage operation, in *Postharvest Physiology, Handling and Utilization of Tropical and Sub-tropical Fruits and Vegetables* (Er. B. Pantastico, ed.) AVI Pub. Co., Westport, CT. 1975, p. 314.
242. Lipton, W. J., Some effects of low oxygen atmosphere on potato tubers, *Am. Potato J.* 44:292 (1967).
243. Nielsen, L. W., Accumulation of respiratory CO_2 around potato tubers in relation to bacterial soft rot, *Am. Potato J. 45*:174 (1968).

244. Butchbaker, A. F., D. C. Nelson, and R. Shaw, Controlled atmosphere storage of potatoes, *Trans ASAE 10*:534 (1967).
245. Workman, M. N., and J. Twomey, The influence of oxygen concentration during storage on seed potato respiratory metabolism and on field performance, *Proc. Am. Soc. Hortic. Sci. 90*:268 (1967).
246. Workman, M. N., and J. Twomey, The influence of storage atmosphere and temperature on physiology and performance of Russet Burbank Seed potatoes, *J. Am. Soc. Hortic. Sci. 94*:260 (1969).
247. Yamaguchi, M., W. J. Flocker, F. D. Howard, and M. Timm, Changes in the CO_2 levels with moisture in fallow and cropped soil and susceptibility of potatoes to black spot, *Proc. Am. Soc. Hortic. Sci. 85*:446 (1968).
248. Van Niekerk, B. P., Potatoes can be stored for eight months, *Farming S. Afr. 36*:20 (1960).
249. Salunkhe, D. K., and M. T. Wu, Development in technology of storage and handling of fresh fruits and vegetables, in *Storage, Processing and Nutritional Quality of Fruits and Vegetables* (D. K. Salunkhe, ed.), CRC Press, Boca Raton, FL, 1974, p. 121.
250. Arteca, R. N., Calcium infiltration inhibits greening in Katahdin potatoes, *HortScience, 17*(1):79(1982).
251. Kumar, P., B. Baijal, and D. Alka, Effects of some growth regulators on ascorbic acid and total sugar content of potatoes associated with sprouting during storage, *Acta Bot.* (Indica) *8*:235 (1980).
252. Pederson, S., The effects of ionizing radiations on sprouts prevention and chemical composition of potatoes, *Food Technol. 10*:532 (1956).
253. Panalaskas, T., and O. Pellefier, The effect of storage on ascorbic acid content of gamma radiated potatoes, *Food. Res. 25*:33 (1960).
254. Mikaelsen, K., and L. Roger, Improved storage ability of potatoes exposed to gamma radiation, *Acta Agric. Scand.* 6:145 (1956).
255. Ogata, K., Y. Tatsumi, and K. Chachin, Studies on the browning of potato tubers by gamma radiation. I. Histological observation and the effect of the time of irradiation after harvest, low temperature and polyethylene packaging, *J. Food Sci. Technol. 17*:298 (1970).
256. Tatsumi, Y., K. Chachin, and K. Ogata, Studies on the browning of potato tubers by gamma radiation. II. The relationship between the browning and the changes of o-diphenol ascorbic acid and activities of polyphenol oxidase and peroxidase in irradiated potato tubers, *J. Food Sci. Technol. 19*:508 (1972).
257. Smith, O., *Potatoes: Production, Storing and Processing,* AVI Pub. Co., Westport, CT. 1968.
258. Workman, M., M. E. Patterson, N. K. Ellis, and F. Heilligman, The utilization of ionizing radiation to increase the storage life of white potatoes, *Food Technol. 14*:395 (1960).
259. Salunkhe, D. K., Gamma radiation effects on fruits and vegetables, *Econ. Bot. 15*:28 (1961).
260. Chachin, K., and T. Iwata, Respiratory metabolism and potassium release of irradiated potatoes, Paper read at Seminar on Food Irradiation for Developing Countries in Asia and Pacific, Tokyo, November 9 to 13, 1981, IAEA/Food and Agriculture Organization, Rome, 1981.
261. Palmer, A. K., D. D. Cozens, D. E. Prentice, J. C. Richardson, and D. H. Christopher, Production and longevity of rats fed an irradiated potato diet, Intl. Project in the Field of Food Irradiation, Final Tech. Rep. IF IP-R 25, 1975.
262. Palmer, A. K., D. D. Cozens, D. E. Prentice, D. E. Richardson, D. H. Cristopher, H. M. Gottschaik, and P. S. Elias, Reproduction and longevity of rats fed an irradiated potato diet, *Toxicol. Lett. 3*:163 (1979).
263. Willis, R. H. H., T. H. Lee, D. Graham, W. B. McGlasson, and E. G. Hall, *Postharvest: An Introduction to Physiology and Handling of Fruits and Vegetables,* South Wales University Press, Kenington, Australia, 1981.
264. Booth, R. H., and F. J. Procter, Considerations relevant to the storage of ware potatoes in the tropics, PANS (*Pest Artic. News Summ.*) *18*:409 (1977).
265. Research and development potato mid-week review, *Econ. Times* (India), April 15, 1981.
266. Rustowski, A., Potato storage and storage environment in Survey Papers European Assoc. for Potato Research 7th Triennial Conf. Warsaw, June 26-July 1, 1978. Bonin (Poland), *Field Crop Abstr. 35*:5433 (1979).

267. Metlitskii, L. V., Biological aspects of protection of yield of potatoes, vegetables and fruits from losses during storage, *Izv. Akad. Nauk. SSSR Ser. Biol. 1*:73 (1980).
268. Von Gierke, K., *Potato Storage in Potato Research in Pakistan* (M. A. Shah, ed.), Pakistan Agriculture Research Council, Islamabad, 1978, p. 69.
269. CIP Annual Report, 1981, International Potato Centre, Lima, Peru, 1981.
270. Howard, H. W., The production of new varieties, in *The Potato Crop: The Scientific Basis for Improvement* (P. M. Harris, ed.), Chapman and Hall, London, 1978, p. 607.
271. Harris, J. R., Potato processing industry—what the market requires, in Agriculture Group Symp.: Factors Influencing Storage Characters and Cooking Quality of Potatoes, *J. Sci. Food Agric. 32* (Abstr.):104 (1981).
272. French, W. M., Varietal factors influencing the quality of potatoes, in Agriculture Group Symp.: Factors Influencing Storage Characteristics and Cooking Quality of Potatoes, *J. Sci. Food Agric. 32* (Abstract):126 (1981).
273. Talburt, W. F., and O. Smith, *Potato Processing,* AVI Pub. Co., Westport, CT, 1967.
274. Gray, D., and J. C. Hughes, Tuber quality, in *The Potato Crop: The Scientific Basis for Improvement* (P. M. Harris, ed.), Chapman and Hall, London, 1978, p. 504.
275. McComber, D. R., E. M. Osman, and R. A. Robert, Factors related to potato mealiness, *J. Food Sci. 63*:1423 (1988).
276. Sterling, C., and F. A. Bettelheim, Factors associated with potato texture III. Physical attributes and general conclusions, *Food Res. 20*:130 (1955).
277. Reeve, R. M., Pectin, starch and texture of potatoes: Some practical and theoretical implications, *J. Texture Stad. 8*:1 (1977).
278. Nonaka, M., and H. Timm, Textural quality of cooked potatoes II. Relationship of steam cooking time to cellular strength of cultivars with similar and differing solids, *Am. Potato J. 60*:685 (1983).
279. Loh, J., W. M. Breene, and E. A. Davis, Between species differences in fracturability loss: microscopic and chemical composition of potato and chinese waterchestnut, *J. Texture Stad. 13*:325 (1982).
280. Bretzloff, C. W., Some aspects of cooked potato texture and appearance II. Potato cell size stability during cooking and freezing, *Am. Potato J. 47*:176 (1970).
281. Briant, A. M., C. J. Personius, and E. G. Cassel, Physical properties of starch from potato of different culinary quality, *Food Res. 10*:437 (1945).
282. Barrios, E. P., D. N. Newsom, and J. C. Miller, Some factors influencing the culinary quality of Irish potatoes: physical character, *Am. Potato J. 40*:200 (1963).
283. Unrau, A., and R. Nytund, The relation of physical properties and chemical composition to mealiness in the potato. I. Physical properties, *Am. Potato J. 34*:245 (1957).
284. Heinze, P. H., M. E. Kirkpatric, and E. F. Dochtermann, Cooking quality and compositional factors of potatoes of different varieties from several commercial locations, U.S. Dept. Agric. Tech. Bull. Washington, DC, 1106, 1955.
285. McClendon, J. H., Evidence for the pectic nature of middle lamella of potato tuber cell walls based on chromatography of macerating enzymes, *Am. J. Bot. 51*:628 (1963).
286. Warren, D. S., D. Gray, and J. S. Woodman, Relationship between chemical composition and breakdown in cooked potato tissues, *J. Sci. Food Agric. 26*:1689 (1975).
287. Bettelheim, F. A., and C. Sterling, Factors associated with potato texture I. Specific gravity and starch content, *Food. Res.* 20:71 (1955).
288. Bettelheim, F. A., and C. Sterling, Factors associated with potato texture, II. Pectic substances, *Food Res. 20*:118 (1955).
289. Gray, D., Some effects of variety, harvest date and plant spacing on tuber breakdown on canning in tuber dry matter content and cell wall surface area in the potato, *Potato Res. 15*:317 (1972).
290. Warren, D. S., and J. S. Woodman, Distribution of cell wall components in potato tubers: a new titrimetric procedure for the estimation of total polyuronide (pectic substances) and its degree of esterification, *J. Sci. Food Agric. 25*:129 (1974).
291. Hughes, J. C., R. M. Faulks, and A. Grant, The texture of cooked potatoes: relation between the

comprehensive strength of cooked potato discs and release of pectic substances, *J. Sci. Food Agric. 26*:731 (1975).

292. Hughes, J. C., and R. M. Faulks, Texture of cooked potatoes of different maturity in relation to pectic substances and cell size (Abstract), Proc. Fifth Triennial Conf. Eur. Assoc. Potato (Norwich, 1972), 1973.
293. Reeve, R. M., A review of cellular structure, starch and texture qualities of processed potatoes, *Econ. Bot. 21*:294 (1967).
294. Reeve, R. M., Histological survey of conditions influencing texture in potatoes. II. Observation on starch in treated cells, *Food Res. 19*:333 (1954).
295. Ophuis, B. G., J. C. Hesen, and F. Krosbergen, The influence of the temperature during handling on the occurrence of blue discoloration inside potato tubers, *Eur. Potato J. 1*:48 (1958).
296. Salunkhe, D. K., E. J. Wheeler, and S. T. Dexter, The effects of environmental factors on suitability of potatoes for chip making, *Agron. J. 46*:195 (1954).
297. Salunkhe, D. K., *The Influence of Certain Environmental Factors on the Production and Quality of Potatoes for Potato Chips Industry,* Ph.D. thesis, Michigan State University, East Lansing, 1953.
298. Salunkhe, D. K., S. H. Wittwer, E. J. Wheeler, and S. T. Dexter, The influence of pre-harvest foliar spray of maleic hydrazide on the specific gravity of potato chips, *J. Food Res. 18*:191 (1953).
299. Burton, W. G., The potato, in *A Survey of Its History and of the Factors Influencing Its Yield, Nutritive Value, Quality, and Storage* (H. Veenman and N. V. Zonen, eds.), Wageningen, European Association of Potato Research, The Netherlands, 1966.
300. Smith, O., Environmental factors, in *Potatoes: Production, Storing, Processing,* AVI Pub. Co., Westport, CT, 1968, p. 259.
301. Killick, R. J., and N. W. Simmonds. Specific gravity of potato tubers as character showing small genotype environmental interactions, *Heredity 32*:109 (1974).
302. Kunkel, R., and N. Holstad, Potato chip colour, specific gravity and fertilization of potatoes with N.P.K., *Am. Potato J. 49*:43 (1972).
303. Mazza, G., J. Hung, and M. J. Dench, Processing/nutritional quality changes in potato tubers during growth and long term storage, *Can. Inst. Food Sci. Technol. J., 16*(1):39 (1983).
304. Mazza, G., Correlations between quality parameters of potato during growth and long term storage, *Am. Potato J. 60*:145 (1983).
305. Mazza, G., *Selected Factors Affecting Frying Quality of Potatoes,* Paper presented at the Eleventh Annual Meeting of the Pairie Potato Council, Winnepeg, Manitoba, Feb. 12–16, 1983.
306. Salunkhe, D. K., and L. H. Pollard, Microscopic examination of starch grains in relation to maturity of potatoes, *Proc. Am. Soc. Hortic. Sci.* 64:331 (1954).
307. Reeve, R. M., Suggested improvements for microscopic measurement of tissue cells and starch granules in fresh potatoes, *Am. Potato J. 44*:41 (1967).
308. Sharma, K. N., and N. R. Thompson, Relationship of starch grain size to specific gravity of potato tubers, *Mich. Agric. Exp. Stn. Q. Bull. 38*:559 (1955).
309. Poarrios, E. P., D. W. Newsom, and J. C. Miller, Some factors influencing the culinary quality of southern and northern grown Irish potatoes. I. Chemical composition, *Am. Potato J. 38*:182 (1961).
310. Barrios, E. P., D. W. Newson, and J. C. Miller, Some factors influencing the culinary quality of Irish potatoes. II. Physical characters, *Am. Potato J. 40*:200 (1963).
311. Unrau, A. M., and R. E. Nylund, The relation of physical properties and chemical composition of mealiness in the potato I. Physical properties, *Am. Potato J. 34*:245 (1957).
312. Unrau, A. M., and R. E. Nylund, The relation of physical properties and chemical composition of mealiness in the potato II. Chemical composition, *Am. Potato J. 34*:303 (1957).
313. Esau, K., *Plant Anatomy,* John Wiley and Sons, New York, 1953.
314. Reeve, R. M., Histological survey of conditions influencing texture in potatoes. I. Effects of heat treatments on structure, *Food Res. 19*:323 (1954).
315. Shewfelt, A. L., D. R. Brown, and K. D. Troop, The relationship of mealiness in cooked potatoes to certain microscopic observations of raw and cooked product, *Can. J. Agric. Sci. 35*:513 (1955).

316. Dastur, R. H., and S. D. Agnihotri, Study of the pectic changes in potato tubers at different stages of growth and in storage, *Indian J. Agric. Sci. 4*:430 (1934).
317. Sharma, K. N., D. R. Isleib, and S. T. Dexter, The influence of specific gravity and chemical composition on hardness of potato tubers after cooking, *Am. Potato J. 36*:105 (1959).
318. Whittenberger, R. T., and G. C. Nutting, Observations on sloughing of potatoes, *Food Res. 15*:331 (1950).
319. Mapson, L. W., and T. Swain, Production and Application of Enzyme Preparation in Food Manufacture SCI Monograph II, Society of Chemical Industry, London, 1961, 121.
320. Mapson, L. W., T. Swain, and A. W. Tomalin, Influence of variety, cultural conditions and temperature on storage on enzymatic browning of potato tubers, *J. Sci. Food Agric. 14*:673 (1963).
321. Clark, W., N. Mondy, K. Bedrosan, R. Ferrari, and C. Michon, Polyphenolic content and enzyme activity of two varieties of potato, *Food Technol. 11*:297 (1957).
322. Mulder, E. G., Mineral nutrition in relation to the biochemistry and physiology of potatoes. I. Effect of nitrogen, phophate, potassium, magnesium and copper nutrition on the tyrosine content and tyrosinase activity with particular reference to blackening of the tubers, *Plant Soil.* 2:49 (1949).
323. Welte, E., and K. Muller, The influence of potassium manuring on the darkening of raw potato pulp, *Eur. Potato J. 9*:38 (1966).
324. Mondy, N. I., E. O. Mobley, and S. B. Gedde-Dahl, Influence of potassium fertilization on enzymatic activity, phenolic content and discoloration of potatoes, *J. Food Sci. 32*:378 (1967).
325. Hoff, J. E., C. M. Jones, G. E. Wilcox, and M. D. Castro, The effects of nitrogen fertilization on the composition of the free amino acid pool of potato tubers, *Am. Potato J. 48*:390 (1971).
326. Mulder, E. G., and K. Bakema, Effect of the nitrogen, phosphorus, potassium and magnesium on the content of free amino acids and on the amino acid composition of protein of the tubers, *Plant Soil 7*:135 (1956).
327. Grewal, S. S., and N. P. Sukumaran, *Effect of Maleic Hydrazide (MH) Sprout Suppressant on Various Biochemical Constituents,* Ann. Sci. Rep. 1979 of the Central Potato Res. Inst. Simla, India, 1980, p. 40.
328. Hodge, J. E., Chemistry of browning reactions in model systems, *J. Agric. Food Chem. 1*:928 (1953).
329. Shallenberger, R. S., O. Smith, and R. H. Treadway, Role of the sugars in the browning reaction in potato chips, *J. Agric. Food Chem.* 7:274 (1959).
330. Schwimmer, S., C. E. Hendel, W. O. Harrington, and R. L. Olson, Interrelation among measurements of browning of processed potatoes and sugar components, *Am. Potato J. 34*:119 (1957).
331. Denny, F. E., and N. C. Thornton, Factors for colour in the production of potato chips, Contrib. Boyce *Thompson Inst. 11*:291 (1940).
332. Burton, W. F., and A. R. Wilson, The apparent effect of the latitude of the place of cultivation upon sugar content of potatoes grown in Great Britain, *Potato Res. 13*:269 (1970).
333. Sowokinos, J. R., Maturation of *Solanum tuberosum* I. Comparative sucrose and sucrose synthetase levels between several good and poor processing varieties, *Am. Potato J. 50*:234 (1973).
334. Carlsson, H., Production of potatoes for chipping, *Vaxtodling 26*:9 (1970).
335. Pressey, R., Role of invertase in the accumulation of sugars in cold storage potatoes, *Am. Potato J. 46*:291 (1969).
336. Schwimmer, S., R. W. Makower, and E. S. Rorem, Invertase and invertase inhibitor in potato, *Plant Physiol. 36*:313 (1961).
337. Hughes, J., and T. Swain, After cooking blackening in potatoes. II. Core experiments, *J. Sci. Food Agric. 13*:229 (1962).
338. Hughes, J. C., and T. Swain, After cooking blackening in potatoes III. Examination of interaction of factors by *in vitro* experiments. *J. Sci. Food Agric. 13*:358 (1962).
339. Hughes, J. C., and J. L. Evans, Studies on after cooking blackening in potatoes, 4. Field experiments, *Eur. Potato J. 10*:16 (1967).
340. Smith, O., and C. L. Davis, *Potato Processing in Potatoes: Production, Storing, and Processing* (O. Smith, ed.), AVI Pub. Co., Westport, CT, 1968, p. 558.

341. Smith, T. J., and C. C. Huxsoll, *Peeling Potatoes for Processing in Potato Processing* (W. F. Talburt and O. Smith, eds.), Van Nostrand Reinhold, New York, 1987, p. 333.
342. Commercial infrared peeling process, *Food Proc. 31*:28 (1970).
343. Graham, R. P., C. C. Huxsoll, M. R. Hart, M. L. Weever, and A. I. Morgan Jr., Dry caustic peeling of potatoes, *Food Technol. 23*:61 (1969).
344. Weaver, M. L., H. Timm, M. Nonaka, R. N. Sayre, K. C. Ng, and L. C. Whitehand, Potato composition, III. Tissue selection and its effect on total nitrogen, free amino acid nitrogen and enzyme activity, *Am. Potato J. 55*:319 (1978).
345. Huxsoll, C. C., M. L. Weaver, and K. C. Ng, Double-dip caustic peeling of potatoes, Laboratory scale development, *Am. Potato J. 58*:327 (1980).
346. Burton, W. G., Postharvest behaviour and storage of potatoes, in *Applied Biology*, Vol. 2 (T. H. Coaker, ed.), Academic Press, New York, 1978.
347. Weaver, M. L., K. C. Ng, and C. C. Huxsoll, Sampling potato tubers to determine peel loss, *Am. Potato J. 56*:217 (1979).
348. Finglas, D. M., and R. M. Fawks, Nutritional composition of UK retail potatoes both raw and cooked, *J. Sci. Food Agric. 35*:1347 (1984).
349. True, R. H., J. M. Hogan, J. Augustin, S. R. Johnson, C. Teitzel, R. Toma, and P. Orr, Changes in nutrient composition of potatoes during home preparation. III. Minerals, *Am. Potato J. 56*:339 (1979).
350. Zarneger, L., and A. E. Bender, The stability of vitamin C in machine peeled potatoes, *Proc. Nutr. Soc. 30*:94 (1971).
351. Gorun, E. G., Effect of mode of potato peeling on content of B group vitamins (in Russian), *Izu. Vyssh Uchebn Zaved Pishch Tekhnol. 6*:154 (1973).
352. Mondy, N. I., and G. Barry, Effect of peeling on total phenols, total glycoalkaloids, discoloration and flavor or cooked potatoes, *J. Food Sci. 53*:576 (1988).
353. Marquez, G., and M. C. Anon, Influence of sugars and amino acids in colour development in fried products, *J. Food Sci. 51*:157 (1986).
354. Sullivan, J. F., M. F. Kozempel, M. J. Egoville, and E. A. Talley, Loss of amino acids and water soluble vitamins during potato processing, *J. Food Sci. 50*:1249 (1986).
355. Mishkin, M., I. Saguy, and M. Karel, A dynamic test for kinetic models for chemical changes during processing: ascorbic acid degradation in dehydration of potatoes, *J. Food Sci. 49*:1267 (1984).
356. Warner, K., C. D. Evans, G. R. List, B. K. Boundy, and W. I. Kwolet, Pentane formation and rancidity in vegetable oils and in potato chips, *J. Food Sci. 39*:761 (1974).
357. Moore, M. D., L. D. Van Blaricom, and T. L. Senn, The effect of storage temperature of Irish potatoes on the resultant chip colour, *Clemson Univ. Cell For Recreat. Resour. Dep. For Fer. Res. Ser. 43*:1 (1963).
358. Agle, W. M., and G. W. Woodbury, Specific gravity dry matter relationship and reducing sugar changes affected by potato variety, production area and storage, *Am. Potato J. 45*:119 (1968).
359. Clegg, M. D., and H. W. Chapman, Postharvest dicoloration of chips from early summer potatoes, *Am. Potato J. 39*:176 (1962).
360. Dexter, S. T., and D. K. Salunkhe, Improvement of potato chip colour by hot water treatment of slices, *Mich. State Coll. Exp. Stn. Q. Bull. 34*:399 (1952).
361. Dexter, S. T., and D. K. Salunkhe, Chemical treatment of potato slices in relation to the extraction of sugars and other dry matter and quality of potato chips, *Mich. State Coll. Exp. Stn. Q. Bull. 35*:102 (1952).
362. Dexter, S. T., and D. K. Salunkhe, Control of potato chip color by treatment of slices with glucose solution following an acid treatment, *Mich. State Coll. Agric. Exp. Stn. Q. Bull. 35*:156 (1952).
363. Smith, O., *Improving the Colour of Potato Chips,* Natl. Potato Chip Inst. Potatoes, Article 10, 1, 1950.
364. Smith, O., Factors affecting the methods of determining potato chip quality, *Am. Potato J. 38*:265 (1961).
365. Smith, O., Potato chip research in 1962, Proc. Prod. Tech. Div. Meeting, Potato Chip Inst. Intern, Ithaca, NY, 1963, p. 2.

366. Smith, O., Changes in the manufacture of snack foods. *Cereal Sci. Today 19*:306 (1974).
367. Smith, O., *Potatoes: Production, Storing, Processing,* 2nd Ed., AVI Pub., Westport, CT, 1977.
368. Quast, D. G., and M. Karel, Effects of environmental factors on the oxidation of potato chips, *J. Food Sci.* 37:584 (1972).
369. Reeves, A. F., Potato chip colour ratings of advanced selections from Maine potato breeding program, *Am. Potato J. 59*:389 (1982).
370. Sherman, M., and E. E. Ewing, Temperature cyanide and oxygen effects on the respiration, chip colour, sugars and organic acids of stored potatoes, *Am. Potato J. 59*:165 (1982).
371. Singh, R. P., D. R. Heldman, and B. F. Cargill. The influence of storage time and environmentals on potato chip quality, in *Potato Storage Design Construction, Handling and Environmental Control* (B. F. Cargill, ed.), Michigan State University, East Lansing, 1976.
372. Whiteman, T. M., Improvement in the colour of potato chips and French fries by certain precooking treatments, *Potato Chipper 11*:24 (1951).
373. Young, N. A., *Potato products: Production and Markets in European Communities,* Information on Agriculture, No. 75 Office for the Official Publications of the European Communities, Luxembourg, 1981.
374. Fitzpatric, T. J., E. A. Talley, and W. L. Porter, Preliminary studies on the fate of sugars and amino acids in chips made from fresh and stored potatoes, *J. Agric. Food Chem. 13*:10 (1965).
375. Fitzpatric, T. J., and W. L. Porter, Changes in sugars and amino acids in chips made from fresh stored and reconditioned potatoes, *Am. Potato J. 43*:238 (1966).
376. Jaswal, A. S., Effects of various processing methods on free and bound amino acid content of potatoes, *Am. Potato J. 50*:86 (1973).
377. Bucko, A., K. Obonova, and P. Ambrova, Effects of storage and culinary processing on vitamin C losses in vegetables and potatoes, *Nahrung 21*:107 (1977).
378. Pelletier, O., C. Nantel, R. Leduc, L. Tremblay, and R. Bressani, Vitamin C in potatoes prepared in various ways, *J. Inst. Can. Sci. Technol. Aliment. 10*:138 (1971).
379. Deutsch, R. M., Science looks at potato chips, *Chipper Snacker 35*:15 (1978).
380. Shaw, R., C. D. Evans, S. Munson, G. R. List, and K. Warner, Potato chips from unpeeled potatoes, *Am. Potato J. 50*:424 (1973).
381. Feustel, I. C., and R. W. Kueneman, Frozen French fries and other frozen products, in *Potato Processing* (W. F. Talburt and O. Smith, eds.), AVI Pub. Co., Westport, CT, 1967.
382. Weaver, M. L., R. M. Reeve, and R. W. Kueneman, Frozen potato products, in *Potato Processing,* 3rd ed. (W. F. Talburt and O. Smith, eds.), AVI Pub. Co., Westport, CT, 1975.
383. Zobel, M., Nutritional aspects of potato peeling in the DDR and from the international viewpoint, *Ernaehrungsforschung 24*:74 (1979).
384. Murphy, E. W., A. C. Marsh, K. E. White, and S. N. Hagan, Proximate composition of ready to serve potato products, *J. Am. Diet. Assoc. 49*:122 (1966).
385. Augustin, J., B. G. Swanson, C. Teitzel, S. R. Johnson, S. F. Pometto, W. F. Artz, C. P. Huang, and C. Shoemaker, Changes in the nutrient composition during commercial processing of frozen potato products, *J. Food Sci. 44*:807 (1979).
386. Oguntuna, T. E., and A. E. Bender, Loss of thiamin from potatoes, *J. Food Technol. 11*:347 (1976).
387. Gorun, E. G., Changes in the vitamin activity of potatoes during production of quick frozen fresh fries, *Kholod Tekh. 10*:15 (1978).
388. Feustel, I. C., and R. W. Kueman, Frozen French fries and other frozen potato products, in *Potato Processing,* 2nd ed. (W. F. Talburt and O. Smith, eds.), AVI Pub. Co., Westport, CT, 1967.
389. Tressler, D. K., W. B. Van Arsdel, and M. J. Copley, *The Freezing Preservation of Foods,* Vol. 4, 4th ed., AVI Pub. Co., Westport, CT, 1963.
390. Talburt, W. F., Canned white potatoes, in *Potato Processing,* 2nd ed. (W. F. Talburt and O. Smith, eds.), AVI Pub. Co., Westport, CT, 1967.
391. Markakis, P., T. M. Freeman, and W. H. Harte, Dough containing vital gluch amylopectin and inert starches, U.S. Patent 3027258 (1962).

392. Fast, R. B., C. F. Spotts, and R. A. Morck. Process for making a puffable chip-type snack food product, U.S. Patent 3451822 (1969).
393. Reinertsen, B. J., Method of making a chip-type food product, U.S. Patent 3361573 (1968).
394. Singer, N. S., and E. G. Beltran, Process of making a snack product, U.S. Patent 3502479 (1968).
395. Liepa, A. L. Potato food product, U.S. Patent 3396036 (1968).
396. Hitton, B. W., Potato products and process for making same, U.S. Patent 3109739 (1963).
397. Murray, D. G., N. G. Marota, and R. M. Boettger, Novel amylose coating for deep fried potato products, U.S. Patent 3597227 (1971).
398. Slakis, A. J., W. K. Kubr, R. L. Hughes, and A. J. Neilson, Cyclohexylsulfamic acid as a taste improver of raw potato products, U.S. Patent 3353962 (1970).
399. Olson, R. L., and W. D. Harrington, *Dehydrated Mashed Potatoes—A Review,* U.S. Department of Agriculture, Western Regional Research Laboratory, Albany, CA, A C-297, 1951.
400. Kueneman, R. W., Dehydrated mashed potatoes, *Proc. Potato Utiliz. Conf. 8*:64 (1957).
401. Feustel, I. C., C. F. Hendel, and M. E. Juilly, *Potatoes in Food Dehydration,* Vol. II, *Processes and Products,* AVI Pub. Co., Westport, CT, 1964.
402. Talburt, W. F., F. P. Boyle, and C. F. Hendel, Dehydrated mashed potatoes—potato granules, in *Potato Processing* (W. F. Talburt and O. Smith, eds.), Van Nostrand Reinhold Company, New York, 1987, p. 12.
403. Ooraikul, B., Processing of potato granules with the aid of freeze-thaw technique, Ph.D. Thesis, University of Alberta, Edmonton, Canada, 1973.
404. Jadhav, S., L. Steele, and D. Hadziyev, Vitamin C losses during production of dehydrated mashed potatoes, *Food Sci. Technol. 8*:225 (1975).
405. Jadhav, S., L. M. Berry, and L. F. L. Clegg, Extruded French fries from dehydrated potato granules processed by freeze-thaw techniques, *J. Food Sci. 41*:852 (1976).
406. Potter, A. L. Jr., Changes in physical properties of starch in potato granules during processing, *J. Agric. Food Chem. 2*:516 (1954).
407. Burton, W. G., Mashed potato powder IV. Deterioration due to oxidative changes, *J. Soc. Chem. In. 68*:119 (1949).
408. Willard, M. J., V. M. Hix, and G. Kluge, Dehydrated mashed potatoes—potato flakes, in *Potato Processing* (W. F. Talburt and O. Smith, eds.), Van Nostrand Reinhold Company, New York 1987, p. 13.
409. Pedersen, D. C., and P. M. Sautier, Vitamin enriched potato flakes, U.S. Patent 3,833,739 (1974).
410. Sapers, G. M., O. Panasuik, F. B. Talley, S. F. Ogman, and R. L. Shaw, Flavour quality and stability of potato flakes: Volatile components associated with storage changes, *J. Food Sci. 37*:579 (1972).
411. Augustin, J., B. G. Swanson, S. F. Pometto, C. Teitzel, W. F. Artz, and C. P. Huang, Changes in nutrient composition of dehydrated potato products during commercial processing, *J. Food Sci. 44*:216 (1979).
412. Augustin, J., G. A. Marousek, W. F. Artz, and B. G. Swansco, Vitamin retention during preparation and holding of mashed potatoes made from commercially dehydrated flakes and granules, *J. Food Sci. 47*:274 (1982).
413. Kluge, G., F. S. Y. Appoldt, G. Seiler, and K. Petutsonig, Production of dried potato flakes, British Patent 1473036 (1977).
414. Sapers, G. M., O. Panasuik, S. B. Jones, E. B. Kalan, F. B. Talley, and R. L. Shaw, Iron fortification of dehydrated mashed potatoes, *J. Food Sci. 39*:552 (1974).
415. Glabau, C. A., Cookie production with potato flour as an ingredient, *Bakers Wkly. 178*:40 (1958).
416. Kim, H. S., and H. J. Lee, Development of composite flours and their products utilizing domestic raw material IV. Effect of additives on the bread making quality of composite flours, *Korean J. Food Sci. Technol. 9*:106 (1977).
417. Rivoche, E. J., Food products and method of making the same, U.S. Patent 2791508 (1957).
418. El-Samathy, S. K., A. M. Elias, and M. M. Morad, Effect of potato flour on the rheological properties of the dough, bread quality and staling, *Cereal Microbiol. Technol. Lebensm. 4*:186 (1976).
419. Ceh, M., Baking bread with potato flour addition, *Hrana Ish rana* (Yugoslavian) *15*:113 (1974).
420. Yanez, E., D. Ballester, H. Wuth, W. Orrego, V. Gattas and S. Estay, Potato flour as a partial

replacement of wheat flour in bread: baking studies and nutritional value of bread containing graded levels of potato flour, *J. Food Technol. 16*:291 (1981).

421. Zahana, T., P. Sutescu, F. Popescu, and L. Gontea, The quality of bread supplemented with maize flour and other ingredients, *Igiena* (Romanian) *22*:11 (1972).
422. Bushway, A. A., P. R. Belyea, R. H. True, T. M. Work, D. D. Russell, and D. F. McGann, Potato starch and flour in frankfurters: effect on chemical and sensory properties and total plate counts, *J. Food Sci. 47*:402 (1982).
423. Potter, A. L., and M. L. Belote, New potato snack item, *Bakers Wkly. 197*:42 (1963).
424. Baczkowicz, M., and P. Tomasik, A novel method of utilization of potato juice, *Starch 37*:241 (1985).
425. Cordon, T. C., R. H. Treadway, M. D. Walsh, and M. F. Osborne, Lactic acid from potatoes, *Ind. Eng. Chem. 42*:1833 (1950)
426. Eskew, R. K., P. W. Edwards, and C. S. Redfield, *Recovery and Utilization of Pulp from White Potato Starch Factories,* U.S. Department of Agriculture, Eastern Regional Research Laboratories, AIC-ZO4, 1948.
427. Highlands, M. E., Potatoes and potato pulp for livestock feed, in *Potato Processing* (W. F. Talburt and O. Smith, eds.), AVI Pub. Co., Westport, CT, 1967, p. 540.
428. Treadway, R. H., in *Potato Processing* (W. F. Talburt and O. Smith, eds.), Van Nostrand Reinhold, New York, 1987, p. 647.
429. Szkilladziowa, W., B. Secomska, I. Nadolna, I. Trzebska-Jeska, M. Wartanowicz, and M. Rakowska, Results of studies on nutrient content in selected varieties of edible potatoes, *Acta Aliment Acad. Sci. Hung. 3*:87 (1977).
430. Augustin, J., R. B. Toma, R. H. True, R. L. Shaw, C. Teitzel, S. R. Johnson, and P. Orr, Composition of raw and cooked potato peel and flesh: Proximate and vitamin composition, *J. Food Sci. 44*:805 (1979).
431. Bretzloff, C. W. Calcium and magnesium distribution in potato tubers, *Am. Potato J. 48*:97 (1971).
432. Johnston, F. B., I. Hoffonan, and A. Petrosovits, Distribution of mineral constituents and dry matter in potato tubers, *Am. Potato J. 45*:287 (1968).
433. Kubisk, A., J. TomKowiak, and M. Andrzejewska, The control of some trace elements in different parts of the tuber in five potato varieties, *Hodowla Rosl. Aklim. Nasienn. 22*:81 (1978).
434. Mondy, N. I., and R. Ponnampalam, Effects of baking and frying on nutritive value of potato minerals, *J. Food Sci. 48*:1475 (1983).
435. Chick, H., and E. B. Slack, Distribution and nutritive value of the nitrogenous substances in the potato, *Biochem. J. 45*:211 (1949).
436. Talley, E. A., R. B. Toma, and P. H. Orr, Composition of raw and cooked potato peel and flesh: amino acid content, *J. Food Sci. 48*:60 (1983).
437. Herrera, H., *Potato Protein: Nutritional Evaluation and Utilization,* Ph.D. Thesis, Michigan State University, East Lansing, 1979.
438. Toma, R. B., J. Augustin, P. Orr, R. H. True, J. M. Hogan, and R. L. Shaw, Changes in the nutrient composition of potato during home preparation. I. Proximate Composition, *Am. Potato J. 55*:639 (1978).
439. Hellendoorn, E. W., M. Vanden Top, and J. E. M. Vander Weide, Digestibility in vitro of dry mashed potato products, *J. Sci. Food Agric. 21*:71 (1970).
440. Englyst, H., H. S. Wiggins, and J. H. Cummings, Determination of non-starch polysaccharide in plant foods by gas chromatography of constituent sugars as alditol acetates, *Analyst. 107*:307 (1982).
441. Jone, G. P., D. R. Briggs, M. L. Wahlquist, and L. M. Flentge, Dietary fiber content of Australian foods, I. Potato, *Food Technol. Aust. 37*:81 (1985).
442. Choudhari, R. N., A. A. Joseph, D. Ambrose, V. Narayan Rao, M. Swaminathan, A. Sreenivasan, and V. Subramanyan, Effect of cooking, frying, baking and canning on nutritive value of potato, *Food Sci.* (Mysore) *12*:253 (1963).
443. Jaswal, A. S., Effect of various processing methods on free and bound amino acids content of potatoes, *Am. Potato J. 50*:86 (1973).

444. Johnston, D. E., and W. T. Oliver, The influence of cooking technique on dietary fiber of boiled potato, *J. Food Technol. 17*:99 (1982).
445. Pau, A. A., and D. A. T. Southgate, *McCance and Widdowson's The Composition of Foods,* 4th ed., MRC Serial Report, No. 297, London, 1978.
446. Swaminathan, K., and B. M. L. Gangwar, Cooking losses of vitamin C in Indian potato varieties, *Indian Potato J. 3*:86 (1961).
447. Leichsering, J. M., L. M. Norris, and H. L. Pilcher, Ascorbic acid contents of potatoes: I. Effect of storage and of boiling on the ascorbic, dehydroascorbic and diketoglyconic acid contents of potatoes, *Food Res. 22*:37 (1957).
448. Woolfe, J. A., Effect of storage, cooking and processing value of potato, in *Potato in Human Diet,* Cambridge University Press, London, 1984, p. 4.
449. Fenton, F., Vitamin C retention as a criterion of quality and nutritive value in vegetables, *J. Am. Diet. Assoc. 16*:524 (1940).
450. Richardson, J. E., R. Davis, and H. L. Mayfield, Vitamin C content of potatoes prepared for table use by various methods of cooking, *Food Res. 2*:35 (1937).
451. Domah, A., J. Davidek, and J. Velisek, Changes of L. ascorbic acid and L. dehydroascorbic acids during cooking and frying of potatoes, *Z. Lebensm Unters Forsch. 154*:272 (1974).
452. Page, E., and F. M. Hanning, Vitamin B_6 and niacin in potato, *J. Am. Diet. Assoc. 42*:42 (1963).
453. Kahn, R. M., and E. G. Halliday, Ascorbic acid content of white potatoes as affected by cooking and standing on steam table, *J. Am. Diet. Assoc. 20*:220 (1944).
454. Pelletier, O., C. Nantel, R. Le duc, L. Tremblay, and R. Bressani, Vitamin C in potatoes prepared in various ways, *Can. Inst. Food Sci. Technol. J. 10*:138 (1977).
455. Korschgen, B., R. Baldwin, and S. Snider, Quality factors in beef, pork and lamb cooked by microwaves, *J. Am. Diet. Assoc. 69*:635 (1976).
456. Klein, L. B., and N. I. Mondy, Comparison of microwave and conventional baking of potatoes in relation to nitrogenous constituents and mineral composition, *J. Food Sci. 46*:1874 (1981).

3

Sweet Potato

P. M. KOTECHA AND S. S. KADAM
Mahatma Phule Agricultural University, Rahuri, India

I. INTRODUCTION

Sweet potato is an important and leading vegetable crop of tropical and subtropical countries. It is considered a native of tropical America. In the United States, it is grown extensively throughout the southeastern and south central states and is shipped to the northern markets. China is the largest producer of sweet potato. Other sweet potato–producing countries are Vietnam, Uganda, India, Rwanda, and Brazil (Table 1) (1).

TABLE 1 Major Producers of Sweet Potatoes

Country	Production (1000 MT)
World	131,707
Africa	6,204
North America	1,396
South America	1,538
Asia	121,885
Europe	77
China	112,220
Vietnam	2,000
Uganda	1,670
India	1,300
Rwanda	900
Brazil	760
Madagascar	485
Tanzania	340

Source: Ref. 1.

II. BOTANY

Sweet potato (*Ipomoea batatas* Poir) belongs to the Convolvulaceae (morning glory) family. The specific name "potato" originated with the Indian word *batatas*, which later was also applied to the white or Irish potato. Sweet potato is a swollen, fleshly rooted perennial with prostrate or slender stems. The blossoms resemble those of the common morning glory. Sweet potato is extremely heterozygous and is a hexaploid with a somatic chromosome number of 90 (2). Sweet potato requires a long and warm growing season.

A. Morphology

Sweet potato produces tubers, which develop as a result of secondary growth of a few roots. Most fleshy roots develop from the initial fibrous root system of the plant. The structure of the fleshy roots (Fig. 1A) is similar to that of the primary root. The principal tissues are the periderm (Fig. 1B), the ring of secondary vascular bundles, the tracheids, sieve tubes, and the laticifers (3).

B. Cultivars

Thompson and Beattle (4) classified sweet potato cultivars into eight well-defined food types primarily used for human consumption. "Feed types" are of inferior quality. "Food types" are subdivided into "dry-fleshed" and the "moist-fleshed" cultivars. These terms refer to characteristics of the cooked sweet potato and have no relation to the water content of the raw vegetable. According to Thompson and Kelly (5), in fact, the so-called dry cultivars commonly contain a higher percentage of moisture than the moist cultivars. Boswell (6) suggested the terms "soft" and "firm" to describe cooked potatoes. The soft-fleshed cultivars are of greater commercial importance in the United States than the firm-fleshed cultivars. Nancy Hall is an old cultivar, important in some areas. Nancy Gold is a deep orange-fleshed mutant of Nancy Hall. The new cultivars developed by selection of mutations and by crossing include Ranger, Earlyport, Goldrush, Heart-O-Gold, and Australian Canner. The important firm-fleshed cultivars are Big Stem Jersey, Yellow Jersey, and Maryland Golden (a mutation of Little Stem Jersey). Pelcian Processor and Whitestar are nonfood cultivars with larger yields and higher starch content, principally used for processing. The reported sweet potato lines include W-13, W-78, W-71, W-115, W-125, W-119, and W-154. A new cultivar of sweet potato, Pope, developed at North Carolina State University, is early maturing, high yielding, and has high resistance to *Fusarium* wilt, root-knot nematodes, and flood damage in the field (7). The Travis sweet potato cultivar, released by the Lousiana Agricultural Experiment Station in 1980, has been reported to be early maturing, high yielding, and resistant to soil rot (8). Another new cultivar, Eureka, has been released by Lousiana State University. It has high quality and is resistant to soil rot (9). Other cultivars released include Carmoex, Oklamex Red, and Rojo Blanco.

The most popular cultivars of sweet potato grown in India are V-B (FB. 4004), V-12 (TST-White) or Tie Shen Tun. The Indian Agricultural Research Institute, New Delhi, released three improved varieties of sweet potato in 1967: Pusa Safed, Pusa Sunehari, and Pusa Lal (10). At the Central Tuber Crops Research Institute, Trivendrum, India, the following promising genotypes of sweet potato have been identified (11): open pollinated progenies, OP-1, 2, 34, 217, and 219, Selection S-43, 73, 107, and 162. Pentaploids (2n = 75) of sweet potato have been developed, and few among them have shown tuber development comparable with cultivated hexaploid sweet potato (11).

(A)

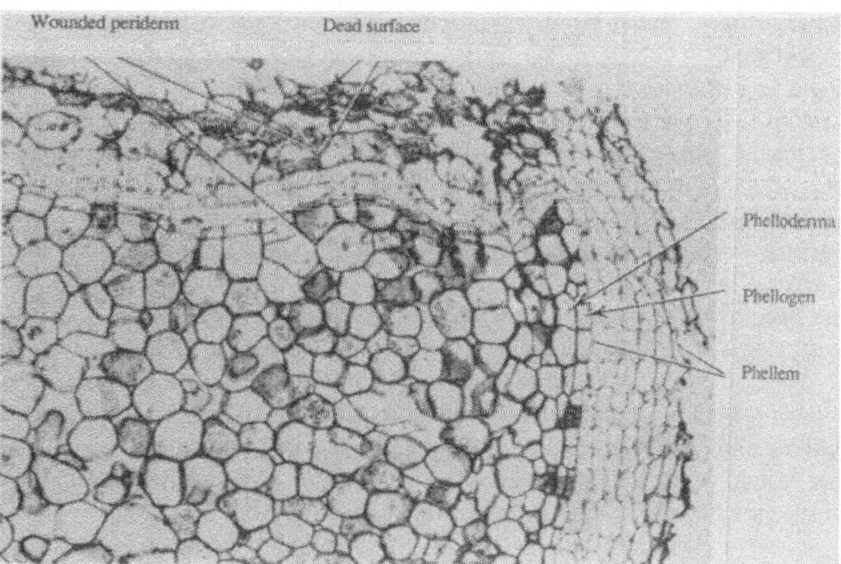

(B)

FIGURE 1 (A) Fleshy roots of sweet potato. (B) Periderm of sweet potato root. Note natural periderm to right and wounded periderm (arrows). (From Ref. 3.)

C. Fleshy Root Development

Fleshy root development in sweet potato is largely governed by soil-environment interaction. It is associated with rate of cell division as well as starch density in the cells and accumulation of starch granules. Starch deposition in the cells of the cortex may contribute to tuber bulk. Starch deposition is mostly confined to the cortical region and is normally observed on the 8th day after planting, when anomalous cambia arise around individual vessels in the vascular region. The intensity of starch deposition appears to be high in the cortical region by the 14th day, when the cortical cells are almost filled with starch grains. By the 20th day, the deposition of starch grain extends to the vascular region.

III. PRODUCTION

A. Soil and Climate

Soil texture, drainage, and aeration have tremendous influence on the growth and productivity of sweet potato vines. The ideal soil for sweet potato must be light, well drained, and aerated. Sandy or loamy sand soil with a clay subsoil are ideal for its growth (10). Sandy loam soils favor maintenance of proper balance between vegetative (aerial) growth and fleshy root development. In contrast, clayey soils or soils that are rich in organic matter produce a great deal of vegetative growth at the cost of fleshy root development, the result being that productivity is adversely affected. In heavy soils, roots become rough, small, irregular in shape and stringy. These soils, due to their sticky nature, cause harvesting difficulties leading to fleshy root injury. In contrast, very light soils without a relatively compact subsoil and deep soils produce long slender roots, which are not commercially desirable (10–12).

In its native environment, sweet potato is a tropical plant. Like other warm climate crops, it thrives better in warm humid weather, requires ample sunshine, and is vulnerable to frost injury. It can withstand drought conditions very well and can yield fairly well even in a semi-arid climate without supplemental irrigation. The optimum temperature for obtaining excellent vine growth as well as tuber yield is 24°C. Temperatures lower than 10°C generally cause chilling injury. The optimum temperature range was found to be 20–30°C. Higher soil temperatures (>30°C) affect tuberization and tend to encourage vegetative growth of aerial plant parts at the cost of tuber formation and growth. Lower soil temperatures (<10°C) also affect tuber development. Fleshy root development in the sweet potato is governed by photoperiod: short days promote it, whereas long days favor vine growth and inhibit fleshy root development.

For the best growth of sweet potato crops, rainfall should be moderate and should occur during the early stages of growth. The optimum rainfall is considered to be around 500–675 mm, but sweet potato can thrive with up to 1000 mm rainfall. The growing season is usually determined by the amount of rainfall received at a particular place and the availability of supplemental irrigation. For example, in heavy rainfall areas, sweet potato is grown mainly as a *rabi* (October-March) crop, while in places where rainfall is moderate, but abundant during the early part of the growing season, crops can be grown throughout the year.

B. Propagation

The most common method of propagation in sweet potato is from cuttings taken from vines of the previous crop. In certain parts of India where the growing season is relatively short, the tubers are planted in nursery beds spaced at 45 × 30 cm and 5–6 cm deep. After sprouting, the shoots are cut and planted in another nursery bed spaced at 60 × 30 cm in order to encourage rapid vegetative growth. When these vines are of the desired length, cuttings are made for further planting in the main field. The cuttings, about 30 cm long, are taken from healthy vines. When tubers are used, it is necessary to recognize the effect of apical dominance—only one terminal shoot emerges from one shoot. To break apical dominance, the tubers must be dipped in warm water (40–42°C) for 10 minutes or in 10 ppm 2,4-D (12). The soil for planting of cuttings is brought to a fine tilth by repeated shallow sowing, harrowing, and leveling.

C. Cultural Practices

1. Planting

The cuttings, taken either from the vines of the old crop or newly planted crop, are planted in flat beds or ridges and furrows. The recommended spacing for rooting of cuttings is 15 × 30 cm for flat beds or 60 × 22 cm for ridges and furrows (10). The land must thoroughly be irrigated to have the cuttings planted in muddy soils. When planted in ridges, the cuttings are planted at about one third the height of the ridges in such a way that both ends of the cuttings are free and the central portions are buried in the soil, which yield roots in about a month's time. One hectare of land accommodates about 1,25,000 such cuttings, and the grower must have additional cuttings to make up for plant mortality, which can be as high as 10–25% (11).

Crop quality is influenced by both time of planting as well as agro-climatic conditions. Biswas et al. (13) recorded maximum tuber weight and maximum yield for sweet potatoes planted in September in West Bengal, while May planting increased the vine weight only. This might have been due to high rainfall, fewer hours of sun, and high humidity. In the Konkan region of Maharashtra, Nawale and Salvi (14) observed that sweet potatoes growing during the *rabi* season (October planting) had higher tuber yield (about 77% increase) compared to *Kharif* (May-June) plantings.

Sweet potato is normally planted in ridges, with space between rows of 60 cm and plant-to-plant spacing of 20 cm, thus accommodating 83,000 plants/ha. However, Singh and Mandal (15) recorded the highest tuber yield (27.9 t/ha) at 45 × 15 cm spacing and the lowest (19.7 t/ha) at 90 × 15 cm.

2. Manuring and Fertilization

Sweet potato requires smaller amounts of nutrients than potato, Choudhury (16) reported that a sweet potato crop yielding 6.6 tons of tubers per hectare removes from the soil 30 kg of nitrogen, 9 kg of phosphorus, and 60 kg of potash. Other reports in this regard indicated that a crop yielding 5.5 tons of tubers required an uptake of 15 kg of nitrogen from the soil (10), while a crop producing 15 tons of tubers per hectare removed 70 kg of nitrogen, 20 kg of phosphorus, and 170 kg of potash. However, an excess of nitrogenous fertilizers must be avoided, as it results in stimulation of vegetative growth at the cost of tuberization and tuber development. Fleshy root development is promoted by higher levels of potassium. Therefore, the fertilizer schedule for sweet potato must involve less nitrogen and higher doses of potassium. Choudhury (16) recommended the application of 60 kg of nitrogen, 60 kg of phosphorus and 120 kg of potash per hectare for optimum growth. In addition, 25 tons of farmyard manure should be added at the time of soil preparation. Other fertilizer recommendations include 56 kg of nitrogen, 56 kg of phosphorus, and 28 kg of potash, and 90 kg of N, 90 kg of P_2O_5, and 90 kg of K_2O/ha (17). The amount of individual fertilizers to be applied depends largely on the type of soil and existing fertility status of the soil (18). Analysis of soil and plant parts indicated that approximately 41 kg P, 68 kg K, 22 kg of Ca, and 8 kg of Mg are needed to produce 18 tons fleshy roots/per hectare of sweet potato (11).

3. Irrigation

A total of 500–700 mm of water is required during the growing season for this crop. Whenever the rainfall is inadequate, supplementary irrigation should be performed. In *rabi* (October-March) crops, irrigation should be done at intervals of 10 days or whenever required during hot season.

Supplementing rainfall with irrigation and proper drainage in low-lying areas are the most important aspects of water management in sweet potato. For proper sprouting and establishment of vine cuttings, planting should immediately be followed by irrigation so as to have moist soil for 4–5 days after planting. For dry-season crops, 8–10 irrigations are required.

4. Pruning of Vines

Pruning of vines has been shown to increase root yields in sweet potato. The retardation of vegetative growth of vines has also been shown to reduce rotting of new vigorous shoots capable of photosynthesis and also reducing wastage of photosynthates, which are in turn translocated to roots. The vines are pruned back to 20–30 cm to increase root yields. Some growers prune the vines after one month to improve their productivity.

5. Turning of Vines

The vines, due to their ability to grow rapidly, can create a favorable microclimate for insect pests to hibernate or conditions conducive to bacterial or fungal growth. The vines produce roots at nodes when they come in contact with the soil, thereby affecting the growth of the main root system. Therefore, vines should be turned at least once or twice before they are fully grown. Some workers have suggested turning the vines every week, while others think this unsafe as it may cause injury to the vines (10).

6. Intercultural Operation

Regular hoeing and weeding are necessary to check weed growth and to maintain soil aeration. First weeding should coincide with the onset of rapid growth of vines, which takes place 3 weeks after planting.

D. Diseases and Pests

Among insects, sweet potato weevil (*Cylas formicarius*) and leaf-eating caterpillars (*Horse convolvuli* and *Diacrisia obligua*) are often serious in sweet potato crops. Steinbaur and Kushman (19) identified the following pathogens infecting sweet potatoes: *Diaporthe batatitis* (dry rot), *Diploidia theobromae* (Java black rot), *Ceratocystis fimbriate* (black rot), *Fusarium oxysporum* (surface rot), *Macrophomina phaseoli* (charcoal rot), and *Rhizopus stolonifer* (soft rot). In most cases, the entire root becomes shriveled; blackrot makes the tissues black and bitter (20). *Diploida* and *Fusarium* develop very slowly, and infection is conspicuous only after 6 weeks of storage at 1.7–4.4°C, when the tissue becomes spongy and turns red. Sweet potatoes may develop softrot with a mushy or starchy odor when stored at low temperatures as a result of infections caused by *Penicillium* sp., *Botritis cinerea*, and *Mucor racemosus*.

E. Harvesting, Handling, and Maturity

For early market, sweet potatoes are harvested as soon as they reach marketable size, regardless of stage of maturity. The main crop intended for storage is generally harvested after full maturity (15). Sweet potatoes are harvested by clearing the vines and digging the tubers. In the United States, and elsewhere, where sweet potatoes are grown on an extensive commercial basis, they are harvested when the maximum number of roots are developed; depending on cultivar and environmental conditions, this may vary from 90 to 130 days after transplanting. In this case, the entire crop is harvested mechanically on a onetime basis. A mold-board plough and a sharp-rolling cutter attached to a tractor may be used for cutting and digging. The implement

TABLE 2 Effects of Harvesting Implements Sweet Potato Loss During Storage

Implement	Loss of decay (%)	Total loss (%)
16-inch plow		
Fast	8.08	19.53
Slow	9.05	20.47
Middle buster	13.08	25.88
12-inch plow	11.92	24.35
Rod-wing middle buster	8.58	19.98
Mechanical digger	14.52	29.10

Source: Ref. 21.

used for digging sweet potatoes should be such that physical injury and bruising is kept to a minimum. Lutz et al. (21) demonstrated that a 16-inch turn plough and a rod wing middle buster operated better than a 12-inch plough, a middle buster without rod wings, or a mechanical digger (Table 2) (21). The roots are separated after digging and left on the row to dry. Excessive handling of roots should be avoided by collecting and depositing sweet potatoes directly into storage crates. The harvesting operation is usually performed early in the morning to avoid heat injury (sunscald). According to Thompson et al. (22), in steep areas after the roots are pulled, the soil is commonly returned to allow smaller roots to develop.

There are no definite maturity indices for sweet potatoes (23). A random sampling or trial digging of the area indicates the fleshy root maturity. Austin et al. (24) reported that mature tubers had high starch content. The maximum concentration of carotene and total carotenoid pigments occurred at the time of usual harvest for storage (25). Significant cultivar differences were observed by these investigators. In general, the carotene and total carotenoid pigments of sweet potatoes increased during the first part of the harvest period and then decreased. The ascorbic acid content decreased later in the harvest period. Ramanujam and Indira (26) reported that bulking rate, rather than time of fleshy root initiation, was the more suitable criterion to determine maturity period of sweet potato.

The harvested sweet potatoes are prepared for market by careful grading, cleaning, and packaging. Uniform, medium-sized sweet potatoes, free from cuts, bruises, and rot, are in great demand. In the United States, sweet potatoes are usually washed before they are packed for market, machine-washing sometimes being adopted. Sweet potatoes are packed in various types of containers, such as hampers, round staved baskets, crates, and boxes (5).

IV. CHEMICAL COMPOSITION OF FLESHY ROOTS

Sweet potatoes contain more calories than do potatoes (113 vs. 75/100 g). They are also an exceptionally rich source of vitamin A (7100 IU/100 g). With its appreciable quantities of ascorbic acid, thiamine, riboflavin, niacin, phosphorus, iron, and calcium (27,28), the sweet potato, in combination with legumes, can be an ideal food to combat protein-calorie malnutrition (PCM). The chemical composition of tubers varies widely (Table 3) according to cultivar, climatic condition, degree of maturity, and duration of storage (29).

TABLE 3 Chemical Composition of Sweet Potato Roots

Content	Ref. 38	Ref. 36
Moisture (%)	50–80	72.84
Total carbohydrates (%)	—	—
Starch (%)	10–29	24.28
Protein (%)	0.5–2.4	1.65
Ether extract (%)	1.0–6.4	0.30
Fiber (%)	—	—
Reducing sugars (%)	0.5–2.5	0.85
Nonstarch (%)	1.0–7.5	—
Mineral matter (%)	0.9–1.4	0.95
Carotene (mg/100 g)	1–12	—
Thiamine (mg/100 g)	0.1	0.066
Riboflavin (mg/100 g)	0.6	0.147
Nicotinic acid (mg/100 g)	0.9	0.674
Ascorbic acid (mg/100 g)	20–30	22.7
K (mg/100 g)	373	204
P (mg/100 g)	49	28
Ca (mg/100 g)	30	22
S (mg/100 g)	29	—
Mg (mg/100 g)	24	10
Na (mg/100 g)	13	13
Fe (mg/100 g)	0.8	0.59
Zinc (mg/100 g)	—	0.28
Copper (mg/100 g)	—	0.169
Manganese (mg/100 g)	—	0.355
Vitamin A (IU/100 g)	—	20063
Food energy (kJ/100 g)	—	441

A. Starch

Sweet potato tubers are rich in starch (15–28%); the amylose content of starch varies from 17 to 22%. Much of the starch is converted to maltose during cooking. The starch forms a clear stable gel, a useful product for the confectionary and bakery industry.

B. Sugars

During storage of tubers, some starch is converted into reducing sugars and subsequently into sucrose. The reducing sugar content was 4–25 times higher in baked roots than in raw roots (Table 4) (30). It is known that sweet potatoes contain β-amylase (31). Ikemiya and Deobald (32) reported an α-amylase in sweet potatoes which was sufficiently heat stable to allow it to act during baking after gelatinization of starch granules. This enzyme lowers the viscosity via starch hydrolysis. The amount of α-amylase increases during curing and storage (33). Harvat et al. (34), reporting on the sugar content in raw and cooked samples of three cultivars of sweet potatoes, found that maltose increased the most in baked Jewel and Tainung 57 (Table 5). These results substantiated the findings of others that maltose is the predominant sugar formed during baking (35–37), contributing greatly to the sweet taste. Experiments using microwave baking found that the levels of maltose in Jewel could be reduced by almost a factor of two (Table 5).

TABLE 4 Reducing Sugar and Starch Content of Four Sweet Potato Cultivars, Raw and Baked, at Harvest and Postharvest

	Carbohydrates in raw and baked sample (%)			
	Reducing sugars		Starch	
Cultivar and treatment	Raw	Baked	Raw	Baked
Acadian				
Harvest	0.45	7.58	15.30	7.13
Cured	0.25	6.84	11.70	3.80
Stored (weeks)				
4	0.29	7.33	12.45	1.97
13	0.40	6.85	12.15	1.35
21	0.37	6.30	10.23	2.18
Heartogold				
Harvest	0.68	7.17	12.60	7.05
Cured	1.62	7.50	10.35	3.90
Stored (weeks)				
4	1.16	7.65	11.70	3.17
13	1.60	7.43	10.40	2.09
21	1.35	7.33	10.73	3.26
Unit 1				
Harvest	0.39	6.96	12.15	6.85
Cured	0.38	7.67	12.15	2.44
Stored (weeks)				
4	0.66	6.91	10.50	1.30
13	1.11	7.45	10.57	1.34
21	1.35	7.33	10.73	3.26
Early port				
Harvest	0.29	7.41	14.40	8.70
Cured	0.66	6.51	13.50	6.57
Stored (weeks)				
4	0.68	7.67	13.65	4.10
13	0.95	7.58	14.01	3.78
21	1.01	6.51	11.47	3.28

Source: Ref. 30.

Picha (28) reported that maltose level in microwave-heated roots was about the same as in oven-baked roots. Microwave heat can penetrate through parenchymal tissue much faster than conventional baking, and thus may reduce the time between starch gelatinization and inactivation of amlyases. This could explain the reduced levels of maltose found after microwave heating (Table 5) (34).

C. Organic Acids

The nonvolatile acid contents in sweet potato are presented in (Table 6) (34). These acids probably make a minor contribution to taste.

TABLE 5 Sugar and Nonvolatile Acid Composition of Raw and Baked Sweet Potatoes[a]

Component	Raw			Cooked		
	Jewel	99	Tainung 57	Jewel	99	Tainung 57
Succinic	0.06	0.65	0.05	0.11	0.06	0.06
Malic	0.16	0.12	0.25	0.40	0.12	0.25
Citric	0.05	0.20	0.02	0.08	0.01	0.01
Quinic	0.05	0.20	0.09	0.04	0.15	0.10
Fructose	0.85 ± 0.07	0.25 ± 0.05	0.18 ± 0.02	0.83 ± 0.01	0.21 ± 0.05	0.25 ± 0.01
Galactose	0.02	0.01	0.02	0.02	0.01	0.02
Glucose	1.01 ± 0.06	0.31 ± 0.05	0.21 ± 0.01	0.97 ± 0.01	0.25 ± 0.06	0.39 ± 0.03
Inositol	0.06	0.09	0.07	0.03	0.08	0.05
Sucrose	4.24 ± 0.04	2.92 ± 0.002	2.79 ± 0.32	3.65 ± 0.13	2.38 ± 0.03	3.46 ± 0.53
Maltose	0.03	—	0.04	3.80 ± 0.17	0.07 ± 0.04	5.30 ± 0.54
Total sugars	6.21	3.58	3.27	9.31	3.00	9.45
S/A ratio	19.41	3.65	7.97	14.78	8.82	22.50

[a]Based on fresh weight of parenchyma tissue (g/100 g); average of two extracts. Standard deviation shown for major sugars.

Source: Ref. 34.

TABLE 6 Sugar and Nonvolatile Acid Composition of Cooked Sweet Potatoes (cv. Jewel)[a]

Compound	Microwave	Oven
Succinic	0.11	0.13
Malic	0.12	0.15
Citric	0.23	0.19
Quinic	0.03	0.06
Fructose	1.24 ± 0.03	1.22 ± 0.17
Galactose	0.03	0.04
Fucose	1.39 ± 0.15	1.60 ± 0.24
Sucrose	5.83 ± 0.12	5.96 ± 0.47
Maltose	1.63 ± 0.01	2.91 ± 0.18

[a]Based on fresh weight of parenchyma tissue (g/100 g), average of two extracts ± SD for the major sugars. Potato samples were purchased at a local supermarket.
Source: Ref. 34.

D. Proteins

The protein content of tubers varies from 1.0 to 2.5% (about 5% on dry weight basis). The peel contains more protein than the flesh. The major amino acids present in total protein are valine, leucine, isolucine, arginine, and lysine. The amino acid composition of baked, canned, and flaked sweet potatoes is given in Table 7 (38).

E. Vitamins and Minerals

Some of the yellow-fleshed cultivars are high in carotene (10 mg/100 g), mostly in the form of β-carotene, which is a precursor of vitamin A. Sweet potato is also a good source of ascorbic acid (20–30 mg/100 g) and vitamin B complex. The tubers are rich in potassium and phosphorus.

F. Volatile Compounds

Several volatile compounds have been isolated from baked sweet potatoes (39,40), including aldehydes, alcohols, ketones, aromatic hydrocarbons, heterocyclic compounds (pyridine and furan derivatives), and palmitic acid. Harvat et al. (34) identified some new compounds in cooked potatoes, including limonene, cineole, terpineol, β-cyclocitral, α-cadinene, and palmitic acid (Table 8). Many of these volatiles are known to have distinctive aromatic properties and may, therefore, contribute to the characteristic aroma of the respective cultivars. Tui et al. (41) reported that five volatile compounds were associated with desirable flavor.

G. Pigments

Sweet potatoes are rich in β-carotene. Yellow-fleshed cultivars contain higher amounts of β-carotene than white types. The roots of red sweet potatoes contain anthocyanin pigment. Imbert et al. (42) reported that the pigments in red sweet potatoes were dicaffeoyl derivatives of cyanidin and peonidin-3-glycosylglucoside-5-glycoside. Tsukui et al. (43) reported that cyanide-3-(α-D-glucopyranosyl-α-D-fructofuranoside)-5-α-D-xyloside was the main

Table 7 Amino Acid Composition of Baked, Canned, and Flaked Sweet Potatoes[a]

	Products (g/16 g N recovered)			
Amino Acid	Baked	Canned	Flaked	FAO
Threonine	4.50	4.90	4.82	4.0
Valine	6.83	6.81	6.07	5.0
Methionine	2.69b	2.38ab	2.06a	
Half-cystine	0.56	0.64	0.86	
Total sulfur	3.25	3.02	2.92	3.5
Isoleucine	4.57	4.53	4.31	4.0
Leucine	7.47a	9.01b	7.57a	7.0
Tyrosine	5.81	5.75	5.04	
Phenylalanine	7.32	7.82	6.65	
Total aromatics	13.13	13.56	11.69	6.0
Lysine	6.60b	4.84a	7.74a	5.0
Tryptophan	0.44	0.46	0.75	1.0
Total essential	46.79	47.13	42.87	
Aspartic acid	20.22	16.82	22.19	
Serine	3.83a	5.20b	5.14b	
Glutamic acid	7.41	7.48	7.25	
Proline	3.99	3.74	3.92	
Glycine	4.19	4.57	4.69	
Alanine	6.24a	7.12b	5.63a	
Histidine	2.75a	2.90b	2.36a	
Arginine	4.28	5.11	5.65	
Total nonessential	52.91	53.04	56.83	

[a]Means of three replicates for August, September, and November. Values in horizontal rows with the same or no superscripts are not significantly different at $p = 0.05$.
Source: Ref. 38.

pigment. The presence of acyl groups in anthocyanin was reported to confer stability for pigments from sweet potato (44,45).

V. STORAGE

A. Chemical Changes During Storage

Sweet potatoes undergo substantial changes in carbohydrate metabolism during development and storage (46). Purcell et al. (47) studied changes in dry matter (DM), protein (N × 6.25), and nonprotein nitrogen (NPN) during storage of sweet potatoes. Both DM and protein were lost during storage, and the rate of loss was about twice as fast for DM. Sistrunk (48) found that all carbohydrates decreased during 10 days of storage except for sugar and water-soluble pectin.

B. Curing

Well-matured, thoroughly cured, and carefully handled sweet potatoes, free from physical injury and diseases, keep well in storage for a long time. Curing is one of the most effective means of

TABLE 8 Volatile Compounds Identified in Cooked Sweet Potatoes

Compound[a]	Jewel (area %)	Tainung 57 (area %)	99 (area %)
Toluene[c]	tr	tr	tr
Pyridine[c] + xylene	tr	tr	tr
Furfural[c]	1.1	1.3	1.0
2-Acetylfuran[c]	tr	tr	tr
Benzaldehyde[c]	2.7	1.0	1.5
5-Methyl-2-furfural[c]	10.1	2.1	5.9
Limonene	tr	tr	tr
Cineole	tr	tr	—
Phenylacetaldehyde[c]	2.6	tr	2.2
Linalool[c] + nonanal[c]	4.3	1.2	tr
α-Terpineol	4.4	tr	tr
β-Cyclocitral[b]	tr	tr	tr
α-Copaene	5.8	2.3	13.8
Caryophyllene	1.7	tr	2.7
Sesquiterpene hydrocarbon I + M 204	tr	tr	tr
Sesquiterpene hydrocarbon II + M 204	tr	tr	tr
(E)-β-Farnesene	2.5	tr	1.9
α-Cadinene	1.3	—	1.0
β-Ionone[c]	4.6	1.8	7.4
Sequiterpene hydrocarbon III + M 204	1.2	tr	—
Sequiterpene hydrocarbon IV + M 204	tr	1.3	tr
Palmitic acid	2.5	1.1	1.5

[a]Identified by GC/MS and GC retention times.
[b]Tentative identification based on mass spectrum.
[c]Previously identified.
tr = 1% based on GC peak areas.
Source: Ref. 34.

reducing postharvest water loss and pathological attack of sweet potatoes. Curing is often carried out at the farm level at little or no cost. It is a wound-healing process during which general skin of sweet potatoes gets strengthened. The suberization of the outer tissue involved in the process of curing and subsequent development of a wound periderm act as an effective barrier against infection and reduces water loss considerably (49). Curing, essential for good keeping of sweet potatoes, is accomplished best by holding them at about 29.4°C and 85% relative humidity. Lower temperature or lower humidity prolongs the time required for curing. High humidity promotes healing of wounds by favoring wound-cork formation and also reduces shrinkage of the roots through water loss. During curing, there is loss in weight, varying from 5 to 10%, through loss of water by evaporation and loss of solids through respiration.

During curing and storage, and to a greater extent during cooking, some root starch is converted into sugars and dextrins, which improve the eating quality of sweet potato. Short-chain saturated fatty acid content decreases, and tetracosenoic acid content correspondingly increases (50). Sweet potatoes are also stored in pits between layers of wood ash, earth, and straw (51). The traditional methods of curing and storage are typified by the Indian practice of spreading the root in the sun for one week, providing suitable waterproof cover during the night, and then storing in a well-ventilated room with frequent inspections to eliminate any unhealthy roots (52).

TABLE 9 Effects of Curing on Sensory Evaluation of Baked Sweet Potatoes

Panel score	Days of curing 0	4	7
Flavor			
Sweet aromatic	2.7	4.3	4.7
Sweet basic	2.7	4.3	4.7
Starchy	4.5	3.4	3.4
Caramel	1.4	2.4	2.8
Sweet aftertaste	2.0	3.1	3.3
Texture			
First bite denseness	10.8	9.7	9.4
First bite moistness	5.8	6.9	7.6
Mastication gumminess	10.4	8.6	8.8
Ease of swallow	8.2	9.2	9.3
Chalkiness	5.1	3.8	3.3

Source: Ref. 55.

In the United States, where sweet potato processing is well developed, the common practice is to cure at about 30°C with a RH of 85–90% for 7 days, followed by storage of cured roots at 13–16°C with a RH of 85–90% (53). If the roots are quickly placed under good curing conditions, the wounds develop periderm (Fig. 1B), which is identical to normal (54). Hamann et al. (55) studied the effect of curing on the flavor and texture of sweet potatoes (Table 9) (55). Sucrose content, however, increased significantly in cured and stored potatoes in comparison to the uncured group (56). The highest sucrose content was found in cured roots (Table 10). No significant changes were detected in β-carotene among uncured, cured, and stored sweet potatoes

TABLE 10 Effects of Fresh Storage and Frozen Storage on Sugar Content of Sweet Potatoes

Storage treatment	Glucose (%)	Fructose (wet basis) (%)	Sucrose (%)	Maltose (%)
Fresh				
Uncured	0.51[c]	0.26[c]	2.77[d]	9.85[a]
Cured	0.58[c]	0.27[c]	4.52[c]	9.69[a]
1 mo	1.31[a]	1.01[a]	3.76[b]	10.44[a]
3 mo	1.18[b]	0.91[b]	3.34[b]	9.77[a]
Frozen				
Unfrozen	0.87[a]	0.62[ab]	3.50[a]	10.53[a]
1 mo	0.91[a]	0.72[a]	3.59[a]	9.93[a]
3 mo	0.97[a]	0.58[b]	3.48[a]	9.58[a]
6 mo	0.92[a]	0.56[b]	3.83[a]	9.81[a]
9 mo	0.92[a]	0.46[c]	3.66[a]	9.84[a]

[a-d]Means within columns separated by Duncan's multiple range test, $p = 0.05$. Values with common letter not significantly different ($p = 0.05$).
Source: Ref. 56.

Table 11 Effects of Fresh and Frozen Storage on Sensory Scores of Baked Potatoes

Storage treatment	Color	Flavor	Texture	Overall rating
Fresh				
Uncured	3.86a	2.82c	3.05b	3.07c
Cured	3.95a	3.43b	3.56a	3.59ab
1 mo	3.79a	3.72a	3.66a	3.67a
3 mo	3.84a	3.30b	3.56a	3.45b
Frozen				
Unfrozen	3.70b	3.36a	3.56a	3.49a
1 mo	3.92a	3.42a	3.59a	3.57a
3 mo	4.80a	3.41a	3.35ab	3.44a
6 mo	3.87ab	3.43a	3.62a	3.60a
9 mo	3.68b	2.88b	3.04b	2.99b

[a–c]Means within columns separated by Duncan's multiple range test, $p = 0.05$. Values with common letter are not significantly different ($p = 0.05$).
Source: Ref. 56.

(56). However, there was an increase in β-carotene during frozen storage (56,57). The quantity of vitamin C was significantly different among uncured, cured, and cured/stored roots (56). Vitamin C content was also significantly lower in tubers stored 3 months compared to cured (unstored) roots. An approximate 50% loss of vitamin C was observed during the first month of frozen storage of sweet potatoes, after which no significant difference was detected. There were few significant differences in sensory scores of baked sweet potatoes during the first 6 months frozen storage (Table 11) (56). After 9 months of frozen storage, samples were rated significantly lower in flavor, texture, and overall sensory properties in comparison to 6-month samples (58). However, 9-month samples were still of acceptable sensory quality, with average scores of around 3 or higher. Sweet potatoes normally lose weight by water evaporation and respiration during storage (59).

C. Low-Temperature Storage

Sweet potatoes store best at 12.8–15.16°C (60). Extensive studies by these investigators (60) indicated marked cultivar differences in the storage behavior of sweet potatoes. Storage temperatures above 15.6°C increased the growth rate of decay-causing organisms such as *Rhizopus* and *Fusarium*, and severe chilling injury occurred when the tubers were exposed to temperatures below 10°C (60). Uncured roots are more susceptible to chilling than cured ones. Cultivars differ slightly in their ability to withstand chilling injury, but all cultivars are sufficiently susceptible to make it desirable to avoid chilling conditions. Short periods at temperatures as low as 10°C need not cause alarm, but after a few days at 10°C or shorter periods at even lower temperatures, sweet potatoes may develop discoloration of the flesh, internal breakdown, off-flavors, and hard core when cooked and increased susceptibility to decay (60). Pantastico et al. (61) recommended temperatures of 10–12.8°C and 80-90% relative humidity for storing sweet potatoes for about 13–20 weeks.

D. Controlled-Atmosphere Storage

The usefulness of controlled-atmosphere (CA) storage for sweet potatoes has not been studied extensively. Mattus and Hassan (62) found that roots held in an atmosphere with 2–3% CO_2 and

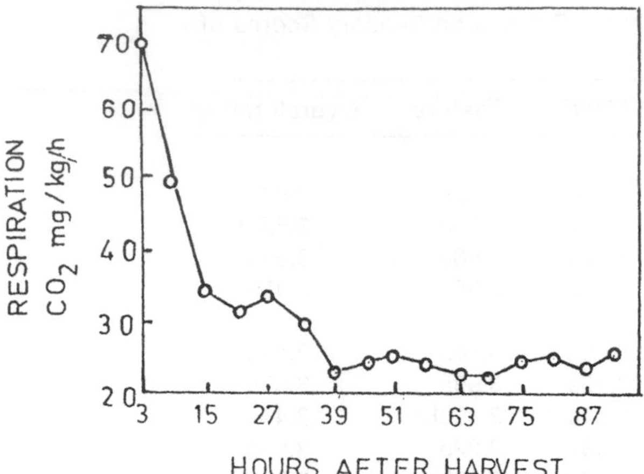

FIGURE 2 Changes in respiratory rate of uncured sweet potato roots during the first 93 hours after harvest when held at 21% O_2. (From Ref. 63.)

7% O_2 suffered fewer losses during storage than air-stored samples. With CO_2 levels above 10% and O_2 concentrations below 7%, the roots developed an unpleasant flavor. Chang and Kays (63) reported the effect of low-oxygen storage on sweet potato roots. The total sugar content accumulated with low oxygen (2.5 and 5.0%) during storage. Protopectin was low in roots stored at low O_2 concentrations, but water-soluble pectin was not significantly affected. The rate of respiration of uncured sweet potatoes was high at harvest but decreased rapidly, stabilizing at about 25 mg CO_2/kg/hr after 38–40 hours (Fig. 2) (63). After cured roots were placed in an atmosphere containing less than 20% oxygen, it took several days before the respiratory rates of roots stabilized. The respiratory rate of roots stored at O_2 concentrations of 5–10% were lower than those of roots held at either 2.5% or 20% oxygen (Fig. 3) (63). The protopectin content of sweet potato also increased with increasing oxygen concentration in the storage atmosphere (Fig. 4) (63).

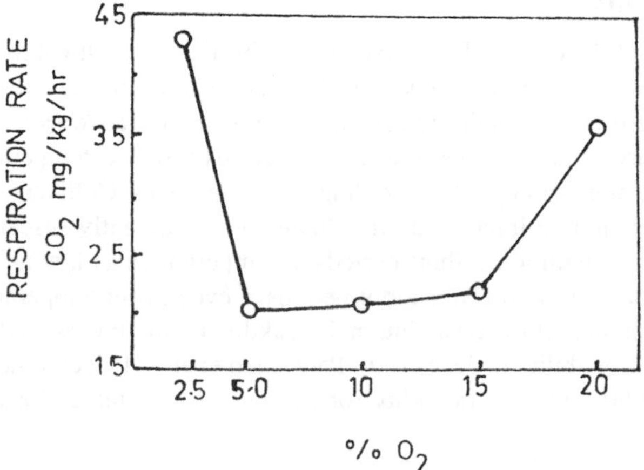

FIGURE 3 Effects of concentration of oxygen in storage atmosphere on the respiratory rate of cured sweet potato root 7 days after beginning of treatment. (From Ref. 63.)

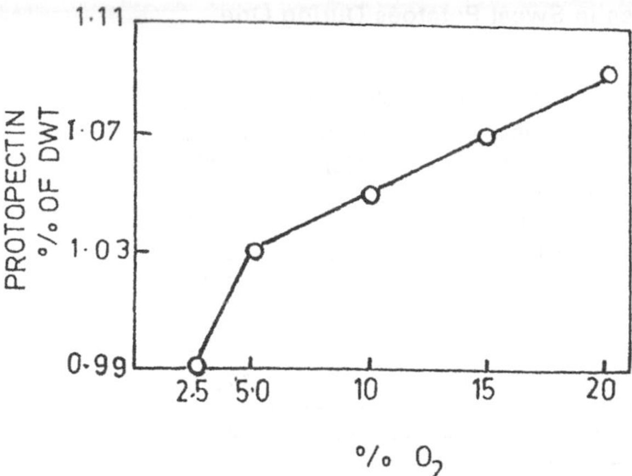

FIGURE 4 Effects of oxygen concentration of storage gas on the concentration of protopectin (% dry wt.) of sweet potato. (From Ref. 63.)

E. Chemical Control of Postharvest Losses

Chemicals such as borax, sodium-*o*-phenylphenate (50 ppm, *p*-acetic acid, dicloran, and thiabendazole (TBZ) have been found to be effective in controlling various rots of sweet potatoes (21). Borax is used in a dipping solution at 0.1–1.0% (64) or in combination with a spray emulsion (65). Kushman (66) found that dipping of sweet potatoes in 0.6% solution of SOPP reduced both blackrot and *Rhizopus* rot. Para-acetic acid is very effective against blackrot and, to a certain extent, *Rhizopus* rot. Daines (65) observed that dicloran dips effectively reduced *Rhizopus* rot in properly treated sweet potatoes, but were not effective against *Penicillium* sp. A 0.4% solution of TBZ effectively controlled the black rot of sweet potatoes (67). Vahab and Kumaran (68) reported that application of foliar spray of Ethrel (150–450 ppm) or CCC (500–1000 ppm) significantly increased the starch and total sugars but decreased the protein content of sweet potatoes. Ceponis et al. (69) reported that sweet potatoes treated with 2,6-dichloro-4-nitroaniline (DCNA) and packaged suffered less waste after one week in a retail store than either treated or untreated roots retailed in bulk. Effective control of *Rhizopus* soft rot of sweet potatoes on sleeve-wrapped and completely overwrapped trays was obtained using DCNA (Table 12) (69). The combination of DCNA and complete overwrap effectively reduced retail losses from store rot and desiccation.

VI. PROCESSING

Although in most developing countries sweet potatoes are consumed fresh after cooking, in several advanced countries they are also canned or dehydrated. Both yellow and white types are used for canning, but yellow types are preferred because of their attractive color. Yellow types are not as sweet as the white cultivars, but they have good flavor and cooking qualities (70). McConnel and Gottschall (71) recommended freshly dug sweet potatoes for canning purposes.

In most developing countries, the sweet potato is consumed immediately after harvest without any intermediate storage. It may be consumed after baking, boiling, steaming, or frying, or it may be candied with syrup, sliced into chips, or pureed. During cooking,

TABLE 12 Decay and Weight Losses in Sweet Potatoes During One Week in a Retail Store

Treatment	Unsalable packages (%)	Soft-rot losses[a] (%)	Moisture losses[b] (%)	Total losses[c] (%)
Sleeve-wrap—DCNA	3.1	0.4d	3.3	3.70
Sleeve-wrap	28.1	6.1ab	3.1	9.2abc
Complete wrap—DCNA	2.1	0.5d	2.3	2.8c
Complete wrap	28.6	6.3a	2.3	8.6bcd
Bulk—DCNA	—	1.0d	8.3	9.3ab
Bulk	—	4.4abc	8.3	12.7a

[a]Means separation within columns by Duncan's multiple range test, 5% level.
[b]Includes weight of culls shriveled by desiccation in bulk lots.
[c]Losses for sweet potatoes during one week in a retail store (combined data for eight tests).
Source: Ref. 69.

some starch is hydrolyzed to maltose and dextrins by β-amylase action. Sweet potato is sometimes sliced and dried in the sun, then ground into flour for baking bread and confectionery. In Indonesia, sun-dried chips are fried and packed in polyethylene for retail sale (72–74). In some African countries, boiled or roasted roots are pounded with peanuts to produce *futali*.

In the United States about 60–70% of the sweet potato crop is consumed by humans in the form of fresh roots or as canned, frozen, or dehydrated foods and used in a variety of products, such as pie fillings, purees, and candied and baby foods. In Japan, about half of the total crop is utilized for the production of starch for use in the textiles, paper, cosmetics, adhesives, glucose syrup, and food-manufacturing industries. During the production of food products or starch, sweet potato peels and other residues may be advantageously extracted for pectins or compounded with vine tops and leaves to serve as animal feed.

A. Canning

Freshly dug roots are preferred over cured and stored ones for canning. To pack fancy whole potatoes, the canner is interested in roots 25–45 mm wide and 50–175 mm long. The essential processing steps are shown in Figure 5 (75). Preheating (63°C for 8 min) drives out intercellular gases and helps in attaining a good can vacuum. In lye peeling, a 1.0% lye solution and a temperature of 93–94°C for 8-10 minutes are commonly used, although higher lye concentrations and temperatures are also used. The addition of surfactants aids in peel removal. Steam peeling uses steam temperatures of 150°C and retention times between a few seconds and 1.5 minutes. Discolored areas, fibrous ends, and damaged parts are removed during inspection, trimming, and sorting. After blanching, the material is immediately packed in cans and covered with syrup to prevent discoloration (75). Although canning is an established method of preserving food, the market for canned sweet potatoes is likely to be limited in developing countries due to its high cost—mainly a result of the cost of the can. Until alternative packaging technology is developed, canned roots can only be manufactured for export.

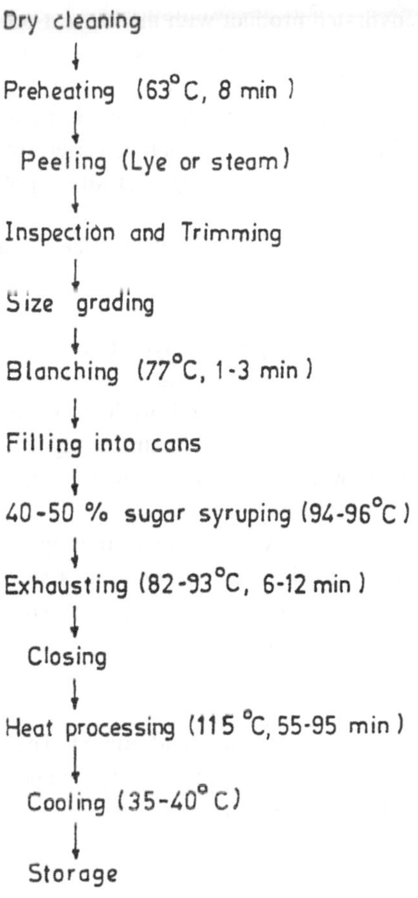

FIGURE 5 Canning of sweet potato. (From Ref. 75.)

B. Dehydrated Products

Dehydration of fresh and precooked sweet potatoes by sun-drying has long been practiced in developing countries to produce chips and flour. Both white and colored varieties are suitable for sun-drying. The roots are cut into 2- to 3-mm slices and submerged in boiling water for 6 minutes. Blanched or unblanched slices, with or without sodium metabisulfite treatment, are sun-dried to 60% moisture. The slices are then brittle and can be ground into flour. Both slices and flour will remain in good condition for a long time if kept under dry conditions (76).

Cruess (77) described in detail the method for dehydrating sweet potatoes. Raw sweet potatoes are peeled by abrasive peelers or lye-peeled in 15–20% sodium hydroxide lye solution. They can also be peeled under steam pressure, as in the case of white potatoes. The sweet potatoes are then washed thoroughly to remove all lye and disintegrated peel. Then they are trimmed, sliced or cut into julienne strips, or diced. After thorough steam-blanching for about 7 minutes, they are dehydrated on trays at about 71°C to a moisture content of less than 7%. Sulfiting of sweet potatoes prolongs their keeping quality. Trays weigh about 1–1½ pounds per square foot, and the drying ratio of fresh to finished product is about 4:1. Arthur and McLemore (78) reported the changes in the chemical composition of sweet potatoes during dehydration.

Goldrush cultivar was recommended for a freshly dehydrated product with the highest possible carotene content. The drying ratios (pounds of raw material required to make pound of dried product) varied from 5:1 to 6:1. The total sugar and reducing sugar contents of the dried sweet potatoes ranged from 28–47% and 14–25%, respectively, on a dry weight basis. The carotene content of cultivars ranged from 172 to 545 ppm, and the ascorbic acid content varied from 40 to 88 mg/100 g, both reported on a moisture-free basis. The loss in weight of sweet potatoes during lye-peeling was in the range of 20-34% (average 27%).

Purcell and Walter (38) related the changes in the amino acid content of sweet potato to the heat-processing treatment. Baking of the roots caused less amino acid loss than either canning in syrup or dehydrating into flakes. The major nutritional change caused by heat processing was the destruction of lysine, which was significantly greater in canning and dehydration than during baking. Roots canned in syrup contained significantly less nitrogen than did roots processed by other treatments. The total protein contents of baked, canned, and flaked products were 7.52, 5.55, and 7.06%, respectively, on a dry weight basis. The type of heat treatment significantly affected the amino acid content (38). The lysine contents were higher and serine contents were lower in the baked samples than in the canned and flaked samples.

Modern sweet potato dehydration practices, however, involve many more unit operations, some also used in the canning process (79). In Indonesia, fresh roots are sometimes soaked in 8-10% salt solution for about one hour before cutting into chips and sun-drying. The brine treatment is reported to inhibit microbial growth during drying (80).

1. Flakes

To produce dehydrated flakes, thicker slices (13 mm) are cut, blanched, and pureed. The puree (20% solids) is passed through steam-heated drums (0.13 mm apart; speed 2 rpm; retention time 20 sec, steam pressure 6 atm). The moisture content of the flakes is around 2–4%.

2. Flour

In India, sweet potato flour is produced by dehydration (usually sun-drying) and grinding of the peeled, sliced roots and is used to supplement cereal flours in bakery products, pancakes, and puddings. It acts as a dough conditioner in bread manufacturing and as a stabilizer in ice cream. It is also consumed in some food preparations along with peanut cake. Sweet potato flour can also be produced by spray-drying or cabinet-drying of the peeled sliced roots. The peeled and sliced roots are dried in an air oven at 60°C for 12 hours and milled in a disc mill to produce flour. Dipping the sliced root in 0.5% sodium metabisulfite solution for 2 minutes produces flour of acceptable color (81,82).

3. Chips

For chips, the peeled roots are rinsed to remove excess starch, and after slicing they are deep-fried and packed (Fig. 6). The high sugar content of sweet potatoes causes chips to turn dark brown and gives a burned flavor when deep-fried. Wiley and Bouwkamp (83) have developed a technique to extract sugars from sweet potatoes and other vegetables and fruits by liquid extraction before they are to be fried.

Dehydration of sweet potato appears to have a great potential as a method of preserving food in developing countries. Since steam is the major energy input, countries having geothermal energy resources, like Indonesia and the Philippines, can operate dehydration plants quite cheaply. In Indonesia, sweet potato is cultivated in the vicinity of some geothermal energy sources.

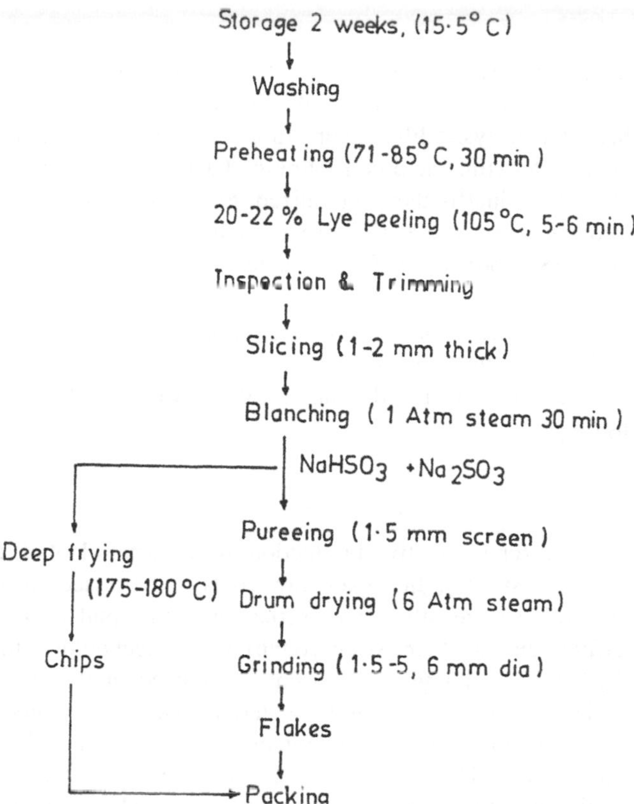

FIGURE 6 Preparation of chips and flakes from sweet potato. (From Ref. 79.)

C. Frozen Products

Sweet potatoes are peeled, rinsed, and sliced into strips 0.6 cm thick × 0.9 cm wide. The strips are blanched 2 minutes, in water (100°C) and frozen. Frozen strips are fried in oil before use. All sweet potato cultivars have an amylolytic enzyme system, which decreases starch content during storage (84). The postharvest history has been shown to have an impact on the textural properties of French fries (85). The softening of a fresh fry-type product prepared from Jewel sweet potatoes decreased linearly when the tissue pH was incrementally lowered from 6 to 3.8 prior to blanching (86).

D. Starch

Sweet potato varieties that have a high starch content and light flesh are used for the production of starch. The processing steps are similar to those for starch making from other roots, except for color removal. The process for sweet potato starch production is kept alkaline throughout (pH 8.6) by using lime, which flocculates the impurities and dissolves pigments. The tubers are ground in lime water, and starch is separated from the pulp by washing over a series of screens. The starch suspension is then blanched with sodium hypochlorite, if required, and centrifuged. Starch is stored wet in concrete tanks or dried to a moisture content of about 12% in a vacuum drier, pulverized, and screened. The yield is 20–26% (79).

E. Glucose

Sweet potato startch can be advantageously converted into glucose by acid or enzymatic hydrolysis. While acid hydrolysis gives a better conversion, the enzymatic hydrolysis product has better flavor and color (87). In acid hydrolysis, dilute hydrochloric acid is added in autoclove to a 35–40% suspension of starch in water (final acid concentration 0.015–0.02 N) and heated to 140–160°C for 15–20 minutes. The product is then neutralized, treated with active carbon, heated again, and filtered hot through a centrifuge. Glucose crystalizes from the filtrate on cooling. Yield is 75–80%. Agra et al. (88) reported satisfactory yields of starch by boiling a 13% suspension of sweet potato starch with 0.2 N hydrochloric acid for 3 hours.

In enzymatic hydrolysis, the roots are cooked in water, pulped, and saccharified by treatment with malt (0.2%) at 60° C. The solution is clarified and concentrated to a 70% solid content. The product is not as sweet as cane sugar and contains 30.1% water, 43.0% maltose, 7.0% sucrose, and 14.0% dextrins (89).

F. Alcohol

Sweet potatoes are a good raw material for fermentive production of industrial alcohol, lactic acid, acetone, butanol, vinegar, and yeast (90). In Japan, raw roots and pressed juice are employed for preparation of alcoholic beverages after saccharification with liquid *Koji*. Similarly, local wines are made in Africa and in other Asian countries. The technology for the production of alcohol from carbohydrates is well known. Sugars are directly converted to alcohol and CO_2 by the action of enzymes in the yeast. But if starchy materials like sweet potatoes and other roots and grains are to be converted into sugar, fermentation must be carried out in dilutions that permit yeast cells to survive and multiply. The resultant product, therefore, is diluted alcohol, which is rectified by fractional distillation. In Indonesia, a large number of alcohol plants are being planned using sweet potato, cassava, and sago as raw materials. Preference is being given to sweet potato, since it can be harvested three times a year as opposed to cassava, which is harvested once every 8–12 months.

G. By-products

Low-grade sweet potatoes are utilized in the United States as a carbohydrate-rich feed for livestock, especially for pigs, after chopping and dehydration. Various uses of sweet potato reported by Ge et al. (91) are outlined in Figure 7. The dried material is equal to corn in total digestible nutrients but low in proteins. However, if cottonseed meal is added to sweet potato meal, it performs satisfactorily as an animal feed.

Fleshy roots were found to be a satisfactory feed for horses, mules, and hogs. Sweet potato meal was found to be 90% efficient as a feed for lactating dairy cows when compared to cornmeal feed and could be substituted up to 25% as poultry feed. Since shredding or slicing was necessary to facilitate drying, several inexpensive machines were developed. Drying was carried out under natural and artificial conditions. However, sweet potatoes have not yet replaced corn as an animal feed in the United States to any appreciable degree, mainly due to high dehydration costs (75).

Sweet potato vines serve as a nutritive and palatable green feed for cattle (92). The feeding value of vines is close to that of alfalfa. Bennett and Gieger (93) studied ensiling of sweet potato vines alone and in combination with molasses and roots. The sweet potato silage was found to be equal in performance to corn silage when fed to dairy cows.

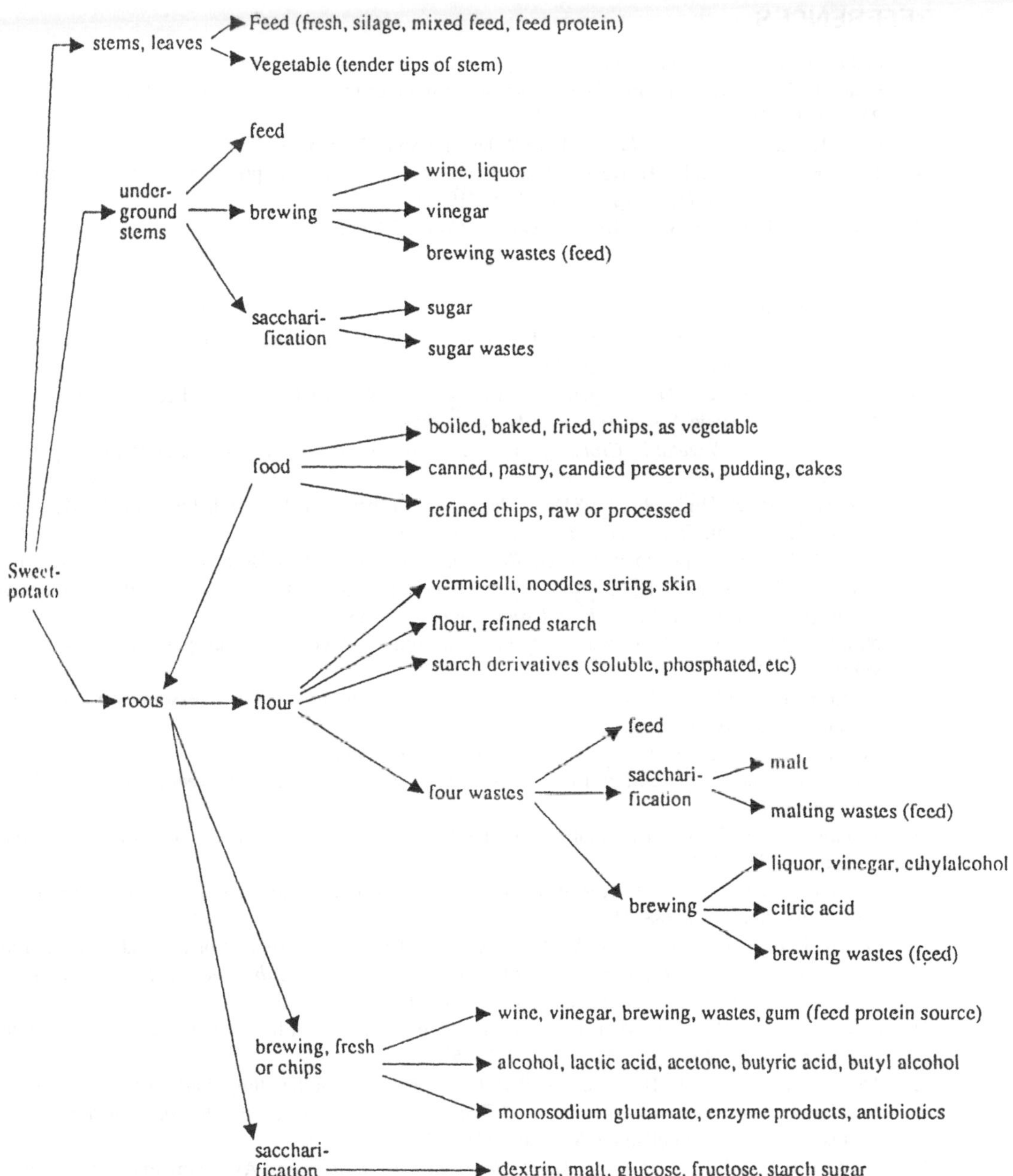

FIGURE 7 Various uses of sweet potato. (From Ref. 91.)

REFERENCES

1. FAO, *Production Year Book*, Food and Agriculture Organization, Rome, 1980.
2. King, J. R., and R. Bamford, The chromosome number in *Ipomoea* and related genera, *J. Hered. 28*:279 (1937).
3. Esau, K., *Anatomy of Seed Plants*, John Wiley and Sons, New York, 1960.
4. Thompson, H. C., and J. H. Beattle, Classification and variety descriptions of American varieties of sweet potatoes, U.S. Dept. Agric. Bull. 1021 (1922).
5. Thompson, H. C., and W. C. Kelly, *Vegetable Crops*, 5th ed., McGraw-Hill, New York, 1957.
6. Boswell, V. R., Commercial growing and harvesting sweet potatoes, U.S. Dept. Agric. Farmers Bull., 2020, (1950).
7. Collins, W. W., and J. W. Moyer, 'Pope' Sweet potato, *HortScience, 17*(2):265 (1982).
8. Hernandez, T. P., R. J. Constantin, H. Hammett, W. J. Martin, C. Clark, and L. Rolston, 'Travis' sweet potato, *HortScience 16*(4):574 (1981).
9. Tigchelaar, E. C., ed., New Vegetable Varieties List XXI. The Garden Seed Research Committee, American Seed Trade Association, *HortScience 15*:565 (1980).
10. Yawalkar, K. S., *Vegetable Crops of India*, 2nd ed., Agriculture Horticultural Publishing House, Nagpur, India, 1980.
11. CTCRI, Research Highlights (1981) of The Central Tuber Crop Research Institute (ICAR), Sreek-ariyam, Trivendrum, India, *Tech. Bull. No. 1*:1 (1982).
12. Mandal, R. C., Sweet potato, in *Tropical Root and Tuber Crops*, Agro Botanical Publisher India, 1993.
13. Biswas, J., H. Sen, and S. K. Mukhopadhyay, Effect of time of planting on tuber development of sweet potato (*Ipomoea batatas* L.), *J. Root Crops, 14*(1):11 (1988).
14. Nawale, R. N., and M. J. Salvi, Effect of season on yield of sweet potato, *J. Root Crops, 9*(1&2):55 (1983).
15. Singh, K. D., and R. C. Mandal, Performance of sweet potato in relation to seasonal variations (Time of planting), *J. Root Crops*, *2*(2):17 (1976).
16. Choudhury, B., *Vegetables*, National Book Trust, New Delhi, 1976.
17. Pande, U. C., and O. S. Jauhari, Effect of N, P, K on yield and quality of sweet potato, *Punjab Hort. J. 10*:265 (1970).
18. Mohankumar P. M., B. S. Kabeerathumma, and P. G. Nair, Fertility management of tuber crop, *Indian Farming 33*:35 (1984).
19. Steinbaur, C. E., and L. J. Kushman, *Sweet Potato Culture and Diseases, U.S. Department of Agriculture Handbook* 388, 1971.
20. Eckert, J. W., P. P. Rubio, A. K. Mattoo, and A. K. Thompson, Diseases of tropical crops and their control, in *Postharvest Handling and Utilization of Tropical and Subtropical Fruits and Vegetables* (E. B. Pantastico, ed.), AVI Publishing, Westport, CT, 1975, p. 415.
21. Lutz, J. M., J. K. Park, and M. T. Deonier, Influence of methods of harvesting sweet potatoes on their storage behaviour, *Proc. Am. Soc, Hortic. Sci.* 57:297 (1951).
22. Thompson, A. K., M. B. Bhatti, and P. P. Rubio, Harvesting and handling: harvesting, in *Postharvest Physiology Handling, and Utilization of Tropical and Subtropical Fruits and Vegetables* (E. B. Pantastico, ed.), AVI Publishing, Westport, CT, 1975, p. 236.
23. Knott, J. E., and J. R. Deanon, Jr., *Vegetable Production in Southeast Asia*, University of Philippines Press, College of Agriculture, Laguna, (1967).
24. Austin, M. E., L. H. Aung, and B. Graves, The use of carbohydrate contents as an index of sweet potato maturity, *Proc. 2nd Int. Symp. on Trop. Root and Tuber Crops 1*:42 (1970).
25. Ezell, B. D., M. S., Wilcox, and J. N. Crowder, Pre-and post-harvest changes in carotene, total carotenoids and ascorbic acid content of sweet potatoes, *Plant Physiol. 27*:355 (1952).
26. Ramanujam, T., and P. Indira, Maturity studies in sweet potato (*Ipomoea batatas* (L.), Lam.), *J. Root Crops* (India) *5*:43 (1979).
27. *Potatoes and Sweet Potatoes, Final Estimate 1978–1982*, Stat Bull. 709, USDA, 1984.
28. Picha, D. H., Crude protein, minerals and total carbtenoids in sweet potatoes, *J. Food Sci. 50*:1768 (1985).

29. Kay, D. E., *Root Crops, Sweet potato*, Tropical Products Institute, London, 1973, p. 144.
30. Ali, M. K., and L.G. Jones, The effect of variety and length of storage of carbohydrate contents and table quality of sweet potatoes, *Pak. J. Sci. Ind. Res. 10*:121 (1965).
31. Balls, A. K., M. K. Malden, and R. R. Thompson, Crystalline betaamylase from sweet potatoes, *J. Biol. Chem. 173*:9 (1948).
32. Ikemiya, M., and H. J. Deobald, New characteristics of alpha-amylase in sweet potatoes, *J. Agric. Food Chem. 14*:237 (1966).
33. Deobald, H. J., T. A. Mclemore, V. C. Hasling, and E. A. Catalona, Control of sweet potato alpha-amylase for producing optimum quality precooked dehydrated flakes, *Food Technol.* 22:93 (1968).
34. Harvat, R. J., R. F. Arrendale, G. G. Dull, G. W. Chapman Jr., and J. J. Kays, Volatile constituents and sugars of three diverse cultivars of sweet potatoes (*Ipomoea batatas* L.), *J. Food Sci. 56*(3):714 (1991).
35. Chapman, G. W. Jr., and R. J. Horvat, Determination of nonvolatile acids and sugars from fruit and sweet potato extracts by capillary GLC and GLC/MS, *J. Agric. Food Chem. 37*:947 (1989).
36. Sistrunk, W. A., J. C. Miller, and L. G. Jones, Carbohydrate changes during storage and cooking of sweet potatoes, *Food Technol. 8*:223 (1954).
37. Walter, R. M., A. E. Purcell, and A. M. Nelson, Effect of amylolytic enzymes on "moistness" and carbohydrate changes of baked sweet potato cultivar, *J. Food Sci. 40*:793 (1975).
38. Purcell, A. E., and M. W. Walter, Jr. Stability of amino acids during cooking and processing of the sweet potatoes, *J. Agric, Food Chem. 30*:443 (1982).
39. Purcell, A. E., D. W. Later, and M. L. Lee, Analysis of the volatile constituents of baked Jewel sweet potatoes, *J. Agric, Food Chem. 28*:939 (1980).
40. Kays, S. J., and R. J. Horvat, Insect resistance and flavor chemistry intergration into future breeding programs, *Proc Amer. Soc. Hort. Sci. Trop. Region 27*(B):97 (1983).
41. Tui, C. S., A. E. Purcell, and W. W. Collins, Contribution of some volatile compounds to sweet potato aroma, *J. Agric. Food Chem. 33*:223 (1985).
42. Imbert, M. P., C. E. Seaforth and D. B. Williams, Anthocyanin pigments of sweet potato: *Ipomoea batatas*, *J. Am. Hort. Sci. 88*:481 (1966).
43. Tsukui, A. K. Kuwana, and T. Metanura, Anthocyanin pigment isolated from purple root of sweet potato, *Kaseiqaku Zasshi 34*:153 (1983).
44. Brouillard, R., Origin of the exceptional color stability of the Zebrina anthocyanin, *Phytochemistry 20*:143 (1981).
45. Bassa, I. A., and F. J. Francis, Stability of anthocyanins from sweet potatoes in a model beverage, *J. Food Sci. 52*:1753 (1987).
46. Edmond, J. B., and G. R. Ammerman, *Sweet Potatoes: Production, Processing, Marketing,* AVI Publishing Co., Westport, CT, 1971.
47. Purcell, A. E., M. W. Walter Jr., and F. G. Giesbrecht, Changes in dry matter, protein and non-protein nitrogen during storage of sweet potatoes, *J. Am. Soc. Hort. Sci. 103*(2):190 (1978).
48. Sistrunk, W. A., Relationship of storage, handling and cooking method to color, handcore tissue, and carbohydrate composition in sweet potatoes, *J. Am. Soc. Hort. Sci. 102*(4):381 (1977).
49. Booth, R. H., Postharvest deterioration of tropical root crops, losses and their control, *Trop. Sci. 16*:49 (1974).
50. Boggess, T. S. I. R. Marion, J. G. Woodroof, and A. H. Demsey, Changes in lipid compostion of sweet potatoes as affected by controlled storage, *Food Sci. 32*(5):554 (1967).
51. Wargiono, H. J., Pedoman bercocok tanam ubijalar (in Indonesian), Central Res. Inst. for Agric. Bogor Indonesia, Bull. 5, (1960).
52. Indian Council of Agricultural Research, Harvesting and storage of sweet potato. Farm News Release No. 173, in *Wealth of India: A Dictionary of Indian Raw Materials*, Council of Scientific and Ind. Res., New Delhi, 1956, p. 244.
53. Covington, H. M., L. G. Wilson, C. W. Averre, J. V. Baird, K. A. Sorenson, E. A. Proctor, and E. O. Beasley, Growing and marketing quality of sweet potatoes, North Carolina State University Ext. Publ. AG-09, 1976.

54. Walter, M. W., Jr., and W. E. Schadel, Structure and composition of normal skin periderm and wound tissue from cured sweet potatoes, *J. Am. Hort. Soc. 108*:909 (1983).
55. Hamann, D. D., N. C. Miller, and A. E. Purcell, Effect of curing on the flavor and texture of baked sweet potatoes, *J. Food Sci. 45*:992 (1980).
56. Wu, J. Q., S. J. Schwartz and D. E. Carroll, Chemical, Physical and sensory stabilities of prebaked frozen sweet potatoes, *J. Food Sci. 56*(3):701 (1991).
57. Schwartz, S. J., W. M. Walter, Jr. D. E. Carroll, and F. G. Giesbrecht, Chemical, physical and sensory properties of sweet potato french-fry type product during frozen storage, *J. Food Sci. 52*:617 (1987).
58. Walter, W. M. Jr., and M. W. Hoover, Effect of preprocessing conditions on the composition, microstructure, and acceptance of sweet potato patties, *J. Food Sci. 49*:1259 (1984).
59. Walter, W. M. Jr., L. K. Hammett, and F. G. Giesbrecht, Wound healing and weight loss of sweet potatoes harvested at several soil temperatures, *HortScience 114*:94 (1989).
60. Cooley, J. S., L. J. Kushman, and H. F. Smart, Effect of temperature and duration of storage on quality of stored sweet potatoes, *Econ. Bot. 8*:21 (1954).
61. Pantastico, E. B., T. K. Chattopadhyay, and H. Subramanyam, Storage and commercial storage operations, in *Postharvest Physiology, Handling and Utilization of Tropical and Subtropical Fruits and Vegetables* (E. B. Pantastico, ed.), AVI Publishing Co., Westport, CT, (1975). p. 314.
62. Mattus, G. E., and F. M. Hassan, Controlled atmosphere storage studies with sweet potatoes, *HortScience, 3*:90 (1968).
63. Chang, L. A., and S. J. Kays, Effect of low oxygen storage on sweet potato roots, *J. Am. Soc. Hort. Sci. 106*:481 (1981).
64. Person, L. H., E. O. Olson, and W. J. Martin, Effectiveness of fungicides in controlling black rot of sweet potatoes, *Phytopathology, 38*:474 (1948).
65. Daines, R. H., Effects of temperature and 2,6-dichloro 4-nitroaniline dip on keeping qualities of 'Yellow Jersey' sweet potatoes during the post storage period, *Plant Dis. Rep. 54*:486 (1970).
66. Kushman, L. J., Fungicidal control of decay in 'Puerto Rico' sweet potatoes during marketing, U.S. Department of Agriculture Marketing Service Report 452, 1961.
67. Martin, W. J., Evaluation of fungicides for effectiveness against the sweet potato black rot fungus, *Ceratocystis fimbriata, Plant Dis. Rep. 55*:523 (1971).
68. Vahab, N. A., and N. M. Kumaran, Starch, sugar and protein content of sweet potato tubers as influenced by Ethrel and CCC, *Proc. Natl. Semin. Tuber Crop Production Technology*, November 21 to 22, 1980, Tamilnadu Agricultural University, Coimbatore, India, 1980, p. 133.
69. Ceponis, M. J., J. K. Kaufman, and W. H. Tietjen, Effects of DCNA and prepackaging on the retail quality of sweet potatoes, *HortScience 8*(1):41 (1973).
70. Rodriguez, R., B. L. Raina, E. B. Pantastico, and M. B. Bhatti, Quality of raw materials for processing, in *Postharvest Physiology, Handling and Utilization of Tropical and Subtropical Fruits and Vegetables*, (E. B. Pantastico, ed.), AVI Publishing Co., Westport, CT, 1975, p. 467.
71. McConnel, E. R., and P. B. Gottschall, Effect of canning of new and familiar sweet potato varieties with particular emphasis on breakdown and firmness, *Food Technol. 11*:209 (1957)
72. Winaro, F. G., W. Soesarsona, and S. Iran, *Agricultural Marketing System in Boger Area*, Inst. of Dev. Econ. Tokyo, Japan, 1980.
73. Onwueme, I. C., *The Tropical Tuber Crops*, John Wiley, New York, 1978, p. 191.
74. Research and development, Potato mid-week review, *Economic Times* (India), April 15, 1981.
75. Edmond, J. B., Production for industrial uses and feed, in *Sweet Potatoes, Production, Processing and Marketing*, (J. B. Edmond and G. R. Ammerman, eds.), AVI Publishing Co. Inc., Westport, CT, 1971, p. 334.
76. Jackson, T. H., and B. B. Mohamed. Sundrying of fruits and vegetables, *FAO Agricultural Service Bulletin* 5, 1969.
77. Cruess, W. V., *Commercial Fruit and Vegetable Products*, 4th ed., McGraw-Hill, New York, 1958, p. 619.
78. Arthur, J. C., and T. A. McLemore, Sweet potato dehydration: effects of processing conditions and variety on properties of dehydrated products, *J. Agric. Food Chem. 3*(9):782 (1955).

79. Sparde, J. T., and E. L. Patton, Precooked dehydrated sweet potato flakes, *Food Eng. 33*(7):46 (1961).
80. Namaken, Sembiring, Alat-alat yang telah dibuat. FTDC Bull. Penelitian dan Pengembangan (1):46 Food Tech. Dev. Center, IPB, Bogor, Indonesia, 1979.
81. Gore, H. C., The value of sweet potato in breadmaking, *Ind. Eng. Chem. 15*:1238 (1923).
82. Hamed, M. G. S., F. Y. Refai, M. F. Hussein, and S. K. El-Sa Smahy, Effect of adding sweet potato flour to wheat flour on physical dough properties of baking, *Cereal Chem. 50*(2):140 (1973).
83. Wiley, R. C. and J. Bouwkamp, Sweet potato chips, *Food Eng. Int. 4*(5):28 (1979).
84. Walter, W. M. Jr., A. E. Purcell, and A. M. Nelson, Effects of amylolytic enzymes on 'moistness' and carbohydrates changes from baked sweet potato cultivars, *J. Food Sci. 40*:793 (1975).
85. Walter, W. M. Jr., and M. W. Hoover, Preparation, evaluation and analysis of a fresh fry-type product from sweet potatoes, *J. Food Sci. 51*:967 (1986).
86. Walter, W. M. Jr., H. P. Fleming, and R. F. Mofeeters, Firmness control of sweet potato French fry-type, product by tissue acidification, *J. Food Sci. 57*(1):138 (1992).
87. Santosa, A. L., Pengaruh kelembaban dan suhu penyimpanan pada sifat fisiko-kimia ubi jolar (*Ipomoea batatas*, Poir) (in Indonesian), Dissertation, Bogor Agricultural University (IPB), Bogor, Indonesia, 1977.
88. Agra, I. D., Sri Warnajati, and R. S. Risadi, Hydrolysis of sweet potato starch a high atmospheric pressure, *Res. I.* (Indonesian) *2*(3):34 (1969).
89. Dawson, P. R., Sweet potato for food and feed, *U.S. Agric. Yearbook 51*:204 (1951).
90. Scott, G. J., and V. Sundrez, Transorming traditional food crops: Product development for roots and tubers, in *Product Development for Root and Tuber Crops Volume III—Africa.* (G. J. Scott, P. I. Ferguson, and J. E. Herrera, eds.), International Potato Centre Apartado 5969, Lima, Peru, 1993, p. 3.
91. Ge, L. W., W. Xiuguin, C. Huiyi, and D. Rong, Sweet potato in China, in *Product Development for Roots and Tubers*, Vol. I—*Asia*, (G. Scott, S. Wisersema, and P. Ferguson, eds.), Proc. Intl. Workshop held April 22–May 1, 1991, at Visayas State College of Agriculture, Baybay, Leyle, Philippines, 1992, p. 41.
92. Mudaliar, Cattle feed from sweet potato vine, *Madras Agric. J. 37*:421 (1950).
93. Bennett, H. W., and M. Geiger, Sweet potato vine silage, Mississippi Agriculture Experiment Station Information Sheet 215, 1960.

4

Cassava

D. B. WANKHEDE, P. N. SATWADHAR, AND A. R. SAWATE
Marathwada Agricultural University, Parbhani, India

I. INTRODUCTION

Cassava (*Manihot esculenta* Crantz), a popular industrial root crop, is one of the main sources of calories for people in many tropical regions of the world. It is mostly cultivated in the tropics for its starchy roots. It is the fourth most important source of calories in human diet in tropical regions of the world (1). In fact, it is native to South America, and was subsequently popularized in other tropical parts of the world. The crop is known by many different popular names—cassava, mandioc, manioc, manihot, yuka, kahoy, etc.—in different countries (2). It is one of the most efficient crops in terms of carbohydrate production (3). The world annual production of cassava roots in 1993 was estimated at 153 million metric tons, about 46% of which was produced in Africa, 33% in Asia, and 21% in Latin America (4). The major producing countries are Nigeria, Brazil, Zaire, Thailand, and Indonesia, which together produce more than two thirds of the world's total production (Table 1). Most of this crop is used as food for humans, with lesser amounts being used for animal feed and in the industrial production of alcohol.

II. BOTANY

Cassava, or manioc (*Manihot esculenta* Crantz), is a herbaceous shurb (Fig. 1) belonging to the family Euphorbiaceae. It grows to a height of about 1.5–3 m, and is cultivated for its tuberous roots measuring 4–10 cm in diameter and more than 1 in length (5). The size and shape of cassava root depends upon the variety and environmental conditions. The roots are generally 15–100 cm long and 3–15 cm wide (Fig. 2). They can be cylindrical, conical, or oval with a coffee-, pink-, or cream-colored peel covered by a thin brown bark (1). The parenchyma is generally white, cream, or yellow. The plants produce 5–10 roots weighing 0.5–2.5 kg each (1).

Cassava root has three distinct layers of tissue: bark (periderm), peel, and parenchyma. The parenchyma is the edible portion of the fresh root and comprises about 85% of the total

TABLE 1 World Production of Cassava

	Production (1000 MT)	
Continent/Country	1989	1993
World	147,500	153,628
Africa	62,098	74,778
North America	941	1,007
South America	29,902	27,460
Asia	54,378	50,174
Oceania	780	208
Brazil	23,247	21,719
Thailand	23,460	19,610
Indonesia	16,681	16,356
Nigeria	16,500	21,000
Zaire	16,300	20,835
India	5,250	5,340
Paraguay	4,000	2,680
Mozambique	3,400	3,511
Guinea	3,327	4,200
China	3,185	3,406
Vietnam	2,900	2,631

Source: Ref. 4.

weight (1). The parenchyma consists of xylem vessels distributed in a matrix of starch-containing cells. The peel layer comprises sclerenchyma, cortical parenchyma, and phloem and constitutes 12% of root weight. The periderm layer contributes 2% to the total weight of the root.

Cassava cultivars are mostly developed in India and elsewhere on the basis of the following attributes: high root yield, high starch content, high protein content, acceptable

FIGURE 1 The cassava plant.

FIGURE 2 Cassava roots.

culinary quality, low hydrocyanic acid and prussic acid content, early maturation, and resistance to pests and diseases, particularly mosaic and bacterial blight. Moreover, they have also been distinguished on the basis of morphological characteristics, shape of the roots, yield, and cyanogenic glucoside content of the roots. They are broadly classified into three groups on the basis of the cyanide content in the edible portions of the roots, namely, sweet or nontoxic (< 50 mg/kg fresh weight), intermediate (50–100 mg/kg fresh weight), and bitter or toxic (> 100 mg/kg fresh weight). The sweet variety is usually used as food, however, the bitter variety is generally used as the raw material for starch production and not as food or feed (6). The roots are usually reddish or brown, but some are white. Several cultivars, including MCol 22, M Mex 59, M Col 113, Popayam, M Mex 15, CMC 40, CMC 76, CMC 92, and M Pan 70, released from Colombia, Bahia Preta, Curvelo, and Manjari, are early maturing cultivars of high starch and dry matter content grown in Brazil. Other currently grown cultivars in Brazil are Grande Preta, Saracura, Brancade Santa Catarina, Rosa, Brava de Padua, Salangor Preta, Graveto, Graveto de Santa Terezinha, Sutinga Preta, Aipim Gigante, Aipim Icara, Aipim Preto, Mandim Branca, Amerela, Mico, Sipdal 1 and 2, Barrinha, Mamao, Cria Menino Preto, Dhoverde, and Itapicura de Serra. In India, cassava is cultivated mainly in Kerala and Tamilnadu for both edible and industrial purposes. Many hybrid cultivars have been developed in India. These include H-388, H-688, H-1423, H-92, H-1253, H-97, H-165, H-226 (Sree Prakash), H-1687 (Sree Visakham), H-2304, (Sree Sahaya), and H-3641 (Mulluvadi) (7,8).

III. PRODUCTION

A. Soil and Climate

Cassava can be grown on many types of soil, but a friable fertile, well-drained sandy loam is considered better than heavier types. A hardpan layer below the surface layer is thought to be desirable because it tends to prevent the development of the tuberous roots at too great depths.

The cassava crop is grown between 30°N and 30°S with more than 750 mm rainfall and an annual mean temperature greater than 18–20°C. A small proportion of cassava is grown near the equator in South America and in Africa at altitudes up to 2000 m (9). It requires a warm climate free from frost for at least 8 months. The plant is tender to frost, and growth is checked by continued cool weather without frost. Frequently, environmental influences exceed cultivar differences. Maximum photosynthetic activity is observed at 20–30°C, while below 18°C and above 40°C photorespiration is 80% lower than the maximum (10). Low light intensity substantially reduces dry matter deposition, which reduces root development and growth more than shoot growth. However, cassava can tolerate prolonged drought coupled with high temperatures and low atmospheric humidity conditions (10).

B. Propagation

Commercial cassava is exclusively established from planting stem cuttings from mature plant. Whole mid-section stem cuttings 25–30 cm long and 2–2.5 cm thick planted vertically give the best growth and yield response. In some cases, the availability of plantable stems has retarded the rate of expansion of cassava production. However, CIAT (9) has developed a rapid propagation technique with the potential of producing 3,00,000–4,50,000 plantable, 30-cm-long cassava stem cuttings from a single mature plant. In vitro culture provides a higher multiplication rate for cassava than traditional propagation methods. Pest/Disease elimination, germplasm maintenance, and somatic embryogenesis have also been reported (10) in cassava.

C. Cultural Practices

1. Planting

Stem cuttings are planted in furrows and covered to the depth of 2–4 inches. Planting the cuttings vertically has been found to be better than slanting or horizontal methods. Dipping the basal end of the cuttings in *Azotobacter* gives higher yield. The ideal spacing for branched-type cassava is 90 × 90 cm and for the nonbranched type 75 × 75 cm. In both types, retaining only two shoots is best for increasing yield. The tuber number per plant and tuber weight is higher under the wider spacing (90 × 90 cm) with 12,340 plants/ha (11).

2. Manuring and Fertilization

For most soils, it is desirable to apply some fertilizer or manure or both. The practice of farmyard manure (FYM) application can be eliminated by green manuring in situ with cowpea. This has been found to improve the nitrogen content from 0.075% to 0.083% (11). It has been recommended to apply 12.5 tons of FYM, 120 kg of N, 60 kg of P, and 180 kg of K/per hectare (12). The application of 100:100:100 kg NPK/ha was also found to be significant with respect to tuber number and weight (11). Sulfur-containing fertilizers were found to be significantly superior to non–sulfur-containing fertilizers. Sulfur at 50 kg/ha gave significantly higher yields (12). Cassava has also shown good response to calcium application (10).

3. Irrigation

In India, cassava is grown as a rainfed crop. However, studies show that irrigation at frequent intervals maximizes the yield. Irrigation at 60% available moisture recorded the maximum tuber yield of 24,591 kg/ha, consuming 1907 mm of water during the crop growth (13).

4. Intercultivation

Two intercultivations, the first one month after planting and a second 30 days later, are required to ensure proper tuber development. Intercropping in cassava is an economical proposition, and crops like groundnut and cowpea can be grown successfully along with cassava at early stages of growth. Growth regulators like CCC at 10,000 ppm and ethephon at 250 ppm have significantly increased cassava yields (14).

D. Diseases and Pests

Cassava is susceptible to bacterial blight, being attacked by *Cercospora henningsii*, *Cercospora viscosae*, *Erwinia carotovora*, and *Phoma* sp. Cassava mosaic virus disease, namely, *Tottikappa* and *Kalikalan*, were reported from India (14). Ingram and Humphrics (15) listed a number of insect pests causing serious losses of cassava products.

E. Harvesting and Handling

The maturity of the cassava is influenced by variety and climatic conditions. The early type matures in about 6 months under favorable climatic conditions. A maturity period of about 9–12 months is required to get the highest yields. Prolonged maturity period may, however, turn the tubers fibrous and poor in quality. The tubers of cassava are harvested as and when needed for home use and local markets. The aboveground portions of cassava are cut into stubs 15–20 cm long (16), which furnish a handhold to pull the roots from the ground. Various handtools are used to loosen the soil. Mechanical harvesters have been developed in Brazil, Venezuela, Nigeria, and Cuba (1).

Cassava is cured at relatively high temperature and humidities. The curing is done at 25–40°C and 80–85% relative humidity. Under these conditions, suberization occurs in cassava in 1–4 days and a new cork layer forms around wounds in 3–5 days. Curing of cassava roots delays the onset of primary deterioration and reduces both secondary deterioration and moisture loss.

IV. CHEMICAL COMPOSITION

Cassava is one of the most important food crops of rural lowland tropical areas. It is consumed as a major source of carbohydrate by poor people living close to the subsistence level. These tubers are rich in carbohydrates (≈ 80–90%) on a dry weight basis, of which starch is a major constituent. However, it is known that a wide variation exists in the contents of starch and dry matter. This may be attributed to variety, age of the plant, and environmental and geographical conditions. The proximate chemical composition of cassava tubers compared to other tubers is presented in Table 2. The distribution of various nutrients in different anatomical parts suggests

TABLE 2 Nutritional Composition of Cassava (g/100 g edible portion)

Constituent	Cassava[a,b]	Sweet potato[a,b]	Yam[a]	Taro[a]
Water (g)	59.4–62.5	68.0–70.0	70.0–72.4	70.0–72.4
Carbohydrate (g)	34.7–38.7	27.3–31.0	24.0–25.0	24.0–25.0
Protein (g)	0.7–1.2	1.3–1.5	2.0–2.4	1.9–2.0
Fat (g)	0.2–0.3	0.4–0.7	0.2	0.2
Food energy (kcal)	144–146	117–121	105–108	104–108
Calcium (mg)	25–50	30–34	15–22	23–30
Phosphorus (mg)	40–50	49–60	30	80–140
Iron (mg)	0.5–0.9	0.7–1.0	0.8–100	1.0–2.1
Vitamin A (IU)	Trace	500–19450	Trace	30
Thiamine (mg)	0.02–0.06	0.09–0.14	0.05–0.09	0.04–0.15
Riboflavin (mg)	0.01–0.07	0.05–0.10	0.03–0.06	0.03
Niacin (mg)	0.3–0.6	0.6–0.7	0.4–0.5	0.4–0.9
Vitamin C (mg)	30–36	16–23	4.5–21.5	5.10

[a]From Ref. 9
[b]From Ref. 18.

that peel has more protein, fiber, and sugars than parenchyma and less dry matter and starch (Table 3). Total carbohydrates, including starch, free sugars, and cellulose and/or hemicellulose components together, make up over 90% of parenchyma dry weight (1).

A. Carbohydrates

Cassava contains very low amounts of free sugars (70% ethanol-soluble sugars) to the extent of 1–6%. Among the free sugars in the root, a nonreducing sugar, sucrose, predominates, with small quantities of glucose, fructose, and maltose (17).

Cassava contains 20–30% starch on a fresh weight basis. The chemical composition of the starch is presented in Table 4. The protein and ash contents of cassava starch are considerably lower than tuber starches (18). These values are in close agreement with the values given by ISI for cassava starch (19). Some of the important physicochemical properties of cassava and tuber starches are given in Table 5. Microscopic examination of cassava starch granules reveal that

TABLE 3 Constituents of Cassava Root Parenchyma and Peel

	% dry weight	
Constituent	Parenchyma	Peel
Starch	70–91	45–59
Total sugars	1.5–5.8	5.2–7.1
Crude fiber	3.0–3.0	5.0–15.0
Ash	1.0–2.5	2.8–4.2
Protein	1.0–6.0	7.0–14.0
Fat	0.3–1.5	1.5–2.8

Source: Ref. 1.

TABLE 4 Chemical Composition (%) of Cassava and Tuber Starches

Composition	Cassava[a,b]	Yam[c] Elephant Tuber	Potato[a]	Cassava starch[b] ISI specifications	
				Edible	Textile/ Industry use
Moisture	13.0	10.55	19.0	13.0	15.0
Ash	0.20	0.58	0.40	0.40	0.40
Fiber	—	—	—	0.2	0.60
Protein (N × 6.25)	0.10	0.27	0.06	—	0.30
Lipid	0.10	0.10	0.05	—	0.20
Viscosity (0.2% paste in seconds in redwood)	—	—	—	—	40.0
pH	—	—	—	4.5–7.0	4.80

ISI = Indian Standard Institutions.
[a]From Ref. 21.
[b]From Ref. 24.
[c]From Ref. 18.

they are mostly round in shape. They measure 4–35 μm in length. However, yam elephant tuber starch granules have a polygonal to round shape. They measure 7–13 μm in length and 5–25 μm in width (18). Sweet potato granules have a polygonal shape, and measure 5–100 μm in diameter. (20,21). The hilum in cassava starch granules is well defined and possesses some fissures. However, yam elephant starch granules exhibit no fissures but faint hilum (18). Other tuber starches also exhibit distinct faint hila. These starch granules examined under polarized light exhibit strong centric polarization crosses (18,21). The gelatinization temperature range of cassava starch is given in Table 5. It compares well with other tuber starches such as sweet potato and yam elephant (18,21). The Brahender pasting temperature is found to be lower (59°C) than that of yam elephant and potato starches (18,21). Cassava starch exhibits peak viscosity to the extent of 1400 B.U. at a concentration level of 7% (W/v), whereas yam elephant tuber and potato starches exhibit 1570 and 2500 B.U. at 10 and 7% (W/v) concentration levels,

TABLE 5 Important Physicochemical Properties of Cassava, Yam Elephant, and Potato Starches

Properties	Cassava[a]	Yam elephant tuber[b]	Potato[a]
Amylose	17.00	24.50	21.00
Average degree of polymerization	18,000	—	14,000
Granule diameter (μm)	4–25	5–30	5–100
Shape	Mostly round	Polygonal round	oval
Gelatinization temperature (C°)/ Brabender pasting temperature	58.5–70	73.0–75.5–80.5	56-66
Brabender peak viscosity (B.U.) at 95°C	1400	1570	2500

[a]From Ref. 20.
[b]From Ref. 18.

respectively (18,21). The amylose content of cassava starch is found to range between 17 and 25% (20,21). These values are well within the range reported for various other tuber starches (18,20,21).

B. Proteins

The protein content of cassava and other roots ranges between 1.6 and 2.6%. The protein content of cassava roots is very low, and the essential amino acid makeup is not very desirable (22). According to Gopalan et al. (23), methionine is the most limiting essential amino acid in cassava. However, Raja et al. (24) reported that arginine, histidine, leucine, isoleucine, lysine, phenylalanine, threonine, and tryptophan are present in substantial amounts in cassava protein. Similarly, Splittstoesser et al. (25) reported that the true protein of cassava contains large quantities of lysine and tryptophan, while methionine, cystine, and cysteine are found in lower amounts.

C. Minerals and Vitamins

The vitamin and mineral contents of cassava tubers and other tubers are given in Table 6. Cassava contains appreciable amount of calcium and phosphorus, but it is deficient in other minerals. Cassava is a rich source of vitamin C, but it lacks other important vitamins (17). Oke (26) reported that cassava roots are a good source of minerals such as calcium, iron, magnesium, and phosphorus.

D. Antinutritional Factors

There is a large variation in total cyanide content between different varieties of cassava (1). The consumption of cassava has been associated with several kinds of pathological disorder (27) due to the presence of cyanogenic glucosides linamarin and lotaustralin (Fig. 3), which on hydrolysis release HCN. However, several million people living in the tropics consume cassava on a daily basis without any obvious sign of intoxication. This suggests that the concentration of toxic

TABLE 6 Nutritional Composition of Cassava (g/100 g edible portion)

	Tuber crops			
Constituents	Cassava	Sweet potato	Yam	Taro
Water (g)	59.4–62.5	68–70	70.0–72.4	70.0–72.4
Carbohydrate (g)	34.7–38.7	27–31	24–25	24–25
Fat (g)	0.7–1.2	1.3–1.4	2.0–2.4	1.9–2.0
Ash (g)	1.0–2.4	0.66–1.98	—	—
Calcium (mg)	25–50	30–40	15–22	104–108
Phosphorus (mg)	40–50	49–60	35	23–30
Iron (mg)	0.5–0.9	0.7–1.0	0.8–1.0	1–2.1
Vitamin A (IU)	Traces	500	Traces	30
Thiamine (mg)	0.02–0.06	0.09–0.14	0.05–0.09	0.04–0.05
Riboflavin (mg)	0.01–0.07	0.05–0.10	0.03–0.06	0.03
Niacin (mg)	0.3–0.06	0.6–0.70	0.4–0.50	0.40–0.90
Vitamin C (mg)	30–60	16–23	4.5–21.5	5.1
Food energy (kcal)	144–146	117–121	105–108	104–108

Source: Refs. 17, 18, 23.

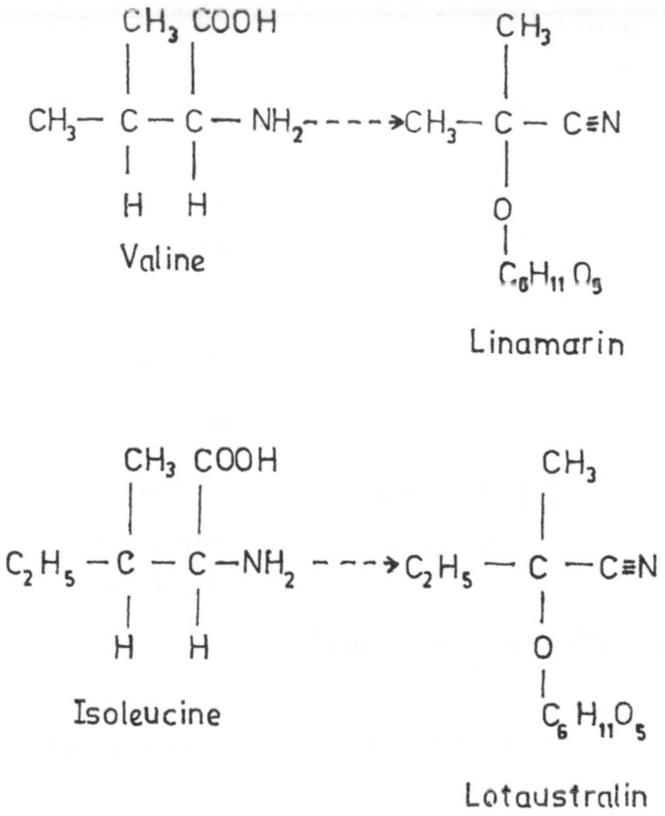

$$CH_3-CH(CH_3)-CH(COOH)-NH_2 \dashrightarrow CH_3-C(CH_3)(O-C_6H_{11}O_5)-C\equiv N$$

Valine — Linamarin

$$C_2H_5-CH(CH_3)-CH(COOH)-NH_2 \dashrightarrow C_2H_5-C(CH_3)(O-C_6H_{11}O_5)-C\equiv N$$

Isoleucine — Lotaustralin

FIGURE 3 Linamarin and lotaustralin and their metabolic precursors valine and isoleucine. (From Ref. 30.)

compounds in cassava-based foods is naturally low. They are sufficiently reduced during the processing of cassava into food, and they do not cause any harm (1). The cassava plant accumulates linamarin and lotaustralin, which are synthesized from the amino acids valine and isoleucine, respectively. These are stored inside vacuoles in the cytoplasm, while the enzyme linamarase, a β-glucosidase that can hydrolyze them, is located on the cell wall (28). It is generally assumed that in intact plant cells, the enzyme and cyanogenic glucosides do not come in contact. When tissues are crushed, the cell structure may be damaged to such an extent that the enzyme comes in contact with the cyanogenic glucosides. A second enzyme, hydroxynitrile lyase, carries out the decomposition of cyanohydrin, producing cyanide (Fig. 4). There is wide variation in the cyanogenic potential of different cassava cultivars, ranging from 0.2 to 62.4 mg HCN eq/100 g (29,30). Cassava roots may be dried and milled into flour. During processing, the structural integrity of the cell is usually lost and cyanogenic glucosides come into contact with linamarase, thus initiating the formation of HCN. During processing, which commonly includes heating, the HCN produced is likely to evaporate completely.

The cyanogenic potential of cassava leaves is 5–20 times higher than that of edible portions of the roots (2). In many countries, cassava leaves constitute a highly priced vegetable. Young leaves are usually pounded and boiled for 15–20 minutes. Various ingredients are then added to improve the taste. When leaves are processed in this manner, the cyanogenic potential is considerably reduced during pounding and is virtually eliminated after boiling (30). A similar phenomenon is observed when cassava roots are ground into a mash, dewatered, and heated

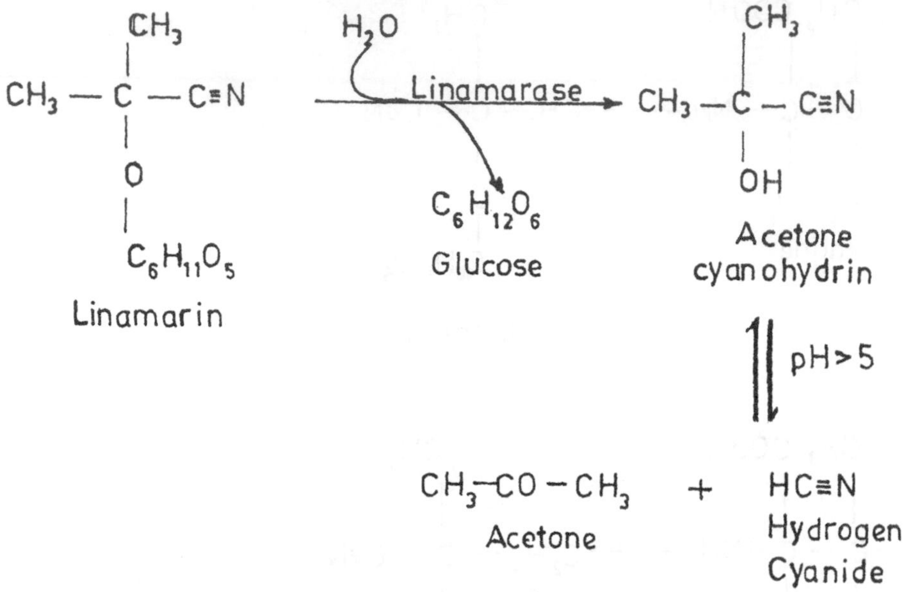

$$\underset{\text{Linamarin}}{CH_3-\overset{\displaystyle CH_3}{\underset{\displaystyle O-C_6H_{11}O_5}{C}}-C\equiv N} \xrightarrow[\text{Linamarase}]{H_2O} \underset{\text{Glucose}}{C_6H_{12}O_6} + \underset{\text{Acetone cyanohydrin}}{CH_3-\overset{\displaystyle CH_3}{\underset{\displaystyle OH}{C}}-C\equiv N}$$

$$\text{Acetone cyanohydrin} \underset{}{\overset{pH>5}{\rightleftharpoons}} \underset{\text{Acetone}}{CH_3-CO-CH_3} + \underset{\text{Hydrogen Cyanide}}{HC\equiv N}$$

FIGURE 4 Enzymatic hydrolysis of linamarin. (From Ref. 30.)

on a flat hot surface to produce roasted granules known as *farinha* in Brazil and *gari* in West Africa (30).

V. STORAGE

Cassava roots have a short postharvest life. Roots become inedible within 1–3 days after harvest due to rapid chemical and biochemical changes in the roots. Phenolic compounds (catechins, coumarins, leucoanthocyanins) are synthesized, and polymerization of these compounds produces blue, brown, and black pigments. There is a rapid accumulation of a coumarin called scopoletin (blue fluorescence) during storage (1).

Cassava roots can be stored in a dry and warm climate for several weeks. Traditional storage techniques include reburial, coating with mud, placing under water, or piling in heaps and giving thorough daily water (31). In most cassava-producing areas, the roots are left on the ground until needed (32). Although they can be harvested over longer periods, older roots tend to become fibrous and woody (33). In areas where refrigeration is not available or not economical, it is a common practice to rebury harvested roots in a cool place for short storage (34).

Prestorage curing followed by storage in simple field clamps or in boxes packed with moist material prolong the marketable life of cassava up to 3 months (34). Recently, packing roots in polyethylene bags (with or without perforation) and protecting against microbial deterioration with a Thiabendazole-based fungicide of low toxicity has been developed as a simple and effective storage technology (16). Neelakanthan and Maniangalai (34) reviewed postharvest deterioration and storage problems of cassava tubers, and techniques such as cold storage, freezing, fungicidal surface treatments, fumigation methods, hot water dip, and storage in specially designed clamps can reduce cassava losses during storage.

Waxing of the cassava can extend the marketable life of the roots up to 1 month by reducing the rate of gas transfer between tissue and atmosphere (35). The freshly harvested roots are packed in moist sawdust in boxes to prolong their postharvest life for 1–2 months (35). Cassava keep well at 0–5°C and 85–90% relative humidity (32).

VI. PROCESSING AND UTILIZATION

A variety of dried cassava products such as chips, flour granules (*gari* and *forinha de mandico*), and starch are prepared in Africa, Asia, and Latin America. According to local needs, tradition, and taste (36–39), parboiled cassava chips made in India are reported to have a longer storage life than ordinary chips or flakes (38). Cassava meal or flour is made by grinding the chips. The granules in the form of *gari* and *farinha* are important cassava products in Western Africa and Brazil. Very recently, Kottapale (40) prepared extruded products containing cassava starch. Cassava starch is used in laundering, for sizing paper, in making glue, and in other industrial products. Roots are prepared for the table by boiling and baking and are served much like potato. The typical chemical composition and nutritive value of cassava and some of its important products is shown in Table 7.

A. Roots

Cassava roots are generally peeled prior to consumption. Peel accounts for 10–20% of the fresh root, while the edible portion accounts for 80–90%. The edible fleshy portion of the cassava root contains moisture (62%), carbohydrate (35%), protein (1–2%), fat (0.3%), and fiber (1–2%). It is known that the unprocessed cassava root contains high amounts of prussic acid. Moreover, these fresh roots cannot be stored for more than a few days after harvesting. These factors limit their utilization in foodstuffs, and therefore, prior to consumption they are peeled and cut into pieces and boiled.

After cooking the processed roots are eaten as such or along with other side dishes, especially fish or coconut or any other curries. These roots are also consumed after toasting and baking in some parts of the world. In Kerala, parboiled cassava slices are used for human consumption. Haranandani and Advani (41) reported that parboiling increases the crispness of

TABLE 7 Nutritional Composition of Cassava and Some of Its Products

Composition	Tubers	Flour	Gari	Fufu	Lafun	Macaroni product
Moisture (%)	59.4	9.5	14.4	15.3	19.5	10.6
Protein (%)	0.7	1.6	0.9	0.6	0.8	11.2
Fat (%)	0.2	0.4	1.1	0.14	0.4	1.9
Fiber (%)	0.60	0.80	0.40	0.20	0.73	0.70
Carbohydrate (%)	38.1	84.9	81.8	75.8	76.4	73.8
Ash (%)	1.0	1.8	1.4	0.5	2.0	1.8
Calcium (mg%)	50	60	70	160	220	30
Phosphorus (mg%)	40	80	40	20	40	140
Thiamine (mg%)	0.05	0.08	—	—	—	0.22
Iron (mg%)	0.9	3.5	2.2	6.2	6.6	2.9
Vitamin C (mg%)	25	—	—	—	—	—
Energy (kcal/100 g)	157	—	323	393	391	—

Source: Ref. 17.

chips and also increases their shelf life. During parboiling, some enzymes are denatured, which may play an important role in extending shelf life (41).

B. Flour

Cassava flour is prepared by chopping the roots into small pieces and then sun-drying and pounding or milling these into a coarse mill or flour (42–44). Traditional Indian foods such as *chappati*, *Uppama*, *Puttu*, *Idli*, and *Dosa* can be prepared using cassava flour (41). Similarly, a porridge can be made by mixing hot water with cassava flour. According to Castillo (45) various delicacies such as *bibingka*, *Suman*, and *Kalamay* can be prepared using cassava flour. It can be used successfully up to a content of 15% in breadmaking. Kasasian and Dendy (46) reviewed the use of cassava flour and starch in breadmaking.

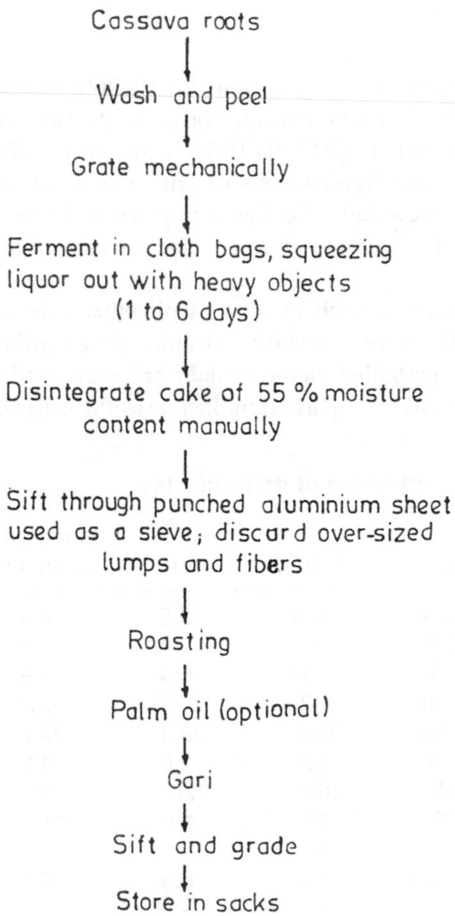

FIGURE 5 Flow diagram of traditional process for production of gari (West Africa). (From Ref. 39.)

C. Gari

Gari is a fermented, cooked and dehydrated cassava meal. It is prepared by fermenting the roots of the cassava plant (47). It has been estimated that 70% of the cassava grown in Nigeria is used for manufacturing gari (47). The preparation of gari is shown in Figure. 5. The corky outer peel and thick cortex are removed from the cassava root and the inner portion is grated. The grated pulp is pressed to express some of the juice. Fermentation takes place for 3–4 days, after which the cassava is sieved to remove any coarse lumps and heated to reduce the moisture content to 10–15%. Gari is usually consumed as a hot rehydrated staple (48).

The fermentation steps crucial to the development of the characteristic aroma and sour flavor are mainly due to lactic acid volatiles such as aldehyde, diacetyl esters, and ethanol (16). Fermentation also helps to detoxify the mash, while grating allows linamarase to come into contact with the cyanoglucoside substrate.

Recently, microorganisms associated with cassava fermentation have been identified (49–51), including *Lactobacillus plantarum*, *Streptococcus faecium*, and *Leuconostoc mesenteroides*. They are responsible for the rapid acidification that characterizes cassava fermentation. Other organisms detected in cassava fermentation include *Bacillus* species and *Corynebacterium,* but these are usually present in small number.

Submerged fermentation is the most efficient process for reducing the levels of cyanogens in cassava. Reductions in the level of cyanogens during fermentation to the extent of 95–100% are often reported (52–55). This is probably due to (a) textural changes in plant tissues that make it possible for vacuole-bound cyanogenic glucosides to diffuse and come in contact with membrane-bound linamarase and for hydrolyzed or intact compounds to leach out, (2) an increase in β-glucosidase activity in cassava tissue, and (3) utilization of cyanogenic glucoside and its breakdown products by fermentation of microorganisms.

A different mechanism has been reported for the detoxification of cassava in mash fermentation (56). The grating of cassava roots to obtain the mass disrupts the structural integrity of the plant cells, allowing cyanogenic glucosides from storage vacuoles to come in contact with linamarase. The subsequent fermentation contributes very little to the breakdown of the glucosides (57). The low pH rapidly achieved during fermentation inhibits linamarase and stabilizes cyanohydrins, thus slowing down linamarin hydrolysis and cyanohydrin breakdown.

D. Farinha

Farinha de mandioca is a major dietary staple in Brazil (58). The process for the production of farinha is presented in Figure. 6. Unlike gari, farinha preparation does not include a fermentation step (16). Large-scale production facilities in Brazil can process 50–200 tons of cassava roots per day. In addition, many thousands of small-scale traditional operations exist that make farinha. This product is an integral ingredient in many traditional dishes. It is consumed with meals and beans.

E. Fufu

Fufu is another fermented product similar to gari. It is prepared by soaking peeled cassava tubers in water for 2–3 days. The tubers are then subjected to crushing and sieving, and the resultant pulp is transferred to a bag or sack from which excess water is extracted by keeping heavy weights on the sacks. The resulting dough is used for cooking (59). Fufu is generally prepared by cooking with water while stirring until a sticky dough is formed. The fufu usually is consumed with a vegetable curry.

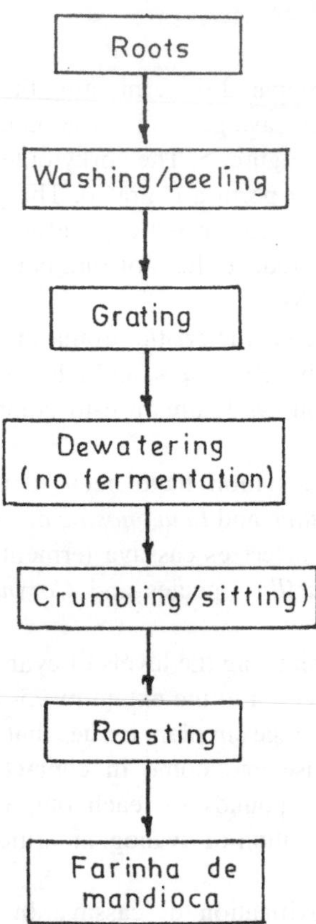

FIGURE 6 Flow diagram of the traditional process for production of farinha de mandioca in Brazil. (From Ref. 39.)

F. Lafun

Lafun is another fermented food product mostly consumed in Nigeria. It is prepared by soaking unpeeled cassava tubers in water for 5 days. These are then subjected to peeling and sun-drying for 3–4 days. The resultant dried product is powdered and sieved into flour.

G. Macaroni and Extruded Products

Macaroni is prepared by blending cassava, groundnut flour, and wheat semolina in a ratio of 60:15:15. The mixture is then subjected to extrusion cooking. The resultant product contains appreciable amounts of protein. It can be consumed after boiling. Kottapalle (40) prepared extruded products using cassava starch to the extent of 10% followed by extrusion cooking. The resultant extruded product can be consumed after frying in oil (Fig. 7).

H. Starch

Cassava roots are utilized for the production of starch in China, Indonesia, Thailand, and many countries of southeast and Latin America (16). Cassava starch is used in local food industries to

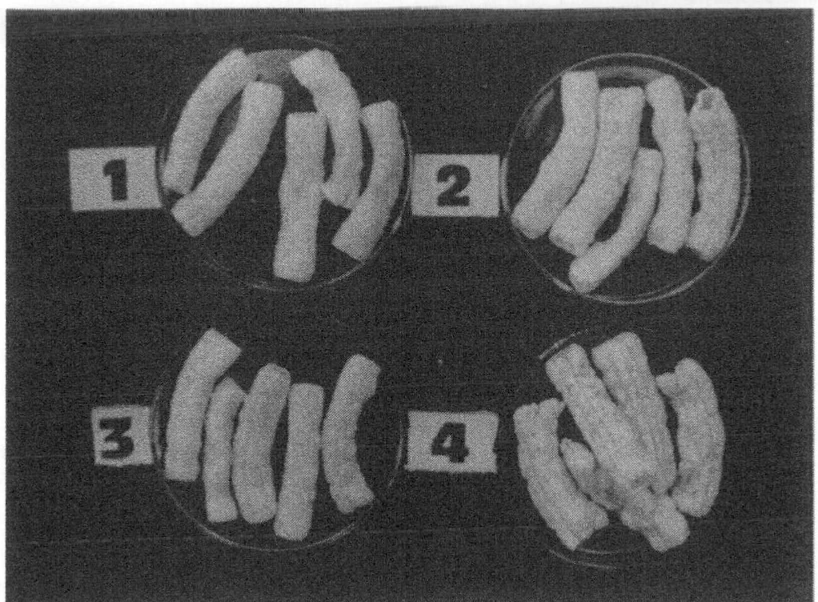

FIGURE 7 Extruded products of blends of various cereal flours (85%) cassava starch (10%), and green gram flour (5%) after frying: (1) wheat flour (85%) + cassava starch (10%); (2) sorghum flour (85%) + cassava starch (10%); (3) rice flour (85%) + cassava starch (10%); and (4) corn flour (85%) + cassava starch (10%). (From Ref. 40.)

make a wide range of traditional products, such as *sago* in India, *Krupuk* in Indonesia, and *Chipa* in Paraguay. In Brazil and Colombia, moist starch is fermented before drying and a naturally modified starch is obtained (16). Subsequently, solid starch is separated from the starch/water slurry by sedimentation (Fig. 8) or centrifugation. The resultant starch is dried to a final moisture content of 12–14%. The starch extraction rate varies from 18 to 25%, depending upon the efficiency of the process.

While making *sago* (Figure 9), partially dried starch is globulated on a vibrating surface, then the resulting small globules are steamed or baked to gelatinize the surface layer of starch (16). The end product is used for many purposes. To obtain sour starch, the sedimented starch is left in tanks for 20–30 days before drying. Sour starch has excellent expansion power, thus, the volume of dough containing this starch increases manyfold during baking.

I. Alcohol

At the Central Tuber Crops Research Institute, Trivendrum, India, a technique of producing alcohol from cassava by using immobilized enzymes has been standardized. Attempts were made to ferment cassava starch under vaccuum, steady state, and tower fermenters for the large-scale production of ethanol. A hand-operated cassava tuber chipping machine has been evolved for producing uniform-sized chips (37). A process to convert cassava starch into alcohol using enzymes and yeast has been developed (37). The process consists of liquefaction of the cassava flour slurry, saccharification, fermentation, and recovery of alcohol. The alcohol thus produced conforms to the specification for potable alcohol. The Central Food Technological Research Institute (CFTRI), Mysore, India, has also developed a solid-state fermentation technique

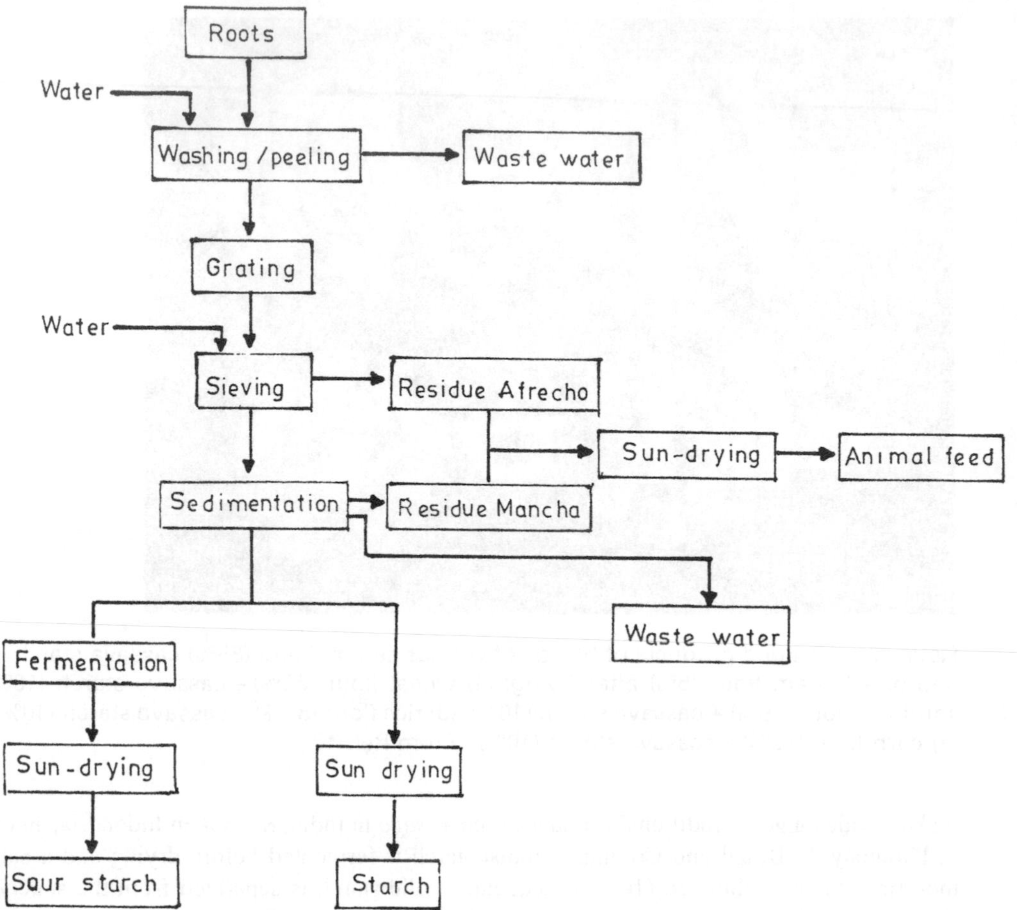

FIGURE 8 Steps in production of fermented and nonfermented cassava starch. (From Ref. 1.)

to manufacture amyloglucosidase, one of the two enzymes required in the process of alcohol production.

Poulsen and Garbeg (44) described a process of producing alcohol and fructose from cassava and maize. They stated that by using a semicontinuous alcohol process (Fig. 10), the fresh cassava can be handled in concentrated solutions up to about 25% distillage. Poulsen and Garberg (44) also demonstrated a continuous process of alcohol production. Both cassava and maize can be converted into alcohol using simple equipment and limited amounts of energy. A fructose syrup of acceptable quality can be made from cassava (Fig. 11), which could be a viable supplement to industrial sugar (44).

J. Miscellaneous Uses

Cassava roots are rich in carbohydrate with small quantities of protein. They are also rich in some of the minerals and vitamins. However, they contain some antinutritional factors, such as prussic acid and cyanogenic glucosides, which limit their utilization. Cassava roots, after processing, can also be used in animal feeds (17).

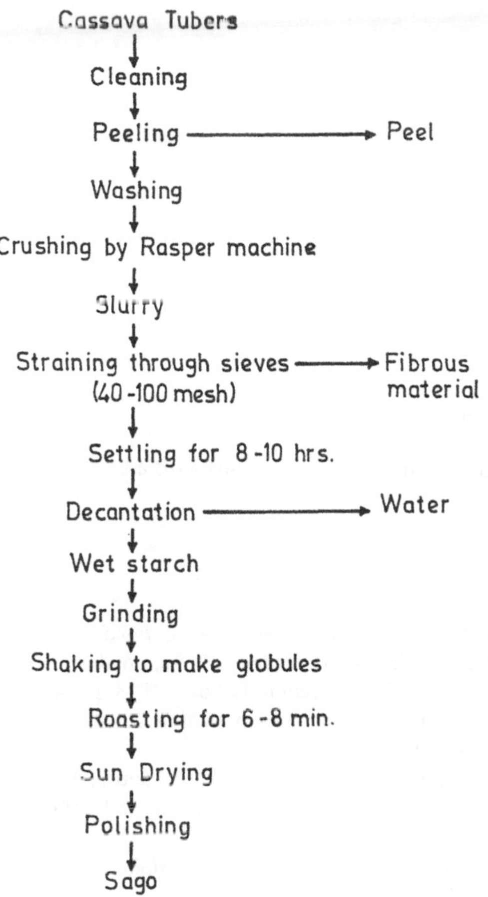

FIGURE 9 Flow diagram of production of sago in India. (From Ref. 16.)

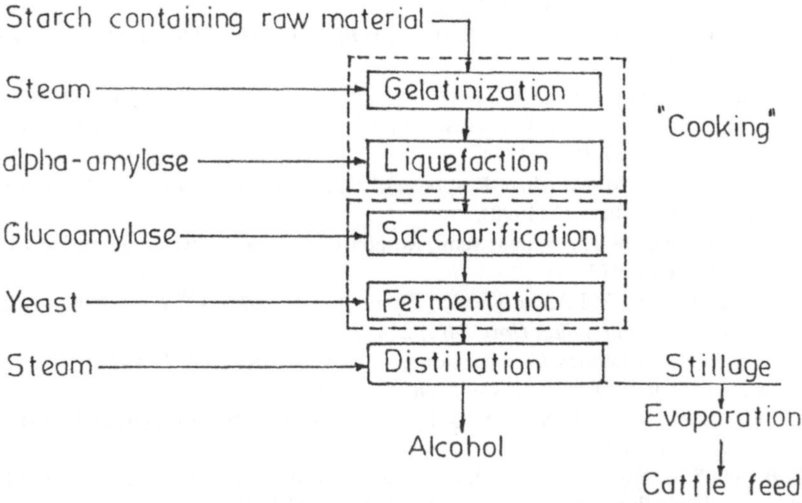

FIGURE 10 Stages in alcohol production from cassava. (From Ref. 44.)

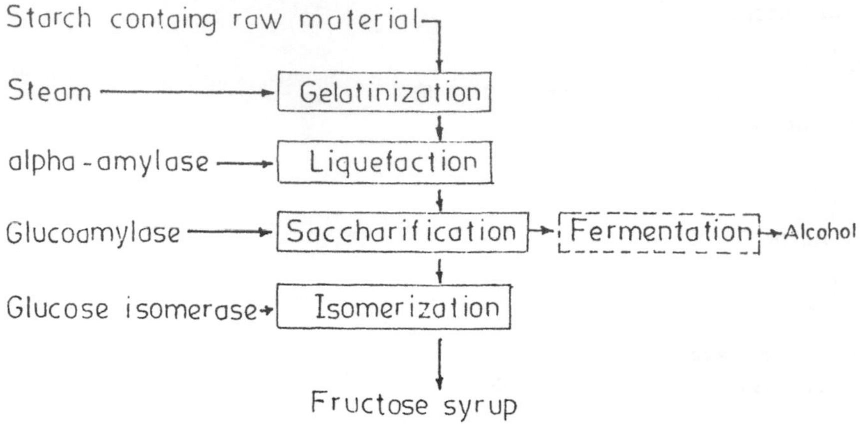

FIGURE 11 A process for fructose production from cassava. (From Ref. 44.)

REFERENCES

1. Wheatley, C. C., and G. Chuzel, Cassava, in *Encyclopaedia of Food Science, Food Technology, and Nutrition* (R. Macrae, R. K. Robinson, and M. J. Sadler, eds.), Academic Press, London, 1993, p. 734.
2. Kay, D. E., *Root Crops, and Crop Product Digest*, TPI Publication, London, 1973, p. 24.
3. De Vries, C. A., J. D. Ferwerda, and M. Flach, Choice of food crop in relation to actual and potential production in the tropics, *Neth. J. Agric. Sci. 15*:241 (1967).
4. FAO, *Production Yearbook*, Food and Agricultural Organization, Rome, 1989 and 1993.
5. Mandal, R. C., Cassava, in *Tropical Root and Tuber Crops*, Agro. Botanical Publisher (India), Calcutta, 1993.
6. Fukuba, H., and E. M. T. Mendoza, *Tropical Root Crops: Post Harvest Physiology and Processing*, 1984, p. 171.
7. Roca, W. M., A. Rodriguez, L. F. Pitena, R. C. Barbara, and J. C. Toto, Improvement of propagation techniques for cassava using single leaf cutting: preliminary report, *Cassava Newsletter 8*:4 (1980).
8. Abraham, T. E., K. C. M. Raja, V. P. Sreedharan, and H. S. Nathan, Some quality aspects of a few varieties of cassava, *J. Food Sci. Techno. 16*:237 (1979).
9. Salunkhe, D. K., and B. B. Desai, Cassava, in *Postharvest Biotechnology of Vegetables*, Vol. II, CRC Press, Boca Raton, FL, 1984, p. 135.
10. Splittstocsser, W. E., and G. O. Tuniya, Crop physiology of cassava, *Hort. Rev. 13*:105 (1992).
11. Mohankumar, C. R., R. C. Mandal, and N. Hrishi, Effect of density, fertility levels and stages of harvest on cassava, *Root Crops 1*(1):59 (1974).
12. Hedge, M., D. P. Kumar, and M. M. Khan, Response of tapioca to different levels of nitrogen and potash, *South Indian Hort. 38*(40):193 (1990).
13. Selvaraj, K. V., R. Kaliappa, and P. P. Ramaswami, Response to tapioca to water and nutrient levels, *South Indian Hort. 25*(2):62 (1977).
14. Vijay Kumar, M., and J. B. M. Md-Abdul Khader, Effect of ethrel and CCC on certain growth and yield attributes of cassava, *South Indian Hort. 34*(4):228 (1986).
15. Ingram, J. S., and J. R. O. Humphries, Cassava storage—a review, *Trop. Sci. 14*:131 (1972).
16. Scherry, R. W., Manioc: A tropical stuff of life, *Econ. Bot. 1*:20 (1947).
17. Ghosh, S. P., T. Ramnujam, J. S. Tos, S. N. Moorthy, and R. G. Nair, *Tuber Crops*, Oxford and IBH Publishing Co., New Delhi, 1988, p. 87.
18. Wankhede, D. B., and S. U. Sajjan, Isolation and physicochemical properties of starch extracted from yam elephant (*Amorphophallus companuclatus*), *Die Starke 33*:153 (1981).
19. Specification for tapioca starch, Indian Standard Institution, New Delhi, 1958.

20. Muhrbeck, P., and A. C. Eliasson, Influence of pH and ionic strength on the viscoelastic properties of starch gels—A comparison of potato and cassava starches, *Carbohydrate Polymer 7*:291 (1987).
21. Whistler, R. L., E. F. Paschall, J. N. BeMiller, and H. J. Roberts, *Starch: Chemistry and Technology*, Vol. II, Academic Press, New York, 1967, pp. 643, 651.
22. Jones, W. O., *Manioc in Africa*, Stanford Press, Stanford, 1959.
23. Gopalan, C., B. V. Ramasastri, and S. C. Balasubrahmanian, *Nutritive Value of Indian Foods*, National Institute on Nutrition, 1971, p. 75.
24. Raja, K. C. M., E. Abraham, H. S. Nathan, and A. G. Mathew, Chemistry and technology of cassava, *Indian Food Packer 33*(3):31 (1979).
25. Splittstoesser, W. E., F. W. Martin, and A. M. Rhodes, The nutritional value of some tropical root crops, *Proc. Trop. Reg. Amer. Soc. Hort. Sci. 17*:290 (1973).
26. Oke, O. L., Cassava as food in Nigeria, *World Rev. Nutr. Dietet. 9*:227 (1968).
27. Rosling, H., Cassava toxicity and food security. A. review of health effects of cyanide expore from cassava and of ways to prevent these effects, *Report for UNICEF AFRICAN Household Food Security Programme*, Truck Kontakt, Uppasala, 1988.
28. Mkpong, O. E. H. Yan, G. Chism, and R. J. Sayre, Purification, characterization and localization of linamarase in cassava, *Plant Physiol. 93*:176 (1990).
29. Bokanga, M., The cyanogenic potential of cassava, in *Root Crops for Food Security in Africa* (M. O. Akoroda, ed.), *Proc. 5th Symp. of Intl. Soc. For Tropical Root Crops - Africa Branch*, Kampals, Ugand, Nov. 22-28, Intl. Inst. Trop. Agric., Ibadan, Nigeria, (1994).
30. Botanga, M., Biotechnology and cassava processing in Africa, *Food Technol. 49*(1):86 (1995).
31. Booth, R. H., Storage of fresh cassava (*Manihot esculenta*) II. Simple Storage technique, *Exp. Agric. 13*:119 (1977).
32. Andrade, A. M., B. V. Rocha, and H. Correa, Storage of cassava roots, *Inf. Agro. Pecuria 5*:94 (1979).
33. Dala, E. S., and F. Q. Arrieda, Village level technology of cassava storage, *Radix 3*(1):17 (1981).
34. Neelakanthan, S., and G. Manianegalai, Postharvest deterioration and storage of cassava tubers, in *National Seminar on Tuber Crops—Production Technology*, Coimbatore, India, 1980.
35. American Society of Heating, Refrigeration and Air-Conditioning Engineers Handbook, Application 59, Atlanta, 1982.
36. Cock, J. H., Cassava, a basic energy source in the tropics, *Science 218*:755 (1982).
37. *Some Recent Developments*, Central Food Technological Research Institute, Mysore, India, 1975.
38. Raman, A. I., Naturally dried cassava chips: A solution to cassava marketing problem, *Cassava Newslet. 7*:5 (1983).
39. Onyekwere, O. O., I. A. Akminerele, O. A., Koleojo, and G. Heys, Industrialization of gari ferementation, in *Industrialization of Indigenous Foods* (K. H. Steinkraus, ed.), Marcel Dekker, New York, 1989, p. 363.
40. Kottapalle, S. L., Studies on protein enrichment in extruded products, M. Tech. (Food Sci.) thesis, Marathwada Agricultural University, Parbhani, India, 1994.
41. Haranandeni, G. J., and K. H. Advani, A report on the marketing of tapioca in India, Mark Sr. No. 88, Directorate of Market Inspection, Ministry of Food and Agriculture, India, 1955.
42. Bradbury, J. H., and W. D. Holloway, Chemistry of tropical roots crops: Significance for nutrition and agriculture in the pacific, Australian Centre for International Agricultural Research, Monograph No. 6, 1988.
43. Rickard, J. E., and D. G. Coursey, Cassava storage, Part I, Storage of fresh cassava roots, *Trop. Sci. 23*:1 (1981).
44. Poulsen, P. P., and P. Garberg, Increasing the diversification of the use of enzymes. A tool for producing a variety of products from biomass with particular reference to the production of alcohol from fructose syrup from cassava, Presented at the AHARA 82, Intl. Food Conf. Bangalore, India, May, 1982.
45. Castillo, L. S., The cassava industry of the Philippines, in *Cassava Processing and Storage*, Araullo, E. V. B. Nestel, and M. Campbell, IDRC, Canada, 1974, p. 63.

46. Kasasian, R., and D. A. V. Dendy, Composite flour technology bibliography, Rep. G. 111, Tropical Products Institute, London, 1977, p. 70.
47. Olayade, S. O., D. Olatunbosun, E. O. Idusogie, and J. D. Abiagom, A quantitative analysis of food requirements, supplies and demand in Nigeria, 1988–1985, Fed Dept. Agric., Lagos, Nigeria, 1988.
48. Beuchat, L. R. Application of biotechnology of indigenous fermented, *Food Technol. 49*(1):97 (1995).
49. Bokanga, M., Microbiology and biotechnology of cassava fermentation, Ph. D. thesis, Cornell University, Ithaca, NY, 1989.
50. Wankwo, N., D. E. Anadu, and R. Usoro, Cassava fermenting organisms, MIRCEN, *J. Appl. Microbiol. Biotechnol. 5*:169 (1989).
51. Okafor, N., B. Ijioma, and C. Oyalu, Studies on microbiology of cassava retting for foo-foo production, *J. Appl. Bacteriol. 56*(1):1 (1984).
52. Bokanga, M., S. K. O'Hair, R. R. Narayanan, and K. H. Steinkraus, Cyanide detoxification and nutritional changes during cassava fermentation, in *Proc. 8th Symp. Int. Soc. Trop. Root Crops*, (R. H. Howeler, ed.), Bangkok, Oct. 30-Nov. 5, 1988, p. 385.
53. Ayernor, G. S., Effect of retting of cassava on product yield and cyanide detoxification, *J. Food Technol. 20*:89 (1985).
54. Ezeala, D. O., and N. Okoro, Processing techniques and hydrocyanide content of cassava based human foodstuffs in Nigeria, *J. Food Biochem. 10*:125 (1986).
55. Westby, A., Importance of fermentation in cassava processing, in *Tropical Root Crops in Developing Economy* (P. Ofori and S. K. Hahn, eds.) *Int. Soc. for Hort. Sci.* Wageningen, 1994.
56. Vasconcelos, A. T., D. R. Twiddy, A. Westby, and P. J. A. Reilly, Detoxification of cassava during preparation, *Int. J. Food Sci. Technol. 25*(2):198 (1990).
57. Balagopalan, C., C. Padmaja, S. K. Nanda, and S. N. Moorthy, *Cassava in Food Feed and Industry*, CRC Press, Boca Raton, FL, 1988.
58. Lancaster, P. A., J. S. Ingram, M. Y. Lim, and D. G. Coursey, Traditional cassava based food: Survey of processing techniques, *Econ. Bot. 36*:12 (1982).
59. Devlo, F. E. Cassava as food in Ghana, *Home Scientist* (Ghana) 2:9 (1973).

5

Carrot

P. M. Kotecha and B. B. Desai
Mahatma Phule Agricultural University, Rahuri, India
D. L. Madhavi
University of Illinois, Urbana, Illinois

I. INTRODUCTION

Carrot is one of the important root crops cultivated throughout the world for its fleshy edible roots. It is grown in spring, summer, and autumn in temperate countries and during winter in tropical and subtropical regions (1). Carrots are used for human consumption as well as animal feed. They are cooked alone or with other vegetables in the preparation of soups, stews, curries, and pies; fresh grated roots are used in salads; tender roots are pickled (2). Carrots possess many medicinal properties and are used in ayurvedic medicine (2). The white varieties of carrots are valued as feed for horses and dairy cattle. Carrots are a rich source of β-carotene and contain appreciable amounts of thiamine and riboflavin. Worldwide production of carrots increased from 10.09 to 13.37 million metric tons (MT) during 1980–1990 (Table 1). The major carrot-producing countries are the former USSR, China, Japan, France, the United States, and the United Kingdom (2).

II. BOTANY

The carrot (*Daucus carota* var. Sativa) is a native of Europe, Asia, and Northern Africa, and possibly North and South America. Carrot is a biennial vegetable and belongs to family Umbelliferae. The genus *Daucus* contains about 60 species, of which very few are cultivated. Carrots are a dicotyledonous herbaceous crop grown for the enlarged taproot. The color of cultivated carrot roots varies from white to yellow, orange, light purple, deep red, or deep violet, and the shape varies from short stumps to tapering cones. The root diameter and length can vary from 2 to 6 cm and 6 to 30 cm, respectively.

The most important bunching varieties of carrot in the United States are Imperator, Gold Spike, and Gold Pak, having long slender roots with a good smooth exterior. Varieties Red Cored Chantenay, Royal Chantenay, and Autmn King have good color and are popular processing

TABLE 1 World Production of Carrots

Continent/Country	Production (1000 MT)
Africa	614
Algeria	160
South Africa	175
North Central America	1,777
United States	1,312
Dominican	378
South America	667
Argentina	185
Colombia	178
Asia	3,993
China	2,749
Japan	700
Turkey	174
Europe	3,930
France	490
Germany	445
Netherlands	330
Poland	740
United Kingdom	470
USSR	2,200
Developed, all	8,700
Developing, all	4,599
World	13,369

Source: Ref. 2.

varieties. The highly flavored French or Chantenay varieties are not suitable for canned juice preparations because of the development of undesirable off-flavors (3). The Royal Chantenay grows about an inch longer and is more cylindrical than Red Cored Chantenay. Nantes is a good home-garden variety (1). Other popular varieties are Robicon, Perfection, Bunching, Hutchinson, Bagley, Touchon, and Coreless. The hybrids Woodland, 13 C-24, and Fanci Pak have been found to be the most promising of the 61 carrot cultivars grown in southwest Texas (4). Many varieties of carrot are grown throughout the world. They may be grouped into two types, namely, (1) the temperate or European cultivars, which are biennial, orange in color, and uniform in thickness with a small core, and (2) the tropical or Asiatic cultivars, which are annual, red in color, more juicy, have a bigger core, and a heavier top (5). Some important commonly grown varieties of carrot are Amsterdam, Barlikuner, Chantenay, Royal Chantenay, Delattya, Danvers, Emperador, Flakker, Goldpack-28, Gonsen Lemer, Gross Karotel, Honey Sweet, Karotina, Pusa Kessar, Pusa Meghalli, Rubica, Berlicum, Rote Rusen, Selection-23, Nantes, No. 29, Nantesca, Saint Valery, Sperton Fancy, Tip-top, UH-AC, and Pusa Yamadagni (6,7).

III. PRODUCTION

A. Soil and Climate

Carrots can be grown in a wide variety of soils. However, the ideal soil should be deep, loose, friable, well drained, and rich in humus. Loamy or sandy loam soils with ample quantities of humus are well suited to the cultivation of carrots. The ideal pH range for obtaining good yield is 5.5–6.5. Soils with pH up to 7.0 can also be used, but too alkaline or acidic soils are unsuitable for this crop (8).

Carrot is a cold weather crop. However, it thrives well in warm climates. The best temperatures for getting excellent growth of the top portion ranges between 16 and 18°C, while temperatures above 28°C drastically reduce top growth. Temperatures lower than 16°C affect the development of color and result in long slender roots, while higher temperatures produce shorter and thicker roots (9). Kale and Kale (10) reported that temperatures between 15 and 20°C result in attractive roots with very good red color and quality.

B. Propagation

The carrot crop uses seeds that are drilled or broadcast in the field. The seeds are small—approximately 800 per gram. They remain viable for nearly 3 years, and up to 85% germination is very common. However, germination of some local varieties may be poor. Therefore, it is necessary to ascertain the germination percentage when calculating the seed requirement. For best results, it is also necessary to procure clean, healthy, and viable seeds from reliable sources. The seeds take approximately 7–21 days for complete germination. The best seed germination occurs at 20–30°C.

C. Cultural Practices

1. Sowing

Carrots can be grown practically throughout the year if the temperature does not go below 3.9°C. The best time for planting depends primarily on the climatic conditions prevailing in the given area. In the plains, carrots are grown during the middle of August to December. The quality of roots produced from late-sown crops is much superior than from early-sown crops. However, some of the varieties have a tendency to bolt at low temperatures. In contrast, some of the European types are more tolerant of low temperatures and do not bolt unless the temperature is very low. Such varieties can be grown until the end of November or the early part of December.

The soil for carrot cultivation should be properly prepared to get the desired tilth. It is necessary that the field provides a loose, friable, deep, and well-drained mellow substrate for seeds to germinate. This can be achieved by repeated deep ploughing at least 20–30 cm deep followed by harrowing, leveling, and cleaning. All old debris or remains of previous crops should be removed completely to obtain the desired seed bed condition (11). Since the seeds are very small and delicate, the seed bed should be very finely prepared. Beds of convenient size should be prepared before sowing.

The seeds are either broadcast, drilled, or dibbled by hand and are mixed with sand, ash, or fine soil to facilitate sowing (12). The seeds are sown either in ridges or in a flat bed. In any case, a shallow furrow is made 30–45 cm in length. When the seeds are dibbled, they are sown with 7.5–10 cm spacing. With such conditions about 4–6 kg of seeds are needed per hectare. The seeds are then lightly covered with soil or sand (13). Some growers irrigate the field about 24 hours before sowing to ensure that adequate moisture is present at the time of sowing. Many varieties germinate in about 10–15 days (14).

2. Manuring and Fertilization

According to Thompson and Kelley (1), carrot crops yielding 28–30 tons of roots require 32 kg nitrogen, 18 kg phosphorus, and 100 kg potash per hectare. Carrot is a heavy potassium feeder.

Potassium deficiency can affect the quality of the roots in addition to disturbing the overall metabolism of the plant. Potassium-deficient roots are less sweet and the flesh does not have the desired luster.

The fertilizer recommendation for carrots should be based on the soil type, variety, and the season (15,16). Most of the recommendations stress potassium fertilization. Kumar and Mathur (17) recommended the application of 23 tons of farmyard manure along with 50 kg nitrogen, 40 kg phosphorus, and 50 kg potash per hectare. In case farmyard manure is not available, the doses of inorganic fertilizers should be increased to 80 kg nitrogen, 60 kg phosphorus, and 60 kg potash per hectare. Choudhury et al. (18) suggested the application of 54 kg nitrogen and 54 kg phosphate per hectare for alluvial soils. Abo-Sedera and Eid (19) reported the best results in terms of vegetative growth, yields, and quality with 60 kg N_2 + 72 kg K_2O with the Red Cored Chantenay cultivar.

The entire quantity of farmyard manure should be mixed thoroughly with the soil, preferably to a depth of 20–25 cm, applying it at the time of land preparation. Half the dose of nitrogen and the full dose of phosphorus and potassium should be applied before sowing. The remaining nitrogen should be given 5–6 weeks after germination.

3. Irrigation and Intercultural Operations

The first irrigation should be light and carried out immediately after sowing. Subsequent irrigations are given as needed. When the seeds are broadcast or drilled, they should be thinned to maintain proper spacing or plant density. The frequency of irrigation depends upon soil type, season, and variety. In general, one irrigation every 4–5 days in summer and 10–15 days in winter provides adequate moisture for the crop. During the rainy season only occasional irrigations are needed. Water stress should be avoided during root development to avoid cracking of the roots, which also become hard. Earthing-up may also be necessary, particularly in the long-rooted types.

D. Disorders, Diseases, and Pests

Calcium deficiency in carrots causes a disorder known as cavity spot (20). Calcium deficiency may be induced by excess potassium uptake during the ontogeny of the plant. Increasing calcium accumulation reduced the incidence of cavity spot in carrots (21). Ryall and Lipton (22) mentioned a "scab spot complex" of carrots, which is related to climatic, nutritional, and genetic factors. This disorder causes considerable market losses in carrots produced in the southwestern United States. This disorder was originally attributed to a bacterium and was called bacterial blight, but it is now known to be a physiological disorder, although a few species of bacteria may be found in the lesions. Carrots are also susceptible to freezing injury at temperatures below freezing (≈ -1.5°C). Splitting or cracking of carrot root is a major problem in many carrot-growing areas. Although cracking is controlled by genetic factors, a number of other factors such as high nitrogen and chloride levels in the soil may be involved (5).

Diseases are not a serious problem in carrots. However, some microbial infections have been reported to occur in different parts of the world. These include leaf blight (*Alternaria radicina, Alternaria dauci*), leaf spot or *Cercospora* spot (*Cercospora carotae*), powdery mildew (*Erysiphe* sp.), and several viral diseases such as yellows, carrot leaf virus, carrot motely dwarf, and carrot mosaic, which may cause extensive damage to the crop.

Some of the common insect pests that cause damage to carrot crop are carrot rust fly, lygus bug, vegetable weevil, carrot beetle, parsley caterpillar, wingless May beetle, and cut and army worms (5). Besides insect pests, nematodes cause galls and distorted roots. The most important nematode damaging carrots is a species of *Pratylenchus*, but *Heterodera carotae* and

Meliodogyne sp. are also of common occurrence. Control of nematodes can be obtained with ASC 66824 900EC. Foliar applications of oxamyl was found to be ineffective (23).

E. Maturity, Harvesting, and Handling

Carrots are harvested when the roots are about 1.8 cm or larger in diameter at the upper end. The soil may be loosened with a special plough (carrot lifter) or an ordinary plough. The field is irrigated once the day before harvesting to facilitate harvest. Carrots to be sold in bunches are hauled to a packing house, where they are washed, iced, and packed for market. Most of the carrots go to the market with the tops cut off and the roots packaged in transparent film bags (1). Topping and bagging greatly reduce the loss of moisture during transportation to the market in addition to enhancing the shelf life of carrots (24). The packaging films are perforated to prevent the development of off-flavors. Prepackaging operations such as topping and washing are done mechanically, and the roots are sorted by size before packaging.

IV. CHEMICAL COMPOSITION

Carrot root is valued as a food component mainly because it is a rich source of β-carotene, the precursor of vitamin A. Carrots are also good sources of carbohydrates and minerals like calcium, phosphorus, iron, and magnesium (Table 2). The moisture content varies between 86 and 89% (25–28). The edible portion of carrots contains about 10% carbohydrates. The soluble carbohydrates ranged from 6.6 to 7.7 g/100 g in four carrot cultivars (26). Kaur et al. (29) reported 1.67–3.35% reducing sugars, 1.02–1.18% nonreducing sugars, and 2.71–4.53% total sugars in six cultivars. Simon and Lindsay (30) reported that reducing sugars accounted for

TABLE 2 Chemical Composition of Carrot Roots

Particulars	Ref. 27	Ref. 33
Moisture (%)	86.0	88.8
Protein (%)	0.9	0.7
Fat (%)	0.2	0.5
Carbohydrates (%)	10.6	6.0
Total sugars (%)	—	5.6
Crude fiber (%)	1.2	2.4
Total ash (%)	1.1	—
Calcium (mg/100 g)	80	34
Iron (mg/100 g)	2.2	0.4
Phosphorus (mg/100 g)	53	25
Sodium (mg/100 g)	—	40
Potassium (mg/100 g)	—	240
Magnesium (mg/100 g)	—	9.0
Copper (mg/100 g)	—	0.02
Zinc (mg/100 g)	—	0.2
Carotenes (mg/100 g)	—	5.33
Thiamine (mg/100 g)	—	0.04
Riboflavin (mg/100 g)	—	0.02
Niacin (mg/100 g)	—	0.2
Vitamin C (mg/100 g)	—	4
Energy value (kJ/100 g)	—	125

6–32% of the free sugars in four hybrid varieties of carrots. The free sugars identified are sucrose, glucose, xylose, and fructose (6). The crude fiber in carrot roots consists of 71.7, 13.0, and 15.2% cellulose, hemicellulose, and lignin, respectively (31). The cellulose content varied from 35 to 48% in four carrot varieties (32). The edible portion of carrots contain 0.8–1.1 g protein/100 g (26). The presence of nitrate and nitrite has also been reported in carrots. The average nitrate content in fresh carrot was reported to be 40 mg/100 g and the nitrite content was 0.41 mg/100 g (5, 34). Kalra et al. (6) suggested that the taste of carrots is mainly due to the presence of glutamic acid and the buffering action of free amino acids. Presence of trace amounts of succinic acid, α-ketoglutaric acid, lactic acid, and glycolic acid has been reported (6). Caffeic acid is the predominant phenolic acid in carrots. Carrots also contain anthocyanins. The anthocyanin content in the roots may vary from trace amounts in pink cultivars to 1750 mg/kg in black carrots (35). The major anthocyanins have been identified as cyanidin 3-(2-xylosylgalactoside), cyanidin 3-xylosylglucosylgalactoside, and cyanidin 3-ferulylxylosylglucosylgalactoside (36).

Carrot roots are a rich source of carotenoids. The total carotenoid content in the edible portion of carrot roots ranges from 6,000 to 54,800 μg/100 g (37). The carotenoid content in carrots is associated with the color of the root. The yellow and orange carrots contain relatively more carotenoids (38,39). The predominant carotenoids in orange-colored carrots are β-carotene, α-carotene, and γ-carotene (40). The proportion of individual pigments reported include β-carotene (45–80%), α-carotene (15–40%), γ-carotene (2–10%), and others (3–6%) (39). Thiamine, riboflavin, niacin, folic acid, and vitamin C are also present in appreciable amounts in carrot roots (5, 26).

Numerous reports have appeared on the volatile constituents in carrots. Buttery et al. (41) reported the presence of α-pinene, camphene, β-pinene, myrcene, α-terpinene, *p*-cymene, limonene, γ-terpinene, terpinolene, caryophyllene, β-bisabolene, γ-bisabolene, heptenol, octanol, nonanol, 2-nonenal, terpinene-4-ol, α-terpineol, bornyl acetate, 2,4-decadienol, dodecanal, myristicin, and falcarinol in steam-distilled oil of carrots. They suggested that 2-nonenal contributed most to the odor of a dilute aqueous solution of the distillate. Heatherbill et al. (42) investigated the volatile constituents in raw, canned, and freeze-dried carrots. They identified 23 compounds in raw carrots, of which diethyl ether, acetaldehyde, acetone, propanol, methanol, ethanol, and β-phellandrene were not reported earlier. Table 3 presents the effects of processing on some of the "higher-boiling" carrot volatiles. Buttery et al. (43) further identified geranyl-2-methyl butyrate, geranyl isobutyrate, β-ionone, geranyl acetone, *p*-cymen-8-ol, elemicin, eugenol, *p*-vinylguaiacol, and 4-methyl isopropenyl benzene in the volatile oil of carrots. McLellan et al. (44) studied the aroma characteristics of raw carrot and reported that the aroma constituents of carrot play a minor role in taste parameters for the acceptance of raw carrot.

Several compounds are known to influence the flavor of carrots (44–51). These include the free sugars glucose, fructose, and sucrose, which contribute to sweetness; volatile mono- and sesquiterpenoids, which contribute to harshness; 2-nonenal, which imparts cooked flavor; isocoumarin and other phenolic compounds, which impart bitterness; 2-methoxy-3-butylpyrazine, which contributes to the carrot aroma; free amino acids, which contribute to the delicate flavor; and ionones, which impart a floral off-flavor. Lund (52) reported the presence of polyacetylenes, which contribute to the unpleasant odor in carrots. It was further shown that the previously reported compound falcarindiol-3-monoacetate consisted of a mixture of the 3-acetate and its allylic isomer, 1-acetoxyheptadeca-2,9-diene-4,6-dien-8-ol. The other polyacetylenic carbonyl previously identified, falcarinolone, was present only as an artifact derived from the autoxidation of falcarindiol during the isolation procedure.

TABLE 3 Effects of Processing on the Higher-Boiling Carrot Volatiles

Compound	Raw carrot (ppm)	Canned carrot (ppm)	Freeze-dried carrot (ppm)
α-Pinene	0.09	0.10	0.04.
Camphene	0.04	0.02	tr
β-Pinene	0.05	0.05	tr
Sabinene	0.10	0.07	0.03
Myrcene	0.35	0.21	0.04
α-Phellandrene	tr	tr	tr
Limonene	0.46	0.22	0.06
γ-Terpinene	tr	tr	tr
p-Cymene	tr	0.45	tr
Terpinolene	6.05	2.5	0.46
Octanol	0.02	0.06	tr
2-Decanal	0	tr	0
α,p-Dimethyl styrene	0	+	0
Bornyl acetate	0.22	0.08	tr
Caryophyllene	10	5	1.95
Terpinene-4-ol	—	—	—
β-Bisabolene	0.18	0.87	0.6
γ-Bisabolene	5.55	3.5	2.15
Carotol	0.22	0.19	0.2
Myristicin	0.30	0.35	0.25
Total	26	14	6

tr, Trace; +, increase in concentration.
Source: Ref. 42.

V. STORAGE

A. Chemical Changes During Storage

Several chemical changes occur in carrots during storage (53–58). Polysaccharides are converted into simple sugars and sucrose into reducing sugars during storage (54). The addition of as little as 2.5% CO_2 to the storage atmosphere retarded the latter conversion (56). The relative sweetness was slightly increased early in storage with air atmosphere but reduced if ethylene was present, even though total sugar levels in roots from either atmosphere were comparable (55). Harse flavor associated with volatile terpenoids does not change upon storage in air, but it does arise with bitterness in carrots exposed to ethylene (55). The occurrence of isocoumarin generally coincides with bitter flavor. Ethylene stimulates development of isocoumarin synthesis. A short anaerobic treatment before storage prevents isocoumarin formation (57). By storing washed carrots in sealed polyethylene bags, CO_2 concentration rises and O_2 concentration drops to retard ethylene synthesis (54). Sarkar and Phan (58) reported that eugenin and isochlorogenic acid are synthesized along with isocoumarin during the development of bitterness in carrot. However, the exact role each of these compounds in the bitterness of carrots is not fully understood.

B. Precooling and Preconditioning

Prompt cooling to 5°C or below after harvest is essential for extended storage of carrots (53). Poorly precooled roots decay more rapidly. Mature carrots can be stored for 7–9 months at 0–1°C with a very high relative humidity of 95–100%. However, even under these conditions, 10–20% may show some decay after 7 months. The highest freezing point for carrots is –1.2°C. Severe tissue injury after freezing results in lengthwise cracking and blistering caused by the formation of ice crystals beneath the surface. After thawing, a darkened (dark brown or black) and water-soaked skin is observed (59). Toivonen et al. (60) suggested that low-temperature preconditioning improved the shelf life of fresh carrots stored at 13°C and 90% relative humidity. Preconditioned carrots showed significantly less weight loss and maintained a brighter orange color than unconditined carrots. Preconditioning enhanced deposition of suberin on the surface of the periderm and lignification of the subsurface cell layers. These changes are suggested as possible mechanisms for the reduction of weight loss and discoloration of carrots.

C. Low-Temperature Storage

Carrots can be stored for about 6 months in good condition at a temperature of 0°C and a relative humidity of 93–98%. They deteriorate in quality during storage owing to a slow loss of sugar in respiration. The loss of sugar can be slowed down under good storage conditions and the carrots will retain high quality for at least 5 months (59). Pantastico et al. (61) recommended a storage temperature of 0°C for topped or bunched carrots. The bunched carrots need slightly higher relative humidity than the topped carrots. The storage life of the latter is considerably longer (about 20–24 weeks) as compared to the bunched carrots (4 weeks). Raghavan et al. (62) reported that the average weight loss per month of carrots stored loose and in plastic bags at –2°C and 96–100% relative humidity was only 0.49 and 0.24%, respectively. In both conditions, carrots remained firm and crisp until the end of the storage period (78 days). With loosely packed carrots, the weight loss during the initial 6-week cooling period amounted to 16% per month and they suffered severe decay; those stored under low temperature subsequently, however, kept well for the remaining 5 weeks with little weight loss. The carrots stored in plastic bags retained their quality throughout the storage period.

D. Controlled Atmosphere Storage

The controlled atmospheres (CA) used for other commodities are not useful for carrots (63). Vandenberg and Lentz (64) found that an atmosphere of 3% O_2 and 6% CO_2 markedly increased decay after several months of storage. Ryall and Lipton (22) suggested 2–4% CO_2 to prevent the development of bitterness. Bohling and Hansen (65) investigated the effects of CA storage (mixtures of 4% CO_2 + 2% O_2 and 2% CO_2 + 2% O_2) on the respiratory behavior of carrots. The respiration was reduced to 50% under either of the CA conditions. The exposure of carrots to plain air at the end of the storage period led to about the same respiration intensity as was observed in the samples stored in air (Table 4). No significant differences were observed with respect to visual appearance due to weight loss and losses caused by rots.

McKeown and Loughee (66) have investigated the potential use of low-pressure storage (LPS) in the storage of carrots. LPS provides a simple means of reducing the effect of ethylene-producing vegetables upon carrots when they are stored together. However, with no ethylene source in the storage environment, LPS has very little benefit. A reduction in the air pressure to 10 kilopascals (0.1 atmosphere) is equivalent to reducing O_2 concentration to about 2% at normal atmospheric pressure. LPS is also expensive to construct because of the low internal pressures required, and the application costs are high, which limits its utilization in the storage of carrots (67).

TABLE 4 Effects of CA on Respiratory Activity of Carrots

	Respiratory activity		% Activity in air	
Atmosphere	mlCO_2/kg/hr	mlO_2/kg/hr	CO_2	O_2
Air	2.83 (± 0.21)	2.8 (± 0.26)	100	100
2% CO_2/2% O_2/rest N_2	1.45 (± 0.18)	1.59 (± 0.27)	51.2	55.5
4% CO_2/2% O_2/rest N_2	9.37 (± 0.21)	1.39 (± 0.34)	49.5	48.6

[a]Cultivar Rothild
Source: Ref. 65.

E. Modified Atmosphere Storage

Packaging of fresh "ready-to-use" grated carrots with films of high permeability (about 22000 cc/m^2/day/atm at 25°C) allows better preservation and improved retention. In packs with high-permeability films stored at 10°C, the atmosphere at equilibrium contained about 20% CO_2 and 5% O_2 (Fig. 1) (68). With these films, however, a reduction in sugar content was observed, especially in sucrose, which is the main component contributing to the taste of carrots (69). Packaging films had a slight influence on the carotene content, an important aspect of the nutritive value of carrots (70, 71).

F. Irradiation

Salunkhe (72) investigated the usefulness of γ-irradiation to minimize sprouting losses of several bulb, tuber, and root crops, including carrots. Canned carrots were irradiated at 1.9, 3.7, 7.4, 9.4, 14.9, 29.8, and 59.5 × 10^3 rad. Following irradiation, the cans were stored for 5 months at 10°C and 85% relative humidity. At the end of the storage period, it was found that the higher doses (> 14.9 x 10^3 rad) either retarded or inhibited sprouting of carrots. A dose of 18.6 × 10^3 rad was found to be effective for practical purposes. Higher doses did not result in additional significant decrease in sprouting. The level of irradiation did not change the sensory quality of carrots during the first month of storage. However, in subsequent storage for 4 months at 10°C and 85% relative humidity, the quality scores increased as the dose of irradiation increased, which may be due to a reduction in sprouting losses.

G. Postharvest Diseases and Their Control

Carrots are susceptible to bacterial soft rot (*Erwinia carotovora*), black rot (*Stemphylium radicinum*), gray mold rot (*B. cinerea*), *Rhizopus* soft rot (*R. tritici*, *R. stolonifer*, and *R. oryzae*), and watery soft rot (*S. sclerotiorum*). *Rhizopus* sp. and *Monilinia* sp. have been reported to cause heavy losses (73). Ryall and Lipton (22) have described these diseases and suggested preventive measures. Treatment of carrots immediately before storage with TBZ-20 (thiabendazole) at 1.10%, quintozene at 60 g/100 g, or formaldehyde at 0.05 liter/m^3 significantly reduced storage rots caused by *A. radicina* and *S. sclerotiorum* (74). Dichloran, an aromatic amine, was very effective in controlling *Rhizopus* and *Monilinia* infections (73).

Certain preharvest disease control treatments may have an effect on the chemical composition and shelf life of carrots. Salunkhe et al. (75) investigated the effects of the nematicides Telone (1,3-dichloropropene and other chlorinated hydrocarbons) at the rate of 10, 20, and 30

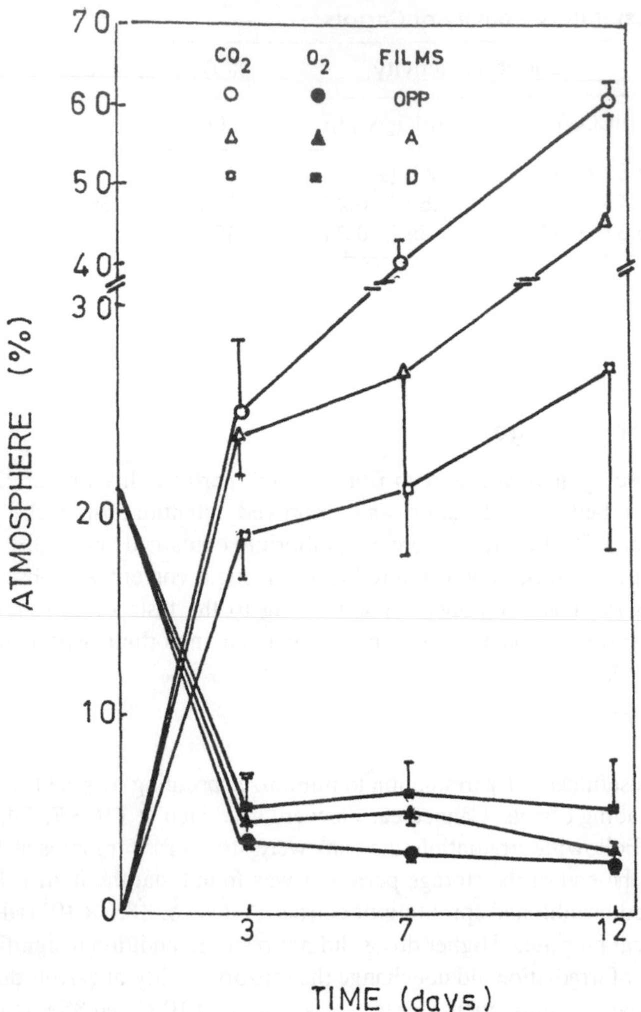

FIGURE 1 Changes in CO_2 and O_2 in grated carrots with different packaging films at 10°C. OPP, Packs made of oriented polypropylene; A, oriented polypropylene A, △, and ▲; D, oriented polypropylene D, □, and ■; Control, ○ and ●. (From Ref. 68.)

gal/acre and Nemagon (1, 2-dichloro-3-chloropropane) at the rate of 1, 2, and 3 gal/acre applied as soil fumigants one week before planting of carrots. The chemicals significantly enhanced the total carotene, β-carotene, and sugar levels, and decreased the rate of respiration (Tables 5 and 6). The chemicals, however, did not change the reducing sugar and total nitrogen contents. The investigators suggested that the reduction in respiratory rate may prolong the shelf life of carrots.

VI. PROCESSING

Carrots are processed into products such as canned carrots, dehydrated carrots, juice, beverages, candy, preserves, intermediate-moisture foods, and halva (6). The bright orange color, absence of fibrous core, tender texture, sweet flavor, and persistence of color throughout the processing

TABLE 5 Effects of Telone and Nemagon on Total Carotenes and β-Carotene

Treatment			
Chemical	Dosage (gal/acre)	Total carotenes (μg/100 g)	β-Carotene (μg/100 g)
Control	—	5359	4927
Telone	10	6361[a]	5881[a]
	20	6746[a]	6270[a]
	30	7790[a]	7315[a]
Nemagon	1	6216[a]	5668[a]
	2	6734[a]	6182[a]
	3	7724[a]	6650[a]

[a]Significant at 0.01.
Source: Ref. 75.

stages are considered desirable characteristics for processing. The cultivar Red Cored Chantenay is exclusively used for the canning of whole carrots. The most suitable cultivars for canning of slices and dices are Red Cored Chantenay, Danvers Half Long, Imperator, and Autumn King. Imperator and Red Cored Chantenay can also be dehydrated, and Chantenay, Long Imperator, and Nantes are used for freezing. Carrots with longer lengths, medium diameter, orange or yellow color, smooth surface, good core, and less bitterness are preferred for processing (76,77).

A. Canning

Firmness is an important quality attribute in canned vegetables. Loss of firmness during processing and subsequent storage could result in an undesirable product. Sistrunk (78) found that variety and time and temperature of blanching are important factors affecting the quality of canned carrots. Blanching at 71°C for 3–6 minutes resulted in a better quality canned prod-

TABLE 6 Effects of Telone and Nemagon on Sugars and Respiration

Treatment					
Chemical	Dosage (gal/acre)	Total sugars (%)	Reducing sugars (%)	Total nitrogen (%)	Respiration (μl O_2/hr/g)
Control	—	4.7	2.27	0.14	108.8
Telone	10	5.95[a]	1.89[ns]	0.13[ns]	94.9[b]
	20	5.42[a]	2.04[ns]	0.14[ns]	88.5[b]
	30	5.72[a]	2.18[ns]	0.16[ns]	79.1[a]
Nemagon	1	5[b]	2.28[ns]	0.15[ns]	83.7[b]
	2	6.0[a]	1.98[ns]	0.14[ns]	82.8[b]
	3	5.95[a]	2.1[ns]	0.15[ns]	75.3[a]

[ns]Not significant at 0.05.
[a]Significant at 0.01.
[b]Significant at 0.05.
Source: Ref. 75.

uct than blanching at 87.5°C for a short time. Tender and small carrots are usually used for canning (76, 77).

Cruess (79) described freezing and canning of carrots. Diced carrots are frozen in retail-size packages and in large containers for use by soup canners. If thoroughly blanched, carrots retain their flavor and freshness very well in freeze storage. The carrots are canned in various forms—diced, halved, quartered, or whole. Improvement in the color and quality of canned carrots by heat treatment and by the use of chemicals has been reported by several workers (80–84).

Chiang et al. (80) studied the effect of water quality on canned carrots. Two chelating agents, calcium disodium ethylene diaminetetraacetic acid ($CaNa_2$-EDTA) and sodium hexametaphosphate (Na-HMP) were used at a 800 ppm level for canning Honey Sweet carrots with water containing 0, 20, 40, 80, and 160 ppm of calcium and 20 ppm of magnesium. The canned carrots were stored at 21.1 or 37.8°C and were evaluated at 60-day intervals for six months. The carrots were found to have better organoleptic acceptability when canned with the chelating agents. Carrots canned with $CaNa_2$-EDTA were firmer and retained color better than those canned with Na-HMP or the controls without any chelating agents. The rate of loss of firmness and color increased with time of storage.

B. Dehydration

Cruess (79) described a process for the dehydration of carrots. The carrots are dried to about 10% moisture and transferred to portable finishing bins to complete dehydration at 44.4°C. Alternately, they may be dried on trays at 68.3°C to about 4–5% moisture and packed in air-tight containers. The method of preparation and improvement in color, taste, and flavor of dehydrated carrots have been reported by a number of workers (85–88).

Ambadan and Jain (76) found that blanching of carrots shreds in 5% sugar solution prior to dehydration not only imparts an attractive color but improves the organoleptic and keeping quality of the product. The quality of dehydrated carrots may be improved by treating the fresh carrots with sodium tetraborate and sodium metabisulfite. Freeze-drying gives a better product with improved taste and texture.

C. Freezing

Large amounts of carrots are frozen in the United States and other developed countries (87). Washed carrots are mechanically separated into small, medium, and large sizes, to be used, respectively, for whole, sliced, or diced carrots. The tips and butt ends are cut off, and trimmed carrots are peeled in abrasive or steam peelers. Large carrots are mechanically diced or sliced. These are transported to an air cleaner to remove small pieces by air blast. Carrots are usually blanched by steam for 2–8 minutes depending upon size, maturity, and texture to inactivate the enzymes. Blanched carrots are subjected to fludized freezing or other IQF technique. Carrots are marketed in containers of various sizes and in bulk containers for mixing with peas, beans, and sweet corn.

Small whole carrots are processed in the same way as large carrots. Some processors cut off the ends of smaller carrots and run these through an abrasive peeler to round off the cut end. The finished product looks like a tiny carrot and is marketed as such. There is growing demand for this product. Hence, varieties of baby carrots for freezing are being developed in developed countries.

D. Intermediate-Moisture Carrots

Sethi and Anand (89) prepared intermediate-moisture carrot slices using a solution containing sugar, glycerol, water, acid, and preservatives. The processed product had good color, flavor, and texture. Treatment with 500 ppm SO_2 and 0.45% potassium sorbate inhibited microbial spoilage at 39.2% moisture (water activity 0.85). At low temperatures (1–3°C), the ready-to-serve product remained acceptable for 6 months in a glass container with a β-carotene retention of 40%.

E. Juice/Nectar

The extraction, canning, and storage of carrot juice has been described (90–92). Carrot beverages have been prepared by mixing carrot juice with other fruit juices or skim milk (93).

F. Pickled Carrots

Generally, carrots are pickled by lactic acid fermentation. Pruthi et al. (94) reported that carrots could be preserved in good condition for 6 months at room temperature even in non–air-tight containers using acidified brine with potassium metabisulfite. The product can be used for the manufacture of good-quality pickles.

G. Toffee

Jagtap (95) prepared carrot toffee by applying sugar, fat, citric acid, and salt to whole carrot shreds or pomace (Figs. 2 and 3). Toffees prepared from Asiatic cultivars were found to have better chemical composition than those prepared from European cultivars. Carrot pomace toffees were relatively inferior to whole carrot toffee in their chemical composition (Table 7).

H. Traditional Products

1. Preserve (Conserve)

Carrot candy or preserve is prepared by covering small whole carrots or slices of large carrots with sugar or heavy sugar syrup so that the total soluble solid content increases to 70–75 Brix (96). Carrots have been processed to obtain intermediate-moisture foods containing about 50% moisture (97–99). The product can be consumed as such without further dehydration or preparation. Carrot *halva* is prepared by cooking shredded carrots with sugar, hydrogenated oil, and milk. Fruits, nuts and spices are added to enhance the flavor (100,101).

2. Puree and Pie

For making carrot puree, carrots are cooked in water (4:1) until soft, processed into a pulp, and sterilized at 121°C for 50 minutes. The use of carrots for the preparation of pie may seem somewhat unusual. Kalra et al. (6) developed an acceptable carrot pie filling of equal or better nutritional value than pumpkin filling.

3. Flakes and Chips

Fresh, unpeeled carrots are cooked, comminuted, made into a puree, and dried for 35 seconds on a drum drier. The flakes thus prepared retain good carrot flavor. Carrot chips can be prepared by frying the slices in oil with constant agitation. The chips contain 35–40% oil and have a flavor distinct from traditionally cooked carrots.

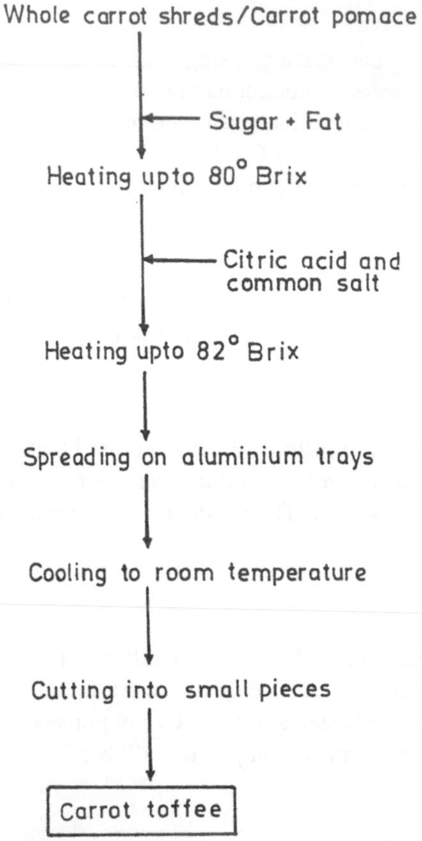

Whole carrot shreds/Carrot pomace
↓ ← Sugar + Fat
Heating upto 80° Brix
↓ ← Citric acid and common salt
Heating upto 82° Brix
↓
Spreading on aluminium trays
↓
Cooling to room temperature
↓
Cutting into small pieces
↓
Carrot toffee

FIGURE 2 Flowchart for preparation of carrot toffee. (From Ref. 95.)

FIGURE 3 Whole carrot toffee of European cultivar. (From Ref. 95.)

TABLE 7 Yield and Chemical Composition of Carrot Toffees

Carrot toffees	Yield (g)	Moisture (%)	T.S.S. (^{o}Brix)	Total sugars (%)	Reducing sugars (%)	Nonreducing sugars (%)	Acidity (%)	Ascorbic acid (mg/100 g)
Whole carrot toffee								
Asiatic cultivar	1570	8.8	82	75.7	34.9	40.8	0.22	1.89
European cultivar	1590	8.6	82.4	73.4	32.5	40.9	0.24	2.04
Carrot pomace toffee								
Asiatic cultivar	2674	8.4	82.7	78.5	38.4	40.1	0.15	0.81
European cultivar	2680	8.3	82.8	78.1	37.6	40.2	0.17	0.86
S.E. ±	—	0.23	0.24	0.81	0.83	0.98	0.03	0.18
C.D. at 5%	—	ns	ns	2.42	2.49	ns	0.09	0.54

ns, Not significant.
Source: Ref. 95.

VII. MEDICINAL PROPERTIES

The use of carrots in herbal medicine has been documented. It is considered a popular remedy for jaundice; the roots and seeds are used as an aphrodisiac and nerve tonic as well as poultice for ulcers, burns, and scalds (102). Carrots have also been reported to have diuretic, nitrogen-balancing properties and are effective in the elimination of uric acid (103). The presence of high concentrations of antioxidant carotenoids, especially β-carotene, may account for the biological and medicinal properties of carrots. Numerous animal experiments and epidemiological studies have indicated that carotenoids inhibit carcinogenesis in mice and rats and may have-anticarcinogenic effects in humans. The carotenoids, and in particular β-carotene, are well known for their antimutagenic, chemopreventive, photoprotective, and immunoenhancing properties. Such effects are mainly linked to the antioxidant properties of these compounds. β-Carotene functions as a free radical–trapping agent and a singlet oxygen quencher in biological systems. β-Carotene has been reported to reduce the risk of cancers of lung, cervix, esophagus, and stomach. It has been found to be helpful in the treatment of patients suffering from photodamaging skin conditions. β-Carotene has also been reported to reduce the risk of cataract formation (104).

The antioxidant properties of carrot extract have been reported. Dietary supplementation of a combination of carrot and orange juice was found to reduce the oxidation of low-density lipoproteins in habitual cigarette smokers (105). Studies by Bishayee and Chaterjee (106) have indicated that carrot aqueous extract may have a hepatoprotective effect against oxidative damage. The increased lipid peroxidation and decreased glutathione levels induced by acute carbon tetrachloride treatment in mice was significantly reversed by pretreatment with an aqueous extract of carrots. The extract also dose-dependently reduced the activity of enzymes linked with antioxidant and free radical defense system, catalase, glutathione peroxidase, and glutathione S-transferase with a simultaneous elevation in the activity of glutathione reductase that were altered remarkably due to carbon tetrachloride induction. The results indicated that the extract affords protection against oxidative stress in the tissue by inhibiting lipid oxidation and exerting membrane stabilization.

Carrot juice has been reported to have antimutagenic, anti–tumor-promoting activity. Dietary carrot was shown to reduce the genotoxic effects of the cytostatic drug cyclophosphamide and its metabolite phosphoamide in rats (107). The authors have suggested that β-carotene, the antioxidant component of carrots, may play an important role in modulating the in vivo metabolism of cyclophosphamide. Carrot extract has been found to induce cancer protective Phase II enzyme quinone reductase in an in vitro assay using murine hepatoma cells. The enzyme is involved in xenobiotic metabolism and detoxification of chemical carcinogens, a well-established protective mechanism against the toxic and neoplastic effects of carcinogens (108).

Carrot seeds have been reported to possess antifertility activity in female albino rats. The effects of carrot seed extract on hormone-regulated uterine changes and its abortifacient properties have been reported by several workers (109–112). The methanolic or chloroform fraction of a petroleum ether extract of carrot seeds inhibited spontaneous uterine mobility and oxytocin-induced contractions in isolated rat uterus (104,105). The investigators have suggested that the antifertility effects of carrot seed extract could be due to its effects on the uterus or on the transport of fertilized egg or inhibition of sperm transport.

The antifungal and antibacterial effects of carrots have been reported. Carrot seed oil and root tissue have been reported to inhibit aflatoxin production by *Aspergillus parasiticus* (113). Carrot root is known to have antiseptic properties in preventing putrescent changes in the tissues (102). Carrot slices and juice have been shown to have significant antilisterial effect, which resulted in a decrease in the number of viable bacteria and in sublethal damage. Five cultivars of carrots were highly effective against nine strains of *Listeria monocytogenes* and single strains of *L. innocua*, *L. ivanovii*, *L. seeligeri*, and *L. welshimeri* (114).

REFERENCES

1. Thompson, H. C., and W. C. Kelly, *Vegetable Crops*, McGraw-Hill, New York, 1957.
2. FAO, Production Yearbook, Food and Agriculture Organization, Rome, Italy, 1990.
3. Rodriguez, G. R., B. L. Raina, E. B. Pantastico, and M. B. Bhatti, Quality of raw material for processing—postharvest physiology: Harvest indices, in *Postharvest Physiology, Handling, and Utilization of Tropical and Subtropical Fruits and Vegetables* (E. B. Pantastico, Ed.), AVI, Westport, CT, 1975, p. 56.
4. Daniello, F. J., and R. R. Heinman, Carrot varieties evaluation in southeast Texas, Texas Agric. Expt. Stn. Publ. No. PR-3871 (Hort. Abst. 52: 2288) 1982, p. 11.
5. Bose, T. K., and M. G. Som, *Vegetable Crops in India*, Naya Prokash, Calcutta, India, 1986.
6. Kalra, C. L., S. G. Kulkarni, and S. K. Berry, The carrot, popular root vegetable, *Indian Food Packer 41*(6):48 (1987).
7. Gill, H. S., K. D. K. Lakhanpal, and S. R. Sharma, Pusa Yamadagni carrot, *Indian Hort. 36*(3):18 (1986).
8. Brar, J. S., and K. S. Nandpur, *Cultivation of Root Crops*, Punjab Agricultural University, Ludhiana, India, 1972.
9. Yamaguchi, M., *World Vegetables: Principles, Production, and Nutritive Values*, AVI, Westport, CT, 1983, p. 204.
10. Kale, P. N., and J. Kale, *Bhajipala Utpadan*, Continental Pub. Co., Pune, India, 1984.
11. Choudhury, B., *Vegetables*, National Book Trust, New Delhi, India, 1987.
12. Nath, P., S. Velayudhan, and D. P. Singh, *Vegetables for the Tropical Region*, ICAR, New Delhi, India, 1987.
13. Singh, B., and M. S. Saimbhi, Influence of sowing methods and spacing on root yield of carrot, *J. Res. PAU 21*:639 (1984).
14. Maurya, K. R., Effect of IBA on germination, yield, carotene, and ascorbic acid content of carrot, *Indian J. Hort. 43*:118 (1986).
15. Kumar, J. C., B. N. Sharma, P. B. Sharma, and Y. Paul, Effect of nitrogen levels and sowing methods on nutritive value of carrot roots, *Indian J. Hort. 31*:262 (1974).
16. Maurya, K. R., and R. K. Goswami, Effect of NPK fertilizer on growth and quality of carrots, *Prog. Hort. 17*:212 (1985).
17. Kumar, V., and M. K. Mathur, Fertilizing turnips pays well, *Fertilizer News 10*(10): 18 (1965).
18. Choudhury, B., M. G. Som, and A. K. Das, *Proc. 3rd Int. Symp. Subtrop. Trop. Hort.*, Bangalore, India, 1972, p. 150.
19. Abo-Sedera, F. A., and S. M. M. Eid, Plant growth, chemical composition, yield, and quality of carrots as affected by NK-fertilization and NAA foliar spray, *Assiut J. Agric. Sci. 23*(2):209 (1992).
20. Maynard, D. M., B. Gersten, E. F. Vlach, and H. F. Vernell, The effects of nutrient concentrations and calcium level on the occurrence of carrot cavity spot, *Proc. Am. Soc. Hort. Sci. 78*:339 (1958).
21. Maynard, D. M., B. Gersten, E. F. Vlach, and H. F. Vernell, The influence of plant maturity and calcium level on the occurrence of carrot cavity spot, *Proc. Am. Soc. Hort. Sci. 83*:506 (1963).
22. Ryall, A. L., and W. J. Lipton, *Handling, Transportation, and Storage of Fruits and Vegetables*, Vol. I: *Vegetables and Melons*, AVI, Westport, CT, 1972.
23. Johnston, S. A., P. R. Probasco, and J. R. Phillips, Evaluation of fumigants and nematicides for the control of root knot nematodes on carrot, Fungicide and Nematicide Tests *48*:208 (1993).
24. Hardenburg, R. E., M. Lieberman, and L. Schomer, Pre-packaging carrots in different types of consumer bags, *Proc. Am. Soc. Hort. Sci. 61*:404 (1953).
25. *The Wealth of India: Raw Materials*, Council of Scientific and Industrial Research, New Delhi, India, 1952, p. 21.

26. Howard, F. D., J. H. MacGillivray, and M. Yamaguchi, Nutrient composition of fresh California grown vegetables, *Bull. No. 788, Calif. Agric. Expt. Stn.*, University of California, Berkley, 1962.
27. Gopalan, C., B. V. Ramasastry, and S. C. Balasubramanian, *Nutritive Value of Indian Foods*, National Institute of Nutrition, Hyderabad, India, 1991, p. 47.
28. Gill, H. S., and A. S. Kataria, Some biochemical studies in European and Asiatic varieties of carrot (*Daucus carota*), *Current Sci. 43*:184 (1974).
29. Kaur, G., S. P. Jaiswal, K. S. Brar, and J. C. Kumar, Physico-chemical characteristics of some important varieties of carrot, *Indian Food Packer 30*(2):5 (1976).
30. Simon, P. W., and R. C. Lindsay, Effect of processing upon objective and sensory variables of carrots, *J. Am. Soc. Hort. Sci. 108*(6):928 (1983).
31. Kochar, G. K., and K. K. Sharma, Fiber content and its composition in commonly consumed Indian vegetables and fruits, *J. Food Sci. Technol. 29*(3):187 (1992).
32. Robertson, J. A., M. A. Eastwood, and M. M. Yeoman, An investigation into the dietary fiber content of normal varieties of carrot at different developmental stages, *J. Agric. Food Chem. 39*(1):388 (1979).
33. Holland, B., J. D. Unwin, and D. H. Buss, *Vegetables, Herbs, and Spices:* Fifth Supplement to McCance and Widdowson's *The Composition of Foods*, London, 1991.
34. Miedzobrodzka, A., E. Cieslik, and E. Sikora, Changes in the contents of nitrate and nitrites in carrot roots during storage in a clamp, *Rocz. Panstw. Zakl. Hig. 43*(1):33 (1992).
35. Mazza, G., and E. Miniati, Roots, tubers, and bulbs, in *Anthocyanins in Fruits, Vegetables, and Grains*, CRC Press, Boca Raton, FL, 1993, p. 265.
36. Harborne, J. B., A unique pattern of anthocyanins in *Daucus carota* and other *Umbelliferae, Biochem. Syst. Ecol.* 4:31 (1976).
37. Simon, P. W., and X. Y. Wolff, Carotene in typical and dark orange carrots, *J. Agric. Food Chem. 35*(6):1017 (1987).
38. Gabelman, W. H., The prospects for genetic engineering to improve nutritional values, in *Nutritional Quality of Fresh Fruits and Vegetables* (P. White, ed.), Futura Pub. Co., New York, 1974, p. 147.
39. Laferriere, L, and W. H. Gabelman, Inheritance of color, total carotenoids, carotene, and β-carotene in carrots (*Daucus carota* L.), *J. Am. Soc. Hort. Sci. 96*(1):408 (1968).
40. Heinonen, M. I., Carotenoids and provitamin A activity of carrot (*Daucus carota* L.) cultivars, *J. Agric. Food Chem. 38*(3):609 (1990).
41. Buttery, R. G., R. M. Seifert, D. G. Guadagni, D. R. Black, and L. C. Long, Characterization of some volatile constituents of carrots, *J. Agric. Food Chem. 16*(6):1009 (1968).
42. Heatherbell, D. A., R. E. Wrolstad, and L. M. Libbey, Carrot volatiles I. Characterization and effect of canning and freeze drying, *J. Food Sci. 36*(1):219 (1971).
43. Buttery, R. G., D. R. Black, W. F. Haddon, L. C. Long, and R. Teranishi, Identification of additional volatile constituents of carrot roots, *J. Agric. Food Chem. 27*(1):1 (1979).
44. McLellan, M. E., J. N. Cash, and J. I. Gray, Characterization of the aroma of raw carrots (*Daucus carota* L.) with the use of factor analysis, *J. Food Sci. 48*(1):71 (1983).
45. Bajaj, K. L., K. Gurdeep, and B. S. Sukhija, Chemical composition and some plant characteristics in relation to quality of some promising cultivars of carrot, *Qual. Plantar. 30*:97 (1980).
46. Alabran, D. M., and A. P. Marbrouk, Carrot flavors, sugars, and free nitrogenous compounds in fresh carrots, *J. Agric. Food Chem. 21*:205 (1973).
47. Atkin, J. D., Bitter flavor in carrots I. Progress report on field and storage experiments, *Bull. N.Y. Agric. Expt. Stn. 774*:1 (1956).
48. Sarkar, S. K., and C. T. Phan, Naturally occurring and ethylene induced phenolic compounds in carrot root, *J. Food Prot. 45*:526 (1979).
49. Cronin, D. A., and P. Stanton, 2-Methoxy 3-sec-butylpyrazine, an important contributor to carrot aroma, *J. Sci. Food Agric. 27*:145 (1976).
50. Otsuka, H., and T. Take, Lipid component in carrots, *J. Food Sci. 34*:392 (1969).
51. Ayers, J. E., M. J. Fishwick, D. G. Land, and T. Swain, Off-flavor of dehydrated carrot stored in oxygen, *Nature (London) 203*:81 (1964).

52. Lund, E. D., Polyacetylenic carbonyl compounds in carrots, *Phytochemistry 31*(10):3621 (1992).
53. Hasselbring, H., Carbohydrate transformation in carrot during storage, *Plant Physiol. 2*:225 (1927).
54. Phan, C.T., H. Hsu, and S.K. Sarkar, Physical and chemical changes occurring in the carrot root during storage, *Can. J. Plant Sci. 53*:635 (1973).
55. Simon, P. W., Genetic effects on the flavor of stored carrot, *Acta Hort. 163*:137 (1984).
56. Denny, F. E., N. C. Thornton, and E. M. Schroeder, The effect of carbon dioxide upon the changes in sugar content of certain vegetables in cold storage, *Contrib. Boyce Thompson Inst. 13*:295 (1944).
57. Carlton, B. C., C. E. Peterson, and N. E. Tolbert, Effect of ethylene and oxygen on production of bitter compounds by carrot roots, *Plant Physiol. 36*:550 (1961).
58. Sarkar, S. K., and C. T. Phan, Naturally occurring and ethylene induced phenolic compounds in the carrot root, *J. Food Prot. 42*:526 (1979).
59. Plantenius, H., Physiological and chemical changes in carrots during growth and storage, Cornell Memoir No. 161 (1934).
60. Toivonen, P. M. A., M. K. Upadhya, and M. M. Gaye, Low temperature preconditioning to improve shelf-life of fresh market carrots, *Acta Hort. 343*:339 (1993).
61. Pantastico, E. B., T. K. Chattopadhyay, and H. Subramanyam, Storage and commercial storage operations, in *Postharvest Physiology, Handling, and Utilization of Tropical and Sub-tropical Fruits and Vegetables* (E. B. Pantastico ed.), AVI, Westport, CT, 1975, p. 314.
62. Raghavan, G. S. V., R. Bovell, and M. Chayet, Storability of fresh carrots in a simulated jacketed storage, *Trans. ASAE* 23:1521 (1980).
63. Hatton, T. T., E. B. Pantastico, and E. K. Akamine, Controlled atmosphere storage, in *Postharvest Physiology, Handling, and Utilization of Tropical and Sub-tropical Fruits and Vegetables* (E. B. Pantastico, ed.), AVI, Westport, CT, 1975.
64. Vandenberg, L., and C. P. Lentz, Effect of temperature, relative humidity, and atmospheric composition on changes in quality of carrots during storage, *Food Technol. 20*:104 (1966).
65. Bohling, H., and H. Hansen, Influence of controlled atmospheres on the respiration activity of achlorophyllous vegetable kinds, *Acta. Hort. 116*:1965 (1980).
66. McKeown, A. W., and E. C. Lougheed, Low pressure storage of some vegetables, *Acta Hort. 116*:83 (1980).
67. Wills, R. H. H., T. H. Lee, D. Graham, W. B. McGlasson, and E. G. Hall, *Postharvest: An Introduction to the Physiology and Handling of Fruits and Vegetables*, South Wales University Press, Kensington, Australia, 1981.
68. Carlin, F., C. Nguyen, G. Hilbert and Y. Charubroy, Modified atmosphere packaging of fresh, ready-to-use grated carrots in polymeric films, *J. Food Sci. 55*(4):1033 (1990).
69. Rym, G. and H. Hansen, Gas Chromatographische Bestimmung löslicher Inhaltstoffe, in *Controlled Atmosphere, Galenbanwissenschaft 38*:281 (1973).
70. Aubert, S., La Carrotte (*Daucus carota* L.) revue de quelques bacterus d'interet dietetique, *Cah. Nutr. Diet. 16*:173 (1981).
71. Bajaj, K. J., G. Kaur, and B. S. Sukhija, Chemical composition and some plant characteristics in relation to quality of some promising cultivars of carrot, *Qual. Plant. 90*:97 (1989).
72. Salunkhe, D. K., Gamma irradiation effects on fruits and vegetables, *Econ. Bot. 15*(1):28 (1961).
73. Eckert, J. W., in *Control of Postharvest Diseases in Antifungal Compounds*, Vol. 1 (M. R. Biegel and H. D. Sisler, eds.) Marcel Dekker, New York, 1977, p. 269.
74. Tasca, G., E. Giurea, and K. Milin, Results obtained in preventing losses of produce caused by microorganisms in vegetables and fruits during storage by applying fungicide, *Hort. Abst. 50*:6899 (1980).
75. Salunkhe, D. K., M. Wu, M. T. Wu, and B. Singh, Effects of Telone and Nemagon on essential nutritive components and the respiratory rate of carrot (*Daucus carota* L.) roots and sweet corn (*Zea mays*) seeds, *J. Am. Soc. Hort. Sci. 96*(3):357 (1971).
76. Ambadan and N. L. Jain, New blanching medium for dehydration of carrot, *Indian Food Packer 25*(4):10 (1971).

77. Girdhari Lal, G. S. Siddappa, G. L. Tandon, *Preservation of Fruits and Vegetables*, Indian Council of Agricultural Research, New Delhi, 1986, p. 207.
78. Sistrunk, W. A., In canning carrots, variety, time, and temperature of blanch are important, *Arkansas Farm Res. 18*:6 (1969).
79. Cruess, W. V., *Commercial Fruit and Vegetable Products*, 4th ed., McGraw-Hill, New York, 1958.
80. Chiang, J. C., B. Singh, and D. K. Salunkhe, Effect of water quality on canned carrots, sweet cherries, and apricots, *J. Am. Soc. Hort. Sci. 96*(3):353 (1971).
81. Jelen, P., and C. S. Chan, Firming of canned vegetables by the addition of lactose, *J. Food Sci. 46*(5):1618 (1981).
82. Edwards, C. G., and C. Y. Lee, Measurement of provitamin A and carotenoids in fresh and canned carrots and green pea, *J. Food Sci. 51*(2):534 (1986).
83. Bourne, M. C., Effect of blanch temperature on kinetics of thermal softening in carrots and green beans, *J. Food Sci. 52*(3):667 (1987).
84. Heil, J. R., and M. J. McCarthy, Influence of acidification on texture of canned carrots, *J. Food Sci. 54*(4):1092 (1989).
85. Feinberg, B., S. R. Schwimmer, R. Reeve, and M. Juilly, Vegetables, in *Food Dehydration*, 2nd ed. Vol. 2 (W. B. Van Arsdel, M. J. Copley, and A. J. Morgan Jr., eds.), AVI, Westport, CT, 1964.
86. Stephens, T. H., and T. A. McLemore, Preparation and storage of dehydrated carrot flakes, *Food Technol. 23*(12):160 (1969).
87. Luh, B. S., and J. G. Woodroof, *Commercial Vegetables Processing*, 4th ed., AVI, Westport, CT, 1982, p. 388.
88. Mudhar, G. S. Jr., and J. J. Jen, Infiltrated biopolymer effect on quality of dehydrated carrots, *J. Food Sci. 57*(4):520 (1991).
89. Sethi, V., and J. C. Anand, Studies on the preparation and storage of intermediate moisture carrot preserve, *J. Food Sci. Technol. 19*(4):168 (1982).
90. Stephens, S. T., G. Saldana, and B. J. Lime, Neutralized juice of acid treated carrots, *J. Food Sci. 41*(5):1245 (1976).
91. Bawa, A. S., and S. P. S. Saini, Effect of method of preservation on the quality of carrot juice, *Indian Food Packer 41*(1):42 (1987).
92. Grewal, K. S., and S. C. Jain, Physico-chemical characteristics of carrot juice beverage, *Indian Food Packer 36*(5):44 (1982).
93. Saldana, G., S. T. Stephens, and B. J. Lime, Carrot beverage, *J. Food Sci. 41*(5):1245 (1976).
94. Pruthi, J. S., A. K. Saxena, and J. K. Mann, Studies on the determination of optimum conditions of preservation of fresh vegetables in acidified sulphited brine for subsequent use in indian style curries, etc. *Indian Food Packer 34*(6):9 (1980).
95. Jagtap, S. A., Studies on the preparation of toffee from carrot (*Daucus carota*, L.), M.Sc. Thesis, Mahatma Phule Krishi Vidyapeeth, Rahuri, India, 1994.
96. Beerh, O. P., A. K. Saxena, and J. K. Manan, Improvement of the traditional method of manufacture of carrot murrabba, *Indian Food Packer 38*(4):59 (1984).
97. Jayaraman, K. S., and D. R. Dasgupta, Development and storage stability of intermediate moisture carrot, *J. Food Sci. 43*(6):1880 (1978).
98. Bhatia, B. S., and G. S. Mudhar, Preparation and storage studies on some intermediate moisture vegetables, *J. Food Sci. Technol. 19*(1):40 (1982).
99. Sethi, V., and J. C. Anand, Studies on the preparation, quality and storage of intermediate moisture carrot preserve, *J. Food Sci. Technol. 19*(4):168 (1982).
100. Sampathu, S. R., S. Chakraborthy, K. Prakash, H. C. Bisht, M. D. Agarwal, and N. K. Saha, Standardization and preservation of carrot halwa, *Indian Food Packer 35*(6):60 (1981).
101. Premavalli, K. S., K. Vidyasagar, and S. S. Arya, Effect of vegetable oil on the stability of carotenoids in carrot halwa, *Indian Food Packer 45*(4):22 (1991).
102. Nadakarni, K. M., *Indian Materia Medica*, Vol. 1, Popular Prokashan, Bombay, India, 1976, p. 442.
103. Shastri, B. N., *The Wealth of India: Raw Materials*, Vol. III, Council of Scientific and Industrial Research, New Delhi, India, 1952, p. 20.

104. Deshpande, S. S., U. S. Deshpande, and D. K. Salunkhe, Nutritional and health aspects of food antioxidants, in *Food Antioxidants—Technological, Toxicological, and Health Perspectives* (D. L. Madhavi, S. S. Deshpande, and D. K. Salunkhe, eds.), Marcel Dekker, New York, 1995, p. 361.
105. Abbey, M., M. Noakes, and P. J. Nestel, Dietary supplementation with orange and carrot juice in cigarette smokers lowers oxidation products in copper-oxidized low-density-lipoproteins, *J. Am. Dietet. Assoc. 65*:671 (1995).
106. Bishayee, A., and M. Chatterjee, Carrot aqueous extract protection against hepatic oxidative stress and lipid peroxidation induced by acute carbon tetrachloride intoxication in mice, *Fitoterapia LXIV*:261 (1993).
107. Darroudi, F., H. Targa, and A. T. Natarajan, Influence of dietary carrot on cytostatic drug activity of cyclophosphamide and its main directly acting metabolite: Induction of sister-chromatid exchanges in normal human lymphocytes, Chinese hamster ovary cells, and their DNA repair-deficient cell lines, *Mutat. Res. 198*:327 (1988).
108. Prochaska, H. J., A. B. Santamaria and P. Talalay, Rapid detection of inducers of enzymes that protect against carcinogens, *Proc. Natl. Acad. Sci. 89*:2394 (1992).
109. Dhar, V. J., Studies on *Daucus carota* seeds, *Fitoterapia* LXI:255 (1990).
110. Dhar, V. J., V. S. Mathur, and S. K. Garg, Pharmacological studies on *Daucus carota* Part I, *Planta Med. 28*:12 (1975).
111. Kaliwal, B. B., R. Nazeer Ahamed, and A. M. Rao, Abortifacient effect of carrot seed (*Daucus carota*) extract and its reversal by progesterone in albino rats, *Comp. Physiol. Ecol. 9*:70 (1984).
112. Kaliwal, B. B., Efficacy of carrot seed (*Daucus carota*) extract in inhibiting implantation and its reversal as compared with estradiol-17β in albino rats, *J. Curr. Biosci. 6*:77 (1989).
113. Batt, C., M. Solberg, and M. Ceponis, Effect of volatile components of carrot seed oil on growth and aflatoxin production by *Aspergillus parasiticus*, *J. Food Sci. 48*:762 (1983).
114. Ngyuen, C., and B. M. Lund, The lethal effect of carrot on *Listeria* species, *J. Appl. Bact. 70*:479 (1991).

6

Other Roots, Tubers, and Rhizomes

S. D. Masalkar and B. G. Keskar
Mahatma Phule Agricultural University, Rahuri, India

I. INTRODUCTION

Cassava, sweet potato, carrot, beet, radish, taro, yam, celeriac, parsnip, and salsify are some of the important root vegetables grown in many parts of the world. Most root crops are hardy to frost. Radish and turnip are susceptible to heat injury. They are grown for markets as well as for domestic consumption but have relatively less market demand. Cassava, sweet potato, and carrots are grown extensively worldwide. For information on these root vegetables, see Chapters 3–5. Information on turnip, horseradish, and rutabaga is presented in Chapters 12–15. In this chapter, an attempt is made to briefly summarize information on beet, radish, taro, yam, and other vegetables, such as amorphophallus, celeriac, parsnip, and salsify.

II. BEET

A. Introduction

Beet (*Beta vulgaris* subsp. *rubra*) is thought to have originated in the Mediterranean regions and near Asia. It is an important home garden and market vegetable crop and is also grown for shipping to distant marketplaces and for processing. The major producers of sugar beets are Russia, France, the United States, Germany, Poland, and Italy (1).

B. Botany

Beet (*Beta vulgaris* subsp. *rubra*) is a member of the Chenopodiaceae family and is probably a native of Europe (2). It behaves as a biennial, producing a thickened root and a rosette of leaves in the first year, flowers and seed in the second year. Under prolonged cold conditions, the plant may produce flowers and seeds in the first year. They are fairly hardy and grow best in cool weather to produce good quality beets with high sugar content and dark internal color throughout the root.

Watts and Watts (3) reported four general classes of beet varieties: root oblate or top shaded, root oval, root half-long, and root long conical. Varieties belonging to the root- oblate class are only of commercial importance. Based on red or yellow color, each of these classes can be subdivided into varieties that also differ in form, earlyness, and vigor.

The most popular early beets for market are Crosby Egyptian and Early Wonder (2,3). The former variety is a flat globe with a small taproot and smooth exterior and dark purple red internal color. Early Wonder is a flattened globe with smooth, dark, red skin and a dark red interior. Detroit Dark Red and strains derived from it are grown widely in the United States. It is globular with dark red smooth skin and a dark blood-red interior color. It is a good all-purpose beet used for bunching and for processing. Another favorite beet variety for processing is Perfected Detroit, whose interior color is darker than that of Detroit Dark Red and which is well adapted to mechanical harvesting. Perfected Detroit roots attain their globe shape at an early age and are also used for sliced and diced beets due to their uniform dark red color, flavor, and tenderness. Detroit Shot Top (strain of Detroit Red) and Halt Long Blood (root long variety) are also grown in some parts of Europe (2). Cylindra is used for canning and for pickling in sliced form. Kind Red and Crosby Egyptian are recommended for freezing and Morse Detroit and Ohio Canner for dehydration (4). Although different color forms exist (white and orange forms are used for their novelty value), only deep red forms are grown commercially (5).

C. Production

1. Soil and Climate

Beet can be grown in a wide range of soils, but light soils produce better crops than heavier ones. A heavy clay soil hinders the development of roots. Therefore, the best quality beet crop can be obtained in sandy loam soils that are deep, well drained, and rich in plant nutrients. The pH range most suitable for production is around 6.0–7.0, but beet can be grown even in more alkaline (pH 9–10) soils.

Beet is a biennial crop but is grown as an annual for its root. The sugar content of the root increases at low temperatures. Beets require low-temperature treatment (vernalization 2 weeks at 4–10°C) for flower bud initiation. This crop can tolerate frost much better than other vegetables but cannot withstand hot dry weather. The ideal temperature for producing excellent quality roots is 16–18°C.

2. Propagation

Beet seed is produced in cold climates. Beet root seeds are dark brown or black in color and contain up to 5000–6000 seeds per 100 g. They are usually sown in October. Under ideal conditions they take about 5–10 days for germination. The best temperature for germination of beet root seed is 20–30°C.

3. Cultural Practices

Sowing

The seeds are sown by broadcasting, drilling, or dibbling. They are sown in rows that are 30–45 cm apart. A plant-to-plant distance of 15–20 cm is maintained, depending upon the variety and soil type. When the seeds are broadcasted, thinning of seedlings may be necessary. In general, about 12–15 kg of seed may be needed per hectare.

The soaking of seeds in water for about twelve hours before sowing improves seed germination. Some of the growth regulators are also used for improving the germination.

Manuring and Fertilization

Beet root does not need too much nitrogenous fertilizer if the land is rich in organic matter. However, it needs higher levels of phosphorous and potassium. According to Kale and Kale (5), a beet crop yielding one ton of roots needs about 2 kg of nitrogen and 4.5 kg each of phosphorus and potash. For better yields of excellent quality roots, they suggested the application of 10–15 cartloads of fully decomposed farmyard manure, 60–70 kg of nitrogen, 100–120 kg of phosphorus, and 60–70 kg of potash per hectare. The entire dose of farmyard manure should be given at the time of land preparation, while the inorganic fertilizers are applied at the time of sowing. Adequate soil moisture is necessary to obtain the best results from fertilizer application (5).

Irrigation and Intercultural Operations

The first irrigation is carried out immediately after sowing. Subsequent irrigations are performed depending upon the soil type, weather conditions, and variety. In general, one irrigation every 6–10 days during winter and 3–4 days during summer should provide adequate moisture. This crop is most sensitive to water stress at the time of root development and during early stages. However, waterlogging should be avoided at all times. Intercultural operations like shallow hoeing and weeding are necessary to check weed growth.

4. Disorders, Diseases, and Pests

Internal black spot of beets is caused by boron deficiency and can be controlled by adequate fertilization with borax (7,8). Hard or corky black spots develop on the roots, especially in the light-colored zones of the young cells and tissues. According to Walker et al. (9), Long Dark Blood is the variety most resistant to internal black spot. Beets are susceptible to freezing injury between −1.0 and −1.5°C (10), hence they should not be held where they may cool below their freezing point for more than a few hours.

Beet roots are affected by black rot caused by *Phoma betae*. The disease is usually confined to topped beets out of storage. The initial brown and water-soaked lesions become black, dry, and spongy later, affecting mostly the tip of the root. Ryall and Lipton (10) stated that close clipping of tops, good air circulation, and storage at 0°C retard the development of black rot.

5. Harvesting, Handling, and Maturity

Beet maturity is generally indicated when the leaves fall away from a vegetable in the upright position (5). Beets for bunching are harvested when they attain a diameter of about 3.75 cm. Larger beets having a diameter of more than 5 cm are not in great demand. Beets are tied in bunches of three to six according to size and market custom. Bunching is done most economically in the field when the beets are pulled. The dead and unsightly leaves are removed, and beets are washed to remove soil. Beets may be harvested mechanically, where the topped and washed beets are packaged in the transparent film bags with mechanical harvesters (2). Topped beets in transparent film bags have a longer shelf life than bunched beets with the tops on. Late beets grown in the North and those grown for processing are generally harvested mechanically (2). The machine lifts the beets, cuts off the tops, and conveys the topped roots into a truck moving along with the harvester. The beets may then be hauled to the processor or to the storage house for packaging.

D. Chemical Composition

The chemical composition of raw beet root is summarized in Table 1. It is a good source of sugar (5). Beets are usually cooked and can be served cold or warm, either alone or in conjugation with leafy salads. Beets are also added to fish and meat salads to add color and flavor.

E. Storage

The best storage temperature for beets is near the freezing point, but beets should not be allowed to freeze. Enough humidity is needed to avoid wilting and withering of roots. Thompson and Kelly (2) stated that beets keep well at 0°C with a relative humidity of about 90%, which is common practice where beets are stored for processing. Chrimes (11) reported that topped beets stored well for 3–5 months and can be held as long as 8 months in cold storage if destined for processing. Under most storage conditions, relatively large roots keep better than small ones because they shrivel more slowly (10). Bunched roots have a shorter storage life because the tops are highly perishable—about 10–15 days at 0°C and only about a week at 4.4°C. Three years

TABLE 1 Nutritional Composition of Beet Root

Constituent (per 100 g fresh weight)	Content
Water (g)	87.1
Total nitrogen (g)	0.27
Protein (g)	1.7
Fat (g)	0.1
Carbohydrate (g)	7.6
Energy value (kJ)	54.0
Starch (g)	0.6
Total sugars (g)	7.0
Dietary fiber (g)	1.9
Sodium (mg)	66
Potassium (mg)	380
Calcium (mg)	20
Magnesium (mg)	11
Phosphorus (mg)	51
Iron (mg)	1.0
Sulfur (mg)	16
Carotene (μg)	20
Thiamine (mg)	0.01
Riboflavin (mg)	0.01
Niacin (mg)	0.1
Vitamin B_6 (mg)	0.03
Folate (μg)	150
Pantothenate (mg)	0.12
Vitamin C (mg)	5

Source: Ref. 5.

of storage research data obtained by Schouten and Van Schaik (12) showed that mechanical harvesting and storage at 0–1°C increased black spots under the skin and rotting of the beet roots, compared with storage at higher temperatures. Storage at 6–7°C tended to prolong the storage period compared to 0–1°C but led to a rapid loss of quality. These workers recommended storage temperatures of 3–4°C and high air humidity (12).

The deterioration of beet roots in store may be of physiological origin (13). The effects of controlled atmosphere (CA) storage of beets have not been investigated thoroughly. Ryall and Lipton (10), however, reported that elevated CO_2 increases beet root decay.

F. Processing

Beets can be preserved by freezing, drying, or canning. Uniform dark red color without zones or rings and tender flesh free of fiber are desirable processing qualities (4). Beets destined for processing must be harvested before they develop woodiness. Beets to be dehydrated are harvested at a slightly later stage of maturity than those used for canning. Canners prefer small beets, which command a premium price in the whole canned form. Small roots are also desirable for freezing except when diced and frozen. Rodriguez et al. (4) further stated that Detroit Dark Red is the best variety for canning whole where small, round, or globular-shaped beets are required. Cylindra is used for canning or for pickling since it gives much less wastage when sliced transversely.

Hardley (5) reported that the market for fresh beets is relatively small. For the supermarket and green grocery trade, beet roots are cooked and usually presented in prepacked form. Processed beet is the major outlet for beet root, where it is either pickled whole, sliced, or to a lesser extent diced into glass jars.

III. RADISH

Radish is probably indigenous to China and middle Asia (14). It is a popular vegetable crop for home gardening and the fresh market (15). Radish crops grow well in cool weather. Radish roots are eaten raw as salad or cooked as a vegetable. In ancient medicine, radish was used for liver and gall bladder ailments (16).

A. Botany

Radish (*Raphanus sativus* L.) is both an annual and a biennial and belongs to the Cruciferae or mustard family.

1. Taxonomy

There are four types of radish commonly cultivated in different continents of the world (1): small, cool season radish, large radish with a wider range of temperature adaptation, rat-tail or mougri radish forming no fleshy roots but forming long slender (20–60 cm) pods, and fodder radish, also producing no fleshy roots. All four types of radish belong to the species *Raphanus sativus* L. (2n = 18), and botanically they are known as *radicula*, *niger*, *mougri* (or *caudatus*), and *oleifera*, respectively. All four types intercross freely and with related wild species (17).

2. Morphology

Radish root (the edible portion of radish) develops from both the primary root and the hypocotyl. Roots vary greatly in size, shape, and other external characteristics as well as in the length of time they remain edible. The shape varies from oblate to long tapering, and the exterior color may be white or various shades of scarlet (17). The length varies from 2.5 to 9 cm. The inflorescence of radish is a typical terminal raceme of Cruciferae. The flowers are small, white, rose, or lilac with purple veins in bractless racemes. The fruit is indehiscent, 3–7 cm long, and 1.5 cm in diameter with 6–12 seeds and a long conical beak. The seeds are globose and about 3 mm in diameter. The seed, when mature, is at first yellowish but turns reddish-brown with age (17). The lack of lignification of the vascular tissue and the pressure of considerable thin-walled parenchyma are responsible for the succulence of the root system. The enlarged roots are available in a wide range of sizes, shapes, and colors, depending on cultivar. Early-maturing European salad types can be round, olive-shaped, or cylindrical red-white or red with a white tip and with crisp whole flesh. Winter and oriental radishes are often white (although red and purple skins are also found) with a large conical shape and white flesh. Early-maturing radishes produce small roots, which are harvested when approximately 1.25 cm in diameter. Winter and oriental radish roots are often 25–40 cm in length and can weigh approximately 2–3 kg (although even larger roots are available) (16).

3. Cultivars

Depending on the season of the year in which the radish is grown, three distinct varieties are known: the quick-growing spring varieties, which include most of the commercially grown varieties, the late-maturing summer varieties, and the long-duration winter varieties, which require about twice as long to mature as the spring varieties (18,19). The most popular market types of spring radish are bright red globular radishes such as Cavelier, Cherry belle, Comet, Sparkler, and Early Scarlet Globe (2). Scarlet Globe is the most important and quick maturing and is a leading market radish in nearly all parts of the United States (3). The long-rooted spring varieties are Cinncinnati Market, White Iscle, and Long Scarlet Short Top. The summer varieties are long-rooted and include Strasburg, Stuttgart, and White Vienna, whereas Golden Globe is a round summer variety. The winter varieties are grown as fall crops for storage; popular ones are White Chinese, China Rose, Long Black Spanish, Round Black Spanish, and Sakurjima. The last grows to an enormous size and has solid, firm flesh of good flavor.

The Indian varieties of radish are Jaunpuri, Bombay Long White, Long Red, and Ganesh Synthetic (Fig. 1). The Indian Agricultural Research Institute, New Delhi, has released Pusa Himani as an improved variety of radish having white skin, crisp flesh, and sweet flavor (20). Several Asiatic or tropical varieties such as Japanese White, Pusa Desi, Pusa Chetki, and Pusa Reshmi (IARI) are also grown extensively in India. Cultivar Co-1 has recently been released by the Tamilnadu Agricultural University, Coimbatore (21).

B. Production

1. Soil and Climate

Like most vegetable crops, radish can be grown on a wide range of soils, but a deep, friable, well-drained, and rich sandy loam or alluvial soil produces the best quality crop. The presence of ample humus and the ability of the soil to retain moisture are the important factors to be considered in radish production. Heavier soils such as clay or silt clay produce small, misshapen, and fibrous roots. Sandy soils lacking in humus are also not suitable for radish production. Radish grows very well over a relatively wide pH range, but soils with near neutral or slightly acidic pH are best.

FIGURE 1 Ganesh Synthetic variety released by Mahatma Phule Agricultural University, Rahuri, India.

Radish is a cool season vegetable but can tolerate warm temperatures. The best quality roots are produced at temperatures between 10.0 and 15.5°C. As the temperature increases, the pungency of the root increases before reaching marketable size. Roots may acquire a repulsive flavor and become more fibrous and mature early at higher temperatures. Varieties also differ in their temperature requirements. Therefore, the selection of variety for a given location is very important to obtain the best quality roots. Radish has a tendency to bolt under long days (more than 8–10 hr/day) coupled with high temperatures.

2. Propagation

Radish is grown from seeds, which are sown directly in the field either by drilling, broadcasting, or dibbling. The seeds are brown to reddish-brown in color; there are about 100 seeds per gram. They remain viable for several years. At 20°C, the seeds take about 4–6 days to germinate. Seeds can be sown at half-weekly or weekly intervals for a continuous supply of roots over a period of time.

3. Cultural Practices

Sowing

As the seeds are sown directly in the field, the land should be thoroughly prepared to provide a proper substrate for them to germinate. This can be done by repeated ploughings, harrowings, and leveling to achieve fine tilth. Old debris or the remains of a previous crop should be removed completely before final preparation.

In the plains, this crop can be grown almost throughout the year except during summer months, as the roots produced in that period will be too pungent, lack pleasant flavor, and be of poor quality. Roots can be continuously supplied over a longer period by staggered planting at intervals of 10–15 days.

Radish seeds are usually sown in ridges 30–45 cm apart. A plant-to-plant distance of 7.5–10 cm is recommended for this crop. Two to three seeds can be sown at each spot in order to ensure that at least one will germinate. When the seeds are drilled, sowing is done in rows at 30–45 cm. The seeds can be mixed with sand or ash to make sowing easy. The seedlings are thinned once the germination is complete and the desired plant population is maintained. Approximately 10–12 kg of seeds are required for each hectare.

Manuring and Fertilization

Radish is considered to be a heavy feeder, particuarly of nitrogen and potassium. A radish crop yielding about 20 tons of roots requires 120 kg of nitrogen, 65 kg of phosphorus, and 100 kg of potassium per hectare (8). However, according to Laske (22), radish removes 50 kg of nitrogen, 20 kg of phosphorus, 50 kg of potash, and 30 kg of calcium oxide (CaO) per 10 tons of roots. The actual fertilizer doses to be applied depend on the variety, season, soil type, method of production, and location. Maurya et al. (23) recommended the application of 150 kg each of nitrogen, phosphorus, and potassium for Jaunpuri radish. On the other hand, Nath (24) stated that 24.0–37.0 tons of well prepared farmyard manure, 18–50 kg of nitrogen, and 50 kg each of phosphorus and potash should be applied to get the best quality roots. It is important not to use fresh farmyard manure, as it causes forking or branching of roots in radish.

The entire dose of farmyard manure should be applied at the time of land preparation so that it is thoroughly mixed with the soil. In addition, half of the nitrogen and the entire dose of phosphorus and potassium are applied at the time of sowing, while the remaining nitrogen is applied after the seedlings have completely emerged.

Irrigation and Intercultural Operation

The first irrigation should be light and given immediately after sowing. It should be applied by sprinklers or hand-sprinklers. It is not necessary to irrigate the field during the rainy season, but during summer and winter watering should be done every 4–5 days and 10–15 days, respectively. Weeding and hoeing are needed particularly in the early stages. When hoeing is done, care must be taken not to injure the roots. The long-rooted varieties need earthing up, which is done about a month after germination to encourage root development.

Intercrops

Since radish crops do not need much space, other vegetable crops such as brinjal, cauliflower, cabbage, *methi*, onion, etc. can be planted as intercrops. Sometimes, radish is grown as an intercrop in fruit orchards or as a border crop in wheat, onion, and chile fields.

4. Diseases and Pests

Radish is a short-duration crop, making it unsusceptible to slow-developing disease (17). Some important diseases of radish include black rot (*Aphanomyces raphani* Kendr.), Alternaria blight (*Alternaria raphani* Groyes and Skolko), crown gall disease (*Agrobacterium tumefaciens*), white rust (*Albugo candida* (Pers. ex. Chav.) Kuntze), and radish mosaic virus. Aphids (*Myzus persicae*) and mustard sawfly (*Athalia proxima*) are important radish pests.

5. Harvesting and Handling

Radishes are harvested as soon as the roots attain edible and marketable size. The quick-maturity spring varieties become strong and pithy if not harvested at the proper time. The summer varieties remain edible much longer than the spring varieties, and the winter varieties remain edible for several months if stored properly (18). Crop maturity varies from about 3–4 weeks for the quick-growing spring varieties to 8–14 weeks for Chinese varieties (25). Watts and Watts (3) stated that at this stage radishes are mild, tender, and crisp, and usually of the proper marketable size. Topping below the growing tip improves the keeping quality of radishes (26–29). Radishes grown for the home garden and with small plantings are pulled by hand and laid in bunches of 6–12 in the field and washed to remove soil. They are usually packed in baskets or crates and iced for transport to market. Tops of winter radish are removed before they are stored. Mechanical harvesters are used for commercially grown crops in the United States.

C. Chemical Composition

The chemical composition of radish is presented in Table 2. Radish contains glucose as the major sugar and smaller quantities of fructose and sucrose. Pectin and pentosan are also reported to be present. It is a good source of vitamin C, containing 15–40 mg/100 g of edible portion and supplies a variety of minerals. Pink-skinned radishes are generally richer in ascorbic acid than white-skinned ones. The vitamin C content of radish roots is greatly influenced by light conditions and fertilizers (17).

The characteristic pungent flavor of radish is due to the presence of volatile isothiocyanates (trans-4-methyl thiobutenyl isothiocyanate), and the color of the pink cultivar is due to the presence of anthocyanin pigments. Thiocyanate content may increase linearly with increasing sulfate level in culture solution (17).

Radish leaves are a good source for extraction of proteins on a commercial scale. The seeds are a potential source of a nondrying fatty oil suitable for soap making (17).

D. Storage

Radish roots are topped and packed in plastic bags. They are cooled quickly to 5°C or below to maintain their crispness. Hydrocooling is an effective method of cooling radishes (30). Radishes can be held for 3–4 weeks at 0°C and a somewhat shorter time at 5°C. When temperatures are higher than 0°C, low oxygen (1%) is beneficial in reducing tops (31). The regrowth of tops can be greatly reduced by trimming off the growing points, which are aggregated within a few millimeters on top of the root (2).

1. Low-Temperature Storage

Radishes can be stored for about 3–4 weeks at 0°C and high relative humidity, but at 7.2°C their shelf life is less than a week (32). The life of bunched radishes is limited by that of the tops. Storage life can be lengthened by trimming the thin roots and by clipping the tops just above the crown. These practices reduce decay, and the latter practice retards regrowth of tops. Winter radishes can be stored much longer than others. Pantastico et al. (33) recommended a tempera-

TABLE 2 Nutritional Composition of Red and White Radish

Constituent (per 100 g fresh weight)	Red radish	White radish
Water (g)	95.4	93.0
Total nitrogen (g)	0.11	9.13
Protein (g)	0.7	0.8
Fat (g)	0.2	0.1
Carbohydrate (g)	1.9	2.9
Energy value (kJ)	49	64
Starch (g)	Traces	Traces
Total sugars (g)	1.9	2.9
Dietary fiber (g)	1.9	—
Sodium (mg)	11	27
Potassium (mg)	240	220
Calcium (mg)	19	30
Magnesium (mg)	5	15
Phosphorus (mg)	20	25
Iron (mg)	0.6	0.4
Sulfur (mg)	38	—
Carotene (μg)	Traces	0
Thiamine (mg)	0.03	0.03
Riboflavin (mg)	Traces	0.02
Niacin (mg)	0.4	0.5
Vitamin B_6 (mg)	0.07	0.07
Vitamin B_{12} (μg)	0	0
Folate (μg)	38	38
Pantothenate (mg)	0.18	0.18
Vitamin C (mg)	17	24

Source: Ref. 5.

ture of 0°C and 88–92% relative humidity for storing topped radishes for about 3–5 weeks. Winter radishes can be stored at 0°C and 98% relative humidity for 6 months (10).

Each (34) placed radishes in boxes covered with plastic or not covered in a cold store at 3–4°C or held at room temperature for 0.5°C for 2–3 days. Cooling reduced radish temperatures from 17 to 3°C in 5–6 hours. Leaf yellowing was greatly reduced due to cooling; even the shortest cooling period was beneficial. Cooling of radishes for 2–3 days without covering gave the best results. The keeping quality of radishes (cv. Cherry Belle) is improved by precooling and storing at 0–1°C in a partial vacuum (26).

2. Controlled Atmosphere Storage

Radishes are generally topped, prepackaged in film bags, and stored. The bags must be ventilated to permit adequate aeration. Ryall and Lipton (10) do not recommend shipping of radishes in sealed bags because of the danger of low O_2 injury, especially at 4.4°C or above. Where O_2 can be controlled, 1–2% CO_2 is beneficial because it reduces root and top growth during prolonged storage.

Holding for 24 hours at 2°C in 9.4–10.0% CO_2 and low O_2 atmosphere prevented regrowth of leaves in topped red radishes of the cultivar Cherry Belle (28). This treatment also partly inhibited the development of *Peronospora parasitica* during the holding period and subsequent shelf life. No such effect was noted when the atmosphere was rich in N_2 or when a high CO_2 content occurred as a result of gradual buildup inside packages. These workers also found that raising the CO_2 level inside polyethylene bags to 25–35% prevented leaf regrowth during storage over 4–5 weeks at 2°C and during subsequent shelf life (29).

3. Physiological Disorders, Diseases, and Their Control

Radishes are susceptible to freezing injury. Ryall and Lipton (10) have described the disorder, stating that freezing followed by thawing causes the injured tissue to appear translucent. In severe cases, roots soften, lose moisture rapidly, and shrivel. Parsons and Day (35) reported that in red radishes the pigment oozes out of the roots with the moisture, leaving them yellowish and bleached.

Bacterial black spot caused by *X. nesicatoria* and downy mildew caused by *Peronospora parasitica* are the important diseases of radish causing severe postharvest losses (2). Segall and Smoot (36) reported that most of the bacterial infections responsible for black spot of radishes started at injuries caused during mechanical harvesting, washing, and sizing.

Satisfactory control of black spot can be obtained by washing radishes with water containing 40–60 ppm of chlorine (36). Zisman and Temkin-Gorodeiski (26) reported that treatment of radishes with maleic hydrazide (1000–3000 ppm) or chloropham (CIPC) in the field had no effect on leaf sprouting during storage. The postharvest dips in growth inhibitors such as ABA and CCC inhibited sprouting and rootlet growth for a certain period but did not prevent texture deterioration. Packing radishes in polyethylene bags with air extracted to form a partial vacuum inhibited leaf sprouting and rootlet growth for considerable periods of time (27). Dipping radishes in natural waxes or in a defoliant ($MgCl_2$) caused severe bleaching (28).

E. Processing and Utilization

Early-maturing radishes are eaten raw in salads, whereas winter and oriental radishes are eaten either raw or as a cooked vegetable rather like turnips. The very characteristic flavor is popular in Japan, the Philippines, and Hawaii, and roots are used to prepare food products such as *takuwan* and *cabaizuku*. Leaves and pods of some cultivars can be boiled and eaten as a vegetable. Roots have been used in the treatment of liver and gall bladder complaints (16).

IV. TARO

Taro is one of the important tuber crops grown in Africa and Asia (Table 3). About 90% of the area under this crop is in Africa with 80% of the total production. It is also an important crop in Hawaii. Taro is recommended for gastric patients and its flour is considered good baby food. In Hawaii, a fermented product, *poi*, is very popular. In Africa, the corm paste prepared from cooked taro is called *fufu*. In India, it is used as a vegetable.

A. Botany

The subfamily *Colocasioideae* of family *Araceae* consists of three edible tuber crops, namely taro (*Colocasia esculenta* Schott), tannia (*Xanthosoma* spp.), and giant taro (*Alocasia* spp.) (37).

TABLE 3 World Production of Taro

Continent/Country	Production (1000 MT) 1989	1993
World	5814	5639
Africa	3780	3459
Ghana	1063	1236
Nigeria	1800	1300
North America	21	22
South America	—	7
Asia	1714	1817
China	1140	1316
Japan	400	330
Philippines	110	110
Oceania	294	333
Papua New Guinea	180	218
Samoa	40	37

Source: Ref. 1.

Among these crops, taro and tannia are cultivated most, while giant taro is not as common as a commercial crop.

1. Morphology

Colocasias are herbaceous plants up to 2 m tall and have underground corms. Leaves are large, with long petiole clasping at the base (Figure 2). Corms are cylindrical with short internodes and a few side tubers (Figure 3). The root system is shallow and fibrous (37).

2. Cultivars

Onwueme (38) described two botanical varieties: viz. *Colocasia esculenta* (L.) var. *antiquorum* (Eddoe types), and *C. esculenta* var. *esculenta* (Dasheen types). The taro cultivars vary in color of leaves and petioles and shape of corms and cormels (38).

Dasheen type
West Indies: Purple
Hawaii: Lehua, Piko Uaua, Piko Kea, and Piko Uliuli
Eddoe type
West Indies: Common
United States: Trinidad

Panch mukhi, Sahasra Mukh, Tamara Kannan, Kovvur, Khasi, Bhunga, C-9, C-135, C-266, and C-149 are some of the cultivars grown in India.

B. Production

1. Soil and Climate

Taro is a tropical crop and can be grown even in heavy soils. It prefers abundant water supply and can withstand waterlogging. The optimum soil pH range is 5.5 – 7.0. Taro will not grow in

FIGURE 2 Plant of *Colocasia*. (Courtesy Prof. A. R. Karnik, Konkan Agricultural University, Dapoli, India.)

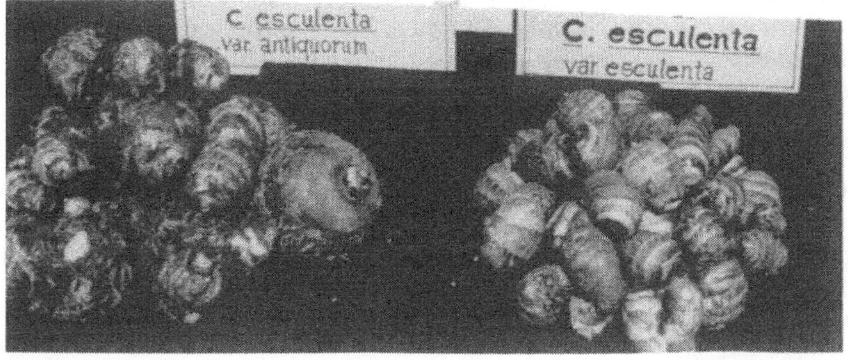

FIGURE 3 Tubers of *Colocasia*. (Courtesy Prof. A. R. Karnik, Konkan Agricultural University, Dapoli, India.)

a cooler climate where frost is a problem, but will grow even in a subtropical climate where the temperature averages around 20°C. It is grown on the hills and plains of northeastern India as a rainfed crop where rainfall is very heavy. Erratic rainfall and intermediate short-term drought during the rainy season affect corm quality.

2. Propagation

Taro is vegetatively propagated. About 1 cm corm tips with 15–20 cm petioles are used in Hawaii, which produce higher yields than corms as planting material (39). Cormels are also used as planting material in many areas. The cormels are preferred in places where winter is severe for the convenience of storing till the next planting season.

3. Cultural Practices

Planting

Land preparation varies from place to place. The land is puddled before planting in Hawaii, whereas it is planted in ridges and furrows and in beds in India (40). The distance of planting ranges from 5 to 10 cm. A spacing of 60 × 45 cm is followed in India, whereas spacing is 90 × 30 cm in Cuba. In Japan, spacing is 60 × 100 cm. Spacing can be closer in lowlands than in uplands because water may not be a limiting factor. The planting is done during the rainy season, but it can be done any time if irrigation facilities are available. In temperate regions, planting is avoided during the winter.

Manuring and Fertilization

Taro is responsive to fertilizer application. It requires more potassium than nitrogen and phosphorus. Under Indian conditions, levels of 103.9, 74.1, and 135.7 kg/ha of N, P, and K were found to be optimum (37). In Hawaii, satisfactory yield was reported by applying 250 kg each of N, P, and K per hectare (37). Fertilize requirements vary with cultivation methods, including spacing, cultivar, and type of soil.

Irrigation

Irrigation increases corm yield with increasing water levels because of very extensive rooting beyond 9 months under flooded conditions (40). Irrigation throughout the life cycle of taro resulted in higher total yield of corms and cormels than for unirrigated plots (41).

Intercultural Operation

Weeding at initial periods of crop growth results in better growth of taro crop. Plucknett et al. (39) reported that herbicides like propanil, Prometon, and Nitrofen were effective in wetlands, while for uplands, they recommended Ametryne. Mulching with dried leaves and other plant material helps in controlling weeds.

4. Diseases and Pests

Leaf blight caused by Phytophthora *Colocasiae* Rac. and Phythium rot caused by *Phythium* spp. are the important diseases of taro. In addition, sclerotium rot and leaf spot (*Cladosporium colocasiae* and *Phytosticta colocasiophila*) diseases also reported (32). A severe mosaic disease, originally reported in the United States, has been reported in India (32).

Taro leaf hopper (*Tarophagus proserpina*) causes losses in most taro-growing areas. Other pests include the egg-sucking bug (*Cyrtorhinus fulvus*), aphid (*Aphis qossypii*), and red spider mite (*Tetranychus cinnabarinus*) (37).

5. Harvesting

Time to maturity depends upon the type, cultivar, and climate. The crop matures in 3 months in Sri Lanka, while it is ready for harvest after 5–7 months in India (37). In Hawaii, it takes 12–15 months. Dasheen matures in 8–10 months, whereas Eddoes take only 5–6 months in Trinidad (42). The crop is harvested by pulling the plants out of the soil with the help of hoes or spades. The leaves become yellow when the crop matures.

C. Chemical Composition

The chemical composition of taro corms, leaves, and stems is given in Table 4. The tropical edible corm has a higher starch content than either potatoes or sweet potatoes. The taro corm is rich in starch, which contains 15–25% amylose. It is a good source of calcium and iron but a poor source of vitamin C (42).

D. Storage

Corms can be stored for 4–5 months at 7–10°C if harvested and handled carefully (43). Taro does not require high-temperature curing as do sweet potatoes. Good air circulation is desirable. Storage rot is caused by a number of fungi like *Aspergillus niger* van Tiegham, *Botryodiplodia theobromae* Pat., *Fusarium solani* (Mart.) Sacc., *Rhizopus Stolonifer*, and *Sclerotium rolfsii* Sacc. (37). Mechanical injury during harvesting is known to predispose the corm to the pathogens in storage. Dipping of taro roots in Difolatan (0.1%) or Mitox (0.20%) solution may check the rotting in storage.

E. Processing and Utilization

Taro is consumed fresh after boiling or baking. It is also processed into chips with a distinct nutty flavor (44). Corms are sliced thinly and cooked in hot oil, resulting in a product similar to potato chips. Taro is processed into flour, noodles, cake, and infant food as well as flaked, canned, or frozen corms.

TABLE 4 Average Composition of Taro Plant (per 100 g of raw product)

Constituent	Leaves	Stem	Corm
Moisture (g)	90	94	73.1
Carbohydrate (g)	5.7	3.6	21.1
Protein (g)	2.4	0.3	3.1
Fat (g)	0.6	0.3	0.1
Calcium (mg)	98	60	40
Iron (mg)	2.0	0.5	1.7
Vitamin A (IU)	1800	104	24
Ascorbic acid (mg)	11	3	—

Source: Ref. 37,38.

V. YAM

Yam is one of the underutilized tuber crops cultivated in tropics particularly in West Africa (Table 5), West Indies, Tropical America and Southeast Asia (45). Yam tubers are consumed after roasting, boiling, or with other vegetables. It is also used as chips, flakes and flour.

A. Botany

The yam belongs to the family Dioscoreaceae and genus *Dioscorea*, which contains about 600 species (45). Of the total species, about 12 are used for edible purposes. The species *Dioscorea alata* L. (Fig. 4), *D. rotundata* (L) Poir (Fig. 5), *D. esculenta* (Lour.) Burk. (Fig. 6), *D. bulbifera* L. (Fig. 7), and *D. trifida* L. are the principal yams of the tropics (Table 6). However, *D. alata* L., *D. rotundata* (L) Poir, and *D. esculenta* (Lour.) Burk are extensively cultivated for edible purposes (46).

1. Morphology

Dioscoreas are deciduous perennial plants. The root system develops from the end of the tuber, thick, unbranched at the beginning, becoming fibrous and branched later. Most species produce underground stem tubers; a few wild species are rhizomatous. Tubers vary in number, size, and form: they may be globular, elongated, or flattened. Some species produce bulbils in leaf axils (45).

2. Cultivars

Martin (47) has described several cultivars of *D. alata*, including Florida, Forastero, Gamelos, Leone Globe, and Veveen Guniea. A white-fleshed cultivar of *D. cayenensia* was popular in

TABLE 5 World Production of Yams

Continent/Country	Production (1000 MT) 1989	1993
World	23459	28126
Africa	22266	26810
Cote Divoire	2370	2480
Nigeria	16000	20000
Togo	400	529
Zaire	265	315
North America	350	480
Haiti	130	200
Jamaica	133	215
South America	397	305
Brazil	210	215
Colombia	142	49
Asia	195	239
Japan	169	210
Oceania	249	290
Tonga	35	31

Figure 4 *Discorea alata* tubers. (Courtesy Prof. A. R. Karnik, Konkan Agricultural University, Dapoli, India.)

Puerto Rico. Several cultivars of *D. alata* having high tuber yields including Da 60, Da 80, and Da 122 were reported by Central Tuber Crops Research Institute, Trivandrum (45). De 11 was the selection in *D. esculenta* reported from the same institute (45).

B. Production

1. Soil and Climate

Yam is mainly grown in tropical and subtropical regions. However, the major yam-growing areas are in the tropical regions. The optimum temperature for yam growth is between 25 and 30°C. High rainfall favors good growth of tubers. Tuber growth depends on the texture and preparation

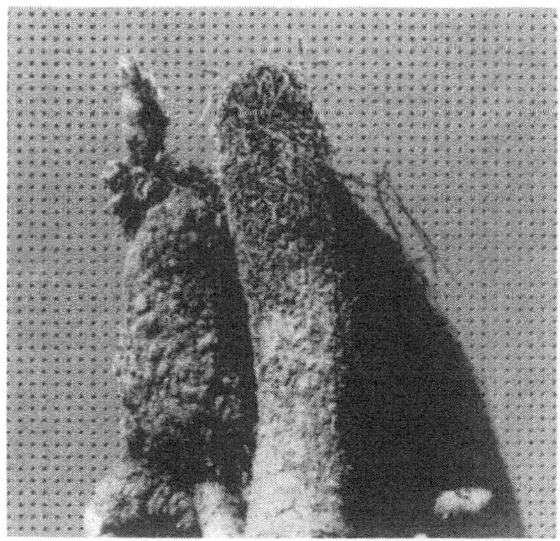

Figure 5 *Dioscorea rotundata* tubers. (Courtesy Prof. A. R. Karnik, Konkan Agricultural University, Dapoli, India.)

FIGURE 6 *Dioscorea esculenta* tubers. (Courtesy Prof. A. R. Karnik, Konkan Agricultural University, Dapoli, India.)

of the soil, Well-drained, loose friable soil containing adequate organic matter is preferred for yam cultivation (48).

2. Propagation

Yam is propagated vegetatively using tuber pieces or small whole tubers as planting material. The size and weight of tuber pieces at planting influence the yield (38). Multiplication of yams by vine cuttings is also possible, but tuber production by this method is low. Tissue culture has been used for clonal propagation of yams.

FIGURE 7 *Dioscorea bulbifera* tubers. (Courtesy Prof. A. R. Karnik, Konkan Agricultural University, Dapoli, India.)

TABLE 6 Description of the Most Common *Dioscorea* Species Growr for Food

Species	Common name	Origin	Tuber number and shape	Flesh color	Leaf shape and phyllotaxy	Stem appearance
Dioscorea alata 1.	Water or greater yam	Southeast Asia	Single, cylindrical	White to purple	Ovate, opposite	Winged
D. cavenensis Lam.	Yellow or yellow guinca yam	Western Africa	Single variable	Yellow	Pointed, opposite, alternate	Spiny
D. esculenta Burk	Lesser yam	China	4–20, ovoid	White	Simple alternate	Spiny
D. ratundata Poir	White or white guinca	Western Africa	Single, cylindrical, spherical	White to yellow	Simple, opposite	Circular, smooth-spiny
D. trifida L.	Cush-cush	Tropical America	Numerous, elongate	White to purple	Lobed, alternate; opposite	Winged

Source: Ref. 68.

3. Cultural Practices

Planting

Yams are planted in flat or raised beds or mounds formed over pits. The seed tubers are planted at a depth of 10–15 cm. Seed tubers weighing 100–250 g are used for planting. A spacing of 75 × 75 cm or 90 × 90 cm is followed for optimum yield.

Training of Vines

Training of vines has been found to be beneficial in obtaining higher yields of yam. The emerging shoots are provided with support to avoid any injury to the tender shoots. This is done within 15 days of sprouting using coir rope attached to an artificial support or tress. Yams are also staked on living or dead trees or other quick-growing plants.

Manuring and Fertilization

Yams require plenty of organic matter to support growth (49–51). Yams are very responsive to the application of fertilizers. Application of N, P, and K at the rate of 80, 60, and 80 kg/ha has been recommended for *D. esculenta* (49). In *D. alata* N, P, and K at 120, 80, and 80 kg/ha was found to be optimum for obtaining higher yields of good-quality tubers. A half-dose of nitrogen and potassium along with a full dose of phosphorus should be applied at the time of first interculturing. The remaining half-dose of N and K should be applied a month later along with the second interculturing. The nutrient uptake in several species of *Dioscorea* varied between 116.1 and 168 kg of N, 25.5 and 30 kg of P, and 108 and 146 kg of K per hectare (50).

Intercultural Operation

The first interculturing operation is performed one week after 50% of tubers have sprouted; the second and third at intervals of 15 days to one month, depending upon weed growth.

4. Diseases and Pests

Blight caused by *Cercospora* sp. and dieback caused by *Colletotrichum* sp. are the major fungal diseases of yam. Mosaiclike symptoms of viral diseases are observed in *Dioscorea* species. The important pests include scale insects (*Aspidiella hartii*), white grubs (*Leucopholis coneophora*), termites, chrysomelids, and hairy caterpillars. Among these, scale insects are the most serious. Several nematode species attack yam tubers, including *Scutellonema bradys*, *Meloidogyne* spp., and *Pratylenchus* spp.

5. Harvesting

Yam crops are ready for harvesting about 8–9 months after planting. When the leaves turn yellow and the vines dry up completely, the crop is ready for harvesting. Care must be taken so that tubers are not cut or damaged, as this causes rotting of tubers.

C. Chemical Composition

The proximate composition of yam tubers is given in Table 7. They are rich in carbohydrates and have a high water content. The carbohydrates consist mainly of starch with less than 1% sugar in the form of glucose, sucrose, and traces of fructose. Yam starch consists mainly of

TABLE 7 Nutritional Composition of Yam and Amorphophallus Corms

Constituent[a]	Yam	Amorphophallus
Moisture (g)	74	78.72
Protein (g)	1.5	1.2
Fat (g)	0.2	0.1
Carbohydrates	24.2	18.4
Thiamine (mg)	0.1	0.06
Riboflavin (mg)	0.01	0.07
Niacin (mg)	0.8	0.7
Ascorbic acid (mg)	15	—
Calcium (mg)	12	50
Phosphorus (mg)	35	34
Iron (mg)	0.08	0.6

[a]Per 100 g edible portion
Source: Ref. 77.

amylopectin; the amylose content ranges from 10 to 25%, depending upon the species. There is considerable variation in the starch composition within the tuber.

D. Storage

The most common method of yam storage is to leave the yam in the field and harvest whenever required, but the tuber quality deteriorates after the onset of rains. In some parts of the world, tubers are kept in a shallow pit covered with soil (52). In some cases, they are heaped in a pyramid and covered with moist soil. Yam barns are constructed using vertical wooden frames to which yams are tied. In Oceania, specially constructed huts are used for stacking the yams.

Yams are sensitive to chilling and will be injured at 12°C and below. Low-temperature injury has been observed after 5 weeks at 5 or 7°C, 3 weeks at 3°C, and 5 days at 2°C (53,54). When chilled, tissues discolor, soften with a waterlogged consistency, and eventually decay.

Storage at 16°C with 70–80% relative humidity and adequate ventilation is recommended for cured yams. Curing is accomplished after harvest by holding the tubers at 29–32°C with 90–95% relative humidity for 4–8 days (53,54). Curing allows suberization of surface injuries and reduces weight loss and rotting in storage. This curing can be accomplished either in a controlled room or under tropical ambient conditions. Properly cured yams should keep 6–7 months at 16°C. Cured yams keep longer than noncured ones.

Yam tubers may be infected with *Botrydiplodia* sp. *Penicillium* sp., *Aspergillus* sp., and *Fusarium* sp. during storage, resulting in considerable losses due to rotting. Dipping of tubers in fungicides like Benlate and Captan and in 2500 ppm thiabendazole is effective in controlling storage rots.

E. Processing

In West Africa, yams are processed into yam flour (44). The production of flour is largely restricted to the northern areas where sun-drying is practiced. A similar product known as *Yam Kokonte* is prepared in some parts of Ghana. To prepare flour, tubers are sliced to a thickness of

Figure 8 *Amorphophallus* plant. (Courtesy Prof. A. R. Karnik, Konkan Agricultural University, Dapoli, India.)

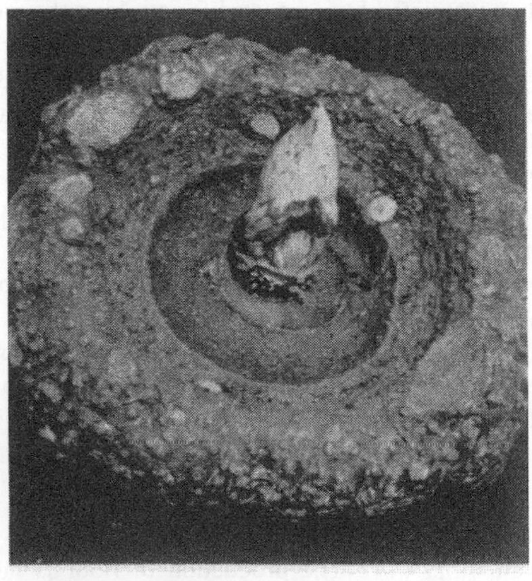

Figure 9 Edible corm of *Amorphophallus*. (Courtesy Prof. A. R. Karnik, Konkan Agricultural University, Dapoli, India.)

about 1 cm, and the slices are peeled and dried in the sun until the moisture content is reduced to 5–10%. In some cases, slices are boiled or parboiled before sun-drying, which softens the tissue considerably and gives more palatable products. The dried pieces are ground in mortars or mills to yield a coarse flour. The raw yams can be sliced and fried. Yams are thus prepared and consumed in various forms, e.g., baked, boiled and fried.

VI. MINOR VEGETABLES

A. Amorphophallus

Amorphophallus or elephant-foot-yam or *Suran* (*Amorphophallus campanulatus* Blume) is a popular vegetable in tropical and subtropical regions. It is commercially grown in India, China, Malaysia, Java, the Philippines, Sri Lanka, and certain Pacific Islands. Its commercial value is greatest in Japan, where it is valued as a food crop.

Amorphophallus belongs to the family of Araceae, having diploid chromosome number 26. The genus *Amorphophallus* consists of 90 species. *A. campanulatus*, *A. oncophyllus*, *A. variabilis* and *A. rivieri* are edible species from India. They are cultivated for their underground modified stem, or corm (56).

Amorphophallus is either a perennial or an annual herb generally bearing one broad, long petiolate leaf. The leaves are simple, the petioles smooth or warty and variously spotted. The blade is simple, 3-parted, and the divisions pinnatifid (Fig. 8). The spathe is funnel or bell shaped at the base, springing from the great bulblike corm in advance of the leaves, the latter, usually pedated, differs from related genera by its botanical characteristics. A flattened edible corm (Fig. 9) is produced at the base of a single compound leaf, enlarging from one year to next if left unharvested. By the end of the first year, the corm can weigh as much as 7 kg. Two cultivars, *Santragachi* and *Kovvur*, are popularly grown in the eastern and southern parts of India.

Amorphophallus needs well-drained and aerated loam or sandy loam soil for successful crop production. It is a tropical and subtropical crop, requiring well-distributed rainfall and humid and warm weather during vegetative growth. It requires reasonably high temperatures with ample moisture during shoot emergence. However, waterlogging is detrimental during crop growth. It is propagated by offsets of corms. These offsets are miniature tubers that grow out of the parent corms and are called buds or daughter corms. Certain types do not produce daughter corms. In this case, the mother corm is cut into pieces and used for planting. The daughter corms are planted in flat beds or ridges. The planting distance varies according to the size of the daughter corms, varying from 50–60 to 90–100 cm. Closer planting increases per hectare yield and suppresses weed growth.

The plant requires a high dose of nitrogen and potassium. The FYM at 30 t/ha is applied during planting. Mohankumar et al. (49) reported removal of 121.9 kg of N, 30.5 kg of P and 176.4 kg of K per hectare from the soil with a yield of 36 t/ha. Mandal and Saraswat (51) found that an economic fertilizer dose was 25 tons of FYM and 40 kg of N, 40 kg of P, and 80 kg of K per hectare. Mohankumar et al. (49) recommended 25 tons of FYM and 80, 60, and 100 kg/ha of N, P, and K, respectively. The application of fertilizers should be done in split doses. A first application at planting with a full dose of P and half-dose of N and K along with FYM encourages rapid development at the early stage. A second application of the remaining N and K should be done 60–70 days after emergence of shoots.

The crop is irrigated first during the emergence of shoots, and subsequent irrigation is required if the soil is dry before the onset of monsoon. When the crop approaches maturity, irrigation should be light and intermittent in the absence of optimum moisture in soil. Waterlogging during the rainy season should be avoided. The crop is harvested after 7–10 months. Yellowing and dropping of leaves are signs of crop maturity. The crop can be harvested before full maturity for early market. It can be stored in ventilated rooms without any damage.

The chemical composition of Amorphophallus is presented in Table 7. It is a good source of carbohydrates, minerals, and vitamins A and B. The main carbohydrate is not metabolized by humans. There is a large variation in the protein content of amorphophallus cultivars. The corms of wild types contain calcium oxalate, making the edible portion highly irritating. Amorphophallus is mainly used as a vegetable in the cooked form. It is also used for preparation of curries and pickles. The plant is employed in certain ayurvedic preparations to treat hemorrhoids (56).

B. Celeriac

Celeriac (*Apium graveotens* L.) is popularly known as celery root (55). It belongs to the same species as celery. However, wild ancestor was bred to produce a root that resembles a turnip. Celeriac is not a true root, but is a stem that has developed with a large mass, measuring up to 15 cm in diameter. It grows half in the ground and half above; a rosette of leaves is produced with a fibrous root system produced below (55).

Cultivation of celeriac is very similar to that of celery, except that plants are not earthened up during growing season. Like celery, it cannot tolerate drought conditions, although it can be grown successfully on a wide range of soil types. Celeriac is usually transplanted, as the seeds are too small to direct-drill. The crop is planted in the spring and is ready to harvest by the middle of September. Roots are usually lifted early in October and stored in dry, cool conditions until required. Roots can remain in the field if the soil is covered with straw in mid-October to provide some winter protection. Roots can then be lifted as required.

The nutritional composition of raw celeriac is presented in Table 8. Celeriac can be boiled and served as a hot vegetable or served cold in salads. It has a celerylike flavor (55).

C. Jerusalem Artichoke

Jerusalem artichoke, also known as girasole, is a herbaceous perennial belonging to the Compositae family. It has edible tuberous roots, which are fleshy and irregular in shape (55). The name artichoke was given to denote a similarity in taste and flavor to the globe artichoke. It is now widely grown in both temperate and subtropical and tropical areas, although it has never been very popular and areas under cultivation are not very large. Only two varieties, Mammoth White French and Sutton's White, are widely grown. Jerusalem artichoke is native to North America, where it grows wild, although it is well adapted to almost any temperate climate.

The edible portions of the Jerusalem artichoke plant are underground tubers similar in shape to white or Irish potato—3–6 cm thick and 7–10 cm long (55). They range in color from white to yellow or from red to blue, depending on the variety.

The crop is grown from sets, which are small tubers or pieces of tuber normally weighing approximately 60 g with at least two or three eye buds. The crop normally reaches maturity in 18–24 weeks, although early-maturing types can be harvested as soon as the leaves begin to die back. The tubers can be lifted by hand or by machine after first cutting back the tops. Tubers are perishable and do not store well.

The chemical composition of boiled Jerusalem artichoke is summarized in Table 8. The tubers are unique in that the major storage carbohydrate is not starch but inulin, a polymer of

Table 8 Nutritional Composition of Celeriac and Jerusalem Artichoke

Constituent	Celeriac	Jerusalem artichoke
Water (g)	88.8	78.01
Protein (g)	1.3	2.0
Fat (g)	0.4	0.01
Carbohydrate (g)	2.3	17.44
Energy value (kJ)	73	318
Dietary fiber (g)	3.7	—
Sodium (mg)	91	—
Potassium (mg)	460	—
Calcium (mg)	40	14
Magnesium (mg)	21	17
Phosphorus (mg)	63	78
Iron (mg)	0.8	3.4
Vitamin A (IU)	26	20
Vitamin B_1 (mg)	0.18	0.2
Vitamin B_2 (mg)	0.02	0.06
Nicotinic acid (mg)	0.2	1.3
Vitamin C (mg)	14	4.0

Source: Ref. 76.

fructose that yields fructose upon hydrolysis. The tubers can be used by diabetics as a carbohydrate source since fructose can serve as a replacement for glucose (55).

Extensive water loss and physical injury can occur due to the thin delicate skin of the Jerusalem artichoke, limiting its storage to a few weeks. Good, sound disease-free tubers, however, can be stored for several months at 0°C and high humidity (57).

Jerusalem artichoke tubers are boiled and eaten in much the same way as potato. Tubers are also grown for stock feeding. The green tops are also used as animal feed. Tubers can be utilized for flour, as a source of fructose, and for the preparation of 5-hydroxymethyl furfurol. They can also be used as a source of industrial alcohol (58).

D. Parsnip

Parsnip (*Pastinaces sativa* L.) is a native plant of Europe and western Asia, and the cultivated form (Subsp. *sativa*) was grown as a food crop by the ancient Greeks (55). Parsnip types are distinguished by the shape of their roots. There are bulbous cultivars, wedge-shaped cultivars, and long, tapering or bayonet cultivars (55). Long, tapering roots are the more traditional cultivars, although wedge and bulbous cultivars are now favored as they are less prone to breakage during mechanical harvesting. Average yields are approximately 20 t/ha. Generally parsnips are grown in the same areas used for carrot production.

Parsnip has traditionally been favored as a winter vegetable, since it is very hardy and resistant to frosts. However, parsnips are now required nearly year round, so that the earliest crops are produced in small quantities from July onwards (55), although the main market for this crop is during the winter months. Frosting of the roots during the winter months improves their sweetness.

Parsnips are grown for the fresh market at a density of 12 plants/m^2, whereas the small roots required for prepacking are grown at a higher density. For early crops, sowing occurs in February, while the main crop is sown from the end of March until the end of April. Parsnip seed has an unusual flat shape and is difficult to drill. In addition, because of its notoriously poor germination (generally no better than 60%), it is one of the few crops for which there is no statutory minimum germination rate (55).

The crop is harvested and stored in unheated barns for 3 weeks prior to marketing. This short-term storage provides stocks if subsequent weather conditions make harvesting difficult.

The chemical composition of raw parsnips is summarized in Table 9. Parsnips are grown for the fresh market, prepackaging, and processing. Small roots (40–65 mm diameter) are required for prepacking, while larger roots are required for the fresh market (40–130 mm diameter). Very large roots are suitable for soup manufacture. A range of plant densities is used to achieve these different root sizes. Parsnips are generally eaten cooked. They are processed for canning, freezing, and diced for incorporation into soups (55).

E. Salsify

Salsify (*Tragopoqon porrifolius* L.) is little known by consumers. It has, therefore, little importance commercially and is only produced by specialized growers. The roots are long and tapering with pale yellowish flesh and long, narrow leaves of a greyish-green color. It is favored as a winter vegetable (55). It is very hardy under moderately temperate conditions and can be

TABLE 9 Nutritional Composition of Salsify and Parsnip

Constituent (per 100 g fr. wt.)	Salsify	Parsnip
Water (g)	83.3	79.3
Total nitrogen (g)	0.22	0.29
Protein (g)	1.2	1.8
Fat (g)	0.3	1.1
Carbohydrate (g)	10.2	12.5
Energy value (kJ)	113	271
Starch	3.5	6.2
Total sugar (g)	1.5	5.7
Dietary fiber (g)	3.2	4.6
Sodium (mg)	110	10
Potassium (mg)	370	450
Calcium (mg)	200	41
Magnesium (mg)	26	23
Phosphorus (mg)	44	74
Iron (mg)	38	0.6
Carotene (μg)	20	30
Thiamine (mg)	0.06	0.23
Riboflavin (mg)	0.01	0.01
Niacin (mg)	0.2	1.0
Vitamin B_6 (mg)	0.07	0.11
Folate (μg)	57	87
Vitamin C (mg)	3	17

Source: Ref. 76.

left in the soil well into the autumn (fall). As with carrots, the seedbed should be deep and friable to ensure straight, smooth roots. Salsify is usually sown at the end of March, and roots can be harvested from the middle of October onwards. They are treated in the same way as parsnips and can be stored in clamps or can be left in the ground to be lifted as required (55). In the spring, the unlifted roots develop young and succulent shoots known as "chards."

The nutritional composition of raw salsify is summarized in Table 9. The stored roots are washed and cleaned and served raw in longitudinal strips, diced, or cubed. The attraction is its distinctive flavor, and it is served raw as a separate dish and dipped in the appropriate dressing and condiments. This vegetable has never gained any degree of popularity in North America, but is enjoyed in Europe (55).

REFERENCES

1. FAO, *Food and Agriculture Organization Production Yearbook*, Rome, 1993.
2. Thompson, H. C., and W. C. Kelly, *Vegetable Crops*, 5th ed., McGraw-Hill, New York, 1957.
3. Watts, R. L., and G. S. Watts, *The Vegetable Growing Business*, Orange Judd Publishing, New York, 1954.
4. Rodriguez, R., B. L. Raina, E. B. Pantastico, and M. B. Bhatti, Quality of raw materials for processing, in *Postharvest Physiology, Handling and Utilization of Tropical and Sub-tropical Fruits and Vegetables* (E. B. Pantastico, ed.), AVI., Westport, CT, 1975, p. 475.
5. Hadley, P., Carrot, parsnip and beet root, in *Encyclopaedia in Food Science, Food Technology and Nutrition* (R. MaCrae, R. K. Robinson, and M. J. Sandler, ed.), Academic Press, London, 1993, p. 4729.
6. Kale, P. N., and S. P. Kale, *Bhajipala Utpadan* (Marathi), Continental Press, Pune, India, 1984.
7. Kale, P. N., and S. D. Masalkar, Agro-techniques for root crops, in *Advance in Horticulture—Vol. 5—Vegetable Crops* (K. L. Chadha and G. Kalloo, ed.), Malhotra Pub. House, New Delhi, 1993, p. 465.
8. Walker, J. C., Internal black spot of garden beet, *Phytopathology 29*:120 (1939).
9. Walker, J. C., J. P. Jolivette, and W. W. Hare, Varietal susceptibility in garden beet to boron deficiency, *Soil Sci. 59*:461 (1945).
10. Ryall, A. L., and W. J. Lipton, *Handling Transportation and Storage of Fruits and Vegetables, Vol. I., Vegetables and Melons*, AVI Pub. Co., Westport, CT, 1972.
11. Crimes, J. R., Proper storage will extend the beet-root supply for processors, in *Handling, Transportation and Storage of Fruits and Vegetables* Vol. I. *Vegetables and Melons*, (Ryall A. L., and W. J. Lipton, eds.), AVI Pub. Westport, CT, 1972.
12. Schouten, S. P., and A. C. R. Van Schaik, The storage of red beets, Proc. Symp. Postharvest Handling of Vegetables, *Acta Hortic. 116*:25 (1981).
13. Tucker, W. C., Observations on the storage performance of individual beetroots, *J. Hort. Sci. 56*:97 (1981).
14. Salunkhe, D. K., and B. B. Desai, *Postharvest Biotechnology of Vegetables*, Vol. II, CRC Press, Boca Raton, FL, 1984, p. 129.
15. Kumar, V., and K. Singh, Agronomy of radish production—a review, *Haryana J. Hort. Sci. 3*(3–4):318 (1974).
16. Hadley, P., Swede, turnip and radish, in *Handbook of Food Science, Food Technology and Nutrition* (R. MaCrae, R. K. Robinson, and M. J. Sadler, eds.), Academic Press, London 1993, p. 4734.
17. Sadhu, M. K., Radish crops, in *Vegetable Crops in India* (Bose T. K., and M. G. Som, eds.), Naya Prakash, Calcutta, 1986, p. 385.
18. Banga, O., Radish, in *Evolution of Crop Plants* (N. W. Simmonds, ed.), Longmans-Green, London, 1974.
19. Simmonds, N. W., ed., *Evolution of Crop Plants*, Longmans-Green, London, 1974.
20. Tamburaj, S., K. G. Shanmugavelu, O. A. A. Pillai, S. Anbu, and C. R. Muthukrishnan, Co-1 radish, a new variety for plains, *South Indian Hort. 29*:152 (1981).

21. Yawalkar, K. S., *Vegetable Crops of India*, 2nd ed., Agril. Horticultural Pub. House, Nagpur, India, 1980, p. 370.
22. Laske, P., *Manuring of Vegetables in Field Cultivation*. Verlagesellsch aft Fur Ackerban, Pub., Hannover, 1962.
23. Maurya, A. N., L. N. Singh, and V. Singh, Nutritional requirement of radish, Proc. Third Int. Symp. Sub-Trop. and Trop. Hort. Crops, Bangalore, 1972, p. 149.
24. Nath, P., *Vegetables for the Tropical Region*, Indian Council of Agricultural Research, New Delhi, 1987.
25. Knott, J. E., and J. R. Deanon, Jr., *Vegetable Production in Southeast Asia*, University of the Philippines, 1967.
26. Zisman, U., and N. Temkin-Gorodeiski, The effect of cultural factors on the keeping quality of red radishes for export, Preliminary Report (No. 771) of the Institute for Technology and Storage of Agricultural Products, Bet, Dagan, Israel, 1977-1978, p. 19.
27. Zisman, U., and N. Temkin-Gorodeiski, Prolonging the storage potential of radishes for export, In Scientific Activities, 1974-1977 of the Institute for Technology and Storage of Agricultural Products Pamphlet No. 184, Bet Dagan, Israel, 1978.
28. Zisman, U., N. Temkin-Gorodeski, S. Grinberg, and A. Daos, Influence of various treatments on the keeping quality of red radishes for export, Preliminary Report (1979/1980), Division of Scientific Publs., Bet Dagan No. 795, 1981.
29. Zisman, U., N. Temkin-Gorodeiski, S. Grinberg, and A. Daos, The effect of CO_2 treatment on the keeping quality of radishes for export, Preliminary Report (1979/1980), Division of Scientific Publg. Bet Dagan No. 798, 1982.
30. Stewart, J. K., and H. M. Couey, Hydro-cooling vegetables—a practical guide to predicting final temperatures and cooling times, U.S. Dept. Agr. Market Res. Report 637, 1963, p. 32.
31. Lipton, W. J., Market quality of radishes stored in low atmosphere, *J. Am. Soc. Hort. Sci.* 164 (1972).
32. Hardenburg, R. E., A. E. Watada, and C. Y. Wang, *The Commercial Storage of Fruits, Vegetables and Florist and Nursery Stock*, U.S. Department of Agriculture, ARS, Agriculture Handbook No. 66, 1986, p. 69.
33. Pantastico, E. B., T. K. Chattopadhyay, and H. Subrahmanyam, Storage and commercial storage operations, in *Postharvest Physiology, Handling and Utilization of Tropical and Sub-tropical Fruits and Vegetables* (E. B. Pantastico, ed.), AVI Publishing, Westport, CT, 1975, p. 314.
34. Esch, H. G., and A. Van, Cool radishes against leaf yellowing, *Greenten en Fruit 35*:37 (1980).
35. Parsons, C. S., and R. H. Day, Freezing injury of root crops; Beets, carrots, parsnips, radishes and turnips, *US Dept. Agric. Marketing Res. Rep.*, 1970, p. 866.
36. Segall, R. H., and J. J. Smoot, Bacterial black spot of radish, *Phytopathology 52*:970 (1962).
37. Parthasarathy, V. A., Taro, in *Vegetable Crops in India* (T. K. Bose and M. G. Som, ed.), Naya Prokash, Calcutta 1986, p. 72.
38. Onwueme, I. C., *The Tropical Tuber Crops*, John Wiley and Son, New York, 1978, p. 199.
39. Plucknett, D. L., R. S. dela Pena, and F. Obrero, Tuber crops, *Field Crops Abst. 23*:413 (1970).
40. The Central Tuber Crops Research Institute, Research Highlights—1981, Sreekariyam, Trivandrum India, Tech. Bull. No. 1, 1982.
41. Ezumah, H. C., and D. L. Plucknell, Effect of irrigation on yield of taro, *J. Root Sci. 7*:41 (1981).
42. Hodge, W. H., The dasheen: A tropical crop for the south, *U.S. Dep. Agric. Circu.*, 1950.
43. Praquin, J. Y., and J. C. Miche, Trials on storage of taros and macabos in cameroon; Preliminary Report, Dschang Station, Institute de Recherches Agronomique Tropicals Ouagadougou Upper Votta, 1971.
44. O'Hair, S. K., and D. N. Maynard, Edible aroids, in *Encyclopaedia of Food Science, Food Technology and Nutrition* (R. MaCrae, R. K. Robinson, and M. J. Sadler, eds.), Academic Press, London, 1993, p. 4755.
45. Bose, T. K., and T. Maharana, Yam, in *Vegetable Crops in India* (T. K. Bose and M. G. Som, eds.), Naya Prokash, Calcutta, 1986, p. 740.
46. Maynard, D. N., and S. K. O'Hair, Root crops of lowlands—vegetables of tropical climate, in

Encyclopaedia of Food Sci. Food Technology and Nutrition (R. MaCrae, R. K. Robinson, and M. J. Sandler, eds.), Academic Press, 1993 p. 4753.
47. Martin, F. W., Dioscorea, Memo. Fed. Expt. Sta. Mayagunez, Puerto Rico, 1974.
48. Janick, J., Tropical root and tuber crops, *Hort. Rev. 12*:157 (1990).
49. Mohankumar, B., S. Kabeeranthumma, and P. G. Nair, Effect of NPK on yield of tuber crops, *Indian Farming 33*:35 (1984).
50. Nair, G. M., C. S. Ravindran, and P. Mangal, Effect of spacing and NPK on yield of Amorphophallus, *Indian Farming 33*:29 (1984).
51. Mandal, R. C., and V. N. Saraswat, Manurial requirement of sweet yam in laterite soils of Kerala, *Indian Agric. 12*:25 (1968).
52. Been, B. O., C. Perkins, and A. K. Thompson, Yam: Curing for storage, *Acta Hort. 62*:311 (1977).
53. Coursey, D. G., Low temperature injury in yams, *J. Food Technol. 3*:143 (1968).
54. Gonzalez, M. A., and A. C. de Rivera, Storage of fresh yam (*Dioscorea alata* L.) under controlled conditions, *J. Agric. Univ. Puerto Rico. 56*:46 (1972).
55. Hadley, P., and R. Fordham, Vegetables of temperate climatc, in *Encyclopaedia of Food Science, Food Technology and Nutrition* (R. MacRae, R. K. Robinson, and M. J. Sadler, eds.), Academic Press, London, 1993, p. 4737.
56. Sen, H., and N. Roychoudhury, Amorphophallus, in *Vegetable Crops in India* (T. K. Bose and M. G. Som, eds.), Naya Prokash Culcutta, 1986, p. 733.
57. Salama, S. B., M. Saad, and I. Rawhija, Studies on artichoke storage, *Agric. Res. Rev.* (Cairo) *40*(3):56 (1962).
58. Sachs, R. M., C. B. Low, A. Vasavada, M. J. Sully, L. A. Willis, and G. C. Ziobro, Fuel alcohol from Jerusalem artichoke, *Calif. Agric. 35*(9/10):4 (1981).

7

Tomato

D. L. Madhavi
University of Illinois, Urbana, Illinois

D. K. Salunkhe
Utah State University, Logan, Utah

I. INTRODUCTION

Tomato is one of the most popular and widely grown vegetable crops in the world. It is also one of the most commonly grown vegetables in the home garden. The estimated worldwide annual production in 1994 was about 77.5 million metric tons (1) (Table 1). Tomato is a very versatile vegetable for culinary purposes. Ripe tomato fruit is consumed fresh and utilized in the manufacture of a range of processed products such as puree, paste, powder, ketchup, sauce, soup, and canned whole fruits. The unripe green fruits are used for pickles and preserves and are consumed after cooking (2). Tomatoes are important sources of lycopene and vitamin C and are valued for their color and flavor.

II. BOTANY

A. Anatomy and Morphology

Tomato belongs to the genus *Lycopersicon* of the family *Solanaceae.* Bailey (3) classified tomatoes as belonging to two species: *L. pimpinellifolium* and *L. esculentum.* The latter species is the parent of commercial tomatoes, and practically all belong to the variety Commune. Other varieties included in this species are Grandifolium, the large-leafed or potato-leafed type, Validum, the upright or dwarf type, with dense, dark green foliage; Cerasiforme, the cherrylike type, with normally two-celled globular fruits and standard foliage; and Pyriforme, with pearlike or oblong fruits and standard foliage. In addition to these two species, Taylor (4) recognized several more species of tomato such as *L. cheesemanii*, *L. peruvianum*, *L. hirsutum*, and *L. glandulosum* (Table 2).

The genus consists of annual or short-lived perennial herbaceous plants. The stems of

TABLE 1 World Production of Tomatoes

Continent/Country	Production (1000 MT)
World	77,540
Continents	
Asia	27,430
Europe	15,537
North Central America	14,874
Africa	8,315
South America	5,335
Oceania	433
Australia	365
Countries	
United States	12,085
China	8,935
Turkey	6,300
Italy	5,259
India	5,059
Egypt	4,600
Spain	3,066
Brazil	2,550
Mexico	1,560
Chile	1,151

Source: Ref. 1.

young seedlings are round, soft, brittle, and hairy, but gradually become angular, hardy, and almost woody when reaching maturity. Three different vine types occur in tomato: indeterminate, determinate, and semi-determinate (5). With the indeterminate types, the primary shoot dominates the side shoot development, resulting in a sprawling growth pattern. The primary shoot continues to grow as long as the plants remain healthy and growing conditions are suitable. In the determinate types, the primary shoot terminates in a flower cluster, forcing side shoots to develop. The branches terminate their growth at approximately the same distance from the crown, resulting in a compact and symmetrically circular growth. The semi-determinate types have vine characteristics that are intermediate between the other two.

Tomato fruits vary in shape from a flat round, to a true round, square round, oblong, pear, or oxheart, and many variations in between. They can be divided into the skin, pericarp, and locular contents. The fruit is a two- to many-celled berry with fleshy placenta containing numerous small kidney-shaped seeds covered with short stiff hairs. The seeds are surrounded by jellylike parenchyma cells, which fill the locular cavities. The color of mature fruits may be red, pink, orange, yellow, or white. Most commercial cultivars are red or pink.

B. Cultivars

Numerous tomato cultivars have been developed both for commercial growers and home gardeners. The cultivars developed for fresh market and processing tomatoes have distinct

TABLE 2 *Lycopersicon* Species

Subgroup	Characteristics
esculentum complex	Hybridize readily
esculentum	Normal tomato
esculentum var. *cerasiforme* (cherry tomato)	Adapted to wet, tropical conditions
cheesemanii	Yellow to orange fruit, diameter 6–9 mm Source of jointless gene, aid to machine harvesting
hirsutum and its forms	Green fruits with purplish stripes, diameter 1.5–2.5 cm Source of cold tolerance, and resistance to several insect pests, root knot nematodes, some fungi and bacteria
parviflorum and *chmielewski*	Yellow-green fruit, diameter 1–1.4 cm, *chmielewski* is a source of high solids
pennellii	Small, green fruit Initially classified in the genus *Solanum* Resistant to drought and some sucking pests
pimpinellifolium (currant tomato)	Resistant to *Fusarium* and bacterial speck
peruvianum complex	Do not hybridize readily with *esculentum* complex
chilense	Green fruit, ciameter 1–2 cm Source of the Tm2a gene Resistant to the tobacco mosaic virus
peruvianum	Green fruit, diameter 1–3 cm Very diverse species that includes coastal and mountain races Potentially valuable source of genes for resistance to many fungal and viral diseases and insect pests

Source: Refs. 4 and 6.

characteristics. Fresh market cultivars include both determinate and indeterminate types. The fruits should be firm, well colored, with an acceptable flavor. Tomatoes for shipment should have fruits that are smooth and firm enough to withstand transportation. Processing cultivars have well-colored, firm fruits and plants with enough foliage to shade the fruits. Machine-harvested fruits should have small determinate vines, concentrated fruit set, and uniform ripening. The incorporation of the jointless character in several modern processing cultivars enables the stems to separate cleanly from the fruit, minimizing puncture damage to the fruits during harvest and handling (5).

Tomato cultivars range widely in fruit shape, size, color, plant type, disease resistance, ripening time, and processing characteristics. Most of the modern cultivars incorporate multiple resistance to many pathogenic diseases and pests (6). Earliana, Bonny Best, Marglobe, Stone, and Pink are some of the important groups of tomato cultivars mentioned by earlier workers.

The varieties of Earliana are characterized by early ripening and bright red fruits. Bonney Best has medium-sized, round to oblate, deep red fruits with a high incidence of cracking and scalding. The varieties of Marglobe are tolerant to *Fusarium* wilt, heavy yielding, with bright red, medium to large fruits. They are considered good for canning in the United States. Stone is grown extensively for canning and has scarlet-red, flattened, and firm fruits. The fruits of Pink are large, globose or heart-shaped to irregular in shape, purplish pink to pinkish red with few seeds, low acidity, and have a pleasant flavor. Table 3 lists the characteristics of some of the tomato cultivars developed in the United States. Several tomato cultivars have been developed in India. These include Pusa Ruby, Pusa Early Dwarf, HS 101, HS 102, HS 10, Sel. 7 (Hissar Arun), Sel. 18 (Hissar Lalima), NRT 8 (Hissar Lalit), HS 4 (Hissar Anmol), CO 3, KS 2, Punjab Chhuhara, Roma, SL 120, Pant T-3, Arka Vikas, and Arka Saurabh (7).

C. Growth and Development

Tomato fruit setting is associated with moderate vegetative growth and a proper balance between nitrogen and carbohydrate level in the plant. When nitrogen supply is abundant, vegetative growth is rapid with a reduction in the concentration of carbohydrates, and plants do not fruit even with abundant flowering (8). Fruit setting in tomato depends on the accumulation of a considerable surplus of carbohydrates above the actual needs of the plant for vegetative growth (2). Another critical factor affecting fruit set is the night temperature. The optimum temperature range is 15–20°C, and fruit setting is not observed at temperatures of 12.8°C or below (9,10). Moore and Thomas (11) observed that high light intensity accompanied by high temperature was harmful to fruit set. Plant growth regulators have been successfully used to increase tomato fruit set both in the greenhouse and in the field. The use of 75–100 ppm of *O*-chlorphenoxyacetic acid and 50–100 ppm of β-naphthoxyacetic acid have been found to increase fruit set significantly (11–15).

Tomatoes take 6–7 weeks from flowering, depending on temperature, to reach full size. Cell division continues for about 2 weeks after flowering, but the bulk of the increase in fruit size is the result of cell expansion. In normal cultivars, the first appearance of red or pink color at the blossom end signals the completion of growth and onset of ripening. Laboratory studies with fruit harvested at mature green stage have shown that ripening actually begins about 2 days before the external color changes. An early indication of ripening is a small increase in ethylene production, which can be measured with a sensitive gas chromatograph (6).

Tomato is a climacteric fruit in which ripening is accompanied by an increase in both respiration and ethylene production (Fig. 1). Ethylene plays a significant role in both the initiation of early biochemical events in ripening and the integration of subsequent changes. The disappearance of starch, degradation of chlorophyll, synthesis of lycopene, flavor components, and polygalacturonase, a cell wall–hydrolyzing enzyme, are highly integrated with the changes in respiration and ethylene production (6). The physiological and biochemical changes associated with ripening have been extensively reviewed by Baker (15), Salunkhe and Desai (2), and Hobson and Grierson (16).

III. PRODUCTION

A. Soil and Climate

Tomatoes can be grown on a wide variety of soils ranging from light-textured sandy or sandy loam to heavier clay soils. Like other vegetable crops, tomato also grows better and yields more

Table 3 Characteristics of Some Tomato Cultivars

Cultivar	Chief use	Season	Plant type	Diseases resistant to	Fruits	
					Size	Shape
Better Boy F_1	Home	Medium	I	FW, VW, N	Medium-large	Globe
Burpee's Big Boy F_1	Market	Medium	I	VF	Medium	Round
Campbell 1327	Market	Medium	D	VW, FW, cracking	Large	Flat, globe
Dombito F_1	Greenhouse	Medium early	I	FW, cracking	Large	Round
Earlypak 707	Shipping, canning	Medium early	D	VW, FW	Medium	Globe
Empire F_1	Market, shipping	Medium	D	VW, FW	Large	Oblate
Glamour	Market, home	Medium early	SI	Cracking	Medium	Flat, globe
Heinz 722	Processing	Medium	D	VW, FW	Small	Pear
Heinz 1350, 1439	Market, home	Early	D	VW, FW, cracking	Medium	Flat, globe
Heinz 2653	Processing	Early	D	VW, FW	Medium	Blocky
Jet Star F_1	Market	Early	I	VW, FW	Large	Flat, globe
Jubilee	Market	Medium	I	—	Large	Globe
Jumbo F_1	Greenhouse	Medium	I	VW, FW	Large	Globe
Morton Hybrid F_1	Market	Early	I	—	Medium	Flat, globe
Mountain Price F_1	Market, shipping	Medium late	D	VW, FW	Large	Globe
New Yorker	Market	Early	D	VW	Medium	Globe
Pik Red F_1	Market, shipping	Early	D	VW, FW, N	Large	Globe
Red Cherry	Home	Early	I		Small	Globe
Roma VF	Processing	Early	I	FW, VW	Small	Pear
Rutgers 39	Market	Medium late	I	VW, FW, cracking	Medium	Flat, globe
Spring Set F_1	Market	Early	D	VF	Large	Round
Supersonic F_1	Market	Medium	I	VW, FW, cracking	Large	Flat, globe
Traveler 76	Shipping	Medium early	I	FW	Large	Flat, globe
VF 134–1–2	Processing	Early	D	FW, VW	Small	Long, globe
UC 82	Processing	Medium early	D	FW, VW	Medium	Blocky
UC 204	Processing	Early	D	FW, VW	Small	Blocky

D, Determinate; I, indeterminate; SI, semi-determinate; FW, *Fusarium* wilt; VW, *Verticillium* wilt; N, nematodes.
Source: Ref. 5.

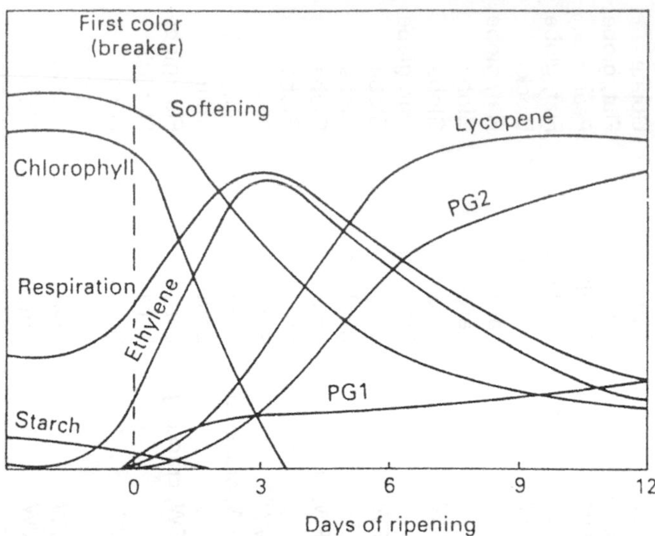

FIGURE 1 Changes in the metabolism and composition during ripening of tomato fruit. PG, polygalacturonase. (From Ref. 6.)

when grown in a rich sandy loam or loamy soil. The soil should be rich in nutrients and organic matter. The ideal soil pH should be near neutral but never below 6.0 or above 7.0.

Tomato is a warm-season crop reasonably resistant to heat and drought and grows under a wide range of climatic conditions. It thrives well when the weather is clear and rather dry, and the temperatures are uniformly moderate in the range of 18–30°C. The crop cannot tolerate low temperatures and is susceptible to frost injury. Varieties vary in their response to frost. The foliage has been shown to be less susceptible to frost damage than the fruits. The fruits are also susceptible to drought and high temperatures, which decrease fruit set and yield. The ripening process needs clear warm days without rainfall.

B. Propagation

Seeds are used for the commercial propagation of tomato. Tomato seeds are dull white in color with minute hairs on the surface of the seed coat, and the seeds count about 30,000/100 g. Under ideal conditions, the seeds remain viable for about 3–4 years, and up to 90% germination is expected. Presowing treatment of tomato seeds with gibberellic acid, 3-indolepropionic acid, β-naphthoxyacetic acid, 2,4-dichlorophenoxyacetic acid, or *p*-chlorophenoxyacetic acid results in quicker germination and healthy seedlings (17).

C. Cultural Practices

1. Planting

Tomato seedlings are usually raised in nursery beds and transplanted to the field. Transplanting is preferred when expensive hybrid seeds are used, for early plantings, and when growing fresh market tomatoes. About 500 g of seeds would provide enough seedlings for one hectare of land. The seeds are sown by broadcasting on well-prepared seed bed lightly covered with soil. The seeds may also be sown in shallow rows. The nursery bed should be irrigated immediately after sowing, and the tender seedlings must be protected from direct sun, heavy rains, and frost.

Watering of the seed bed should be regular until the seedlings are 5–7 cm tall, after which they are irrigated based on need. Excessive irrigation in the seedling stage promotes extraordinary stem elongation, which in turn affects the ability of seedlings to withstand transplanting. The seedlings become ready for transplanting in 3–4 weeks. If the seedlings are allowed to remain in the nursery bed for longer periods, they do not establish well after transplanting and the growth will be relatively slower. Tomatoes are also started by direct seeding in the field, especially when the crop is to be harvested mechanically. Processing tomatoes are direct seeded in many areas.

The best time for planting tomato crops depends largely upon geographic and climatic conditions, transplanting or direct seeding, and mechanical or hand-harvesting. Of these, temperature stability seems to be the most important factor. Tomatoes grow best with average monthly temperatures of 21–24°C, but can be grown with average temperaturs as low as 18°C. The distance of transplanting varies with the cultivar, season, and inherent soil fertility. Plants are subjected to severe shock when transplanted even under the most favorable conditions. When possible, transplanting should be done either during the afternoon or on a still cloudy day. Before removing the plants from the bed, the soil in which they are growing should be thoroughly watered and the plants should have a moisture covering to prevent excess wilting (18). Preplanting treatment with growth regulators has been shown to be beneficial. Several chemicals have been tried to determine their influence on growth and fruit yield (17).

2. Irrigation

Irrigation is employed to maintain soil moisture, and waterlogging should be avoided at all times during the growth of the crop. The frequency of irrigation depends upon soil type, season, and variety. During the rainy season, watering is not necessary. Wherever there is water stagnation, additional drainage should be provided. A winter crop may need watering once every 2–3 weeks, whereas a summer crop may need it once a week. Irrigating the plants during periods of frost helps to maintain the temperature above freezing. Too heavy watering after a long dry spell without prior light irrigation is harmful, as it results in fruit cracking. Similarly, irrigating late in the growing season results in watery fruits, which are poor in quality (19).

3. Manures and Fertilizers

Nutrition plays a major role in increasing productivity of the plants and quality of the fruits. Most of the research work has shown that tomato is a heavy feeder of nitrogen, phosphorus, and potassium. A crop yielding about 40 tons of fruits removes about 93 kg nitrogen, 20 kg phosphorus, and 126 kg potash from one hectare (19). The amount of fertilizer to be applied usually depends on soil fertility, season, and the cultivar (20–25). Some fertilizer recommendations include 40–60 kg nitrogen, 60–80 kg phosphorus, and 100–120 kg potassium per hectare (19); 100 kg nitrogen, 80 kg phosphorus, and 50 kg potassium with 25 tons of farmyard manure per hectare (25).

The entire dose of farmyard manure, phosphorus, and potassium are applied before transplanting. The farmyard manure is mixed before final ploughing, while phosphorus and potassium are applied on both the sides of the rows and mixed with the top 8–10 cm of soil. Nitrogenous fertilizers are applied in two equal split doses during the winter or three equal split doses during the rainy season. The first dose is given before transplanting, the second after one and a half months, and the third at flowering. Nitrogenous fertilizers can also be applied through foliar spray.

Boron is another element that requires greater attention because its deficiency causes cracking of fruits (8). Boron deficiency also affects the formation and utilization of different carbohydrates. This problem is more complex in soils rich in calcium. Other symptoms of boron deficiency in fruits include pitted and corky areas, deformed shape, malformation, and uneven ripening of fruits. The problem can be corrected by spraying 0.3–0.4% borax or soil application of borax (8–12 kg/hectare). Zinc is required for the formation of ascorbic acid in the fruits, but higher levels of zinc may cause a reduction in carbohydrates (8).

4. Training and Pruning

The indeterminate vine type requires support. The plants may be trained on stakes or trellises, or suspended on twine from overhead wires. In general, a considerable amount of labor is required to prune and train the plants. Indeterminate cultivars for the fresh market are used in commercial greenhouses, and yields of 250 tons/hectare have been reported in a full growing season in the United Kingdom. In Europe, some greenhouse tomatoes are grown hydroponically. This system enables precise control of the mineral composition and concentration in the medium (6).

Most of the tomatoes commercially grown outdoors have a determinate habit. The plants have a compact bush habit and set fruit over a brief period of about 1 month. Determinate fresh market tomatoes may be grown as a ground crop in dry climates or may be staked or trellised in regions with relatively high rainfall. Production on stakes or trellises facilitates spraying to control foliar diseases and pests. Tomatoes for processing are universally grown as a ground cover crop on raised beds. Irrigation by trickle is well suited to outdoor tomato production. Pruning is generally done by pinching off the lateral branches as they appear in the leaf axils. The flowering and fruiting are significantly influenced by spacing and pruning (6,26).

D. Disorders, Diseases, and Pests

Blotchy ripening or greenback is one of the major physiological disorders in tomato fruits (27). The affected fruits show areas of yellow or orange discoloration on the surface intermixed with normal fruit color. Internally the parenchyma surrounding the vascular bundles of the outer fruit walls becomes necrotic and disorganized, apparently from a stress reaction (28,29). The affected tissues may either be opaque or brownish in color and are lignified or starchy. Hobson (29) further observed that some biochemical features of the green mature fruits are retained in the blotchy tissue during ripening. Blotchy ripening can de induced by preharvest nutritional status of the crop. Potassium deficiency or excess nitrogen may bring about more blotchy fruits (30,31). Venter (32) was able to produce the symptoms of greenback by localized heating of the fruit. The intensity of greenback depended on length of exposure and degree of temperature. Lipton (33) observed that defective coloration of tomato shoulders could be due to short-wave radiation and proposed the term "solar yellowing" to indicate that the disorder is a solar radiation effect. The short-wave radiation causes a reduction in carotenoid synthesis in blotchy fruits and inhibits ethylene production in fruits affected by greenback (34).

Spurr (35) described blossom-end rot disorder as brown proteinaceous inclusions occurring in the epidermis and pericarp at the stylar end of the fruit. The cell membranes become disorganized and tissue necrosis develops underneath the skin (36,37). The causal factor for blossom-end rot could be calcium deficiency, and water stress may aggravate the symptoms (27,35,38). Cutical cracks or skin cracks are shallow, slightly dark but well-healed cracks on the fruit surface. Since the waxy covering is removed, water loss is rapid, resulting in shriveling and discoloration (39). The disorder fruit tumor, also called waxy blister, results in waxlike, irregular tumor on the fruit surface, which becomes brown, depressed, and cracks as the fruit matures (39,40). Such blisters may be induced by rubbing green tomatoes and storing at 21–35°C. According to Treshow (41), rubbing induces the synthesis of more growth regulators with kininlike activity, causing increased cell division and tumorus growth. Puffiness disorder is generally caused by poor pollination, resulting in poor development of seed bearing tissue. The affected fruit is hollow and light in weight. The puffy tomatoes are downgraded or are rendered unmarketable in serious cases (27).

Growth cracks are characterized by the rupturing or cracking of the surface of the fruit and often result in large losses (8). Two kinds of cracks are generally noticed: one radiates from the stem and the other develops concentrically around the shoulder of the fruit. Growth cracks affect the appearance of the fruit and become sites for the development of infections. Frazier (42) suggested that abundant rainfall and high temperatures favor rapid growth and predispose the fruits to growth cracks. The specific cause of rupturing could be an uncoordinated tissue expansion during growth or simply a turgidity phenomenon.

The tomato crop is susceptible to various microbial infections, which may cause low yields and poor quality fruits. The fungal diseases include anthracnose (*Colletotrichum phomoides*), *Fusarium* wilt (*Fusarium oxysporum*), *Verticillium* wilt (*Verticillium* sp.), early blight (*Alternaria solani*), late blight (*Phytophthora infestans*), and root rot (*Phythium* sp., *Phytophthora parasitica*). The bacterial diseases include bacterial canker (*Cornybacterium michiganense*), bacterial wilt (*Pseudomonas solanacearum*), bacterial spot (*Xanthomonas vesicatoria*), and bacterial speck (*Pseudomonas tomato*). Tobacco mosaic, cucumber mosaic, leaf-roll, bunchy top, and spotted wilt are some of the viruses affecting tomato crops.

The insect pests affecting tomato crops include fruitworm, hornworm, cutworm, flea beetles, cabbage loopers, aphids, leaf miners, white fly, and caterpillars. Nematodes cause a malformation of the roots called root knot, resulting in small swellings to large galls.

E. Harvesting

Tomatoes are harvested at a variety of stages of ripeness, from mature green to light pink, when they are easily separated from the vine by a half turn or twist. The stage of maturity at which tomatoes are picked depends upon the purpose for which they are grown and the distance they are to be transported. The various stages recognized are immature-green, mature-green, breaker, pink, full-ripe, and overripe. For long-distance transporting, fresh market tomatoes are picked at the mature-green or breaker stage, whereas for roadside sales, the fruits are picked at a more mature breaker stage onward. Processing tomatoes are harvested when fully mature in order to optimize various quality parameters such as color, flavor, texture, acidity, and total solids.

Commercially, most of the fresh market tomatoes are harvested by hand. Some harvest machines have been developed, but they have not been widely adopted because of excessive fruit damage and limited cost savings. Mechanical harvest aids are more widely used. These machines transport the pickers along the rows and convey the harvested fruits into bulk bins. Most of the processing tomatoes are harvested mechanically in developed countries. These machines cut the bushes slightly below ground level, lift and shake the plants on a conveyor, separate the fruits from the vines, and deliver the fruits to a side trailer. The fruits are sorted electronically by color, and a sorting crew is also present to discard fruits for other defects. The sorted fruits are transported to the packing house in standard 2-foot-deep bins. Mature-green tomatoes handle well with less bruising in bins than in small containers. Packing house operations include cleaning, grading, waxing, wrapping, and packing. Many packing houses also have ripening rooms where immature tomatoes are exposed to ethylene gas (6). Figure 2 presents the postharvest handling system for fresh market tomatoes.

Since tomato ripening has been closely associated with ethylene, exogenous application of the ethylene-releasing compound ethephon has been commercially used to accelerate ripening and to control the percentage of red-ripe fruits. Users have indicated that the mature-green fruit

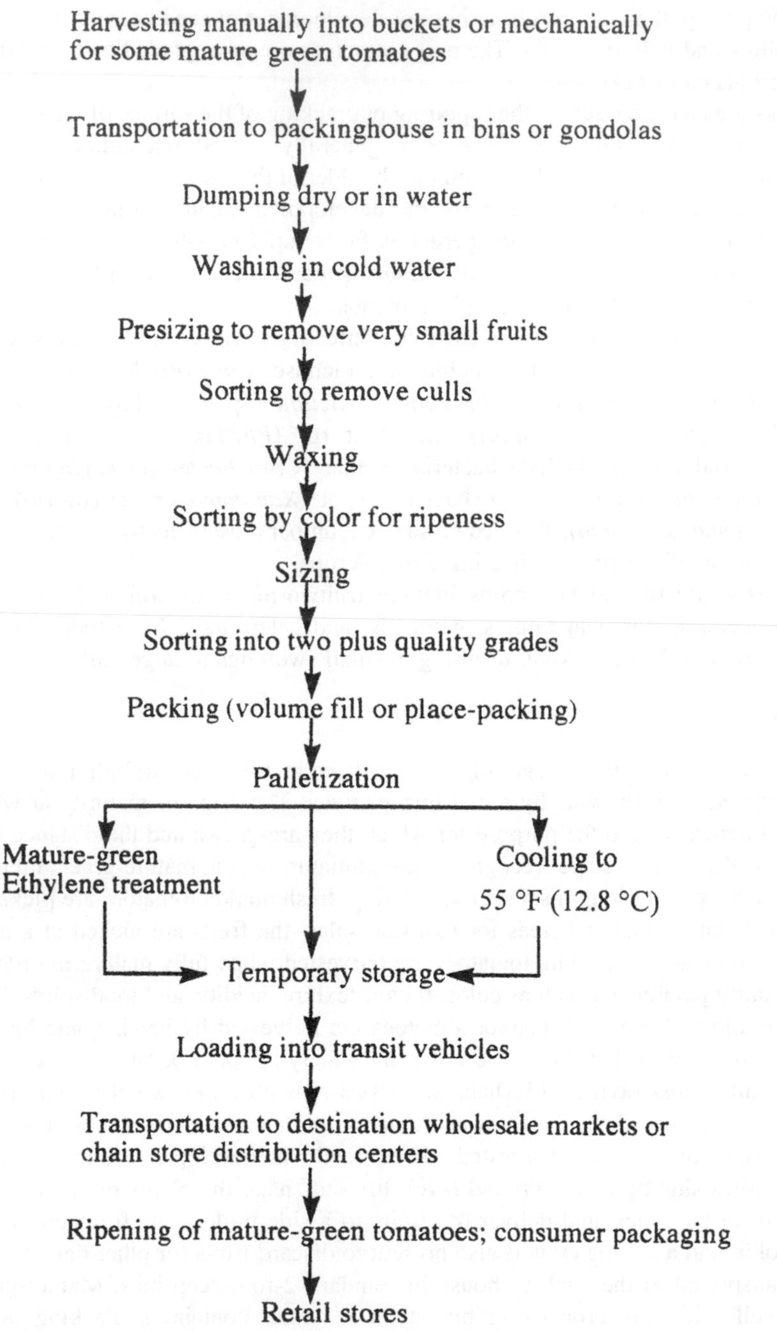

FIGURE 2 Handling system for fresh market tomatoes. (From Ref. 43.)

will turn red in 7–18 days after proper application of ethephon (18). Salunkhe and Wu (43) reviewed the effects of ethylene-releasing compounds and concluded that ethephon had the following advantages: reduction in sorting cost due to uniformity of ripening; reduction in weight loss due to fast ripening rates and actual prolongation of the shelf life of the fruits; reduction in ripening room requirements; increase in yield from once-over harvest; and hastening the maturity of tomatoes early in the season to obtain marketable fruits at premium prices.

F. Quality Improvement

Tomatoes used either fresh or in processing must have distinct quality characteristics. Fresh tomatoes must have acceptable flavor, color, texture, and taste parameters to satisfy consumer demands and handling requirements. Processing tomatoes, on the other hand, must have intrinsic rheological characteristics that make them suitable for various processing applications, such as juice, ketchup, or sauce production (44). Numerous tomato varieties have been developed to meet some of these requirements by breeding methods. However, breeding methods that have been traditionally used are slow and may not impart the desired characteristics. In recent years, genetic engineering methods have been successfully applied to develop varieties with quality characteristics that could not be gained by conventional breeding methods.

In tomatoes, the onset of fruit ripening results in the initiation of various biochemical and physiological processes that affect the quality of the fruit. Some of these changes include modification in the structure and composition of cell walls affecting fruit firmness, metabolism of sugars and acids involved in flavor determination, biosynthesis and deposition of carotenoids determining fruit color, and synthesis of hormones responsible for the rate of ripening (44). In the past 10 years, considerable advances have been made in the understanding of fruit ripening at a molecular level, which resulted in the isolation and identification of a number of genes expressed exclusively in ripening/ripe tomatoes. Table 4 lists some of the identified genes and the fruit characteristics affected by them. Of the various genes listed, genes encoding the enzymes polygalacturonase (PG) and pectinesterase (PE) have been utilized for targeted single-gene modification studies. These enzymes determine the texture, a major paramenter of fresh and processed tomatoes, by affecting the metabolism of pectin in the cell wall. PG hydrolyzes the α-1,4 linkages in the polygalacturonic acid component reducing the chain length, while PE modifies the degree of esterification of pectin. The method involves introduction of whole or part of the gene in an antisense orientation leading to an inhibition of the expression of the gene. Genetically modified tomatoes with reduced levels (1%, 10% PG and 10% PE activity) of these cell wall–modifiying enzymes have been successfully developed (44).

The quality charcteristics of these modified tomatoes have been determined in depth. The average pectin chain length in low-PG fruits was found to be considerably greater than in unmodified fruits (45). The low-PG tomatoes (92% PG reduction) are reported to be firmer than unmodified tomatoes, have better home-grown flavor, last longer on the vines, can ripen more fully before being picked, and are less prone to spoilage on and off the vine, leading to better

TABLE 4 Genes Isolated from Tomatoes and Affected Fruit Characteristics

Fruit Characteristics	Determined by	Gene
Viscosity	Cell wall structure	Polygalacturonase Pectinesterase
Handling characteristics	Cell wall structure	Polygalacturonase Pectinesterase
	Rate of ripening	Ethylene synthase Ethylene oxidase
Soluble solids	Sugars	Invertase
	Pectins	Polygalacturonase Pectinesterase
Color	Lycopene	Phytoene synthase
Taste	Sugars/Acids balance	Invertase

Source: Ref. 44.

postharvest handling (44, 46–48). They ripen without getting overly soft and have enhanced shelf life (2–7 days more) compared to the unmodified tomatoes. It has also been demonstrated that low-PG tomatoes have significantly better processing characteristics compared to unmodified tomatoes (45,49). Bostwick viscosity, a measure of paste yield potential, was increased by more than 80% in low-PG tomatoes. Bostwick viscosity is influenced largely by insoluble cell wall polymers. The increase in viscosity is therefore attributable to the changes in the average molecular weights of cell wall pectins in low-PG tomatoes. These results suggest that a yield increase during the manufacture of tomato paste should be possible using low-PG tomatoes (44).

The low-PE tomatoes have also been assessed. The degree of esterification in the ripe low-PE fruits was found to be comparable to that of control green tomatoes. The major effect of reducing PE was an increase in serum viscosity, largely due to different methylation status of the soluble pectin in the fruit. The increase in serum viscosity enhanced the glossy appearance of the tomato paste (44). Tomato lines in which both low-PG and low-PE traits were combined have also been developed and assessed. These lines showed the quality characteristics of the parental lines, e.g., increases in Bostwick and serum viscosities. In addition, significant increases in soluble solids content were also observed (44). Table 5 presents the potential broad-range benefits of low-PG and low-PE tomatoes.

Based on these studies, Calgene Fresh, a California-based biotechnological company recently developed an improved quality low-PG fresh market tomato named Flavr Savr™. This tomato contains an antisense version of the PG gene as well as a bacterial marker gene that confers resistance to the antibiotic kanamycin. The presence of this gene attracted opposition to Flavr-Savr™ from critics of genetic engineering and consumer-activist groups, who claimed that the gene might jeopardize the use of kanamycin as an antibacterial or might transfer antibiotic resistance to gut-living bacteria (50–52). After years of scrutiny, the U.S. Food and Drug Administration approved the Flavr-Savr™ tomato on May 18, 1994, thus making it the world's

TABLE 5 Potential Benefits of Use of Low-PG and Low-PE Tomatoes

Benefit	Recipient			
	Farmer	Processor	Consumer	Environment
Reduced chemical input	X			X
Reduced water usage on farm	X			X
Reduced production cost	X			
Reduced harvest cost	X			
Reduced harvest waste	X	X		
Reduced transport waste	X	X		
Reduced transport cost		X		
Reduced water usage in processing plant		X		X
Reduced waste water		X		X
Increased yield		X	X	
Reduced product cost		X	X	
Improved flavor		X	X	
Differentiated texture and flavor		X	X	
Improved range of products		X	X	

Source: Ref. 44.

TABLE 6 Representative Chemical Composition of Fresh Market Tomatoes

Constituents	Content (per 100 g edible portion)
Energy (kJ)	56
Gross constituents (g)	
Water	94.7
Protein	1.0
Fat	0.1
Dietary fiber	1.6
Carbohydrates (g)	
Glucose	0.9
Fructose	1.0
Sucrose	0
Starch	0
Organic acids (g)	
Citric	0.43
Malic	0.08
Oxalic	0
Other	0
Vitamins (mg)	
Vitamin C	18
Thiamine	0.04
Riboflavin	0.02
Nicotinic acid	0.7
β-Carotene (equivalent)	0.34
Minerals (mg)	
Potassium	200
Sodium	6
Calcium	8
Magnesium	10
Iron	0.3
Zinc	0.2

Source: Ref. 6.

first genetically engineered whole food to reach U.S. consumers. The Flavr-Savr™ tomato is currently being sold in a few areas in California and Illinois (53).

IV. CHEMICAL COMPOSITION

The chemical composition of tomato fruits depends upon factors such as cultivars, maturity, and environmental and cultural conditions. The chemical constituents are important in assessing the quality in respect to color, texture, appearance, nutritional value, taste, and flavor of the fruits. Table 6 presents the chemical composition for a fresh-market cultivar representative of the modern determinate types grown outdoors. Salunkhe et al. (54) have reviewed the compositional changes associated with tomato ripening. Some of the compositional changes associated with ripening of tomato are presented in Table 7.

The soluble solids in tomatoes are predominantly sugars, which are important contributors

TABLE 7 Some Compositional Changes of Tomato Fruit[a] Associated with Ripening

Composition[b]	Stage of maturity				
	Large green	Breaker	Pink	Red	Red-ripe
Dry matter (%)	6.4	6.2	5.81	5.8	6.2
Titratable acidity (%)	0.285	0.31	0.295	0.27	0.285
Organic acids (%)	0.058	0.127	0.144	0.166	0.194
Ascorbic acid (mg%)	14.5	17	21	23	22
Chlorophyll (mg%)	45	25	9	0	0
β-Carotene (mg%)	50	242	443	10	0
Lycopene (mg%)	8	124	230	374	412
Reducing sugars (%)	2.4	2.9	3.1	3.45	3.65
Pectins (%)	2.34	2.2	1.9	1.74	1.62
Starch (%)	0.61	0.14	0.136	0.18	0.07
Volatiles (ppb)	17	17.9	22.3	24.6	31.2
Volatile reducing substances (μeq.%)	248	290	251	278	400
Amino acids (μmol%)	—	2358	3259	2941	2723
Protein nitrogen (mg N/g dry wt.)	9.44	10	10.27	10.27	6.94

[a]Fireball cultivar, except V. R. Moscow cultivar for amino acid contents.
[b]Expressed on the basis of fresh weight unless specified.
Source: Ref. 54.

to flavor. In general, the flavor of the fruit becomes pronounced when its sugar content peaks. The free sugars, representing more than 60% of the solids in tomatoes, are mainly D-glucose and D-fructose, with traces of sucrose, a ketoheptose, and raffinose. Some species other than *L. esculentum* are known to accumulate a significant proportion of sucrose. *L. chmielewski* has been crossed with *L. esculentum* to produce high-sugar breeding lines. The sugar content increases uniformly from small and green-mature to large and red-pipe tomatoes (6,55–57). Tomato fruit accumulates low levels of starch in the immature stages. Yu et al. (58) reported that the starch accumulation continues up to the large green stage and then rapidly decreases as ripening begins.

The lipid fraction of the tomatoes is composed of triglycerides, sterols, sterol esters, free fatty acids, and hydrocarbons. Stigmasterol, β-sitosterol, α-amyrin, and β-amyrin are the sterols found in tomato fruit. Kapp (59) reported 33 saturated and unsaturated fatty acids in the pericarp of all the varieties tested. Linoleic, linolenic, oleic, stearic, palmitic, and myristic acids comprised the major portion of the fatty acid fraction and increased during the period of greatest color development. Ueda et al. (60) found considerable amounts of total lipids in green tomato fruits on plants and lesser amounts in fruits harvested at the breaker stage. According to Jadhav et al. (61), contents of linoleic and linolenic acids decreased with advancing maturity of the fruits. The unsaponifiable fraction of the tomato skin contains hydrocarbons and sterols, while the alkali soluble fraction contains long-chain organic acids (62,63).

Yu et al. (58) reported that the total nitrogen content gradually decreased to a minimum at the large green stage of maturity and then increased steadily up to the red stage, followed by a rather gradual decline at the red-ripe stage. Nonprotein nitrogen increased with advancing maturity of the fruit. The decrease in protein content was correlated to increased production of volatile components. The ripe tomatoes are reported to contain 20 amino acids in addition to small amounts of tryptamine, 5-hydroxy tryptamine, and tyramine (54). The concentration of individual amino acids varies with different stages of maturity. Significant increases in glutamic acid and aspartic acid and a reduction in the levels of alanine, arginine, leucine, and valine with ripening has been reported (58,64–66). Certain amino acids also serve as precursors of volatile components in tomato fruit. Glutamic acid, aspartic acid, γ-aminobutyric acid, and glutamine comprise about 80% of the free amino acids and contribute to the taste of the tomato fruit (6,54).

Although minerals represent a small fraction of the dry matter of fruit, they play an important role in the nutritional composition of the fruits. Because of the large quanities consumed, tomatoes are important sources of potassium in the diet. The mineral content in general increases during growth and maturation of the tomato fruits.

Pectin contributes significantly to the textural characteristics of the fruit. The pectin content and composition depends on the degree of maturation of the fruits. Dalal et al. (55) showed that protopectin increased up to the large green stage and then progressively decreased. Increases in the soluble pectins during maturation coincided with a progressive softening of the fruit (67).

Tomato fruit is a good source of ascorbic acid. On the basis of fresh weight, vitamin C content averages about 25 mg/100 g (68); however, the values vary with cultivars. Increases in ascorbic acid content during maturation have been reported by Dalal et al. (55), Fryer et al. (69), and Brown and Moser (70). Tomato cultivars ripening at a faster rate were shown to contain higher amounts of vitamin C as compared to those that ripened at a relatively slower rate (71). Fruits exposed to direct sunlight contain a higher concentration of ascorbic acid than those ripening in the shade. Hobson and Davis (72) have reviewed the effects of sunlight on ascorbic acid content in tomato fruits. Tomato also contains folic acid, pantothenic acid, biotin, and vitamin K, in addition to nicotinic acid, riboflavin, and thiamine.

Citric acid and malic acid are the organic acids that contribute to the typical taste of tomato fruit. Other acids such as acetic, formic, trans-aconitic, lactic, fumaric, galacturonic, and α-oxo acids have been detected. The locular juices contain higher concentrations of organic acids than the outer locular walls. Blended fruit tissue has a pH of 4.0–4.7. As the fruit ripens from mature-green to red, acidity increases to a maximum value and then decreases. Maximum acidity was found in the breaker and pink stages (55,57,73).

Presence of flavonoids and other phenolics has been reported in tomato fruits. Wu and Burrel (74) isolated naringenin and quercetin from the skin. *p*-Coumaric, caffeic, ferulic, and chlorogenic acids are the other phenolics reported in the fruit. Walker (75) found increasing concentrations of these compounds except *p*-coumaric acid during ripening.

Tomato fruits contain tomatine (Fig. 3), a glycosidic steroidal alkaloid. Traces of solanine were also found. According to Kajderowicz-Jasosinska (76), the largest amount of tomatine, 0.087%, was found in green tomatoes. In yellowish fruits, the level was reduced to nearly half. Red-ripe tomatoes lost all their tomatine when left on the plants for 2–3 days.

Color is perhaps the most important and reliable index of tomato maturity. Chlorophyll *a* and *b* are the major green pigments of tomato fruit until the mature green stage, while the typical red color of ripe tomatoes is mainly due to lycopene. During ripening, the chloroplasts are transformed into chromoplasts. Chlorophyll disappears completely within about 4 days at 20°C, and lycopene synthesis reaches a maximum by about 6 days. As destruction of chlorophyll progressed during ripening, different shades of color such as green-yellow, yellow-orange with some traces of green, orange-yellow, orange-red, and red develop in sequence (6,54).

Mature-green tomatoes are reported to contain only α-and β-carotenes (77). β-Carotene and lycopene contribute 7 and 87% respectively of the carotenoids in the normal red tomato fruit (78). Curl (79) isolated 22 xanthophylls from the fruit, of which the major proportion consisted of lutein, violaxanthin, and neoxanthin. Carotenoids consist of predominantly carotenes. Trombly and Porter (80) listed 19 carotenes from the red tomatoes. Presence of colorless carotenoids

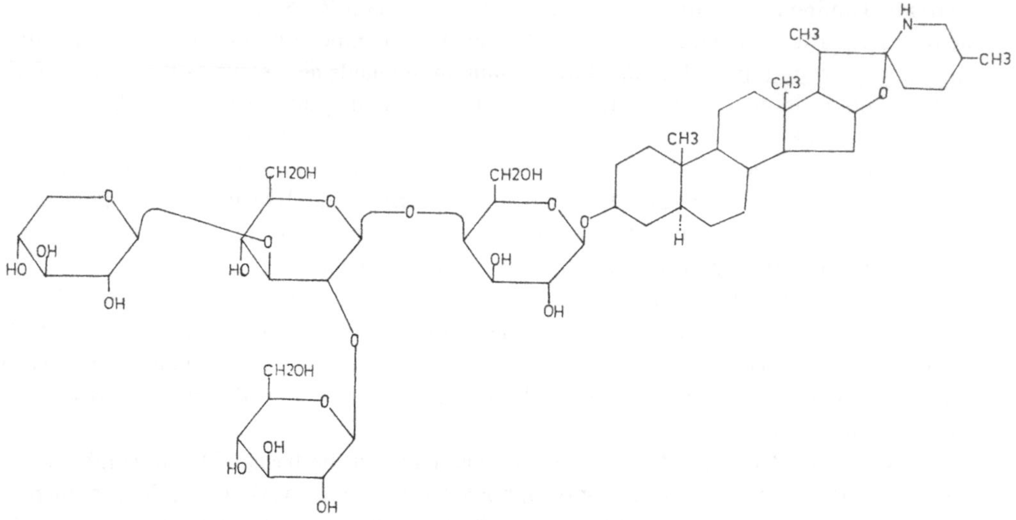

FIGURE 3 Structure of tomatine. (From Ref. 54.)

phytene and phytofluene has also been reported (54). McCollum (81) showed that the color of red tomatoes depended upon the total carotenoids as well as the ratio of the dominant pigments lycopene (red color) to β-carotene (yellow color). Pigment distribution studies have indicated that the concentration of total carotenoids was highest in the outer pericarp, while β-carotene concentration was greatest in the locular region (81). Changes in the carotenoid composition with maturity has been reported. Dalal et al. (55,82) found very rapid decreases in β-carotene content after the pink stage of ripeness. In addition to a large increase in lycopene, other carotenoids such as phytoene, phytofluene, γ-carotene, and ζ-carotene increased during ripening (54). On the basis of structural similarity and high correlation between high-boiling volatiles (terpenoids) of tomato, Stevens (83) hypothesized the production of such volatiles from the oxidation of polyene carotenes.

Numerous volatile or aroma components have been reported in tomato fruits (54,84). More than 200 volatile compounds have been reported in ripe fruits, which include carbonyls, alcohols, esters, acids, hydrocarbons, nitrogen compounds, lactones, acetals, ketals, sulfur compounds, ethers, and chlorine compounds. Of these, alcohols, aldehydes, carbonyls, and sulfur compounds may have a significant influence on flavor quality. Quantitative and qualitative differences in the volatile components have been reported with different cultivars, cultural conditions, degree of maturity, time of harvest, artificial ripening, storage conditions, and processing. Figure 4 represents the degradation of various chemical constituents to volatiles.

The green tomato fruits accumulate low levels of volatile compounds. Yu et al. (85–87) reported that crude enzyme preparations from green tomatoes synthesized short-chain carbonyls when alanine, leucine, and valine were used as substrates. At a later stage of maturation, the enzyme preparations were active with a number of amino acids. This shows that as the fruit ripens, more intricate enzyme systems become operative for synthesizing volatile compounds. Dalal et al. (88) observed that isopentanal and hexanol increased with maturation up to the breaker and large green stage, respectively. The concentration of all volatiles except isopentenal and hexanol increased during ripening.

Shah et al. (89) evaluated volatiles of field-ripened, artificially ripened, and overripened tomatoes (Fig. 5). They concluded that the typical aroma of the field-ripened tomato is due to carbonyls (32%), short-chain (C_3-C_6) alcohols (10%), hydrocarbons, long-chain alcohols, and esters (58%). In field-ripened tomatoes, the concentrations of nonanal, decanal, dodecanal, neral, benzaldehyde, citronellyl propionate, citronellyl butyrate, geranyl acetate, and geranyl butyrate (peaks 21,26,35,34,25,57,58,59,60 in Fig. 5b respectively) were higher compared to artificially ripened tomatoes. In artificially ripened tomatoes, the concentration of butanol, 3-pentanol, 2-methyl-3-hexanol, isopentanal, 2,3-butanedione, propyl acetate, and isopentyl acetate (Peaks 10, 13, 14, 1, 30, 6, 18 in Fig. 5a, respectively) were higher compared to the field-ripened tomatoes. Overripened fruits showed an increase in the concentration of isopentyl butyrate, citronellyl butyrate, and geranyl butyrate (Peaks 18, 58, 60 in Fig. 5c, respectively) compared to field-ripened fruits. Shah et al. (89) proposed that major contributions of long-chain carbonyls and terpene esters were essential for the ripe tomato aroma. Kader et al. (90) reported that tomatoes picked at earlier stages of ripeness and ripened at 20°C were evaluated by panelists as being less sweet, more sour, less tomatolike, and possessing more off-flavor than those picked at the table-ripe stage. Hayase et al. (91) identified 130 compounds using GC-MS in field-grown tomatoes at various stages of ripening and artificially ripened fruits. The concentrations of hexanal, *trans*-2-hexenal, 2-iso-butylthiazole, 2-methyl-2-hepten-6-one, geranylacetone, and farnesylacetone, which were estimated to be important volatile components of fresh tomato aroma by the GC-Sniff method, increased with natural and artificial ripening. However, many volatile compounds showed complicated changes in the case of artificially ripened tomatoes.

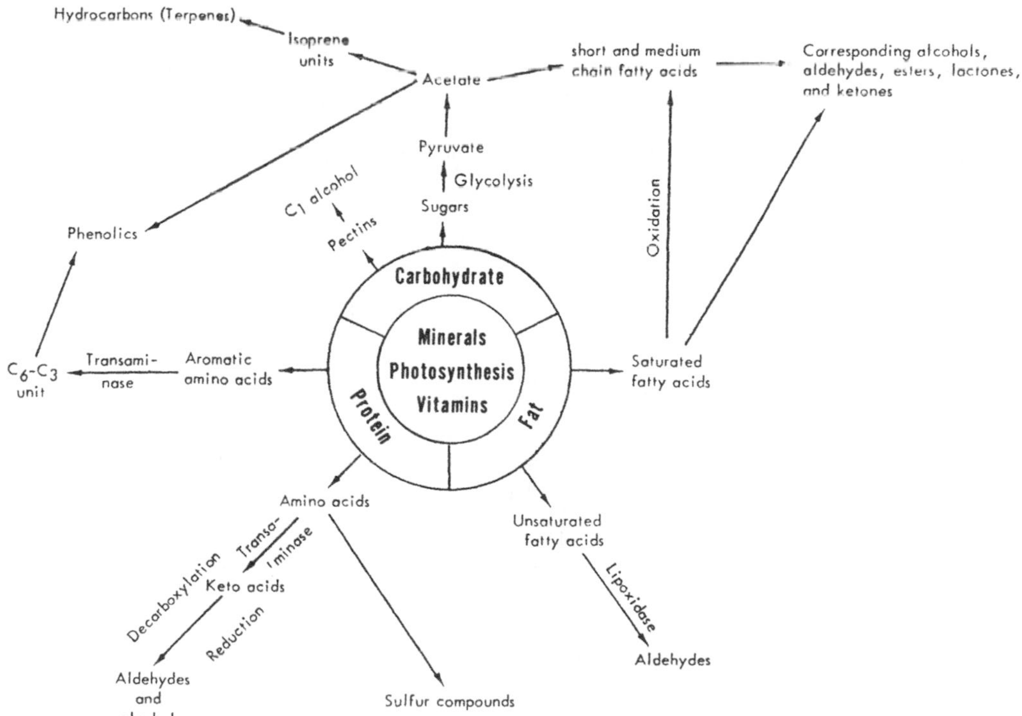

FIGURE 4 Degradation of chemical constituents to volatiles. (From Ref. 54.)

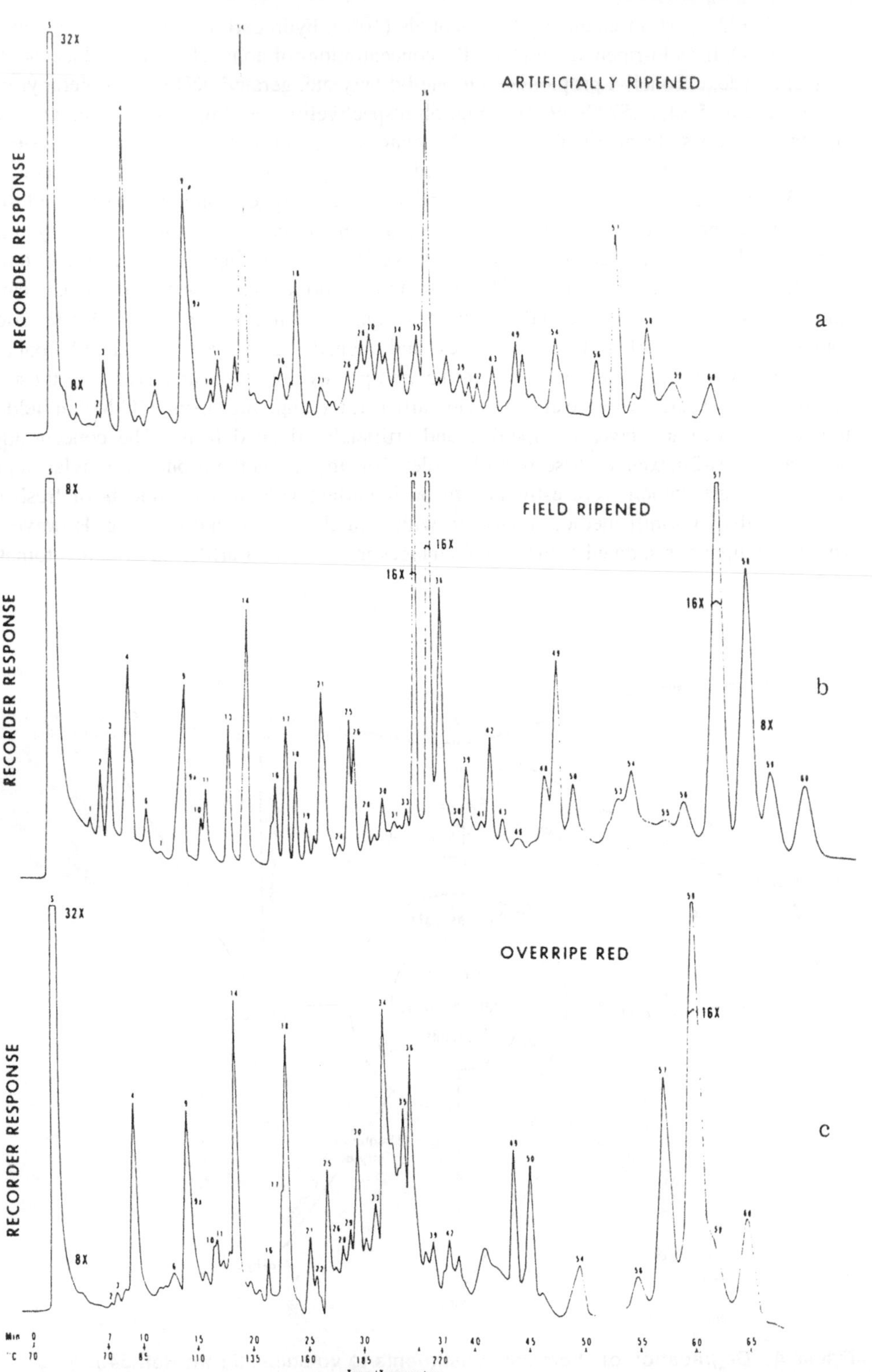

FIGURE 5 Gas chromatograms of volatiles from artificially ripened, field ripened, and overripened tomatoes. (From Ref. 89.)

V. STORAGE

A. Low-Temperature Storage

Since tomatoes are harvested at various stages of ripeness, the recommended storage temperature varies with the degree of ripeness of the fruit. Ryall and Lipton (92) stated that the need for precooling of the mature-green tomatoes depended on their initial temperature and on the timetable of ripening. Rapid precooling is necessary only if the fruit temperatures are above 26.7°C and ripening is to be delayed. Kasmire (93) reported that cherry tomatoes could be cooled from 32.2 to 15.6°C within about 3–5 minutes using hydrocooling with water temperatures of 1.1–4.4°C. Where hydrocooling is not feasible, cooling with an air flow below 4.4°C is not harmful when the fruits are exposed to such temperatures for no longer than 24 hours (94). Holding of tomatoes below 10°C for more than 24 hours must be avioded because various symptoms of chilling injury may seriously affect the market quality of the fruit. Fruits at breaker or turning stage are less sensitive to chilling injury than mature-green fruits. Pink tomatoes are even less sensitive to chilling injury than the turners and can be held at 4.4°C for 4 days without evident injury. Fully ripened red fruits are held between 1.7 and 4.4°C until ready for use. Such holding should be limited to a few days, because ripe tomatoes not only lose color and soften even at low temperatures, they also lose their characteristic aroma and flavor (95). In general, 12°C and 86–90% RH are optimal for tomato storage. In case of most shipments, temperatures of 10–12.8°C are preferable if delay in ripening is not essential (2).

B. Controlled Atmosphere Storage

The effects of controlled atmosphere (CA) storage on the shelf life and quality of tomatoes has been well documented. Parsons et al. (96) found that 3% O_2 with the remainder N_2 was the most promising atmosphere to prolong the shelf life of tomatoes at 12.8°C. Mature-green tomatoes held under these conditions colored normally when transferred to air at 18.3°C. Fewer than 5% of the fruits decayed compared to 90% decay of tomatoes held in air. The tomatoes had an acceptable flavor. Salunkhe and Wu (97) reported that low O_2 atmospheres of 10, 3, and 1% O_2 (the rest being N_2) extended the storage life of Green-Wrap tomatoes for 62, 76, and 87 days, respectively, at 12.8°C. Low O_2 atmospheres (1% or 3%) inhibited the degradation of chlorophyll and starch and synthesis of lycopene, β-carotene, and soluble sugars in the fruits.

Controlled atmosphere containing CO_2 has also been found beneficial in retarding ripening and color change during storage. Hatton et al. (98) noted that breaker and pink tomatoes stored in 4–8% O_2 and 1–2% CO_2 at 12.8°C had a prolonged storage life due to retarded ripening. An atmosphere below 4% O_2 and above 4% CO_2 was found to cause uneven ripening of the fruits. Dennis et al. (99) investigated the use of CAs (3% O_2 + 5% CO_2 + 92% N_2 or 5% O_2 + 5% CO_2 + 90% N_2) on greenhouse-grown and outdoor-grown tomatoes held at 13°C, 93–95% RH for 6–10 weeks. The fruits ripened more uniformly on transfer to air at 20°C than the fruit stored in air. The fruits stored in 5% O_2 and 5% CO_2 had better flavor than fruits stored in 3% O_2 and 5% CO_2. In another study, Bhowmik and Pan (100) stored mature-green tomatoes in 2.5% O_2, 5% CO_2, and N_2 as the balance gas at 85 or 98% RH and 12°C for a period of ≤ 8 weeks. After 40 days, the fruits were allowed to ripen in air at 23°C. The weight loss was significantly reduced at 98% RH. The tomatoes remained green for up to 40 days of storage and changed color gradually during consecutive storage in atmospheric conditions. The appearance of tomatoes was acceptable after ripening.

C. Subatmospheric (Hypobaric) Pressure Storage

Subatmospheric or hypobaric storage has the same sort of effects as conventional CA storage. In subatmospheric storage, however, ethylene and other gases produced by the respiring fruits are removed by a continuous evacuation of air, which delays ripening and extends the shelf life of the produce (101). Wu et al. (102) demonstrated that it was possible to store tomatoes up to 100 days using hypobaric storage at 102 mmHg if the fruits were subsequently transferred to 646 mmHg at 12.8°C and 90–95% RH. The ripening of Green Wrap tomatoes was retarded and the storage life was extended by hypobaric storage. The control fruits ripened in 35 days at 12.8°C. They ripened in 65 and 87 days when stored at 471 and 278 mmHg, respectively, but did not ripen or deteriorate if stored at 102 mmHg beyond 100 days. Tomatoes previously stored at 102 mmHg for 100 days ripened normally in 7 days when transferred to 646 mmHg at 12.8°C and 90–95% RH. The subatmospheric pressure delayed the loss of chlorophyll and inhibited the synthesis of lycopene, β-carotene, and starch degradation.

Tolle (103) found that pressures from 180 to 190 mmHg retarded the ripening of tomatoes due to a reduction in the partial pressure of oxygen. He hypothesized that the retardation of ripening and color development of mature-green tomatoes may be more directly related to the oxygen levels than to the levels of total pressure and reduction in ethylene. Burg (104) observed that the hypobaric method was useful in the shipment or the storage of breaker tomatoes. These fruits could be kept for over 4 weeks under vacuum conditions without advance in ripening beyond a light pink state. This provides an opportunity to ship breaker tomatoes having superior eating quality without excessive ripening while in transit. The normal refrigerated trailers are not satisfactory for this purpose because the fruit becomes so soft in transit resulting in a substantial spoilage followed by further losses during subsequent distribution.

D. Irradiation

Salunkhe (105) reported the effects of γ-radiation on the storage and nutritional quality of tomatoes. The ripening process in green tomatoes was slowed by radiation. The threshold dose was in the vicinity of 1.86×10^5 rad. Lycopene was more susceptible to irradiation than other carotenoids. In another study using ripe tomatoes, the fruits along with small packets of activated charcoal were sealed in cans and the cans were irradiated at 1, 2, and 4×10^3 rad and stored at 21.1°C; they were opened 5 days after radiation and evaluated for quality. The quality of irradiated tomatoes with charcoal was somewhat better, indicating that activated charcoal might have absorbed the volatile substances and other harmful gases. Salunkhe (105) recommended the inclusion of a package of activated charcoal in cans during the process of irradiation and immediately after irradiation. Such cans can be opened to permit normal metabolism of the product provided the product is packaged in a semipermeable film permitting normal respiration of the produce while at the same time prohibiting the entry of microbes. In another study, Larrigaudiere et al. (106) reported that γ-irradiation of early climacteric cherry tomatoes caused a sharp burst of ethylene production during the first hour. The extent of ethylene production was dose dependent and was maximum at about 3 kGy. Application of γ-irradiation to mechanically wounded tissues inhibited the continuous increase of ethylene production, but the initial burst of ethylene caused by the wounding was not inhibited. The investigators reported that γ-irradiation inhibited the activity of ethylene-forming enzymes at doses of >1 kGy.

E. Packaging

The packaging of tomatoes in polyethylene bags provides modified atmospheres and consequently reduces fruit decay, softening, and loss of soluble solids during storage. Reductions in transpiration and respiration losses increase the shelf life and retain quality (54). Ben-Yehoshua et al. (107) investigated the effects of individual seal-packaging of tomato fruits in a 10-mm-thick film of high-density polyethylene on the development of fruit bemishes. Various blemishes and chilling injury of tomatoes were markedly inhibited due to seal-packaging. The beneficial effects of the packaging were related to acceleration of healing of small wounds by the saturated humidity around the fruit and inhibition of red blotch due to prevention of dehydration of the peel. Rawnsley (108) reported that the postharvest losses could be reduced from 15 to 3% by using "upright cone" baskets together with dry grass as a packaging material between layers of fruit in place of "inverted cone" baskets.

F. Postharvest Disease Control

Storage at low-temperature, CA storage, chemical treatments to retard ripening, and application of fungicides are reported to reduce postharvest diseases and decay of tomatoes. Parsons and Spalding (109) observed that tomatoes held in 3% O_2 and 5% CO_2 had smaller soft-rot lesions than fruits stored in air. Fuchs et al. (110) reported that postharvest treatment with gibberellic acid inhibited ethylene production in mature-green tomatoes. Application of calcium and other divalent cations also retarded ripening (111). Addition of Purafil®, a chemisorbent of ethylene (made of aluminum oxide and $KMnO_4$), has been reported to delay the ripening of tomatoes in storage (112).

Chemicals such as Captan®, Dithane®, *o*-phenylphenate, Thiram®, and Benomyl® have the greatest potential as fungicide treatments in tomatoes (113). The blossom-end rot of the tomato can be markedly reduced by the application of $Ca(NO_3)$ and gypsum or by spraying with $CaCl_2$ solution. Field spraying of Ziram®, Maneb®, Captan®, Dyrene®, and Phaltan®, at 3 lb/100 gal water, or Dithane Z-78® (0.2%) effectively controlled the advent of fruit rot in the field and prevented the spread of fungi that caused latent infection leading to anthracnose or ripe-rot (114,115). Among the chemicals effective as dips are 1% Chlorox, 6 oz of Borax in 1 gal water, sodium polysulfide (1 gal/150 gal water), and formaldehyde (116). A wetting agent such as soap and detergent ensures a uniform covering. Aureofungin at 100 and 200 ppm is effective against *Alternaria* (117). Most of these dip treatments leave no residues after washing the fruits in water. Aulakh and Grover (118) found that coating with Mobil® and cottonseed oil at 50 and 75% completely inhibited tomato rotting. Barkai-Golan and Ben-Yehoshua (119) reported that postharvest dipping of tomatoes in sodium hypochlorite was effective immediately after removal of the fruits from storage; a hot (40°C) solution was superior to a cold one. Dipping the fruits in hot water at the same temperature without the addition of any disinfectant was also effective in reducing fruit rot. The investigators noted that dipping in 5% sodium dimethyldithiocarbamate solution was most effective and decreased rotting to 18% as compared to 30% in untreated fruit.

VI. PROCESSING

A. Canned Tomato Products

Tomatoes for canning should be medium to large in size, smooth, of uniform ripeness, with a firm red flesh and good flavor. Fruits of irregular shape or with wrinkled skins are difficult to peel and result in excessive loss in preparation. Those with large seed cavities soften badly in the can, giving an unattractive appearance, and the can appears slackly filled. The important canning cultivars of tomato are Stone, Pearson Moran, Ohio 7870, and San Marzano (2). In addition to whole peeled tomatoes, diced, wedged, and sliced tomatoes have gained considerable acceptance. Stewed tomatoes and tomatoes canned with other vegetables have also become popular. Gould (18) has described the following process for canning tomatoes.

At the processing plant, the tomatoes are washed and sorted to remove off-color and diseased specimens from firm ripe fruits. Tomatoes with a small amount of rot or green area may often be trimmed to remove the defective part. The cores, if present, are removed by hand or machine. The peeling of tomatoes is a high-cost operation in terms of labor. When coupled with coring it accounts for approximately 60% of the labor cost for processing tomatoes. Three major tomato-peeling methods are currently used: Steam peeling, where the fruits are scalded in live steam (98–100°C) long enough to loosen the skin (≈ 30–60 sec) followed by cold water spray to crack the skin; lye peeling, wherein the fruits are immersed in hot (88–93°C) 16–20% caustic soda for 20–30 sec, which selectively dissolves the cuticular tissue (the broken-down outer tissue, soft and gelatinous, is easily washed away with a water spray, leaving an intact inner tissue); and exposure to infrared radiation for 4–20 seconds, when the epidermal cells are exposed to high temperatures of 816–982°C. The water contained in the epidermal cells is changed from liquid to steam and the peel breaks away from the fruit, to be removed by a spray of cold water. Recently cryogenic scalding has been introduced as a peeling process, wherein liquid nitrogen is used to freeze the skin in a few seconds. The fruits are rapidly thawed, leaving them in a loose-fitting skin, which can be easily removed. The whole peeled tomatoes are machine-filled in cans and covered with brine (2% sugar and 1% salt) or 1% salt, and juice from whole tomatoes or from peels and cores are added to fill the cans. Calcium salts (not to exceed 0.08%) are also added to maintain firmness of the canned product. The pH is adjusted to 4.1–4.3 by the addition of citric acid to control flat-sour spoilage of the canned product. The open cans are exhausted to obtain sufficient vacuum before sealing. The cans are then sterilized for 35–100 minutes based on can size with the center temperatures of cans at 82–88°C to ensure commercial sterility.

B. Juice

In manufacturing tomato juice and puree and other products, the tomatoes are washed and sorted similar to preparation for canning. The fruits are chopped to 0.4–0.6 in. for crushing prior to juice extraction. Immediately following chopping they may be subjected to hot-break or cold-break procedures. In the hot-break method, the chopped or crushed tomatoes are rapidly heated to at least 82°C for 15 seconds to inactivate the pectic enzymes. Heat treatment is usually given in rotary coil tanks followed by a heat exchanger and holding tube to achieve 104°C to retain at least 90% of the potential serum viscosity in the original fresh tomato. The hot-break produces a better-quality juice with respect to cooked tomato flavor and body. A heavier-bodied more homogeneous juice is obtained because of the inactivation of pectic enzymes and efficient of pectin. In the cold-break procedure, tomatoes are scalded to loosen the skin before chopping. The fruits are chopped or crushed at temperatures less than 66°C and then fall into a holding tank, where they remain static for periods ranging from seconds to minutes. During this holding period, the pectic enzymes liberated during crushing can catalyze the breakdown of pectins. The cold-break procedure is claimed to give better-colored, better-flavored juice and a better retention of vitamin C. Quick processing of the extracted juice is necessary to produce high-quality juice from the cold-break procedure (18).

Following either the hot-break or the cold-break process, the chopped tomatoes are conveyed to a cyclone for juice extraction. Seventy to eighty percent of the juice is extracted commercially. The juice is deaerated immediately after extraction to prevent loss of vitamin C, acidified with citric acid to enhance the flavor, salted if necessary, and filled in cans or bottles. The fill machines are adjusted to give maximum fill to exclude as much headspace as possible.

The cans are closed at about 82–88°C followed by water cooling, or the juice is presterilized at 121°C for about 0.7 minutes and the hot juice is poured into cans to ensure sterilization of the containers. A minimum closing temperature of 93°C is suggested, after which the cans are held inverted and conveyed for a minimum of 3 minutes at this temperature prior to water cooling (18).

C. Puree and Paste

Tomato puree and paste are prepared from one or any combination of the liquid obtained from mature tomatoes of red or reddish varieties; liquid obtained from the residue consisting of the peels and cores after preparing such tomatoes for canning; and liquid obtained from the residue after partial extraction of juice from such tomatoes. For the preparation of puree, the liquid is separated from seeds, skin, cores, etc. and concentrated in tanks with rotary steam coils or in vacuum pans. The liquid is evaporated to less than or equal to one-half its volume to obtain a puree of specific gravity 1.035–1.05. The puree is then filled into cans at 88°C, sealed, and cooled carefully to prevent loss of flavor, color, and stack burning. Tomato paste is prepared in a similar manner with the addition of salt, spices, flavorings, and sometimes baking soda (18).

D. Ketchup

Tomato ketchup can be made directly from fresh cyclone juice or from concentrated pulp or bulk-stored tomato paste. The other constituents of ketchup are sugar, salt, vinegar, onions, and spices. The ingredients are cooked for 30–45 minutes in steam kettles or tanks with steam coils. The thickness of the bottled ketchup is an important part of its quality. Part of the thickness of the ketchup is due to the pectin from the tomatoes. Some manufacturers prefer a hot-break before the tomatoes are cycloned in order to retain the largest possible amount of pectin. The hot-break method dissolves some of the mucilaginous material from the seeds, again contributing to the final consistency. The ketchup is bottled, deaerated, and sealed at a temperature of 82–88°C and cooled to prevent loss of flavor and stack burning. The total soluble solids content should be 25–29 for Grade C, 29–33 for Grade B, and over 33 for Grade A (18).

VII. MEDICINAL PROPERTIES

Tomato extract has been used to treat various diseases in traditional medicine in different countries such as Japan, Greece, Peru, and Guatemala. Hot water extract of dried fruits has been used in the treatment of ulcers, wounds, hemorrhoids, and burns (120,121). Tomato poultice has been used in the treatment of edema during pregnancy (122). The fresh fruits are reported to be effective as digestive aids and in the treatment of kidney and liver problems (123). In recent years, several epidemiological studies have shown that tomatoes protect against cancers of the digestive tract and prostate (124–126). The biochemical mechanisms or the micronutrients responsible for these cancer-preventing properties are still under investigation. The beneficial influence of tomatoes could be due to the presence of antioxidant compounds such as carotenoids and vitamin C (Fig. 6). Tomatoes are low in β-carotene, but high in lycopene, an effective antioxidant compound. They are also rich sources of vitamin C, known for its therapeutic, antioxidant properties (127).

In a recent epidemiological study, Giovannucci et al. (125) reported that the risk of prostrate cancer was reduced nearly 45% among men who ate at least 10 servings a week of tomato-based foods. The study was conducted over a period of 6 years with 47,894 health professionals initially free of diagnosed cancer. An intake of 1.5 and 10 servings of tomato-based

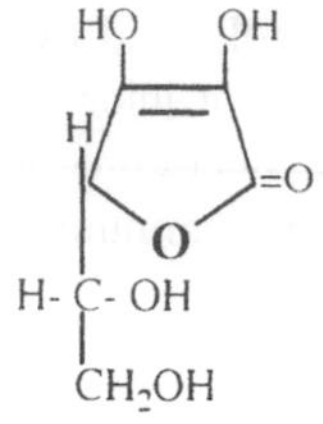

Vitamin C

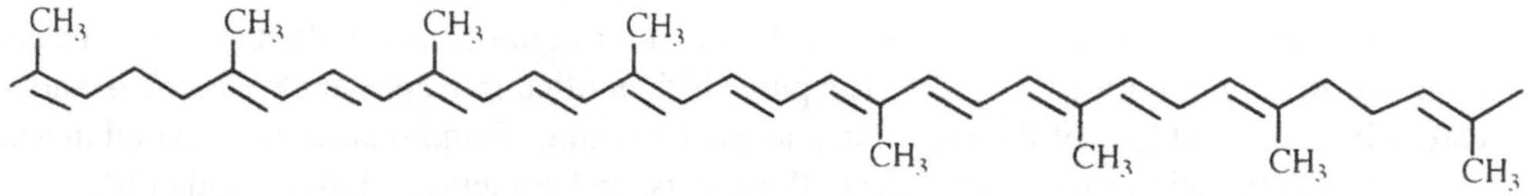

Lycopene

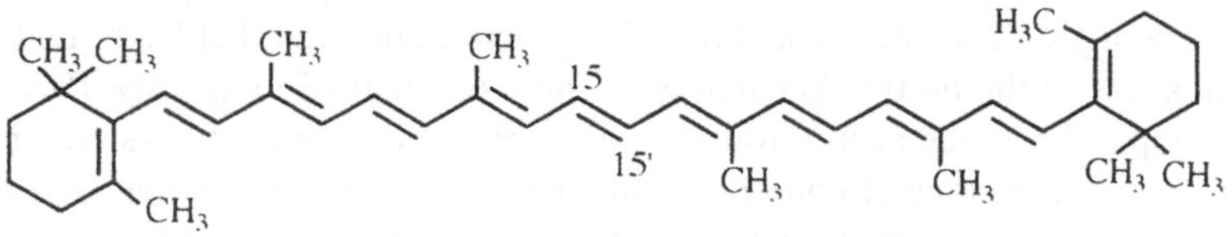

β-Carotene

FIGURE 6 Some antioxidant compounds in tomatoes.

products also seemed to derive some benefits. Of 46 vegetables and fruits or related products, tomato sauce, tomatoes, and pizza, which were primary sources of lycopene, showed significant association with lower prostate cancer risk. Intakes of other carotenoids—β-carotene, α-carotene, lutein, and cryptoxanthin—were not associated with prostate cancer risk. Also, tomato sauce had the strongest association, while tomato juice was not associated with prostate cancer risk. The investigators hypothesized that heating tomato products in oil enhances intestinal absorption of lycopene, which accounts for the observed differences between the two products. This study is supported by earlier reports on the organ distribution of lycopene and effects of processing on the uptake of lycopene in humans. Stahl and Sies (128) studied the absorption of lycopene from processed (boiled with 1% corn oil at 100°C for 1 hr) and unprocessed tomato juice in six healthy adults. Serum lycopene concentration (both *cis* and *trans* isomers) increased significantly only when processed tomato juice was consumed. In addition, studies on the carotenoids in human blood and tissues have revealed that the predominant carotenoids in human tisues are β-carotene and lycopene. They are found in the liver, adrenal glands, fat, pancreas, and lung, but high concentrations of lycopene have been reported in the testes (129,130).

The protective effects of tomatoes against cancers of the digestive tract have been reported from different parts of the world. Cook-Mozaffari et al. (131) reported from Iran a significant 40% reduction of esophageal cancer risk in men following consumption of tomatoes at least once a week. In another study in Israel, Modan et al. (132) reported a significant inverse relationship between tomato consumption and gastrointestinal cancers. Colditz et al. (133) reported a significant 50% reduction in mortality from cancers of all sites among elderly Americans reporting high tomato intake. Macquart-Moulin et al. (134) in France and Benito et al. (135) in Spain have included tomatoes in a subgroup of low-fiber vegetables significantly protective against cancer of the colon and rectum. In a recent epidemiological study, Franceschi et al. (126) reported the positive effects of tomatoes on the risk of digestive tract cancers. The data were obtained from a series of hospital-based case-control studies conducted between 1985 and 1991 in northern Italy, where tomato intake is high and also heterogeneous. The study showed a consistent pattern of protection by high intake of raw tomatoes in all examined cancer sites such as upper digestive tract, stomach, colon, rectum, and most notably gastrointestinal neoplasms.

Compared with carotenoids and other antioxidant compounds including vitamin E, lycopene has been reported to be a more efficient quencher of singlet oxygen in vitro (136,137). A recent study found that carotenoids are effective protectors of blood lymphocytes from nitric oxide radical damage and lycopene is at least twice as effective as β-carotene (138). Lycopene is also as effective as β-carotene in inhibiting low-density lipoprotein oxidation (139). Lycopene has been associated with lowered risk of cardiovascular diseases. One study found that Lithuanians have higher coronary heart disease mortality rates than Swedes because Lithuanians have significantly lower levels of lycopene than the people in Sweden (138). According to one analysis, foods that revealed the strongest association to reduced cardiovascular disease mortality were tomato ketchup and chili sauce, which according to the study are better sources of lycopene than β-carotene (138,140).

Lycopene has been reported to enhance resistance against total body x-ray irradiation and survival rates of mice exposed to x-ray irradiation (141,142). Recently, Ribaya-Mercado et al. (143) have reported that skin lycopene is destroyed preferentially over β-carotene during ultraviolet irradiation in humans, suggesting a role of lycopene in mitigating oxidative damage in tissues. This observation also supports the in vitro studies that indicate that lycopene may have superior antioxidant properties compared to β-carotene. The investigators have suggested that further studies are necessary to define the role of lycopene in the prevention of antioxidant damage in the skin and other body tissues.

REFERENCES

1. *FAO Production Yearbook*, Vol. 48, Food and Agriculture Organization, Rome, Italy, 1994, p. 129.
2. Salunkhe, D. K., and B. B. Desai, *Postharvest Biotechnology of Vegetables*, Vol. I, CRC Press, Boca Raton, FL, 1984.
3. Bailey, L. H., *Manual of Cultivated Plants*, 2nd ed., Macmillan, New York, 1949.
4. Taylor, I. B., Biosystematics of tomato, in *The Tomato Crop: A Scientific Basis for Improvement* (J. G. Atherton and J. Rudich, eds.), Chapman and Hall, London, 1986.
5. Swaider, J. M., G. W. Ware, and J. P. McCollum, *Producing Vegetable Crops*, Interstate Publishers, Inc., Danville, IL, 1992.
6. McGlasson, B., Tomatoes, in *Encyclopedia of Food Science, Food Technology, and Nutrition* (R. MaCrea, R. K. Robinson, and M. J. Sandlers, eds.), Academic Press, New York, 1993, p. 4579.
7. Kalloo, G., Utilization of wild species of crop plants, in *Distant Hybridizaton of Crop Plants* (G. Kalloo and J. B. Choudhury, eds.), Springer-Verlag, New York, 1991.
8. Thompson, H. C., and W. C. Kelly, *Vegetable Crops*, McGrawHill, New York, 1957.

9. Went, F. W., Plant growth under controlled conditions. V. The relation between age, light, variety, and thermoperiodicity of tomatoes, *Am. J. Bot. 32*:469 (1945).
10. Went, F. W., and L. Coser, Plant growth under controlled conditions. VI. Comparison between field and air conditioned greenhouse culture of tomatoes, *Am. J. Bot. 32*:643 (1945).
11. Moore, E. L., and W. O. Thomas, Some effects of shading and *para*-chlorophenoxyacetic acid on fruitfulness of tomatoes, *Proc. Am. Soc. Hort. Sci. 60*:289 (1952).
12. Wittwer, S. H., Control of flowering and fruit setting by plant growth regulators, in *Plant Regulators in Agriculture* (H. B. Tukey, ed.), Wiley and Sons, New York, 1954.
13. Wittwer, S. H., and W. A. Schmidt, Further investigations of the effect of hormone sprays on the fruiting response of outdoor tomatoes, *Proc. Am. Soc. Hort. Sci. 55*:335 (1950).
14. Parsons, C. S., and E. W. Davis, Hormone effect on tomatoes grown in nitrogen rich soil, *Proc. Am. Soc. Hort. Sci. 62*:371 (1953).
15. Baker, J. E., Morphological changes during maturation and scnescence, in *Postharvest Handling and Utilization of Tropical and Subtropical Fruits and Vegetables* (E. B. Pantastico, ed.) AVI, Westport, CT, 1975, p. 128.
16. Hobson, G., and D. Grierson, Tomato, in *Biochemistry of Fruit Ripening* (G. B. Seymour, J. E. Taylor, and G. A. Tucker, eds.), Chapman and Hall, London, 1993, p. 405.
17. Adlakha, P. A., and S. K. Verma, Use of plant growth regulators for transplantings to notice their effect on growth and yield, *Punjab Hort. J. 4*:107 (1964).
18. Gould, W. A., *Tomato Production, Processing, and Technology*, CTI Publications, Inc., Baltimore, 1992.
19. Shukla, V., and L. M. Naik, Agro-techniques for solanaceous vegetables, in *Advances in Horticulture*, Vol. 5, (K. L. Chadha and G. Kalloo, eds.), Malhotra Publishing House, New Delhi, 1993, p. 365.
20. Anand, N., and C. R. Muthukrishnan, Effect of potassium on growth, yield, and quality of tomato, *Potash Rev. 8*:8 (1974).
21. Arora, S. K., M. L. Pandita, and S. C. Pandey, Effect of PCPA and micronutrients on the fruit set, earlyness, and total yield of tomato variety, HS 102, *Haryana J. Hort. Sci. 12*:212 (1983).
22. Das, R. C., and G. Das, Effect of micronutrients with and without urea on growth and yield of tomato cv Pusa Ruby, *Orissa J. Hort. 9*:22 (1981).
23. Wittmeyer, E. C., *Summary of Practices Followed by Growers in the Ohio Top Ten Club*, Ohio State Univ. Dept. Hort., Columbus, OH, 1971.
24. Zobel, M. B., Mechanization of tomato production, Proc. Natl. Conf. Tomatoes, Dept. Hort. Purdue Univ., Natl. Food Proc. Assoc., Lafayette, IN, 1966.
25. Kamalnathan, C., and A. Thamburaj, Response of tomato to N, P, K, and plant spacing, *Madras Agric. J. 57*:525 (1979).
26. Rajeswar, S. R., and U. K. Patil, Flowering and fruiting of some important varieties of tomato as affected by spacing and pruning, *Indian J. Agric. Sci. 49*:358 (1979).
27. Pantastico, E. B. and F. Venter, Physiological disorders other than chilling injury. Part 2. Tomato, in *Postharvest Handling and Utilization of Tropical and Subtropical Fruits and Vegetables* (E. B. Pantastico, ed.) AVI, Westport, CT, 1975, p. 376.
28. Sadik, S., and P. A. Minges, Symptoms and histology of tomato fruits affected by blotchy ripening, *Proc. Am. Soc. Hort. Sci. 88*:532 (1966).
29. Hobson, G. E., Phenolase activity in tomato fruit in relation to growth and to various ripening disorders, *J. Sci. Food Agric. 18*:523 (1967).
30. Winsor, G. W., and M. I. E. Long, The effects of nitrogen, phosphorus, magnesium, and lime in factorial combination on ripening disorders of glasshouse tomatoes, *J. Hort. Sci. 42*:391 (1967).
31. Winsor, G. W., J. N. Davies, and M. I. E. Long, Liquid feeding of glasshouse tomatoes: the effects of K concentration on fruit quality and yield, *J. Hort. Sci. 36*:254 (1961).
32. Venter, F., Investigations on greenback of tomatoes, *Acta Hort. 4*:99 (1965).
33. Lipton, W. J., Effects of high humidity and solar radiation on temperature and color of tomato fruit, *J. Am. Soc. Hort. Sci. 95*:680 (1970).

34. McCollum, J. P., Effects of light on the formation of carotenoids in tomato fruits, *Food Res. 19*:182 (1954).
35. Spurr, A. R., Anatomical aspects of blossom-end rot in the tomato with special reference to calcium nutrition, *Hilgardia, 28*:269 (1959).
36. Van Goor, B. J., The role of calcium and cell permeability in the disease blossom-end rot of tomatoes, *Physiol. Plant. 21*:110 (1968).
37. Wiersum, L. K., Calcium content of fruits and storage tissues in relation to the mode of water supply, *Acta Bot. Neerl. 15*:406 (1966).
38. Evans, H. J., and R. V. Troxler, Relation of calcium nutrition to the incidence of blossom-end rot in tomatoes, *Proc. Am. Soc. Hort. Sci. 32*:519 (1953).
39. McColloch, L. P., H. T. Cook, and W. R. Wright, *Market Diseases of Tomatoes, Peppers, and Eggplants*, U.S. Dept. Agric. Handbook, 28, Washington D.C., 1968.
40. Treshow, M., The etiology, development, and control of tomato fruit tumor, *Phytopathology 45*:132 (1955).
41. Ramsey, G. B., J. S. Wiant, and L. P. McColloch, *Market Diseases of Tomatoes, Peppers, and Eggplants*, U.S. Dept. Agric. Handbook, 28, Washington D.C., 1952.
42. Frazier, W. A., A study of some factors associated with the occurrence of cracks in the tomato fruit, *Proc. Am. Soc. Hort. Sci. 32*:519 (1935).
43. Salunkhe, D. K., and M. T. Wu, Developments in technology of storage and handling of fresh fruits and vegetables, in *Storage, Processing, and Nutritional Quality of Fruits and Vegetables*, (D. K. Salunkhe, ed.), CRC Press, Cleveland, OH, 1974, p. 121.
44. Schuch, W., Improving tomato quality through biotechnology, *Food Technol. 48*(11):78 (1994).
45. Smith, C. J. S., C. F. Watson, P. C. Morris, C. R. Bird, G. B. Seymour, J. E. Gray, C. Arnold, G. A. Tucker, W. Schuch, S. Harding, and D. Grierson, Inheritance and effect on ripening of antisense polygalacturonase gene in transgenic tomatoes, *Plant Mol. Biol. 14*:369 (1990).
46. Gray, J., S. Picton, J. Shabbeer, W. Schuch, and D. Grierson, Molecular biology of fruit ripening and its manipulation with antisense genes, *Plant Mol. Biol. 19*:69 (1993).
47. Kramer, M., R. A. Sanders, R. E. Sheehy, M. Melis, M. Kuehn, and W. R. Hiatt, Field evaluation of tomatoes with reduced polygalacturonase by antisense RNA, in *Horticultural Biotechnology* (A. D. Bennett and S. D. O'Neil, eds.), Wiley-Liss, New York, 1990, p. 347.
48. Kramer, M., R. A. Sanders, H. Bolkan, C. Waters, R. E. Sheehy, and W. R. Hiatt, Postharvest evaluation of transgenic tomatoes with reduced levels of polygalacturonase: Processing, firmness, and disease resistance, *Postharvest Biol. Technol. 1*:241 (1991).
49. Schuch, W., J. Kanczler, D. Robertson, G. Hobson, G. Tucker, D. Grierson, S. W. J. Bright, and C. Bird, Fruit quality characteristics of transgenic tomato fruits with altered polygalacturonase activity, *HortSci 26*:1517 (1991).
50. Shapiro, L., A tomato with a body that just won't quit (Flavr Savr bioengineered tomato), *Newsweek 123*:80 (1994).
51. Young, E., Altered tomato faces ban from British shops, *New Scientist 143*:9 (1994).
52. Holden, C., Tomato of tomorrow, *Science 264*:512 (1994).
53. Thayer, A., FDA gives go-ahead to bioengineered tomato (Calgene's Flavr Savr), *Chem. Eng. News 72*:7 (1994).
54. Salunkhe, D. K., S. J. Jadhav, and M. H. Yu, Quality and nutritional composition of tomato fruit as influenced by certain biochemical and physiological changes, *Qual. Plant. 24*:85 (1974).
55. Dalal, K. B., D. K. Salunkhe, A. A. Boe, and L. E. Olson, Certain physiological and biochemical changes in the developing tomato fruit (*Lycopersicon esculentum* Mill.), *J. Food Sci. 30*:504 (1965).
56. Lambeth, V. N., M. L. Fields, and D. E. Huecker, The sugar-acid ratio of selected tomato varieties, *Mo. Agr. Exp, Sta. Bull. 850*:1 (1964).
57. Winsor, G. W., J. N. Davies, and D. M. Massey, Composition of tomato fruit III. Juices from whole fruit and locules at different stages of ripeness, *J. Sci. Food Agric. 13*:108 (1962).
58. Yu, M. H., L. E. Olson, and D. K. Salunkhe, Precursors of volatile components in tomato fruit-I. Compositional changes during development, *Phytochemistry 6*:1457 (1967).

59. Kapp, P. P., Some effects of variety, maturity, and storage on fatty acids in fruit pericarp of *Lycopersicon esculentum* Mill., *Dissertation Abst. 27*:77B (1966).
60. Ueda, Y., T. Minamide, K. Ogata, and H. Kamata, Lipids of fruits and vegetables and their physiological and qualitative role Part I. Method of quantitative determination of neutral lipids and changes in lipids during maturation of tomato fruit, *J. Jap. Soc. Food Technol. 17*:49 (1970).
61. Jadhav, S. J., B. Singh, and D. K. Salunkhe, Metabolism of unsaturated fatty acids in tomato fruit: Linoleic and linolenic acid as precursors of hexanal, *Plant Cell Physiol. 13*:449 (1972).
62. Brieskorn, C. H., and H. Reinartz, Composition of tomato peel I. The extractable components, *Z. Lebensm. U. -Forsch. 133*:137 (1967).
63. Brieskorn, C. H. and H. Reinartz, Composition of tomato peel II. Alkali soluble fraction, *Z. Lebensm. Unters. Forsch. 135*:55 (1967).
64. Freeman, J. A., and C. G. Woodbridge, Effect of maturation, ripening, and truss position on the free amino acid content of tomato fruits, *Proc. Am. Soc. Hort. Sci. 76*:515 (1960).
65. Davies, J. N., Changes in the non-volatile organic acids of tomato fruit during ripening, *J. Sci. Food Agric. 17*:396 (1966).
66. Hamdy, M. M., and W. A. Gould, Varietal differences in tomatoes: A study of α-keto acids, α-amino compounds, and citric acid in eight tomato varieties before and after processing, *J. Agric. Food Chem. 10*:499 (1962).
67. Luh, B. S., F. Villarreal, S. J. Leonard, and M. Yamaguchi, Effect of ripeness level on consistancy of canned tomato juice, *Food Technol. 14*:635 (1960).
68. Oliver, M., Occurrence in foods, in *The Vitamins*, (W. H. Sebrell and R. S. Harris, eds.), Vol. I., Academic Press, New York, 1967, p. 359.
69. Fryer, H. C., L. Ascham, A. B. Cardwell, J. C. Frazier, and W. W. Wills, Effect of fruit cluster position on the ascorbic acid content of tomatoes, *Proc. Am. Soc. Hort. Sci. 64*:360 (1954).
70. Brown, A. P., and F. Moser, Vitamin C content of tomatoes, *Food Res. 6*:45 (1941).
71. Clutter, M. E., and E. V. Miller, Ascorbic acid and ripening time of tomatoes, *Econ. Bot. 15*:218 (1961).
72. Hobson, G. E., and J. N. Davies, The tomato, in *The Biochemistry of Fruits and their Products* (A. C. Hulme, ed.), Academic Press, New York, 1971, p. 437.
73. Jane, B. E., Some chemical differences between artificially produced parthenocarpic fruits and normal seed fruits of tomato, *Am. J. Bot. 28*:639 (1941).
74. Wu, M., and R. C. Burrell, Flavonoid pigments of the tomato (*Lycopersicon esculentum* Mill.), *Arch. Biochem. Biophys. 74*:114 (1958).
75. Walker, J. R. L., Phenolic acids in 'cloud' and normal tomato fruit wall tissue, *J. Sci. Food Agric. 13*:363 (1962).
76. Kajderowicz-Jarosinska, D., Content of tomatine in tomatoes at different stages of ripeness, *Acta Agr. Silvestria*, Ser. Roln. *5*:3 (1965).
77. Meredith, F. I., and A. E. Purcell, Changes in the concentration of carotenes of ripening Homestead tomatoes, *Proc. Am. Soc. Hort. Sci. 89*:544 (1966).
78. Ferrari, R. A., and A. A. Benson, The path of carbon on photosynthesis of lipids, *Arch. Biochem. Biophys. 93*:185 (1961).
79. Curl, A. L., The xanthophylls of tomatoes, *J. Food Sci. 26*:106 (1961).
80. Trombly, H. H., and J. W. Porter, Additional carotenes and a colorless polyene of *Lycopersicon* species and strains, *Arch. Biochem. Biophys. 43*:443 (1953).
81. McCollum, J. P., Distribution of carotenoids in the tomato, *Food Res. 20*:55 (1955).
82. Dalal, K. B., D. K. Salunkhe, and L. E. Olson, Certain physiological and biochemical changes in greenhouse-grown tomatoes (*Lycopersicon esculentum* Mill.), *J. Food Sci. 31*:461 (1966).
83. Stevens, M. A., Relationship between polyene-carotene content and volatile compound composition of tomatoes, *J. Am. Soc. Hort. Sci. 95*:461 (1970).
84. Johnson, A. E., H. E. Nursten, and A. A. Williams, Vegetable volatiles: A survey of components identified. Part II., *Chem. Ind. 43*:1212 (1971).

85. Yu, M. H., L. E. Olson, and D. K. Salunkhe, Precursors of volatile components in tomato fruit-II. Enzymatic production of carbonyl compounds, *Phytochemistry 7*:555 (1968).
86. Yu, M. H., D. K. Salunkhe, and L. E. Olson, Production of 3-methylbutanal from L-leucine by tomato extract, *Plant Cell Physiol. 9*:633 (1968).
87. Yu, M. H., L. E. Olson, and D. K. Salunkhe, Precursors of volatile components in tomato fruit-III. Enzymatic reaction products, *Phytochemistry 7*:561 (1968).
88. Dalal, K. B., D. K. Salunkhe, L. E. Olson, and J. Y. Do, Volatile components of developing tomato fruit grown under field and greenhouse conditions, *Plant Cell Physiol. 9*:389 (1968).
89. Shah, B. M., D. K. Salunkhe, and L. E. Olson, Effects of ripening processess on chemistry of tomato volatiles, *J. Am. Soc. Hort. Sci. 94*:171 (1969).
90. Kader, A. A., M. A. Stevens, M. Albright-Holton, L. L. Morris, M. Algazi, Effect of fruits ripeness when picked on flavor and composition in fresh market tomatoes, *J. Am. Soc. Hort. Sci. 102*:724 (1977).
91. Hayase, F., T. Y. Chung, and H. Kato, Changes of volatile components of tomato fruits during ripening, *Food Chem. 14*:113 (1984).
92. Ryall A. L. and W. J. Lipton, *Handling, Transportation, and Storage of Fruits and Vegetables*, Vol. I. *Vegetables and Melons*, AVI, Westport, CT, 1972.
93. Ryall, A. L., and W. J. Lipton, in *Handling, Transportation, and Storage of Fruits and Vegetables*, Vol. 1. *Vegetables and Melons*, AVI, Westport, CT, 1972.
94. Srivastava, H. C., N. V. N. Moorthy, and N. S. Kapur, Refrigerated storage behavior of pre-cooled tomatoes (Var. Marglobe), *Food Sci. 11*:252 (1962).
95. Hall, C. B., The effect of low storage temperatures on the color, carotenoid pigments, shelf-life, and firmness of ripened tomatoes, *Proc. Am. Soc. Hort. Sci. 78*:480 (1961).
96. Parsons, C. S., R. E. Anderson, and R. W. Penney, Storage of mature green tomatoes in controlled atmosphere, *J. Am. Soc. Hort. Sci. 95*:791 (1970).
97. Salunkhe, D. K., and M. T. Wu, Effects of low oxygen atmosphere storage on ripening and associated biochemical changes in tomato fruits, *J. Am. Soc. Hort. Sci. 98*:12 (1973).
98. Hatton, T. T., E. B. Pantastico, and E. K. Akamine, Individual commodity requirements, in *Postharvest Handling and Utilization of Tropical and Subtropical Fruits and Vegetables* (E. B. Pantastico, ed.) AVI, Westport, CT, 1975, p. 201.
99. Dennis, C., K. M. Browne, and F. Adamicki, Controlled atmosphere storage of tomatoes, *Acta Hort. 93*:75 (1979).
100. Bhowmik, S. R., and J. C. Pan, Shelf-life of mature green tomatoes stored in controlled atmosphere and high humidity, *J. Food Sci. 57*:948 (1992).
101. Salunkhe, D. K., and M. T. Wu, Subatmospheric storage of fruits and vegetables, in *Postharvest Biology and Handling of Fruits and Vegetables* (N. F. Haard and D. K. Salunkhe, eds.), AVI, Westport, CT, 1975, p. 153.
102. Wu, M. T., S. J. Jadhav, and D. K. Salunkhe, Effects of subatmospheric pressure storage on ripening tomato fruits, *J. Food Sci. 37*:952 (1972).
103. Tolle, W. E., *Hypobaric Storage of Mature Green Tomatoes*, U.S. Dept. Agric. Market Res. Rep. No. 842, Washington D.C., 1969.
104. Burg, S. P., Hypobaric storage and transportation of fruits and vegetables, in *Postharvest Biology and Handling of Fruits and Vegetables* (N. F. Haard and D. K. Salunkhe, eds.), AVI, Westport, CT, 1975, p. 172.
105. Salunkhe, D. K., Gamma radiation effects on fruits and vegetables, *Econ. Bot. 15*:28 (1961).
106. Larrigaudiere, C., A. Latche, J. C. Pech, and C. Triantaphylides, Short-term effects of γ-irradiation on 1-aminocyclopropane-1-carboxylic acid metabolism in early climacteric cherry tomatoes, *Plant Physiol. 92*:577 (1990).
107. Ben-Yehoshua, S., I. Kobiler, and B. Shapiro, Effects of individual seal-packaging of fruit in film and high density polyethylene (HDPE) on various postharvest blemishes of citrus and tomatoes, *HortSci* (Abst.) *15*:93 (1980).

108. Rawnsley, J., *Crop Storage*, Tech. Rep. No. 1, Food Research and Development Unit, Ministry of Agriculture, Accra, Ghana, Food and Agriculture Organization, Rome, Italy, 1969.
109. Parsons, C. S., and D. H. Spalding, Influence of controlled atmosphere, temperature, and ripeness on bacterial soft rot of tomatoes, *Am. Soc. Hort. Sci. 97*:297 (1972).
110. Fuchs, Y., R. Barkai-Golan, and N. Aharoni, Postharvest studies with fresh market tomatoes intended for export, in *Scientific Activities of the Institute for Technology and Storage of Agricultural Products, 1974-1977*, Pamphlet No. 184, Bet Dagan, Israel, 1978, p. 33.
111. Wills, R. B. H., and S. I. H. Tirmazi, Effect of calcium and other minerals on ripening of tomatoes, *Aust. J. Plant. Physiol. 6*:221 (1979).
112. Ben-Yehoshua, S., I. Kobiler, B. Shapiro, and I. Gero, The effect of a new packing method on the rate of deterioration of tomatoes for export, in *Scientific Activities of the Institute for Technology and Storage of Agricultural Products, 1974-1977*, Pamphlet No. 184, Bet Dagan, Israel, 1978, p. 32.
113. Domenico, J. A., A. R. Rahman, and D. E. Westcott, Effects of fungicides in combination with hot water and wax on the shelf-life of tomato fruit, *J. Food Sci. 37*:957 (1972).
114. Sharma, S. L., and B. R. Verma, Residual effect of field sprays on storage rots of tomato, *Himachal J. Agric. Res. 1*:55 (1971).
115. Linn, M. B., and R. G. Emge, Development of anthracnose and secondary rots in stored tomato fruits in relation to field spraying with fungicides, *Phytopathol. 39*:898 (1949).
116. Tisdale W. B., and S. O. Hawkins, Experiments for the control of *Phoma* rot of tomatoes, Fla. Agric. Exp. Stn. Bull. 308, Gainesville, FL, 1937.
117. Kaushik, C. D., J. N. Chand, and D. O. Thakur, Parasitism and control of *Alternaria tenuis* causing ripe rot of tomato fruits, *J. Appl. Sci. Ind. 1*:15 (1969).
118. Aulakh, K. S., and R. K. Grover, Use of oils for controlling ripe fruit rots of tomatoes caused by *Phoma destructiva* and *Curvularia lycopersici*, *Plant Prot. Bull. FAO 17*(4):90 (1969).
119. Barkai-Golan, R., and S. Ben-Yehoshua, Cold storage and disinfection treatments of export grade tomatoes (cultivar S-5), in *Scientific Activities of the Institute for Technology and Storage of Agricultural Products, 1974-1977*, Pamphlet No. 184, Bet Dagan, Israel, 1078, p. 57.
120. Caceres, A., L. M. Giron, S. R. Alvarado, and M. F. Torres, Screening of antimicrobial activity of plants popularly used in Guatemala for the treatment of dermatomucosal diseases, J. *Ethnopharmacol. 20*:223 (1987).
121. Ramirez, V. R., L. J. Mostacero, A. E. Garcia, C. F. Mejia, P. F. Pelaez, C. D. Medina, and C. H. Miranda, Vegetales empleados en medicina tradicional norPeruana, Banco Agrario Del Peru Nacl. Univ. Trujillo, Peru, 1988:54 (1988).
122. Velazco, E. A., Herbal and traditional practices related to maternal and child health care, *Rural Reconstruction Review*, 1980, p. 35.
123. Liebstein, A. M., Therapeutic effects of various food articles, *Am. Med. 33*:33 (1927).
124. A tomato a day for preventing prostate cancer? Diet may be key, *Geriatrics 51*:21 (1996).
125. Giovannucci, E., A. Ascherio, E. B. Rimm, M. J. Stampfer, G. A. Colditz, and W. C. Willett, Intake of carotenoids and retinol in relation to risk of prostate cancer, *J. Natl. Cancer Inst. 87*:1767 (1995).
126. Franceschi, S., E. Bidoli, C. L. Vecchia, R. Talamini, B. D'Avanzo, and E. Negri, Tomatoes and risk of digestive-tract cancers, *Int. J. Cancer 59*:181 (1994).
127. Deshpande, S. S., U. S. Deshpande, and D. K. Salunkhe, Nutritional and health aspects of food antioxidants, in *Food Antioxidants—Technological, Toxicological, and Health Perspectives*, (D. L. Madhavi, S. S. Deshpande, and D. K. Salunkhe, eds.), Marcel Dekker, New York, 1995, p. 361.
128. Stahl, W., and H. Sies, Uptake of lycopene and its geometrical isomers is greater from heat processed than from unprocessed tomato juice in humans, *J. Nutr. 122*:2161 (1992).
129. Stahl, W., W. Schwarz, A. R. Sundquist, and H. Sies, Cis-trans isomers of lycopene and β-carotene in human serum and tissues, *Arch. Biochem. Biophys. 294*:173 (1992).
130. Kaplan, L. A., J. M. Lau, and E. A. Stein, Carotenoid composition, concentrations, and relationships in various human organs, *Clin. Physiol. Biochem. 8*:1 (1990).
131. Cook-Mozaffari, P. J., F. Azordegan, N. E. Day, A. Ressicaud, C. Sabai, and B. Aramesh, Esophageal

cancer studies in the Caspian Littoral of Iran: Results of a case-control study, *Br. J. Cancer 39*:293 (1979).

132. Modan, B., H. Cuckle, and F. Lubin, A note on the role of dietary retinol and carotenes in human gastrointestinal cancer, *Int. J. Cancer 28*:421 (1981).
133. Colditz, G. A., L. G. Branch, and R. J. Lipnick, Increased green and yellow vegetable intake and lowered cancer deaths in an elderly population, *Am. J. Clin. Nutr. 41*:32 (1985).
134. Macquart-Moulin, G., E. Riboli, J. Cornee, B. Charnay, P. Berthezene, and N. Day, Case-control study of colorectal cancer and diet in Marseilles, *Int. J. Cancer, 38*:183 (1986).
135. Benito, E., A. Obrador, A. Stiggelbout, F. X. Bosch, M. Mulet, N. Munoz, and J. Kaldor, A population-based case-control study of colorectal cancer in Majorca. I. Dietary factors, *Int. J. Cancer 45*:69 (1990).
136. Di Mascio, P., S. Kaiser, and H. Sies, Lycopene as the most efficient biological carotenoid singlet oxygen quencher, *Arch. Biochem. Biophys. 274*:532 (1989).
137. Di Mascio, P., M. E. Murphy, and H. Sies, Antioxidant defense systems: the role of carotenoids, tocopherols, and thiols, *Am. J. Clin. Nutr. 53*:194S (1991).
138. Lycopene as potent an antioxidant as β-carotene, FDA told, *Food Labeling Nutr. News 4*(23):6 (1996).
139. Lynch, S. M., J. D. Morrow, L. J. Roberts II, and B. Frei, Formation of non-cycloxygenase-derived prostanoids in plasma and low-density lipoprotein exposed to oxidative stress in vitro, *J. Clin. Invest, 93*:998 (1994).
140. Verlangieri, A. J., J. C. Kapeghian, S. el-Dean, and M. Bush, Fruit and vegetable consumption and cardiovascular mortality, *Med. Hypotheses, 16*:7 (1985).
141. Lingen, C., L. Ernster, and O. Lindberg, The promoting effect of lycopene on the non-specific resistance of animals, *Exp. Cell Res. 16*:384 (1959).
142. Forssberg, A., C. Lingen, L. Ernster, and O. Lindberg, Modification of x-irradiation syndrome by lycopene, *Exp. Cell Res. 16*:7 (1959).
143. Ribaya-Mercado, J. D., M Garmyn, B. A. Gilchrest, and R. A. Russel, Skin lycopene is destroyed preferentially over β-carotene during ultraviolet irradiation in humans, *J. Nutr. 125*:1854 (1995).

8

Capsicum

J. C. Rajput and Y. R. Parulekar
Konkan Agricultural University, Dapoli, India

I. CHILI

Chili, also known as hot pepper, is an important vegetable and condiment crop grown in the tropical and subtropical regions of the world. It is an indispensible commodity in every home in the tropics. Due to their medicinal properties, vitamin contents, and the demand for chilies has been increasing all over the world. They are grown commercially in China, Korea, Indonesia, Pakistan, Sri Lanka, Turkey, Japan, Mexico, Ethiopia, Nigeria, Uganda, Yugoslavia, Spain, Italy, Hungary, and Bulgaria. India is the largest exporter of chili followed by China, Indonesia, Japan, Mexico, Uganda, and Kenya and Nigeria (1,2).

A. Botany

The genus *Capsicum* is a member of the Solanaceae family. Five major cultivated species of *Capsicum* have been recognized (Table 1). Heiser (3) discussed distinguishing varieties of *Capsicum annuum*. All cultivated as well as wild species are diploid ($2n = 24$).

Seven botanical varieties of *C. annuum* have been described (4):

1. *C. annuum* var. *abbreviatum,* Fingerh: wrinkled peppers; fruits generally ovate, wrinkled, about 5 cm long or even less
2. *C. annuum* var. *acuminatum,* Fingerh: fruits linear, oblong, over 9 cm long, usually pointed, pungent, widely grown in India, e.g., Bydagi
3. *C. annuum* var. *cerasiformae* (Miller) Irish: cherry peppers; fruits globose with firm flesh, 1.2–2.5 cm in diameter, red, yellow, or purple, pungent
4. *C. annuum* var. *conoides* (Miller) Irish: Cone pepper; tabasco type; fruit erect, conical, about 3 cm long, pungent
5. *C. annuum* var. *fasciculatum* (Stuart) Irish: cluster peppers; fruits clustered, erect, slender, about 7.5 cm long, very pungent (As the fruits are not born singly, it probably belongs to *C. frutescens.*)
6. *C. annuum* var. *grossum* (L.) Sendt.: sweet peppers, paprika; fruit large with basal

TABLE 1 *Capsicum* Species

Species	Synonyms	Distribution	Characteristics
C. annuum L.	*C. purpureum* *C. grossum* *C. cerasiformae*	Columbia to southern United States, throughtout Latin America, Asia	Blue anthers, milky white corolla, inconspicuous calyx, lobing and solitary peduncle
C. baccatum L.	*C. pendulum* C. microcarpum C. angulosum	Argentina, Bolivia, Brazil, Columbia, Equador, Peru, Paraguay, etc.	Yellow or brown spots and prominent calyx teeth, peduncle erect at anthesis
C. frutescens (Tabasco pepper)	*C. minimum*	Columbia, Costa Rica, Guatemala, Mexico, Puerto Rico, Venezuela	Blue anthers, milky greenish or yellowish, white corolla, two or more peduncles in a node
C. chinese L.	*C. luteum* *C. umbilicatum* *C. sinense*	Bolivia to Brazil, Belize, Costa Rica, Mexico, Nicaragua, West Indies	Constriction below the calyx, two or more flowers in each node
C. pubescence	*C. eximium* *C. tovari* *C. cardenasii*	Bolivia to Columbia, Costa Rica, Guatemala, Honduras, Mexico	Thick flesh, dark rugose seeds, purple corolla

depression inflated, red or yellow flesh thick and mild, e.g., California Wonder, Yellow Wonder, Ruby Giant, etc.

7. *C. annuum* var. *longum* (DC) Sendt: long peppers; fruits mostly dropping, tapering at apex

After conducting flavonoid analysis of three *Capsicum* species, Lopes et al. (5) suggested that there exists an affinity between *C. annuum* and *C. frutescens* and between *C. pendulum* and *C. frutescens.* Protein electrophoretic studies by Pradeepkumar (6) revealed species-specific protein bands in *C. chinense, C. baccatum,* and *C. chacoense.* Among five species, namely, *C. annuum, C. fruitescense, C. chinense, C. baccatum,* and *C. chacoense,* a close relationship was established between *C. chinense* and *C. fruitescens.*

1. Morphology

The chili (*Capsicum annuum* L. var. *acuminatum*) is an annual herbaceous plant. The branches turn woody or brittle with age. Branching is greatly influenced by genotypic difference. The main shoot is radial, but lateral branches are cincinnate (Fig. 1). The leaves are simple, alternate, exstipulate, elliptic, lanceolate, and glabrous. Leaf size is variable and differs according to cultivar as well as management practices. The petiole is 0.5–2.5 cm long. Chili plants possess a strong tap root, which is usually broken or arrested during transplanting, resulting in the development of profusely branched laterals as much as 1 m long. The

FIGURE 1 Chili plant, cv. Phule Jyoti.

entire root system remains restricted to the upper 30 cm of the soil layer. It has been observed that more active feeding roots are found 10 cm from the base of the plant both laterally and vertically (7).

2. Floral Biology

The number of days required for flowering mainly depends on variety and meterological parameters. However, in most varieties flowering commences about 40 days after transplanting. Flowers are usually solitary, but in some cultivars they occur in clusters. They are bracteolate, pedicellate (1.5 cm long), bisexual, and hypogynous. Most flowers open at 5 a.m. Stigma becomes receptive from the day of anthesis and remain so for 2 days after anthesis. Another dehiscence takes place between 8 and 10 a.m. Pollen grains become fertile a day before anthesis, with maximum fertility on the day of anthesis. Chili is basically a self-pollinated crop, but cross-pollination to an extent of 16% has been reported (12). Bees, ants, and thrips are the principal pollinating agents. About 40–50% fruit set is observed in chili (2).

3. Fruit Development and Ripening

The chili fruit is a berry with a short thick peduncle, which varies in shape, color, and pungency. The pericarp is leathery or succulent, and changes in color from green or purple to red (Fig. 2). The berry develops from a bicarpellary ovary with axile placentation. The placenta carries

FIGURE 2 Ripe chilies, cv. Phule—C-7.

numerous seeds. The pericarp begins to dry on full ripening. The major fruit components are seed, pericarp, placenta, and pedicel. All of these components vary greatly, depending on the variety and climatic conditions.

It takes about 30–35 days from fruit set to complete development of fruit for harvest at the green stage. The fruit starts ripening 80–90 days after fruit set. The carotenoid content increases to about 120% at the time of full ripeness as compared to the green stage (2). The sugar content in fruit increases up until ripening, but it declines during and after ripening.

B. Production

1. Soil and Climate

The chili crop requires well-drained soil. Water stagnation or saturated conditions for just a few days will lead to death of the plants. However, chili can be grown in alluvial, red loamy, and sandy, soils. Highly acidic or alkaline soils are not suitable for chili growth. Crops grown in entisols, utisols, inceptisols under rainfed conditions mature earlier than in vertisols. This may be due to moisture stress created in light soils. The optimum pH range for satisfactory growth is 5.0–7.5, although chili can be grown in soils with a pH as high as 9.0.

Chili is grown in both tropical as well as subtropical regions from sea level to 2000 m altitude. A frost-free period of 4 months with a maximum temperature ranging between 20 and 35°C and a minimum temperature of 10°C is optimum for successful cultivation. Soil temperatures above 30°C will retard root development.

High temperatures and low relative humidity during flowering increase the transpiration pull, resulting in abscission of buds, flowers, and fruits. Higher temperatures were found to be responsible for high capsaicin content in Capsicum. Low night temperatures (8–10°C and 15°C) during the flower-opening period reduces pollen viability, leading to formulation of parthenocarpic fruits or fruits with fewer seeds.

2. Propagation

Traditionally and commercially, chili plants are propagated through seeds. In recent years, grafting and anther culture have been developed for propagation of this crop, but their use is restricted for testing resistance under laboratory conditions (2). Thomas (9) reported that rubbing seeds between the palms followed by washing reduced the pungency and facilitated germination as well as helped to repel ants. Similarly, seeds are dressed with thirum (2–3 g/kg seeds) to avoid damping-off disease during the nursery stage (10).

About 35–40 raised beds of convenient size (3 × 1 m) are required to raise chili seedlings sufficient for 1 ha. According to Thomas (9) 6-week-old seedlings 15–20 cm tall are ideal for transplanting. Sometimes, hardening of seedlings is done in the nursery by inducing moisture stress for one week.

3. Cultural Practices

Preparation of fine tilth is necessary for successful cultivation of chili crop. Compost or farmyard manure (FYM) is thoroughly mixed with the soil during tilth preparation. Chili seedlings are generally planted in ridges and furrows in medium and heavy soils. However, dwarf and compact growing varieties can be successfully grown in flat beds in light soils. Spacing of plants varies according to the cultivar, soil type, season, and fertility of the soil (Table 2).

Planting

Chili is transplanted mainly in June-July if grown under rainfed conditions. Planting in June-July resulted in the highest green as well as red chili yields (11). During *kharif* season, planting is done in June-July, whereas the *rabi* crop is planted in October-November to obtain maximum yield and minimize leaf curl complex (15). Seedlings are transplanted in February-March. Soil type also influences the time of planting. In light soils, transplanting must be done early in the monsoon, as the retention of soil moisture is low. Watering should be done immediately after transplanting.

Manuring and Fertilization

Chili responds well to applied fertilizers. Usually, 10–20 tons of FYM per hectare is incorporated in the soil before transplanting. A complete dose of phosphatic and potassic fertilizers along with a one-third dose of nitrogen should be applied just before transplanting. The remaining dose of nitrogen should be given in two or three equal splits at intervals of 30 days.

Heavy application of nitrogenous fertilizers may increase vegetative growth and delay maturity. Chili also responded to the potassium up to 50 kg per hectare, beyond which there was a decrease in yield (17,18). Potassium increases the concentration of nucleoproteins in the leaves and enables the plant to synthesize more carbohydrates (18). Application of phosphatic fertilizers induces early flowering in chili (16). Rajput and Wagh (1) reported that the balanced application of organic manures and inorganic fertilizers is essential to maximize the yield of green chillies

TABLE 2 Recommended Spacing for Chili in Different Soils

Soil type	Plant Stature	Spacing	Ref.
Heavy soil	Dwarf	60 × 60 cm	12
	Tall	90 × 90 cm or 75 × 75 cm	12
Medium soil	Dwarf	60 × 45 cm	11
		60 × 60 cm	13
	Tall	75 × 75 cm	13
Light soil	Dwarf	60 × 30 cm	1
		60 × 45 cm	14

and to improve the keeping quality of fruits. Application of ≈ 2 tons of neem cake and 75:25:25 kg NPK/ha resulted in a high yield with good-quality fruits.

A combination of 60 kg of N, 30 kg of P_2O_5, and 25 kg of K_2O, when combined with pretreatment of seedlings with *Azospirillum* culture through root dip, resulted in the maximum yield of 2.62 tons of dry chilies per hectare compared with 1.65 tons for controls (9). The use of weak starter solutions at the time of transplanting was reported to be beneficial to chili crops. A starter solution containing urea, single superphosphate, and muriate of potash in 2:1:1 proportions in liquid form at a rate of 1300 ml/hill improved the seedling survival (94.36%) after transplanting and increased the yield of the red chili variety Konkan Kirti significantly (19–21). Chili crops also responded very well to foliar application of nutrients. Spraying 75 kg of N as 1.5% urea on the 30th day of transplanting and subsequently at 2-week intervals accelerated flowering in G-4 cultivars and resulted in higher yields (22).

Irrigation

Chili can be grown as a rainfed crop where 80–100 cm of rainfall is evenly distributed throughout the life span of the crop. Irrigation requirements depend on the season, weather factors, soil type, and type of irrigation system used. About 50–60% of irrigation water was saved when the drip method was used when compared to surface irrigation; at the same time there was a significant increase in yield (23).

In general, the crop is irrigated immediately after transplanting and thereafter at weekly intervals (24). Irrigation at 12-day intervals along with spraying of 20 ppm Alachlor solution as an antitranspirant was most economical for the cultivar Pusa Jwala (26). Chili cannot withstand water stagnation and excess moisture. If saturated conditions exist for 24 hours, the plants will die. Saturated conditions inhibit plant growth, leading to reduced yield. The most critical stages for moisture stress are the initial establishment of transplanted seedlings and just prior to flowering. Moisture stress at blossoming leads to flower and fruit drop. Similarly, a long dry spell followed by heavy irrigation will also result in flower drop.

Weed Management

Weed control is one of the most important operations in the successful raising of chili crops. Usually weeds are controlled manually or mechanically. Three to four weedings are required during the life span of the crop. In the initial stages of crop growth, hoeing is done to minimize the weeds and to loosen the soil for improvement of aeration.

The use of weedicides is becoming popular for weed control in chili (Table 3). Weedicides were observed to be effective against *Cynadon dactylon, Cyperus rotundus, Convolvulus arvensis, Dinebra rotroflexa, Panicum isachne, Digetaria emarginata, Echinochloa* sp., *Trianthema portulagastrum, Cynotis* sp., *Commelina bengalensis, Acalypha indica,* and *Phyllanthus* sp.

Growth Regulators

Various growth regulators (31–35) have been found to be effective for reducing flower drop and increasing fruit set and yield in chili (Table 4).

4. Diseases and Pests

Powdery mildew, caused by *Leveilluta tauica,* is the major disease of chili in tropical climates. Spraying of a fungicide like wettable sulfur (0.25%) or benylate or carbendezim (0.1%) three times at 3-week intervals minimizes the disease incidence (7). Other important diseases of chili are anthracnose and dieback caused by *Colletotrichum gloesporioides.* These can be controlled by spraying difoliation (0.2%) or Bavistin (0.2%), benylate (0.1%), or Dithane M-45 (38). Fruit

TABLE 3 Control of Weeds in Chili Using Weedicide

Weedicide	Dose	Time	Mode of application	Ref.
EPTC followed by Nitrofen Alachlor	3.75 kg/ha 1–2 kg ai/ha 2.5 kg ai/ha	10 days before transplanting	Soil incorporation	22
TOK-E-25	2 kg ai/ha + one hand-weeding	10 days before transplanting	Soil incorporation	23
Trifluralin	1 kg ai/ha	10 days before transplanting	Soil incorporation	24
Trifluralin	0.9–1.5 kg ai/ha	7–10 days before transplanting	Soil incorporation	25
Diphenamid	4.8 kg/ha	Preemergence	Soil incorporation	26

TABLE 4 Growth Regulators Used in Chili

Growth regulator	Concentration (ppm)	Stage of application	Effects
NAA (naphthelene acetic acid)	50	Full bloom stage	Control of fruit drop and 41% yield increase
NAA (Planofix)	10	First at flowering, second 5 weeks later	Increased fruit set
NAA (Planofix)	60	Bloom stage	Increase in yield
MH (maleic hydrazide)	3000	Preflowering	Suppression of flowering
Ethrel (Ethephon)	200	Preflowering	Improved fruit set and flowering
GA (gibberellic acid)	50	Fruit setting	Decrease in flower shedding and increased fruit set
Tricontanol	2	First spray at 30 days after transplant; second at bloom stage	Reduction in flower drop and increased fruit yield
Ethephon	500	At maturity	Turning color, increase in yield

rots caused by the fungus *Colletotrichum* spp., *Botryodiplodia palmarum*, and *Phytophthora capsici* cause damage to red ripe fruits. Yellow spots develop on the fruit, which subsequently become black, greenish, or dirty grey in color. Fruits maturing during the rainy season are affected worse than those maturing in dry weather conditions. Spraying of bordeaux mixture (1%) and Dithane M-45 (0.2%) four times at weekly intervals commencing 30 days after transplanting effectively controlled the disease. Bacterial leaf spot caused by *Xanthomonas vesicatoria* is another major disease. Leaf curl, chili mosaic, and yellow mosaic are the important viral diseases of chili. The important pests include thrips (*Scirtothrips dorsalis*), aphids (*Aphis gossypii*), green peach aphids (*Myzus persicae*), mites (*Hemitasonemus latus*), and white flies (38).

5. Maturity Indices and Harvesting

Chili fruits are harvested at two stages: one for green vegetables and other for dry chilies. The green chilies are harvested when they are fully mature but before they change from green to red. Chili crops take about 40 days to flower after transplanting and a further 30 days to develop fruit suitable for green harvest. They are picked at frequent intervals, sometimes twice a week. The frequent harvesting of green fruits stimulates further flowering and fruit set. Red chilies are harvested about 80–90 days after transplanting. Red fruits are harvested in two or three pickings. Reddy and Murthy (39) reported that six pickings gave significantly higher yield as compared to three to four pickings.

Chilies are usually picked when ripe and are then dried and allowed to equalize in moisture content in covered piles. Water is usually added to the chilies after drying toreduce brittleness. They are then packed tightly into sacks and are generally stored in nonrefrigerated warehouses for up to 6 months. The temperature of the warehouses depends to some extent on their construction and the way in which they are managed but chiefly on the outside temperature (10–25°C).

The average yield of green chilies ranges from 10 to 12.5 t/ha, whereas the dry chili yield ranges from 1 to 1.5 t/ha.

6. Grading and Packaging

Rajput and Wagh (1) reported that grade III (i.e., about 2.5 g pod weight) constituted the maximum percentage both by number (48.08%) and by weight (55.81%) and was considered the most common grade of the chili variety Konkan Kirti. Prepackaging of green chilies in 200 gauge ventilated polyethylene bags increased the shelf life of chili fruits (cv. Konkan Kirti) up to 11 days (1). Similarly, Anandaswamy et al. (40) found 150 gauge polyethylene film to be best for prepackaging green chilies.

C. Chemical Composition

Both ripe and green chilies are an important constituent used to impart pungency, flavor, and color to food. The pungency is due to an active principle known as capsaicin or capsicutin mainly present in the pericarp and placenta of the fruit. Capsaicin ($C_{18}H_{27}NO_3$) is a crystalline, pungent, colorless compound. It is a condensed product of 3-hydroxy-4-methoxybenzyl amine and decyclenic acid. The pungent principle is chiefly a mixture of capsaicin (69%) and dehydrocapsaicin (22%). Other minor constituents are homocapsaicin and homodehydrocapsaicin (41). The amount of capsaicin varies from 160 to 210 mg/100 g in green chilies, whereas the red ripe fruits contain around 113–160 mg/100 g (41). The variation in the

TABLE 5 Composition of Green Chili (per 100 g edible portion)

Constituent	Content
Moisture (g)	85.7
Protein (g)	2.9
Fat (g)	0.6
Minerals (g)	1.0
Fiber (g)	6.8
Carbohydrate (g)	3.0
Calcium (mg)	3.0
Magnesium (mg)	24
Riboflavin (mg)	0.39
Oxalic acid (mg)	67
Nicotinic acid (mg)	0.9
Phosphorus (mg)	80
Iron (mg)	1.2
Sodium (mg)	6.5
Potassium (mg)	217
Copper (mg)	1.55
Sulfur (mg)	34
Chlorine (mg)	15
Thiamine (mg)	0.19
Vitamin A (IU)	292
Vitamin C (mg)	111

composition of fruits depends on the stage of maturity. Green chilies are rich in vitamin A and C. They are also a good source of energy, protein, and minerals (Table 5).

D. Storage

1. Fresh Chilies

The storage of chili fruits in cold temperatures (10 ± 2°C) increased the shelf life up to 10 days as compared to storage in cool chambers or under ambient temperatures (42). Red fruits appear to be more susceptible to decay during storage than yellow ones (43). After 12 days of storage, percentage decay was lowest in red and yellow fruits dipped in Milton® (10 and 20%, respectively) compared with 76 and 68% decay, respectively, in undipped red and yellow control fruits. (43). Both cultivars stored best at 10°C (Table 6). Decay caused by *Erwinia carotovora* was hastened by high relative humidity (44). The shelf life of hot peppers in paper bags was due to a longer delay in fungal decay, which started 4 days later than in pods kept in low-density polyethylene bags.

2. Dry Chilies

The moisture content of chili and other hot peppers when stored should be low enough (10–15%) to prevent mold growth. A relative humidity of 69–70% is desirable. With a higher moisture content, the pods may be too pliable for grinding and may have to be redried. With lower moisture content (under 10%) pods may be so brittle that they shatter during handling, causing

Table 6 Effects of Packaging on Percentage Decay-Free and Percentage Marketable Hot Pepper Pods at 10, 20, and 30°C After 12 Days of Storage

	Decay-free[a] (%)		Marketable[b] (%)	
Temperature (°C)	LDPE	Control	LDPE	Control
10	50.0	75.0	66.7	100.0
20	16.7	25.0	16.7	58.3
30	0.0	16.7	16.7	33.3

LDPE, Low-density polyethylene bag storage; control, paper bag storage.
[a]Error variance, 3.27 d.f. 60.
[b]Error variance, 3.77, d.f. 60.
Source: Ref. 44.

losses and a release of dust that is irritating to the skin and respiratory system. The use of polyethylene film liners within bags allows longer storage and reduces the dust problem. The liners ensure that the pods maintain a constant moisture content during storage and until the time of grinding; thus, they permit successful storage or shipment under a wide range of relative humidities (45).

E. Processing

Dry chili powder and green chilies are commonly used in various culinary preparations. They are also widely used in the preparation of sauces, soup, ketchup, and salads. Green chili pickles are popular in the Indian subcontinent.

1. Drying

Chili pods are spread on trays either as whole pods or slices. Sliced pods not only dry faster but give a superior initial color. A temperature of 80°C has been used for drying chilies. Lease and Lease (46) reported that a drying temperature of 65°C results in improved initial color, color retention, and pungency. At 65°C, drying time for whole pods averages about 12 hours and for slices 6 hours. Chilies are dried to a final moisture content of 7–8%. In California, some processors dry the whole or cut pods down to 12–15% moisture and store at 0°C. Then when the chilies are to be ground, they are spread on trays again and dried to 7–8%. This method is reported to give maximum color and pungency retention. Drying can be done using two-stage tunnel-drying, drying the chilies to 20% moisture in the first stage and to 3% in the second stage, followed by powdering and refrigerated storage until shipment.

2. Chili powder

Dry chilies can be stored in powder form. The important properties of chili powder, such as color, pungency, and free-flowing character, have been shown to be adversely affected by moisture, light, and oxidation, and deterioration is accelerated by storage temperature. Mahadeviah et al. (47) studied the effect of packaging and storage on chili powder in flexible consumer packages. Severe deteriorative effects of sunlight include bleaching with a reduction in extractable color of 30–60% (Table 7). High temperature and high humidity reduced color by 25–45%. The

TABLE 7 Changes in Moisture and Color of Chili Powder in Different Packaging After Storage for 90 days

	Storage conditions					
	27°C and 65 ± 2% RH		38°C and 90 ± 2% RH		Sunlight, ambient	
Packaging	Moisture (%)	Color (g/kg)	Moisture (%)	Color (g/kg)	Moisture (%)	Color (g/kg)
Glass	4.50	1.23	4.50	1.16	4.50	0.90
	8.80	1.13	8.80	1.10	8.80	0.87
	11.5	1.00	11.50	1.0	11.50	0.81
LDPE, 300 G	6.80	0.85	11.10	0.70	6.20	0.52
HDPE, 300 G	6.50	1.14	9.0	0.59	6.10	0.44
LDPE, black, 300 G	—	—	11.70	0.70	6.0	0.59
LDPE, amber 200 G	—	—	12.70	0.71	6.80	0.68
Saran/Cello, Saran/PE	7.20	1.14	13.10	0.62	6.70	0.57
Aluminum foil laminate	5.60	1.14	5.50	0.89	5.50	0.88
Kraft paper	13.90	0.82	Highly moldy	—	7.50	0.70
Grease-proof paper	13.70	0.92	Highly moldy	—	8.70	0.58

Source: Ref. 47.

laminate had no advantage over plain high-density polyethylene film packs. Chili powders are generally packed in small quantities in simple packages that can be used within 30 days.

3. Oleoresins

Oleoresins are desolventized total extracts or products obtained by beneficiation of one or more functional components of total extracts by some amount of fractionation (48). They are used as food additives. Oleoresins contain the components that constitute color and flavor (pungency, aroma, and related sensory factors) and truly recreate upon dilution the sensory qualities of the original fresh mateiral (48). Various solvents are used for the extraction of oleoresins. The yield and quality of oleoresins depends upon the solvent used (Table 8). Ethylene dichloride is a commonly used solvent for the extraction of oleoresin. The yield of oleoresin ranges from 8.0 to 17.4% (Table 9). The use of oleoresins proves economical as well as convenient since con-

TABLE 8 Yield of Oleoresin, Color, and Capsaicinoids with Different Solvents from Chili (*C. annum* var. *acuminatum* L.)

Solvent	Yield of oleoresin (%)	Color recovery (%)	Capsaicinoids (%)
Alcohol	17.5	19.7	0.523
Acetone	15.6	67.7	0.515
Chloroform	16.4	61.1	0.587
Ether	16.1	66.5	0.701
Hexane	15.0	58.1	0.605

Source: Ref. 48.

Table 9 Capsaicinoid Content and Oleoresin Yield of Some Chili Varieties

Variety	Capsaicinoids in chili (%)	Oleoresin yield (%)	Capsaicinoids in oleoresin (%)
Mombasa, Uganda	0.9–0.85	12.0–12.5	6.8–6.9
Mombasa, African	0.42	13.1	3.2
Small chilies			
African	0.82	13.3	6.2
Bahamian	0.51	12.5	4.1
Bird chilies	0.36	8.7	4.1
India	0.56	13.0	4.3
Santaka, Japan	0.30	11.5	2.6
Sannam, India	0.33	16.5	2.0
Mundu, India	0.23	16.0	1.4
Jwala, India	0.63	9.0	7.0
Coimbatore, India	0.45	17.4	2.6

Source: Ref. 48.

centrated forms save on transport and storage over bulky inventory of powdered spices. They are more stable than the whole or powdered forms. The United States is the largest consumer of oleoresins.

II. BELL PEPPERS

Bell pepper, *C. annuum* var. *grossum* (Syn. Simla mirch, sweet pepper, Bullnose capsicum), is widely cultivated in almost all parts of India, Central and South America, Pery, Bolivia, Costa Rica, Mexico, Hong Kong, and almost all European countries (49). The major countries producing bell peppers are listed in Table 10. Sweet peppers, both green and red, are eaten raw

Table 10 World Production of Green Peppers

Continent/Country	Production (1000 MT)
World	10630
China	2977
Turkey	965
Nigeria	900
Mexico	760
Spain	673
Indonesia	440
Italy	314
Korea Republic	270
United States	260
Algeria	215
Egypt	190
Netherlands	171
Hungary	130

Source: Ref. 49.

or after cooking. They are used in salads, in stews for imparting flavor, pizza, meatloaf, dehydrated processed meat, and for canning. They are also used in pickles, brine with cucumbers, baking, and stuffing.

A. Botany: Classification

The classification system for peppers was developed by Dr. P. G. Smith, Department of Vegetable Crops, University of California, Davis. The system is based on grouping cultivars that are horticulturally similar in major characteristics such as fruit shape, size, color, texture, flavor, and pungency as well as uses. It also provides alternate sources for various processed products of pepper industry. Accordingly, the following classification has been proposed.

Bell group: The varieties belonging to this group bear large fruits (7–12 cm × 5–10 cm) with a smooth surface, thick flesh, and blocky, blunt shape having three to four lobes or square to rectangular or tapering longitudinally. Color is usually green at the immature stage, turning red at full maturity. Most of the cultivars are nonpungent, although a few pungent forms are known. They are mostly used for fresh market, salads, pizza, meatloaf, processed meat, canning, etc. (e.g., California Wonder and Yellow Wonder).

Pimiento group: The cultivars of this group produce large green fruits (5–10 cm × 5–7 cm) turning red, smooth, and thick-walled. They are nonpungent and commonly used for fresh market, salds, soups, processed meat, canning, meatloaf, etc. Cultivars such as Pimiento and Pimiento Select belong to this group.

Squash or cheese group: In this group, the genotypes bear small to large fruits (2.5–5 cm × 5–10 cm), which are wider than deep, scalloped or rounded flat or semipointed, smooth or rough. Fruits are medium- to thick-walled and nonpungent, medium green or yellow to mature red. These are used for processing, canning, freezing, pickling, as well as for culinary purposes and as salads. Cultivars like Cheese, Gambo, and Antibois belong to this group.

Ancho group: Fruits from this group are large (10–12 cm × 5–7 cm), heart-shaped, smooth, thin-walled, cup-shaped, mildly pungent, and used as dried whole powdered, fresh roasted, and peeled. Cultivars like Mild California, New Mexican Chili, and Big Jam are examples of this group.

Anaheim chili group: Pods of this group are long, slender (12–15 cm × 2–3 cm), tapering toward the stylar end, smooth, with medium thick flesh, medium to dark green in color, turning red at maturity, sweet to moderately pungent in taste. The fruits are normally dehydrated as whole pods or processed into powder. They are also used for canning and preparation of sauces. The cultivars California Chili and Paprika belong to this group.

Cayenne group: Fruits of this group are long, slender (12–22 cm × 1.5–6 cm), thin-walled, and medium green turning red at maturity. Fruits are irregular in shape with wrinkles and are highly pungent. They are commonly used for fresh market, powder, hot sauce, pickling, salads, and culinary purposes.

Cuben group: Fruits of this group are long (10–20 cm × 1.5–5 cm), yellowish-green turning red at maturity, thin-walled, irregular, blunt, and mildly pungent. They are used for fresh market, salads, pickles, and frying.

Jalapeno group: Fruits of this group are elongated (5–7 cm × 2–5 cm), round, cylindrical in shape, smooth, thick-walled, dark green turning red at maturity. They are irregular in shape with or without a corky network on the skin of mature fruits. They are highly pungent and used as fresh green peppers or as mature dried pods. They are also used for canning and in the preparation of sauces. One cultivar belonging to t his group is the Jalapeno.

Small hot group: Fruits of this group are slender (4–7 cm × 1–2 cm), medium- to thin-walled, green becoming red with maturity, and highly pungent. They are used as fresh green pods or in the form of dried powder and sauces. The cultivars belonging to this group include Red Chili and Sontaka.

Cherry group: Fruits of this group are small, spherical (2–5 cm in diameter), somewhat flat, thick, and green becoming red at maturity. Fruits are pungent and used for pickling. The cultivars Red Cherry Large and Red Cherry Small belong to this group.

Short wax group: Fruits of this group (5–7 cm × 2–5 cm) are yellow and turn orange red at maturity. Fruits are smooth, medium- to thick-walled, and tapered. They are used in fresh pickling, processing, sauce, cooking, etc. The cultivars Floral Jam, Cascabella, and Caloro are examples of this group.

Long wax group: Fruits of this group (8–12 cm × 1–4 cm) are yellow turning red, pointed, or blunt. Fruits are used for fresh market, pickle, sauces, canning, etc. The cultivars Hungarian Yellow Wax and Sweet Banana belong to this group.

Tabasco group: Fruits of this group are slender (2–5 × 0.5 cm), yellow or yellowish to red, and highly pungent in taste. They are used for vinegar, sauce, and pickles. Cultivars belonging to this group include Tabasco and Green Leaf Tabasco.

B. Production

1. Soil and Climate

Bell peppers can be grown on almost any type of soil, but well-drained clayey loam soil is considered ideal. On sandy loam soils, crops can be grown successfully provided that heavy manuring is done. A soil pH between 6.0 and 6.5 is considered ideal. However, peppers can be grown in soils with a pH range of 5–9. Peppers are grown in tropical as well as subtropical regions. In general, sweet peppers are grown under lower temperature conditions than hot peppers. Heavy rainfall during flowering and fruiting causes heavy flower drop and fruit rotting. Similarly, moisture stress during flowering flavors flower bud abscission, causing heavy flower drop. Temperatures higher than 37.8°C affect fruit development adversely (50).

2. Planting

Transplant spacing varies according to soil type and locality. In heavy soils, more space is required than for medium and light soils. Seedlings are transplated at 60 × 45, 45 × 45, or 60 × 30 cm. The highest yield of bell pepper was recorded by Gunewardhane and Pereira (17) with 90 × 30 cm spacing. Planting of the cultivar California Wonder at 45 × 45 cm resulted in the highest yield (17).

3. Manuring and Fertilization

Bell pepper responds well to applied fertilizers. Nutrient uptake by the plant is very high. Hence, correct manuring practices with both organic and inorganic nutrients were found to increase crop growth as well as yield (50). Shrinivas and Prabhakar (51) reported that applicaiton of approximately 150 kg of N per hectare in three splits increased the yield of the cultivar California Wonder. Shukla et al. (52) recorded maximum yield of bell pepper when the crop was supplied with 180 kg of N and 50 kg of P_2O_5 per hectare. Application of phosphorus is essential for early and increased yield. Similarly, potassium is necessary, as it increases uptake of both P and K. Application of potash in the form of potassium chloride was found more beneficial for increasing yield than as the potassium sulfate (52,53).

4. Irrigation

The first irrigation should be given immediately after transplanting. Then light irrigation at an interval of 2-3 days should be given till the establishment of seedlings. Later the field should be irrigated at an interval of 6 to 8 days depending on soil type. In summer, irrigation at an interval of 4 to 6 days is essential. In winter, irrigation at an interval of 8 to 10 days is optimum. Frequent irrigations decrease the yield significantly.

5. Weed Management

The field should be kept as weed-free as possible. Three hand-weedings at 3, 6, and 12 weeks after transplanting increase the yield significantly. Weeds can be effectively controlled by using grass mulch or black polyethylene mulch.

6. Harvesting

Fruits are usually harvested when they are of market size and still green in color. The fruits are picked by snapping off the brittle stem. Then they are generally graded and packed. Fruits will show shriveling if they are picked at an immature stage. However, they will remain in good condition on the plant for a few days. Boswell (54) stated that bell peppers harvested at the immature green stage should be well shaped, waxy, firm, and shiny. Peppers for canning are allowed to attain full size. Fruits used for sauce are allowed to ripen on the plant before harvesting.

7. Packaging

Bell peppers are generally not prepacked, although the practice certainly would retard wilting (55). Fruits with a large surface-to-volume ratio, such as long or small tgypes, are particularly susceptible to water loss (31). Adequate ventilation for the packages is necessary to avoid moisture condensation inside the package (56–61).

Peppers rapidly lose water after harvest, limiting longevity. New Mexican-type peppers become flaccid in 3–5 days (57,58) at 20°C (7–10% weight loss). Lownds et al. (56) reported that placing pepper fruit in perforated polyethylene packages reduced water loss rates 20 times or more so that water loss no longer limited postharvest storage. Packaging also eliminated flaccidity (Table 11) and reduced color development across cultivars at 14 and 20°C (Table 12). Packaged fruit, however, developed diseases that limited postharvest longevity.

Table 11 Effect of Storage Temperature on Flaccidity Rating for Unpackaged Fruit of Nine Pepper Cultivars After Storage for 14 days at 8, 14, or 20°C

		Firmness		
Type	Cultivar	8°C	14°C	20°C
Bell	Keystone	1.1	7.3	7.7
	Mxibell	1.0	7.5	8.2
New Mexican	NuMex R Naky	1.0	9.0	9.0
	NuMex Conquistador	1.3	8.7	9.0
	New Mexico 6–4	1.0	9.0	9.0
Yellow wax	Santa Fe Grande	2.0	8.8	9.0
	Cascabella	1.8	8.5	9.0
Jalapeno	TAM Jalapeno	1.4	6.7	6.5
Serrano	TAM Hidalgo	1.7	9.0	8.7
LSD 0.05		0.6	0.6	0.4

Firmness rating based on a 0-to-9 scale, where 0 = firm and 9 = soft.
Source: Ref. 54.

TABLE 12 Effect of Storage Temperature and Packaging on Color Rating for Nine Pepper Cultivars After 14 Days of Storage at 8, 14, or 20°C

		Color rating[a]					
		8°C		14°C		20°C	
Type	Cultivar	NP	P	NP	P	NP	P
Bell	Keystone	1.3	1.6	6.3	2.3	7.7	3.9
	Mexibell	3.0	2.0	7.8	5.2	8.3	9.0
New Mexican	NuMex R Naky	3.7	2.0	9.0	7.7	9.0	6.5
	NuMex Conquistador	2.2	2.0	6.8	2.3	9.0	2.7
	New Mexico 6–4	2.5	2.0	9.0	5.8	9.0	4.5
Yellow wax	Santa Fe Grande	2.8	1.8	8.8	3.0	9.0	4.2
	Cascabella	2.3	2.1	8.5	3.2	9.0	5.2
Jalapeno	TAM Jalapeno	2.0	1.7	4.7	1.8	5.3	2.1
Serrano	TAM Hidalgo	2.7	2.3	6.0	2.2	9.0	4.2
LSD 0.005		0.7		2.4		1.4	
Significance							
Cultivar		0.058		0.001		0.001	
Package		0.001		0.001		0.001	
Cultivar × package		0.015		0.230		0.001	

NP = Nonpackaged; P = packaged.
[a]Color rating based on 0 = 100% green and 9 = 1300% red.
Source: Ref. 54.

C. Chemical Composition

Sweet peppers are very rich in vitamins, especially vitamins A and C (Table 13). They are also a good source of β-carotene. Capsaicin as a pungent principle is oxidized by peroxidase present in the fruit (62). Green fruits contain chlorophyll A and Chlorophyll B, which are probably synthesized de novo during chloroplast development (63).

D. Storage

1. Low-Temperature Storage

Sweet, or bell, peppers are subject to chilling injury at temperatures below 7°C, whereas temperatures above 13°C encourage ripening and spread of bacterial soft rot. Bell peppers should not be stored longer than 2–3 weeks even under the most favorable conditions. At 0–2°C, peppers usually develop pitting in a few days. Peppers held below 7°C long enough to cause serious chilling injury also develop numerous lesions due to *Alternaria* rot (64,65). Holding at 4.5°C and below predisposes peppers to *Botrytis* decay (66).

Sweet peppers prepackaged in moisture-retentive films, such as perforated polyethylene, have a storage life at 7–10°C up to a week longer than nonpacked peppers (67). The use of film crate liners can help in reducing moisture loss.

Bell pepper fruits are commonly waxed before shipment to reduce moisture loss and scuffing. Strict sanitation of the packing line and drying of surface moisture on the stem and

TABLE 13 Composition of Bell Pepper per 100 g Edible Portion

Constituent	Content
Energy	48 kcal
Proteins	2.0 g
Fats	0.8 g
Carbohydrates	10 g
Fiber	2.6 g
Calcium	29 mg
Phosphorus	61 mg
Iron	2.6 mg
Thiamine	0.12 mg
Riboflavin	0.15 mg
Niacin	2.2 mg
Ascorbic acid	140 mg

Source: Ref. 2.

calyx are essential to avoid increasing decay due to waxing treatment (67). Waxing provides some surface lubrication, which not only reduces chafing in transit but also reduces shrinkage; the result is longer storage and shelf life (68). Senescence of sweet peppers is hastened by ethylene. Therefore, it is not a good practice to store peppers with apples, pears, tomatoes, or other ethylene-producing fruits.

Rapid precooling of harvested sweet peppers is essential for reducing marketing losses, and it can be accomplished by forced air cooling, hydrocooling, or vacuum cooling. Properly vented cartons are recommended to facilitate forced-air cooling. Hydrocooling has been found to increase the incidence of decay (69,70). A blast of cool air directed at the fruits after they leave the hydrocooler may dry them sufficiently to check decay (70). Fruits must be kept at high relative humidity of about 95%, otherwise they will quickly become flabby. Drying conditions also accentuate symptoms of chilling injury by accelerating hydration of injured tissue and thus pitting. Wetting of fruits for 1 or two hours for retail display is harmless at 4.4–7.2°C but increases decay as temperatures approach or exceed 10°C (55).

2. Controlled-Atmosphere Storage

Low-oxygen atmospheres retard ripening and respiration during transit and storage (71). High concentrations of carbon dioxide delay the loss of green color. However, high carbon dioxide also causes calyx discoloration (72). Ryall and Lipton (55) reported that oxygen below 2% combined with carbon dioxide at 10% caused injury at 12.7°C. Hatton et al. (72) observed that the transport life of fruits would be extended when they were held at 4–8% O_2 and 2–8% CO_2. The storage life of fruits of the cultivar California Wonder was prolonged for 38 days compared to 22 days in air by holding fruits at 5% O_2 and 10% CO_2 at 8.8°C (73).

Hughes et al. (73) studied bell pepper storage under controlled atmosphere, modified atmosphere, and hypobaric conditions. The green cultivar Bellyboy stored at 8°C in various controlled atmospheres showed greater weight loss and no increase in storage life as compared to air controls. Moreover, fruit decay increased when stored at 6% CO_2 + 2% O_2. Fruits wrapped

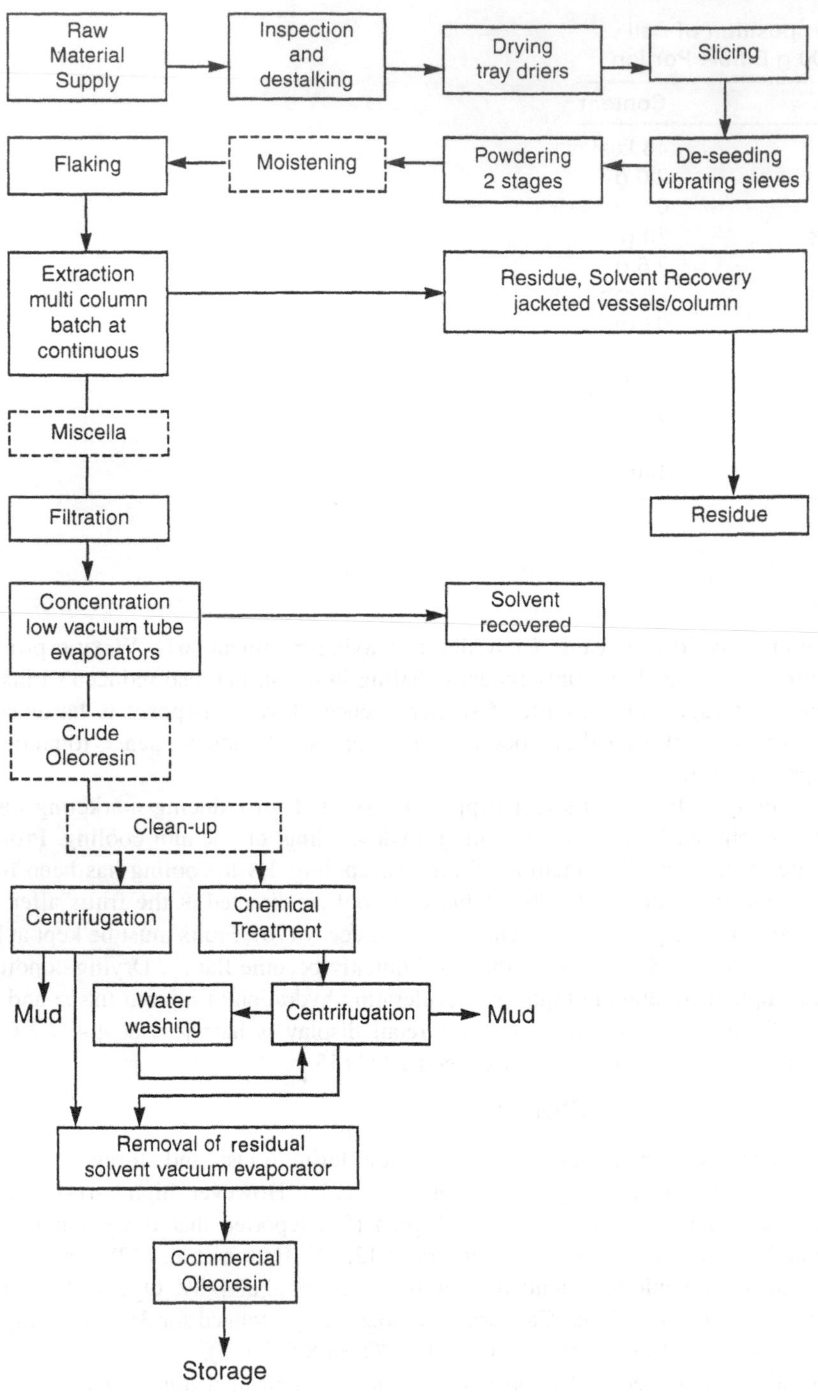

FIGURE 3 Flow sheet for paprika oleoresin production. (From Ref. 48.)

in plastic film showed a significant increase in storage life and had less weight loss than in other treatments. Hypobaric storage did not improve storage life and greatly increased weight loss. Hughes et al. (73) concluded that plastic wraps improved the storage life of peppers.

E. Processing

Sweet peppers are important components of mixed pickles. Commercially mixed vegetable pickles are usually made from vegetables temporarily preserved by fermenting and stored in brines varying in salt content from about 7 to 15%. This product is known as a salt stock. It is further processed by desalting followed by finishing in suitably spiced, sweetened vinegar. The finished product is then generally subjected to mild thermal processing or pasteurization (74).

1. Freezing

Bell peppers freeze very well (75). Frozen peppers are sold both red (mature) and green. The harvested peppers are washed, inspected and graded for color and size, and then fed to a coring machine, which removes and ejects the stem, pithy placenta, and seeds and cuts the peppers in halves. The halves are washed under strong water sprays to remove any remaining seeds or pithy material. Green and red peppers freeze very well unblanched but tend to develop an objectionable flavor after 6 months in storage. To achieve low bacterial and mold counts, it is common to blanch peppers in water or steam for about 2 minutes. Diced bell peppers are packed in cartons, while whole or halved peppers are packed in polyethylene bags. Most frozen bell peppers are destined for food service use or remanufacturing use in mixed frozen or canned vegetables (75).

2. Dehydration

Dehydrated peppers are sold primarily in diced form to remanufacture as an ingredient in canned stewed tomatoes and dry soup mixes. The fruits are diced and sprayed with a sulfite solution to give a final SO_2 content of 1000–2500 ppm (75). The diced peppers are dried to 6% moisture and ground. The powder is rehydrated to 12%, the optimum moisture level for storage under refrigeration.

3. Oleoresins

The production of paprika oleoresins in Spain and Hungary is handled by a few exclusive plants on a large scale. The process is summarized in Figure 3. Oleoresin is characterized by very high color, low or no pungency, and a characteristic sweet odor. It is available in two forms: free from pungency and pungent. The first is used as a coloring agent in prepared meat, salad dressing, and margarines, the pungent form as a flavoring and coloring for snack foods.

REFERENCES

1. Rajput, J. C., and Wagh, R. G., *Maharashtratil Bhajipala Lagwad* (book in Marathi), 1994.
2. Shukla, V., and L. B. Naik, Chilli, in *Advances in Horticulture* (K. L. Chadha and G. Kalloo, eds.), Maholtra Pub. House, New Delhi, 1993, p. 380.
3. Heisen, B., Jr., Peppers; *Capsicum* (Solaneceae), in *Evoluation of Crop Plants* (N. W. Simmonds, ed.), Longman, London, 1976, p. 265.
4. Redgrove, H. S., *Spices and Condiments*, Pitman, London, 1930.

5. Lopes, C. R., M. M. Viera, and N. Silva, Chemotaxanomic study of genus *Capsicum*, *Ciencia C. Culture 30*: 526 (1928).
6. Pradeepkumar, J., Inter-specific hybridization in capsicum, M. Sc. thesis, Kerala Agricultural University, Vellanikara, 1990.
7. Reddy, K. M., B. A. Sathe and P. V. Reddy, Studies on active root distribution by employing 32p soil injection technique under irrigated conditions, *Proc. Nat. Seminar on Chillies, Ginger and Turmeric*, Hyderabad, India, 1988.
8. Hawthorn, L. R., and H. Pollard, *Vegetable and Flower Seed Production*, Blackiston Company, Inc., New York, 1954.
9. Thomas, K. M., No need to be chery of chillies, *Indian Farming 11*(7):27 (1961).
10. Singh, A., and Singh, H. N., Inheritance of quantitative characters in chilli (*Capsicum annuum* L.), *Plant Sci. 10*:17 (1976).
11. Lawande, K. E., S. B. Raijadhav, P. N. Kale, and K. G. Choudhary, Status of chilli research in Maharashtra: A review, *Proceeding of Nat. Seminar on Chillies, Ginger and Turmeric*, Hyderabad, India, 1988, p. 13–16.
12. Tikloo, S. K., Crop management of soleneaceous vegetables: Tomato, brinjal and chillies, in *Recent Advances in Tropical Vegetable Production*, Kerala Agricultural University, Vellanikara, India, 1991.
13. Purseglove, J. W., *Tropical Crops*, Vols. I and II, Longman, London, 1977, p. 524.
14. Gopalkrishnan, T. R., Inheritance of clusterness, destalkness and deep red colour in chilli (*Capsicum annuum*. L.), Ph.D. thesis, Kerala Agricultural University, Vellanikara, India,
15. Premnath, S. Velayudhan, and D. P. Singh, *Vegetables for the Tropical Region*, ICAR Publication, Pusa, New Delhi, 1994.
16. Rajput, J. C., S. B. Palve, B. P. Patil, and M. J. Salvi, Recent research on chilli in Konkan region of Maharashtra, *Proc. of Nat. Seminar on Chillies, Ginger and Turmeric*, Hyderabad, India, 1988, pp 6–12.
17. Gunewardhane, S. D. I. E., and P. D. Pereira, Agronomic investigations with chilli in West Zone studies on interaction between fertilizer levels and plant density, *Trop. Agric. 52*:134 (1975).
18. Rajput, J. C., B. M. Jamadagni, P. A. Fugro, and R. C. Wagh, Response to the applied potassium in vegetable crops in Maharashtra, Workshop on Use of Potassium in Maharashtra Agriculture, organized by Potash and Phosphate Institute of Canada-Indian Programme, Gurgaon (Haryana) and Konkan Krishi Vidyapeeth, Dapoli, March 20, 1993.
19. Veeraraghavatham, D., S. Sundararajan, S. Jayashankar, E. Vadivel, and K. G. Shanmugavelu, Effect of nitrogen, phosphorus and *Azospirillum* inoculation of chilli (*Capsicum annuum* L.), *Proc. Nat. Seminar on Chillies, Ginger and Turmeric*, Hyderabad, India, 1988, pp. 65–70.
20. Rajput, J. C., P. A. Fugro, G. D. Joshi, and C. V. Nimkar, Vegetable production for export—research strategies, *Proc. Vasantrao Naik Memorial Nat. Seminar on Export Oriented Horticultural Production—Research and Strategies*, held at College of Agriculture, Nagpur, Dec. 5–6, 1993.
21. Pawar, N. G., M. G. Desai, and M. J. Salvi, Studies on effect of application of different starter solutions on flowering and fruiting in chilli (*Capsicum annuum* L.), *South Indian Hort. 33*:240 (1985).
22. Annual Report, Tamilnadu Agricultural University, Coimbatore, 1977.
23. Annual Report, Tamilnadu Agricultural University, Coimbatore, 1979.
24. Scientific Workers Conference Horticulture, Tamilnadu Agricultural University, Coimbatore, 1980.
25. Saimbhi, M. S., and K. S. Randhava, Herbicide control of weeds in transplanted chilli, *Punjab Hort. J. 16*:146 (1976).
26. Annual Report, Tamilnadu Agricultural University, Coimbatore, India, 1981.
27. Ignatov, B., The effect of some herbicides on different capsicum varieties grown on calcereous chernozem soils, *Gradinarska Loxarsk Nauba 9*(2):61.
28. Filipov, G. A., and F. P. Inshakov, The use of herbicides in crops of pepper and eggplants, *Khimira v. Selskom Khozyaistre 17*(10):38 (1979).
29. Whiting, F. C., C. F. Lippert, and J. M. Lyons, Chemical weed control in peppers, *Calif. Agric. 24*(7):8 (1970).

30. Dempsey, A. H., and T. A. Fretz, New herbicides for weed control in pepper, *Research Report, Georgia Agric. Expt. Sta.*, No. 153, 1973.
31. Yamgar, V. T., U. T. Desai, P. N. Kale, and K. G. Choudhari, Fruit and seed yield in chilli (*Capsicum annuum* L.) as influenced by naphthalene acetic acid, *Veg. Sci. 13*:83 (1986).
32. Patil, V. B., P. B. Sangale, and B. K. Desai. Chemical regulation of yield and composition of chilli (*Capsicum annuum* L.) fruits, *Curr Res. Rep. 1*:39 (1986).
33. Usha, P., and K. V. Peter, Control flower fall through stimulants and regulators in chillies, *Veg. Sci. 15*:185 (1988).
34. Gorcha, M. S., Effect of ethephon on ripening and quality of chillies (*Capsicum annum* L.), M.Sc. thesis, Punjab Agricultural University, Ludhiana, 1990.
35. Kanade, V. M., J. C. Rajput, C. B. Gondhalekar, and R. G. Wagh, Effect of plant growth regulators on growth and yield of chilli (*Capsicum annum* L.) var. Konkan Kirti, *Pestology 16*(4):14 (1992).
36. Sivaprakasam, K., R. Jaganathan, K. Pillayarswamy, and L. Anavaradham, Control of powdery mildew of chillies. *Madras Agric. J. 63*:52 (1976).
37. Eswarmurthy, S., C. M. Papaiah, M. Muthuswamy, S. Muthuswamy, and P. Gomathinayagam, Chemical control of die-back of chillies, *Pesticides 22*(3):38 (1988).
38. Eswarmurthy, S., S. Chelliah, and S. Jayaraj, *Aphidus platensis* Brethes. A potential parasite on *Myzus persicae* Suly, *Madras Agric. J. 63*:182 (1976).
39. Reddy, K. M., and N. S. Murthy, Effect of number of pickings on yield of chillies under irrigated conditions, in *Proc. Nat. Seminar on Chillies, Ginger and Turmeric*, Hyderabad, India, 1988, pp. 52–53.
40. Anandaswamy, B., H. B. N. Murthy, and N. V. R. Iyengar, Prepaking studies on fresh produce, *C. grossum Sendt.* and *C. acuminatum* Fingerh, *J. Sci. Indust. Res. India* Sect. A. *18*(a):274 (1959).
41. Kosuge, S., and P. Furita, Chemical composition of the pungent principles, *Agric. Biol. Chem. 34*:248 (19XX).
42. Platenius, H., F. S. Jamison, and H. C. Thompson, Studies on cold storage of vegetables, *Cornell Bull.*:602 (1934).
43. Mohammed, M., Effects of polyethylene bags, temperature and time on storage of two hot pepper (*Capsicum frutescens* L.) cultivars, *Trop. Agric.* (Trinidad) *67*:194 (1990).
44. Mohammed, M., Storage of sweet peppers and tomatoes in polymeric films, M.Sc. thesis, University of Guelph, Canada, 1984.
45. Worthington, J. T., Moisture content and texture of dried chilli peppers stored in polyethylene lined burlap bags, U.S. Dept. of Agric. Market. Serv. AMS, 664, 1961, p. 6.
46. Lease, J. G., and E. J. Lease, Effect of drying conditions on initial color, color retention and pungency of peppers, *Food Technol. 16*(11):104 (1962).
47. Mahadeviah, B., K. S. Chang, and NB. Balasubrahmanyam, Packaging and storage studies on dried whole and ground chillies (*Capsicum annuum*) in flexible consumer packages, *Indian Food Packer 30*(6):33 (1976).
48. Govindarajan, V. S., Capsicum: Production, technology, chemistry and quality, Part II—Processes products, standards world production and trade, *CRC Crit. Rev. Food Sci. Nutr. 23*(3):207 (1984).
49. FAO, *Production Year Book*, Food and Agriculture Organization, Rome, Italy, 1993.
50. Rastogi, K. B., B. N. Karla, and S. S. Saini, Effect of different levels of nitrogen and spacing on fruit yield of bell pepper, *Punjab Hort. J. 20*:88 (1980).
51. Srinavas, M., and B. S. Prabhakar, Response of capsicum to nitrogen fertilization, *Veg. Sci. 9*:71 (1982).
52. Shukla, V., K. Srinavas, and B. S. Prabhakar, Response bell pepper to nitrogen, phosphorus and potash fertilization, *Indian J. Hort. 44*:81 (1987).
53. Thompson, H. C., and W. C. Kelly, *Vegetable Crops*, 5th ed., McGraw-Hill, New York, 1957, p. 148.
54. Boswell, V. R., *Pepper Production*, USDA, Inf. Bull. No. 276, 1964.
55. Ryall, A. L., and W. J. Lipton, *Handling, Transportation and Storage of Fruits and Vegetables*. Vol. 1. *Vegetables and Melons*, The AVI Publishing Co. Inc., Westport, CT, 1972, p. 473.
56. Lownds, N. K., M. Banaras, and P. W. Bosland, Postharvest water loss and storage quality of nine pepper (Capsicum) cultivars, *HortScience 29*(3):191 (1994).

57. Lownds, N. K., and P. W. Bosland, Studies on postharvest storage of pepper fruits, *HortScience 23*:71 (1988).
58. Miller, W. R., L. A. Risse, and R. E. McDonald, Deterioration of individually wrapped and non-wrapped bell peppers during long-term storage, *Trop. Sci. 63*:1 (1986).
59. Ben-Yehoshau, S., Individual seal-packaging of fruit and vegetables in plastic film—a new postharvest technique, *HortScience 20*:32 (1985).
60. Ben-Yehoshua, S., B. Shapiro, J. E. Chen, and S. Lurie, Mode of action of plastic film in extending life of lemon and bell pepper fruits by alleviation of water stress, *Plant Physiol. 73*:87 (1983).
61. Bussel, J., and Z. Kenigsberger, Packaging green bell pepper in selected permeability films, *J. Food Sci. 40*:1300 (1975).
62. Bernal, M. A., A. A. Calderon, M. A. Pedreno, R. Munoz, A. Ros Barcelo, and F. Merino De Caderes, Capsaicin oxidation by peroxidase from *Capsicum annuum* (var. annuum) fruits, *J. Agric. Food SChem. 41*(7):1041 (1993).
63. Newman, L. A., N. Hadjeb, and C. A. Price, Synthesis of two chloroplast specific proteins during fruit development in *Capsicum annuum, Plant Physiol. 91*(2):455 (1989).
64. McColloch, L. P., Chilling injury and alternaria rot of bell peppers, U.S. Dept. of Agri Market Res. Rep. 536, 1962, p. 16.
65. Morris, L. L., and H. Platenius, Low temperature injury to certain vegetable after harvest, *Proc. Am. Soc. Hort. Sci. 36*:609 (1938).
66. McColloch, L. P., and W. R. Wright, *Botrytis* rot of bell peppers, U.S. Dept. Agric. Market. Res. Rept. 754, 1966, p. 9.
67. Hardenburg, R. E., A. E. Watada, and C. Y. Wang, *The Comercial Storage of Fruits, Vegetables and Florist and Nursery Stocks*, U.S. Department of Agriculture, Handbook No. 68, 1986, p. 65.
68. Hartman, J. D., and F. M. Isenburg, Waxing vegetables, N.Y. State Col. Agric. Cornell Ext. Bull. 1956, p. 1.
69. Morris, L. L., and A. A. Kader, Commodity requirements and recommendations for transport and storage—selected vegetables (D. H. Dewey, ed.), *Proc. 2nd Natl. Controlled Atmos. Res. Conf. Mich. State Univ. Hort. Rpt.*, 28, 1977, p. 266.
70. Johnson, H. B., Effect of hot water treatment and hydrocooling on postharvest bacterial soft rot in bell peppers, USDA, AMS-517, 1964.
71. Pantastico, E. B., T. K. Chattopadhyay, and H. Subramanyam, Storage and commercial storage operations, in *Postharvest Physiology Handling and Utilization of Tropical on Subtropical Fruits and Vegetables*, The AVI Publishing Co. Inc., Westport, CT, 1975, p. 314.
72. Hatton, T. T., Jr., E. B. Pantastico, and E. K. Akamine, Controlled atmosphere storage; 3-individual commodity requirements, in *Postharvest Physiology, Handling and Utilization of Tropical and Sub-tropical Fruits and Vegetables* (E. B. Pantastico, ed.), The AVI Publishing Co. Inc., Westport, CT, 1975, p. 201.
73. Hughes, P. A., A. K. Thompson, R. A. Pumbley, and G. B. Seymour, Storage of capsicuum (*Capsicum annuum* L.) under controlled atmosphere, modified atmosphere and hypobaric conditions, *J. Hort Sci.* 261 (1981).
74. Wang, C. Y., Effect of CO_2 treatment on storage and shelf-life of sweet pepper, *J. Am. Soc. Hort. Sci. 102*:808 (1977).
75. Jones, L. D., Effects of processing by fermentation on nutrients, in *Nutritional Evaluation of Food Processing* (R. S. Harris and E. Karmas, eds.), The AVI Publishing Co. Inc., Westport, CT, 1975, p. 324.

9

Eggplant (Brinjal)

K. E. LAWANDE AND J. K. CHAVAN
Mahatma Phule Agricultural University, Rahuri, India

I. INTRODUCTION

Eggplant or brinjal is thought to have originated in tropical India (1). It is a staple vegetable in many tropical countries. In the United States, the South Atlantic States, New Jersey, and California grow considerable quantities of eggplant for shipment, although the fruit is of minor importance in the United States. It is of great importance in the warm areas of the Far East and is grown more extensively in India, China, and the Philippines (2). The world production of eggplant increased from 3.48 MT in 1969–71 to 8.68 MT in 1992 (Table 1). The major producers of eggplant are China, India, Japan, and Turkey. Eggplant is known to have some ayurvedic medicinal properties and is said to be good for diabetic patients. It has also been recommended as an excellent remedy for those suffering from liver complaints (1).

II. BOTANY

A. Types

Eggplant (*Solanum melongena* L.) belongs to the Solanaceae (or nightshade) family. There are three main botanical varieties under the species *melongena*. The round or egg-shaped cultivars are grouped under var. *esculentum*. The long slender types are included under var. *serpentinum*, and the dwarf brinjal plants are categorized under var. *depressum* (4). The common eggplant, to which the large-fruited forms (e.g., New York Improved) belong, is known under the name *S. melongena* var. *esculentum*. The plant is bushy and grows to a height of 60–120 cm; the leaves are large and arranged alternately on the stems. The flowers are large, violet-colored, and either solitary or in clusters of two or more. The serpentine or snake eggplants are placed under the variety *Serpentinum*. The fruits of this group are long and slender, 2.0 cm or less in diameter, and 30–40 cm long. The dwarf eggplants are known under the variety *depressum*, which produce small pear-shaped fruits that are purple in color.

TABLE 1 Worldwide Production of Eggplant

Continent/Country	Production (1000 MT)
World	8682
Africa	532
Algeria	25
Cote Divoire	40
Egypt	350
Sudan	70
North and Central America	86
Dominican RP	14
Mexico	30
USA	37
Asia[a]	7501
Indonesia	175
Iraq	160
Japan	520
Philippines	112
Europe	535
France	27
Greece	71
Italy	261
Spain	130

[a]Production figure not available for India.
Source: Ref. 3.

B. Morphology

Eggplant is a very tender plant requiring a long, warm growing season for successful production (2). The aboveground portion of the plant is erect, compact, and well branched (Fig. 1). The leaves are large, simple, and lobed (Fig. 2). The underside of the leaves of most cultivars are covered with dense wool-like hairs. The flowers are large with purple corolla (4). The flowers are hermaphroditic, and the stamens dehisce when the stigma becomes receptive. Thus, self-pollination is the rule, although there is some cross-pollination by insects. The pendant fruit is a fleshy berry, borne singly or in clusters in the axils of leaves. The color of the mature fruits varies from purple, purple-black, white or yellowish-white to green and striped, depending upon the cultivar. The seeds are borne on the fleshy placenta, which together completely fill the cavity (4).

C. Cultivars

The white, yellow, brown, green, and striped eggplant varieties and elongated or pear-shaped forms are not popular in the United States, where all commercially important varieties are dark purple, large-fruited, and more or less round. Black Beauty and Fort Mayers Market are the preferred varieties among market especially in North America. Florida High Bush is a vigorous variety, upstanding in growth and well adapted to the climatic conditions of the principal shopping districts. Early Long Purple is an extra early sort for use where the warm season is too short for the larger varieties. The fruits are oblong, slender, and not popular in the markets. Some other early varieties are New Hampshire Hybrid, Bountiful, and Blackie.

Extensive work has been done in India to develop improved eggplant cultivars (4–11). The

FIGURE 1 Brinjal plant.

FIGURE 2 Leaf, flower, and fruit of eggplant.

FIGURE 3 Fruits of "Krishna" cultivar of eggplant.

popular cultivars in India are Black Beauty, Long Black, Muktakeshi, Round Purple, Pusa Purple Long, Pusa Purple Round, Pusa Purple Cluster, Long Purple, Banaras Giant, Pusa Kranti, Arka Navneet, and several other hybrids such as Vijay, S-1, S-4, S-5, S-8, S-16 (Ludhiana), H-4 (Hissar), and A-61 (Pantnagar). The high-yielding varieties grown extensively in Maharashtra include Manjri Gota, Vaishali, Pragati, and Krishna (Fig. 3) (13–15). A hybrid between Pusa Purple Long and Hyderpur developed by the Indian Agricultural Research Institute, New Delhi, has been found to be a promising variety with very attractive, long to oblong, purple fruits.

III. PRODUCTION

A. Soil and Climate

Eggplant can be grown on a wide variety of soils. The major soil considerations are texture, drainage, fertility, and depth. Rich and well-drained loamy soils are ideal for eggplant cultivation, but it also grows very well in silt-loam and clay-loam soils. The ideal pH range is between 5.5 and 6.0. Since the crop remains in the field for nearly 4–5 months, soils rich in organic matter and nutrients are preferred.

Like other warm season crops, eggplant is also susceptible to frost. It needs a long warm season for optimum growth and fruiting. The ideal growing temperature is about 22–30°C, while growth is inhibited below 17°C. At the same time, eggplant does not tolerate climates that are too warm and dry, where the high transpiration rate will cause the plant to wilt. Poor fruit set either due to dessication of pollen or stigmatic fluid is commonly associated with high temperature. Anthocyanin synthesis in purple or bicolor varieties is affected due to high temperature and low humidity. Hot and direct sun rays during fruiting can cause sunburn damage in young fruits (14).

B. Propagation

Eggplant seeds are light (1000 seeds weigh about 4 g). The seeds retain up to 80% viability under controlled conditions for 3 years and take about a week for proper germination. About 500–750 g of seeds are required to produce seedlings to cover 1 ha. However, seed requirements can be reduced up to 375–500 g/ha with proper nursery management.

C. Cultural Practices

1. Planting

The seedlings are raised in nurseries on raised beds. The seeds are sown in shallow rows across the length of the bed. The first irrigation is given immediately after sowing, with subsequent irrigations given as necessary. In general, regular irrigations are performed until seedlings are about 5 cm tall, after which frequent irrigations should be avoided to harden the seedlings. This results in checking of excessive stem elongation and allows the seedling to withstand transplanting shock. Chougule and Pandey (18) revealed that irrigating the nursery bed every 11th or 12th day increases the productivity of the plants when transplanted. The response of plants to hardening depends largely on the variety. It can be safely said that seedling hardening definitely improves the plant's ability to withstand adverse conditions and also increases fruit yield.

The best time for sowing depends largely on the location and the season. In the plains, eggplant can be grown almost throughout the year. The first crop is sown during February-March, the second crop in May-June, and the third during October-November. The ideal sowing time in the hills is April-May.

This crop will stay in the field nearly 4–5 months, therefore, the soil needs to be well prepared by repeated ploughing and harrowing. Beds of convenient length and width or ridges and furrows are prepared and seedlings about 10–15 cm tall are transplanted. Seedlings should be lifted along with the moist soil adhering to the roots. The spacing depends on the variety, soil fertility, and location. In general, the most common spacing used is 60–90 × 45 cm and tall varieties are planted further apart. Seedlings take about 8–10 days for reestablishment. Some growers transplant the seedlings in raised beds.

2. Irrigation

The first irrigation is given immediately after transplanting, and subsequent irrigations are given as necessary. Generally, irrigating the field every 3rd or 4th day in summer and every 10th to 14th day in winter is considered adequate to maintain proper soil moisture. Best results can be obtained when the eggplant crop is irrigated at 75 mm cumulative pan evaporation. Eggplant responds well to drip irrigation, which not only increases yield but also decreases weed population.

3. Manuring and Fertilization

Eggplant is a heavy feeder. Therefore, nutrient depletion from the soil occurs quickly. Also, eggplant remains in the field for a relatively long period of time. There are several recommendations regarding manurial and fertilizer requirements of this crop: nitrogen varying from 56 to 300 kg, phosphorus from 45 to 200 kg, and potassium from 28 to 253 kg per hectare (4). Yawalkar (17) suggested a combination of 56 kg of nitrogen, 56 kg of phosphorus, and 28 kg of potassium per hectare along with 20 cartloads of farmyard manure. The entire dose of farmyard manure as well as phosphatic and potassic fertilizers are applied before transplanting. Farmyard manure is thoroughly mixed with soil at the time of final ploughing, while phosphorus and potassium fertilizers are applied in furrows at the time of transplanting. Nitrogenous fertilizers are applied as a top dressing in three equal split doses: the first at 45 days, the second at 75 days, and the last at 105 days after transplantation. Other recommendations include 45–50 tons of farmyard manure, 75–100 kg each of nitrogen and phosphorus, and 50 kg potassium (13) or 40–60 kg of nitrogen, 60–80 kg of phosphorus and 100–120 kg of potash.

Several reports (19–31) discuss fertilizer requirements for obtaining higher yields for eggplant crops. The application of nutrients through foliar sprays has been shown to have several advantages over fertilization through the soil (32,33). The application of micronutrients has been shown to have a positive effect on the yield and composition of fruits in intensive vegetable-growing areas (32).

4. Interculture Operation

Interculture operations like earthing up and weeding are regularly needed to check the weed growth. Precautions to avoid root damage during earthing up are necessary. Mulching with black cover reduces weed growth and increases crop growth and yield. Intercrops/mixed crops can be taken in spring. The plants suitable for mixed crops are pumpkin, melons, gourds, green onion, and radish.

5. Use of Growth Regulators

Growth regulators can be used to improve fruit set. Under natural conditions, four types of flowers having variable styled lengths are produced in every variety. Only long-styled flowers set fruits, while short-styled and pseudo-styled flowers do not set fruits (33). When flowers are sprayd with 2,4-D (2 ppm), naphthoxyacetic acid (NAA, 75 ppm) at weekly intervals for 75–90 days the fruit set was increased by 50% (34). Similar results were reported when seeds were soaked in these solutions for 24 hours before sowing. Other chemicals shown to be effective are NAA and *para*chlorophenoxy acetic acid (34).

D. Diseases and Pests

Eggplant is subject to attack by many diseases affecting roots, leaves, stems, and fruits (35,36). The fungal, bacterial, and viral diseases are listed in Table 2. Pests that damage eggplant crops include the shoot and fruit borer (*Leucinodes orbonalis*), *Epilachna* beetle (*Epilachna vigintioctopunctata*), jassids, red spider mites, leaf rollers (*Eublemma divacea*), mealybugs (*Centrococcus insolitus*), cotton aphids, bud worm, lacewing bugs and termites (37–40). Root knot nematodes (*Meloidogyne* spp.) also cause heavy damage to eggplant crops (41).

Table 2 Diseases of Eggplant

Disease	Causal organism(s)
Damping-off	*Phythium* spp.
	Phytophthora spp.
	Rhizoctonia spp.
Phomopsis blight	*Phomopsis vexans*
Leaf spot	*Alternaria* spp.
	Cercospora spp.
Wilt	*Verticillium* spp.
	Fusarium solani
	Sclerotium rolfsi
	Macrophomina phaseolina
Bacterial wilt	*Pseudomonas solanacearum* E.F. Sm.
Little leaf	Mycoplasma

Source: Ref. 4.

E. Harvesting, Handling, and Maturity Indices

The fruits of eggplant are edible from the time they are one-third grown until they are ripe (42). They remain in an edible condition for some time after they are fully grown and become colored. Eggplant fruits are best if their diameter does not exceed about 10 cm, although some up to 15 cm remain tender and may not be overly strong-flavored (43). The small and elongated Japanese cultivars have exceptional quality when they are about 10 cm long. Fruits are usually cut from the plants since the stems are hard and woody. The large calyx and a short piece of the stem are left on the fruit, but care must be taken to prevent the stem from injuring nearby fruits. The eggplant fruits should be handled with care even though they are not as perishable as, say, tomatoes. The fruits are sometimes put in paper bags or wrapped in paper before being packed for shipping. They are paked in various kinds of containers, including special crates, bushel baskets, and sometimes berry crates. Before packing they are usually graded somewhat to separate by size and cull out inferior fruits (42). The changes in physicochemical properties with maturity in some varieties of eggplant have been documented (44). Pantastico et al. (45) reported that the fruits must be picked as soon as they have attained the desired size before they harden or show streaks of unusual color. The skin should be bright and glossy. Overmature fruits are dull, seedy, and fibrous.

IV. CHEMICAL COMPOSITION

Eggplant fruits are quite high in nutritive value and can justifiably be compared with tomato (4). The chemical composition of fruits of different eggplant varieties is presented in Table 3.

The percentage of nitrogen and ether extract values were similar for purple, green, and white eggplant cultivars (47). The physical characteristics of fruits, such as shape and color, presence of spines on the calyx, or foliage color, were found to influence the composition (Table 4). The white cultivar contained twice as much crude fiber as the purple and green cultivars. A report by Sherman (48) indicates that there is a negative correlation between the incidence of heart disease and dietary fiber. The amino acid contents were higher in the purple cultivar and lowest in the white (Table 5). Flick et al. (50) identified three distinctive isoenzyme patterns among the three cultivars. Eggplant fruit tissues were analyzed for relative activities of specific and nonspecific phosphatases. NaCN completely inhibited the enzyme in all cultivars.

Bajaj et al. (51) analyzed fruits of 19 eggplant cultivars for dry matter, amido protein, total water-soluble sugars, free reducing sugars, anthocyanins, phenols, and glycoalkaloid (as solanin) (Table 6). On average, white cultivars were rich in crude fiber, whereas the long-fruited cultivars contained large amounts of the other components (Table 7).

Bitterness in eggplant fruits is due to the presence of glycoalkaloids which occur widely in plants belonging to Solanaceae family. Generally, high amounts of alkaloid (20 mg/100 g fresh weight) produce a bitter taste and off-flavor. Flick et al. (50) analyzed mature purple, green, and white eggplants for organic and mineral elements and the activities of three enzymes. All three varieties had high concentrations of potassium, chlorine, magnesium, and calcium. Potassium and chlorine are highest in the green and lowest in the purple variety. The dry matter content ranged from 5.53 to 10.09% (52). For processing purposes, the varieties with high dry matter content are desirable. Earlier Sidhu et al. (53) reported dry matter contents varying from 6.41 to 8.21%. The white cultivars, long white and round white, lacked anthocyanins. A wide variation in the anthocyanin content of different cultivars of eggplant fruits has been reported (53,54).

TABLE 3 Chemical Composition of Eggplant (per 100 g edible portion)

Constituent	Content
Moisture content	92.7%
Carbohydrates	4.0%
Protein	1.4%
Fat	0.3%
Fiber	1.3%
Oxalic acid	18 mg
Calcium	18 mg
Magnesium	16 mg
Phosphorus	47 mg
Iron	0.9 mg
Sodium	3.0 mg
Copper	0.17 mg
Potassium	2.0 mg
Sulfur	44 mg
Chlorine	52 mg
Vitamin A	124 IU
Vitamin B	
Thiamine	0.4 mg
Riboflavin	0.11 mg
β-Carotene	0.74 μg
Vitamin C	12 mg
Energy	24 kcal

Source: Ref. 46.

TABLE 4 Effects of Physical Characteristics of Fruits/Plants on Composition (dry weight basis)

Fruit/plant characteristic[a]	Dry weight (%)	Crude protein (%)	Crude fiber (%)	Total minerals (%)	Total sugars (%)	Total phenolics (%)	Ascorbic acid (%)
Fruit shape							
Round (27)	7.8	13.3	11.9	7.0	12.1	0.84	51.1
Oblong (30)	7.6	13.3	11.3	6.7	12.6	0.97	63.3
Fruit color							
Full purple (33)	7.8	13.8	11.2	7.2	12.5	0.92	57.0
Partial purple (15)	7.5	12.7	12.3	6.4	12.1	0.94	54.9
Green (9)	7.7	12.5	11.9	7.0	15.6	0.69	66.8
Spininess							
With spines (24)	7.7	12.6	11.5	6.9	11.8	0.95	56.5
Without spines (33)	7.7	13.3	11.5	6.9	13.6	0.80	58.3
Foliage color							
Green (45)	7.8	13.2	11.7	7.2	12.8	0.89	58.9
Violet (12)	7.6	13.5	11.3	6.6	13.5	0.86	59.9

[a]Figures in parentheses indicate number of cultivars analyzed.
Source: Ref. 47.

TABLE 5 Amino Acid Composition of Proteins in Eggplants (dry weight basis)

	Content (mg/g)		
Amino acid	Purple	Green	White
Lysine	0.769	0.541	0.541
Histidine	0.475	0.338	0.332
Ammonia	0.558	0.744	0.401
Arginine	1.206	0.724	1.033
Aspartic acid	3.274	2.666	1.969
Threonine	0.776	0.527	0.493
Serine	0.815	0.568	0.562
Glutamic acid	3.582	2.992	2.405
Proline	0.784	0.585	0.534
Glycine	0.776	0.542	0.548
Alanine	0.995	0.658	0.677
Valine	1.212	0.807	0.795
Isoleucine	0.722	0.655	0.638
Leucine	1.266	0.950	0.944
Tyrosine	0.419	0.287	0.313
Phenylalanine	0.869	0.617	0.617

Source: Ref. 49.

There appears to be a correlation between the amount of copper present and polyphenoloxidase activity (purple, green, and white in decreasing order) and between the amounts of iron in each variety and catalase activities (green, purple, and white in decreasing order). Both the zinc content and alcohol dehydrogenase activity are higher in the white than in the purple; zinc contents in the two types were 5.8 and 6.1 ppm, respectively (55). Brinjal polyphenoloxidase activities were highest in the purple variety (Fig. 4), lower in the green, and lowest in the white. This activity was most evident during peeling. Darkening of the fruit when it is exposed to air occurs faster in the purple variety (50). Green fruits discolored less than did purple ones (53). Both catalase and alcohol dehydrogenase (Figs. 5 and 6) activities were reported to be highest in the green eggplant but catalase activity in the purple variety was only slightly lower than in the green. Peroxidase activity was significantly higher in the white than in the other two varieties (Fig. 7); it was only slightly higher in the green variety than in the purple.

Peroxidation of lipids catalyzed by lipoxygenase is considered a primary cause of quality deterioration in unblanched vegetables. Lipoxygenase in eggplant was isolated and partially characterized (50). Both the activity and the reaction rate for purple eggplant were substantially greater than for the green and white varieties (Fig. 8).

Ramaswamy and Rege (58,59) isolated polyphenoloxidase from eggplant fruit and characterized its phenolic compound substrates. These authors also isolated 11 polyphenolic compounds from the fruit pulp. The eggplant contained 7.6 mg α-solanine per 100 g fresh weight (60).

TABLE 6 Anthocyanin, Glycoalkaloid, and Phenol Contents in Eggplant

Variety	Dry matter (%)	Anthocyanin content (mg/100 g) (fresh wt. basis)	Glyco-alkaloids (mg/100g)	Total phenols (%)	Orthodihydroxy Phenols (mg/100g) (dry wt. basis)	Crude fiber (%)
ARU-1-C	7.35	540.50	1.55	0.52	75.00	10.00
ARU-2-C	5.53	260.10	0.48	0.70	140.00	13.00
K-209-9	6.36	752.10	1.03	0.86	212.50	13.00
Azad Hybrid	7.68	368.00	1.73	0.72	122.50	9.90
Pusa Hybrid-1	7.85	593.40	0.98	1.08	167.50	10.75
PBR-91-2	8.17	501.40	0.56	0.74	90.00	10.10
Pusa Hybrid-2	6.35	134.64	1.82	0.92	212.50	11.00
SM-17-4	7.29	223.38	0.91	0.64	137.50	16.00
Pb. Chamkila	6.57	544.68	1.19	1.00	210.00	13.00
Pusa Selection-1	6.38	229.50	1.79	1.20	285.00	16.50
Long White	10.00	ND	1.64	0.86	242.00	27.40
Kat-4	7.77	566.10	1.94	0.98	330.00	10.00
Azad Kranti	6.05	632.50	2.15	0.52	92.50	17.50
PBR-129-5	6.00	422.28	1.42	1.06	282.50	16.00
J Long × R-34	8.85	409.40	ND	0.86	87.50	ND
K-2910	8.78	6.90	2.36	0.58	100.00	32.60
RHR-58	7.32	159.12	0.64	1.00	255.00	19.00
Round White	7.44	ND	1.21	0.84	132.50	23.40
MEAN VALUES	7.35	396.06	1.38	0.84	176.40	15.90
LSD 5%[a]	0.60	107.94	0.30	0.10	40.64	3.48

ND = No data.

V. STORAGE

Eggplant fruits are sensitive to low temperature (5≤10°C) and deteriorate rapidly at warm temperatures, so they are not adapted to long storage. Pitting, surface bronzing, and browning of seeds and pulp are symptoms of chilling injury. The sensitivity of eggplant to chilling varies with cultivar, maturity, size of fruit, and season of harvest (61–63). Wrapping eggplant in film reduces weight loss and maintains firmness due to high relative humidity. However, wrapped eggplants decay rapidly if the film is not perforated (64).

A. Low-Temperature Storage

Pantastico et al. (65) recommended temperatures of 10–12.8°C and a relative humidity of 92% to store eggplant for 2–3 weeks. Ryall and Lipton (43), however, stated that eggplant fruits can be stored at 10–12.8°C and 95% relative humidity for 10–14 days. Gadakh et al. (47) observed maximum weight loss in unpacked fruits and minimum loss in polyethylene packed fruits. The mean cumulative weight loss on the 9th day was highest (53.2%) in Manjri Gota and lowest (22.8%) in Poona Selection × Pragati under open conditions, while the losses were reduced to 7.2% in the variety Ruchira when the fruits were packed in polyethylene bags. Weight loss varied for different varieties at different stages. The hybrid Pragati × Dorli kept to the 8th day with a 24% weight loss, whereas Ruchira fruits remained acceptable to the 9th day with

TABLE 7 Mineral Content (mg/100g) in Eggplant (estimated on dry weight basis)

Variety	Zn	Ca	Fe	Mn
ARU-1-C	2.35	1.00	8.50	3.25
ARU-2-C	2.80	1.08	6.80	1.60
K 202-9	1.65	0.70	5.95	1.28
Azad Hybrid	2.35	1.35	5.00	3.25
Pusa Hybrid-1	2.80	1.60	8.60	2.00
PBR 91-2	2.35	1.35	8.50	1.45
Pusa Hybrid-2	2.35	1.00	10.00	2.00
SM-17-4	2.55	1.35	5.00	1.38
Pb. Chamkila	2.80	1.35	8.50	2.00
Pusa Selection-1	3.50	1.60	12.50	2.38
Long White	2.35	1.00	10.00	2.00
Kat-4	2.55	1.35	10.00	1.38
Azad Kranti	2.80	1.35	8.50	2.00
PBR-129-5	3.96	3.22	9.90	2.67
J Long × R-34	2.80	1.35	12.50	2.00
K-29210	1.95	0.65	5.00	3.25
RHR-58	2.20	1.00	5.00	2.00
Round White	4.00	1.60	8.50	1.38
Mean	2.67	1.33	8.26	2.08
LSD 5%[a]	0.32	0.28	1.23	0.33

[a]Least significant difference at 95% level of confidence.
Source: Ref. 51.

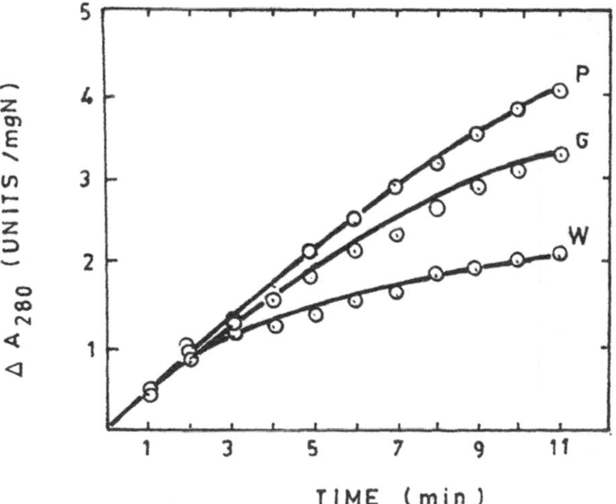

FIGURE 4 Polyphenoloxidase activity in purple, green, and white eggplants. P = purple; G = green; W = white. (From Ref. 56.)

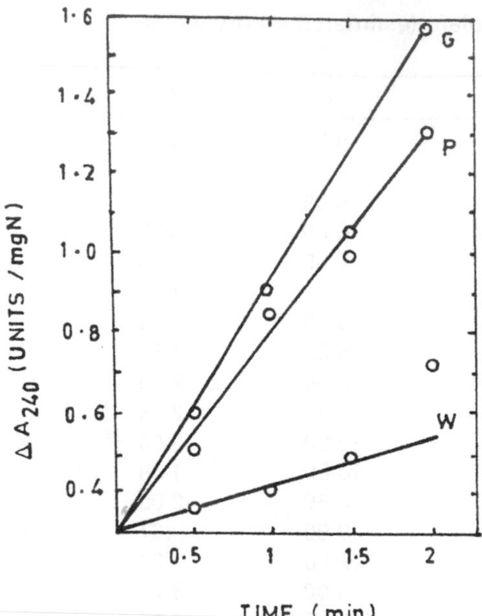

FIGURE 5 Catalase activity in purple, green, and white eggplants. P = purple; G = green; W = white. (From Ref. 56.)

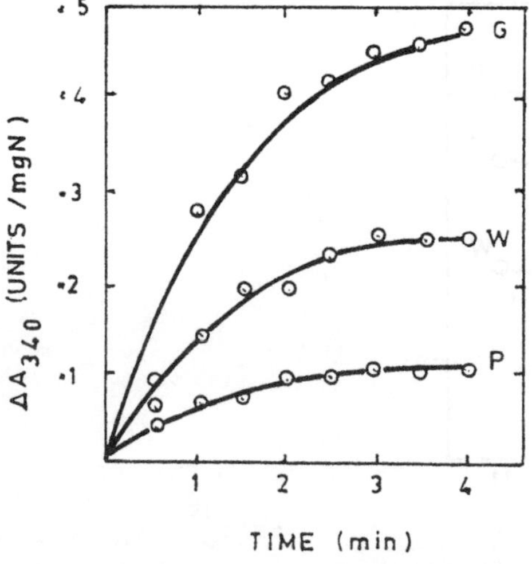

FIGURE 6 Alcohol dehydrogenase activity in purple, green, and white eggplants. P = purple; G = green; W = white. (From Ref. 56.)

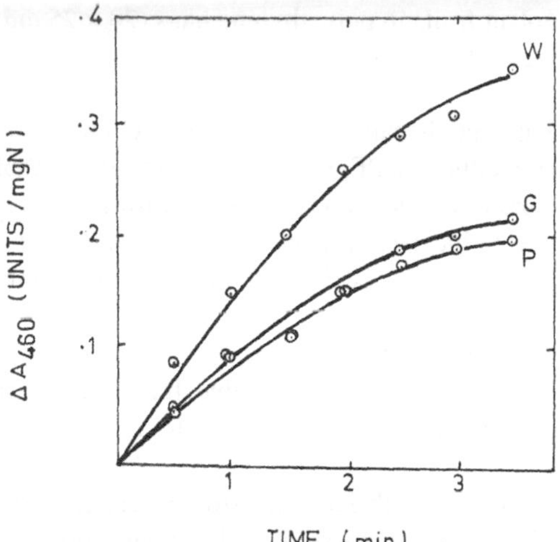

FIGURE 7 Peroxidase activity in purple, green, and white eggplants. P = purple; G = green; W = white. (From Ref. 56.)

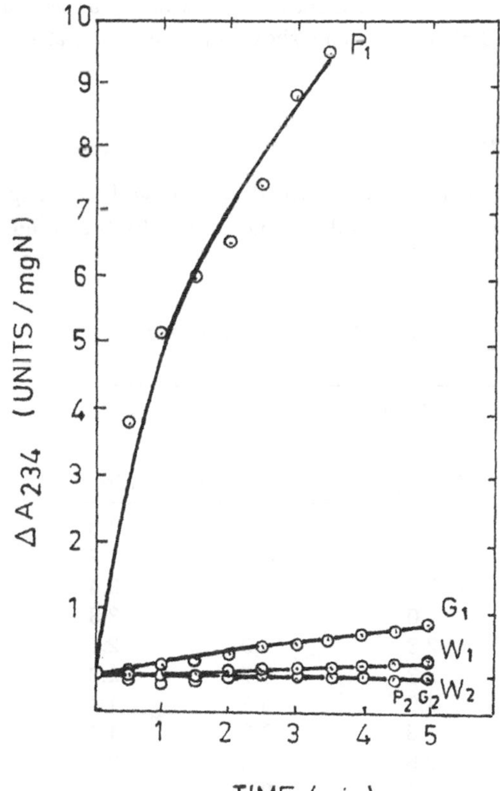

FIGURE 8 Lipoxygenase activity in purple, green, and white eggplants with and without added KCN. (P_1G_1 and W_1 are activity curves for purple, green, and white eggplants without added KCN; P_2G_2 and W2 are curves for extract with added KCN.) (From Ref. 56.)

minimum weight loss of 7.2%. Packing of eggplant fruits in polyethylene bags (20 × 25 cm, 200 guage with 1% vent) prolonged shelf life compared with unpacked fruits in all seven varieties evaluated (48).

Mohammed and Sealy (64) observed that brinjals can be stored at 8–9°C up to 10 days with excellent marketable quality when packed in either LDPE, HDPE, or shrink-wrap (Table 8). Storing fruits up to 17 days followed by exposure to ambient conditions even for a day resulted in rapid quality deterioration. Polyethylene packing has been found to extend shelf life up to 20 days at room temperature, and storage of samples at 3–8°C and 85–90% RH extended their shelf life by 40–50 days (Table 9); storage at −11°C caused freezing injury. Storage at ambient conditions caused a marked decrease in dry matter, starch, total, reducing, and nonreducing sugars, and chlorophyll. The usefulness of hydrocooling is questionable for large eggplants. Their low surface-to-volume ratio would result in slow cooling, and the water also might result in water-spotting of the fruits.

Eggplant fruits should never be kept in contact with ice. The storage qualities of nine eggplant cultivars were studied by Abe et al. (58). Fruits of marketable size when stored at 10°C showed slight pitting after 5 days, which gradually became severe over the next 10 days, however, smaller fruits did not show pitting in the same period. When fruits were stored at 20°C, wilting and decay of the calyx were the main causes of deterioration, and when the fruits were packed in polyethylene and stored at 1, 10, 15, or 20°C, the smaller fruits deteriorated more rapidly than the larger ones and the latter showed less tendency to calyx decay as well as no pitting injury at 15°C. The fruits having high diphenol contents showed pitting injury at 1°C and developed peel browning at 20°C (58).

B. Controlled-Atmosphere Storage

Controlled-atmosphere storage so far has not been shown to extend the storage life of eggplant fruits. At more than 7% CO_2, injury to fruits has been noticed in some cases (66). If continuously

TABLE 8 Effects of Packaging on the Marketable Quality of Two Eggplant Cultivars Stored at 8–9°C

		Storage[a]			
Packaging	Cultivar	10 days	17 days	17 + one day at ambient	Mean
LDPE	La Pastora	4.0	3.0	1.3	2.8
	Long Purple	4.0	1.0	0.0	1.7
HDPE	La Pastora	4.0	2.8	3.5	3.4
	Long Purple	4.0	4.0	2.0	3.3
Shrink-wrap	La Pastora	4.0	2.8	1.0	2.6
	Long Purple	4.0	2.2	0.7	2.3
Control	La Pastora	2.2	1.5	0.5	1.4
	Long Purple	1.0	2.3	0.0	1.1
Mean	La Pastora	3.6	2.5	1.6	2.6
	Long Purple	2.5	2.4	0.7	2.1

[a]0 = Unmarketable; 1 = poor, 2 = fair, 3 = good, 4 = excellent.
Source: Ref. 64.

Table 9 Effect of Storage Temperature on Chemical Constituents in Eggplant Cultivar Arka Shirish

Constituent	Storage treatments[a]						
	1	2	3	4	5	6	7
Marketable storage period (days)	—	20	35	10	40	50	40
Weight loss (%)	—	13.5	3.6	—	1.9	4.8	10.8
Dry matter (%)	9.4	9.9	7.0	9.9	8.5	8.1	8.8
Alcohol-insolubles (%)	5.9	6.2	5.2	7.2	5.5	4.9	4.7
Starch (%)	0.58	0.33	0.22	0.72	0.41	0.47	0.24
Total sugars (%)	2.32	1.67	1.11	1.57	1.53	1.65	1.89
Nonreducing sugars (%)	0.39	0.10	0.06	0.06	0.05	0.19	0.22
Reducing sugars (%)	1.91	1.59	1.04	1.51	1.47	1.45	1.66
Total chlorophyll in skin (mg/100 g)	13.0	7.52	3.43	11.48	5.09	9.28	12.17

[a]1 = Fresh sample; 2 = ambient, without polythene cover; 3 = ambient, in polythene cover; 4 = stored at −11 ± 1°C and 85–90% RH; 5 = stored at 3 ± 1°C and 85–90% RH; 6 = stored at 5 ± 1°C and 85–90% RH; 7 = stored at 8 ± 1°C and 85–90% RH.
Source: Ref. 65.

held at about 10°C and prepackaged, eggplant can be stored for 2–3 weeks without serious deterioration (67).

C. Postharvest Diseases and Their Control

Fruit rot and wilt are the most serious diseases of the eggplant. Fruit rot is caused by the fungus *Phomopsis vexans*, which attacks all parts of plants above ground. Spots on fruits start with light brown blotches, which develop into a soft rot, frequently covering the entire fruit (2). Wilt caused by *Verticillum alboatrum* is characterized by yellowing of the foliage and gradual defoliation. Flea beetles, Colorado potato beetle, eggplant lace bug, and aphids also attack eggplant. *Alternaria* rot, bacterial soft rot, and *Phomopsis* rot caused by *Alternaria tenuis*, *Erwinia carotovora*, and *Phomopsis vexans*, respectively, have been described (43).

Eckert et al. (68) stated that the most common postharvest diseases of eggplant include fruit rot and anthracnose. *Phytophthora parasitica* causes brown rot, which starts as circular, dark, water-soaked lesions. Superficial growth of the fungus appears on the infected fruit. Dry rot is caused by *Diaporthe vaxans*. Anthracnose infection is characterized by yellowish and sunken parts of the fruit with moist pink spore masses of *Gloeosporium melongense* (69).

Eckert et al. (68) recommended using dithiocarbamates (Zineb) against anthracnose disease. *Alternaria* rot, bacterial soft rot, and *Phomopsis* rot do not develop when eggplant fruits are held at 10°C. Postharvest dipping of eggplant fruits in cold Propineb® suspensions had little effect on anthracnose caused by *C. gloeosporiodes* or *F. melongenae*. Hot water (50°C), however, even with fairly low concentrations of fungicide, gave good control (67).

Bajaj and Mahajan (70) investigated the effects of nematicides on the chemical composition of eggplant fruits grown in a field naturally infested with nematodes. The five nematicides applied were aldicarb (as Temik) at 4 kg/ha, carbofuran (as Furadan) at 4 kg/ha, phorate (as

Table 10 Effects of Processing on Vitamin C Content in Eggplant

Processing	Vitamin C (mg/100 g)
Unprocessed	15.6
Washing	13.1
Cutting	11.2
Cutting and washing	6.6
Washing and cutting	10.4
Cooking in open pan	0.6
Cooking in closed pan	3.0

Source: Ref. 73.

Thimet) at 10 kg/ha, fensulfothion (as Desanit) at 10 kg/ha, and Nemagon (1,2-dibromo-3 chloropropane, 60% e.c.) at 20 liters ha. The treated plants had reduced proteins and increased solasodine contents. The nematicides also had varied effects on the biosynthesis of total phenolics and orthodihydroxy phenolic compounds. Dry matter contents in fruits from the treated plants did not differ from the controls.

VI. PROCESSING

Fresh eggplant can be satisfactorily dehydrated using a hot air dryer (71). The resultant powder retained good organoleptic qualities, and its nutritive value appeared to be intact after 5 months of storage. Eggplant also can be sun-dried and preserved. Detailed information on the grading, preparation, and drying of eggplant has been reviewed (72). Bajaj et al. (52) reported that for pickling and drying purposes, eggplant should have a high dry matter content and a low level of phenolics. High anthocyanin and a low glycoalkaloid content are considered desirable regardless of how the fruit is to be used.

The eggplant fruits are widely used to prepare curried vegetable by washing, cutting, mixing with spices and cooking. Such processes often cause significant losses in vitamin C (Table 10). Among the processes, cutting followed by washing and cooking in open pan are more

Table 11 Effects of Grilling and Processing into *Bharta* on Chemical Composition of Eggplant (cv. Round Green Sel, on dry weight basis, %)

Composition	Fresh	Roasted	*Bharta*[a]
Water	93.7	91.8	79.7
Total sugars	49.7	40.0	22.0
Reducing sugars	17.3	17.5	17.2
Ether extract	7.0	7.7	42.2
Crude fiber	17.0	12.7	8.4
Total ash	5.2	5.0	8.4

[a]Recipe: Roasted eggplant 100g + 30 g tomatoes, 30 g chopped onion, 5 g ginger, 1.2 g salt, 200 mg chili powder, and 8 g peanut oil.
Source: Ref. 74.

detrimental. An indigenous product called *Bhartha* is prepared by roasting of eggplant and mixing with spices and oil (74,75). Out of six recipes tested, a product with 30% tomatoes, 30% onion, 5% ginger, 1.2% salt, 0.2% chilli powder and 8% oil was found to be excellent. Roasting caused a loss in total sugars and crude fiber (Table 11), but imparted excellent color, aroma, taste, and mouthfeel to the product. The product was canned and stored for 6 months with acceptable quality.

REFERENCES

1. Shukla, V., and L. B. Naik, Agro-techniques of solanaceous vegetables, in *Advances in Horticulture*, Vol. 5, *Vegetable Crops*, Part 1 (K. L. Chadha and G. Kalloo, eds.), Malhotra Pub. House, New Delhi, 1993, p. 365.
2. Thompson, H. C., and H. C. Kelly, *Vegetable Crops*, 5th ed., McGraw-Hill Book Co. Inc., New-York, 1957.
3. Food and Agriculture Organization, *Production Year Book*, Rome, Italy, 1992.
4. Som, M. G., and T. K. Maity, Brinjal, in *Vegetable Crops in India* (T. K. Bose and M. G. Som, eds.), Naya Prokash, Calcutta, 1986, p. 293.
5. Choudhury, J. B., Exploiting hybrid vigour in vegetables, *Indian J. Hort. 10*:56 (1966).
6. Chadha, M. L., and C. M. Sharma, Inheritance of yield in brinjal, *Indian J. Hort. 46*(4):485 (1989).
7. Chadha, M. L., C. M. Sharma, and K. L. Bajaj, Inheritance of bitterness in brinjal, *Indian J. Hort. 47*;244 (1990).
8. Chadha, M. L., R. K. Hegde, and K. L. Bajaj, Heterosis and combining ability studies of pigmentation in brinjal, *Veg. Sci. 15*:64 (1988).
9. Choudhary, D. N., and G. M. Mishra, Heterosis for morphological characters in inter-varietal crosses of brinjal (*Solanum melongena* L.), *Haryana J. Hort. Sci. 17*:221 (1988).
10. Dahiya, M. S., B. S. Dhankar, and G. Kalloo, Hybrid performance in egg plant (*Solanum melongena* L.), *Haryana J. Hort. 13*:147 (1984).
11. Dixit, J., and N. C. Gautaum, Studies on hybrid vigour in egg plant (*Solanum melongena* L.,) *Indian J. Hort. 44*:74 (1987).
12. Gill, H. S., K. A. Balakrishnan, A. S. Kataria, D. Singh, S. A. Memane, K. M. Pal, B. A. Tomer, and R. Verma, Project Directorate of Vegetable, Annual Report, Part-I, Presented at XIth Workshop held at Dr. Y. S. Parmar University of Horticulture and Forestry, Solan, June 4–7, 1990.
13. Sonone, H. N., B. P. Deore, and S. K. Patil, Vaishali (RHR51)—a high yielding variety of brinjal in Maharashtra, *J. Mah. Agril. Univ. 9*:34 (1984).
14. Kale, P. N., K. E. Lawande, S. B. Raijadhav, K. G. Choudhary, and H. N. Sonone, Pragati—a new brinjal cultivar with high yield and excellent fruit quality, *Mah. J. Hort. 5*:154 (1989).
15. Lawande, K. E., S. P. Gadakh, P. N. Kale, C. D. Badgujar, and V. R. Joshi, Krishna—a promising F_1 hybrid in brinjal for western Maharashtra, *Mah. J. Hort. 6*:51 (1992).
16. Choudhary, B., *Vegetables*, National Book Trust of India, New Delhi, 1979.
17. Yawalkar, K. S., *Vegetable Crops in India*, Agricultural Horticultural Pub. House, Nagpur, 1969.
18. Chougule, B. A., and S. G. Pandey, Hardening of seed for moisture in relation to growth, yield, and irrigation requirement of brinjal fruit, *Indian J. Agron. 3*:1 (1958).
19. Ghanakumari, and G. Satyanarayan, Effect of N, P and K fertilizers at different rate of flowering on yield and composition of brinjal (*Solanum melongena* L.), *Indian J. Agric. Sci. 41*:554 (1971).
20. Patnaik, B. P., and A. Farooqui, Effect of nitrogen, Single and in combination with P_2O_5 and K_2O in brinjal, *Fertilizer News 9*(9):21 (1964).
21. Fajardo, P. S., Ammonium sulfate doubles yield of eggplant, *Agric. Indust. Life. 16*(10):27 (1954).
22. Singh, K., and D. S. Sandhu, Effect of soil and foliar application of nitrogen on growth and yield of brinjal (*Solanum melongena* L.), *Punjab Hort. 10*:103 (1970).
23. Seth, J. N., and D. G. Dhiandor, Effect of fertilizers and spacing on the seed yields and quality in brinjal variety Pusa Purple Long, *Prog. Hort. 1*(4):45 (1969).

24. Lal, S., and D. N. Singh, Response of hybrid brinjal to nitrogen fertilization, *Indian J. Hort. 33*:258 (1976).
25. Umrani, N. K., and B. D. Khot, Response of brinjal (*Solanum melongena* L.) to nitrogen, phosphorus and irrigation intervals, *Indian J. Agric. Sci. 43*:786 (1973).
26. Gupta, A., V. Shukla, K. Srinavas, and J. V. Rao, Response of brinjal (*Solanum melongena* L.) to nitrogen, phosphorus and potassium fertilization with different plant spacing, *Indian J. Hort. 35*:352 (1973).
27. Mishra, I. P., Effect of starter solution on growth and yield of brinjal (*Solanum melongena* L.) *J.N.K.V.V. Res. J. 16*:303 (1982).
28. Singh, S., N. Singh, and J. Mangal, Effect of nitrogen and phosphorus application on brinjal (*Solanum melonguna* L.) productivity under rainfed conditions, *Haryana J. Hort. Sci. 17*:237 (1988).
29. Subbaiah, K., S. Sundarajan, and S. Muthuswamy, Effect of varying levels of organic and inorganic fertilizer on yield and nutrient uptake of brinjal, *South Indian Hort. 31*:287 (1983).
30. Reddy, K. B., and E. N. Reddy, Effect of plant nutrient on growth and yield of brinjal (*Solanum melongena* L.), *South Indian Hort. 34*:100 (1986).
31. Ramu, N., and S. Muthuswamy, Preliminary studies on foliar application of nitrogen on brinjal, *Madras Agric. J. 51*:80 (1964).
32. Nagarajan, S. S. and S. Shanmugasundaram, Studies on micronutrients on crop plants. I. Foliar nutrient studies on brinjal, *Madras Agric. J. 418*(2):22 (1961).
33. Krishnamurthy, S., and D. Subramanian, Some investigations on types of flower in brinjal (*Solanum melongena* L.) based on style length and their fruit set under natural conditions and in response to 2,4 D as a plant growth regulator, *Indian J. Hort. 11*:63 (1954).
34. Chauhan, D. V. S., *Vegetable Production in India*, Ramprasad and Sons, Agra, 1981.
35. Kale, P. B., S. W. Mankar, D. D. Wangikar, and V. S. Gonge, Chemical composition of little leaf disease affected leaves of different genotypes of brinjal, *Veg. Sci. 13*:199 (1986).
36. Peter, K. V., and M. Joseph, Disease resistance in tomato, chilli and brinjal, *Veg. Sci. 13*:250 (1986).
37. Kale, P. B., U. V. Mohod, V. N. Dod, and H. S. Thakare, Screening of brinjal germplasm for resistance to shoot and fruit borer (*Leucinodes orbonalis* Guen) under field conditions, *Veg. Sci. 13*:376 (1986).
38. Mote, U. N., Chemical control of brinjal (eggplant) jassids (Amrasca devastens) and shoot and fruit borer (*Leucinodes orbonalis* (Guen), *Pesticides 12*:20 (1977).
39. Mote, U. N., Varietal resistance of brinjal to fruit borer, *Leucinodes morbonalis* Guen. I. Screening under field conditions, *Indian J. Entymol. 43*:112 (1981).
40. Mote, U. N., Varietal susceptibility of brinjal to jassid, *J. Mah. Agric. Univ. 7*:59 (1982).
41. Shetty, K. D., and D. D. R. Reddy, Resistance in *Solanum* species to root-knot nematodes *Meloidogyne incognila, Indian J. Nematol. 15*:230 (1985).
42. Thompson, A. K., M. B. Bhatti, and P. P. Rubio, Harvesting and handling, in *Postharvest Physiology, Handling, and Utilization of Tropical and Sub-tropical Fruits and Vegetables* (E. B. Pantastico, ed.), AVI Pub. Co., Westport, CT, 1975, p. 236.
43. Ryall, A. L., and W. J. Lipton, *Handling, Transportation, and Storage of Fruits and Vegetables*, Vol. 1, *Vegetables and Melons*, AVI Pub. Co., Westport, CT, 1972.
44. Singh, B. P., N. K. Sharma, and G. Kalloo, Physico-chemical changes with maturity in some promising varieties of brinjal (*Solanum melongena* L.), *Haryana J. Hort. Sci. 19*(3–4):318 (1990).
45. Pantastico, E. B., A. K. Mattoo, T. Murata, and K. Ogata, Physiological disorders and diseases: Chilling injury, in *Postharvest Physiology, Handling, and Utilization of Tropical and Sub-tropical Fruits and Vegetables* (E. B. Pantastico, ed.), AVI Pub. Co., Westport, CT, 1975, p. 339.
46. *Nutitive Value of Indian Food*, National Institute of Nutrition, Indian Council of Medical Research, Hyderabad, 1980.
47. Dighe, A. H., Studies on biochemical and nutritional composition of promising brinjal (*Solanum melongena* L.) cultivars, M.Sc. thesis, Mahatma Phule Agricultural University, Rahuri, India, 1995.
48. Sherman, W. C., Eggplant, *Food Nutr. News 46*:3 (1974).

49. Flick, G. J., Jr., F. S. Burnette, L. H. Aung, R. L. Ory, and A. H. Angelo, Chemical composition and biochemical properties of mirlitons (*Sechium edule*) and purple, green and white eggplants (*Solenum melongena*), *J. Agric. Food Chem. 26*:1000 (1978).
50. Flick, G. J., Jr., R. L. Ory, and A. J. Angelo, Comparison of nutrient composition and enzyme activity in purple, green and white eggplants, *J. Agric. Food Chem. 25*:117 (1977).
51. Bajaj, K. L., G. Kaur, and M. L. Chadha, Glycoalkaloid content and other chemical constituents of the fruits of eggplant, *J. Plant Foods 3*:199 (1979).
52. Bajaj, K. L., K. D. Bansal, M. L. Chadha, and P. P. Kaur, Chemical composition of some important varieties of eggplant (*Solanum melongena* L.), *Trop. Sci. 30*:255 (1990).
53. Sidhu, A. S., G. Kaur, and K. L. Bajaj, Biochemical constituents of varieties of eggplant, *Veg. Sci. 9*(2):112 (1982).
54. Bajaj, K. L., G. Kaur, and M. L. Chadha, Glycoalkaloid content and other chemical constituents of the fruits of some eggplant varieties, *J. Plant Foods 33*:215 (1979).
55. Ramaswamy, S., and D. V. Rege, Polyphenoloxidase of *Solanum melongena* and its natural substrate, *Hort. Abstr. 47*:1565 (1977).
56. Ramaswamy, S., and D. V. Rege, Polyphenolic compounds in tissues of brinjals (*Solanum melongena*), *Hort. Abstr. 47*:1566 (1977).
57. Jones, P. G., and G. R. Fenwick, The polyalkaloid content of some edible solanaceous fruits and potato products, *J. Sci. Food Agric. 32*:419 (1981).
58. Abe, K., K. Chachin, and K. Ogata, Chilling injury in eggplants. VI. Relationship between storability and contents of phenolic compounds in some eggplant cvs. *J. Jpn. Soc. Hort. Sci. 49*:269 (1980).
59. Uncini, L., F. L. Gorini, and A. Sozzi, Cultural value and reaction to cold storage of the first Italian eggplant hybrids in comparison with other foreign varieties, *Ann. Dell. VTPA 17*:191 (1976).
60. Warshavski, S., N. Temkin, Gorodeiski, and M. Schiffonan-Nadal, The storage of eggplant at various temperature of cooling and ripening of fruits in relation to quality, *Bull. Int. Inst. Refrig. Supp.* 239 (1973).
61. Risse, L. A., and W. R. Miller, Film wrapping and decay of eggplants, *Proc. Fla. State Hort. Sci. 96*:350 (1983).
62. Pantastico, E. B., T. K. Chattopadhyay, and H. Subramanyam, Storage and commercial storage operations, in *Postharvest Physiology, Handling, and Utilization of Tropical and Sub-tropical Fruits and Vegetables*, AVI Pub. Co., Westport, CT, 1975, p. 314.
63. Badgujar, C. D., K. E. Lawande, and P. N. Kale, Polyethylene packaging for increasing shelf-life in brinjal fruits, *Current Research Reporter, Mahatma Phule Agril. University. 3*(2):33 (1987).
64. Mohammed, M., and L. Sealy, Extending the shelf-life of melongena (*Solanum melongena* L.) using polymeric films, *Trop. Agril. 63*:36 (1986).
65. Selvaraj, V., E. R. Suresh, and N. G. Divakar, Studies on the effect of storage temperature on biochemical constituents in brinjal varieties, *Indian Food Packer. 38*(1):28 (1974).
66. Viraktamath, C. S., Pre-packaging studies on fresh produce. III. Brinjal (eggplant) (*Solanum melongena*), *Food Sci.* (Mysore) *12*:326 (1963).
67. Hatton, T. T., E. B. Pantastico, and E. K. Akamine, Controlled atmosphere storage: 3. Individual commodity requirements, in *Postharvest Physiology, Handling, and Utilization of Tropical and Sub-tropical Fruits and Vegetables* (E. B. Pantastico, ed.), AVI Pub. Co., Westport, CT, 1975, p. 201.
68. Eckert, J. W., P. P. Rubio, A. K. Mattoo, and A. K. Thompson, Diseases of tropical crops and their control, in *Postharvest Physiology, Handling, and Utilization of Tropical and Sub-tropical Fruits and Vegetables*, (E. B. Pantastico, ed.), AVI Pub. Co. Westport, CT, 1975, p. 415.
69. Fournet, I., Eggplant anthracnose in the French West Indies. I. Characteristics and specificity of the pathogen. II. Control, *Ann. Phytopathol. 5*:1 (1973).
70. Bajaj, K. L., and R. Mahajan, Effects of nematicides on the chemical composition of the fruits of eggplants (*Solanum melongena* L.), *Qual. Plant. Pl. Foods Hum. Nutr. 30*:69 (1980).
71. Dei-Tutu, J., Preservation of garden eggs by dehydration, *Hort. Abstr. 44*:7720 (1974).

72. Kalra, C. L., S. K. Berry, and R. L. Sehgal. Resume on brinjal (*Solanum melongena* L.)—a most common vegetable, *Indian Food Packer 42*(2):46 (1988).
73. Tapadia, S. B., A. B. Arya, and P. Rohini Devi, Vitamin C contents of processed vegetables, *J. Food Sci. Technol. 32*:513 (1995).
74. Berry, S. K., C. L. Kalra, O. P. Beerh, S. G. Kulkarni, S. Kaur, and R. C. Sehgal, Screening of different varieties of brinjal (*Solanum melongena* L.) for preparation and canning of *bharta*, *Indian Food Packer 42*(2):52 (1988).
75. Bhupinder, K., and K. Harinder, An improved process for ready-to-eat curried brinjal (*Solanum melongena* L.) preparation (*Bharta*), *Indian Food Packer 46*(3):9 (1992).

10

Cucumber and Melon

A. M. Musmade and U. T. Desai
Mahatma Phule Agricultural University, Rahuri, India

I. INTRODUCTION

Cucumber and melon are important fruit vegetables belonging to the family Cucurbitaceae, which consists of 90 genera and 750 species (1). Cucurbits, being warm season crops, are of tropical origin and grown mostly in Africa, tropical America, and Asia, mainly Southeast Asia (2). Most cucurbits are climbing or prostrate dicotyledonous plants of the tropics, subtropics, and milder regions of the temperate zones. Most are herbaceous annuals and some are perennials, but all are frost-sensitive. Cucurbits are grown mostly for their fruits, however, the shoot and flowers of some species are used as food. They play an important role in supplying fresh fruit vegetables during the summer season in many tropical countries. This group consists of a wide range of vegetables used as salad (cubumber), for cooking purposes (all the gourds), as dessert fruits (muskmelon and watermelon), and as candied or preserved products (ash gourd). They are mostly seed-propagated, although a few are vegetatively propagated like pointed gourd (*Parwal*) and a few perennials like chow-chow and dry gourd (3). The cucurbits are grown for their ripe or unripe fruits, which are a good source of carbohydrates, vitamin A, ascorbic acid, and minerals.

II. CUCUMBER

Cucumber (*Cucumis sativus* L.) is an important and popular vegetable belonging to Cucurbitaceae family. It is one of the oldest cultivated vegetable crops and has been found in cultivation for 3000–4000 years (4). It is a native of Asia and Africa. Some authorities claim that it originated in India and from there spread to Asia, Africa, and Europe.

A. Botany

The cucumber belongs to the genus *Cucumis*, which includes 50 or more species from Asia and Africa. Of these, only two species—*Cucumis sativus* (cucumber) and *Cucumis melo* (muskmelon)—are of economic importance.

1. Morphology

The cucumber is characterized by monoecious sex expression but andromoncecious, gynoecious, and trimonoecious forms are also found. The cucumber is a trailing or climbing plant with hairy, angular stems, leaves with three to five lobes (Fig. 1) and long petioles. The flowers are axillary, staminate flowers, being more numerous than the pistillate. Under long days and high temperatures, the plant produces more staminate than pistillate flowers. However, under cooler conditions, the ratio of pistillate to staminate flowers increases. It is a common experience that cucumber yield depends upon the number of female flowers produced by the vine, and growers are naturally interested in a variety that gives more female flowers or adopting cultural practices to increase the number of female flowers on the vine. Pollination is usually by bees. Some European cultivars bear parthenocarpic fruits and do not require pollination for fruit set. Gynoecious lines are used in the production of hybrid seeds, and it becomes easier because monoecious plants need not be roughed. When ethephon is sprayed on cucumber plants, only pistillate flowers are produced (5).

2. Cultivars

A number of cultivars have been developed in many parts of the world differing in the size and shape of fruit, thinness, spininess, and color of rind, ranging from whitish-green to dark green, others turning yellow or rusty brown when mature. Cucumber cultivars are usually classified as either pickling or slicing types (fresh market cucumber). The fruits of slicing cucumbers are larger than pickling cucumbers, and they develop a dark and thin skin with a uniform cylindrical shape. Present-day pickling cultivars tend to develop shorter vines than slicing ones and are very prolific (3). The fruits of pickling cultivars are smaller and the seed coat develops more slowly. In pickling cultivars, a thin skin with a lighter color is preferred, and these are used mainly for small pickles. Slicers are often pickled when they are somewhat immature for dills or larger pickles. The important cultivars developed in the United States are Sharmrock, Maine No. 2, P.R. 39, Wiscosin SMR 12, Pixie, Summer, Wisconsin 2757, Midget, Burpee Hybrid, MSU 713-5, Spartan Dawn, Little Leaf, and Castlepik (7).

FIGURE 1 Cucumber vine with leaves, flowers, and fruits.

Cucumber cultivars are also classified according to spine color—either white or black—on the fruit. White spines are usually market types as they keep longer in immature conditions and become dull yellow when fully ripe. Black spine types are almost exclusively used for pickling because of the better color retention when kept in brine (7). Important slicing varieties grown in the United States are Marketer, Colorado, Burpee Hybrid, Surecrop Hybrid, Sensation Hybrid, Santee, Palmetto, and Niagara. The varieties grown exclusively for pickling are National Pickling, Chicago Pickling, Model, M.R. 18, and York State Pickling.

Nath (9) classified a large number of indigenous varieties available in India into four groups: Balam Khira type, long spines, green type, spineless long green, and Sikkim cucumber with reddish-brown fruits. Significant progress has been made in recent years in producing hybrid cultivars on a commercial scale. These hybrids have high yield potential, prolonged harvests, multiple disease resistance in both pickling and slicing types for greehouse as well as open field cultivation. All present-day hybrids are gynoecious, parthenocarpic, and multiple disease resistant. Some of the cucumber cultivars released in India include Straight Eight, Pusa Sanyog, Poinsette, Japanese Long Green, and Himangi (Fig. 2) (10,11).

B. Production

1. Soil and Climate

Cucumber can be grown successfully on many kinds and classes of soil from sandy to heavy clay. Sandy loams are considered best for early crops, whereas heavy clay soils produce high yields. Soil should be well drained. Cucumber can be grown successfully on slightly acid soils. Optimum soil pH varies from 5.5 to 6.8

Cucumber, a warm season crop grown mainly in tropical and subtropical regions, grows best in a temperature range of 18–30°C. The plants suffer from chilling injury at temperatures below 10°C (8). Cucumber can be grown under milder conditions but is easily killed by frost. High humidity and short day length promotes female flower production (13). Cucumber seed does not germinate at 11°C, but it may remain in cold soil for a long time and then germinate when temperature becomes favorable. The speed and percentage of germination are greater at

Figure 2 Himangi—a white cultivar of cucumber.

25–30°C than at 18°C (14). High atmospheric humidity promotes the incidence of disease and insect pests, especially those affecting the foliage.

2. Cultural Practices

Sowing

Sowing can be done on raised beds, in furrows, or in pits according to the system followed. Two seeds per hill are generally sown on both sides of the beds. In garden soil, furrows are made at 1–1.5 meters spacing (row-to-row distance 1–1.5 m; plant-to-plant 50–60 cm). In case of a pit system, pits 45 × 45 cm (about 60 cm deep) are dug at proper spacing and filled with farmyard manure (FYM) or cattle manure and earth in equal proportions (15). Three to four seeds per pit are sown. About 2.5–3.5 kg of seed would be sufficient to sow one hectare. In Punjab, recommended plant spacing is 2.5–3 m × 60 cm (16), whereas 1.5–2 m × 45–60 cm has been recommended by Seshadri (2). Singh (17) has recommended a spacing of 1.2–1.5 m × 60–90 cm.

Manuring and Fertilization

Both animal manure and commercial fertilizers are used in cucumber growing, but in most of the large producing areas commercial fertilizer (18–20) and soil-improving crops are recommended depending upon the nutrient status of the soil (Table 1). Some growers follow the practice of applying a small quantity (4–5 t/acre) of manure in furrows over which the rows are made. The manure is covered with soil and the seed planted over the manure. If large applications of manure are made, broadcasting is better than applying in the furrow. Nath et al. (12) recommended 37–49 tons FYM per hectare, which may be mixed with soil or applied to each pit. About 70 kg of N, 110 kg of P, and 70 kg of K per hectare have been recommended for cucumbers by Yamaguchi (8). About half of the nitrogen is applied at the time of sowing, with the remaining used as a side dressing after plants are thinned at the three- to four- true leaf stage. Other fertilizer recommendations include 100 kg of ammonium sulfate, 100 kg of superphosphate, and 55 kg of potassium sulfate (21). It has been shown to be beneficial to apply 50% of total N at the time of sowing, followed by 25% when growth starts and the remaining 25% at the time of fruiting.

Interculture Operations

The land should be kept clean and aerated by two to three shallow cultivations. Frequent weeding is necessary in order to keep weeds in check. A first weeding may be performed 15–20 days after

TABLE 1 Recommended Levels of Nutrients for Cucumber in India

	Nutrients (kg/ha)		
State	N	P	K
Punjab	100	50	50
Himachal Pradesh	100	50	50
Karnataka	60	0	50
Tamil Nadu	35	0	0
Maharashtra	50	40	0
Orissa	50	30	75

Source: Refs. 18–20.

sowing. The root system must not be disturbed by hoeing or weeding operations. Weedicides can be used for weed control (22).

Irrigation

Cucumber needs adequate amounts of water during growth. Therefore, summer crops require frequent irrigation every 5–7 days depending upon the soil type. There should be regular irrigation at the time of fruit development. The rainy season crop requires less irrigation if rainfall is well distributed. A minimum of 40 cm of water is necessary to grow cucumber in a dry climate (8). Precautions are needed to keep the foliage from coming in contact with water to avoid foliage damage.

3. Diseases and Pests

Powdery Mildew

This disease often becomes very severe in warm, rain-free growing areas. It affects leaves and stems. The disease covers both surfaces of the leaf with powdery growth, some attacked leaves become brown and shriveled, and defoliation may occur. Fruits of the affected plants do not develop fully and remain small. The fungus has been identified as *Erysiphe cichoracearum* D.C. Spraying with Karathane (6 g/10 liters water) or Bavistin (10 g/10 liters water) at 5- to 6-day intervals controls the disease if given when the initial symptoms appears. In cucumber, PI 197087 and PI 120815 are resistance sources. Resistance is governed by one gene (Palmetto and Ashley), two genes (Poinsette and Cheerckee), and three genes (PI 197087) (23,24).

Downy Mildew

This disease is caused by the fungus *Pseudoperonospora cubensis*. It is prevalent in high-humidity areas, especially when summer rains occur regularly. The disease is characterized by the formation of yellow, more or less angular spots on the upper surface of leaves. The disease spreads rapidly, killing the plant quickly through rapid defoliation. Hence, control by fungicidal spray must be undertaken early. Spraying with Dithane M-45 (Maneb), if done early and repeated two or three times, can control the disease effectively, though not completely.

Anthracnose

Eckert et al. (25) describe how anthracnose caused by *Colletotrichum lagenarium* produces black lesions with dry spots beneath. The disease can be controlled by repeated spraying at 5- to 7-day intervals with Dithane M-45 0.2% or Dithane Z-78 (Zineb) 0.2%. Poinsette is one highly resistant cultivar, and a moderately resistant source is PI 195111. In the United States, a high degree of resistance to all the types occurring in cucumber has been found in PI 197087 with some kind of multigene inheritance.

Other Diseases

Several other fungal and bacterial diseases affect cucumber, although they are not as important in India as in other countries. Cucumber scab disease is one such disease, which is prevalent in subtropical and temperate countries. Pickling varieties are frequently attacked by *Cladosporum cocumerinum* E K and Arth, causing scab, whereas a shallow decay is produced under humid conditions. Ryall and Lipton (26) described bacterial spot, bacterial soft rot, and black rot diseases of cucumber. Bacterial spot is caused by *Pseudononas lachrymans*, whereas the pathogen involved in black rot is *Mycosphaerella citrullina*.

Viral Diseases

A large number of viruses can cause much damage to cucumber. Slightly affected vines show normal green leaves or have a roughened surface. Viruses are transmitted through seeds, aphids, and mechanically. Chemical control of insect vectors by spraying Malathion at 5- to 7-day intervals may partially check viral spread. Viruses infecting cucumber include cucumber mosaic virus (cucumber virus-1), tobacco virus group, Cucumis virus 2B, 2C, and cucumis virus-3. Squash mosaic virus and muskmelon mosaic virus can also infect cucumber.

Red Pumpkin Beetle

These are brightly colored small beetles orange-red in color. The beetles (*Aulacophora foveicollis* Lucas) attack at the seedling stage, especially at the cotyledon leaf stage. They make holes in cotyledon leaves and cause severe damage. They can be controlled by spraying Carbaryl (sevin) 0.1–0.2% or Rogor 0.1%.

Aphids

Aphids damage the plant by sucking the leaf sap in the young stage. The leaves turn yellow and the plant loses its vigor and yield. Aphids can be easily controlled by spraying Malathion 0.1% or Rogor 0.1–0.2%.

Fruit Fly

Maggots of this fly (*Dacus cucurbitae, Dacus dorsalis*) cause severe damage to young developing fruits. The adults lay eggs below the skin of the young ovaries. The eggs hatch into maggots, which feed inside the fruits and cause rotting. The fly attack is severe, especially after summer rains when the humidity is high. Adult flies can be controlled using high traps at night and poison baits.

4. Harvesting and Handling

Generally the crop is ready for harvest about 60–70 days after sowing. The fruit takes about 7–10 days from setting to reach marketable stage (2). In salad or slicing cucumbers, dark green colored fruits are picked when they are 15–25 cm long; for pickles they are harvested when they are 5–15 cm long (14). Overmature fruits turn brownish-yellow and show carpel separation in transverse section (2). The fruits should be picked at 2-day intervals (17).

Cucumbers are picked by hand to avoid injury to the vine. The stems are left attached to the fruit. Slicing cucumbers fetch better prices when they are long, moderately slender, and dark green in color. Watts and Watts (27) recommended regular picking two to four times a week according to growing season to secure maximum yields and high grades. Pruning of all secondary shoots up to the 5th node gave significantly high-graded fruit yield compared to unpruned controls (28). Careless or delayed picking reduces yields, as one maturing cucumber exhausts the vine more than many young ones. Vines should be handled gently, and it is advisable to cut or clip away the fruits rather than twist them off.

Cucumbers for slicing are graded on the basis of size, shape, and general appearance. Usually field-grown cucumbers should be separated into two grades. For pickles, they are also graded according to size, shape, and general appearance. Specifications for U.S. standard grades for slicing cucumbers are available from the Bureau of Agricultural Economics. Grades for pickling cucumbers are generally stated in the contract between the pickling company and the grower, and grading usually is done at a sorting table in the field. Slicing cucumbers are generally washed and graded at the packing shed (27).

C. Chemical Composition

The cucumber is rich in vitamin B and C and minerals such as calcium, phosphorus, iron, and potassium (Table 2). The fruit is also eaten raw as salad. It is known to prevent constipation,

TABLE 2 Chemical Composition of Cucumber (per 100 g edible portion)

Constituent	Content
Edible portion	83 g
Moisture	96.3 g
Proteins	0.4 g
Fats	0.1 g
Minerals	0.3 g
Fibers	0.4 g
Carbohydrates	2.5 g
Energy	13 kcal
Calcium	10 mg
Phosphorus	25 mg
Iron	1.5 mg
Thiamine	0.037 mg
Riboflavin	0.2 mg
Vitamin C	7.0 mg

Source: Ref. 3.

jaundice, and indigestion (29). It is a popular vegetable in developed countries where there is excessive calorie intake (30).

Takei and Ono (31) attributed the flavor of cucumber to two compounds: 2,6-nonadienal and 2,6-nonadienol. The pleasant aroma of cucumber is said to derive mainly from the 2,6-nonadienal, with assistance from 2-hexenal, the more astringent taste being contibuted by 2-noenal. Kemp et al. (32) identified some volatile compounds in cucumber which were not reported earlier: inonanol, *trans*-2-nonen-1-ol, *cis*-3-nonen-1-ol, *cis*-6-nonen-1-ol, *trans,cis*-2-6 nonadien-1-ol, *cis,cis*-3,6-nonadien-1-ol, *cis*-6-nonenal, and C_{10}–C_{15} saturated straight-chain aldehyde. The presence of 3-alkyl-2-methoxypryrazine in volatiles of raw cucumber was reported, and a possible biogenesis of these compounds with amino acids and glyoxal as precursors has been suggested. A large increase in carbonyl takes place when cucumbers are blended with water in the presence of oxygen (33). This increase can be prevented by blending at pH 1, by blending in an oxygen-free atmosphere, or by heating the cucumbers to 77°C before blending.

Cucurbits are characterized by the presence of bitter principles called cucurbitacins, which are tetracyclic triterpenes (2). They occur in nature as free glycosides or in bound form. A high concentration of the bitter principle is found in fruits and roots, with leaves being normally or slightly bitter. Bitter principles found in roots differ from those in fruits. The pollen also contains bitter principles, and hence when bitter pollen fertilizes nonbitter ovules, the resulting fruit becomes bitter. This phenomenon is called metaxenia (2). Several changes occur in fruit during maturation, including changes in concentration of bitter principles, ascorbic acid, minerals, and enzymes such as superoxide dimutase (34) and polygalacturonase.

D. Storage

1. Low-Temperature Storage

Cucumbers are susceptible to chilling injury at 10°C or less and to yellowing at 16.4°C or higher, restricting desirable storage temperatures to a narrow range (26). The optimum temperature for cucumbers is 10–13°C (26). However, for short storage periods of 1–2 weeks, 10°C may be preferable because chilling is minimal and yellowing is retarded. Apeland (36) found that yellowing of cucumber is slower at 13°C than at higher or lower temperatures. Ryall and Lipton (26) stated that susceptibility of cucumbers to chilling injury does not preclude their exposure to temperatures below 4.4°C as long as they are utilized immediately after removal from cold storage because symptoms develop rapidly only at higher temperatures. Thus, 2 days at 0°C, 4 days at 4.4°C, or 8 days at 7.2°C are harmless under these conditions. Kapitsmiadi et al. (37) reported that the optimum storage temperature was 11°C for Plura, 11–12°C for Rhensk Druv, and 13–14°C for Kokard and Spangbergers. Mechanical stress induces accelerated aging. Cucumbers usually are not precooled, but are cooled after packing and loading into rail car or truck trailer. Hydrocooling is useful for pickling and slicing cucumbers harvested during hot weather. Cucumbers quickly become flaccid unless they are kept at about 95% RH. At such high RH, softening and pitting are retarded. Water loss can be effectively guarded against by waxing the cucumbers, which is a widespread practice. Packaging of cucumbers in ventilated films reduces water loss considerably (38,39).

2. Controlled-Atmosphere Storage

About 5% CO_2 or 5% O_2 retard yellowing of cucumbers, although a combination of both is much more effective (26). These authors (26) have cautioned about use of chilling temperatures in controlled atmospheres, because high CO_2 and to a lesser degree low O_2 aggravate chilling injury. Even at desirable temperatures, CO_2 should not exceed 10% and O_2 should not fall below 2%. Controlled atmosphere storage with 3% O_2 or 5–10% CO_2 is particularly useful to inhibit ethylene in mixed-storage cucumbers with melons, tomatoes, apples, or pears. Fellers and Pflug (40) found that storage life of SMR-15 could be extended by 2–3 weeks with a combination of 5% O_2 and 5% CO_2.

3. Chemical Treatments

Sharma and Wahab (41) found that dipping cucumbers in aureofungin (50–200 ppm) in dimethyl formamide protected the fruit for 50–70 hours of storage against *Phythium*. Untreated fruits developed decay within 15 hours of storage. Atwa et al. (42) studied the effects of some chemical substances on keeping quality of cucumbers (cv. Beit Alpha). The fruits were dipped in boric acid (1%), borax (4%), and a mixture of boric acid plus borax at these rates. The treated fruits were stored at room temperature (31 ± 2°C) or at 7°C for 16 days and then at room temperature. The fruits treated with borax alone showed less decay than those treated with boric acid alone. Atwa et al. reported a decrease in chlorophyll, sugar, ascorbic acid, and carbohydrate contents and fruit firmness, all of which decreased as storage progressed. Sciumbato and Hegwood (13) studied the effects of various fungicides to control fruit rot caused by *Rhizoctonia solani*. Several treatments controlled fruit rot 7 days after spraying (Table 3). Quintozene plus ethazol controlled fruit rot best at 7 days after treatment, but none of these treatments gave acceptable levels of control of fruit rot at 13-, 21-, and 29-day intervals (Table 3).

E. Processing

Cucumber is the most important vegetable for commercial and home preservation for pickling. Most cucumber pickles are made by fermenting the cucumber with salt (6). It is estimated by Pickle Packers International that 40% of cucumbers processed in the United States are brined, 40% are fresh packed, and 20% are refrigerated (7). Cucumber pickles such as hamburger dill chips, sweet gherkins, and relish are made and preserved by lactic fermentation in salt brine.

TABLE 3 Infection of Cucumber by *Rhizoctonia solani* as Influenced by Fungicide Treatments

Fungicide	Rate (kg ai/ha)	Infected cucumber fruit (%)			
		Days after spraying			
		7	15	21	29
Benomyl	1.1	83ab*	83ab	92ab	100a
Captan	5.6	92ab	69bd	100a	92a
Captan + lignin	5.6	100a	69bd	92ab	92a
Captafol	3.6	832ab	83ab	92ab	92a
Captafol + lignin	3.6	100a	75ac	100a	100a
Carboxin	2.2	100a	83ab	100ab	100a
Chlorothalonil	2.5	83ab	92a	83ab	92a
Chlorothalonil + lignin	2.5	83ab	92a	83ab	100a
Folpet	9.0	67ac	69bd	100a	100a
Folpet + lignin	9.0	100a	69bd	100a	100a
Quintozene	11.2	33c	92a	83ab	58a
Quintozene + lignin	11.2	33c	83ab	50b	42a
Quintozene	22.4	92ab	92a	100a	92a
Quintozene + lignin	22.4	33c	47d	83ab	50a
Quintozene + CGA 48988	22.4 + 0.6	42bc	50d	67ab	50a
Quintozene + ethanol	22.4 + 0.6	8c	56cd	75ab	100a
Lignin	11.4	100a	75ac	100a	100a
Nontreated	—	100a	75ac	100a	100a

*Mean of three and four cucumber replications. Mean separation in cucumber fruit by Duncan's multiple range test, 5% level.
Source: Ref. 13.

About 40% of commercially produced pickles are produced by fermentation, while the remainder are processed by pasteurization [fresh pack style or pickled and stored refrigerated (44,45)].

1. Natural Fermentation

Cucumbers are preserved by brining, which may or may not be accompanied by fermentation. Cucumbers undergo lactic acid fermentation during storage, which offers the advantages of acid formation and removal of fermentable sugars, which serve to prevent growth of pathogenic microorganisms and to stabilize the products. This offers the potential for flavor enhancement in the products (46).

Cucumbers are brined initially in solutions containing 5–8% sodium chloride during fermentation (46). After fermentation, applications of dry salt might be sufficient for the cover brines to reach 16–18° Brix (47). High salt concentration is used to prevent softening spoilage during storage and to prevent freezing of brines. Before pasteurization, preservation of pickles depends upon the conversion of carbohydrates into organic acid and/or addition of sufficient amounts of vinegar, sugar, and other ingredients to fully cured and packed cucumbers in order to minimize microbial growth. The preservation of genuine dill pickles, which are not pasteurized, depends upon the added salt, the acid formed, and by being fully cured (complete removal of sugars and change in flesh from an opaque to a translucent appearance). Processed dill pickles and sour pickles are also prepared from fully cured brine stock, but the final products are acidified further with acetic acid. However, such pickles are subject to softening during primary fermentation. These yeasts are acid tolerant and continue to grow as long as fermentable carbohydrates are present. During the postfermentation stage, microbial growth is restricted to the surface of brines exposed to air. The principal bacteria involved are *Lactobacillus brevis*, *Pediococcus cerevisiae*, and *Lactobacillus plantarum* (50). The movement of cucumbers from field to finished pickles is shown in Figure 3 (7).

Bloating, softening, bleaching, and development of off-flavors and odors are the major defects that occur during brine storage of cucumbers. Bloating is a most serious defect, which occurs in large fruits (over 39 mm diameter) and has been attributed to the production of excess CO_2 in cucumber tissue (7). Softening is commonly observed in dill pickles. The firmness of brined cucumber can be retained if cucumbers are washed prior to brining and the temperature of the brine stock is maintained at 15.5°C or lower. The addition of about 1% calcium chloride to brine results in firmer cucumbers. Purging of CO_2 from the brine tank by injecting bubbling N_2 gas has become a widely accepted practice (49–51). Softening has been related to fungi carried on the surface of cucumber fruits. Bleaching is caused by exposure of brine stock to sunlight. Off-flavor results from growth of undesirable microorganisms. They can be controlled by maintaining favorable conditions, including good sanitation.

2. Controlled Fermentation

In recent years a process for controlled fermentation has been developed (52,54). Attempts have been made to develop a closed-top tank for fermentation and storage of brined cucumbers (52). Thompson et al. (52) found that good-quality pickles were obtained by a controlled fermentation process utilizing brines with 5.5% NaCl and 0.1% $CaCl_2$ and a storage temperature of 15.5°C. Fleming et al. (53) found that excellent-quality pickles could be produced by maintaining 2% NaCl and 0.2% $CaCl_2$ during fermentation and storage. The use of pure culture of lactic acid bacteria (LAB) for brine-fermented cucumbers has been recommended as part of the controlled fermentation procedure (53). Two species of lactobacillus, *Lactobacillus plantarum* and *Pediococcus pentosaneus*, are predominantly used in the pickle industry (43). Calcium chloride and potassium sorbate reduce sodium chloride used during natural cucumber fermentation and storage. The presence of $CaCl_2$ helps to maintain cucumber firmness. A synergistic action between NaCl, $CaCl_2$, and potassium sorbate was noticed, which allowed good-quality pickles to be produced when moderate amounts of all three components (5% NaCl, 0.2% $CaCl_2$, and 0.2% potassium sorbate) were present in the brine (43).

3. Canned Pickles

Cucumber pickles of all kinds and mixed pickles are successfully canned in heavily lacquered cans. The pickles are sometimes packed into the cans carefully by hand. Brine vinegar or spiced sweet vinegar is added. The cans are treated at 93°C (about 8–10 min), then they are sealed and no further sterilization is given. Heat treatment removes the air, expands the contents, and thus creates a vacuum in the cans. Pasteurization and processing variables can influence the final product texture.

4. Refrigerated Dills

These are also called overnight dills and half-sours. Refrigerated dills are essentially nonheated, well-acidified, low-salt, refrigerated green cucumbers containing one or more preservatives with spices and flavoring agents (7). The cucumbers are washed and packed by hand into containers. The cucumbers are then covered with a brine consisting of water, acetic acid (vinegar), and salt.

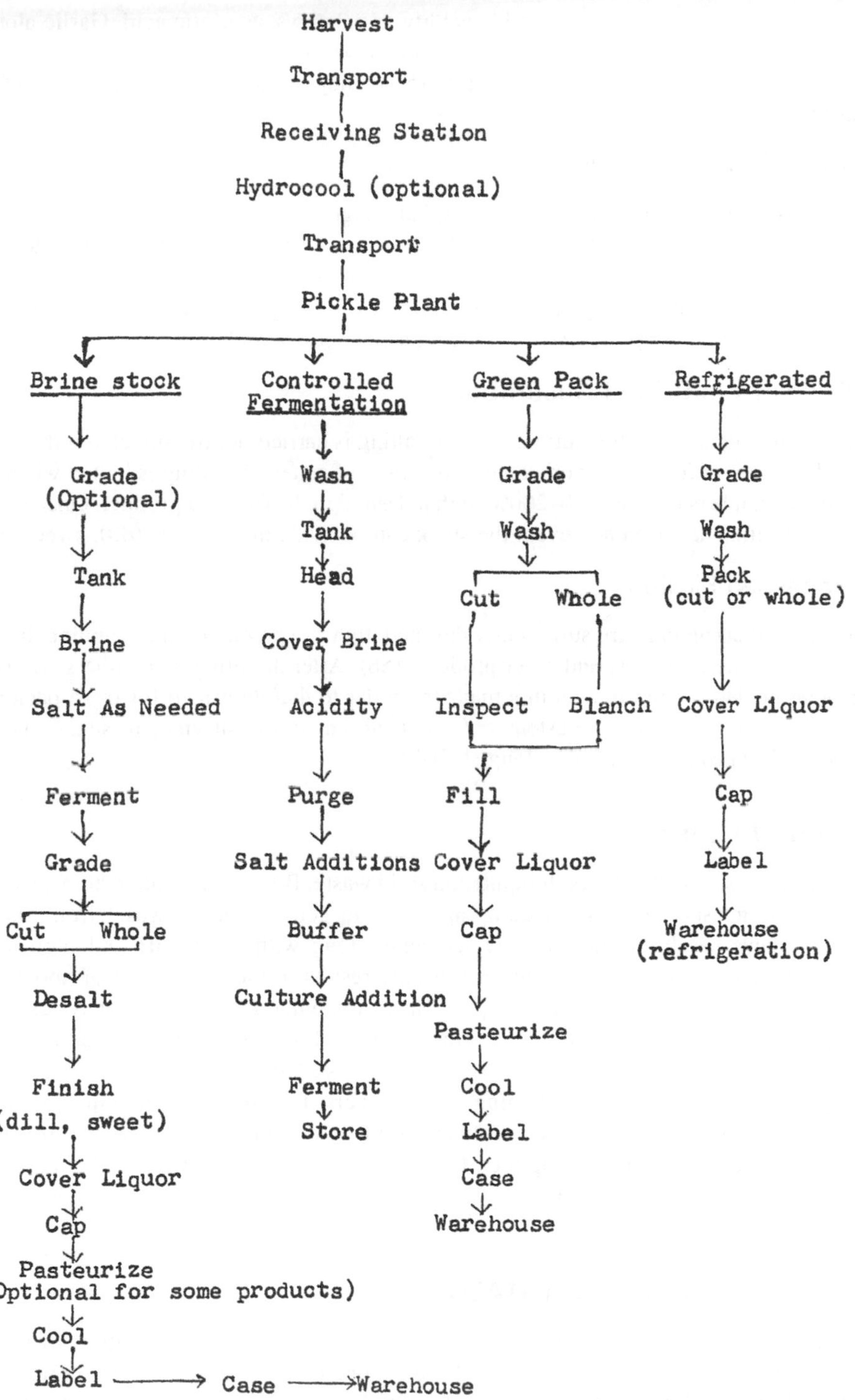

FIGURE 3 Flow chart depicting the movement of cucumbers from field to finished pickles. (From Ref. 7.)

The pH desired is 4.2–4.3 with a titratable acidity of 0.3–0.5% as acetic acid. Garlic along with flavoring agents and sodium benzoate as a preservative are added. The containers are closed, cased, and moved to refrigerated warehouses (55). They are kept refrigerated (7–10°C) until purchased by consumers.

5. Fresh Cucumber Pickles

Several types of pickles including whole dills, hamburger, slices (cross-cut), spears (cut lengthwise), bread and butter, and others are made from fresh cucumbers. Spices are added to the container before it is filled with cucumbers. When containers are filled with cucumbers, a cover liquor consisting of water, vinegar, salt, and other flavoring agents is added. The containers are then capped, sent through pasteurizers, cooled, labeled, cased, and stored.

6. Pickles from Brined Cucumbers

Cucumbers are cured in a salt solution. Then desalting is carried out by immersing the salt stock in several changes of fresh water or a continuous flow of water. Desalting is halted when the salt level in the cucumbers reaches 150–200% of that desired in the finished product. Once the proper degree of desalting has been achieved, the stock can be made into pickles (dill, sweet, or sour).

7. Processed Pickles

Once the brined cucumbers are sufficiently desalted, they are processed into genuine dills, sweet pickles, sour pickles, relishes, and other products (56). After desalting, cucumbers are placed in jars or other containers and treated in a manner similar to that described for fresh-pack pickles. Some of these products are not pasteurized provided that sufficient vinegar, sugar, and preservatives have been added to ensure a long shelf-life.

F. Waste Utilization

Cucumber processing results in both liquid and solid waste. Broken cucumber stems, leaves, and other debris is transported to local dumping areas. In general, liquid waste from cucumber-processing plants has high NaCl and O_2 contents along with a low pH and high total and suspended solid content (57). Fermentation usually results in the production of about 40% by volume of spent brine. For disposal of liquid waste, a popular and satisfactory system involves the use of lagoons (7). The use of pure culture results in lowering the use of salt and generally higher yields of products than natural fermentation procedures (58). The spent brine contains softening enzymes, which cause softening of cucumbers if reused. Heat treatment of 80°C for 30 seconds was sufficient to inactivate pectinase from molds common on fruits and flowers (58). Chemical treatment involving raising the pH of brine to 11.0 or higher for 36 hours resulted in 90% inactivation of pectinase (59).

III. MUSKMELON AND CANTALOUPE

Muskmelon and cantaloupe are important desert fruits grown in many countries of the world. Muskmelon is said to be native to tropical Africa, more specifically the eastern region south of the Sahara Desert. Whitaker (60) considered Central Asia, comprising some parts of southern Russia, Iran, Afghanistan, Pakistan, and Northwest India, as a secondary center of muskmelon. The muskmelon is a popular crop, although it is not an easy one to grow in most regions. It is grown in market gardens and as a commercial crop in many parts of the world (61).

A. Botany

1. Taxonomy and Morphology

The place of origin of muskmelon is not known with certainty, but as the wild species of *Cucumis* occur in Africa, it is likely that it originated on that continent. Secondary centers of variation occurred in India, China, Persia, and southern Russia.

The species *Cucumis melo* is a large polymorphic, encompassing a large number of botanical and horticultural varieties. Horticulturally, the muskmelon and cantaloupe differ somewhat in physical characteristics and regional adaptation. Today cantaloupe simply refers to cultivars that are highly uniform in overall netting (corky tissue on the rind) with relatively indistinct vein tracts (2). Internally, the cantaloupe flesh is thick, salmon-orange in color, with a characteristic flavor, and the seed cavity is small and dry. Muskmelon cultivars, on the other hand, have a stronger aroma, juicier flesh, and larger seed cavity. In the United States, most cantaloupe cultivars have been developed to meet requirements for packing in crates and long-distance transport. Wintermelons conversely require a long growing season at relatively higher temperatures under semi-arid conditions and are suited to limited storage and shipment after harvest.

Cantaloupes belong to *Cucumis melo* var. *reticulata*. Most American cantaloupes and Persian melons belong to this group. European cantaloupes belongs to *C. melo* var. *cantaloupensis*. The fruits are warty, scaly and rough, without netting. These are still cultivated in some parts of Europe. The winter melons of the United States comprising Honeydew (green-fleshed), Casaba, and Crenshaw have fruits with little musky odor. They ripen late. Their leaves are light green or medium green and the fruit surface is usually smooth. The less important cultivated species *Cucumis anguria* L is called the West Indian Gherkin and has been known in a semi-wild state in West Indies. *Cucumis melo* var. *momordica* is the Phoot of north and eastern India, also called the snap melon, whose fruit skin bursts or cracks on maturity. The rind is yellow and the flesh is fluffy and sour (2).

Muskmelon vines have palmate leaves. A mostly andromonoecious sex form is common in most of the varieties grown in India. Male flowers appear first as a cluster on main and secondary branches, but hermaphroditic flowers appear alone on secondary branches. The shape of the ovary varies from ovoid to elongated. After pollination, the ovary wall rapidly expands and develops into pericarp with an exocarp, mesocarp, and endocarp. The edible portion is mostly mesocarp. The number of fruits to develop per vine ranges from one to several.

2. Cultivars

Two important classes of muskmelon have been recognized, both belonging to *Cucumis melo* but of different subspecies. Most muskmelons grown in the United States are *Cucumis melo* var. *reticulatus* (Naudin), called netted melons. The other class of melons belongs to *C. melo* var. *inodorous* and include the so-called wintermelons (14).

Among the leading cantaloupe cultivars are Hale's Best, Edisto, Campo, Jacumba, Top Mark, Perlita and Planters Jumbo. In the Persian group, Persian Large, and Persian Small. Among long-duration winter melons, Honey Dew, Casaba, Golden Beauty, Crenshaw, and Santa Claus are important. Melons known as Charentias in France are grown in greenhouses.

There are several local varieties of musklemon grown in different regions of India (61). Among the fully netted desert cultivars are Lucknow Safeda, Kanpuria, Jogia, Mathuria, Batti, Kajra, Jaunpuri, Bhagpat, and Mahaban in Uttar Pradesh; Sankheda and Balteshwar in Gujrat; Kharri in Jalgaon; Kabri Gurbeli, and Kavita in Madhya Pradesh; Sanganer Hara Gola and Motea in Rajasthan; Sharbat-e-Anar, Bathesa, Chiranji, and Ladoo in Andhra Pradesh; Kadapa in

Karnataka; Batti, Goose, and Jam in Maharashtra; Kutana and Bagpat in Haryana; Har Dhari and Musa in Punjab; and F-1 hybrid and M-3 from New Delhi. Some private seed companies also market hybrid muskmelons like MHC-5 and MHC-6 from Jalna and Swarna from Bangalore, India.

B. Production

1. Climate and Soil

Muskmelon is a warm season crop, requiring a hot and dry climate: the optimum temperature requirement is 27–30°C (12). The seed does not germinate at temperatures lower than 18°C. High atmospheric humidity coupled with clouds adversely affects the sugar formation, texture, and flavor in fruits. Short day length promotes female flower production (13).

For proper ripening and a high sugar content, high temperature, low humidity, and plenty of sunshine are essential. Plants are sensitive to low temperature and frost. The muskmelon can be brown on all types of well-drained and fertile soils. Sandy loams are considered best for early crops, while loams are good for high yield. Sandy soils can also be treated with humus or compost. A very good crop can also be raised on river beds, provided sufficient organic manure is applied.

2. Propagation

Seed propagation is commercially used for multiplying muskmelon. Muskmelon seeds are creamish in color and retain their viability for nearly 5 years. On average, one can get up to 85–90% germination. About 3000–3500 seeds can be obtained from 100 g. The best temperature for seed germination is 23–24°C.

3. Cultural Practices

Planting

Land should be prepared by repeated ploughing to get fine tilth and should be perfectly level before sowing. If the soil is too dry, a light irrigation should be given at the time of final ploughing so that sufficient moisture is available for seeds to germinate.

Seeds are planted 1.5–4.0 cm deep in rows or hills, 6–7 feet (180–210 cm) apart. The depth of planting depends on soil type and moisture. Space between plants should be 30–60 cm (single plants, 30 cm; two plants, 60 cm) (8). In case of river beds, trenches about 0.5–1.0 m deep are dug at recommended row spacing and beds are treated with FYM and fertilizers. About 3–7 kg of seeds are sufficient to plant one hectare. The seeds take 4–6 days to germinate. Seedlings are thinned to keep two to three good seedlings per hill.

Prabhakar et al. (63) reported the highest test yield of cultivar Hara Madhu with 3 m x 60 m spacing. Seshadri (2) recommended spacing of 2–2.5 m between rows and 50–75 cm between plants, whereas Nath et al. (12) recommended 1.5–2.5 m spacing between rows and 60–120 cm between hills. The recommended spacings for different varieties grown in India are presented in Table 4 (61).

Manuring and Fertilization

About 25–45 tons of FYM per hectare has been recommended for muskmelon (12,16). It may be mixed with the soil at the time of ploughing or applied to each pit. In addition, fertilizer application is also required. The recommended doses vary from 50 to 100 kg of N, 30 to 50 kg of P, and 50 to 75 kg of K per hectare (64–67). Yamaguchi (8) recommended about 65–135 kg N/ha

TABLE 4 Recommended Spacing for Different Varieties of Muskmelon in India

Crop variety	Region	Spacing Rows	× Plants
Hara Madhu	Punjab	4 m	75 cm
Punjab Sunehri and Punjab Hybrid	Punjab	3 m	60 cm
Arka Rajhans	Karnataka	2.4 m	1 m
Arka Jeet	Karanataka	2 m	1 m
Durgapura Madhu	Western India	3.5 m	60 cm

Source: Ref. 61.

and 28–66 kg P/ha. About half of the fertilizer is placed about 6 cm to the side and 6 cm below the seed row; the rest of the fertilizer is side-dressed shortly after thinning when the seedlings are at the four- to six-true-leaf stage. Potash may be added to the fertilizer mix if levels of this nutrient are low in the soil. Potash may be applied through soil or as foliar sprays. Randhawa and Singh (67) reported a significant increase in muskmelon yield with foliar sprays of potash at 1.5%. Foliar sprays of boron at 4 mg/liter or Ca at 20 mg/liter (67) or molybdenum at 3 mg/liter (68) have also been found to increase the number of female flowers and yield.

Interculture

A first weeding may be started about 15–20 days after sowing; two or four weedings should be performed before the vine covers the whole ground. Herbicides have also been tested for weed control in the muskmelon (69,70).

Irrigation

A first irrigation should be given immediately after sowing to facilitate quick and better germination. Regular irrigation at 5- to 7-day intervals are required up to the fruit-setting stage (2). The frequency of irrigation may be reduced at the time of fruit ripening. In general, when the crops are grown in flat beds, about five or six irrigations are needed for the whole duration of crop growth. Precautions should be taken not to flood the field at any time to avoid infection of leaves and stems.

4. Diseases and Pests

Powdery mildew caused by *Sphaerotheca fuliginea*, downy mildew caused by *Pseudoperonospora cubensis*, and Fusarium wilt caused by *Fusarium oxysporium* are important fungal diseases of muskmelon. Viral diseases include mosaic and cantaloupe latent virus (aphid transmitted). Most of the pests damaging cucumber also attack melons.

5. Harvesting and Handling

Crops will be ready for harvest about 70–90 days after seed sowing, depending upon variety and season. The fruits of muskmelon take about 28–30 days from fruit setting to reach maturity (71). Thompson and Kelly (14) stated that, for local markets, muskmelons should be left on the vines until they are fully ripe but still solid; for shipment, they should be picked before they are fully ripe, but not so immature that they never develop good edible quality. Correct harvest maturity of melons is difficult to determine. As melons approach maturity, the netting becomes fully rounded out, whereas in immature fruits, the netting is flattened and has a slight crease along the top. The color changes from a dark green to a grayish-green to a yellowish-green as the melons near maturity. As ripening advances, a crack develops around the peduncle at the base of the fruit, and when fully ripe, the fruit slips easily from the stem, leaving a large scar. The soluble solids, the refractive index, and the sucrose content of the juice of muskmelons increase, and the percentage of starch in the seeds decreases as the melons ripen (72). Edible quality of melons depends on texture, flavor, and sweetness. Sugar content is the most important factor; it may be estimated by determining the total solids in the juice of edible portion (73,74).

Purewal (73) described the characteristics of a ripe muskmelon for determing harvesting stage:

1. Softening of rind, which can be felt by pressing the rind with hand. Ripening in muskmelon starts at the blossom end.
2. Change in the color of the base of the pedicel from green to waxy.
3. Ripened fruits have nutty aroma.
4. Formation of abscission layer. If the fruit is in "full slip stage," the fruit separates from the stem easily without much force required, leaving a depression on the stem end of fruit. If the abscission layer is not formed or is incomplete, the fruit does not separate from the stem easily and is said to be in half-slip stage. The fruit must be picked at "full-slip stage," at which time it will have good flavor with pulp of excellent color and consistency. Fruits at the half-slip stage are good for distant markets only. Harvesting is generally done in the morning.

Careful and prompt postharvest handling of melons is very important. Rough handling causes bruises, which make fruit more susceptible to decay. Delay in getting the melons packed and loaded into refrigerator cars may result in too rapid ripening with consequent loss. Muskmelons (cantaloupes) are graded in the United States according to USDA specifications and packed in crates of different sizes.

C. Chemical Composition

The chemical composition of muskmelon is presented in Table 5. According to Aykrod (75), fruits contain 0.6% protein, 0.2% fats, 3.5% carbohydrates, 32 mg calcium, 14 mg phosphorus, 1.4 mg iron, 16 mg carotene, and 26 mg vitamin C per 100 g fresh weight. The total soluble solid content varies from 8 to 17%. The vitamin contents in muskmelon and other melons are presented in Table 6. Color, thickness, and texture (juicy, smooth) of the flesh and absence of fibrous material along with characteristic aromatic flavor and sweetness determine the quality

TABLE 5 Proximate Composition of Melons (per 100 g edible portion)

	Muskmelon			
Constituent	Casaba	Honeydew	Netted	Watermelon
Water (g)	92.00	89.70	89.80	91.50
Energy (kJ)	109	147	147	134
Protein (g)	0.90	0.46	0.88	0.62
Fat (g)	0.10	0.10	0.28	0.43
Fiber (g)	0.50	0.60	0.36	0.30
Total carbohydrates (g)	6.20	9.18	8.36	7.18
Minerals (g)	0.80	0.60	0.71	0.26

Source: Refs. 5, 119.

TABLE 6 Vitamin Content of Some Melons (per 100 g edible portion)

Vitamin	Muskmelon			Watermelon
	Casaba	Honeydew	Netted	
Ascorbic acid (mg)	16.00	24.80	42.20	9.60
Thiamine (mg)	0.06	0.08	0.04	0.08
Riboflavin (mg)	0.02	0.02	0.02	0.02
Nicotinic acid (mg)	0.40	0.60	0.57	0.20
Pantothenic acid (mg)	—	0.21	0.13	0.21
Vitamin B_6 (mg)	—	0.06	0.12	0.14
Total carotene (mg)	0.05	0.07	5.37	0.61

Source: Refs. 5, 119.

of muskmelons. Increase in total sugars until full maturity and during ripening of melons on the vine has been reported (76). Softening of fruits is accompanied by changes from insoluble to soluble forms of pectin. Total solids of about 8–10° Brix are generally recommended for fruits to be shipped (76).

D. Storage

Precooling of melons, especially when they are picked during hot weather, is important. Melons need precooling soon after harvest to reduce high temperature, and this can be done with cold water, cold air, or ice. Pentzer et al. (77) showed that melons picked in the morning took much less time to precool, and the quantity of ice used in precooling was less than for melons picked during the hottest part of the day. Hydrocooling is the most efficient rapid cooling method for cantaloupes, which should be cooled to at least 10–15°C.

1. Low-Temperature Storage

Melons can be stored for a short time at 0°C and 80–90% relative humidity (RH) (78). Low-temperature breakdown is noticed in Honey Dew but not on cantaloupes, especially at 0–1.1°C (79). Package icing can be used for cantaloupes packed in moisture-proof cartons (80). Ryall and Lipton (26) recommended storage temperatures of 3.3–4.4°C with 95% RH for muskmelons. Under these conditions cantaloupes can be stored for 10–14 days. Pantastico et al. (39), however, reported that 1.7–3.3°C with 85–90% RH is suitable to store muskmelons for 11 days. Honey Dew, Casaba, and Persian melons are sensitive to chilling and should not be held below 5°C. The best holding temperatures are 7°C for Honey Dew and Persian melons and 10°C for Casaba melons (Table 7) (81).

2. Controlled-Atmosphere Storage

A slight to moderate benefit of transporting PMK-45 cantaloupe in 2% O_2 and 10–20% CO_2 atmosphere at 5°C was noticed (82). Development of decay-causing organisms was inhibited and ripening was delayed under controlled-atmosphere storage. A level of 1% O_2 appeared to be injurious. Hatton et al. (82) indicated that a level of 2–5% O_2 atmosphere at 0.6°C extended transport life of Honey Dew melons. The effects of temperature, relative humidity, modified CO_2 and O_2 levels, ethylene, and other factors on the quality of Honey Dew melons in van containers have been reported (83).

TABLE 7 Recommended Storage Temperature and Relative Humidity and Approximate Shelf-life of Some Melons

Type	Temperature (°C)	Relative humidity (%)	Shelf-life
Muskmelon			
Casaba	10	90	3 weeks
Honeydew	7	90	3 weeks
Netted			
Near-ripe	2–5	95	15 days
Ripe	0–2	95	5–14 days
Watermelon	10–15	90	2–3 weeks

Source: Ref. 81.

3. Chemical Contol

Muskmelons, including cantaloupes, Honey Dews, Persians, Crenshaws, Casabas, and their hybrid derivatives are all infected by *Alternaria* rot (*A. tenuis*), bacterial soft rot (*Erwinia* spp.), bacterial spot (*P. lachrymans*), blue rot (*Penicilium* spp.), green mold rot (*Cladosporium cucumerinum*), *Fusarium* rot (various species of *Fusarium*), and *Rhizopus* soft rot (*R. stolonifer*).

Carter (84) dipped muskmelon (cv. Perlita) fruits in sodium dimethyldithiocarbamate (SDMC) at 4000 mg/liter, sodium phenyl phenoate (SOPP) at 2500 mg/liter, or water at 24 or 57°C for 0.5 minutes and stored at 90% RH at 5°C for 5 days followed by 4 days at 26°C. SDMC at 57°C was the most effective treatment, reducing the number of *Fusarium roseum* and *Diaporthe melonis* lesions per fruit from 21.8 and 3.0 with water at 57°C to 2.9 and 0.9, respectively. Water alone at 57°C was no more effective than water at 24°C, but fungicide efficacy was increased at 57°C compared with 24°C.

Temkin-Gorodeiski and Zisman (76) found that waxing Haogen melons with Tag 16 wax with the addition of the fungicide sodium dimethyl dithiocarbamate (SDMC) reduced the rate of decay and maintained fruit firmness during shipping at 6°C. Imazalil alone or in combination with wax prevented development of decay in Galia melons longer than SDMC wax treatment. Aharoni and Zauberman (85) reported that En Dor melons coated with Tag wax containing SDMC and stored for 18 days at 10°C were in good condition without any signs of decay when transferred to 20°C for 2 days. The control melons (not treated with wax or fungicide) suffered 10% decay caused by the fungus *Rhizopus stolonifer*.

IV. WATERMELON

Among cucurbitacious crops, watermelon is one of the most important and widely cultivated all over the world. Tropical Africa is considered to be the place of origin of watermelon (86). Watermellon [*Citrullus lanatus* (Thumb.) Matsumura and Nakai] is an important cash crop and yields good profits in a short period (87). It is highly appreciated by consumers for its sweet juice. It is also used to treat hepatic congestion and intestinal disorders (88).

A. Botany

1. Taxonomy

There are four species, *Citrullus vulgaris* Schrad, which includes cultivated watermelon [now renamed *C. lanatus* (Thunb.) Matsumura and Nakai], *C. colocynthis* L., *C. ecirrhosus* Cogn., and *C. naudinianus* (Sond.) Hook. The species classification of *Citrullus* and their relationship have been studied in relation to cucurbitacin content or bitter principles (2). Shimotsuma (89) studied morphological and cytogenetic aspects of the four species. It was observed that all four species were cross-compatible with each other and several factors like geographical isolation, differences in flowering habit, genetic differences and structural changes in chromosomes would have contributed to the maintenance of their identity. According to Fursa (86), cultivation of watermelon began in ancient Egypt and India and spread from there to different countries via the Mediterranean, Near East and Asia. As a result of prolonged cultivation and selection, new forms of table watermelon have evolved, the varieties grown today bearing little resemblance to the ancestral African forms. Cultivation of forage watermelon *C. lanatus* var. *citroides*, is comparatively recent and Russian varieties grown today still have the shape of their African ancestors (*C. lanatus* var. *caffer*).

2. Morphology

The watermelon plant resembles the muskmelon plant, but the stems are angular in cross section and the leaves, which are cordate at the base, and pinnately divided into three or four pairs of lobes. The root system is very extensive but shallow, consisting of a taproot and many lateral roots growing within the top 2 feet of soil. Early destruction of the taproot may be advantageous in getting superior yields (91). The plant typically is a trailing (vine) annual, but recently dwarf forms described as "bush" have been located and are being used to develop cultivars suitable for growth in a limited space. Dwarfing is primarily related to shortened internodes (92).

Fruit shape varies ranging from long and cylindrical to spherical (Fig. 4). Rind colors vary from white to various shades of green, with some cultivars having striped rind and some mottled rind. Flesh color may be white, yellow, orange, pink, or red (Fig. 4); texture varies from fine grained and melting to firm coarse and fibrous; seeds range in size from very small to large, usually black, white, and red.

3. Floral Biology

Watermelon is monoecious and female flowers appear on the main branches. The carolla becomes compact one day before and on the day of anthesis (93). Anthesis continues from 6:00

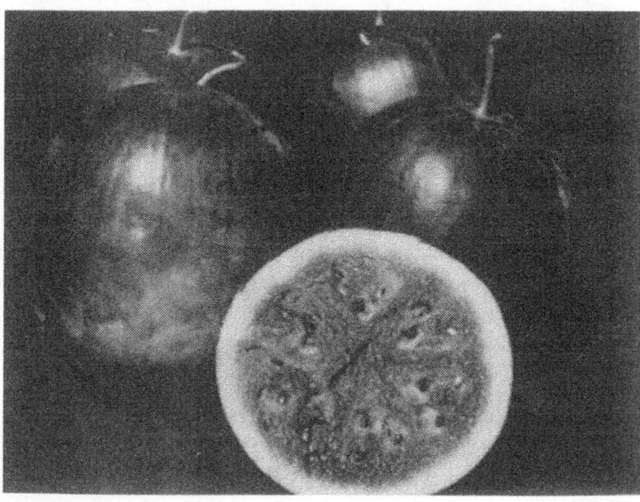

FIGURE 4 Watermelon fruits.

to 7:30 a.m., with the peak between 6:30 and 7:00 a.m. The anthesis is completed in a short time. Dehiscence starts one hour before anthesis and continues up to 6:30 a.m. (93). At high temperature, stigmatic fluid starts drying, and the stigma becomes nonreceptive by 3:00 p.m. (93).

4. Cultivars

Several watermelon cultivars are available for commercial cultivation. These include Sugar Baby, Asahi Yamato, Earliest New Hampshire, Midget, Arka Manik, Arka Jyoti, Shipper, Special No. 1, Durgapura, Meetha, Kesar, and 10-6-1 (94,95). Seedless watermelons are also produced in different parts of the world (94–100).

B. Production

1. Soil and Climate

Watermelon can be grown on a variety of soil types. Sandy loam to loam soils with good drainage are preferred for the cultivation of watermelon. Heavier soils with adequate drainage or silt soil could also be used. When sandy soils are used, they should be supplemented with humus or compost. The crop can withstand acid conditions better than other cucurbits. The optimum pH range is 6.5–7.0 (101). Continuous cropping of the same land should be avoided. Rotation once every 4–6 years is desirable and once every 10 or more years if *Fusarium* and nematodes are a problem (8).

Watermelon can be grown in subtropical as well as in hot arid regions. It is a warm season crop and cannot withstand cold climates, especially frost. High humidity adversely affects the yield and quality of fruits and increases the incidence of diseases. Seeds germinate best when temperatures are higher than 20°C.

2. Propagation

Watermelon is propagated through seed. The seeds of watermelons are large, flat, and count about 500–600 seeds per 100 g. If properly stored, the seeds remain viable for 1.5–2 years without any appreciable reduction in germination.

3. Cultural Practices

Planting

The land is prepared by repeated ploughing, harrowing, and leveling to obtain desired tilth. The sowing may be done on raised beds, furrows, trenches, or pits. Spacing of 2.5 × 3.5 m, 2.0 × 4.0 m (12), 2.5 × 1.5 m (102), 3.5 × 1.2 m (103), and 2.0 × 1.0 or 2 × 1.2 m (104) has been recommended. Presowing treatment of seeds with fungicides prevents the decay of seeds and damping-off of seedlings. Soaking seed in lukewarm water for 12–24 hours, followed by wrapping in moist cloth and storing in baskets in a cool, dry place for several days improved sprouting (21). Sprouting and seedling emergence can also be improved by a presowing soaking of seeds in CEPA (240–480 ppm). The optimum soil temperature for seed germination is 25–35°C. At 20°C, germination is satisfactory, but it takes about 12 days for emergence (8).

Manuring and Fertilization

Watermelons require adequate quantities of manure and fertilizers (105). Watermelon crops require 15–22.5 tons of farmyard manure (FYM) per hectare (102), which may be mixed with the soil before sowing or applied to the pits. Purewal et al. (73) recommended the application of 25 tons of FYM mixed into the soil before the final ploughing. In addition, they have suggested a supplementary dose of 139–187 kg of ammonium sulfate at the fruit-setting stage for better fruit set and better quality fruits. Bhosale et al. (106) found 75 kg of N, 30 kg of P, and 75 kg of K per hectare to be optimum. Deswal and Patil (107) reported 70 kg of N, 50 kg of P, and 50 kg of K per hectare to be optimum. Foliar sprays of boron 3 mg or calcium 3 mg or molybdenum 3 mg per liter of water at two- and four-leaf stages of growth were found to result in considerable increase in number of fruits per plant as well as in fruit yield (87).

Pruning and Training

In light textured soils, watermelon requires a minimum of 380 mm water during growth. Less irrigation is required for watermelon than for other melons. Crops should be irrigated when the irrigation water/cumulative pan evaporation (IW/CPE) ratio is one (110). Drip irrigation (with one emitter per two plants) gave higher fruit yield than furrow irrigation (111). During ripening, excess soil moisture reduces the sugar content and adversely affects the development of flavor. Therefore, irrigation should be completely stopped during fruit ripening. Crops planted in river beds do not need any irrigation unless the soil is too dry.

Interculture

Herbicides have been tested and recommended for weed control in watermelon. Saimbhi and Randawa (69) used sprays of pre- and postemergent herbicides like simazine, alachlor, dichlormate, and propanil one day after sowing of seeds. Leela (112) found butachlor at 2.0 kg ai/ha an effective herbicide for watermelon both in terms of reduced dry matter production of weeds and enhanced fruit yield. Khurana et al. (113) reported that preplant application of bensulfide at 4 kg/ha or trifluralin at 1–2 kg/ha can be used for weed control in watermelon. Other intercultural operations followed in watermelon include protection of young seedlings from frost, thinning seedlings, earthing up of seedlings to protect them from strong winds, timely trailing of vines to encourage them to grow straight, and protecting crops against pests and diseases.

Use of Growth Regulators

The number of female and hermophroditic flowers in watermelons can be increased by treatment with plant growth regulators. Tri-iodobenzoic acid (TIBA) at 25–50 ppm, boron and molybdenum at 3 ppm, and calcium at 20 ppm proved to be effective in inducing more female flowers and increasing fruit set and fruit yield (87). Similarly, gibberellic acid (GA) when sprayed at 25, 50, and 100 ppm significantly increased the number of female flowers, while naphthalene acetic acid (NAA) and indol acetic acid (IAA) increased the proportion of male flowers (87). Maleic hydrazide at 40,250 ppm inhibited the growth of leader shoots, but increased the number of laterals and hermophordite flowers and improved the fruit set (87).

4. Diseases and Pests

Several fungal diseases cause damage to watermelon crops. Powdery mildew is a predominant disease in warm rain-free growing regions. It is caused by *Sphaerotheca fuliginea* (Schlecht Fr.) Poll. Fusarium wilt caused by *Fusarium oxysporium* is also a common disease of watermelon. The fungus *Mysosphaerella citrullina* causes gummy stem blight of watermelon, particularly in warm humid climates. Several varieties resistant to powdery mildew *Fusarium* wilt and gummy stem blight (Citron) have been developed (87). Watermelon is attacked by the same insects, including aphids and beetles. The melon aphid is more injurious to watermelon than to either cucumber or muskmelon. Crops can be set back in the earlier stages of growth due to an attack of red pumpkin beetles.

5. Harvesting, Handling, and Grading

The maturity of watermelon is based on the following parameters:

1. Dull sound when the fruit is thumped, in contrast to metallic sound.
2. Withering of tendrils at the fruit axil.
3. Ground spot (where the fruit touches the ground) turning from white to yellow.

The fruit should be cut from the vine with at least 5 cm of stalk so that Bordeaux paste can be applied to the stalk to prevent stem end rot during transit.

The USDA has established grade standards for watermelon, and a large percentage of shipments are inspected to certify the grade by licensed inspection. Trucks or railroad cars are used for transportation. Watermelon boxes should be lined with paper to protect fruit from bruises resulting from rubbing against the sides of the car. The floor of the car can also be covered with about 7.5 cm of straw or similar material. In most shipments, watermelons are placed in box cars, cattle cars, or refrigerated cars (14).

C. Chemical Composition

Watermelons are high in carbohydrates but are only a moderate source of vitamin A and vitamin C (Table 5). Elmstrom and Davis (114) reported a wide variation in the ratio of total reducing sugars to nonreducing sugras. Most cultivars have deep pink or pale pink flesh color with a slightly reddish tinge, containing largely lycopene and anthocyanin pigments. The color of the flesh has no relationship to the degree of sweetness; pale pink flesh can be sweeter than deep pink flesh in some cultivars.

D. Storage

Watermelons, being chill sensitive, must be held at moderate temperatures. Lutz and Hardenburg (115) recommended 7.2–10°C as an acceptable storage temperature because neither chilling injury nor decay is serious in this stage. The color and flavor of watermelon improve during storage at or above room temperature, and the color fades at 10°C or below (116). Ryall and Lipton (26) stated that since properly handled watermelons do not decay readily, temperatures of 12.7–15.4°C may be preferable for marketing periods of 2 weeks or less. If watermelons are to be stored longer than 2 weeks, the humidity requirements are less exacting than for some other melons, because they do not readily lose moisture. Relative humidity in the range of 80–85% may be satisfactory. Ryall and Lipton (26) further stated that the storage life of watermelons is difficult to assess, because they remain edible for 2–3 months at moderate temperatures, even though their quality may be low by most standards when they are held beyond 2 weeks.

Temkin-Gorodeiski and Zisman (117) studied the relationship between storage temperature and fruit quality in Noy Amid melons. Temperature of 4–10°C did not affect the rate of flesh deterioration, but caused chilling injury and increased rate of decay. It was suggested (117) that melons of cultivar Noy Amid be stored at temperatures above 10°C and below 17°C.

E. Processing

The potential uses of watermelon in products such as juice and fruit residues have been reported by Crandall and Kesterson (118). Fruits of the cv. Charleston Gray were processed by removing the rind with a knife, separating the juice from the flesh, and concentrating the juice. The yield was 41% juice, 8% pulp, 1% seeds, and 50% rind on a wet weight basis. The rind was further chopped, treated with lime, and pressed to remove water (63% on a wet weight basis). The residue after juice extraction was made into an animal feed with a nutrient composition of 13% protein, 7% ether extract, 20% fiber, and 5% ash. Watermelon rind is sometimes pickled as a condiment, and candied citron rind is used in making fruit cakes and other confections (119).

V. OTHER MELONS

A. Long Melon

Long melon or Oriental pickling melon (*Cucumis* melo L. var. *conomon*) is also called Chinese white cucumber. Its origin is obscure. In Japan, it is grown mainly for making pickles called *tsukemono* or *koko*.

The fruits of long melon are eaten raw as salad throughout India. The fruits have a cooling effect and are easily digestible if taken with salt and pepper. The varieties are classified according to rind color into light green and green. The improved varieties include IIHR-4 and Karnal Selection (21). Long melon is a warm season crop but can tolerate cooler temperatures better than other melons. It resembles other cucurbitaceous crops in soil and climatic requirements. There are several cultivars of oriental pickling melon characterized by color: White Shirouri, Green Gouri, and Stripped Shima Uri. A trailing form of long melon is grown in parts of Southeast Asia. The optimum temperature for growth is 25–30°C; at 13°C growth is very much retarded.

The early crop is sown in November-December, while the late crop is sown from January to April. It requires 10 tons of FYM, 40 kg of nitrogen, and 80 kg of phosphorus per hectare (2). Other cultural practices are similar to those for other melons. The fruits are picked when they are still tender—about one-third or one-fourth their full size. The fruits do not keep well for more than one day, so they should be disposed of promptly. For pickling, fruits are harvested when full size but still immature. The seed cavity is removed and only the rind used in the making of pickles. The fruits are used as summer squash in varius ethnic dishes.

B. Squash Melon

Squash melon (*Praccitrullos fistuloses*, Pong) is also called round melon, Indian squash, and round gourd. It is commonly called *tinda* in India (2). It is an important summer vegetable cultivated extensively in northern India and less extensively in the South. Local cultivars have been selected for specific areas. Arka Tinda was developed by the Indian Institute of Horticultural Research, Bangalore, and Mahyco Tinda by a private seed company at Jalna, India.

For Squash melon, a warm and dry climate is best for vegetative growth. The optimum temperature for growth is 27–30°C (2). Seed germination falls in low temperature. It resembles other cucurbitaceous crops in soil requirements. Mahakal et al. (120) reported the highest fruit yield with 50 kg of N, 50 kg of P, and 100 kg of K per hectare. The first irrigation may be given at the time of seed sowing, subsequent intervals, irrigations depend upon the soil type and climatic conditions (IW/CPE ratio should be about 0.9) (121). There will be total of 10–11 irrigations at 5- to 8-day intervals. Drip irrigation was also found to give higher fruit yield (122). Frequent weeding should be performed. The first weeding may be carried out about 15–20 days after seed sowing and later weedings when required. Saimbhi and Randhawa (69) found nitrofen at 1.25 liters/ha and alachor at 2.5 liters/ha effective in controlling weeds.

The crop will be ready for harvest about 60–90 days after seed sowing, depending upon the variety and season. The fruit of squash melon will take about 7–8 days after fruit set to reach edible maturity (2). Picking should be done when fruit are still small and at 3- to 4-day intervals.

The immature fruits are cooked as vegetables. The fruits contain 1.4% protein, 0.4% fat,

3.4% carbohydrates, 13 mg of carotene, and 18 mg of vitamin C per 100 g fresh weight (75). The fruits are also considered good for dry cough and improving blood circulation (29).

REFERENCES

1. Chakravarty, M. L., Monograph on Indian Cucurbitaceae (Taxonomy and Distribution), *Records Bot. Survey India 17*(1):6 (1959).
2. Seshadri, V. S., Cucurbits, in *Vegetable Crops in India* (T. K. Bose and M. G. Som, eds.), Naya Prokash, Calcutta, 1986, p. 128.
3. Bose, T. K., and M. G. Som, *Vegetable Crops in India*, Naya Prokash, 206 Bidhan Sarani, Calcutta, India, 1986.
4. de Candolle, A. *Origin of Cultivated Plant*, New York, 1982.
5. Chadha, M. L., and T. Lal, Improvement of cucurbits, in *Advances in Horticulture* (K. L. Chadha and G. Kalloo, eds.), Malhotra Pub. House, New Delhi, 1993, p. 137.
6. Work, P., and J. Carew, *Vegetable Production and Marketing*, John Wiley and Sons, New York, 1955, p. 537.
7. Miller, C H, and T C Wehner, Cucumbers, in *Quality and Preservation of Vegetables* (N. A. Michael Eskins, eds.), CRC Press, Boca Raton, FL, 1989, p. 245.
8. Yamaguchi, M., *World Vegetables: Principles, Production and Nutritive Value*, AVI Pub. Co., Westport, CT, 1983.
9. Nath, P., Cucurbitaceous Vegetables in Northern India, *Univ. Udaipur Extn. Bull. 7*:64 (1965).
10. Gill, H. S., J. P. Singh, and D. C. Pachuri, Pusa Sanyog out yields other cucumbers, *Indian Hort. 18*:11 (1973).
11.. Musmade, A. M., P. N. Kale, S. R. Gadakh, and K. E. Lawande, Himangi—a new high yielding cucumber with excellent shelf life, Golden Jubilee Symposium Organized by Hort. Soc. India, Bangalore, May 24–28, 1993.
12. Nath, P., S. Velayudhan, and D. P. Singh, *Vegetable for Tropical Region*, ICAR, New Delhi, 1987.
13. Bose, T. K., and M. S. Ghosh, Effect of photoperiod on growth and sex expression on some cucurbits, *Indian J. Agric. Sci, 45*:487 (1970).
14. Thompson, H. C., and W. C. Kelly, *Vegetable Crops*, 5th ed., McGraw-Hill Book Co. Inc., New York, 1957, p. 611.
15. Dutta, S., *Horticulture in Eastern Regions of India*, Ministry of Food and Agriculture, New Delhi, 1966.
16. Anonymous, Package of practices for vegetables and fruit crops, Punjab Agricultural University, Ludhiana, 1987.
17. Singh, S. P., *Production Technology of Vegetable Crops*, Agric. Res. Comm. Centre, Karnal, 1989, p. 219.
18. Verma, J. P., Effect of nutrients and growth regulators on growth and sex expression of cucumber (*Cucumis sativus* L.), *South Indian Hort. 27*:114 (1975).
19. Kadam, K. G., Effects of NPK fertilizers levels on yield of cucumber (*Cucumis sativus* L.), M. Sc. thesis, Mahatma Phule Agricultural University, Rahuri, India, 1983.
20. Maurya, K. R., Effect of nitrogen and boron on sex ratio protein and ascorbic acid content of cucumber (*Cucumis sativus* L.), *Indian J. Hort. 44*:239 (1987).
21. Yawalkar, K. S., *Vegetable Crops of India*, 2nd ed., Agril. Horticultural Pub. House, Nagpur, 1980, p. 370.
22. Khurana, S. K., J. L. Mangal, and G. R. Singh, Chemical weed control in cucurbits: A review, *Agric. Rev. 9*:69 (1988).
23. Barnes, W. C., Multiple disease resistance in cucumbers, *Proc. Am. Soc. Hort. Sci. 77*:417 (1961).
24. Barnes, W. C., Development of Multiple disease resistant hybrid cucumber, *Proc. Am. Soc. Hort. Sci. 89*:390 (1966).
25. Eckert, J. W., P. P. Rubio, A. K. Matto, and A. K. Thompson, Postharvest pathology, II. Diseases of tropical crops and their control, in *Postharvest Physiology Handling and Utilization of*

Tropical and Subtropical Fruits and Vegetables (E. B. Pantastico, ed.) AVI Pub. Co. Inc. Westport, CT, 1975, p. 415.

26. Ryall, A. L., and W. J. Lipton, *Handling, Transportation and Storage of Fruits and Vegetables*, Vol. I, *Vegetables and Melons*, AVI Pub. Co. Inc., Westport, CT, 1972, p. 473.
27. Watts, R. L., and G. S. Watts, *The Vegetqable Growing Business*, War Dept. Education Manual EM 885, Orange, Judd. Pub. Co. Inc., Washington, DC, 1940.
28. Mangal, J. L., and A. C. Yadav, Effect of pruning on growth yield and quality of cucumber, *Punjab Hort. J. 19*:194 (1979).
29. Nadkarni, K. N., *Indian Materia Medica*, Nadkarni and Company, Bombay, 1927.
30. Grubben, G. J. H., *Tropical Vegetables and Their Genetic Resources*, International Board for Plant Genetic Resources, FAO, Rome, 1977, p. 197.
31. Takei, S., and M. Ono, Leaf alcohol. III. Fragrance of cucumber, *J. Agric. Chem. Soc. Jpn. 15*:193 (1939).
32. Kemp, T. R., D. E. Knavel, and L. P. Stoitz, Identification of some volatile compounds from cucumber, *J. Agric. Food Chem. 22*:717 (1974).
33. Fleming, H. P., W. Y. Cobb, J. L. Etchells, and T. A. Bell, The formation of carbonyl compounds in cucumbers, *J. Food Sci. 32*:572 (1968).
34. Rabinwitch, H. D., and D. Skaan, Superoxide dismutase activity in ripening cucumber and pepper fruit, *Physiol. Plant. 52*:380 (1981).
35. Saltveit, M. E., Jr., and R. F. McFeeter, Polygalacturonase activity and ethylene synthesis during cucumber fruit development and maturation, *Plant Physiol. 66*:1019 (1980).
36. Apeland, J., Factors affecting non-parasitic disorders of the harvested produce of cucumber, *Tech. Commun. Intl. Soc. Hort. Sci. 4*:102 (1966).
37. Kapitsimadi, C. M., O. Roeggen, and H. Hoftun, Growth of four cucumber cultivars at sub-optimal temperature and storage behavior of their fruit at different temperature, *Acta Hort. 287*:375 (1991).
38. Miller, C. H., and W. C. Kelly, Mechanical stress, stimulate peroxidase activity in cucumber fruit, *Hort. Sci. 24*(4):650 (1989).
39. Pantastico, E. B., T. K. Chattopadhyay, and H. Subramanyam, Storage and commercial storage operations, in *Postharvest Physiology, Handling and Utilization of Tropical and Subtropical Fruits and Vegetablews* (E. B. Pantastico, ed.), AVI Pub. Co. Inc., Westport, CT, 1975, p. 314.
40. Fellers, P. J., and I. J. Pflug, Storage of pickling cucumbers, *Food Technol. 24*:74 (1967).
41. Sharma, B. B., and S. Wahab, Efficacy of actinodione and auerofungin in the control of postharvest decay of some cucurbitaceous fruits due to *Pythium aphanidermatum*, *Hindustan Antiobiotic Bull. 13*:8 (1976).
42. Atwa, A. A., T. E. Sheikh, S. M. Dessouky, and W. Y. Riad, Effect of chemical substances on keeping quality of cucumber fruits, *Hort. Abstr. 51*:5479 (1981).
43. Sciumbato, G. L., and C. P. Hegwood, Jr., A rapid method of evaluating and determining length of activity of surface applied fungicide for the control of cucumber fruit rot, *HortScience 15*(3):254 (1980).
44. Guillou, A. A., J. D. Floros, and M. A., Cousin, Calcium chloride and potassium sorbate reduced sodium chloride during natural cucumber fermentation, *J. Food Sci. 57*(6):1364 (1992).
45. Chavasit, V., J. M. Hudson, J. A. Torres, and M. A. Daeschel, Evaluation of fermentative bacteria in a model low salt cucumber juice brine, *J. Food Sci. 56*(2): 462 (1991).
46. Fleming, H. P., R. F. McFeeters, R. L. Thompson, and D. C. Sanders, The storage stability of vegetables fermented with pH control, *J. Food Sci. 48*:975 (1983).
47. Etchells, J. L., and L. H. Hontz, The use of equilibrated brine strength based on the average brine-cucumber and temperature, *Pickle Pack Sci. 2*:18 (1972).
48. Pederson, C. S., Pickles and sauerkraut, in *Commercial Vegetable Processing* (B. S. Luh and J. G. Woodroof, eds.), AVI Pub., Westport, CT, 1975, p. 457.
49. McFeeters, R. F., and H. P. Fleming, pH effect on calcium inhibition of softening of cucumber mesocarp tissue, *J. Food Sci. 56*(3):730 (1991).

50. McFeeters, R. F., Cell wall monosaccharide changes during softening of brined cucumber mesocarp tissue, *J. Food Sci. 57*(4):937 (1992).
51. Fleming, H. P., Fermented vegetables, in *Economic Microbiology, Fermented Foods*, Vol. 7 (A. H. Rose, ed.), Academic Press, New York, 1982, p. 227.
52. Thompson, R. L., H. P. Fleming, and R. J. Monroe, Effects of storage conditions on firmness of brined cucumbers, *J. Food Sci. 44*(3):843 (1979).
53. Fleming, H. P., R. E. McFeeters, and R. L. Thompson, Effects of sodium chloride concentration on firmness retention of cucumbers fermented and stored with calcium chloride, *J. Food Sci. 52*(3):653 (1987).
54. Etchells, J. L., T. A. Bell, H. P. Fleming, R. E. Kelling, and R. L. Thompson, Suggested procedure for the controlled fermentation of commercially brined pickling cucumbers—the use of starter and reduction of CO_2 accumulation, *Pickle Pack Sci. 3*:4 (1973).
55. Etchells, J. L., T. A. Bell, R. N. Costilow, C. E. Hood, and J. E. Anderson, Influence of temperature and humidity on microbial enzymatic and physical changes of stored, pickling cucumbers, *Appl. Microbiol. 26*:943 (1973).
56. US Standards for Grades of Pickles, F.R. Doc. 31, FR 10235, U.S. Department of Agriculture, Washington, DC, 1963.
57. Little, L. W., J. C. Lamb, III, and L. F. Horney, Characteristics and treatment of brine waste waters from the cucumber, pickle industry, Report No. 99, University of North Carolina Water Research Institute, Raleigh, 1976, p. 1.
58. Little, L. W., J. G. Wendle, J. Davis, R. M. Harrison, and S. J. Dunn, Reducing waste from cucumber pickling process by controlled culture fermentation EPA-60012-80-046, U.S. Environmental Protection Agency, Cincinnati, OH, 1980, p. 1.
59. McFeeters, R. F., W. Coon, M. P. Palnitkar, M. Velting, and N. Fehringer, Reuse of fermentation brines in cucumber pickling industry, EPA-600/2-78-207, U.S. Environmental Protection Agency, Cincinnati, OH, 1978, p. 1.
60. Whitaker, T. W., *Evolution of Crop Plants*, E.D.H.W. Simmonds, Longman, London, 1978.
61. Nandpuri, K. S., Muskmelon (*Cucumis melo* Linn), *Indian Hort. 33*:38 (1989).
62. Nath, P., and K. S. Khangarot, Physico-systematic studies of muskmelon under alkaline soil conditions, *Rajsthan Agric. 8*:33 (1968).
63. Prabhakar, B. S., K. Srinavas, and V. Shukla, Yield and quality of muskmelon cv. Hara Madhu in relation to spacing and fertilization, *Prog. Hort. 17*:51 (1985).
64. Padda, D. S., B. S. Malik, and J. C. Kumar, Response of muskmelon to nitrogen, phosphorus and potassium fertilization, *Indian J. Hort. 26*:173 (1968).
65. Randhawa, K. S., D. S. Cheema, and K. S. Sandhu, Effect of nitrogen phosphorus and potassium on growth, yield and quality of new muskmelon varieties, *Haryana J. Hortic. Sci. 10*:88 (1981).
66. Jaswal, N. S., K. S. Nandpuri, and K. S. Randhawa, Studies on the effect of irrigation and certain doses of N, P and K on weight of fruit and yield of muskmelon, *Punjab Hort. J. 10*:143 (1970).
67. Randhawa, K. S., and K. Singh, Influence of foliar application of certain chemicals on sex behavior, fruit set and quality of muskmelon (*Cucumis melo* L.), *Plant Sci.* 2:118 (1979).
68. Hooda, R. S., M. L. Pandita, and A. S. Sidhu, A note on the foliar application of calcium and boron on growth and yield of unpruned and pruned muskmelon (*C. Melo*), *Haryana J. Hort. Sci. 10*:111 (1981).
69. Saimbhi, M. S., and K. S. Randhawa, A note on the response of four cucurbits to pre- and post emergence application of different herbicides, *Haryana J. Hort. Sci. 6*:91 (1977).
70. Saimbhi, M. S., Agro-techniques for cucurbits in *Advances in Horticulture*, Vol. 5, *Vegetable Crops* (K. L. Chadha and G. Kalloo, eds.), Malhotra Pub. House, New Delhi, 1993, p. 402.
71. Srinavas, K., D. M. Hegde, and S. D. Doijode. Studies on fruit development in muskmelon (*Cucumis melo* L.), *South Indian Hort. 31*:82 (1985).
75. Chase, E. M., C. G. Church, and F. E. Denny, Relation between composition of California cantaloupes and their commercial maturity, USDA Bull., (1924), p. 1250.
73. Purewal, S. S., *Vegetable Cultivation in North India*, ICAR Farm Bull, 36, 1957.

74. Mutton, L. L., B. R. Cullis, and A. B. Blackeney, The objective definition of eating quality of rock melons (*C. melo*), *J. Sci. Food Agric. 32*:385 (1981).
75. Aykroyd, W. R., V. N. Patwardhan, and S. Ranganathan, The nutritive value of Indian Foods and planning of satisfactory diet, Health Bull. 23, Govt. India, 1951, pp. 1–79.
76. Temkin-Gorodeiski, N., and U. Zisman, Physiological behavior of Haogen melons in the field and during storage in various growing seasons, in *Scientific Activities 1974-1977 of the Institute for Technology and Storage of Agricultural Products*, Pamphlet No. 184 Bet Dagon, Israel, 1974, p. 73.
77. Pentzer, W. T., J. S. Wiant, and J. H. MacGillivray, Maturity, quality and condition of California cantaloupes as influenced by maturity, handling and precooling, USDA Tech. Bull. No. 730, (1940).
78. Plantenius, H., F. S. Jamison, and H. C. Thompson, Studies on cold storage of vegetables, New York Cornell Stn.-Bull. No. 602, 1934.
79. Wiant, J. S., Market storage of Honey Dew melons and cantaloupes, USDA Tech. Bull. No. 613, 1938.
80. Kashmire, R. F., Determining cantaloupe sizes by volume: weight relationship, *Calif. Agric. 22*:13 (1968).
81. Hardenburg, R. E., A. E. Watada, and C. Y. Wang. The commercial storage of fruits, vegetables and florist and nursery stocks, USDA Handbook No. 66, 1986.
82. Hatton, T. T., Hr., E. B. Pantastico, and E. K. Akamine, Controlled atmosphere storage: 3 Individual commodity requirements, in *Postharvest Physiology, Handling and Utilization of Tropical and Subtropical Fruits and Vegetables* (E. B. Pantastico, ed.), The AVI Pub. Co. Inc., Westport, CT, 1975, p. 204.
83. Harvey, R. B., Hardening process in plants and developments from frost injury, *J. Agric. Res. 15*:83 (1918).
84. Carter, W. W., Re-evaluation of heated water dip as a postharvest treatment for controlling surface and decay fungi of muskmelons fruits, *HortScience 16*:334 (1981).
85. Aharoni, Y., and I. Zauberman, Postharvest research on En Dor melons, in *Scientific Activities 1974-1977 of the Institute for Technology and Storage of Agricultural Products*, Bet Dagon, Pamphlet No. 184, 1978, p. 30.
86. Fursa, T. B., History of introduction of watermelon into cultivation, *Trudy po Prikiadnoi Botanike Genetike* i *Selektsii 49*:62 (1973).
87. Choudhary, B., *Vegetables*, National Book Trust, New Delhi, India, 1967, p. 198.
88. Chauhan, D. V. S., *Vegetable Production in India*, Ram Prasad and Sons, Agra, 1981.
89. Shimotsuma, M., Cytogenetic and evolutionary studies in the genus *Citrullus, Seiken Jiho 15*:24 (1963).
90. Mohr, H. C., H. T. Blackhurst, and E. R. Jensen, F1 hybrid watermelons from open pollinated seed by use of genetic marker, *Proc. Am. Soc. Hort. Sci. 65*:399 (1955).
91. Elmstrom, G. W., Watermelon root development affected by direct seeding and transplanting, *HortScience 65*:399 (1955).
92. Mohr, H. C., Mode of inheritance of the bushy growth characteristics in watermelon, *Proc. Assoc. South Agric. Workshop 53*:174 (1956).
93. Seshadri, V. S., Cucurbits, *Indian Hort. 33*:28 (1989).
94. Seshadri, V. S., J. C. Sharma, and B. Choudhary, Pusa watermelon varieties assure good yield and quality, *Indian Hort. 16*:21 (1972).
95. Pal, B. P., M. S. Sikka, and H. B. Singh, Improved Pusa vegetables, *Indian J. Hort. 13*:64 (1956).
96. Kihara, H., Triploid watermelon, *Proc. Am. Soc. Hort. Sci. 58*:217 (1951).
97. Feher, B., and A. Kiss, Seed production in triploid watermelon, *Ag. Rartud Korl. 28*:481 (1969).
98. Whitaker, T. W., and G. N. Davis, *Cucurbits, Botany Cultivation and Utilization*, Interscience Pub., New York, 1962.
99. Walls, V. M., Use of market genes in producing triploid watermelons, *Proc. Am,. Soc. Hort. Sci. 76*:577 (1960).
100. Andrus, C. F., V. S. Seshadri, and P. C. Grimball, Production of seedless watermelons, USDA Tech. Bull. 1425, 1971, p. 12.

101. Nath, P., and S. P. Sachan, Physico-systemic studies of muskmelons under alkaline and saline conditions, *Rajsthan Agric. 7*:1 (1967).
102. Singh, S. P., Production technology of vegetable crops, Agric. Res. Comm. Centre, Karnal, 1989, p. 219.
103. Biswas, D., Standarization of Agro-Techniques on growth, yield and quality of watermelons (*Citrullus vulgaris* Schrad), Ph.D. thesis, Bidhan Chandra Krishi Vishwa Vidyalaya, Kalyani, 1983.
104. Singh, R. V., and K. B. Naik, Response of watermelon (*Citrullus lanatus* Mansf.) to plant density nitrogen and phosphorus fertilization, *Indian J. Hort. 46*:80 (1989).
105. Srinivas, K., Response of watermelon to drip and furrow irrigation under different nitrogen and plant population levels, *Indian J. Agric. Sci. 22*:559 (1987).
106. Bhosale, R. S., M. Khanvilkar, and R. K. Adhale, Yield and performance of watermelon (*Citrullas lanatus* Schrad) variety sugar baby to graded levels of nitrogen, phosphorus and potassium, *J. Mah. Agric. 3*:93 (1978).
107. Deswal, I. S., and V. K. Patil, Effect of N and P on fruit yield of watermelons, *J. Mah. Agric. Univ. 9*:308 (1984).
108. Mangal, J. L., M. L. Pandita, and G. Singh, Effect of pruning and spacing on watermelon, *Haryana J. Hort. Sci. 10*:216 (1981).
109. Virupaksha, M., Studies on effect of nitrogen potassium and fruit thinning on growth, yield and quality of watermelon (*Citrullus lanatus* Thumb Mansf) variety Asahi Yamato, *Mysore J. Agric. Sci. 22*:231 (1988).
110. Desai, J. B., and V. K. Patil, Effect of irrigation and sowing dates on expression of sex ratios and yield of watermelon Cv. Sugar Baby, *Indian J. Hort. 42*:271 (1985).
111. Hegde, D. M., Response of watermelon to nitrogen and irrigation, *J. Mah. Agric. Univ. 14*:70 (1989).
112. Leela, D., Weed controlled by herbicides in cole crops and cucurbitaceous crops, *Trop. Pest Manage. 31*:219 (1985).
113. Khurana, S. K., J. L. Mangal, and G. R. Singh, Chemical weed control in cucurbits: A review, *Agric. Rev. 9*:69 (1988).
114. Elmstorm, G. W., and P. L. Davis, Sugars in developing mature fruits of several watermelon cultivars, *J. Am. Soc. Hort. Sci. 106*:330 (1981).
115. Lutz, J. M., and R. E. Hardenburg, The commercial storage of fruits, vegetables and florists and nursery stocks, USDA Handbook No. 66, 1968.
116. Showalter, R. K., Watermelon color as affected by maturity and storage, *Proc. Fla. State Hort. Sci., 73*:289 (1960).
117. Temkin-Gorodeiski, N., and U. Zisman, Control of storage diseases in melon fruit during storage, in *Scientific Activities 1974-1977 of Institute for Technology and Storage of Agricultural Products, Pamphlet No. 184*, Bet Dagon, Israel, 1978, p. 39.
118. Crandall, P. G. and J. W. Kesterson, Components of processed watermelon fruit, *J. Am. Soc. Hort. Sci. 106*:493 (1981).
119. Lingle, S. E., Melons, squashes and gourds, in *Encyclopedia of Food Science, Food Technology and Nutrition*, Vol. 5 (R. MaCrae, R. K. Robinson, and M. J. Sandlers, eds.), Academic, Press, New York, 1993, p. 2960.
120. Mahakal, K. G., A. J. Joshi, D. P. Deshmukh, and R. P. Puner, Effect of N, P and K on tinda (*Citrullus vulgaris* var. *Fistulosus, Orissa J. Hort. 5*:62 (1977).
121. Mangal, J. L., B. R. Batra, and G. R. Singh, Studies on nitrogen fertilization under various soil moisture regimes on growth and production of round melon (*Citrullus vulgaris* var. *fistulosis*), *Haryana Hort. Sci. 14*:232 (1985).
122. Singh, S. P., and P. Singh, Value of drop irrigation compared with conventional irrigation for vegetable production on a hot arid climate, *Agron. J. 70*:945 (1978).

11

Pumpkins, Squashes, and Gourds

U. T. Desai and A. M. Musmade
Mahatma Phule Agricultural University, Rahuri, India

I. PUMPKINS AND SQUASHES

Squashes and pumpkins belong to the genus *Cucurbita* and species *moschata* and *pepo*. Whitaker and Bemis (1) reported that the genus *Cucurbita* is indigenous to the Americas. There is good archaeological proof that *C. moschata* and *C. pepo* were widely distributed in both North and South America. Cucurbita was generally thought to have been first domesticated in Central America or Mexico. As these two species tolerate hot conditions better than other cultivated species of *Cucurbita*, they are widely grown throughout the tropics of both hemispheres.

A. Botany

1. Types

The terms squash and pumpkin have been used indiscriminately, which has created confusion in the classification of varieties of these vegetables. Squashes are classified as summer or winter types. The summer types, such as *C. pepo* (L), are eaten when immature and do not store well. Winter squashes, such as *C. maxima* (Duchesne) and *C. moschata* (Duchesne) Poir, are eaten when mature and can be stored for several months. The pumpkin *C. pepo* var. *pepo* can also be considered a winter squash.

Squash cultivars are commonly divided into two classes: bush and vining. Bush cultivars produce stems with greatly shortened internodes and set fruits in close succession; the fruits are ready for harvest in about 6 weeks. Bush squashes do not store well and are commonly called summer squashes. Vining cultivars produce large plants with one or more long stems covering the ground to a distance of 6 m. Usually, the fruits are allowed to reach full maturity and are stored for extended periods. Winter squashes included in this group include the orange, pale orange, or yellow-fleshed *Kashiphal* of India. Winter squashes belong to *C. maxima*. There are also vining cultivars in *C. pepo*, and all the pumpkin *C. moschata* cultivars are vining types. The characteristics of the five cultivated species are given in Table 1.

TABLE 1 Characteristics Differentiating Cultivated Species of Cucurbits

Species	Foliage	Stem	Androecium	Penduncle	Fruit flesh	Funicular	Seed margin color
C. peo (annual)	Spiculate	Hard, angular	Short, thick, conical	Hard, angular, ridged	Coarse-grained	Obtuse, symmetrical	Smooth, obtuse, white, buff, or brown
C. moschata (annual)	Nonspiculate	Moderately hard, smoothly angled	Long, slender, columnar	Hard, smoothly angular, flared	Fine-grained or coarse with gelatinous fibers	Obtuse, slightly asymmetrical	Scalloped, obtuse, white, buff, or brown
C. mixta (annual)	Nonspiculate	Hard, angular	Long, slender, columnar	Hard basically angular, but enlarged by hard cork	Coarse-grained	Obtuse, slightly asymmetrical	Barely scalloped, acute, white, buff, or brown
C. maxima (annual)	Moderately spiculate	Soft, round	Short, thick, columnar	Soft basically round, but enlarged by soft cork	Fine-grained	Acute asymmetrical	Smooth, obtuse white, buff, or brownish
C. ficifola	Moderately spiculate	Hard, smoothly angled	Short, thick, columnar	Hard, smoothly angled, slightly flaring	Coarse, tough, fiberous	Obtuse, slightly asymmetrical	Smooth, obtuse, black or tan

Source: Ref. 4.

2. Cultivars

Popular summer, bush-type cultivars of *C. pepo* species are White Scallop or Patty Pan, Early Prolific Straight Neck, Golden Summer Straight Neck, Cocozelle, Zucchini, and Italian Vegetable Marrow's. Some autumn and winter pumpkins include Winter Luxury, Small Sugar or New England Pie, Table Queen or Des Moines (called squash), and Connecticut Field, all of which are of trailing or vining types. The least important species, namely *C. moschata*, is represented by the Cushaw or Winter Crookneck cultivars as well as Large Cheese, Japanese Pie, Tennessee Sweet, Potato, and Kentucky Field cultivars. *C. maxima* consists of true squashes with long keeping qualities and thick, rich, fine-textured flesh. They have soft and spongy fruit stalks. *C. maxima* varieties are generally listed as autumn and winter squashes. Those of the Hubbard group are the most widely planted, high quality, and best keepers. Watts and Watts (2) and Thompson and Kelly (3) mentioned Green Hubbard, Blue Hubbard, Warted Hubbard, Golden Hubbard, Delicious, and Golden Delicious as the popular varieties of slightly earlier types, small but with high-quality fruits. Boston Marrow is the earliest of the large-fruited squash, bright orange in color, very prolific, extensively planted for canning, but inferior to most other varieties in quality and thickness of flesh. Warren or Essex Hybrid is an orange-red, turban-shaped variety that is most popular in New England. Mammoth Chilli, the giant among squashes (weighing 100 pounds or more), has somewhat coarse and stringy flesh. Newly developed and/or released varieties in India include Arka Suryamukhi (*C. maxima*), Pusa Biswas (*C. moschata*), Early Yellow Prolific (*C. pepo*), Australian Green (*C. pepo*), Patty Pan (*C. pepo*), Pusa Alankar (*C. pepo*), and Punjab Chappan Kaddo (*C. pepo*).

B. Production

1. Soil and Climate

Pumpkins grow best and produce excellent quality fruits in rich, light-textured soil with a pH range of 6.5–7.5. Sandy loams or loamy fertile soils are ideal for both pumpkins and squashes. However, heavier soils can also be used as long as the drainage is adequate. Both crops require adequate amounts of essential nutrients and hence only fertile soils should be used. If such soils are not available, then soil should be supplemented with ample quantities of well-rotted farmyard manure and inorganic fertilizers.

Pumpkins and summer squashes are warm climate crops requiring a temperature range of 18–27°C for growth, the ideal being 18–20.5°C. Pumpkins take nearly 3–4 months from seedling emergence to fruit maturity. Therefore, a prolonged warm season is essential to obtain quality pumpkins. Summer squash, on the other hand, require only 40 to 50 days. Both pumpkin and squash tolerate low temperatures better than melons.

2. Propagation

Pumpkin and summer squashes are commercially propagated through seeds, which are sown directly in the field. The seeds of these crops are large, creamish-white in color, and have 75–85% germination under ideal conditions. When conditions are optimum, seeds can be stored for 4 years without any appreciable loss in viability. The best temperature for germination is 23.7–26.5°C (4).

3. Cultural Practices

Planting

The soils for pumpkin and summer squash should be thoroughly prepared by repeated ploughing and leveling to achieve complete pulverization. This facilitates better contact between seed and the soil moisture required for rapid and better germination. Pumpkin seeds are sown directly in the field on raised beds or in furrows, trenches, or pits. Two seeds per hill are sown on both sides of the ridges. In the pit method, 60 × 60 cm pits are dug 60 cm deep at recommended spacing and filled with farmyard manure (FYM) and topsoil. Spacing of 2.5–3.0 m between rows and 1–1.5 m between hills is recommended. The optimum plant spacing is 3 m × 60 cm (6). Four to five seeds are sown per pit. The recommended seed rate is 6–8 kg/ha

Manuring and Fertilization

About 20–25 tons of FYM per hectare should be applied at the time of soil preparation or to the individual pit (5). Sharma and Shukla (7) suggested using 103 kg of nitrogen, 106 kg of phosphorus, and 40 kg of potash per hectare for summer season pumpkin crops and 96 kg of nitrogen, 88 kg of phosphorus, and 40 kg of potash for rainy season crops. Yawalkar (8) recommended the application of 25–30 tons of well-rotted FYM, added in basins at the rate of 4–5 kg per basin, before planting. Pumpkin crops yielding 50 tons of fruit per hectare required 75 kg of nitrogen, 80 kg of phosphorus, and 80 kg of potash, and 30 kg of calcium (8).

For summer squash, Gill (9) recommended a dose of 60 kg of nitrogen, 30 kg of phosphorus, and 30 kg of potash in addition to 25–30 tons of FYM. Yamaguchi (10) suggested a fertilizer dose consisting of 110 kg of nitrogen, 40 kg of phosphorus, and 90 kg of potash per hectare. Half of the total dose of these fertilizers should be applied at the time of sowing, and the remaining half-dose applied at the four- to six-leaf stage.

Irrigation

The first irrigation should be given immediately after sowing, with the second and subsequent irrigations given at weekly intervals or even more frequently, depending upon need. Waterlogging should be avoided at all times. However, no irrigation is usually required for rainy season crops.

Interculture

Pumpkins and squashes require frequent weeding. The first weeding may be performed 15–20 days after seed sowing. A total of three weeding operations will be required. Herbicides can also be used for this purpose. According to Khurana et al. (11) preplant incorporation of bensulide 4–6 kg/ha or Alachlor 2.5 kg/ha preemergence can be used for weed control in pumpkins.

Harvesting

Pumpkin crops reach maturity about 75–180 days after seed sowing depending, on variety, season, and other conditions. The mature fruit is often brown, but in some varieties the mature fruit may not turn brown. Pumpkin fruits, when harvested at full maturity, have a long storage life. They can also be transported easily to distant markets.

Summer squash is ready for harvest about 60–80 days after seed sowing. Fruits can be harvested 7 days after fruit set (12). The picking should be done at 2- to 3-day intervals in order to get highest fruit yield. Tender fruits are preferred by consumers.

Pumpkins and squashes are usually cut from the vines with a portion of the stem attached. It is best to remove stems from squash that is to be stored. Careful handling of all varieties of pumpkins and squashes during storage and transport is essential to avoid bruises and injuries, through which decay organisms may enter. Pumpkins and squashes are seldom graded to any particular specifications. With winter pumpkins and squashes, some grading is done based on size, shape, and color (3).

Work and Carew (13) stated that pumpkins and winter squashes should be allowed to attain full size and maturity of seeds and rind. Hubbard squash, picked while young and soft-skinned, are very delicious. Summer squashes are edible any time before the skin begins to harden and the seeds attain full size. In Europe and North America, they are harvested when very small: 10–15 cm long for Straight Necks and 7.5 cm in diameter for Scallops. This probably means little gain in quality over harvesting at three quarter maturity, when the skin is tender, seeds immature, and the yield is greater. Early shipments are made in baskets, hampers, and crates. Summer squash from southern United States are sometimes wrapped in tissue paper to safeguard against damage and blemishes. Some careful growers even require their workers to wear cotton gloves to avoid fingernail cuts, and trucks may be lined with quilts and rugs.

C. Chemical Composition

Summer squashes are low in energy value (Table 2) but have a slightly sweet flavor and are moderately high in vitamin A (Table 3). Winter squashes supply more energy and have a high carbohydrate content. Their flavor is sweet and nutty, and they have a very high vitamin A content (Table 3). The chayote is higher in energy than is summer squash (Tables 2, 3).

D. Storage

A storage temperature of about 7–10°C with ≤ 70% RH is suitable to store late or winter squashes for an extended period. Only fully mature fruits with stems and without the smallest mechanical injury are suitable for storage (13). Platenius et al. (15) showed that low-temperature storage reduces weight loss. Squashes should be subjected to a curing or drying period of 2–4 weeks at 20.9–26.4°C.

Thompson and Kelly (3) stated that, after squashes and pumpkins are harvested, they should undergo a ripening or curing process to harden the cell. Although a curing period of about 2 weeks at 24–30°C with good air circulation followed by a reduction in temperature to 10–12.7°C is desirable, much better fruits are obtained when the crop is stored as soon as harvested (16). Winter pumpkins and winter squashes keep best at relatively higher temperatures (10–13°C). Blue Hubbard squash does not keep well stored at temperatures below 7°C even with humidities as low as 20% (16). Platenius et al. (15), however, observed that squashes remained in excellent condition for 5 months at about 4.4°C and 50–70% RH. Under these conditions, shrinkage was less than 3% per month, and the conversion of starch to sugar proceeded much more slowly than at higher temperatures.

In general, summer squashes do not store well at temperatures below 5°C, whereas winter

TABLE 2 Proximate Composition of Some Squashes and Pumpkins (per 100 g edible portion)

Constituent	Pumpkin	Squash winter	Squash summer	Chayote
Water (g)	91.60	88.70	93.70	93.00
Energy (kJ)	109	155	84	101
Protein (g)	1.00	1.45	1.18	0.90
Fat (g)	1.00	0.23	0.21	0.30
Total carbohydrates (g)	6.50	8.80	4.35	5.40
Fiber (g)	1.10	1.40	0.60	0.70
Minerals (g)	0.80	0.80	0.58	0.40

Source: Ref. 14.

TABLE 3 Vitamin Contents of Some Squashes and Pumpkins (per 100 g edible portion)

Vitamin	Pumpkin	Squash winter	Squash summer	Chayote
Ascorbic acid (mg)	9.00	12.00	14.80	11.00
Thiamine (mg)	0.05	0.10	0.06	0.03
Riboflavin (mg)	0.11	0.03	0.04	0.04
Nicotinic acid (mg)	0.60	0.80	0.55	0.50
Pantothenic acid (mg)	—	0.40	0.10	0.48
Vitamin B_6 (mg)	—	0.08	0.11	—
Total carotene (mg)	2.67	6.77	0.33	0.09

Source: Ref. 14.

squashes need temperatures greater than 10°C. The fruits are also sensitive to ethylene and should not be stored or shipped with ethylene producing products. Summer squashes have a short storage life of 1–2 weeks (17) at 5–10°C (Table 4). Winter squashes store well for several months if kept relatively dry (50–70% RH) and cool (10–13°C) to reduce decay.

Temkin-Gorodeiski (18) found *Sclerotinia sclerotirum* and *Botrytis* spp. to be the main causative organisms of style and side rot of *Zucchinis* squash. *Alternaria* and *Fusarium* were secondary factors.

Pumpkins are mainly affected by *Fusarium* rot and *Rhizopus* soft rot, as in the case of muskmelons. Winter squashes are, however, more susceptible to *Alternaria* rot (*A. tenuis*), bacterial soft rot, and black rot caused by *Mycosphaerella citrullina*. Cottony leak, *Rhizopus* soft rot, and scab are of minor importance (19). A 0.2% dichloran dip is very effective in controlling squash decay caused by *S. sclerotiorum* and *Botrytis* spp. (18). Wax coatings also reduced decay considerably.

E. Processing and Utilization

Summer squashes are generally boiled and served as vegetables, but they can be sliced and eaten raw. Winter squashes can be baked, boiled, or steamed. Canned pumpkin may be a mixture of pumpkin and winter squash. The cooked mashed flesh of pumpkin and other winter squashes can also be used as a pie filling. Pumpkin pie is a favorite dessert for autumn and winter holidays in the United States.

TABLE 4 Recommended Storage Temperature, Relative Humidity, and Approximate Shelf Life for Some Squashes

Type	Temperature (°C)	Relative humidity (%)	Shelf life
Winter squash			
Acorn	10	50–70	5–8 weeks
Butternut	10	50	2–3 months
Hubbard	10–13	70	6 months
Summer squash	5–10	95	1–2 weeks
Chayote	7	85–90	4–6 weeks

Source: Ref. 17.

II. GOURDS

Gourds consist of several genera indigenous to Africa, India, Asia, and the Americas. These include *Benincasa hispida* (Thunb.) Cogn., wax gourd or Chinese winte melon; *Lagenaria siceraria* (Mol.) Standl., calabash, bottle, or white-flowered gourd; *Coccinia grandis* (L.) Voigt, scarlet grourds; *Momordica charantia* (L.), bitter gourd, *Luffa cylindrica* (syn. *Luffa aegyptiaca* Mill), and sponge or dishtowel gourd. The buffalo gourd (*Cucurbita foetidissima*) is related to the squashes native to North America. There is increased interest in this plant because it is a long-lived perennial, tolerant of drought, and capable of overwintering. Interest in this plant has focused not on the fruit, which develop 32–28 days after pollination and have fibrous flesh, but on the starchy storage root (50–65 g starch/100 g dry weight) and the seeds that are high in protein (about 30 g/100 g dry weight) and oil (25–36 g/100 g dry weight).

Gourds are generally eaten when immature; some become inedible as they ripen. Gourds can be cylindrical, discoid, or bottle-shaped. Seeds are found in the central cavity. Edible when immature, the mature sponge gourd is very fibrous and is used as a rough sponge or scrub pad. A dried mature calabash gourd makes an acceptable watertight container. Gourds are grown for food mostly in Asia and parts of Africa. Since they are not generally commercially grown or used in industrialized countries, less is known about them. The immature fruits are used like summer squash or cucumbers—eaten raw or cooked. Slices can also be dried for storage.

A. Bottle Gourd

The bottle gourd, *Lagenaria siceraria*, is a commonly used vegetable in India. It is also commonly found growing in Ethiopia, Africa, Central America, and other warmer regions of the world.

Bottle Gourd fruits come in different shapes but resembling a bottle. The most common shapes are cylindrical, round-oval, and oblong. They can be yellowish-green or cream colored, relatively soft in texture, with white pulp and large white seeds. They can be used as a vegetable or for making sweets (e.g., *halva*, *kheer*, *petha*, *burfi*, *rayata*) and pickles. As a vegetable, it is easily digestible, even by patients. A decoction made from the leaf is a very good medicine for curing jaundice. The pulp is good for overcoming constipation, cough, nightblindness, and as an antidote against certain poisons (20).

1. Botany

Bottle gourd is a monoecious annual vine with very large oxalate oval leaves (Fig. 1) and branched tendrils spreading or climbing 3–15 m. The foliage is pubsecent and emits a characteristic somewhat musky and unpleasant odor when bruised. The characterstic large white flowers, borne on slender peduncles, open in the evening and may remain open until the middle of the following day. Cultivars are characterized by the shape and size of their fruits, varying from 10 to 90 cm.

There are two main types of bottle gourds in India, i.e., long and round. Yawalkar (8) recommended the round ones for early crops and the long ones for the rainy season. Several improved varieties have been shown to have better quality and productivity, including Pusa Summer Prolific Long, Pusa Summer Prolific Round, Pusa Meghdut, Pusa Manjari, Punjab Round, Punjab Long, Punjab Karnal, Arka Bahar, Pusa Naveen, and Samrat (Fig. 2) (Phule BTG-1).

FIGURE 1 Bottle gourd vine with long tendrils and white flowers.

FIGURE 2 Samrat cultivar of bottle gourd.

2. Production

Climate and Soil

The bottle gourd is a typical tropical plant requiring a hot and humid climate for best growth. The optimum temperatures for growth are 24–27°C (4). It is highly sensitive to photoperiod. High rainfall amounts with prolonged cloudiness results in a higher incidence of disease and may drastically affect yield. Short days and humid climate promote femaleness (21).

Bottle gourd can be grown on all types of soils, but sandy loam soils with high organic matter content are considered best. Soil should be well drained and the pH between 6 and 7. Bottle gourd can be raised successfully on riverbeds.

Propagation

Bottle gourd is propagated through seeds. The seeds are large, white in color, and seed count is about 450–500/100 g. The seeds can be stored for 3–4 years without any appreciable loss in viability.

Cultural Practices

The best time for planting bottle gourd depends upon the location. The land should be thoroughly prepared in order to get fine tilth for seed bed. The seeds are sown directly on raised beds, in furrows, or trenches or pits. Two seeds per hill are sown in both sides of raised beds or furrows. In Punjab, India, the recommended spacing is 2.5 × 2 m (22), whereas Seshadri (23) has recommended spacing of 3–4 × 2 m. Singh (5) recommended a spacing of 2–3 m between rows and 1–1.5 m between plants. The same spacings are followed in river bed plantings. In Maharashtra, bottle gourd is trained on bowers and sown at 3 × 1 m intervals. In West Bengal and South India, sowing is done in pits. Pits 90 × 90 cm are dug 60 cm deep and filled with well-rotted FYM, followed by topsoil (24). Three or four seeds are sown per pit. The recommended seed rate is 3–6 kg/ha (4). Presowing treatment of seeds with 600 ppm of succinic acid for 12 hours improves germination and seedling growth as evidenced by more leaves per seedling (5).

Nath et al. (4) has recommended 45 t/ha of FYM. This can be mixed with the soil applied individually to the pits. Malik (25) has reported the highest fruit yield with 56 kg of N, 28 kg of P, and 28 kg of K per hectare. Prem Nath (26) suggested application of 25–55 kg of N, 20–60 kg of P, and 17–44 kg of potash per hectare. FYM, phosphorus, and potash fertilizers should be applied at the time of land preparation with half of the nitrogen given at the vining stage and the remaining half top-dressed at the time of fruit set. Singh (5) has recommended 40–60 kg of N, 40–60 kg of P, and 60–80 kg of K per hectare. When grown in pits, Mehta (27) recommended that each plant should be given a soil mixture containing good topsoil and well-rotted FYM (1:1 v/v) along with 50 kg each of ammonium sulfate and single superphosphate.

Various pruning methods failed to exert favorable effects on fruit yield of bottle gourd (28). In Maharashtra, India, bottle gourd is trained on bower systems. This increases yield four to five times over that obtained from untrained vines.

Bottle guard is a shallow-rooted crop, thus only shallow interculture operations are needed. Hand-weeding should be done carefully so that the root system is not damaged. Two or three weedings should be necessary to keep the crop free from weeds. Herbicides can also be used in this crop. Saimbhi and Randhawa (29) found that preemergent application of linuron 0.5 kg/ha, alachlor 2.5 kg/ha, dichlormate 0.4 kg/ha, and diuron 1.25 kg/ha was useful for bottle gourd. Leela (30) obtained better weed control with fluchloralin 2.0 kg ai/ha and alachlor and butachlor 2.5 kg ai/ha in bottle gourd cv. PSPL.

The first irrigation may be given after seed sowing in order to improve seed germination. The crop requires frequent irrigation as high humidity is needed for prolific bearing. The rainy season is very good for bottle gourd crops. During hot and warm weather, irrigation every third or fourth day is needed to maintain proper soil moisture level. Drip irrigation has been reported to yield 48% more fruit than furrow irrigation (31).

Fruit set can be increased by spraying the plants twice at the two- and four-leaf stages with maleic hydrazide (400 ppm), triiodobenzoic acid (50 ppm), boron (3 ppm), and calcium (20 ppm) (32). Yields can also be increased by maintaining adequate soil fertility levels, particularly when growing hybrids. Maleic hydrazide at 400 ppm along with 100 kg N/ha promotes production of female flowers and increases fruit set and in turn the yield.

The crop is ready for harvest approximately 60 days after seed sowing, depending upon the variety and season. Bottle gourd fruit takes about 12–15 days after fruit setting to reach the marketable stage. Precautions should be taken while cutting the fruits so that the vines are not injured. Fruits should be picked every 3–4 days.

3. Chemical Composition

Bottle gourd is a good source of minerals. The edible portion of bottle gourd fruit contains 96.3% moisture, 0.1% fat, 2.9% carbohydrates, 0.5% mineral matter, 0.044 mg thiamine 0.023 mg riboflavin, 0.33 mg niacin, and 13 mg ascorbic acid per 100 g edible portion (33).

4. Processing

Bottle gourd fruits are consumed in various forms. The fruits may be processed into *Doodhi Halva*, *Kofta*, and jelly. The jelly prepared from bottle gourd pectin possesses good strengthening properties. Recently, tutti-frutti was prepared from bottle gourd fruits using different methods (Table 5). Good quality tutti-frutti can be prepared by slow syruping process and can be preserved in polyethylene bags for at least 3 months without affecting the quality.

B. Bitter Gourds

Bitter gourd is commonly cultivated in tropical Africa, Central America, China, Malaya, and other tropical Asian countries. It is a native of tropical America and Asia. Its name derives from

TABLE 5 Chemical Composition of Tutti-Frutti Prepared by Different Methods

Method/ Treatment	Moisture (%)	Acidity (%)	T.S.S. (°Brix)	Ascorbic acid (mg/100 g)	Reducing sugars (%)	Nonreducing sugars (%)	Total sugars (%)
T_1	20.23	0.25	68.00	6.25	29.40	36.00	65.40
T_2	20.30	0.23	68.66	7.01	29.55	35.90	65.45
T_3	20.26	0.22	68.33	6.76	29.50	36.00	65.50
T_4	20.30	0.20	69.00	7.00	29.60	26.00	65.60
S.E. ±	0.033	0.004	0.270	0.061	0.036	0.033	0.025
C.D. at 5%	0.108	0.015	0.880	0.199	0.120	0.108	0.081

T_1 = Papaya slow syruping; T_2 = bottle gourd slow syruping; T_3 = bottle gourd single operation (boiling i syrup); T_4 = bottle gourd slow syruping + 1% $CaCl_2$.
Source: Ref. 34.

the bitter taste of its pulp. The plant as a whole has many medicinal uses. The fruit has germicidal qualities and is used for treating blood diseases and diabetes (20). Bitter gourd leaves are known to act as galactogogs, while the roots are astringent (33). Juice from the leaves is used against lysmenorrhea as well as for external eruptions, burns, or boils. A powder prepared from the plant is good for treating intractable ulcers.

1. Botany

Taxonomy

Bitter gourd belongs to the genus *Momordica*, comprising nearly 23 species in Africa alone. The cultivated species are *Momordica charantia*, *M. cochinchinesis*, *M. dioica*, as well as *M. balsamina* (Balsam apple) and *M. cymbalaria* (syn. *M. tuberosa*). There has been no systematic study on the origin of this gourd and its allied species, which led Zeven and Zhukovsky (35) to regard this genus as one of unidentified origin.

Floral Biology

Bitter gourd is monoecious, and its flowers start opening at 5:00 a.m. and are completely open at 9:30 a.m. They wither away by 7:00 p.m. A lowering of atmospheric temperature delays flower opening and shedding by one hour. Anthers dehisce about 2 hours before blooming, i.e., between 7:00 and 8:00 a.m. The pollen becomes nonviable as the day advances, and after noon, not a single pollen grain is found to germinate; the stigma retains its receptivity for a much longer period. There is normal fruit setting after crossing between varieties.

Cultivars

Arka Harit, Coimbatore Long, VK-1 (Priya), MDU-1, Pusa Do Mausmi, Punjab 14, Hirkani (Fig. 3), and Pusa Vishesh are some important bitter gourd cultivars released for commercial cultivation in India.

FIGURE 3 Hirkani, a new cultivar of bitter gourd.

2. Production

Climate and Soil

Bitter gourd can be grown in either tropical or subtropical climates, but warm and hot climates are best for its growth. This crop is more resistant to low temperatures than other cucurbits. The optimum temperature requirement is 24–27°C (4). Similarly, it takes longer for the seeds to germinate at low temperatures (< 18°C). Short days help increase female flower production (21).

Bitter gourd grows fairly well in a wide variety of soils, but for optimum growth and productivity, it needs rich loamy or sandy loam soils. Nutrients in the soil and drainage are very important considerations in commercial bitter gourd production. Optimum soil pH is 6.00–6.7.

Propagation

Bitter gourd seeds are commercially used for propagation. They have a hard seed coat, which is relatively impermeable to water. The color of the seed coat changes from creamish in immature seeds to brownish in matured seeds. The seeds are heavy and count about 6000/kg. They remain viable for nearly 4–5 years under good storage conditions.

Cultural Practices

Soil used to grow bitter gourd should be prepared well to obtain the desired tilth. The time of sowing varies according to climatic conditions (24,36). The seeds are sown in raised beds, about 1.2–1.5 m wide, with 45- to 60-cm wide furrows separating them. Three to four seeds are sown in each spot 90 cm apart on both edges of the bed. The recommended spacing for Punjab is 1.8 m × 45 cm (22). Nath et al. (4) recommended a spacing of 1.5–2 m between rows and 60–120 cm between plants. In West Bengal, pits are dug and plants spaced 1.8–24 cm apart (24). Pits 60 × 60 cm large are dug 45 cm deep and filled with FYM and topsoil. Two seeds per hill are sown on both sides of the raised beds. In the case of pits, four or five seeds are sown per pit. The recommended seed rate is 4.5–6 kg/ha. Seedlings require protection against strong sun, wind, or chilling temperatures. Water should not be allowed to settle near the seedlings. Rainy season crops usually need support, particularly when they start trailing.

Bowers are erected 6 feet from ground level using wooden poles. Strong poles 10 feet in length and 4 cm in diameter are fixed 45 cm deep in the soil 5 m apart (Fig. 4). The border poles are fastened with wire at the top of the pole and the other end of the wire is fastened to stakes fixed in the soil. Later, all the poles are connected by wire. Then 16-gauge wires are stretched over the 10-gauge wires 60 cm apart, crosswise. Jute string is used for support.

Kniffens are prepared with the help of wooden poles and wire. The wooden poles are fixed in the soil 45 cm deep 5 m apart along the length of the rows. To strengthen each kniffen, 16-gauge wire is stretched and fastened to the poles 45–90 and 135 cm above the ground. Jute string is used to support the vines. Discarded dry cotton plants are placed near the plants upon which vines are allowed to grow without disturbing.

Joshi et al. (37) reported that thc bower system of training was most efficient in increasing the productivity of bitter gourd. There was a 143 and 73% increase in yield due to the bower and kniffen systems, respectively, over the ground (untrained plants). The increase in yield was mainly due to vigorous vine growth, maximum number of branches, proper distribution and exposure of fruiting area to sunshine, which resulted in the production of more fruits with greater length and diameter. There was also less incidence of fruit fly.

Dhesi et al. (8) reported that nitrogen was essential for the growth of vines and productivity, while phosphorus was useful in improving the yield. They recommended fertilizing with

FIGURE 4 Support system for bitter gourd vines.

56 kg each of nitrogen and phosphorus. The chemical fertilizer ranges recommended were 60–80 kg of nitogen, 70–90 kg of phosphorus, and 60–70 kg of potassium per hectare. It was indicated that the entire quantity of FYM, phosphorus, and potassium should be applied before sowing, with half the nitrogen given at the initial fruit-set stage (4).

Several workers reported various recommendations for fertilizer doses. Nath et al. (4) recommended 50–56 kg of N and 56–60 kg of P per hectare in coastal regions of Karnataka, India. Lingaiah et al. (39) reported the highest fruit yield with 80 kg of N, 30 kg of P, and 30 kg of K per hectare. Seed treatment with boron at 3–4 mg/kg or foliar sprays of boron at 3–4 mg/liter gave significantly higher fruit yield when compared with control (40).

The first irrigation should occur immediately after seed sowing, followed by irrigation at weekly intervals or even more often, depending upon the requirements. Summer season crops need regular irrigation and may be irrigated at 5- to 7-day intervals (5). During the rainy season, less irrigation is required, except when no rains occur.

Crops need regular weeding, and intercultural operations are needed before the spread of the crop. Leela (30) observed an increase in the yield of bitter gourd with alachlor and butachlor at 2.5 kg ai/ha, flunchloralin at 0.75 ai/ha, and exadiazen at 0.5 ai/ha.

The bitter gourd crop takes about 55–110 days from seed sowing to reach first harvest. Fruits are usually picked while still tender, as such fruits are generally preferred by consumers. Thus, picking should be done at every 2–3 days as bitter gourd fruits mature very quickly. If harvest is delayed to 15–16 days from anthesis, the seed coat becomes hard and the seed kernel becomes well developed. Fruits start turning yellow 18–20 days from anthesis and become unfit for consumption.

3. Chemical Composition

Bitter gourd fruits are a good source of carbohydrates, proteins, vitamins, and minerals (8). According to Aykrod (41), the chemical composition of bitter gourd depends upon the variety (Table 6). The small-fruited types have been shown to contain higher amounts of proteins, fats, carbohydrates, and minerals.

TABLE 6 Chemical Composition of Bitter Gourd Fruits

Components	Small-fruited types	Large-fruited types
Water (%)	83.2	92.4
Carbohydrates (%)	9.8	4.2
Proteins (%)	2.9	1.6
Fat (%)	1.0	0.2
Fiber (%)	1.7	0.8
Mineral salts (%)	1.4	0.8
Calcium (mg/100 g)	50	20
Phosphorus (mg/100 g)	140	70
Iron (mg/100 g)	9.4	2.2
Vitamin A (IU/100 g)	210	210
Vitamin C (mg/100 g)	88–92	88–92

Source: Ref. 41.

4. Storage

Fruits can be stored for 3–5 days under cool and shady conditions; under cold storage they can be stored longer.

5. Processing

Immature fruits can be treated like vegetables. The fruits are cut into pieces, boiled, and the boiling water discarded. The cooked pieces are then fried in oil along with onion, garlic, and spices. Bitter gourd juice is used by diabetic patients.

C. Luffa Gourds

The ridge gourd (*Luffa acutangula* Rob) and the sponge gourd (*Luffa cylindrica* Roem.) belong to the genus *Luffa* and are cultivated both on a commercial scale and in kitchen gardens throughout India. The fruits of ridge gourd (*Kali tori*) are ribbed (Fig. 5), whereas fruits of sponge gourd (*Ghia tori*) are smooth (Fig. 6). The genus derives its name from the product loofah, which is used in bathing sponges, scrubber pads, doormats, pillows, mattresses, and also for cleaning utensils. Both species contains a gelatinous compound called luffein. Both sponge gourds and ridge gourds have medicinal uses. The oils from the seeds are known to cure cutaneous complaints, the roots have laxative effects, and the juice from the leaves is used to cure granular conjunctivities of the eye, adrenal type diabetes, and hemorrhoids (20).

All but one species of the genus *Luffa* is indigenous to the tropical countries of Asia and Africa. *Luffa* is an essentially old world genus, consisting of two cultivated species, *Luff cylindrica* and *L. acutanglula*, and two wild species, *Luffa graveolens* and *Luffa wechinate.* Wild forms of the ridge gourd (angled loofah) have extremely bitter fruits, but the domesticated types are less bitter.

1. Botany

Both species (*Luffa cylindrica* and *Luffa acutangula*) are annual climbing vines with tendrils and are monoecious in flowering habit. However, perfect ridge gourd flowers have been reported. The two species are easily distinguished by the shape of their fruits. Smooth loofah (sponge gourd) have long, smooth, cylindrical fruits with 10 shallow unribbed furrows 30–60 cm in length (Fig. 6). Angled loofah fruits have 10 distinct, angled prominent ribs 15–60 cm in length

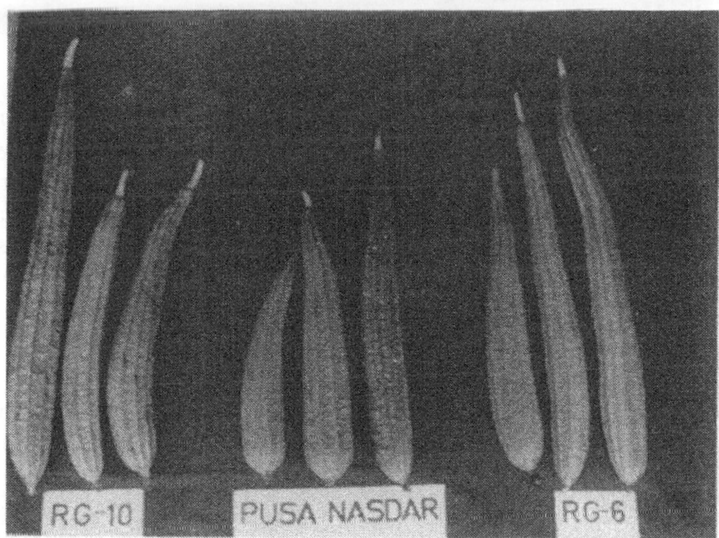

FIGURE 5 Fruits of different cultivars of ridge gourd.

(42). Staminate flowers are produced in a ratio ranging from 25 to 40 per pistillate flower. This rate can be reduced by a factor of two by treatment with indole acetic acid. The smooth gourd stigma is reported to be receptive to pollination for 60 hours after anthesis, the angled loofah for 36 hours.

The Indian Agricultural Research Institute developed some ridge gourd varieties that have enjoyed good consumer acceptance. Pusa Nasdar (42), a ridge gourd variety, and Pusa Chikkani (42), a sponge gourd variety, have been developed (42). Tamil Nadu Agricultural University, Coimbatore, released the CO-1 variety of ridge gourd (44). Punjab Sadabahar is a variety of ridge gourd developed by Punjab Agricultural University, Ludhiana.

FIGURE 6 Sponge gourd fruits.

2. Production

Soil and Climate

Both crops can be grown on a variety of soil types. Loamy soils are considered best if there is a good amount of organic matter and if it is well drained. Optimum soil pH is 5.5–6.7.

Both crops are well adapted to a fairly wide range of climatic conditions. Luffa requires a long warm season for best production. It grows best during the rainy season. Optimum temperature requirement is 25–27°C (4). Due to its hard seed coat, there is a problem with seed germination when the temperature is low. Long day promotes femaleness. Excessive rainfall during the flowering and fruiting period reduces yield.

Propagation

Both crops are propagated through seeds. The seeds of sponge gourd are white, whereas ridge gourd seeds can be black or white. About 3–4 kg of seeds should be enough for one hectare of land.

Cultural Practices

The plot should be prepared to obtain the tilth desirable to facilitate rapid and better germination. The time of planting usually depends upon the season, location, and market demand. There are two methods of sowing: raised bed and pit. In North Indian plains, seed is sown on raised beds 2–3 m × 60 cm apart, and in West Bengal sowing is done in pits. Two seeds per hill in the case of raised beds and three to four seeds per pit are sown. In Maharashtra, a spacing of 1.5 × 1.5 m is followed. The recommended spacing in Punjab is 3 × 2 m (22). Nath et al. (4) recommended spacing of 1.5–2.5 m within rows and 60–120 cm between plants. With pit systems, spacing is 1.5–2.0 m between rows and 1–1.5 m between plants (5).

Chauhan (20) recommended the application of 8–10 tons of FYM at the time of land preparation. In addition, he suggested application of 37 kg of ammonium sulfate per hectare. Nath et al. (4) recommended 22–23 tons of FYM in addition to 17–20 kg of nitrogen per hectare for both sponge gourd and ridge gourd. In Punjab, the recommendation is 100 kg of nitrogen, 60 kg of phosphorus, and 60 kg of potash per hectare (22). Siyag and Arora (44) found 50 kg of nitrogen and 20 kg of phosphorus per hectare to be the optimum dose for sponge gourd. Singh (5) has recommended 10–15 tons of FYM in addition to 40–60 kg of N, 30–40 kg of P, and 30 kg of K per hectare for both these crops.

Although kitchen garden crops are trained on fences or walls, commercial crops are trained on a kniffen system. Weeding should be started 15–20 days after sowing and another weeding may be done 20–25 days later (11). The crop is trained when the seedlings are about 10–15 cm tall. The early crop can be allowed to trail on the bed itself.

The early crop must be irrigated regularly. The first irrigation should be given immediately after sowing. Summer crops require more frequent irrigation than rainy season crops. Irrigation may be carried out according to the soil moisture status and season. During the rainy season, irrigation may be required during the early growth period. Drip irrigation was not found to be better than furrow irrigation (45).

In these crops, the female and male flowers are borne separately on the same plant. The sex ratio can be regulated by growth regulator spray. NAA (200 ppm) promoted the production of female flowers, and the yield was significantly increased in ridge gourd (23).

The crop is ready for harvest about 60–90 days after seed sowing, depending upon season and variety. The fruits attain a marketable maturity about 5–7 days after anthesis. Fruits should be harvested when they are still immature. Picking should be done every 3 or 4 days. If there is

TABLE 7 Proximate Composition of Some Gourds

Constituent	Wax gourd	Calabash gourd	Scarlet gourd	Luffa gourd
Water (g)	96.10	95.30	93.30	93.80
Energy (kJ)	55	59	—	84
Protein (g)	0.40	0.62	1.50	1.20
Fat (g)	0.20	0.02	0.10	0.20
Total carbohydrates (g)	3.00	3.39	3.10	4.36
Fiber (g)	0.50	0.56	1.60	0.50
Mineral (g)	0.30	0.43	0.50	0.40

Source: Ref. 14.

delay in harvesting, the fruits become more fibrous and are unfit for human consumption. The fruits are packed in bamboo baskets with proper padding to prevent injury in transit.

3. Chemical Composition

The chemical composition of ridge gourd and sponge gourd is presented in Table 7. Compared to ridge gourd, sponge gourd contains more proteins, carotene, and fibers (Table 7). However, ridge gourd contains higher amounts of minerals (Table 8).

4. Storage

Fruits harvested at the marketable stage can be stored for 3–4 days in a cool place without any adverse effects. The shelf life of ridge gourd fruits can be improved by dipping the fruits in benzyl adenine (50 mg/liter) (46).

5. Processing

Ridge gourds and sponge gourds are mainly consumed as vegetables. They are cooked and then fried or prepared as a curry.

TABLE 8 Vitamin Content of Some Gourds (per 100 g edible portion)

Vitamin	Wax gourd	Calabash gourd	Scarlet gourd	Luffa gourd
Ascorbic acid (mg)	13.00	10.10	15	12.00
Thiamine (mg)	0.04	0.03	0.07	0.05
Riboflavin (mg)	0.11	0.02	0.08	0.06
Nicotinic acid (mg)	0.40	0.32	0.70	0.40
Pantothenic acid (mg)	—	—	—	—
Vitamin B6 (mg)	—	0.04	—	—
Total Carotene (mg)	0	0	0.16	0.68

Source: Ref. 14.

D. Other Gourds

1. Ash Gourd

Ash gourd (*Benincasa hispida* Thunb. Cogn) believed to have originated in Java and Japan, is widely cultivated in India and several tropical countries. Ash gourd or waxgourd or white gourd is known as *Petha kaddu.* It is mainly grown in North India, especially in Uttar Pradesh, where it is used in the preparation of *Petha* (33).

There are two types of ash gourd: purple-green and green. They are also classified as round and oblong types. Ash gourd is an annual hispid (rough with bristlelike hairs), climbing, herbaceous monoecious plant varying several meters in length. The leaves are broad with 5–11 angular lobes. The staminate flowers have long penduncles; the pistillate flowers have densely haired ovaries on short penduncles. The corolla has five large yellow petals (10).

All types of soil are good for ash gourd. Well-drained sandy loam soils with high organic matter are considered best. Optimum pH is 6.0–7.0. This crop requires a warm and humid climate. It can be grown in milder climates, but it is sensitive to frost. The optimum temperature requirement is 24–30°C (4). Short days, low night temperatures, and humid climate are good for production of female flowers (47).

Crops are sown on raised beds or in furrows (in riverbeds, sowing is done in trenches). The recommended spacing is 1.5–2.5 m between rows and 60–120 cm between plants (4,5). With the pit system, pits are dug with 1.5–2.5 m spacing between plants and rows (24). Two seeds per hill are sown on both sides of a raised bed, whereas three to four seeds are sown per pit. The recommended seed rate is 5–7 kg/ha.

Nath et al. (4) recommended 22–23 tons of FYM mixed with the soil or into each pit in addition to 18–27 kg of nitrogen per hectare. Singh (15) recommended 45–60 kg of N, 50–60 kg of P, and 60–80 kg of K per hectare for ash gourd.

Hand-weeding may be started 15–20 days after seed sowing. A total of three or four hand-weedings is needed for the whole crop season (11). For summer crops, regular and frequent irrigation is needed during the vegetative phase. Intervals between irrigations may be maintained at 6–7 days, depending upon season and soil type. When crops mature, irrigation intervals may be lengthened. Less irrigation is needed for rainy season crops.

Ash gourd starts to flower about 60–80 days after planting and can be harvested in about one week for immature fruits and 30–40 days for mature fruits. The fruits at maturity have a white waxy surface. Aykrod et al. (41) reported that ash gourd fruits contained 0.4% protein, 1.9% carbohydrates, and 0.3% minerals, and traces of vitamins A, B, and C.

The fruits have good storage quality and can be transported long distance via truck. The fruit is mainly purchased by the confectioners, who use it to prepare *petha.* Ash gourd can be stored longer than any other cucurbits. Bruises and cuts on mature fruits are capable of healing by suberization. When stored at 13–15°C and 70–75% RH, fruits maintain good quality for over 6 months. It is best to store fruit individually on shelves, not stacked (10).

2. Pointed Gourd

Pointed gourd (*Trichosanthes dioica* Roxb), commonly called *Parval,* is a perennial cucurbit cultivated in many parts of India. Immature fruits of this plant are a popular vegetable.

Trichosanthes is a large genus principally of Indo-Malayan distribution with about 44 species, 22 of which occur in India (48). Most authors agree that India or the Indo-Malayan region is its original home. The Bengal-Assam area is the primary center of origin.

Some improved selections are commonly grown in various states, e.g., Green Oval, Green

Long Striped, and White Oval in Madhya Pradesh (49), Kazil, Bombay, and Damodar in West Bengal (50), and Dandli, Kalyani, Guli, Bihar Sharit, FP-5, and FP-4 in Uttar Pradesh.

Pointed gourd can be grown successfully on both sandy loam and loamy soils. The soil should be well drained and rich in organic matter. Riverbeds can also be used for this crop, which is a warm season crop. A hot and humid climate is best for its development. During winter, it becomes dormant and sprouts again in summer. Rainfall is considered very good for this crop, but a coastal climate is not suitable.

Pointed gourd is vegetatively propagated through vine cuttings and root suckers. Seed propagation is avoided because of poor germination. Also, 50% of plants may be nonfruiting (i.e., males). In order to ensure maximum fruit set, 10–12% male plants is considered adequate. The varieties should be early maturing with fewer nodes at which the first pistillate flower sets.

Vine cuttings may be transplanted in August upland and in November on riverbeds, spaced at 2 × 2 m (5). In West Bengal, the optimum time for transplanting root cuttings is the seconed half of October with a spacing of 3 × 0.6–9 m (51). Vine cuttings may be planted in a number of ways. In the first method, cuttings 1 or 1.5 m long are folded in the shape of a figure eight and planted in pits filled with a mixture of FYM and soil. The middle portion is buried 10 cm deep. In the second method, vine cuttings are folded in the shape of a ring and planted in pits, keeping the ends aboveground. In the third method, straight vine cuttings are planted in a furrow filled with FYM and soil, keeping the cutting ends aboveground. The fourth method is called the moist-lump method. In this method, cuttings 60–90 cm long are encircled with lumps of moist soil and planted 10 cm deep in furrows.

Plants may also be propagated using root cuttings. In this method, roots from old vines are dug up and planted either in the nursery or in the field. Seedlings would be ready in January-February for transplanting. Since the crop is dioecious, after every 10 female plants, a cutting from a male plant may be planted.

In West Bengal, Das et al. (51) found 90 kg of N, 60 kg of P, and 40 kg of K per hectare to be the optimum dose of fertilizer for the pointed gourd. Singh (5) recommended 20–25 tons of FYM mixed with soil in the furrows or in pits. In addition, he also suggested 90 kg of N, 60 kg of P, and 60 kg of K per hectare.

Weeding should be done only when plants have sprouted and are well set. Shallow interculture operations should be followed. Water should be given as and when required. Early irrigation should be done more frequently. During the rainy season, less irrigation is required. Training the crops over bowers gives high yields. Picking also becomes very easy.

The pointed gourd is ready for harvest about 80–90 days after planting. Harvesting should be done when the fruits are immature and still green. Picking should be done frequently as more pickings means more fruit. Yield varies from 60 to 80 q/ha.

The fruits contain 2% protein, 0.3% fat, 2.2% carbohydrates, 153 IU vitamin A, and 2 mg vitamin C per 100 g edible portion (5). The fruits have a poor shelf life. Spraying with maleic hydrazide (MH) at 1000 mg/liter (5) or GA at 1–10 mg/liter or boric acid at 1000 mg/liter improves the keeping quality of fruits up to 5 days. Postharvest dipping of fruits in 1% salicylic acid, 4% citric acid, and 10% waxol also promotes the keeping quality of the fruits. Pointed gourd in the form of immature fruits is mostly used as a cooked vegetable.

3. Snake Gourd

Snake gourd (*Trichosanthes cucumerina*) occurs in the wild form in India, Southeast Asia, and tropical Australia. The Indian Archipelago is thought to be its place of origin (32). It is known by various names in different regions: e.g., Chichina, Pallakaya, and Attlakaya. The fruits are long, narrow, cylindrical, and pointed at both ends (Fig. 7). The green or white fruits are rich in carbohydrates, proteins, minerals, and vitamins A, B, and C (41).

Snake gourd is a monoecious annual vine with large, angular five- to seven-lobed leaves, climbing by means of branched tendrils. The five or more male flowers, forming a pubescent florescence and a single sessile pistillate flower, are borne in the same leaf node. Both flower types have white, delicately fringed petals, which open late in the afternoon and are quite fragrant.

Snake gourd plants are highly cross pollinated. Certain selections in local strains have been made at several places (52). At Coimbatore, three selections—H-8, H-371, and H-375—have been found promising. Selection 30-2-2 recommended by the Department of Agriculture, Maharashtra, and a promising selection made by the Indian Horticultural Research Institute—16-A—have been reported from Bangalore. The early-maturing cultivars CO-4 and CO-1 were developed by the Tamil Nadu Agricultural University, Coimbatore. A new selection—TA-19—has been recommended by the Kerala Agricultural University in India.

The most ideal soil for snake gourd is rich loamy or sandy loam soil. However, other soils can be used, as long as they are well drained and rich. Like other cucurbitaceous crops, snake gourd needs a warm, humid climate for best growth. However, it can be grown in other climates also.

The seeds of snake gourd are white and larger than some of the other cucurbitaceous seeds. The seed rate varies from 2 to 4 kg per hectare. After the land is thoroughly prepared, pits are dug 60 cm apart and filled with a mixture containing topsoil and well-rotted FYM (1:1 v/v). The pits are irrigated before dibbling the seeds. Generally three to four seeds are put in each pit. Yawalkar (8) recommended sowing of snake gourd seeds in raised beds 1 m wide separated by water channels. In this method, seeds are sown 30–45 cm apart on both the edges of the bed. Usually two to four seeds are dibbled at each spot. The recommended depth of sowing is 1.25–1.5 cm. Under normal conditions, seeds germinate within a week.

The first irrigation is given immediately after sowing. For summer crops, regular irrigation at 3- to 5-day intervals is necessary to maintain the desired soil moisture level. This includes regular hoeing, weeding, staking, and training of vines. The bower system of training is

FIGURE 7 Snake gourd fruits and support system used in commercial cultivation.

recommended for this crop (Fig. 7). To obtain straight fruits, some growers tie weights to the bottom end of fruits.

Yawalkar (8) recommended 35 tons of FYM be applied during land preparation. When the vines start trailing, a first dose of NPK in the ratio of 1:2:0.5, at the rate of 50 kg/pit/spot, is given. A second dose of NPK is given at the time of fruit set (26). Frequent irrigations are to be given after manuring.

Fruit set in snake gourd starts at the beginning of second month. Fruits are hand-picked when they are still tender and about one quarter to one third their full size. Fully mature fruits are lighter, fibrous, and hard and as such as not preferred in the market. If properly managed, a vine yields about 20–25 fruits and can yield as many as 40–50 fruits in about 2–3 months. As the fruits do not keep too well after harvest, they should be sent to market promptly. The fruits need to be properly packed in baskets or in other containers to protect them from injuries during transport.

4. Tondli

Tondli (*Coccinia grandis* (L) Voigt; syn. *C. indica* Wright and Arn, family Cucurbitaceae) is a vegetable crop believed to have had its origin in India. It is grown in most parts of India, tropical Africa and Central America, China, Malaya, and other tropical countries. The fruit is smooth, light green in color and 4–7.0 cm in diameter. It is most commonly used as a vegetable in southern and central India. The pulp is creamish-white or yellowish-green in color when immature, but becomes bright pink to dark pink when ripe. Both immature and mature fruits are used for cooking. The fruits are a rich source of carbohydrates and contain 1.2% protein, 260 I.U. of vitamin A, and 15.0 mg vitamin C per 100 g of fruit. The leaves are also used in some parts of India for their medicinal value.

Tondli is a perennial climbing herb with cordate-type leaves. The fruits are ovoid or elliptical in shape, have white stripes when immature, and become smooth and light or yellowish-green when ripe. The shape and size of fruits vary with the strain. The numerous seeds are soft, dull white or creamish in color, and are edible with the flesh. Fruits can be stored for a week or two.

Although there are no established varieties, various strains are being cultivated in different parts of India. The strain commonly grown in southern states has smaller fruits with white longitudinal stripes. In contrast, the type grown in west India has long, thin fruit with white longitudinal stripes. Some of the types grown in northern and eastern India have larger and bolder fruits, which are excellent in quality and yield.

Tondli is a warm season vegetable and bears excellent crop during summer and the rainy season when the climate is warm. It can also tolerate cold during winter months, producing few fruits. For best growth, it needs a warm and moist climate.

Any soil that is well drained and fertile is suitable for growing the crop. However, sandy loam or loamy soils with perfect drainage and ample nutrient supply favor better growth and production of excellent quality fruits. The roots are very susceptible to water stagnation.

Tondli is propagated through stem cuttings or roots. The ideal cuttings are about 25 cm long and 2 cm thick. Cuttings from relatively older shoots root better and produce an early crop. The tuberous roots can also be used for propagation. These cuttings are planted directly in the field.

Stem or root cuttings are taken from healthy plants during the rainy season and planted in basins prepared after deep ploughing and harrowing. Usually, basins 60–75 cm in diameter are made 2 m apart on both sides. Two cuttings are planted in each basin to obtain one plant. It is necessary to keep at least one bud below the soil surface for producing roots. In places where the winter is not severe, planting can be extended up to January or February. Cuttings usually take about 10–15 days to sprout and start growing. The use of rooting hormone can ensure rapid and better rooting of cuttings. Young plants require support to keep their shoots growing upright. Since the new shoots are delicate, they should be protected from frost or strong sun. The application of two basketsful of well-rotted FYM per vine, three times a year, is adequate to provide the needed quantity of nutrients (8).

For a commercial crop, plants are trained on a bower system constructed to suit the growth of vine. Usually bowers 1.5 m high are erected with pillars placed 3–4 meters apart. The top is strung with galvanized iron wire with 1 × 1 m spacing. The plants are supported by stalks so that they grow straight to the bowers and spread their branches. The branches need to be properly trained on the bower. The central shoots growing from the main stem below the bower should be removed in order to encourage shoots on the bower. Dead or diseased shoots are removed as and when necessary. No regular pruning is necessary after the vines are established.

Weed growth should be checked by regular hoeing and weeding without disturbing the growth of vines. The hoeing should be shallow so as not to disturb the feeding roots. Crops should be irrigated once a week during hot months. No irrigation is needed during the rainy season. Water stagnation should be avoided at all times. Green manuring crops can be grown and buried beneath the soil at a proper stage. This practice improves the fertility of the soil.

Because tondli is a dioecious plant, the male and female flowers are borne on separate plants. It is difficult to recognize the difference between these two types in the vegetative stage. However, at the flowering stage they can be easily differentiated. The female flowers have a conical bulged base, while the male flowers are thin and narrow at the base. It has been estimated that a maximum of 10% of the total population should be male plants to provide enough pollen. Any excess male plants should be removed and replaced. It is common practice to keep a tondli crop on a site for 5 years, after which a different site is used. This ensures good yield and high-quality fruits.

Tondli fruits are produced almost throughout the year. In places where the winter is relatively severe, fruit yield during this period decreases. The stage of harvesting is determined by change in color from dark green to bright or light green. The tender and fully grown fruits are harvested carefully so as not to cause any injury to the vines. These fruits can be shipped long distances or stored for up to 2 weeks at room temperature. An estimate by Prem Nath (26) indicated that under ideal conditions, approximately 8–10 kg of fruit can be harvested per vine per year. However, on average, 200–300 fruits weighing 3–4 kg can be harvested per vine per year.

The fruits are a rich source of carbohydrates and contain 1.2% protein, 260 IU vitamin A, and 15 mg vitamin C per 100 g fruit. The leaves are also used for medicinal purposes.

III. CHAYOTE

Chayote (*Sechium edule* Swartz) is an herbaceous perennial climbing monoecious vine indigenous to southern Mexico and Central America. It grows up to 15 m or more in length and has large shallow-lobed leaves producing both male and female flowers in the same leaf axil. It is grown principally for its pear-shaped fruits, which are cooked many ways.

Chayote cultivars are classified by fruit shape and color. Some cultivars have prickly spines or thorns on the fruit. Local forms have been selected for specific areas. This crop can tolerate high temperatures, although relatively cool night temperatures appear to promote fruit development. Chayote grows well in loose, well-drained soil high in organic matter. The optimum pH is 5.5–6.5. A mature fruit containing the germinating seed is planted to a depth of two thirds of its length in prepared planting holes, with the widest end downwards. Fruit used for seed should be carefully handled, not stored at < 10°C, and planted without much delay. Plants or seeds are placed in hills spaced 2 m and rows 3.5 m apart. Plants require support by poles or a trellis. Nutrients such as NPK are applied before planting, followed by a top-dressing with nitrogenous fertilizer at intervals until fruit formation.

Irrigation should be frequent, and the soil moisture content should be kept high. Chayote is attacked by striped and spotted cucumber beetles and by nematodes in the soil. In high-rainfall areas, the plant can become infested with fungal diseases. Immature fruits may be harvested 100–120 days after sowing. Fruits are produced over a long period. Harvested fruit should be handled carefully to avoid cuts and bruises.

Chayote fruits are very nutritious. They contain 94% water, 0.7% protein, 0.1% fats, 5% carbohydrates, 17 mg calcium, 14 mg phosphorus, 0.4 mg iron, 0.01 mg thiamine, 0.02 mg riboflavin, and 14 mg ascorbic acid per 100 g edible portion (14). Fruits may be maintained in good conditions up to 30 days at temperatures of 10–15°C and relative humidity of 85–90%.

REFERENCES

1. Whitaker, T. W., and W. P. Bemis, Cucurbits, *Bull Torrey Bot. Club. 102*:362 (1978).
2. Watts, R. L., and G. S. Watts, *The Vegetable Growing Business*, War Dept. Education Manual, Em. 855, Orange Judd Pub. Co. Inc., Washington, DC, 1940.
3. Thompson, H. C., and W. C. Kelly, *Vegetable Crops*, 5th ed., McGraw-Hill Book Co. Inc., New York, 1957.
4. Nath, P., S. Velayudhan, and D. P. Singh, *Vegetables for Tropical Region*, ICAR, New Delhi, 1987.
5. Singh, S. P., *Production Technology of Vegetable Crops*, Agric. Res. Comm. Centre, Karnal, 1989, p. 219.
6. Singh, K., S. P. S. Madan, and M. S. Saimbhi, Effect of nitrogen and spacing on growth, yield and quality of pumpkin (*Cucurbita moschata* Duch ex. Poir), *Punjab Agric. Univ. J. Res.* (in press).
7. Sharma, C. B., and V. S. Shukla, Response of pumpkin to nitogen and phosphorus application and their optimum levels, *Indian J. Hort. 29*:179 (1972).
8. Yawalkar, K. S., *Vegetable Crops of India*, 2nd ed., Agric. Horticultural Pub. House, Nagpur, 1980, p. 370.
9. Gill, H. S., Improved varieties of summer squash, *Indian Farmers Digest 12*:9 (1979).
10. Yamaguchi, M., *World Vegetables: Principles, Production and Nutritive Value*, AVI Pub. Co., Westport, CT, 1983.
11. Khurana, S. K., J. L. Mangal, and G. R. Singh, Chemical weed control in cucurbits: A review, *Agric. Rev. 9*:69 (1988).
12. Kumar, J. C., and K. S. Nandpuri, Effect of picking intervals on yield and bearing of marrow (*Cucurbita pepo* L.), *Punjab Hort. J. 18*:99 (1978).
13. Work, P., and J. Carew, *Vegetable Production and Marketing*, John Wiley and Sons, New York, 1957, p. 537.
14. Lingle, S. E., Melons, squashes and gourds, in *Encyclopedia of Food Science, Food Technology and Nutrition*, Vol. 5 (R. McCrae, R. K. Robinson, and M. J. Sandlers, eds.), Academic Press, New York, 1993, p. 2960.
15. Plantenius, H., F. S. Jamison, and H. C. Thompson, Studies on cold storage of vegetables, *New York Cornnel Sta. Bull.*, No. 602 (1934).
16. Yeager, A. F., The storage of hubbard squash, *N.H. Bull.* 356 (1945).
17. Hardenburg, R. E., A. E. Watada, and C. Y. Wang, *The Commercial Storage of Fruits, Vegetables and Florists and Nursery Stocks*, USDA Handbook Number 66, 1986.
18. Temkin-Gorodeiski, N., Storage rots of Zucchini squash (*C. pepo*) and their control, *Israel J. Agric. Res. 20*:978 (1970).

19. Ryall, A. L., and W. J. Lipton, *Handling, Transportation and Storage of Fruits and Vegetables*, Vol. I. *Vegetables and Melons*, AVI Pub. Co. Inc., Westport, CT, 1972, p. 473.
20. Chauhan, D. V. S., *Vegetable Production in India*, Ram Prasad and Sons, Agra, 1972.
21. Bose, T. K., and M. S. Ghosh, Effect on photoperiod growth and sex expression on some cucurbits, *Indian J. Agric. Sci. 45*:487 (1970).
22. Package practices for vegetable fruit crops, Punjab Agricultural University, Ludhiana, 1987.
23. Seshadri, V. S., Cucurbits, in *Vegetable Crops in India*, (T. K. Bose and M. G. Som, eds.), Naya Prokash, Calcutta, 1986, p. 128.
24. Dutta, S., *Horticulture in Eastern Regions of India*, Ministry of Food and Agriculture, New Delhi, 1966.
25. Malik, B. S., Response to compost and fertilizers on development and yield of bottle gourd, *Indian J. Agron. 10*:266 (1965).
26. Prem Nath, *Vegetables for the Tropical Region*, ICAR Low Priced Book Series No. 2, ICAR, New Delhi, 1976.
27. Mehta, Y. R., *Vegetable Growing*, Uttar Pradesh, Bureau of Agricultural Information, Lucknow, 1959.
28. Sharma, H. K., B. S. Dankhar, and M. L. Pandita, Effect of pruning in bottle gourd cv. Pusa Summer Prolific Long, *Haryana J. Hort. Sci. 17*:89 (1988).
29. Saimbhi, M. S., and K. S. Randhawa, A note on the response of four cucurbits to pre- and post-emergence application of different herbicides, *Haryana J. Hort. Sci. 6*:91 (1977).
30. Leela, D., Weed controlled by herbicides in cole and cucurbitaceous crops, *Trop. Pest Manage. 31*:219 (1985).
31. Gupta, J. P., and S. D. Singh, Hydrothermal environment of seed and vegetable production with drip and furrow irrigation, *Indian J. Agric. Sci. 53*:138 (1983).
32. Choudhary, B., *Vegetables*, National Book Trust, New Delhi, India, 1967, p. 198.
33. Nadkarni, K. N., *Indian Materia Medica*, Nadkarni and Company, Bombay, 1927.
34. Babar, V. D., Preparation of tuti-friuty from bottlegourd, M.Sc. thesis, Mahatma Phule Agricultural University, Rahuri, India, 1996.
35. Zeven, A. C., and P. M. Zhukovsky, *Dictionary of Cultivated Plants and Their Centers of Diversity*, Wageningen, The Netherlands, 1975.
36. Srinivasan, C., A note on the optimum time of sowing of bitter gourd (*Momordica charantia* L.), *Prog. Hort. 6*:77 (1974).
37. Joshi, V. R., K. E. Lawande, and P. S. Pol, Effect of different training system on yield of bitter gourd, *Mah. J. Hort. 8*(1):67 (1994).
38. Dhesi, N. S., D. S. Padda, J. C. Kumar, and B. S. Malik, Response of bitter gourd to nitrogen, phosphorus and potash fertilization, *Indian J. Hort. 23*:169 (1966).
39. Lingaiah, H. M., D. C. Uthaiah, R. S. Herle, and K. D. Rao, Influence of nitrogen and phosphorus on yield of bitter gourd in the coastal regions of Karnataka: A preliminary study, *Current Res. Univ. Agric. Sci. 17*:116 (1988).
40. Verma, V. K., P. S. Sirohi, and B. Choudhury, Chemical modification and effect on yield in bitter gourd (*Momordica charantia* L.), *Prog. Hort. 16*:52 (1984).
41. Aykrod, W. R., V. N. Patwardhan, and S. Ranganathan, The nutritive value of Indian foods and planning of satisfactory diet, Health Bull. 23, Government of India Press, 4, 1951, p. 1.
42. Pal, B. P., M. S. Sikka, and H. B. Singh, Improved Pusa vegetables, *Indian J. Hort. 13*:64 (1956).
43. Sundararajan, S., and C. R. Muthukrishnan, Co-1: the delicious ridge-gourd, *Indian Hort. 23*:29 (1978).
44. Siyag, S., and S. K. Arora, Effect of nitrogen and phosphorus on fruit yield and quality of sponge gourd (*Luffa aegyptica* Roxb.), *Indian J. Agric. Sci. 58*:860 (1988).
45. Singh, S. P., and P. Singh, Value of drip irrigation compared with conventional irrigation for vegetable production on a hot arid climate, *Agron J. 70*:945 (1978).
46. Kumar, J., S. K. Arora, and R. Mehra, Effect of antisenescent regulators on shelf life of ridge gourd, *Crop Res.* (India) *1*(1):124 (1988).

47. Kamalnathan, S., Studies on sex expression and sex ratio in ash gourd (*Benincasa hispida*), *Madras Agric. J. 59*:486 (1972).
48. Chakravarty, M. L., *Monograph on Indian Cucurbitaceae: Taxonomy and Distribution*, Records of the Botanical Survey of India, Vol. 17 1959, p. 56.
49. Dubey, K. C., and R. P. Pandey, Parwal cultivation in Madhya Pradesh holds promise, *Indian Hort. 18*:15 (1973).
50. Mukhopadhyay, G. K., and T. K. Chattopadhyay, Studies on propagation of pointed gourd (*T. dioica* Roxb.), *Prog. Hort. 7*:65 (1976).
51. Das, M. K., T. K. Maily, and M. G. Som, Growth and yield of pointed gourd (*Trichosanthes dioica* Roxb) as influenced by nitrogen and phosphorus fertilization, *Veg. Sci. 14*:18 (1987).
52. Nath, P., and M. Seenappa, The south like snake gourd, *Indian Hort. 18*:19 (1973).

12

Cabbage

S. P. Ghosh
Indian Council of Agricultural Research, New Delhi, India
D. L. Madhavi
University of Illinois, Urbana, Illinois

I. INTRODUCTION

Cabbage is one of the most important vegetable crops worldwide. It is grown by market gardeners, vegetable growers, and general farmers in hot tropical and subtropical regions mainly during the winter. It is commercially grown in the United States, China, Japan, several European countries, and India. World production during the year 1994 was 40.24 million metric tons (1). The major cabbage-producing countries are the Russian Federation, China, Korean Republic, Japan, Poland, and India (Table 1).

II. BOTANY

Cabbage (*Brassica oleracea* L. var Capitata) belongs to the family of Cruciferae. It appears in the wild state in England, Denmark, and France. Wild cabbage is herbaceous and usually perennial. Cultivated cabbage differs in the size, shape, and color of its leaves as well as head (2).

A. Morphology

Cultivated cabbage has an unbranched stem, which generally remains less than 30 cm long, its growth being arrested at an early stage. As the growing point continues to form leaf primordia, a rosette of sessile leaves arises (3). The first leaves unfold normally and often reach a length of over 30 cm. After some time, leaves are produced that unfold only partially and form a "skin," embracing the later-formed leaves, which do not expand. By the continuous segregation and growth of young leaves, a compact head is eventually formed, though its solidity varies. Sometimes the pressure of the inner leaves causes the "skin" to crack, which also happens in spring when the plant starts to bolt. The shape of the head varies from flat-topped to long-oval, and it is often more than 20 cm in diameter. According to the leaf characteristics, a distinction

Table 1 Estimated World Production of Cabbage

Continent/Country	Area (1000 HA)	Production (1000 MT)
World	1713	40250
Asia	928	22299
Europe	265	6685
North America	106	2139
Africa	33	841
South America	25	569
Oceania	4	127
Leading countries		
China	419	9850
Russian Fed	180	4680
India	200	3300
Japan	68	2700
Korean Republic	44	2600
Poland	57	162
United States	76	1650
Indonesia	61	1332
Romania	32	800
Uzbekistan	20	800
Ukraine	69	893
Germany	13	659
United Kingdom	20	633

Source: Ref. 1.

can be made between white cabbage (with red-purple leaves) and Savoy cabbage (4). The latter has "blistered," "puckered" leaves because of the veins that grow less rapidly than the other leaf tissues (3).

B. Cultivars

Cabbage cultivars differ in head shape, varying from pointed to round, and the leaves may be green or red and smooth or wrinkled. Savoy types, which are particularly tolerant of cold conditions, have deeply wrinkled dark green leaves (5). Tigchelaar (6) described several new cultivars of cabbage released before 1980, including Blue Haven (early, blue-green, with spherical head, fine quality, white interior), Blue Oak (late, well-wrapped, blue-green, short core, tight interior), Cole Cash (large-framed, late vigorous, blue-green plants with slightly flattened firm heads), Defender (medium size, open plant, smooth round head, with solid interior and short core), Excel (main season blue-green, Danish type for storage), Grand Slam (highly uniform, tight dome-shaped head), Guardian (medium-sized, vigorous plant with smooth head, excellent solid interior, and short core), Jack Pot (early, midseason, uniform, round well-wrapped solid head), Regal (early, blue-green market cabbage with spherical head and fine interior quality), Satellite (compact, uniform, blue-green plants with solid round heads), Shamrock (midseason, blue-green, round slow bolting heads with short core and stem), Sunup (uniform, blue-green round, well-wrapped market-type head), and Venus (early, small uniform round heads with short core). In the United States, popular cultivars grown for the fresh market include Bravo Headstart,

Little Rock, Market Prize, Princess, Rio Verde, and Roundup. Cultivars grown for cold storage and the winter market include Bartolo, Green Winter, Hinova Polinius, and Supergreen. More popular sauerkraut cultivars include King Cole, Krantpacker, Little Rock, Roundup, and Sanibel (2). Chaudhury (7) classified the commonly grown Indian cultivars of cabbage into four groups:

1. Round head or ball head (e.g., Golden Acre, Pride of India, Copenhagen Market, Mammoth Rock Red, and Express)
2. Flat-head or drum head type (Pusa Drum Head)
3. Conical head (Jersey Wakefield)
4. Savoy (Chieftain).

Of these cultivars, Golden Acre, Pusa Drum Head, and Pride of India are grown most extensively in India.

III. PRODUCTION

A. Soil and Climate

Cabbage grows best in relatively cool and moist weather. It can tolerate frost to a great extent. In drier atmospheres leaves tend to be more distinctly petioled. Also, the quality of the head is impaired and much of its delicate flavor is lost.

Cabbage exhibits good vegetative growth at temperatures of 15–20°C. Temperatures below 0°C hinder growth. In regions where the average winter temperature goes below 0°C, it is difficult to overwinter the young tender seedlings in the open for early spring production. Supply in these regions is made by growing storage varieties, which are usually harvested in autumn and stored for some time. Type or variety also determines the appropriate sowing time, i.e., pointed headed and late types bolt less readily while the reverse is true for the round-headed and early types. When early varieties are sown early for harvesing in winter, generally few plants bolt. However, with late varieties sown early, losses may be much greater, because these varieties do not mature until spring when conditions become favorable for bolting.

Both sandy and heavy clay soils are suitable, but the former are preferred for early varieties while the latter are preferred for late ones. A provision for good drainage is a must, particularly for autumn or winter crops, as the young seedlings that overwinter in the field are very sensitive to an excess of soil moisture. The plants are also sensitive to soil reactions that should be between pH 6.0 and 6.5. In saline soils with pH above 6.5, the foliage becomes dark but the leaf margins die back and the plants are highly susceptible to blackleg diseases in such soils. These soils are especially unsuitable for summer crops since the salt level rises under drought conditions and high temperatures. As often as possible, the soil should be tilled shallow to maintain good tilth. The soil for cabbage crops should have a uniform texture to its 30-cm depth. It is also seen that moisture supply and cultivation practices influence the root development.

B. Propagation

About 400–500 g seeds are required for raising seedling sufficient for planting over an area of one hectare. Seeds are usually sown in lines in well prepared raised seedbeds. The seedbeds should be prepared in an open, sunny environment. These should be raised 15 cm above the surrounding surface to ensure good drainage. The soil should be friable and fairly rich in organic matter. Sufficient quantities of well-rotted farm yard manure or well-decayed leaf mould should be incorporated into the first 5-cm layer of topsoil. The surface of these beds should be leveled smooth before seed sowing.

Seedlings should be dipped in mercuric chloride solution (1 in 500 parts of water) before transplanting to guard against club root disease. Chhonkar and Jha (8) reported that application of starter solution of urea 1 part, potassium sulfate 1 part, and single superphosphate 2 parts (100 g of this mixture dissolved in 23 liters of water) at ½ pint per seedling after transplanting and again 15 days after the first application at the same rate increased cabbage yield by 18 tons/ha. Harvests 12–15 days earlier were achieved.

Four or 5-week-old seedlings that have produced three or four leaves and are 10–15 cm in height are best. They successfully resume their growth after transplanting. Older and taller seedlings are not desirable since their roots suffer badly when uprooted. Seedlings of late varieties can be directly transplanted in the field. However, early ones require pricking out or potting, which causes roots to branch and seedlings to become stocky and hardier.

It has been observed that topping the main leaves of seedlings reduces transpiration losses, but it also considerably reduces the photosynthetic surface. The assimilates formed in the retained foliage are just enough to be consumed by foliage and the roots usually starve. Therefore, topping is not advised (4).

C. Cultural Practices

1. Transplanting

The field should be laid out in either flat beds or ridges and furrows. Ridges and furrows are preferable in that they permit adequate irrigation water, drainage, and soil aeration. The soil must remain drier for winter crops.

Plant spacing varies with the variety (9,10). Early varieties are usually closely spaced and late ones widely spaced. Closer planting is conducive to the production of small heads, which are often preferred by consumers. The richer the soil, the closer the plants may be set to produce heads of good marketable size. Planting seedlings 40 cm apart in rows 75 cm apart appears to be optimum. A series of holes are prepared at marked places and then the seedlings are planted either in dry soil or in wet soil and then watered. The seedlings are inserted in the ground up to the cotyledons or up to the first two true leaves. Then the soil around is well firmed to establish close contact with the roots.

2. Irrigation

It has been observed that young cabbage plants take in about 0.3 liters of water from the soil daily, whereas adult plants take in about 0.4–0.5 liters (4). This is due to high evapotranspiration through the leaves. It has been estimated that the evapotranspiration per kg of dry matter ranges from 203–843 liters (11).

In light to medium-heavy soil, the moisture content—ranging from 60–100% field capacity or 80% average field capacity—appears to induce optimum plant growth. At 50% field capacity, the yield is found to be reduced by 20–30%. In loamy sand soil, the plants wither at a soil moisture content of 24% of field capacity. At excessive soil moisture content (i.e., at 100% field capacity), the crop matures late and yield is also slightly reduced (4).

The best yields are obtained if the crop is irrigated as soon as the moisture content of soil layer in which the plant produces its roots drops below 50% field capacity. This can be accomplished by first irrigating the third day after transplanting and applying subsequent irrigations at intervals of 10–12 days in heavy soils or 8 days in light soil. The irrigation interval should be longer for early-maturing varieties and shorter for late-maturing ones. This schedule should be followed until the heads are fully developed and are quite firm. Subsequent irrigation should be practiced sparingly and at longer intervals until harvest. Otherwise, many heads are likely to burst or split within 24 hours due to excess moisture.

3. Manuring and Fertilization

Cabbage is a heavy feeder (12,13). Therefore, adequate and timely application of farmyard manure or compost and nitrogenous, phosphatic, and potassic fertilizers is a must.

Once uptake by a crop and also the quantities of the nutrients available in the soil have been determined by chemical analysis, a rough estimation of the doses to be applied can be made (14–17). It has been estimated that in tropical and subtropical regions a good crop yielding 50 tons of cabbage heads uses 220 kg N, 100 kg P_2O_5, and 220 kg K_2O. It is, therefore, recommended that one apply 50 cartloads of farmyard manure or compost, 50 kg N, 100 kg P_2O_5, and 100 kg K_2o before transplanting and 50 kg N as a top dressing about 3–4 weeks after transplanting. In temperate regions, the dose of nitrogen may be increased to 150 kg per hectare.

It has been observed that a deficient nitrogen supply results in depressed yield, delayed crop maturity, and impaired keeping quality of heads (18–20). The excess supply of water causes the heads to turn coarser and looser, produces more outer leaves, promotes internal tipburn, impairs the keeping quality of heads, and also causes cracking of stored heads (4).

A deficient supply of P_2O_5 delays head formation and increases dry matter content, while excess P_2O_5 depresses yield, produces loose heads, and also adversely affects color intensity (20). An adequate supply of K_2O promotes winter hardiness and increases color intensity. A deficiency of potassium results in loose heads, an unpleasant strong odor, and poor keeping quality. However, excess potassium results in loose heads and reduced yield, probably due to high salt concentration in soil. The crop is sensitive to chloride injury. Hence, potassium sulfate is preferred to muriate of potash as a source of K_2O.

Fertilizer practices have a significant effect on the overall quality of cabbage. Nitrogen fertilization increases the total dry weight and nitrate nitrogen in cabbage but reduces the percent dry weight due to rapid, lush growth. Excess nitrogen increases the size of heads, which burst due to extremely rapid head enlargement particularly as maturity is approached. Excess nitrogen also reduces keeping quality (21–24). Increased synthesis of glucosinolates occurs with higher nitrogen levels (25).

4. Intercultural Operations

All intercultural operations in a cabbage field should be shallow—not deep enough to disturb the root system. It has been observed that before the plant is half grown, the roots cross in the center of the rows, and, therefore, deeper cultivation is likely to do more harm than good by destroying the roots near the surface. In order to produce solid heads, the plants should be earthed up about 5–6 weeks after transplanting. They should be top-dressed as soon as head formation commences.

D. Disorders, Diseases, and Pests

1. Disorders

Tipburn, black speck, and black petiole (midrib) are nonpathogenic internal disorders of cabbage.

Tipburn

This condition is characterized by tan or light brown tissue, which may later appear dark brown or even black. The affected tissue loses moisture and takes on a papery appearance. Tipburn is increased by high nitrogen (26) and high relative humidity (27,28). Lack of adequate calcium, particularly at the margins of the inner leaves, has been implicated in initiating tipburn development (29,30). Recent evidence suggests that tipburn is the result of a calcium imbalance caused by localized calcium deficiency in one or a group of leaves (31,32). Plants undergoing rapid growth, such as those exposed to long hours of daylight, high temperatures, or high levels of nitrogen, tend to be more susceptible to tipburn development. These plants are unable to deliver sufficient calcium to young, actively growing inner head leaves at a critical point in their development (33). An adequate supply of calcium in the soil will not prevent tipburn, nor is foliar spraying with calcium effective because the calcium is fixed by the outer leaves and not translocated to the interior of head. Incidence of tipburn can be reduced by the use of resistant or tolerant cultivars (34,35).

Black Petiole

Black petiole is an internal disorder of cabbage that has been noted in recent years (35). As heads approach maturity, the dorsal side of the internal leaf petiole or midribs turn dark gray or black at or near the point where the petiole attaches to the core. This is a complex physiological disorder in which environment plays an important role in symptom expression.

Black Speck

Black speck is characterized by dark spots occurring on outer leaves or sometimes throughout the head. Symptoms may not be present at harvest but the initial damage or predisposition likely occurs in the field with the typical symptoms developing during storage at low temperature (36,37). Although the casue of black speck is unknown, high rates of fertilizers, cultural conditions promoting vigorous growth, and temperature fluctuation have been reported to increase plant susceptibility (38). High rates of potassium in soil have been shown to reduce the severity of the disease. Some commercial cultivars have an acceptable level of tolerance (39–41).

2. Diseases

Many diseases reduce the overall quality of cabbage either as fresh or stored product. Diseases such as club root (*Plasmodiophora brassicae* Wor) and Fusarium yellows [*Fusarium oxysporum*, F. *conglutinans* (Wr.) Snyder and Hansen] are due to soilborne organisms (42) that prevent normal head development in cabbage. Soil fumigation and a soil pH of about 7.2 or higher aid in limiting the development of clubroot (42). Since yellow fungus is long-lived in soil, the most effective control over the disease in infected fields is the use of resistant cultivars. Black rot [*Xanthomonas campestris* (Pam.) Dows], black leg [*Phoma lingam* (Tode ex. Fr.) Desm.] and downy mildew (*Peronospora parasitica* Pers. Fr.) are field diseases that cause visible damage to leaves or make heads more susceptible to other diseases in storage (42,43). Black rot and black leg are seedborne diseases. The use of disease-free seed along with sanitation and rotation are the only methods of practical control (44). Fungicides are also avaialble for control of downy mildew and black blight.

3. Pests

The important pests of cabbage include cabbage worm (*Pieris rapae* L.), diamond black moth (*Plutella xylostella* L.), cabbage looper (*Trichoplusia ni* Hubner), cabbage aphids (*Brevicoryne brassicae* L.), and onion thrips (*Thrips tabaci* Lindeman) (45). Timely application of insecticide or biological control agents is required to minimize insect damage.

E. Maturity and Harvesting

Cabbage grown for early market is harvested as soon as it attains sufficient size, since earliness is more important than size (46). Midseason and late cabbage is not harvested until the heads are full and hard.

Crops are ready for harvest after about 2–4 months depending upon the type and variety. At maturity, the heads attain their full size and become firm and hard but tender. Pantastico et al. (47) reported that solidity and firmness of heads are the usual signs of maturity, and the color of the head is used as an added index of maturity. The fully developed head attains a lighter shade of green, but if intended for pickling purposes, the cabbage should be allowed to reach full maturity, as indicated by a curling back of the cover leaves and exposure of the white leaves beneath (48). Delay in harvesting causes the heads to crack and rot. Harvesting can be delayed somewhat by slowing the plant growth. This can be achieved by loosening the soil or raising plants without actually digging them up or by deeper cultivation whereby the roots are broken and the intake of soil, water, and nutrients—and thus loss from bursting—can be lessened. Harvesting at this stage gives maximum yield per unit area.

The heads are usually harvested with a knife or sickle, leaving a couple of leaves below the head intact, which provide a kind of protection. The cut heads are placed in a row to be collected by others or tossed to someone on a wagon or truck. Due care is exercised to prevent injuries due to bruising, especially for late cabbages means for prolonged storage. Ryall and Lipton (49) reported that most fresh market cabbage is harvested by hand. Varieties that are intended for sauerkraut or storage are generally harvested by machine all at once. Other varieties are often harvested on several occasions over a period of time, particularly if the heads crack and the crops have grown unevenly. Savoys are harvested with a large proportion of the wrapper leaves, which are, in part, consumed.

F. Grading and Packaging

A uniform product is obtained by sorting out the heads according to variety, size of head, and quality. The outer loose leaves of each head are then removed (50–52). This assures convenience in handling and a better appearance. These heads are then packed and sent to market. The healthy heads of cabbage should be packed in perforated polyethylene bags and stored at low temperature immediately. Disease-producing pathogens can be controlled if the affected heads or leaves are excluded. Close trimming removes most of the infection. Refrigeration at 1.6°C or lower and low relative humidity are needed to control the growth of many pathogens (50).

IV. CHEMICAL COMPOSITION

A. Proximate Composition

The chemical composition of cabbage leaves is presented in Table 2. Cabbage leaves are low in calories, fat, and carbohydrates. They are good sources of protein, which contains all of the essential amino acids, particularly sulfur-containing amino acids. Cabbage proteins compare very favorably to pea proteins in biological value, digestibility, and NPU (54). Cabbage is an excellent source of minerals such as calcium, iron, magnesium, sodium, potassium, and phosphorus (Table 3), most of which are in an available form. Cabbage has substantial amounts of β-carotene (provitamin A), ascorbic acid, riboflavin, niacin, and thiamine

TABLE 2 Nutritional Composition of White and Savoy Cabbage

Constituent (per 100g)	White cabbage	Savoy cabbage
Water (g)	90.7	88.1
Total nitrogen (g)	0.23	0.33
Protein (g)	1.4	2.1
Fat (g)	0.2	0.5
Carbohydrate (g)	5.0	3.9
Energy value (kJ)	113	114
Starch (g)	0.1	0.1
Total sugars (g)	4.9	3.8
Dietary fiber (g)	2.1	3.4

Source: Ref. 53.

(Table 4). The level of ascorbic acid varies from 30 to 65 mg/100 g fresh weight (55–60). Cooking or shredding of cabbage and leaving it exposed to air can result in substantial loss of ascorbic acid.

B. Anthocyanins

Red cabbage is a rich resource of anthocyanin pigments. Both mono- and diacylated cyanidin derivatives have been identified in red cabbage. Some of the anthocynanins characterized are: 3-*O*-(6-*O*-*E*-ferulyl-β-D-glucopyranosyl)-5-*O*-(β-D-glucopyranosyl) cyanidin, 3-*O*-(2-*O*-β-D-glucopyranosyl)-6-*O*-(4-*O*-β-D-glucopyranosyl-*E*-*p*-coumaryl)-β-D-glucopyranosyl-5-*O*-(β-D-glucopyranosyl) cyanidin, 3-*O*-(6-*O*-*E*-sinapyl-2-*O*-β-D-glucopyranosyl)-β-D-glucopyranosyl-5-*O*-(β-D-glucopyranosyl) cyanidin, 3-*O*-(6-*O*-*E*-ferulyl-2-*O*-(2-*O*-*E*-sinapyl-β-D-glucopyranosyl)-5-*O*-(β-D-glucopyranosyl) cyanidin, 3-*O*-(6-*O*-*E*-*p*-coumaryl-2-*O*-(2-*O*-*E*-sinapyl-β-D-glucopyranosyl)-5-*O*-(β-D-glucopyranosyl) cyanidin, and 3-*O*-(6-*O*-*E*-sinapyl-2-*O*-(2-*O*-*E*-sinapyl-

TABLE 3 Mineral Composition of White and Savoy Cabbage

Mineral (mg/100 g)	White cabbage	Savoy cabbage
Sodium	7	5
Potassium	240	320
Calcium	49	53
Phosphorus	29	44
Iron	0.5	1.1
Copper	0.01	0.03
Zinc	0.2	0.3
Sulfur	54	88
Chloride	40	48
Manganese	0.2	0.2
Iodine (μg/100 g)	2	2

Source: Ref. 53.

TABLE 4 Vitamin Contents in White and Savoy Cabbage

Vitamin (per 100 g)	White cabbage	Savoy cabbage
Carotene (μg)	40	995
Vitamin D (μg)	0	0
Thiamine (mg)	0.12	0.15
Riboflavin (mg)	0.01	0.03
Niacin (mg)	0.3	0.7
Vitamin B_6 (mg)	0.18	0.19
Vitamin B_{12} (μg)	0	0
Folate (μg)	34	150
Pantothenate (mg)	0.21	0.21
Biotin (μg)	0.1	0.1
Vitamin C (mg)	35	62

Source: Ref. 53.

β-D-glucopyranosyl)-5-*O*-(β-D-glucopyranosyl) cyanidin (61). Since the mid-1970s, there has been an increased interest in red cabbage pigments as a possible source of natural food colorants. The pigment preparations have been reported to impart a color superior to that obtained with the synthetic colorant Red No. 40 and had a satisfactory color stability in beverages. Presently, a red cabbage colorant is marketed under the name San Red RC, recommended for use in beverages, chewing gums, candies, sherbets, dressings, yogurt, and other fermented products.

C. Flavor

Flavor is one of the important properties of cabbage (62,63). The flavor of cabbage depends upon the cultivar, maturity, season, horticultural practices, storage handling, and method of cooking. Jensen et al. (62), using paper chromotography, showed the presence of 2-propenyl isothiocyanate in both white and red cabbage and 3-butenyl isothiocyanate in red cabbage. Dateo et al. (64) noted that cabbage contains thermolabile sulfur compounds, which evolve sulfurous odors. The major volatile sulfur compounds (dimethyl disulfide and hydrogen sulfide) of cooked cabbage are derived from a precursor L-*S*-methylcysteine sulfoxide, a free amino acid. Maruyama (65) indicated that dimethyl trisulfide was a major aroma component in cooked brassicaceous vegetables and proposed that the strong, unpleasant aroma characteristic of overcooked brassicaceous vegetables is due to the gradual loss of pleasant volatile components and resultant unmasking of unpleasant sulfur components.

MacLeod and MacLeod (66), assessing the similarities among flavors of cabbage, cauliflower, and Brussels sprouts, found that isothiocyanates with allyl isothiocyanate comprised over 6% of the total volatiles. In general, cabbage contains high amounts of objectionable sulfur compounds (67) and has a mushy texture after extended boiling (68). Such vegetables also undergo a significant loss of color from prolonged heating in restaurants or dormitory kitchens and are not preferred by the consumers despite their being excellent sources of proteins, vitamins, and minerals with few calories (69).

D. Toxic Constituents

Cabbage contains goitrogens, which cause enlargement of the thyroid glands. Natural thioglucosides are the source of goitrogens. The same tioglucosides with their associate enzyme(s) impart the

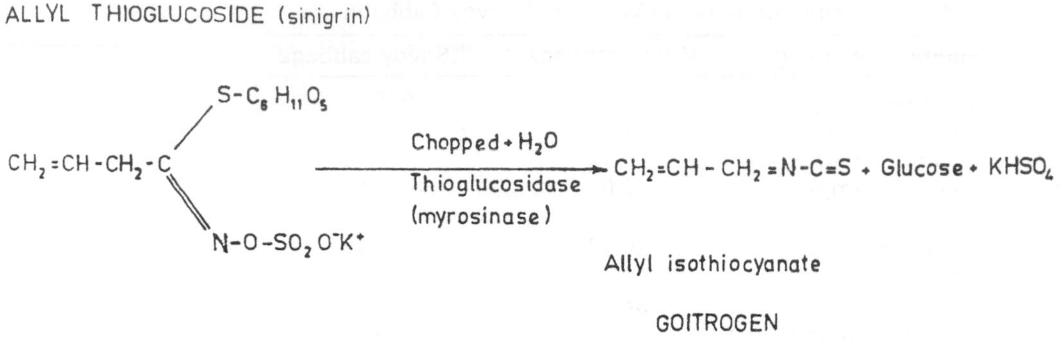

ALLYL THIOGLUCOSIDE (sinigrin)

$CH_2=CH-CH_2-C(-S-C_6H_{11}O_5)(=N-O-SO_2O^-K^+)$ —Chopped + H_2O / Thioglucosidase (myrosinase)→ $CH_2=CH-CH_2=N-C=S$ + Glucose + $KHSO_4$

Allyl isothiocyanate

GOITROGEN

FIGURE 1 The enzymatic hydrolysis of allyl thioglucoside (sinigrin) to allyl isothiocyanate. (From Ref. 54.)

desirable culinary flavor to cabbage, broccoli, and cauliflower. The thioglucosides (sinigrin), nongoitrogenic, upon enzymatic hydrolosis, yield glucose and bisulfate and allylthiocyanate—a goitrogen (Fig. 1). The thioglucosides progoitrin and epi-orogoitrin are the precursors of an antithyroid compound named goitrin (5-vinyloxazolidine-2-thione). Goitrin is formed subsequent to hydrolysis and then through cyclization of an unstable isothiocyanate-containing hydroxyl group (Fig. 2). The total glucosinolate content of a large number of white cabbage cultivars ranged from 299 to 1288 ppm (70). The goitrin content varied from 1.2 to 26.00 ppm. Although the enzymatic hydrolytic products from the glucosinolates are goitrogenic, some of these breakdown products (which include isothiocyanates, thiocyanates, nitrile, and goitrin) also

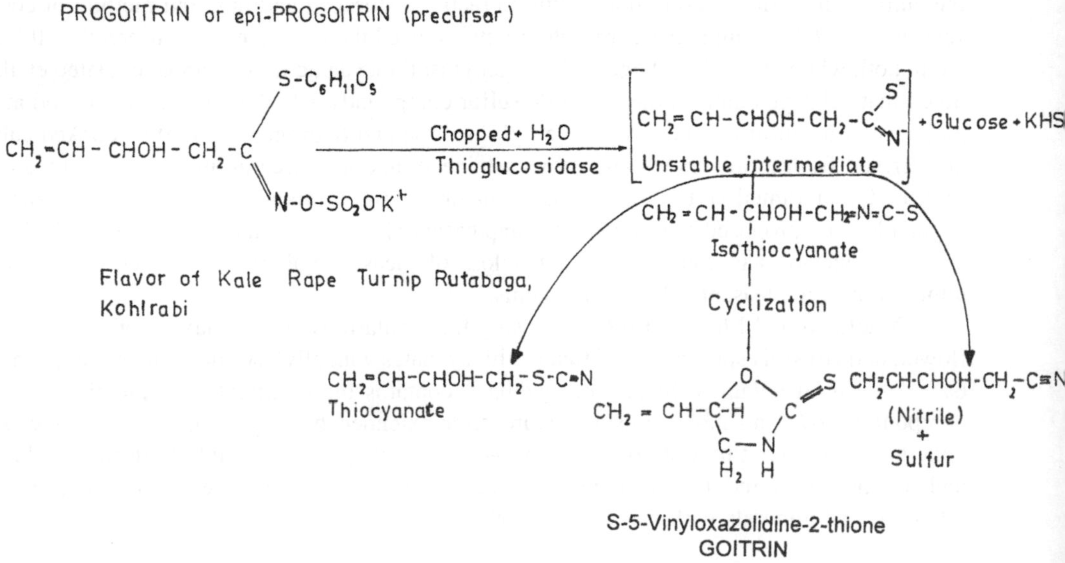

PROGOITRIN or epi-PROGOITRIN (precursor)

$CH_2=CH-CHOH-CH_2-C(-S-C_6H_{11}O_5)(=N-O-SO_2O^-K^+)$ —Chopped + H_2O / Thioglucosidase→ [$CH_2=CH-CHOH-CH_2-C(S^-)(=N^-)$ Unstable intermediate] + Glucose + KHS

$CH_2=CH-CHOH-CH_2-N=C=S$

Isothiocyanate

Cyclization

Flavor of Kale Rape Turnip Rutabaga, Kohlrabi

$CH_2=CH-CHOH-CH_2-S-C\equiv N$

Thiocyanate

$CH_2=CH-CH$ (ring: O, C=S, N-H, CH_2)

$CH_2=CH-CHOH-CH_2-C\equiv N$

(Nitrile) + Sulfur

S-5-Vinyloxazolidine-2-thione

GOITRIN

FIGURE 2 The enzymatic hydrolysis of progoitrin-a thioglucoside to goitrin. (From Ref. 54.)

exhibit the typical cabbage flavor. Greer (70) found that most of the goitrogen properties of the product could be lost during cooking.

V. STORAGE

Cabbage should be handled carefully from field to storage, and only solid heads with no yellowing, decay, or mechanical injuries should be stored. Before the heads are stored, all loose leaves should be trimmed away. Adequate ventilation is essential for successful storage. Upon removal from storage, the heads should be trimmed again to remove loose and damaged leaves. Cabbage should not be stored with fruits emitting ethylene. Concentrations of 10–100 ppm of ethylene cause leaf abscission and loss of green color within 5 weeks.

Thompson and Kelly (46) mentioned the following essentials for successful storage of cabbage:

1. Good storage variety
2. Freedom from disease or injury of any kind
3. Relatively uniform temperature near the freezing point
4. Moderate degree of humidity—enough to prevent wilting, but not so moist as to cause condensation

A. Low-Temperature Storage

Cabbage is stored in mechanically refrigerated rooms or ventilated common storage. The lower the temperature, the longer the cabbage will keep provided the heads do not actually freeze. Ryall and Lipton (49) reported that slight freezing is harmless, but that severe freezing damages quality; solid heads are more susceptible to freezing injury than merely firm heads. A storage temperature of 0°C with 98% relative humidity for storage up to 6 months is recommended (71). Prevention of excessive loss of moisture is essential during storage. Cabbage can be successfully held in storage at 0–1.5°C. Storage facilities should be insulated to prevent freezing of the cabbage. Wilting occurs if storage conditions are too dry. Hence, humidity should be high enough to keep the leaves fresh and turgid. Cabbage stored at 0°C has less decay when relative humidity is maintained at or near saturation (98–100%) than at 90–95% (49). Well-trimmed heads can be stored in nearly water-saturated air without danger of excessive decay if temperature is below 4°C. Respiratory heat must be removed by providing adequate air circulation.

Parsons (71) found that cabbage heads can be stored in excellent condition for at least 8 weeks in bins or crates lined with perforated polyethylene. Perforations with one ¼-inch hole per square foot of liner or bag is sufficient. Cabbage varieties that store well have compact hard heads, such as Danish Ballhead. Chinese cabbage can be stored at 0°C with 95% relative humidity for 1–2 months (49).

B. Controlled-Atmosphere Storage

Slight reductions in weight loss (mainly due to less trimming) and retardation of general senescence, such as yellowing, toughening, and loss of flavor are the major advantages of controlled-atmosphere storage of cabbage (72). White cabbage seems to store well in 5% O_2 plus 2.5–5% CO_2 (49). Suhonen (73), however, found that if O_2 is near zero or if CO_2 is maintained at 15% or higher for one month or longer, the internal leaves of cabbage discolor even though the external leaves may appear to be normal. A concentration of 1.0–2.5% O_2 and 5.5% CO_2

retarded senescence in terms of toughening, loss of flavor, and yellowing and reduced the incidence of pepper spot virus (74). By increasing the CO_2 level to 10%, Suhonen (73) found a reduction in root growth, decay, and sprouting even at near water saturation provided the temperature was O°C. Cabbage stored at 0–2°C in controlled atmosphere (3% O_2, 5–6% CO_2, and the remainder N_2) showed a lower incidence and severity of storage rots caused by *B. cinerea* than did cabbages stored at the same temperature in air. Weight loss was reduced, little or no trimming was needed, and the flavor, texture, and appearance were better in controlled-atmosphere–stored than in air-stored cabbage (74).

C. Postharvest Diseases and Their Control

Black rot is the most serious market disease of the cabbage (75–77). This vascular disease, which blackens the veins, is caused by *Xanthomonas campestris*. The bacteria produce acid mucopolysaccharide, which disorganizes xylem (76). The melanins accumulate among the collapsing cells, causing the veins to blacken. Gray mold rot, caused by *Botrytis cinerea*, is another serious market disease of the cabbage. This pathogen germinates in most packages at 2–4°C, and the infection leads to rotting. Other organisms causing cabbage rots are *Sclerotinia sclerotioria*, *Fusarium roseum*, *Phytophthora*, spp., *Rhizoctonia solani*, and *Alternaria* spp., which develop during transit and storage, producing blemishes (50). Ryall and Lipton (49) have given a detailed account of various diseases, the pathogens involved, development, and possible preventive measures for cabbage and other cruciferous vegetables.

Geeson and Browne (74) reported that postharvest drenches of winter white cabbage with benomyl, thiabendazole, or iprodione (50–60% w.p., 1 g/liter) gave good control of rotting caused by *B. cinerea*, the main cause of spoilage in untreated cabbage stored at 0.1°C and 95% relative humidity. All three fungicides controlled rots caused by *Mycosphaerella brassicicola*, while iprodione also controlled *Alternaria* spp.

Salunkhe et al. (77) reviewed the effects of chemical treatments on storage behavior of several vegetable crops including cabbage and other crucifers. Stewart (78) found that during storage leaf drop in cauliflower and cabbage was reduced by an application of 100–500 ppm of 2,4-D 1–7 days before harvest. These vegetables when wrapped with shredded paper containing 50–100 mg NAA and stored at 0°C showed a reduction in leaf abscission and weight loss. Drenching harvested cabbage with Benlate (benomyl), Tecto RPH (tiabendazole), or Rovral (ibrodione) at 0.45 kg/450 liters reduced the percentage of heads affected by fungal rots and elminated rots caused by *Mycophaerella* spp. (79). Immediate dipping after harvest of the cabbage in 7.5% borax solution gives the best control of soft rot caused by *Erwinia carotovora* (80).

D. Irradiation

Salunkhe et al. (81–83) reported the effects of radiation dose, rate of radiation, and temperature on physiological and chemical changes of cabbage, cauliflower, and asparagus. Medium-sized whole heads of cabbage were irradiated in sealed cans at 0, 0.93, 2.79, and 5.58×10^5 rad/hr. Satisfaction with the taste of irradiated cabbage heads was found to decrease with the increase in radiation dose, whether evaluated 24–28 hours or 10 days after irradiation (Table 5). In general, the intermediate rate, i.e. 0.465×10^6 rad/hr, gave a product with superior taste than did both the fast (0.930×10^6 rad/hr) and slow (0.093×10^6 rad/hr rates of radiation. The taste of irradiated cabbage was improved after 10 days at 10°C. The off-flavor change following irradiation diminished after 10 days of storage at 0°C. Spoilage after 10 days of storage was less

TABLE 5 Effect of Radiation Dose, Radiation Rate and Storage Period on Taste Preference of Cabbage

Rate ($\times 10^6$ rad/hr)	Dose ($\times 10^6$ rad)	Taste preference score: Evaluated 24–48 hrs after radiation	Taste preference score: Evaluated 10 days after radiation
Control	Control	6.9	5.9
0.093	0.093	5.4	7.2
	2.79	3.5	4.8
	5.58	3.6	3.7
	Mean	4.2	5.2
0.045	0.93	6.4	6.7
	2.79	6.4	7.6
	5.58	6.5	6.8
	Mean	6.4	7.0
0.93	0.93	5.5	6.5
	2.79	5.7	6.1
	5.58	4.2	5.9
	Mean	5.1	6.2
Dose mean	Control	6.9	6.9
	0.93	5.8	6.8
	2.79	5.2	6.2
	5.58	4.8	5.5
LSD	Individual	1.5–1.2	1.5–1.2
	Dose means	0.7	0.7
	Rate means	0.7–0.6	0.7–0.6

Source: Ref. 81.

severe at the higher doses than at the lower doses. Mold growth was suppressed by gamma radiation of $> 2.5 \times 10^5$ rad.

VI. PROCESSING

A. Sauerkraut

Cabbage is commonly used to prepare sauerkraut (84–89). Sound, firm cabbage heads are selected for sauerkraut production. The outer leaves are removed by hand and the core is reamed by a rapidly revolving conical knife. The cored cabbage is cut into shreds by thin curved knives attached to the revolving metal disc. The shredded cabbage falls into a fermenting tank or is placed in special carts and moved to the vats or tanks. The conversion of cabbage into sauerkraut is accomplished by lactic acid fermentation. The presence of a moderate concentration of salt is necessary to reduce the growth of spoilage microorganisms and to promote the growth of lactic acid bacteria. The usual proportion of salt is 2.5% by weight, which is well mixed with the shredded cabbage. Bacteria and yeasts develop rapidly, and gas evolution is vigorous during the first stage of fermentaion. Yeasts are present in considerable numbers and may produce small amounts of alcohol, particularly during the initial stages of the fermentation.

The predominant organisms are lactic acid producers (84). The *Leuconostoc* group is always present and is responsible for considerable gas formation as well as production of lactic acid, alcohol, and manitol. If the development of yeasts, aerobacter, and butyric organisms is prevented or at least minimized, the quality of product is improved correspondingly. A temperature of 19°C is optimum for the quality in kraut fermentation. If the cabbage is cold shredded, it is desirable to warm it to 19–20°C in the tanks. The acidity rapidly increases during fermentation and frequently reaches 1–8%. There is a demand for canned sauerkraut since the canned product is convenient for shipment. Sauerkraut used for canning is ordinarily not cured as the canner desires a product of lighter color and lower acidity.

Cabbage is a good source of vitamin C (85); the vitamin C content of sauerkraut during the active fermentation period was equal to that of the original cabbage (86). A slow progressive loss of vitamin C, however, occurs during storage. Nabors and Salunkhe (90) found that low-boiling compounds of sauerkraut were destroyed faster with conventional or microwave drying than with freeze-drying. A cheeselike flavor was associated with carboxylic acids, and greater amounts of butyric and hexanoic acids were found in off-flavored samples (91). White cabbage varieties with a compact head, fine leaves, thin veins, and white color are used for sauerkraut manufacture. Prematurely harvested cabbage give soft kraut of a gray color and inferior taste, and heads with internal abnormalities (intenal tipburn) are not fit for sauerkraut.

B. Freeze-Dried and Canned Products

Cabbage can be freeze-dried or canned after shredding, but these products have a lower quality than fresh products. Loss of vitamin C following canning of cabbage has been reported (89).

Srisangam et al. (92–94) studied the effect of blanching, freezing, and freeze dehydration on the acceptability and nutrient retention of cabbage and attempted to improve the flavor of blanched cabbage. Cabbage was blanched with steam and chemical solution (or water) for 3 minutes at 96°C. In steam blanching, cabbage was impregnated with chemical solution for under 25 in. vacuum and then exposed to steam for 3 minutes. Of solution and steam blanched (without chemical treatment) cabbage, the former had better organoleptic qualities (Table 6), whereas the latter had higher ascorbic acid retention (Table 7). Cabbage blanched in the presence of 0–5% malic acid, 0.05% $NaHCO_3$, and 1.2% NaCl was preferred for texture, color, and overall quality, respectively (Table 8). The results from chemical (vitamin C and chlorophyll contents), physical (firmness), and sensory evaluation of processed and stored (at –19°C up to 24 months) cabbage indicated blanching in the presence of 1.2% NaCl solution to be the preferred treatment.

C. Pickled Products

Pickles, relishes, and condiments are important products of the food industry. Kim Chee pickle or pickled nappa is a well-known delicacy native to Japan and Korea. Kim Chee is the Korean term for this type of pickle, and nappa is the Japanese term for a type of celery cabbage (95). Kim Chee pickle is a fermented product and is prepared by combining an intermediate juice with an intermediate portion of nappa or with fresh nappa, in which case the nappa is combined with intermediate juice and allowed to ferment to the state desired. The intermediate juice is not subjected to further activation in the can, so it does not become spoiled or cause the can to swell or explode (96).

VII. THERAPEUTIC PROPERTIES

Cabbage and other members of genus *Brassica* are well known for their medicinal properties. In ayurvedic medicine, cabbage leaves are prescribed for cough, fever, skin diseases, peptic ulcers, urinary discharges, and hemorrhoids. The seeds are diuretic, laxative, stomachic, and

TABLE 6 Taste Panel Scores for Cabbage Samples Blanched in Solution or in Steam

	Flavor		Color		Texture		Overall quality	
Chemical treatment	Solution blanching	Steam blanching	Solution blanching	Steam blanching	Solution blanching	Steam blanching	Solution blanching	Steam blanching
Control	6.4 ± 1.1[b]	5.8 ± 1.2[b]	6.5 ± 0.9[b]	6.1 ± 1.0[b]	6.2 ± 1.1[b]	5.9 ± 1.2[b]	6.5 ± 1.0[b]	6.0 ± 0.7[b]
EDTA 0.025%	6.5 ± 1.0	6.0 ± 0.9	6.3 ± 1.3	6.0 ± 1.1	6.7 ± 0.9	6.4 ± 1.2	6.3 ± 0.8	6.2 ± 1.3
Malic acid 0.5%	6.0 ± 1.0	5.0 ± 1.2	5.5 ± 1.2	5.8 ± 1.0	6.5 ± 0.9	6.0 ± 1.3	5.3 ± 0.9	6.0 ± 0.7
NaCl 1.2%	7.3 ± 0.8	7.0 ± 1.2	7.1 ± 1.0	6.5 ± 0.9	7.3 ± 0.9	7.0 ± 0.9	7.1 ± 0.9	6.9 ± 1.0
$NaHCO_3$ 0.05%	6.8 ± 1.1	6.1 ± 1.1	7.0 ± 1.0	6.5 ± 1.1	6.0 ± 1.1	6.0 ± 0.7	6.3 ± 1.1	6.4 ± 1.2
LSD at 0.05 level	0.8		0.8		0.8		0.8	

[a]Represents evaluation by 12 trained judges.
[b]Mean ± S.D.
Source: Ref. 93.

TABLE 7 Effects of Chemicals on Chlorophyll and Ascorbic Acid Content of Water Blanched and Steam Blanched Cabbage

Blanching method	Chemical treatment	Chlorophyll (mg/100 g)	Ascorbic acid (mg/100 g
Solution	Control	8.2	30.1
	EDTA 0.025%	9.6	32.9
	Malic acid 0.5%	12.1	33.3
	NaCl 1.2%	9.4	34.1
	$NaHCO_3$ 0.05%	11.0	31.9
Steam	Control	12.2	31.9
	EDTA 0.025%	10.5	33.2
	Malic acid 0.5%	9.9	36.2
	NaCl 1.2%	10.6	34.0
	$NaHCO_3$ 0.05%	13.0	37.7

Source: Ref. 93

antihelminthic (97). The chemoprotective components include ascorbic acid, tocopherols, carotenoids, isothiocyanates, indoles, and flavonoids. Fresh cabbage juice is reported to contain a heat-labile anti-peptic ulcer component, vitamin U (98–100). A clinical study indicated that concentrated cabbage juice is significantly effective in healing peptic ulcers (99,100). Cabbage is one of the largest sources of sulfur-containing amino acids. Among these, *S*-methylcysteine sulfoxide, a nonessential amino acid, has potent antihypercholesterolemic effects in experimental hypercholesterolemic rats (101,102). Petering et al. (103) have reported the presence of antistiffness factors in cabbage.

Fresh cabbage juice is reported to have antibacterial activity and has been shown to inhibit the growth of various strains of lactic acid bacteria (104–106). Among cabbage components, glucosinolate hydrolysis products have been reported to be antimicrobial (107,108). Fresh cabbage is reported to contain 300–1070 μg/g of total glucosinolate compounds (109,110), which upon hydrolysis produce isothiocyanates, nitriles, and thiocyanates. The antibacterial activity of fresh cabbage juice was reported to be higher and more consistent than heat-treated cabbage juice. The pH of the juice was also reported to play a significant role in the inhibitory activity. Fresh cabbage juice adjusted to a higher initial pH (6.5–7.5) showed greater activity than at pH 5.0–6.0 (106). The activating factor may be an enzyme that converts a precursor to the inhibitory compound (106).

Cabbage is also reported to have significant anticancer activity. Both animal model studies and epidemiological data in humans have tended to confirm the protective role of cabbage on the development of cancer. Dietary cabbage has been reported to inhibit chemically induced mammary tumorigenesis in rats (111,112), reduced the formation of colon adenomas in mice (113), and depress liver cancer in rats after aflatoxin B_1 administration (114). Cabbage-containing diets also provide a protective effect against the tumorigenicity of 1,2-dimethylhydrazine in mice (115).

In an epidemiological study involving 256 white male colon cancer patients, 330 white male rectal cancer patients, and 1222 controls with nonneoplastic diseases of systems other than the gastrointestinal tract, Graham et al. (116) reported that the colon cancer risk was two to three times higher in individuals who either rarely or never consumed certain vegetables, including cabbage. A similar decreased risk of colon cancer in cabbage-consuming Japanese was reported by Haenzel et al. (117). Cabbage extract is reported to increase the level of xenobiotic-

TABLE 8 Effects of Blanching on the Mineral Contents[a] (mg/100 g) in Cabbage Samples

	Solution blanching			Steam blanching		
Chemical treatment	Mg	Ca	Fe	Mg	Ca	Fe
Control	262.9 ± 0.8[b]	973.8 ± .0.7[b]	4.7 ± 0.1[b]	237.5 ± 4.4[b]	569.0 ± 1.3[b]	5.0 ± 0.2[b]
EDTA 0.025%	286.3 ± 3.7	866.4 ± 8.7	5.0 ± 0.1	241.2 ± 2.2	583.7 ± 0.6	5.3 ± 0.2
Malic acid 0.5%	202.6 ± 12.6	547.2 ± 8.4	3.9 ± 0.2	193.6 ± 0.2	639.0 ± 6.4	4.2 ± 0.2
NaCl 1.2%	201.7 ± 3.0	748.6 ± 11.16	5.5 ± 0.3	229.3 ± 6.3	766.9 ± 12.6	5.9 ± 0.1
$NaHCO_3$	204.6 ± 0.8	998.3 ± 15.98	3.7 ± 0.3	232.9 ± 0.8	699.7 ± 13.0	5.0 ± 0.2
Fresh and untreated cabbage	221.0	534.8	4.9	221.0	534.0	4.9
LSD at 0.05 level	8.5	16.2	0.4	8.5	16.2	0.4

[a]The contents are expressed for 100 g of moisture-free cabbage.
[b]Mean ± S.D.
Source: Ref. 93.

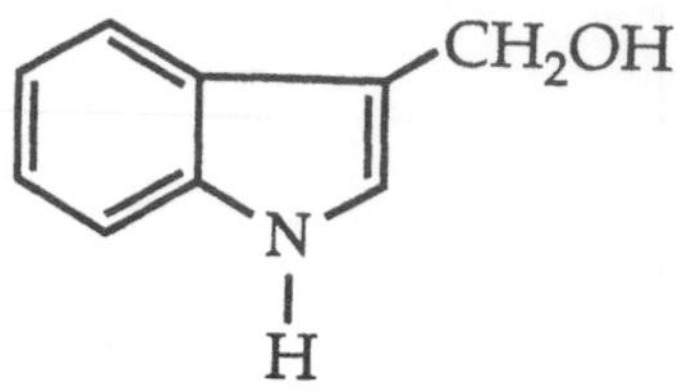

FIGURE 3 Indole-3-carbinol.

metabolizing enzymes in experimental animals and humans (118,119). This may explain the inhibitory effects on chemically induced carcinogenesis observed in rodents and may contribute to a decreased cancer incidence of the colon and rectum in humans (120). The constituents tentatively responsible for these activities are indole-3-carbinol, indole-3-acetonitrile, and 3,3′-di-indolylmethane, of which indole-3-carbinol (Fig. 3) was the most potent enzymne inducer (121). These indoles are generated in crushed plant material via a thioglucosidase mediated autolysis of a parent glucosinolate, commonly referred to as glucobrassisin (122,123). The flavonoids such as apigenin and robinetin showed a potential to inhibit the promotion stage of carcinogenesis (124). Cabbage juice also exhibited a neutrophil-inducing activity in mice indicating the potential as an immunopotentiator (125).

Cabbage extract has been reported to be antimutagenic against various mutagens in the Ames assay (121). The extract has been shown to be active against mutagenesis caused by tryptophan pyrolyzate, oxidized linolenic acid, and nitrate + sorbic acid (121). Munzer (126) demonstrated that some antimutagenic compounds in the extract act by stimulating native detoxification systems in *Salmonella typhimurium* and thus are desmutagenic. There is considerable agreement that the active components include ascorbic acid, cysteine, or other compounds acting as reducing agents (121). In addition, Lawson et al. (127) have reported that methanol, methylene chloride, and petroleum ether extracts of dried cabbage have antumutagenic activity against specific mutagens, *N*-methyl-*N*-nitrosourea and 2-aminoanthracene, in the Ames assay and in the V79 mammalian mutagenesis system. Nonacosane, 15-nonacosanone, pheophytin a, and β-sitosterol were identified as the active compounds in these extracts.

REFERENCES

1. Food and Agriculture Organization, *Production Year Book*, Vol. 48, Rome, Italy, 1994.
2. Pritchard, M. K., and R. F. Becker, Cabbage, in *Quality and Preservation of Vegetables* (N. A. M. Eskins, ed.), CRC Press, Boca Raton, FL, 1989, p. 265.
3. Yamaguchi, M., *World Vegetables: Principles, Production and Nutritive Values*, AVI Pub. Co. Inc., Westport, CT, 1983, p. 219.
4. Chatterjee, S. S., Cole crops, in *Vegetable Crops in India*, (T. K. Bose and M. G. Som, eds.), Naya Prokash, Calcutta, 1986, p. 165.
5. Fordham, R., Cabbage and related vegetables, in *Handbook of Food Science, Food Technology and Nutrition*, (R. McRae, R. K. Robinson, and M. J. Sandler, eds.), Academic Press, London, 1993, p. 4719.
6. Tigchelaar, E. C., New Vegetable Varieties List XXI, The Garden Seed Research Committee, American Seed Trade Association, *HortScience 15*(5):565 (1980).
7. Chaudury, B., *Vegetables*, 4th ed., National Book Trust, New Delhi, 1976, p. 214.
8. Chhonkar, V. S., and R. N. Jha, The use of starters and plant growth regulators in transplanting of cabbage and their response on growth and yield, *Indian J. Hort. 20*:123 (1963).

9. Vaidya, S. J., and A. V. Patil, Influence of date and transplanting and age of seedings on growth, yield and quality of cabbage, *Indian J. Agron. 10*:420 (1965).
10. Mangal, J. L., A. H. Jassin, and S. K. Mehmood, Effect of plant spacing and age of seedling on yield and quality of cabbage, *Haryana J. Hort. Sci. 16*:121 (1987).
11. Chatterjee, S. S., and R. R. Sharma, All about cole crops, *Indian Farmers Digest 3*:37 (1970).
12. Bothar, V. F., Characteristics of mineral nutrition and yield formation in white head cabbage cultivar as dependent on fertilization, Part 3. Nitrogen, phosphorus and potassium content in cabbage plants, *Agrokhimiya 7*:57 (1990).
13. Ghanti, P., S. G. Sounda, P. K. Jana, and M. G. Som, Effect of levels of nitrogen phosphorus, spacing on yield characters of cabbage, *Veg. Sci. 9*:1 (1982).
14. Singh, R. V., and L. B. Naik, Response of cabbage to plant spacing, nitrogen and phosphorus levels, *Indian J. Hort. 45*:325 (1988).
15. Khokar, N. S., R. P. Singh, and M. Prashad, Effect of different levels of nitrogen, phosphorus, potassium and FYM on the yield of cabbage, *Indian J. Agron, 15*:9 (1970).
16. Som, M. G., P. K. Jana, and S. S. Pal, Soil moisture regimes and nitrogen levels on the yield and consumptive use of water by cabbage, *Veg. Sci. 3*:131 (1976).
17. Choudhury, B., *Vegetables*, 8th ed., National Book Trust, 1990.
18. Gupta, A., Effect of nitrogen and irrigation on cabbage production, *Indian J. Hort. 44*:241 (1987).
19. Srinivas, K., Note on response of cabbage to plant density and fertilizers, *Indian J. Hort, 41*:277 (1984).
20. Thakur, P. C., and H. S. Gill, Effect of nitrogen, phosphorus, and potash fertilization on seed yield of cabbage (*Brassica oleracea* var. *Capitata*), *Indian J. Hort. 33*:262 (1976).
21. Janes, B. E., The relative effect of variety and environment in determining the variations of percent dry weight, ascorbic acid and content of cabbage and beans, *Proc. Am. Soc. Hort. Sci. 45*:387 (1944).
22. Jones, B. E., Composition of Florida grown vegetables III. Effects of location, season, fertilizer level and soil moisture on the mineral composition of cabbage, beans, collards broccoli, and carrots, *Agric. Exp. Stn. Fla. Bull.* 488 (1951).
23. Peck, N. H., Cabbage plant responses to nitrogen fertilization, *Agron. J. 73*:679 (1981).
24. Jackson, W. A., J. S. Steel, and V. R. Boswell, Nitrates in edible vegetables and vegetable products. *Proc. Am. Soc. Hort. Sci. 90*:349 (1967).
25. Mengel, K., Influence of exogenus factors on the quality and chemical composition of vegetables *Acta Hort. 93*:133 (1979).
26. Peck, N. H., M. H. Dickson, and G. E. MacDonald, Tipburn susceptibility in semi-isogenic inbred lines of cabbage influenced by nitrogen, *HortScience 18*:726 (1983).
27. Dickson, M. H., Inheritance of resistance to tipburn in cabbage, *Euphytica 26*:811 (1977).
28. Palzkill, D. A., T. W. Tibbitts, and B. E. Sturkmeyer, High relative humidity promotes tipburn on young cabbage plants, *HortScience 15*:659 (1980).
29. Maynard, D. N., B. Gersten, and H. F. Vernell, The distribution of calcium as related to internal tipburn, variety and calcium nutrition in cabbage, *Proc. Am. Soc. Hort. Sci. 86*:392 (1965).
30. Walker, J. C., L. V. Edington, and M. V. Nayuda, Tipburn of cabbage: Nature and control, *Univ. Wisc. Res. Bull.* 230 (1961).
31. Palzkill, D. A., T. W. Tibbits, and P. H. Williams, Enhancement of calcium transport to inner leaves of cabbage for prevention of tipburn, *J. Am. Soc. Hort. Sci. 101*:645 (1976).
32. Wiebe, H. T., Relationship between plant water status and the occurrence of internal tipburn in white cabbage, *Gartenbau Wissenschaft 40*:134 (1975).
33. Palzkill, D. A., and T. W. Tibbitts, Evidence that root pressure flow is required for calcium transport of head leaves of cabbage, *Plant Physiol. 60*:854 (1977).
34. Walker, J. C., P. H. Williams, and G. S. Pound, Internal tipburn of cabbage: Its control through breeding, *Univ. Wisc. Res. Bull.* 258 (1965).
35. Becker, R. E., Tipburn and other internal disorders, *NY Food Life Sci. Bull.* 7 (1971).
36. Standberg, J. O., J. F. Darby, J. C. Walker, and P. H. Williams, Black speck a non-parasitic disease of cabbage, *Phytopathology 59*:1879 (1969).

37. Walsh, J. R., E. C. Lougheed, and P. M. A. Toivonen, The effect of benomyl, sodium hypochlorite and controlled atmosphere upon the incidence of black speck of stored cabbage, *J. Am. Soc. Hort. Sci. 108*:533 (1983).
38. Cox, F. F., Pepper-spot in white cabbage—a literature review, *A-DAS Q. Rev. 25*:81 (1977).
39. Darby, J. F., R. B. Forbes, and J. C. Walker, Cause and control of black speck of cabbage, *Fla. Agric. Exp. Stn. Ann. Rep.* 223 (1967).
40. Strandberg, J. O., and R. B. Forbes, Cause and control of black speck of cabbage, *Fla. Agric. Exp. Sta. Ann. Rep.* 223 (1967).
41. Cook, R. J., Black speck on cabbage in kent, *Plant Pathol. 25*:181 (1976).
42. Chupp, C. and A. F. Sherf, *Vegetable Diseases and Their Control*, Ronald Press Co., New York, 1960, p. 237.
43. Sherf, A. F., Black rot of crucifer crops, *NY State Agric. Exp. Sta.* (1972).
44. Sherf, A. F., Black-leg of crucifers, *NY State Agric. Exp. Sta.* (1972).
45. Andalaro, J. T., and A. M. Shelton, Onion thrips, *NY State Agric. Exp. Sta. 750*:75 (1983).
46. Thompson, A. K., and W. C. Kelly, *Vegetable Crops*, 5th ed., McGraw-Hill Book Co. Inc., New York, 1957, p. 611.
47. Pantastico, E. B., H. Subramanyan, M. B. Batti, N. Ali, and E. K. Akamine, Postharvest physiology: Harvest indices, in *Postharvest Physiology, Handling and Utilization of Tropical and Sub-tropical Fruits and Vegetables* (E. B. Pantastico, ed.), AVI, Westport, CT, 1975, p. 56.
48. Mack, W. B., R. E. Larson, D. G. White, and H. O. Sampson, *Vegetable and Fruit Growing*, Lippincott Publishing Co., New York, 1956.
49. Ryall, A. L., and W. J. Lipton, *Handling, Transportation and Storage of Fruits and Vegetables*, Vol. I, *Vegetables and Melons*, AVI Pub. Co., Inc., Westport, CT, 1972.
50. Eckert, J. W., P. P. Pubio, A. K. Mattoo, and A. K. Thompson, Postharvest pathology: Diseases of tropical crops and their control, in *Postharvest Physiology, Handling and Utilization of Tropical and Sub-tropical Fruits and Vegetables* (E. B. Pantastico, ed.,) AVI Pub. Co. Inc., Westport, CT, 1975, p. 415.
51. Work, P., and J. Carew, *Vegetable Production and Marketing*, 2nd ed., John Wiley and Sons, New York, 1955.
52. Omary, M. B., R. F. Testin, S. F. Barefoot, and J. W. Rushing, Packaging effects of growth of *Listeria innocua* in shredded cabbage, *J. Food Sci. 58*(3):623 (1993).
53. Holland, B., I. D. Unwin, and D. H. Buss, *Vegetables, Herbs and Spices*, Fifth Supplement to McCance and Widdowson's *The Composition of Foods*, London, HMSO, 1991.
54. Salunkhe, D. K., and K. Salunkhe, The evaluation of the nutritive value and quality in fresh brassicaceous vegetables after harvest, during preparation and subsequent to storage, *Proc. IV. Int. Congress Food Sci. Technol. 5*:274 (1974).
55. Poole, C. F., P. C. Grimball, and M. S. Kanapaux, Factor affecting the ascorbic acid content of cabbage lines, *J. Agric. Res. 68*:325 (1944).
56. Pyke, M., The vitamin content of vegetables, *J. Soc. Chem. Ind. London 61*:149 (1942).
57. Burrell, R. C., H. D. Brown, and V. R. Ebright, Ascorbic acid content of cabbage as influenced by variety, season and soil fertility, *Food Res. 5*:247 (1940).
58. Poole, C. F., P. H. Heinze, J. E. Welch, and P. C. Grimball, Differences in stability of thiamin, riboflavin and ascorbic acid in cabbge varieties, *Proc. Am. Soc. Hort. Sci. 45*:396 (1967).
59. Branion, L. D., J. S. Roberts, C. R. Cameron, and A. M. McCready, The ascorbic acid content of cabbage, *J. Am. Diet Assoc. 24*:101 (1948).
60. Truscott, J. H. L., W. M. Johnstone, T. G. H. Drake, J. R. Van Haarlem, and C. L. Thompson, *A Survey of the Ascorbic Acid Content of Fruits, Vegetables and Some Native Plants Grown in Ontario*, Canada, Department of National Health Welfare, Ottawa, Canada, 1945.
61. Mazza, G., and E. Miniati, Cole crops, in *Anthocyanins in Fruits, Vegetables, and Grains*, CRC Press, Boca Raton, FL, 1993, p. 283.
62. Jensen, K. A., J. Conti, and A. Kjaer, Isothiocyanates II. Volatile isothiocyanate in seeds and roots of various Brassicae, *Acta Chem. Scand. 7*:1267 (1953).

63. Clapp, R. C., L. Long Jr., C. P. Dateo, F. H. Bissett, and T. Hasselstrom, The volatile isothiocyanates in fresh cabbage, *Am. Chem. Soc. J. 81*:6278 (1959).
64. Dateo, G. P., R. C. Clapp, D. A. M. MacKay, E. J. Hewitt, and T. Hasselstrom, Identification of the volatile sulphur components of cooked cabbage and the nature of precursors in the fresh vegetable, *Food Res. 22*:440 (1957).
65. Maruyama, F. T., Identification of dimethyl trisulfide as a major aroma component of cooked *Brassicaceous* vegetables, *J. Food Sci. 35*:540 (1970).
66. MacLeod, A. J., and G. MacLeod, Effects of variations in cooking methods on the flavor volatiles of cabbage, *J. Food Sci. 35*:744 (1970).
67. Hing, F. S., and K. G. Weckel, Some volatile components of cooked rutabaga, *J. Food Sci. 29*:149 (1964).
68. Chin, H.-W., and R. C. Lindsay, Volatile sulphur compounds formed in disrupted tissue of different cabbage cultivars, *J. Food Sci. 58*(4):835 (1993).
69. Van Etten, C. H., M. E. Daxenbichler, W. F. Kwolek, and P. H. Williams, Distribution of glucosinolates in the pith cambial-cortex and leaves of head cabbage (*Brassica oleracea*), *J. Agric. Food Chem. 27*:648 (1979).
70. Greer, M. A., Isolatin from rutabaga seed of progoitrin, the precursor of naturally occurring antithyroid compound, goitrin (1-5-vinyl-2-thiooxazolidone), *Am. Chem. Soc. J. 78*:1260 (1956).
71. Parsons, C. S., Effect of temperature and packaging on the quality of stored cabbage, *Proc. Am. Soc. Hort. Sci. 74*:616 (1959).
72. Isenberg, F. M. R., and R. M. Sayles, Modified atmosphere storage of Danish cabbage, *J. Am. Soc. Hort. Sci. 94*:447 (1969).
73. Suhonen, I., On the storage life of white cabbage in refrigerated stores, *Acta Agric. Scand. 19*:8 (1969).
74. Geeson, J. D., and K. M. Browne, Effect of postharvest fungicide drenches on stored winter white cabbage, *Plant Pathol. 28*:161 (1979).
75. Sulton, J. C., and P. H. Williams, Relation of xylem plugging to black-rot lesion development in cabbage, *Can. J. Bot. 48*:391 (1970).
76. Geeson, J. D., and R. W. Kear, Fungicide dip for cabbage storage, *Hort. Ind.* (Oct.):38 (1978).
77. Salunkhe, D. K., J. Y. Do, E. B. Pantastico, and K. Chachin, Regulation of ripening and senescence: Chemical modifications, in *Postharvest Physiology: Handling and Utilization of Tropical and Sub-tropical Fruits and Vegetables* (E. B. Pantastico, ed.), The AVI Pub. Co. Inc., Westport, CT, 1975, p. 148.
78. Stewart, W. S., Maturity and ripening as influenced by application of plant growth regulators, in *Plant Regulators in Agriculture* (H. B. Turkey, ed.), John Wiley and Sons, New York, 1956, p. 132.
79. Geeson, J. D., and R. W. Kear, Fungicide dip for cabbage storage, *Hort. Ind.* (Oct.):38 (1978).
80. Tirtosoaktjo, M. S., and A. J. Quimio, Control of cabbage soft rot by postharvest dip in borax, *Bul. Penelitian Hort. 7*:31 (1979)
81. Salunkhe, D. K., Gamma radiation effects on fruits and vegetables, *Econ. Bot. 15*(1):28 (1961).
82. Salunkhe, D. K., R. K. Gerber, and L. H. Pollard, Physiological and chemical effects of gamma radation on certain fruits, vegetables and their products, *Proc. Am. Soc. Hort. Sci. 74*:423 (1959).
83. Salunkhe, D. K., L. H. Pollard, and R. K. Gerber, Effects of gamma radiation dose, rate and temperature on the taste preference and storage life of certain fruits, vegetables and their products, *Proc. Am. Soc. Hort., Sci. 74*:414 (1959).
84. Jones, I. D., Effects of processing by fermentation on nturients, in *Nutritional Evaluation of Food Processing*, 2nd ed. (R. S. Harris and E. Karmas, eds.), The AVI Pub. Co. Inc., Westport, CT, 1975, p. 324.
85. Penderson, C. S., and M. M. Albury, Sauerkraut fermentation, *NY State Agric. Exp. Sta. Bull.* 273 (1969).
86. Penderson, C. S., G. L. Mack, and W. L. Athawes, Vitamin C content of sauerkraut, *Food Res, 4*:31 (1939).

87. Penderson, C. S., J. Whitcombe, and W. B. Robinson, The ascorbic acid content of sauerkraut, *Food Technol. 10*:365 (1956).
88. Gangopadhyay, H., and S. Mukherjee, Effect of different salt concentrations on the microflora and physiochemical changes in sauerkraut fermentaion, *J. Food Sci. Technol. 8*:127 (1971).
89. Nieuwhof, M., *Cole Crops: Botany, Cultivation and Utilization*, World Crops Books (N. Polunin, ed.), Leonard Hill, London, 1969.
90. Nabors, W. T., and D. K. Salunkhe, Pre-fermentation inoculation with *Leuconostoc mesenteroides* and *Lactobacillus plantarum* on physiochemical properties of fresh and dehydrated sauerkraut, *Food Technol. 23*:67 (1969).
91. Vorbeck, M. L., L. R. Mattick, F. A. Lee, and C. S. Pederson, Volatile flavor of sauerkraut: Gas chromoatographic identification of volatile acidie off-flavor, *J. Food Sci. 26*:569 (1961).
92. Srisangam, C., D. K. Salunkhe, N. R. Reddy, and G. G. Dull, Effects of blanching, freezing and freeze dehydration on the acceptability and nutrient retention of cabbage (*Brassica oleracea* L.), *J. Food Quality 3*:217 (1980).
93. Srisangam, C., D. K. Salunkhe, N. R. Reddy, and G. G. Dull, Physical, chemical and biochemical modification in processing treatments to improve flavor of blanched cabbage (*Brassica Oleracea* L.), *J. Food Quality 3*:233 (1980).
94. Srisangam, C., D. K. Salunkhe, N. R. Reddy, and G. G. Dull, Effects of blanching on the nutritional quality of cabbage (*Brassica Oleracea* L.) proteins, *J. Food Quality 3*:251 (1980).
95. Hanson, L. P., *Commercial Processing of Vegetables*, Noyes Data Corporation, London, 1975, p. 309.
96. Lee, J. K., Production of Kim Chee (Pickled Nappa), U.S. Patent 3,295,994, January 3, 1967.
97. Salunkhe, D. K., K. I. Bhonsle, V. D. Salunkhe and R. N. Adsule, Anticancer agents of plant origin, *CRC Crit. Rev. Plant Sci. 1*:203 (1983).
98. Cheney, G. S., H. Waxler and I. J. Miller, Vitamin U therapy of peptic ulcer, *California Med. 84*:39 (1956).
99. Steigmann, F., and B. Shulman, The time of healing of gastric ulcers: Implications as to therapy, *Gastroenterology 20*:20 (1952).
100. Strehler, U. E., and K. Hunziker, Behandlung von Magnedarmgeschwuren mit Kohlsaft und Bananenfrapps (Antiulcusfakter, Vitamin U), *Schweiz Med. Wochschr. 84*:198 (1954).
101. Fujiwara, M., Y. Itokawa, H. Uchino, and K. Inoue, Anti-hypercholesterolemic effect of sulfur containing amino acid, *S*-methylcysteine sulfoxide, isolated from cabbage, *Experientia 28*:254 (1972).
102. Itokawa, Y., K. Inoue, S. Sasagawa, and M. Fujiwara, Effect of *S*-methylcysteine sulfoxide, *S*-allylcysteine sulfoxide and related sulfur containing amino acids on lipid metabolism of experimental hypercholesterolemic rats, *J. Nutr. 103*:88 (1973).
103. Petering, H. G., L. Stubberfield, and R. A. Delor, Studies on the guinea pig factor of Wulzen and van Wangtendonk, *Arch. Biochem. Biophy. 18*:487 (1948).
104. Yildiz, F., and D. Westhoff, Associative growth of lactic acid bacteria in cabbage juice, *J. Food Sci. 46*:962 (1981).
105. Liu, J. Y., T. Teraoka, D. Hosokawa, and M. Watanabe, Bacterial multiplication and antibacterial activities in cabbage leaf tissue inoculated with pathogenic and non-pathogenic bacterium, *Ann. Phytopathol. Soc. Japan 52*:669 (1986).
106. Kyung, K. H., and H. P. Fleming, Antibacterial activity of cabbage juice against lactic acid bacteria, *J. Food Sci. 59*:125 (1994).
107. Virtanen, A. I., Some organic sulfur compounds in vegetables and fodder plants and their significance in human nutrition, *Angew. Chem. 6*:299 (1962).
108. Zsolnai, V. T., Die antimikrobielle Wirkung von Thiocyanaten und Isothiocyanaten. 1. Mitteilung, *Arzneim. Forsch. 16*:870 (1966).
109. Van Etten, C. H., M. E. Daxenbichler, P. H. Williams, and W. F. Kwolek, Glucosinolates and derived products in cruciferous vegetables. Analysis of the edible part from 22 varieties of cabbage, *J. Agric. Food Chem. 24*:452 (1976).

110. Van Etten, C. H., M. E. Daxenbichler, H. L. Tookey, W. F. Kwolek, P. H. Williams, and O. C. Yoder, Glucosinolates: Potential toxicants in cabbage cultivars, *J. Am. Soc. Hort. Sci. 105*:710 (1980).
111. Wattenberg, L. W., Inhibition of neoplasia by minor dietary constituents, *Cancer Res. 43*:2448 (1983).
112. Bresnick, E., D. F. Birt, K. Wolterman, M. Wheeler, and R. S. Markin, Reduction in mammary tumorigenesis in the rat by cabbage and cabbage residue, *Carcinogenesis 11*:1159 (1990).
113. Temple, N. J., and T. K. Basu, Selenium and cabbage and colon carcinogenesis in mice, *J. Natl. Cancer Inst. 79*:1131 (1987).
114. Boyd, J. N., J. G. Babish, and G. S. Stoewsand, Modification by beet and cabbage diets of aflatoxin B_1-induced rat plasma α-foetoprotein elevation, hepatic tumorigenesis, and mutagenicity of urine, *Food Chem. Toxicol. 20*:47 (1982).
115. Srisangam, C., D. G. Hendricks, R. P. Sharma, D. K. Salunkhe, and A. W. Mahoney, Effects of dietary cabbage (*Brassica oloeracea* L.) on the turmoigenicity of 1,2-dimethylhydrazine in mice, *J. Food Safety 4*:235 (1980).
116. Graham, S., H. Dayal, M. Swanson, A. Mittleman, and G. Wilkinson, Diet in the epidemiology of cancer of the colon and rectum, *J. Natl. Cancer Inst. 61.*:709 (1978).
117. Haenzel, W., F. B. Locke, and M. Segi, A case-control study of large bowel cancer in Japan, *J. Natl. Cancer Inst. 64*:17 (1980).
118. Pantuck, E. J., K-C. Hsiao, W. D. Loub, L. W. Wattenberg, R. Kuntzman, and A. H. Conney, Stimulatory effect of vegetables on intestinal drug metabolism in the rat, *J. Pharmacol. Exp. Ther. 198*:278 (1976).
119. Pantuck, E. J., C. B. Pantuck, W. A. Garland, B. H. Min, L. W. Wattenberg, K. E. Anderson, A. Kappas, and A. H. Conney, Stimulatory effects of Brussels sprouts and cabbage on human drug metabolism, *Clin. Pharmacol. Ther. 25*:88 (1979).
120. National Research Council, *Diet, Nutrition, and Cancer*, National Academy of Science, Washington, DC, 1982.
121. Beecher, C. W. W., Cancer preventive properties of varieties of *Brassica oleracea*: a review, *Am. J. Clin. Nutr. 59*(Suppl):1166S (1994).
122. Virtanen, A. I., Studies on organic sulfur compounds and other labile substances in plants, *Phytochemistry 4*:207 (1965).
123. Bradfield, C. A., and L. Bjeldanes, Dietary modification of xenobiotic metabolism: Contribution of indolylic compounds present in *Brassica oleracea*, *J. Agric. Food Chem. 35*:896 (1987).
124. Birt, D. F., Anticarcinogenic factors in cruciferous vegetables in *Horticulture and Human Health* (B. Quebedeaux and F. A. Bliss, eds.), Prentice-Hall, Englewood Cliffs, NJ, 1988, p. 160.
125. Yamazaki, M., and T. Nishimura, Induction of neutrophil accumulation by vegetable juice, *Biosci. Biotech. Biochem 56*:150 (1992).
126. Munzer, R., Modifying action of vegetable juice on the mutagenecity of beef extract and nitrosated beef extract, *Food Chem. Toxicol. 24*:847 (1986).
127. Lawson, T., J. Nunnally, B. Walker, E. Bresnick, D. Wheeler, and M. Wheeler, Isolation of compounds with antimutagenic activity from savoy chieftain cabbage, *J. Agric. Food Chem. 37*:1363 (1989).

13

Cauliflower

D. L. MADHAVI
University of Illinois, Urbana, Illinois

S. P. GHOSH
Indian Council of Agricultural Research, New Delhi, India

I. INTRODUCTION

Cauliflower is grown for its tender white head (curd) formed by shortened flower parts. It consists of the repeatedly branched terminal portions of the main axis of the plant. It includes a shoot system with short internodes, branch apices, and bracts. Cauliflower curds are cooked by boiling or split into segments and pickled. They are also used raw in salads. The cultivation of cauliflower is more or less similar to that of cabbage. The world production of cauliflower was 10.88 million metric tons in 1994. The major cauliflower-producing countries are China, India, France, Italy, and the United Kingdom (Table 1).

II. BOTANY

The cauliflower (*Brassica oleracea* type Botrytis) is of European origin, probably developed from broccoli (2,3). It is considered to be the most refined and delicate vegetable crop of the cabbage family. It belongs to the Cruciferae family.

A. Morphology

Cauliflower is a biennial crop, with a short stem that thickens to about the same extent as that of cabbage. The leaves are large, generally oblong, longer and narrower than those of cabbage, with the younger ones being almost sessile (4). The buds usually do not arise in the leaf axils. During transition to the generative phase, which is accomplished in the first year, the peduncles in the axils of the bracts formed by the main growing point branch repeated so that branches of even the fifth order can arise. At first, the numerous peduncles do not grow lengthwise but become thick and fleshy. Thus, a colorless, roughly spherical, terminal compact head arises of which the upper surface consists of a vast number (some millions) of naked apical meristems.

TABLE 1 Estimated World Production of Cauliflower

Continent/Country	Area (1000 ha)	Production (1000 MT)
World	606	10,888
Asia	407	7,704
Europe	148	2,445
North and Central America	28	383
Africa	9	160
Oceania	5	107
South America	5	67
Leading countries		
India	270	4,800
China	88	2,265
France	44	553
Italy	23	443
United Kingdom	26	414
United States	22	294
Spain	14	271
Poland	14	220

Source: Ref. 1.

The peduncles are composed of thin-walled parenchyma and vascular bundles, which do not become woody (5). The young curd is at first entirely covered by the foliage. On becoming visible, it already has a diameter of more than 5 cm. After a time, the flower stems elongate and a number of the apices develop into normal flowers. But no part of flower is apparent in the curd.

B. Cultivars

Cauliflower cultivars differ in their cold requirement for floral intiation leading to curd formation (6). Snowball and Erfurt, available in several types depending upon earliness, plant size, head size, leaf length, and other characteristics, are the most widely grown varieties. Super Snowball is one of the earliest varieties of cauliflower to mature; other similar types like Early Snowball A, Super Junior, Snowcap, Earfurt or Snowball types are mostly late and larger than Super Snowball types, with more rounded, thicker, and heavier curds. Early Snowball, Catskill, Dwarf Erfurt, Snowdrift, White Mountain, and Snowball X, Snowball M, Snowball Imperial, and Improved Holland Erfurt are included in this group. Late Pearl, Cossa, and April are late-maturing varieties (in the spring) preferred in England and California. Early Purple Head carries purple heads and does not require blanching; it is popular in home gardens but is not grown commercially.

Tigchelaar (7) reported that Snowball 42 and Snowball 76 were introduced as new cultivars of cauliflower. The former is an open-pollinated cultivar which has a more concentrated maturity than Snowball M. It has deep white curds, which hold well. Snowball 76 is also open pollinated, having a more concentrated maturity than Snowball X. The Indian Council of Agricultural Research, New Delhi, has recommended Pusa Snowball 1, Pusa Snowball 2, and

Early Kumari for general cultivation in India. Pusa Katki, Pusa Deepali, and Showball 16 are some of the varieties released by the Indian Agricultural Research Institute, New Delhi.

III. PRODUCTION

A. Soil and Climate

Like cabbage, cauliflower is also exacting in its climatic requirements, temperature being the most influential factor (8,9). The best curds are produced in cool and moist weather, particularly when days are short. Cauliflower cannot withstand extremities in temperature as cabbage does. Dry weather and low humidity are harmful and cause ricy, fuzzy, loose, leavy, and yellow curds. It is, therefore, of the utmost importance that a particular variety be planted at the right time, otherwise the crop will produce seeds without forming curds. The optimum monthly average temperature required for growth is 15–20°C with an average maximum of 25°C and a minimum of 8°C. Early varieties require higher temperatures and longer days. Temperatures below the optimum during growth delay maturity and produce undersized, unmarketable, hard heads or button heads. For growth of young seedlings, the initial temperature should be around 23°C and later reduced to 20–17°C. Winter or late cauliflower plants grow more rapidly at 23°C. Hence, they should be grown at temperatures of around 14°C in their young stage. During later stages of growth, the temperature should range from 15 to 20°C. At higher temperatures, excessive leaf production takes place at the cost of curd formation, the generative phase is retarded, harvesting is delayed, but a few curds grow loosely and mature simultaneously. For best quality curd formation, the temperature should be around 10°C. Higher temperatures at this stage lead to excessive vegetative growth at the cost of curd formation, and bracts on flower stalks grow strongly and become clearly visible, thereby impairing the curd quality.

Cauliflower does well on a wide range of soils provided they are fertile, loose, and well drained with adequate moisture retention. Medium to heavy fertile soils are ideally suited for cauliflower production. Sandy loams are good for early crops, while loams and clay loams are ideal for late crops.

On light soils, curds grow loosely with a sudden increase in temperature and the plants are susceptible to damage. In contrast, on heavy soils growth may at first be slow, but if plants have sufficient foliage, top-quality curds can be obtained. However, torrential rains may cause extensive damage. Good drainage is important, particularly for winter cauliflowers, as they grow very rapidly and are not hardy enough to withstand adverse conditions.

Cauliflowers grow best in neutral to slightly acidic soils with pH between 6.0 and 7.0. In alkaline soils, boron becomes unavailable, and borax should be applied at 10–15 kg/ha. In soils with pH values below 5.5, molybdenum becomes unavailable, necessitating the application of sodium molybdate at the rate of 1.0–1.5 kg/ha and lime at 0.5–1.0 ton/ha. Cauliflower plants, particularly the early varieties and summer types, are sensitive to total salt content in the soil.

B. Propagation

Cauliflower is commercially propagated through seeds (10). The seeds are sown in raised nursery beds. Cauliflower seed is slightly bigger than that of cabbage, and it remains viable for about 4 years. Under ideal conditions, 85% seed germination is very common (11). Seedbed preparation, seed treatment, method and depth of sowing, moisture supply, layout, and planting are similar to that of cabbage. Seedlings with three to four true leaves become ready for transplanting within

3–4 weeks of sowing. They can be transplanted until they are 7 weeks old. If the seedlings are allowed to remain in the seedbed too long, they usually produce curds prematurely on transplanting in the field (12).

C. Cultural Practices

1. Planting

Seedlings of early sown crops should be planted on low ridges, otherwise, heavy rains can easily destroy all the tender plants. The seedlings should be hardened, spacing them at 6.5 × 6.5 cm both row to row and plant to plant for final transplanting (12). They should be transplanted in the fields according to the soil, season, and variety.

The seedlings should be transplanted in the bottom or on the side of the furrow according to the soil type in such a way that they will not be flooded by irrigation water. The plants produce heads prematurely when kept too long in seed beds or under moisture stress. Therefore, timely transplanting and keeping the soil moist are very important for a successful cauliflower crop. As in cabbage, gaps resulting from the death of seedlings should be filled in as promptly as possible.

2. Irrigation

As in cabbage, for cauliflower growing to be a success, plant growth should at no time during the growing period be interrupted or hindered. This requires a constant supply of irrigation to ensure optimum soil moisture level (13–16). This can be accomplished by irrigating the soil every 5–6 days in the case of early planting and every 10–12 days in the case of late ones. At the time of head formation, there should be adequate moisture in the field requiring more frequent irrigation during this period.

3. Manuring and Fertilization

Like cabbage, cauliflower is a heavy feeder (17–27). A crop producing 17 tons of curd yield per hectare has been shown to remove from soil about 45 kg of N, 17 kg of P_2O_5, and 56 kg of K_2O. Choudhary (28) recommended that 12–20 tons of farmyard manure (FYM), compost, or sludge, 60 kg of N, 80 kg of P_2O_5, and 40 kg of K_2O be applied before transplanting and again 50–60 kg of N per hectare as top-dressing 6 weeks after transplanting. Cauliflower is a relatively heavy feeder of magnesium, boron, and molybdenum, and because deficiencies of these elements are often observed in normal cultivated soil, unless they are supplemented adequately, the crop will suffer from various physiological disorders. It is recommended to mix 400 kg of magnesium oxide, 23.5 kg of boric acid, and 2.3–4.5 kg of sodium or ammonium molybdate and then evenly spread over the field before transplanting. Molybdenum deficiency can also be corrected by applying lime at the rate of ½ to 1 ton/ha, which increases soil pH. Boron can be supplemented as a foliar spray. Two sprays of 0.3–0.4 boric acid solution, the first about 2 weeks after transplanting followed by another spray 2 weeks before head formation, are adequate to correct the deficiency (29,30). This deficiency can also be corrected by giving plants molybdenum in the seedbed before transplanting at 1 g of molybdate/m^2. Even distribution is ensured by dissolving the molybdate in water, which is then sprayed on plants a few days before transplanting. In severe cases of deficiency, spraying a 0.01–0.1% molybdate solution at the rate of 250 g of molybdate per hectare gives satisfactory recovery (31).

4. Intercultural Operations

Weed growth can be checked and the soil around the plants loosened by two or three shallow tillings; the soil can be easily worked after regular irrigation. Weeding should be started as soon as the plants are set in the field. It should be shallow so as not to disturb the root system. In order to produce solid curds, plants should be earthed up about 5–6 weeks after transplanting.

5. Special Cultural Practices

Certain physiological disorders occur during both the vegetative and reproductive phases of plant development. The most common of these are ricy curd, furry curd, and leafy curd (32).

After maturity if the curds are exposed directly to intense sunlight, they lose some of their flavor and turn yellow, thereby reducing market quality and appearance. Only perfect, milky white or snow white curds fetch a good price. Blanching consists of tying with twine or securing with a rubber band the tips of the outer leaves at the top of the curd. The curds should be left covered this way for a maximum of 4–5 days. This period can be extended up to a week in severe winter or reduced to 2–3 days during hot weather. Self-blanching varieties include Austral (96-D), Early Purple Head, Kalimpong Damia, and Kalimpong Showball (32).

D. Disorders, Diseases, and Pests

Like cabbage, cauliflower also suffers serious losses due to several disorders, diseases, and pests (33–35).

1. Disorders

Thompson and Kelly (3) have described the important physiological disorders of cauliflower such as whiptail, browning, buttoning, and blindness, which cause heavy crop losses. Whiptail results from a deficiency of molybdenum occurring when cauliflower is grown in acidic soil (pH < 5.5) (32). Waring et al. (36) reported that whiptail can be prevented by liming the soil to a pH of 6.5 or by applying one pound of sodium or ammonium molybdate per acre. Molybdenum is not available in very acidic soils, and its availability is increased by reducing the acidity with lime.

Browning

Brown rot or red rot results from a deficiency of boron. The disorder first appears as water-soaked areas in the stem and in the center of the branches of the curd. Later these areas change to a rusty brown color, leading to a brown-rot or red-rot condition. Browning is associated with hollow stems, but hollow stems can occur without browning. Curds affected with browning are bitter. Other symptoms of boron deficiency are changes in foloage color, thickening, brittleness, and downward curling of the older leaves. The application of about 10–15 lb. of common borax (sodium tetraborate) per acre controls browning and other symptoms (37).

Buttoning

This term is used to denote the development of small heads or buttons while the plants are small. It is considered by many workers as premature heading. Carew (38) showed that the curd of normal plants begins to develop as early as that of the button. It has also been reported that nitrogen deficiency may result in buttoning (38). Maintenance of rapid vegetative growth through an adequate supply of nutrients and good control of weeds and diseases and pests, especially the cabbage maggot, controls the disorder to a large extent (35).

Blindness

Cauliflower blindness is characterized by plants without terminal buds and large, dark green, thick, leathery foliage. Axillary buds develop in some cases, but plants do not bear marketable heads. This disorder is very common in England on overwintered cauliflower plants and is believed to be due to the effect of low temperature on the small growing plants. It occurs in the United States in all cauliflower regions to some extent and is due to different causes (35).

2. Diseases

Black leg, caused by *Phoma lingam* (Tode ex Fr.) Desm, is one of the important diseases of cauliflower. The root system of the plant decays from the bottom upward and the affected plant falls over in the field. The infection occurs through seeds. The infection can be prevented by spraying seeds or plants with copper oxychloride or with an organomercuric compound. Downy mildew, caused by *Peronospora parasitica* (Pers. ex. Fr.) Fr., is another disease frequently observed in areas where high humidity prevails. The disease can be controlled by spraying with a 1% Bordeaux mixture of Dithane M-45. Rhizoctonia rot caused by *Rhizoctonia solani* Kuehn, black spot by *Alternaria brassicicola* or *A. brassicae*, yellowing by *Fusarium oxysporum* conglutinans (Wr.) Snyder and Hansen, club root by *Plasmodiophora brassicae* Wor., watery soft rot by *Sclerotinia sclerotiorum* (Lib.) de Bary, powdery mildew by *Erysiphe polygoni* DC, damping-off by *Phythium* sp., black rot by *Xanthomonas campestris* (Pam.), and bacterial soft rot by *Erwinia carotovora* (L. R. Jones) Hollander are other important cauliflower diseases. Chatterjee (8) has described control measures for these diseases. In addition to chemical treatments, some of the diseases can be controlled or prevented by using healthy seeds, resistant cultivars, and improved cultural practices.

3. Pests

Cauliflower caterpillar or large white butterfly (*Pieriq brassicae*) is an important cauliflower pest. It can be controlled by a mixture of insecticides. Aphids are another important pest of cauliflower. For the chemical control of this pest, various nonsystemic as well as systemic insecticides are used (33).

E. Maturity, Harvesting, and Handling

Like cabbage, the cauliflower crop becomes ready for harvest in about 2–4 months depending upon the type and variety. When mature, the curds attain their full size and become compact. Loosening of curds or their breaking into segments indicates overmaturity accompanied by loss of flavor and market appearance. The condition of the curd is a better criterion for deciding the proper stage of harvest than the size of the curd, since stunted plants never produce large curds. A good curd should be compact, regular in shape, globular, firm, and pure white in color.

The size of the curds varies widely with the growing conditions, the largest curd attaining a diameter of more than 30 cm. Early cauliflower types generally produce smaller curds than do late ones. Frequently preference is given to medium-sized curds, which are usually about 15–25 cm in diameter. Those smaller than 10 cm in diameter are not acceptable in the market.

The plants are cut off using a long, sharp sickle well below the curds (heads) leaving the leaves intact for protection during handling. The intact leaves should be cut off 1.25–2.5 cm above the curds. The stubs are left to protect the curd from injury during handling in transit and in the market. As all curds do not mature at the same time, several cuttings may be necessary. The entire field should be inspected every other day to see if the curds are ready for harvesting.

A uniform product fetches a good market price. This is achieved by properly sorting the curds. Off-quality and injured or damaged curds should be sorted out of the harvested lot and discarded. They are grouped according to size, color, and condition or visual appearance.

The leaves left intact while harvesting are trimmed to 1.25–2.5 cm above the top surface of the curd. Produce so prepared should be protected from direct sunlight to prevent them from turning pale yellow to blackish-brown in reaction to ultraviolent rays. In Europe and the United States, the curds are hydro-cooled immediately after harvest to remove field heat (Fig. 1). The harvested and precooled curds are usually packed securely in nets, shallow bamboo baskets,

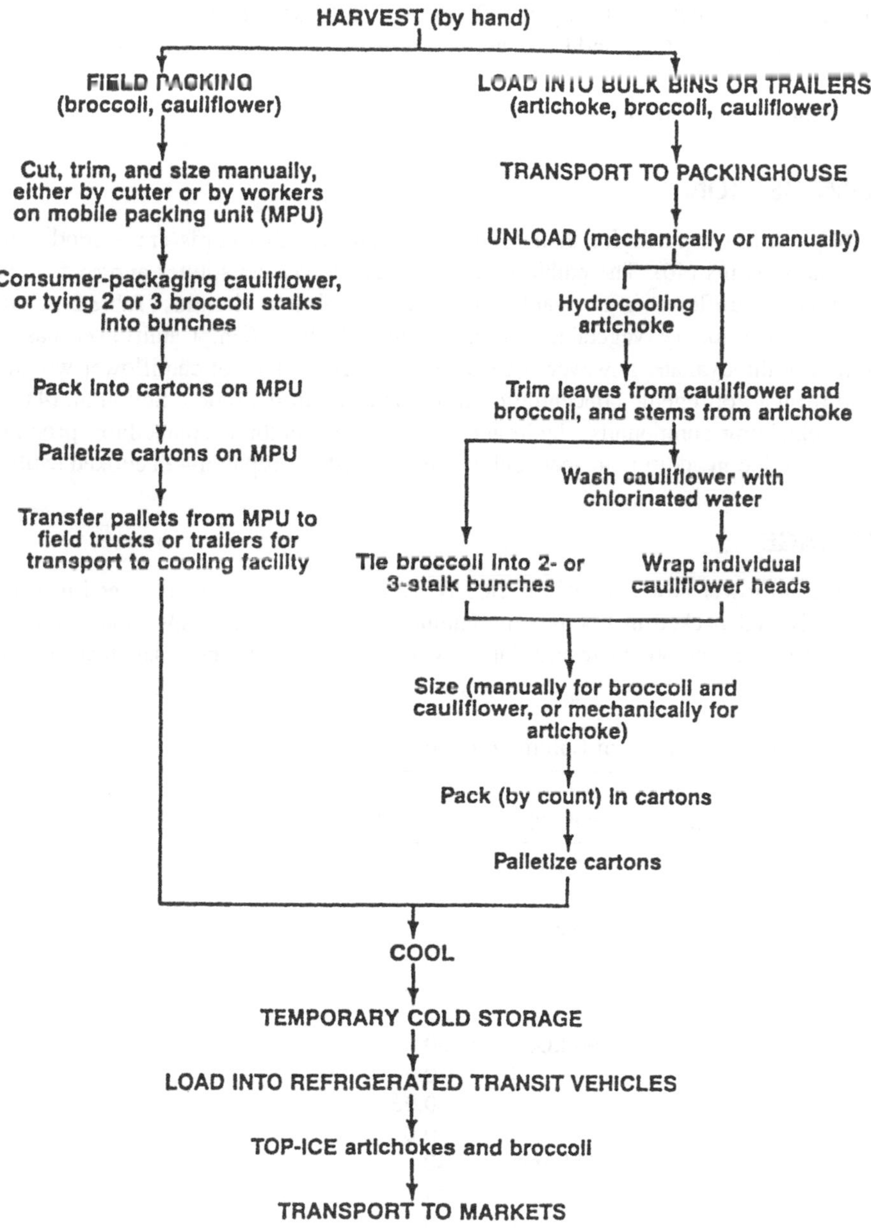

FIGURE 1 Harvest and postharvest operations for floral vegetables (artichoke, broccoli, and cauliflower). (From Ref. 40)

wooden crates, or wire-bound boxes. Tight packing is necessary to prevent displacement and bruising of curds during transit.

F. Packaging and Transportation

Cauliflower sealed in high-density polyethylene (10 μm thick) maintained a fresh flavor longer than that packaged in thicker polyethylene (20 or 30 μm thick), which developed an off-flavor. The concentrations of O_2, CO_2, and C_2H_4 within the sealed packages were 18–20%, 1%, and < 0.1 ppm, respectively (39). The harvested produce must reach the marketplace in a fresh condition as quickly as possible.

IV. COMPOSITION

Analysis of cauliflower curds for their food value indicates that curds are a good source of vitamin B and proteins (40). The edible portion of cauliflower constitutes approximately 45% of the vegetable. Cauliflower is a valuable source of minerals and vitamins (Table 2). Compared to cabbage, it is a luxury vegetable and is consumed less often. Cauliflower has a high concentration of thiocyanate, however, a daily intake of 20 pounds of cauliflower would be required to furnish a goitrogenic concentration of thiocyanate in the blood. Self et al. (41) identified sulfur-containing compounds—hydrogen sulfide, methanethiol, ethanethiol, propanethiol, and dimethyl sulfide in addition to acetaldehyde and 2-methyl propanol—in cooked cauliflower.

V. STORAGE

Most of the cauliflower now marketed is closely trimmed of leaves, prepackaged in perforated film overwards, and packed in fiberboard containers (42). The overwraps should have four to six 5-mm holes to allow adequate ventilation. Cauliflower can be held satisfactorily for 3–4

TABLE 2 Nutritional Analysis of Cauliflower Curd

Contents (per 100 g edible portion)	Ref. 40	Ref. 6
Water	91.7 g	88.4
Protein	2.4 g	3.6
Fat	0.2 g	0.9
Carbohydrates	4.9 g	3.0
Fiber	0.9 g	1.8
Calories	—	142
Vitamin A	90 I.U.	50
Thiamine	—	0.17
Riboflavin	—	0.05
Nicotinic acid	—	0.6
Vitamin C	69.0 mg	43
Phosphorus	—	64
Iron	1.1 mg	0.7
Sodium	16.0 mg	9
Potassium	400.0 mg	380
Calcium	22.0 mg	21
Magnesium	7.0 mg	17
Sulfur	29.0 mg	55

weeks at 0°C. Its storage life is about 15 days at 3°C, 7–10 days at 5°C, 5 days at 10°C, or 3 days at 15°C (43). Slightly immature compact heads keep better than more mature ones. A high relative humidity of at least 95% is necessary to prevent wilting. Van den Berg and Lentz (44) reported that a humidity of 98–100% was satisfactory for cauliflower mainly because it allowed even less weight loss to occur than that at 90–95%. Freezing causes a grayish-brown discoloration and softening of the curd accompanied by water-soaked conditions (45).

A. Low-Temperature Storage

Cauliflower curds cannot ordinarily be stored for long periods. They can barely be kept in good condition for a day or two at room temperature. But good, sound curds with an intact whorl of full leaves can be kept for one month in cold storage at 0°C and 85–95% relative humidity (45). Cauliflower can be hydro-cooled or vacuum-cooled equally successfully (7); the final temperature should be below 7.2°C, preferably below 4.4°C, followed by storage as close to 0°C as feasible.

Stewart and Barger (42) found that cauliflower can be hydro-cooled from 21° to 4.4° C within about 20 minutes if the water is at 1.0°C, and that vacuum-cooling requires about 30 minutes for equivalent cooling if the curds are wet; if dry, they cool to only slightly below 10°C, which is not desirable. Wilting of cauliflower during storage can be avoided by maintaining more than 95% relative humidity; this condition can be reached easily by wrapping each head in perforated film. Ryall and Lipton (7) stated that four to six ¼-inch holes provide adequate ventilation for heads of any size. The storage life of cauliflower heads can be extended for at least one month and possibly 6 weeks under ideal conditions of initial quality, temperature, and humidity (46,47). Usually, however, 2 weeks is the maximum storage time for good final quality. Herregods (48) reported that at 4.4°C, the storage life of cauliflower is about a week to 10 days; at 10°C it is about 5 days; and at 15.4°C it is only 3 days. Pantastico et al. (49) recommended temperatures of 0–1.7°C with 85–95% relative humidity for storing Snowball cauliflowers for about 7 weeks. Damen (50) found that cooling cauliflowers at 2°C for 1 or 3 days did not significantly affect the color of the curds. Cooling for 3 days, however, did reduce weight loss compared with cooling for 1 day or no cooling. Cooling for 3 days also retarded degreening up to 7 days. The storage of cauliflower at low temperature retards (51) loss of vitamin C (Table 3).

B. Controlled-Atmosphere Storage

The reports on controlled-atmosphere (CA) studies for cauliflower are conflicting. Ryall and Lipton (7) stated that the maximum level of CO_2 that should be used at about 0°C is 10%; at higher temperatures (4.4–10°C) even 5% CO_2 causes severe injuries. Such CO_2 injury is evident only after cooking, taking the form of off-flavors and excessively soft curds (52). Low O_2 levels (< 2%) may also produce off-flavors prior to cooking. Aeration completely counteracts CO_2-induced injury, but not low-O_2 damage (53). Lipton and Harris (52) found that curds stored up to 3 weeks at 2.5°, 5°, or 7.5°C in atmospheres with 2, 4, or 6% O_2, respectively, were not

Table 3 L-Ascorbic Acid Content of Cauliflower at Various Times During Storage

	L-Ascorbic acid content (mg/100 g fresh weight)					
Storage temperature (°C)	1 day	3 days	1 week	2 weeks	4 weeks	8 weeks
0	67	62	68	60	60	61
20	67	46	45	34	—	—

Source: Ref. 51.

superior to curds stored in air. The curds stored in 1% O_2 almost always developed off-flavors and off-odors when cooked.

C. Modified-Atmosphere Storage

Bohling and Hansen (54) studied the extent to which modified atmospheres influence the respiration of cauliflower in commercial cauliflower heads without leaves and inflorescences without leaves. They found that inflorescences without leaves exhibited a greater respiration intensity than heads without outer leaves. The respiration-inhibiting effect of the CA, on the other hand, was more pronounced in the inflorescences than in the outer leaves. These studies also confirmed the relatively small effect of an atmosphere in which only the O_2 content was lowered (Table 4).

D. Chemical Control of Postharvest Losses

Treatment of cauliflower with NAA or 2,4-D prevents leaf abscission and increases storage life (53). A combination of one of these chemicals with low-temperature storage resulted in cauliflower remaining in marketable condition for about 6 weeks (46). Stewart and Barger (55) reported that cauliflower wrapped with shredded paper containing 50–100 mg NAA and stored at 0°C had reduced leaf abscission and weight loss (53). The application of 2,4-D, especially if combined with BA, retarded yellowing of cauliflower during storage (53).

VI. PROCESSING

Cauliflower curds are cooked by boiling or are split into segments and used in pickled products (6). Curd segments are also processed by freezing and can be used raw in salads.

A. Freezing

Cauliflower is extensively grown for fresh market and for freezing. Varieties that produce large white heads with relatively smooth surfaces, tender texture, and not too thick floret stalks are desirable for freezing. Varieties suitable for freezing include Early Snowball, Super Snowball, Snowdrift, and St. Valentine.

After harvesting, cauliflower is transported to the processing plant. Rapid transportation from field to freezer is crucial. Cauliflower held for more than a few hours should be kept under refrigeration at 0–1°C and 85% relative humidity. The cauliflower heads are fed to machines that

TABLE 4 Respiration Activity of Cauliflower

Year	Temperature (°C)	Atmosphere	With outer leaves: ml CO_2/kg/hr	With outer leaves: ml O_2/kg/hr	Without outer leaves: ml CO_2/kg/hr	Without outer leaves: ml O_2/kg/hr
1979	0	Air	6.17	6.02	7.43	7.25
	0	3% O_2/rest N_2	4.83	4.23	6.97	8.06
	0	5% O_2/3% CO_2/rest N_2	3.60	5.26	4.68	6.93
1980	3.5	Air	11.28	11.89	14.94	16.1
	3.5	3% CO_2/3% O_2/rest N_2	6.04	6.98	7.68	9.68

Source: Ref. 54.

cut off the base. Then heads are driven to mechanical coring devices, which remove the core and detach the bud cluster in one operation. The curd is then broken into individual florets, which are spected, sorted, and graded. These are subsequently thoroughly washed and blanched for 4–5 minutes in steam, cooled in water, and placed on a conveyor with cold air blowing to cause rapid evaporative cooling. The cooled florets are either IQF-frozen or mechanically packed into cartons to be frozen in a plate freezer or blast freezer (57). The cartons are then sealed, automatically overwrapped with a lithographed wrapper and plate, or air-blast frozen. The frozen product is subsequently used whenever required.

B. Canning

Cauliflower heads are broken apart and handled similar to broccoli. Discoloration of canned cauliflower has been a serious problem. Various recommendations to prevent discoloration include choosing good raw materials, using citric acid and SO_2 to eliminate metal pickup, using lined cans, and avoiding excessive heat treatment (58).

C. Pickling

Cauliflower is often used as a component of mixed pickles and is given the same preservation and processing treatment as cucumber salt stock. Curds are stored for several days in 5–10% brine and afterwards in a 15–16% brine until they are used (59).

VII. THERAPEUTIC PROPERTIES

The therapeutic effects of cauliflower have been documented in folk medicine. The inflorescence extract has been used in the treatment of scurvy, as a blood purifier, and as an antacid (60). The seeds have contraceptive properties (61). Cauliflower extract has been reported effective in the inhibition of initiation and promotion of carcinogenesis in in vitro assays and animal studies. Potential chemopreventive agents include ascorbic acid, carotenoids, tocopherols, isothiocyanates, indoles, and flavonoids.

Koshimizu et al. (62) tested cauliflower extracts for possible anti–tumor-promoting activity in an in vitro short-term assay system of inhbiition of Epstein-Barr virus activation induced by a phorbol-ester promoter, 12-*O*-hexadecanoylphorbol-13-acetate. The methanol extract of cauliflower strongly inhibited the activation, indicating an anti–tumor-promoting activity. Prochaska et al. (63) tested cauliflower extract for its ability to induce xenobiotic metabolizing phase II enzyme quinone reductase in an in vitro assay system using murine hepatoma cells. The acetonitrile extracts of both florets and leaves showed considerable inducing activity, indicating that the extracts may have an inhibitory effect on the initiation stage of carcinogenesis.

Dried cauliflower powder has also been reported to induce cytochrome P-450–dependent monooxygenase activity in mice (64). The indoles indole-3-carbinol, indole-3-acetonitrile, and 3,3′-diindolylmethane are implicated in the induction of these enzymes. Studies by Bradfield and Bjeldanes (65) in rat and mice indicated that these indoles are not the only inducing agents present in cauliflower. They reported the isolation of 1-methoxyindole-3-carbaldehyde, a more potent inducer of the monooxygenase activity than any of the 3-substituted indoles tested. Cauliflower-containing diets inhibited hepatic toxicity induced by food contaminants, polybrominated biphenyls, and aflatoxin B_1 in rats, which could be due to the induction of the detoxification system (66). The effects of dried cauliflower powder on 7,12-dimethylbenz[*a*]anthracene–induced mammary tumorigenesis was determined by Wattenberg (67). When incorporated at 10% level in the diet, cauliflower powder reduced tumor yield to about 30–35% of the control group. Cauliflower extract has been reported to be antimutagenic against various mutagens in the Ames assay (68). The extract has been shown to be active against mutagenesis caused by oxidized linolenic acid, nitrate + methylurea, nitrate + aminopyrene, and nitrate + sorbic acid (68).

REFERENCES

1. *Production Year Book*, Vol. 48, Food and Agriculture Organization, Rome, 1994.
2. Huxley, A. J., M. Grifffiths, and M. Levy (eds.), *The New Royal Horticultural Society, Dictionary of Gardening*, 4th Vol., MacMillan, London, 1992.
3. Swarup, V., and S. S. Chatterjee, Origin and genetic improvement of Indian cauliflower, *Econ. Bot. 26*:381 (1972).
4. Crisp, P., The use of an evolutionary scheme for cauliflowers in screening of genetic resources, *Euphytica 31*:725 (1982).
5. Sidki, S., Morphology of the curd of cauliflowers, *Am. J. Bot. 49*:290 (1962).
6. Fordham, R., Cabbage and related vegetables, in *Encyclopaedia of Food Science, Food Technology and Nutrition* (R. McRae, R. K. Robinson, and M. J. Sandler, eds.), Academic Press, London, 1993, p. 4719.
7. Tigehelaar, E. C., New vegetable varieties List XXI. The Garden Seed Research Committee, American Seed Trade Association, *HortScience 15*(5):565 (1989).
8. Chatterjee, S. S., Cole crops, in *Vegetable Crops in India* (T. K. Bose and M. G. Som, eds.), Naya Prokash, Calcutta, 1986, p. 165.
9. Sidki, S., Factors involvedd in curd and flower formation in cauliflower, *Proc. Am. Soc. Hort. Sci. 90*:252 (1967).
10. Arora, D. N., B. S. Joshi, and S. L. Pandey, Tips for raising the yield of cauliflower, *Indian Hort. 15*:19 (1970).
11. Hari, O., S. P. Tripathi, and R. S. Mishra, Effect of age and seedlings and spacing on the growth, yield and other characteristics of cauliflower variety, Snowball-16, *Prog. Hort. 17*(2):129 (1985).
12. Chatterjee, S. S., and R. R. Sharma, Cultural practices for a food crop of cauliflower, *Indian Hort. 15*:14 (1970).
13. Patel, R. M., and R. P. Jyotishi, Effect of paddy husk used as mulch on cauliflower under two intervals of irrigation and two levels of nitrogen, *Madras Agric. J. 56*:181 (1969).
14. Singh, S. B., and R. S. Mishra, Effect of various mulches on the growth and yield of cauliflower (*Brassica oleracea* var. *Botrytis* L.), *Prog. Hort. 7*:65 (1975).
15. Singh, S. D., Effect of planting configuration on water use and economics of drip irrigation system, *Agron. J. 70*:951 (1978).
16. Singh, U. P., N. B. Syamal, and R. Kumar, Effect of different dates of transplanting on growth and yield of cauliflower (*Brassica oleracea* var. *Botrytis* L.), *Indian J. Hort. 35*:138 (1978).
17. Balean, D. S., B. S. Dhankar, D. S. Rubal, and K. P. Singh, Growth and yield of cauliflower variety Snowball-16, as influenced by nitrogen, phosphorus and zinc, *Haryana J. Hort. Sci. 17*:247 (1988).
18. Gill, H. S., P. C. Thakur, and B. S. Bhullar, Effect of nitrogen and phosphorus fertilization on seed yield of late cauliflower (*Brassica oleracea* var. *Botrytis*), *Indian J. Hort. 32*:94 (1975).
19. Nagda, C., and K. S. Chauhan, Response of cauliflower to nitrogen and sulphur application, *Indian J. Hort. 44*(1-2):62 (1987).
20. Randhawa, K. S., and A. S. Bhail, Growth, yield and quality of cauliflower as influenced by nitrogen, phosphorus and boron, *Indian J. Hort. 33*:83 (1976).
21. Randhawa, K. S., and D. S. Khurana, Effect of nitrogen, phosphorus and potassium fertilization of the yield and quality of cauliflower, *Veg. Sci. 10*:1 (1983).
22. Roy, H. K., Effect of fertilizers on the yield of cauliflower (*Brassica oleracea*), *Indian J. Hort, 7*:1 (1981).
23. Saimbhi, M. S., K. Singh, and D. S. Padda, Influence of nitrogen, phosphorus fertilization on the yield and curd size of cauliflower, *Punjab Hort. J. 9*:198 (1969).

24. Sharma, R. P., and P. N. Arora, Response of mid-season cauliflower to irrigation, nitrogen and age of seedling, *Veg.Sci. 14*:1 (1987).
25. Sharma, R. P., and K. S. Parashar, Response of cauliflower to soil moisture regime, nitrogen and phosphorus levels, *Veg. Sci. 9*:75 (1982).
26. Singh, R. D., S. N. Tiwari, B. B. Singh, and J. N. Seth, Effect of nitrogen and potash on growth and curd yield of cauliflower variety Snowball-16, *Prog. Hort. 7*(4):20 (1976).
27. Yadav, B. R., and K. V. Palieal, Response of cauliflower to nitrogen and phosphorus fertilization on irrigation with saline water, *Veg. Sci. 17*:1 (1990).
28. Chaudury, B., *Vegetables*, 4th ed., National Book Trust, New Delhi, 1976, p. 214.
29. Mehlotra, O. N., N. S. Singh, K. B. Lal, and U. K. Pandey, Foliar fertilization of vegetable crops, *Prog. Hort. 1*:51 (1969).
30. Melrotra, O. N., R. D. L. Srivastava, and H. R. Verma, Response of cauliflower to boron and molybednum, *Prog. Hort. 7*(27):59 (1975).
31. Singh, K. P., and C. B. S. Rajput, Effect of molybdenum on cauliflower (*Brassica oleracea* var. *Botrytis*) in sand nutrient culture, *Exp. Agric. 12*:195 (1976).
32. Thompson, A. K., and W. C. Kelly, *Vegetable Crops*, 5th ed., McGraw-Hill Book Co. Inc., New York, 1957, p. 611.
33. Ramsey, G. B., and M. A. Smith, *Market Diseases of Cabbage, Cauliflower, Turnips, Cucumbers, Melons and Related Crops*, USDA Agric. Marketing Hand Book No. 184, 1961.
34. Ryall, A. L., and W. J. Lipton, *Handling, Transportation and Storage of Fruits, and Vegetables*, Vol. I, *Vegetables and Melons*, AVI Pub. Co. Inc., Westport, CT, 1972.
35. Lee, S. H., and R. L. Carolus, Foliar abscission of stored cauliflower and cabbage to the effects of certain growth regulating substances, *Michigan Agric. Exp. Sta. Tech. Bull.* 216 (1949).
36. Waring, E. J., R. D. Wilson, and N. S. Shirlow, Whip tail of cauliflower, *Agric. Gaz. New South Wales 60*:21 (1949).
37. Purvis, E. R., and W. J. Hanna, Vegetable crops affected by boron deficiency in Eastern Virginia, *Virginia Truck Expt. Sta. Bull.* 105 (1940).
38. Carew, H. J., A study of certain factors affecting buttoning of cauliflower, Cornell University Thesis, 1947.
39. Ben-Ychoshua, S., I Kobiter, and B. Sharpio, The effect of seal packaging in plastic films on deterioration of vegetables, *Hassadeh 61*:63 (1980).
40. Salunkhe, D. K., and B. B. Desai, *Postharvest Biotechnology of Vegetables*, Vol. I, CRC Press, Boca Raton, FL, 1984, p. 157.
41. Self, R., J. C. Casey, and T. Swain, The low boiling volatiles of cooked foods, *Chem Ind.* (London) (21):863 (1963).
42. Stewart, J. K., and W. R. Barger, Effect of cooling method on the quality of asparagus and cauliflower, *Proc. Am. Soc. Hort. Sci. 78*:295 (1961).
43. Herregods, M., The storage of cauliflower, Tuinbouwberichten 28:486 (1971).
44. Vanden Berg, L., and C. P. Lentz, High humidity storage of vegetables and fruits. *HortScience 13*:565 (1978).
45. Platenius, H., F. S. Jamison, and H. C. Thompson, Studies on cold storage of vegetables, *NY Agric. Exp. Sta. Bull.*, 602 (1934).
46. Hruscha, H. W., and J. Kaufman, Storage tests with Long Island cauliflower to inhibit leaf abscission by using plant growth regulators, *Proc. Am. Soc. Hort. Sci. 63*:409 (1949).
47. Karch, G., The practical use of polyethylene film during cold storage of cauliflower, *Deut. Gartenh. 12*:214 (1965).
48. Herregods, M., The storage of cauliflower, *Tuinbouwberichten 28*:486 (1964).
49. Pantastico, E. B., T. K. Chattopadhyay, and H. Subramanyam, Storage and commercial storage operations, in *Postharvest Physiology, Handling and Utilization of Tropical and Sub-Tropical Fruits and Vegetables* (E. B. Pantastico, ed.), The AVI Pub. Co. Inc., Westport, CT, 1975, p 314.
50. Damen, P. M. M., Is the cooling of cauliflower worthwhile? *Groenten on Fruit 36*(18):49 (1980).
51. Persson, A. R., Quality studies in cauliflower, *Acta Hort. 93*:443 (1979).

52. Lipton, W. J., and C. M. Harris, Response of stored cauliflower (*B. oleracea Botritis* group) to low O_2 atmosphere, *J. Am. Soc. Hort. Sci. 101*:208 (1976).
53. Hattoon, T. T., E. B. Pantastico, Jr., and K. K. Akamine, Controlled atmopshere storage, Part 3, Individual commodity requirements, in *Postharvest Physiology Handling and Utilization of Tropical and Sub-tropical Fruits and Vegetables* (E. B. Pantasico, ed.), The AVI Pub. Co. Inc., Westport, CT, 1975, p. 201.
54. Kaufman, J., and S. M. Ringel, Tests of growth regulators to retard yellowing and abscission of cauliflower, *Proc. Am. Soc. Hort. Sci. 78*:349 (1961).
55. Bohling, H., and H. Hansen, Influence of controlled atmosphere on the respiration activity of achlorophyllus vegetables, *Proc. Symp. on Postharvest Handling of Vegetables, Acta. Hortic. 116*:165 (1980).
56. Stewart, J. K., and W. R. Barger, Effects of cooling method on quality of asparagus and cauliflower, *Proc. Am. Soc. Hort. Sci. 78*:295 (1961).
57. Luh, B. S., B. Feinberg, and J. J. Meehan, Freezing preservation of vegetables, In *Commercial Vegetable Processing* (B. S. Luh and J. G. Woodroof, eds.), AVI Pub. Co. Inc., Westport, CT, 1982, p. 317.
58. Chandler, B. V., Discoloration in processed cauliflower, *Food Res. Quart. 24*:11 (1964).
59. Pederson, C. S., Pickles and Sauerkraut in *Commercial Vegetable Processing* (B. S. Luh and J. G. Woodroof, eds.), AVI Pub. Co. Inc., Westport, CT, 1982, p. 457.
60. Liebstein, A. M., Therapeutic effects of various food articles, *Am. Med. 33*:33 (1927).
61. El-Dean Mahmoud, A. A. G., *Study of Indigenous (Folkways) Birth Control Methods in Alexandria*, MS thesis, University of Alexandria Higher Institute of Nursing, Alexandria, Egypt, 1972.
62. Koshimizu, K., H. Ohigashi, H. Tokuda, A. Kondo, and K. Yamaguchi, Screening of edible plants against possible antitumor promoting activity, Cancer Lett. *39*:247 (1988).
63. Prochaska, H. J., A. B. Santamaria, and P. Talalay, Rapid detection of inducers of enzymes that protect against carcinogens, *Proc. Natl. Acad. Sci. 89*:2394 (1992).
64. Bradfield, C. A., Y. Chang, and L. Bjeldanes, Effects of commonly consumed vegetables on hepatic xenobiotic-metabolizing enzymes in the mouse, *Food Chem. Toxicol. 23*:899 (1985).
65. Bradfield, C. A., and L. Bjeldanes, Dietary modification of xenobiotic metabolism: Contribution of indolylic compounds present in *Brassica oleracea*, *J. Agric. Food Chem. 35*:896 (1987).
66. Stoewsand, G. S., J. B. Babish, and H. C. Wimberly, Inhibition of hepatic toxicities from polybrominated biphenyls and aflatoxin B_1 in rats fed cauliflower, *J. Environ. Pathol. Toxicol. 2*:399 (1978).
67. Wattenberg, L. W., Inhibition of neoplasa by minor dietary constituents, *Cancer Res. 43*:2448s (1983).
68. Beecher, C. W. W., Cancer preventive properties of varieties of *Brassica oleracea*: A review, *Am. J. Clin. Nutr. 59* (Suppl):1166S (1994).

14

Broccoli

N. Rangavajhyala and V. M. Ghorpade
University of Nebraska, Lincoln, Nebraska

S. S. Kadam
Mahatma Phule Agricultural University, Rahuri, India

I. INTRODUCTION

Broccoli (*Brassica oleracea* var. *Italica*) is a compact, rapidly developing floral vegetable that is usually harvested when the flowering heads are immature. The name broccoli, therefore, refers to the young shoots that develop in spring on some species of the genus *Brassica* (*brocco* is Italian for shoot). The sprouting broccoli (*Brassica oleracea* var. *Italica*) is believed to be the ancestor of the present quick-growing cauliflowers (1). The usual types of sprouting broccoli first develop a central head of green color, and a number of smaller axillary stalks bear smaller heads, which are sold in bunches. The sprouting broccoli is a vegetable of the highest quality (2). Broccoli is a popular vegetable of commercial importance in the United States (3). It was determined from an acceptibility poll that broccoli and cauliflower are America's favorite vegetables. More women rated broccoli as their favorite vegetable than men (4). Broccoli was one of the top vegetables exported from the United States in 1995 (5). Broccoli and asparagus accounted for about 90% of $102.6 million in exported U.S. vegetables sent to Japan in 1994, and broccoli imports into Japan for that year from the United States were projected to amount to 100,000 t, worth about $200 million (6). In 1989, 60% of the frozen broccoli consumed in the United States was imported. A large part of the U.S. produce goes to the freezing industry. The top broccoli-producing states in 1992 were California, Arizona, and Texas (7). Broccoli is gradually becoming a popular vegetable in other parts of the world.

II. BOTANY

Sprouting broccoli is sometimes called broccoli, although this name is also used to refer to broccoli raab (the sprouts that develop on turnips) and for winter cauliflower (winter broccoli or heading broccoli). It is best to confine the name broccoli to the appropriate sprouting types of *Brassica oleracea* (1). In the United States, it is also known as Italian broccoli, pointing to its

Italian origin, Calabrese, referring to the district of Calabria in southern Italy, was an initially widely used varietal name.

A. Morphology

The broccoli plant forms a kind of head consisting of green buds and thick fleshy flower stalks. The terminal head is rather loose and green in color, and the flower stalks are longer than those of cauliflower. The internodes are long, and the plant produces axillary flowering shoots in addition to the terminal inflorescence. The color of the flower buds, forming the edible portion, varies from white to green depending on cultivar. The inflorescence is made of fully differentiated flower buds, although it is considered to be a good quality fruit if these reach the open stage (8). The sprouts in the leaf axils develop strongly, especially after removal of the terminal head. Both the terminal head and the sprouts with bud clusters are consumed as food.

B. Cultivars

Italian Green Sprouting or Calabrese are the most popular varieties of broccoli. Strains of Italian Green Sprouting—Midway, Green Mountain, and Grand Central—are of the early- the medium-maturing type, which are dark bluish-green and produce large compact heads and many lateral shoots after the central head is cut (3). The cultivar De Cicco is light green in color and produces abundant lateral shoots; it is the main winter broccoli grown in Texas. Texas 107 is similar to De Cicco but matures earlier and more uniformly; it also tends to turn yellow sooner.

A popular and widely adapted variety is Waltham 29, which is late maturing with uniform maturity and plant characteristics. Its color is dark bluish-green, and it produces large compact heads with smaller flowers buds and a moderate number of large lateral shoots. Purple and green variants occur, although in the United States only the green types are grown. The purple types are fairly hardy and can be harvested during the winter. They are grown to a limited extent in England. Cultivars like De Cicco, Green Bud, and Spartan Early are early cultivars that can be harvested 6–7 weeks after transplanting. However, these cultivars may not perform well under unfavorable conditions, especially lower temperatures. Cultivars like Waltham 29, Green Mountain, and Coastal Atlantic are less sensitive to buttoning and can be grown as early or late crops. Medium strains like Green Sprouting Medium take about 100 days to maturity, while late strains like Green Sprouting Late are biennials and may be harvested at the onset of winter or after its end. Several P1 hybrids including Southern Comet, Premium Crop, Clipper, Laser (extra early and early), Corsair, Excalibur, Cruiser Emerald Corona (midseason) and Late Corona, Stiff, Kayak, and Green Surf are marketed by seed companies in Japan, the United States, and Europe (9).

III. PRODUCTION

A. Soil and Climate

The soil should be sufficiently fertile for the rapid development of broccoli seedings after transporting. A good moisture supply is also necessary.

Broccoli is somewhat sensitive to temperature. In warm weather, bud clusters become loose quickly. Some cultivars of broccoli are to a certain extent resistant to frost depending on the size of the plant. These are generally biennial, while the annual cultivars are somewhat sentitive to frost. In such cultivars, even a light frost causes considerable damage to the buds (7).

B. Cultural Practices

Broccoli is generally sown on seedbeds like other cole crops. Seeds are drilled directly to the soil in the United States. Spacing is maintained as 12 × 30 inches between both plants and rows. Crops grown in such close spacing yield more although the heads are smaller and mature slightly later than when wider spacing is followed. In the United States, broccoli crops are grown mainly for harvest in autumn and spring. Crops are sown in June-July, and seedlings are planted out from July to September. Cultivars are chosen on the basis of period and intensity of frost (10).

C. Maturity and Harvesting

Broccoli is harvested shortly before the buds begin to open while their heads remain compact and cut to a length of about 8–10 inches. The axillary shoots, as they develop, are cut at the same stage as the central head. Thompson and Kelly (8) observed that the central head or cluster, with 8–10 inches of stem, should be harvested before the flower buds open. Lateral shoots develop in the axils of the leaves into marketable heads 1–3 inches in diameter, producing a continuous harvest for several weeks.

Harvested broccoli heads are further trimmed to a length of approximately 6 inches, and bunches weighing about 1½ pounds are tied together for packaging. The bunches are further packaged in crates holding up to 28 bunches. Since fresh broccoli continues to develop and the flowers begin to open or turn yellow, the bunches are hydrocooled to 4.4°C, packed with ice, and sotred under refrigeration.

D. Pests, Diseases, and Physiological Disorders

1. Pests

Cabbage Maggot

The North American cabbage maggot, which originated in Europe in the nineteenth century, is considered to be a serious pest of cruciferous crops such as canola, cabbage, radishes, cauliflower, and broccoli. The adult form of this insect looks similar to a common housefly, with a dark stripe along the back of the abdomen. Maggots are the damaging life stage of this pest, which are known to move distances of at least a few kilometers in search of host plants. Cabbage maggots are more likely to be a problem in cool areas and in winter and spring crops. Infestation occurs from the roots, particularly of seedlings, causing the crop to appear pale green and stunted and to wilt on hot and dry days (12). In addition to root damage, the plant becomes vulnerable to diseases as pathogens enter through the lesions created by maggots. Reeve beetles (ground beetles) are considered a natural enemy of cabbage enemy. They eat the eggs and parasitize the pupae. Satisfactory control of cabbage maggot may be obtained by broadcast furrow treatment at the time of planting or by drenching at transplanting. Covering the seed beds with a thin cloth offers additional protection from these flies as well as from flea beetles.

Aphid

Aphids feed on most crops and also infest cole crops. Presence of aphids in the heads of broccoli or cauliflower makes a crop unmarketable. Aphids damage plants by feeding on plant sap, spreading viruses, and excreting a sticky "honeydew" that coats the plant. Sooty molds grow in this honeydew, reducing photosynthesis and making the produce unattractive. Several natural parasites and predators of aphids can offer protection to the crop, unless they are killed by insecticides. Use of insecticidal soap and inundative release of chrysoperia carnea to control aphids on broccoli have been unsuccessful (13). Aphid infestation should be treated when 1–2% of the crop is found infested, and the treatments should be repeated if the aphid infestation reappears.

Diamond-back Moth

The very small caterpillars of this insect attack practically all the cruciferae, eating from the underside of the leaves and producing holes all over the leaves. Moths are about ⅓ in. long, with the folded wing flaring outward and upward toward their tips and, in the male, forming a row of three diamond-shaped yellow spots where they meet down the middle of the back. The hind wings have a fringe of long hairs (14). The small larvae wriggle actively when touched. This insect is notorious for resistance to all types of insecticides. The adult can be monitored with pheremone traps. Infestation can be controlled by employing resistance cultivars, destroying crop debris, and using chemical insecticides, which should be rotated to decrease the opportunity for resistance to develop.

2. Diseases

Black Rot

Black rot infection occurs near the margins of the leaves and is caused by the bacterium *Xanthomonas campestris* pv. *campestris.* This microorganism may attack cruciferous plants at all stages of growth.

Fusarium Yellow

Caused by the fungus *Fusarium oxysporium* f.sp. *conflutinans*, this disease is manifested by a yellowish-green coloration of the foliage. The fungus, which lives in the soil for a number of years without a host plant, can infect plants through rootlets and wounds caused by transplanting or insects.

Downy Mildew

This fungus (*Peronospora parasitica*) is noticeable on the underside of leaves as a gray, fluffy, downy growth in well-defined spots. It is difficult to control this fungus during periods of high humidity and cool temperatures.

Soft Rot

The bacterium *Erwinia corotovora* causes a slimy decay with foul odor to develop during storage and transit. Soft rot also commonly occurs when fields become water saturated.

3. Physiological Disorders

Internal Tipburn

Tipburn causes leaf margins to turn brown and leaves to be burried in the head. This has been ascribed to poor water movement within the plant.

Broccoli Heat Injury

In many parts of the world, high temperatures cause the heads of broccoli to be rough, with uneven bud sizes. Several parameters for screening of heat-sensitive varieties have been developed (15).

E. Integrated Pest Management

Integrated pest management (IPM) is a set of practices designed to avoid economic losses from pests while at the same time minimizing use of pesticides and their possible detrimental effects (13). Rotation with cover crops, grains, and vegetables from different families is the most effective way to prevent insects, diseases and weeds, The greater the difference between plants in rotation, the more effective they are in suppressing pest populations. Planting disease-free seed and seed that has been treated with hot water to eliminate seedborne bacteria helps. In case of fungal infection, fungicide application or planting healthy transplants are the most effective means of control. Also, keeping fields and surrounding areas free of weed hosts will help reduce some viral diseases.

Long distance transportation of horticultural commodities offers the opportunity to control common insect pests by prolonged exposure to low temperature in combination with insecticidal atmosphere. Moderate to high CO_2 treatments were studied in packaged broccoli for postharvest insect control (16). The data indicated that under moderate CO_2 conditions, thrips (*Thysanoptera*) could be killed in 7 days whereas aphid mortality could be achieved in 14 days.

IV. CHEMICAL COMPOSITION

A. Composition of Broccoli

The chemical composition of broccoli (Table 1) resembles that of cauliflower. Several flavor compounds have been shown to be present in broccoli. The concentration of ethanol, 3-hydroxy-2-butanone, and 2,3-butanediol increased and that of C5-C7 aldehydes and alcohols decreased in cooked broccoli that had been stored fresh under nitrogen (17).

B. Glucosinolates

Glucosinolates (thioglucosides) are sulfur-containing compounds present in *Brassica* vegetables (Table 2). A generalized structure of glucosinolate is shown in Figure 1. The glycosyl component of β-D-glucopyranose and all glucosinolates probably has the anti configuration with respect to the sulfate and R groups (18). The glucosinolates are anions and occur in plants mostly as potassium salts (19). Breakdown products of glucosinolates possess important sensory properties such as odor and flavor, and they may induce physiological changes in humans, including carcinogenesis inhibition and goiter formation (20). At high intake levels, certain glucosinolates are associated with toxic effects, especially goiter development (21).

Glucosinolates are readily hydrolyzed under moist conditions by the coexisting endogenous enzyme myrosinase (thioglucoside glucohydrolase, EC 3.2.3.1). The products of enzyme decomposition of glucosinolates are β-D-glucose, sulfate, and an organic aglucon moiety. Depending on conditions, such as the proteins present, pH, trace metals, etc., the aglucon can undergo intramolecular rearrangement and/or fragmentation yielding products such as thiocyanates, isothiocyanates, nitriles, cyanides, and ozazolidine-2-thiones (19,22). Water blanching of broccoli resulted in a significantly higher reduction in major glucosinolates when compared to steam blanching (Table 3) (23). This was ascribed to increased leaching of glucosinolates during water blanching.

The glucosinolate content of purple-headed broccoli was found to be in the range of 72–212 mg/100 g (24). Goodrich et al. (25) observed a number of quantitative changes in specific glucosinolate levels in broccoli harvested in municipal sewage sludge-amended soil, whereas the total glucosinolate content remained unchanged. Glucosinolate concentrations of

TABLE 1 Nutritional Value and Chemical Composition of Cauliflower and Green Broccoli (Calabrese)[a]

Constituent	Cauliflower	Green broccoli (Calabrese)
Water (g)	88.4	88.2
Total nitrogen (g)	0.58	0.71
Protein (g)	3.6	4.4
Fat (g)	0.9	0.9
Carbohydrate (g)	3.0	1.8
Energy value (kJ)	142	138
Starch (g)	0.4	0.1
Total sugars (g)	2.5	1.5
Dietary fiber (g) (Englyst method)	1.8	2.6
Sodium (mg)	9	8
Potassium (mg)	380	370
Calcium (mg)	21	56
Magnesium (mg)	17	22
Phosphorus (mg)	64	87
Iron (mg)	0.7	1.7
Cooper (mg)	0.03	0.02
Zinc (mg)	0.6	0.6
Sulfur (mg)	55	130
Chloride (mg)	28	100
Manganese (mg)	0.3	0.2
Selenium (μg)	Tr	Tr
Iodine (μg)	Tr	2
Carotene (μg)	50	575
Vitamin D (μg)	0	0
Vitamin E (mg)	0.22	(1.3)[b]
Thiamine (mg)	0.17	0.10
Riboflavin (mg)	0.05	0.06
Niacin (mg)	0.6	0.9
Vitamin B_6 (mg)	0.28	0.14
Vitamin B_{12}	0	0
Folate (μg)	66	90
Pantothenate (mg)	0.60	N
Biotin (μg)	1.5	N
Vitamin C (mg)	43	87

[a]Values based on 100 g raw flesh.
[b]Value in parentheses is estimated form other related data.
N = No analysis available, Tr = trace.
Source: Ref. 3.

broccoli with reference to varying environmental conditions at three Ontario sites were investigated by Shelp et al. (26). Their results indicated that boron treatments had no consistent influence on the glucosinolate concentrations over a period of 2 years. Further, the differences observed in total glucosinolate concentrations of each cultivar among the growing sites were greater than differences between cultivars at the same site.

Table 2 Glucosinolates Present in Brassica Vegetables: Structures and Trivial Nomenclature

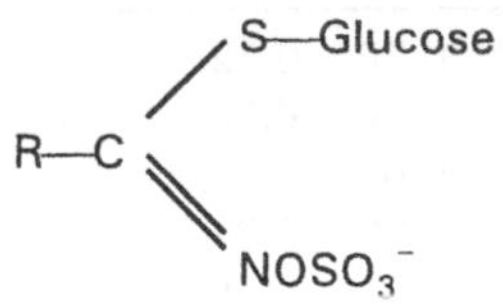

Glucosinolate skeleton

	R	Trivial name
I	Prop-2enyl	Sinigrin
II	2-Hydroxybut-3-enyl	Progoitrin
III	2-Hydroxypent-4-enyl	Gluconapoleiferin
IV	3-Methylthiopropyl	Glucoiberverin
V	3-Methylthiobutyl	Glucoerucin
VI	3-Methylsulfinylpropyl	Glucoiberin
VII	4-Methylsulfinylbutyl	Glucoraphanin
VIII	2-Phenethyl	Gluconasturtiin
IX	Indolyl-3-methyl	Glucobrassicin
X	4-Hydroxyindolyl-3-methyl	4-Hydroxyglucobrassicin
XI	2-Methoxylndolyl-3-methyl	4-Methoxyglucobrassicin
XII	1-Methoxyindolyl-3-methyl	Neoglucobrassicin

Source: Ref. 24.

C. Therapeutic Properties of Broccoli

1. Anticarcinogenic Properties

The National Research Council Committee on Diet, Nutrition, and Cancer has recommended increased consumption of vegetables of the *Brassica* genus as a measure to decrease the incidence of human cancer (27). The American Cancer Society suggested inclusion of crucifer-

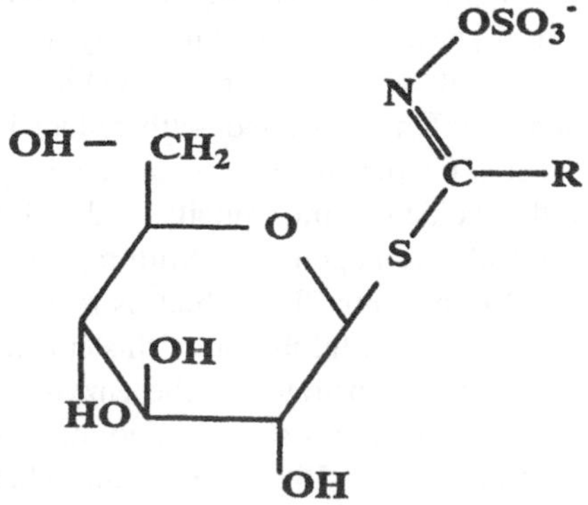

Figure 1 Structure of glucosinolate.

TABLE 3 Effects of Blanching Treatments on Glucosinolate Contents

Glucosinolate[2]	Unblanched (mg/g)	Water blanched (mg/g)	Steam blanched (mg/g)
Glucoiberin[3]	1.27 ± 0.01^{a1}	0.25 ± 0.01^{c}	0.70 ± 0.12^{b}
Progoitrin[4]	0.06 ± 0.00^{a}	0.09 ± 0.01^{a}	0.36 ± 0.08^{a}
Sinigrin[5]	0.05 ± 0.00^{a}	0.0^{b}	0.0^{b}
Glucoraphanin[6]	4.45 ± 0.07^{a}	0.91 ± 0.08^{c}	2.68 ± 0.21^{b}
Glucosinalbin[7]	0.0^{b}	0.08 ± 0.01^{b}	0.38 ± 0.09^{a}
Gluconapin[8]	0.0^{a}	0.0^{a}	0.17 ± 0.17^{a}
Glucobrassicanapin[9]	0.11 ± 0.00^{b}	0.0b	0.0^{b}
Glucobrassicin[10]	4.96 ± 0.07^{a}	0.60 ± 0.34^{c}	2.43 ± 0.50^{b}
Neoglucobrassicin[11]	0.70 ± 0.04^{a}	0.05 ± 0.01^{b}	0.24 ± 0.07^{b}
Total	11.59 ± 0.12^{a}	1.97 ± 0.30^{c}	6.97 ± 1.25^{b}

[1]Mean values are on a freeze-dried basis. Different letter superscripts indicate significant ($p \leq 0.05$) differences of each glucosinolate between treatments.
[2]Trivial names.
Glucosinolate side chains are: [3]3-methylsulfinylpropyl; [4]2-hydroxybut-3-enyl; [5]prop-2-enyl; [6]4-methylsulfinylbutyl; [7]*p*-hydroxybenzyl; [8]but-3-enyl; [9]pent-4-enyl; [10]3-indolylmethyl; [11]1-methoxy-3-indolylmethyl.
Source: Ref. 23.

ous vegetables in the diet to reduce the risk of developing cancer (28). Epidemiological evidence (29) and animal experiments (30,31) indicate that these vegetables possess cancer-inhibiting properties due to the presence of nutritive and nonnutritive constituents known to inhibit chemically induced carcinogens in animals.

Wattenberg (31) suggested that tumorigenesis is inhibited by carcinogen-metabolizing systems induced by compounds in *Brassica* plants. Sulfur-containing phytochemicals of two different kinds are present in broccoli and other cruciferous vegetables: glucosinolates and S-metholcysteine sulfoxide. Numerous studies have indicated that the hydrolytic products of some of the glucosinolates have anticarcinogenic activity (32).

Although the cancer-preventive properties of broccoli seem evident, the mechanism by which the course of carcinogenesis is altered has not been identified. It has been suggested that feeding of vegetables induces enzymes of xenobiotic metabolism and thereby accelerates the metabolic disposal of xenobiotics. Sulforaphane [1-isothiocyanato-4-(methyl-sulfinyl)butane] isolated from broccoli was found to be a potent inducer of phase 2 detoxification enzymes (e.g., glutathione transferase, NAD(P)H:quinone reductase, UDP-glucuronosyltransferases) in murine hepatoma cells in culture. Phase 2 enzyme induction is often associated with reduced susceptability of animals and their cells to the toxic and neoplastic effects of carcinogens. Administration of sulforaphane significantly reduced the incidence, multiplicity, and weight of mammary tumors in rats when challenged with a chemical carcinogen (33). Sulforaphane was shown to inhibit the phase 1 enzyme cytochrome P-450 isoenzyme 2E1, which is responsible for activation of several carcinogens, including diakylnitrosamines. Inhibition of this enzyme by sulforaphane may offer chemoprotection against carcinogenic substrates of the enzyme (34). Indole-3-carbinol, a metabolite of the glucosinolate glucobrassicin, has shown inhibitory effects in studies of human breast and ovarian cancers (32). Chen et al. (35) observed an enhanced activity of colon and deodenum mucosal glutathione levels in a dose-dependent manner in rats fed lyophilized broccoli and challenged with 1,2-dimethyl hydrazine—a chemical carcinogen.

Extract of broccoli has been shown to exert an antimutagenic effect on several mutagens, namely, ethidium bromide, oxidized linoleic acid, etc. (36).

2. Other Nutritional Properties

Cruciferous vegetables contain high concentrations of carotenoids, which are believed to be chemopreventive and associated with a decreased risk for various human cancers in epidemiological studies. In a human feeding trial, participants consuming broccoli and cauliflower had the highest plasma carotenoid concentrations. Hence it was suggested that plasma carotenes may be useful biomarkers of carotenoid-rich food intake (37). Goodman et al. (38) studied the effect of various vegetables in the diet on the survival of lung cancer patients and observed that increased consumtion of broccoli, particularly among women, appeared to improve survival in lung cancer.

Broccoli may also play a role in reducing levels of serum cholesterol. Broccoli was reported to contain about 3.5 g/kg of D-glucaric acid. Purified diets containing calcium D-glucarate or potassium hydrogen D-glucarate markedly lowered serum levels of cholesterol in female Sprague-Dawley rats (39). It has been suggested that dietary calcium requirements can be met in part by the intake of broccoli, among other vegetables (40). Intake of cruciferous vegetables during breast feeding has, however, been associated with colic symptoms in young infants (41).

V. STORAGE AND PACKAGING

A. Storage

The respiration rate of freshly harvested broccoli is very high—comparable to that of asparagus, spinach, or sweet corn. Therefore, brocccoli must be cooled immediately after harvest to rapidly lower the respiration rate and it must be kept at low temperature for maximum shelf life (42).

Italian or sprouting broccoli is highly perishable but is usually stored for only a brief period as needed (43). Broccoli should be hydrocooled or packed in ice immediately after harvest and kept at 0°C to maintain good salable condition, fresh green color, and vitamin C content. If in good condition and stored with adequate air circulation and spacing between containers to avoid heating, broccoli should keep satisfactorily for 10–14 days at 10°C. Longer storage is undesirable because leaves may discolor, buds become yellow and drop off, and tissues soften (43,44).

Film wraps can be beneficial in maintaining high relative humidity and extending storage life (45–54). Such wraps should be perforated or be sufficiently permeable. Otherwise, an atmosphere that causes injury and/or off-odors, particularly at 0°C, may develop (52). Broccoli should not be stored with other fruits such as apples or pears that produce substantial quantities of ethylene, because this gas accelerates yellowing of the buds.

King and Morris (48) studied postharvest senescence of broccoli immediately after harvest at 20°C. Changes in respiration (Fig. 2), ethylene production, and color were determined for florets, branchlets, and heads of three cultivars of field-harvested broccoli: Green Beauty, Dominator, and Shogun. Carbon dioxide produced from heads of container-grown broccoli and from heads, branchlets, and florets of field-harvested broccoli decreased markedly during the first 12 hours of postharvest storage before stability. The maximum storage life at 20°C was 72 hours. Tian et al. (50) reported that the reproductive structures stamens and pistil may have a role in determining the rate of sepal degreening since removing them from florets reduced the yellowing rate (Table 4). Propylene stimulated respiration and ehtylene production and accelerated yellowing (Table 5).

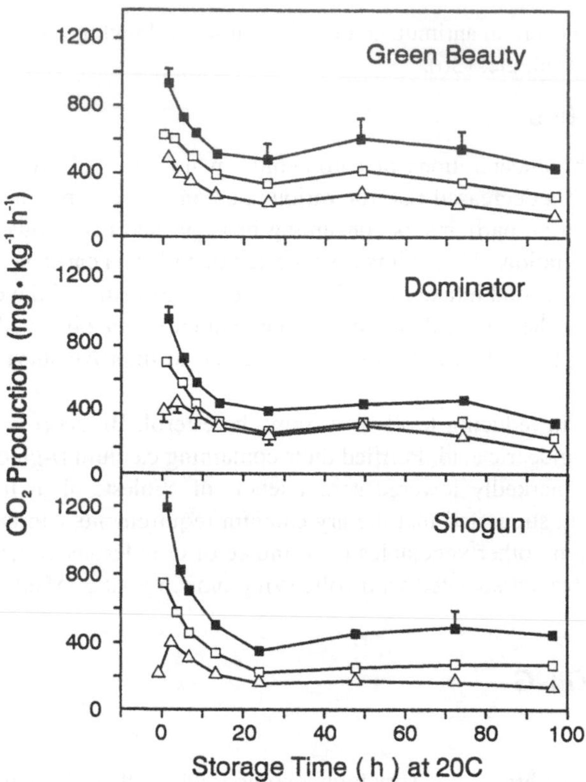

FIGURE 2 Changes in CO_2 production of three broccoli cultivars during storage at 20°C. (■) Florets, (□) branchlets, (▲) heads. Bars are SES and are contained within symbol when not shown. (From Ref. 68).

1. Low-Temperature Storage

The yellowing of broccoli, which generally occurs in 3 days at room temperature as a result of the production of ethylene, is the main problem in its prepackaging. Lieberman and Hardenberg (54) reported that yellowing can be delayed by storing broccoli in an oxygen-free atmosphere. Flushing the packages with oxygen-free gas retarded yellowing appreciably, suggesting that ethylene accumulation in the package may be one of the causes of rapid yellowing.

The important factors of postharvest spoilage of broccoli such as production of ethylene (55) and consequent yellowing can be controlled by refrigeration of packaged broccoli. Thompson and Kelly (8) observed that packaging does not eliminate the need for refrigeration but may

TABLE 4 Chlorophyll Content of Broccoli Sepals from Florets With or Without Stamens and Pistils

Days at 20°C	Floret treatment	Total chlorophyll (μg/g)
0	Intact	677.9
3	Intact	237.7
3	Without stamens and pistils	331.0

Source: Ref. 50.

TABLE 5 Chlorophyll Content of Florets of Harvested Broccoli Head Treated with Propylene (0.5%) for 3 Days at 20°C in the Dark

Days in dark at 20°C	Treatment	Total Chlorophyll (μg/g)
0	—	908
3	In air	305
3	In C_3H_6	90

Source: Ref. 50.

actually accentuate the need for refrigeration to keep broccoli cold constantly. Further, Leberman et al. (56) found that the green color and tenderness of broccoli can be retained and mold growth retarded by holding broccoli at 5–20% CO_2.

2. Controlled-Atmosphere Storage

Broccoli is a highly perishable product whose visual and organoleptic qualities greatly depend on the storage conditions (57–59). The more commonly encountered storage disorders are yellowing of inflorescence, opening of flower buds, toughening of stems, development of undesirable odors, soft rot, and mold (57,58). A controlled atmosphere (CA) of 10% CO_2 and 1% oxygen can increase the shelf life of good-quality broccoli held above 5°C (45). An atmosphere with 10% CO_2 retards yellowing and toughening, whereas one with 15% CO_2 has the same retarding effect but can induce persistent off-odors. A 1% oxygen atmosphere retards yellowing, but 0.1–0.25% oxygen can cause severe injury and result in off-odors and off-flavors in cooked broccoli (52). Makhlouf et al. (60) assessed the effect of long-term storage of broccoli under CA storage. Broccoli was stored for 6 weeks at 1°C under N_2 containing different percentages of CO_2 and O_2: chlorophyll and color retention was better under CA than in air (Fig. 3). This was mainly due to increased CO_2 concentration. The storage of broccoli also delayed the development of soft rot and mold. However, after 6 weeks of storage under an atmosphere containing 10% or more CO_2, the rate of respiration increased simultaneously with the development of undesirable odors and physiological injury. Of the atmospheres tested, 6% CO_2 and 2.5% O_2 was the best for long-term (> 3 weeks) maintenance of broccoli quality while avoiding physiological injury.

3. Modified-Atmosphere Storage

The modified-atmosphere (MA) packaging of broccoli to delay senescence has met with some success, and modified atmosphere has been shown to be potentially useful (60–62). Forney and Rij (62) studied the effects of temperature of broccoli florets at the time of packaging on package atmosphere and quality of broccoli. Warm (20°C) broccoli had 40–50% more CO_2 and 25–30% less O_2 than cold (3°C) broccoli packages 6 hours after sealing (Table 6). Warm florets were rated lower for color turgidity and general appearance than initially cold ones after 7 days at 5°C (Table 7). Objectionable off-odors were formed from florets held in TPM 87 packages that had an O_2 concentration of 1.5%. Broccoli has been traditionally cooled with package ice. In recent years, there has been an increasing market for broccoli florets packaged without ice. The temperature of broccoli florets at the time of packaging also influences the quality of stored broccoli, affecting the respiration rate and therefore the rate of atmosphere modification in the package.

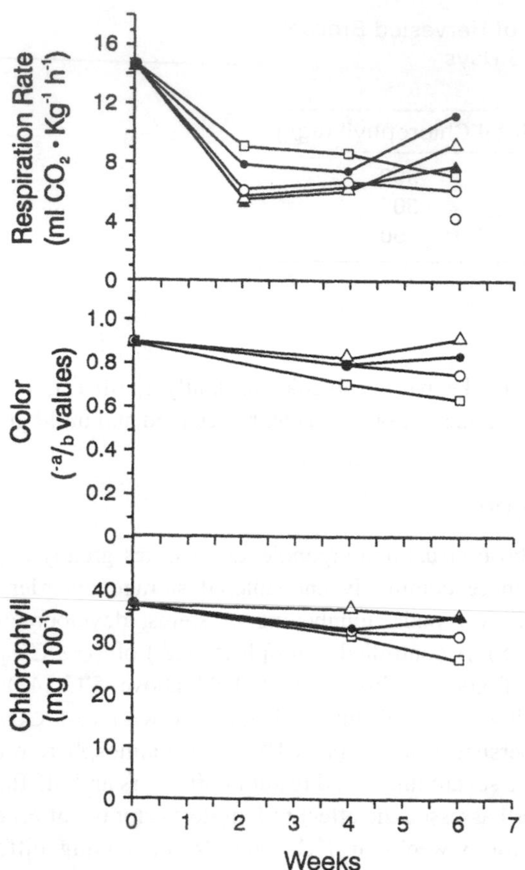

FIGURE 3 Respiration rate of broccoli (m CO_2/kg/hr), color (-a/b values), and chlorophyll content (mg/1100 g fresh weight) of broccoli during storage at 1°C under the following CO_2/O_2 combinations: 0%/20% CO_2 (□); 6%/2.5% (○); 10%/20% (■); 10%/2.5% (▲); 15%/2.5% (△). (From Ref. 48).

B. Packaging

Prepackaged broccoli florets are popular in the United States. Postharvest applications of cytokinins has been shown to delay senescence of whole stalks of broccoli (63–68). Compared to whole stalks, broccoli florets appear to represent a different set of physiological conditions because of the large relative increase in cut surface. Due to wounding, plant tissues generally exhibit an increase in the rates of both respiration and ethylene production (65). Rushing (68) observed broccoli florets treated postharvest with either benzyladenine or zeatin at 10 or 50 ppm before packaging in perforated polyethylene bags and storage at 16°C compared to controls. The application of benzyladenine at 50 ppm reduced the respiration rate 50% (Fig. 4) and ethylene production increased 40% throughout the first 4 days of storage. The total chlorophyll content had dropped 60% in controls but was unchanged in cytokinin-treated florets, which had a 90% longer shelf life than controls.

TABLE 6 Changes in Atmosphere Composition in Packages of Broccoli Florets Over Time as Affected by Package Film Type and Temperature at Time of Packaging

		Time after sealing (hr)					
		6		48		168	
Film type	Temperature (°C)	O_2	CO_2	O_2	CO_2	O_2	CO_2
TPM 87	20	11.1	5.9	2.3	14.6	3.0	15.2
	3	15.7	3.9	3.8	14.1	4.6	14.4
RMF 61	20	12.0	4.8	5.0	6.0	6.7	5.0
	3	16.0	3.4	10.1	6.2	10.7	5.0
Significance							
Film		NS	**	*	***	**	***
Temp		***	***	*	NS	NS	NS
F×T		NS	NS	NS	NS	NS	NS

NS, *, **, *** Nonsignificant or significant at $p = 0.05$, 0.01, or 0.001, respectively, according to F-test.
Source: Ref. 62.

Polyunsaturated fatty acid (PUFA) degradation by lipoxygenase may contribute to postharvest senescence of vegetables and fruits. Both lipid peroxidation and hydroperoxides promote losses of membrane integrity, protein, and chlorophyll, resulting in yellowing (69). Zhuang et al. (70) reported that modified-atmospheric packaging resulted in increased chlorophyll and C-18 polyunsaturated fatty acids in broccoli florets by 96 hours. In comparison, reduction in

TABLE 7 Effect of Film Type and Temperature at Time of Packaging on Quality of Broccoli Florets After 7 Days at 5°C[a]

		Rating				
Film type	Temperature (°C)	Odor	Decay	Color	Turgidity	General appearance
TPM 87	20	7.8	8.6	7.8	7.3	7.3
	3	7.6	8.8	9.0	8.7	8.8
RMF 61	20	8.8	8.3	7.5	7.4	7.0
	3	8.8	8.8	8.8	9.0	8.4
Significance						
Film		*	NS	NS	NS	NS
Temp		NS	NS	**	**	**
F×T		NS	NS	NS	NS	NS

[a]Values represent the average ratings of the means of three replications.
Rating scale: odor—9 = normal, 1 = nauseating; decay—9 = none, 1 = severe; color—9 = dark green, 1 = yellow; turgidity—9 = turgid, 1 = dry; general appearance—9 = excellent, 1 = very poor.
NS, *, ** Nonsignificant or significant at $p = 0.001$ or 0.001, respectively, according to F-test.
Source: Ref. 62.

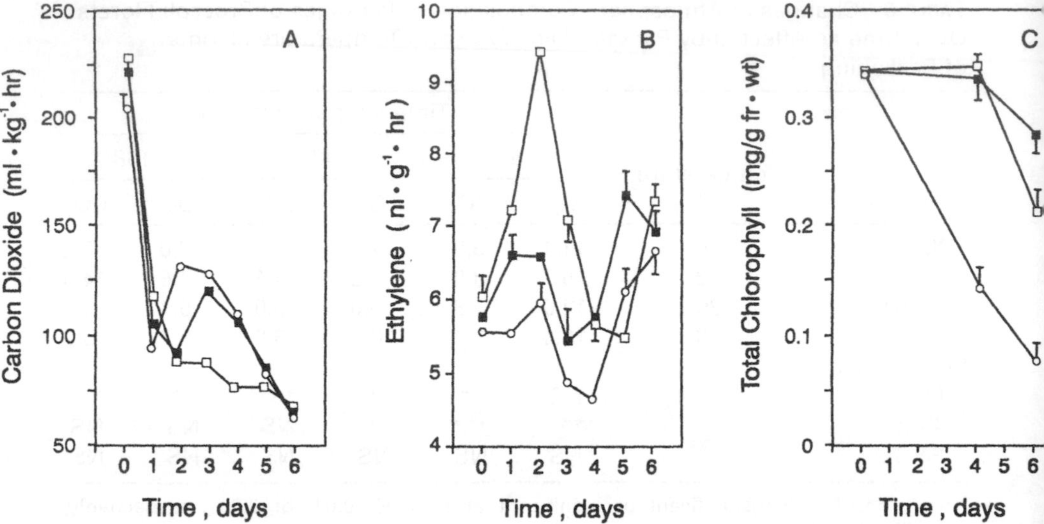

FIGURE 4 Effect of benzyladenine (BA) and zeatin treatments on (A) respiration rate, (ethylene production, and (C) total chlorophyll content of broccoli florets at 16°C. Each point the graph represents the mean of six measurements, and the experiment was conducted twi with similar results each time. Bars represent SE and, when absent, fall under the symbol (C control; □= 50 ppm BA; ■ = 50 ppm zeatin). (From Ref. 60.)

chlorophyll, PUFA, and soluble protein contents were observed in nonpackaged samples. Barth et al. (71) have found that packaging of broccoli spears resulted in greater ascorbic acid retention, providing better vitamin C content for the consumer, and greater chlorophyll retention, contributing to greener appearance. Packaging of broccoli with polymeric films has been therefore found acceptable for postharvest storage.

Microflora changes in MA-packaged broccoli florets stored at refrigerated temperatures were investigated (72). The level of O_2 in the packages decreased while that of CO_2 increased after 94 hours of storage. Although similar growth trends in aerobic plate count, coliform, yeast/mold counts were reported between the packaged and nonpackaged broccoli samples, the packaged samples had significantly lower microbial counts than the nonpackaged samples.

VI. PROCESSING

In the United States sprouting broccoli is frozen on a large scale. After thorough washing, broccoli sprouts are blanched for 3–5 minutes until the catalase and peroxidases are inactivated. This is followed by cooling to 10–15°C to prevent color and flavor losses. The sprouts are then packed into cartons, frozen, and stored at -20°C (73). Consumer purchases are based primarily on external characteristics such as visual appearance and texture. An important component of visual appearance is color, and loss of green color is a major limiting factor in the shelf life of broccoli (74). Color can be stabilized by blanching (75).

TABLE 8 Effects of Chemicals on Hunter Color Values, Shear Press, and Composition of Steam-Blanched Broccoli

Chemical treatment	L	a	b	Shear press	Chlorophyll (mg/100 g)	Ascorbic acid (mg/100 g)
Control	–35.5[a]	–12.1[a]	16.2	560[b]	454.3[a]	80.5[a]
EDTA	36.1[b]	–12.8[b]	16.5	581[c]	452.1[a]	81.6[b]
Malic acid	3.56[a]	–12.0[a]	16.2	556[b]	450.2[a]	82.5[c]
$NaHCO_3$	37.2[c]	–13.8[c]	16.0	542[a]	460.5[b]	81.1[b]
NaCl	36.4[b]	–14.0[c]	16.4	573[c]	466.1[c]	83.0[c]

Means of 4 determinations (2 reps. × 2 samples).
Means with different superscripts are significantly different ($p \leq 0.05$) by LSD.
Source: Ref. 76.

Salunkhe et al. (76) evaluated the effects of blanching, freezing, and freeze-dehydration on the acceptability and nutrient retention of broccoli. The samples were blanched with steam and chemical solution (or water) for 3 minutes at 96°C. In steam blanching, broccoli was immersed in chemical solution for 5 minutes under 25 in. vacuum and then exposed to steam for 3 minutes. The total chlorophyll content of steam-blanched broccoli was higher with NaCl and $NaHCO_3$ treatments but lower with EDTA and malic acid treatments as compared to controls (Table 8). In water-blanched broccoli, all the chemical treatments resulted in higher chlorophyll content than in controls. Taste panel results showed that the color of steam-blanched broccoli treated with NaCl was liked best; the ascorbic acid content of the water-blanched product generally was lower than that of the steam-blanched product (Tables 8, 9). Sodium chloride and malic acid treatment seemed to retain ascorbic acid better when compared with controls. A distinct preference for broccoli cooked by nonconventional method (stir-fry) with incorporation of nucleotide flavor potentiator over one cooked by conventional method without nucleotide flavor potentiator was expressed by taste panelists

TABLE 9 Effects of Chemicals on Hunter Color Values, Shear Press, and Composition of Water-Blanched Broccoli

Chemical treatment	L	a	b	Shear press	Chlorophyll (mg/100 g)	Ascorbic acid (mg/100 g)
Control	35.2[a]	–12.0[a]	16.0	540[b]	446[a]	75.1[a]
EDTA	35.8[b]	–12.1[a]	15.9	545[b]	450[b]	79.1[b]
Malic acid	35.4[a]	–13.5[b]	16.2	551[c]	455[c]	80.0[c]
$NaHCO_3$	36.1[b]	–13.8[b]	16.3	528[a]	458[c]	78.5[b]
NaCl	36.0[b]	–13.6[b]	16.2	545[b]	451[b]	80.1[c]

Means of 4 determinations (2 reps. × 2 samples).
Means with different superscripts are significantly different ($p \leq 0.05$) by LSD.
Source: Ref. 76.

TABLE 10 Effects of Cooking Methods and Nucleotide Flavor Enhancers on Acceptability of Cabbage and Broccoli

	Flavor		Color		Texture		Overall	
Cooking methods	Cabbage	Broccoli	Cabbage	Broccoli	Cabbage	Broccoli	Cabbage	Broccoli
Conventional without nucleotide	5.0[a]	4.8[a]	5.1[a]	5.0[a]	6.2[a]	6.6[a]	5.5[a]	6.0[a]
Conventional with nucleotide	6.7[b]	6.9[b]	5.1[a]	5.2[a]	6.2[a]	6.6[a]	6.2[b]	6.5[b]
Stir-fry with nucleotide	8.0[c]	8.2[c]	7.0[b]	8.5[b]	8.2[b]	8.4[b]	7.8[c]	8.5[c]

Means of 20 judgments (10 judges × 2 reps.).
Means with different superscripts are significantly different ($p \leq 0.05$) by LSD.
Source: Ref. 76.

(Table 10). This method tends to minimize discoloration, softening, and strong flavor development, resulting in cooked broccoli with an attractive green color and firm texture without undesirable strong flavor. Incorporation of the nucleotide flavor potentiator not only brought out the natural flavor of broccoli but also masked the undesirable strong flavor caused by sulfurous compounds.

Blanching of vegetables before processing reduces microbial load, improves color, aids in package fill, and assists in the development of desirable texture. However, blanching broccoli spears in water resulted in a loss of 8–9% solids, compared with approximately 2% solid loss when blanched with steam (77). Blanching may also affect levels of ascorbic acid, one of the most heat-labile nutrients. In a recent study (78), microwave-blanched broccoli was found to have retained the greatest amount of reduced ascorbic acid and had an appearance, color, and texture score equivalent to steam-blanched broccoli. Application of systems eliminating leaching and oxidation will decrease the loss of water-soluble vitamins. Processing of broccoli florets was reported to retain the highest concentrations of vitamins (Table 11) and resulted in better acceptability than boiled broccoli (79). Processing can lead to reduced folate in food, either by degradation or by leaching into processing effluents (80). Folate, being water soluble, is lost during operations such as blanching and canning. DeSouza and Eitenmiller (81) observed a retention of 91% of total folate activity (TFA) in steam-blanched broccoli. Steam blanching led to higher TFA retention than water blanching. A significant loss was observed during the canning process, while no significant loss occurred during canned or frozen storage.

Carotenoids in fruits and vegetables have been considered stable to cooking procedures. Martin et al. (82) reported near or complete retention of carotene with microwave or conventional heating in broccoli. Park (83) investigated the effects of freezing, thawing, cooking, and drying on carotene retention of carrots, broccoli, and spinach. Dehydration, regardless of drying method, significantly reduced carotene in these vegetables. The carotene contents of fresh, vacuum, and microwave dried broccoli samples were 459, 43, and 325 μg/g, respectively (83).

The woody stem parts of broccoli spears are generally considered unfit for consumption, and usually they are trimmed off and discarded. This portion amounts to nearly one third of the weight of broccoli purchased from the market. The waste could be changed to profit by proper utilization of this unwanted vegetable portion. An instant soup ingredient–type product has been developed (76). The woody stem section of broccoli was first skinned, then diced into 5-mm cubes, steam-blanched for 3 minutes, vacuum-cooled in a solution of 0.2% nucleotide flavor potentiator (monosodium glutamate, 5' inosinate, and 5' guanylate in the proportion of 95:2.5:2.5) and 0.5% NaCl to achieve simultaneous impregnations, quick-frozen and freeze-dried and vacuum-packed. The product is an excellent ingredient for soup as well as salad. The healthful uses for broccoli are many (84,85). Broccoli has been used to prepare packaged dinners in combination with other vegetables or meat for infants and young children (86).

TABLE 11 Theoretical 5-Minute Losses After Different Heat Treatments of Broccoli

Treatment	Ascorbic acid (%)	Vitamin B_6, total (%)	Folate (%)
Boiling	36	55	39
Steaming	0	17	12
Sous vide processing	0	3[a]	3[a]

Values are calculated assuming first-order kinetics during the first part of the heat treatment and assuming a constant temperature of 100°C.
[a]The 5-minute loss is based on nonsignificantly different values.
Source: Ref. 79.

REFERENCES

1. Gray, A. R., Taxonomy and evolution of broccoli, *Econ. Bot. 36*:397 (1982).
2. Watts, R. L., and G. S. Watts, *The Vegetable Growing Business*, Orange Judd Publishing, New York 1954.
3. Fordham, R., Broccoli-type brassica, in *Encyclopaedia of Food Science, Food Technology, and Nutrition* (C. R. McCrae, R. K. Robinson, and M. J. Sandler, eds.), Academic Press, London, 1993, p. 4722.
4. American School Food Service Association, Survey shows which vegetables Americans prefer to eat. *School Food Serv J 40*(4):31 (1986).
5. World Horticultural Trade and U.S. Export Opportunities, No. 1:9 (1996).
6. World Horticultural Trade and U.S. Export Opportunities, No. 10:18 (1994).
7. *Vegetables and Specialties: Situation and Outlook Report* (1987-1993), USDA Economic Research Service, Quarterly, U.S. Dept. of Commerce, Bureau of Census, Washington, DC.

7a. Peet, M., *Crop Profiles: Sustainable Practices for Vegetable Production in the South*, Focus Publishing, Newburyport, MA, 1996.

8. Thompson, H. K., and W. C. Kelly, *Vegetable Crops*, 5th ed., McGraw-Hill Book Co. Inc., New York, 1957, p. 611.
9. Salunkhe, D. K., and B. B. Desai, *Postharvest Biotechnology of Vegetables*, Vol. I, CRC Press Inc., Boca Raton, FL, 1984, p. 171.
10. Nieuwhof, M., Cole crops: botany, cultivation, and utilization, in *World Crops* (N. Polunin, ed.), Leonard Hill, London, 1969.
11. Work, P., and J. Carew, *Vegetable Production and Marketing*, 2nd ed., John Wiley and Sons, New York, 1955.
12. Cooperative extension service, *Illinois Pest Control Handbook*, University of Illinois at Urbana–Champaign Urbana, IL, 1991.
13. Peet, M., *Integrated Pest Management: Sustainable Practices for Vegetable Production in the South*, Focus Publishing, Newburyport, MA, 1996.
14. Illinois Agr. Exp. Sta., Urbana, IL, Cir. 437, 1939.
15. Thomas, B. J., and K. Pearson, The heat-sensitive stage of broccoli flower development, Meeting of the American Society for Horticultural Sciences, Montreal, Quebec, 1995.
16. Cantwell, M. R., M. S. reid, A. Carpenter, X. Nie, L. Kushwaha, R. Serwatowski, and R. Brrok, Short-term and long-term high carbon dioxide treatments for insect disinfestation of flowers and leafy vegetables, in *Harvest and Postharvest Technologies for Fresh Fruits and Vegetables*, Proceedings of the International Conference, Guanajuato, Mexico, 1995, p. 287.
17. Hansen, M., C. E. Olsen, L. Poll, and M. I. Cantwell, Volatile constituents and sensory quality of cooked broccoli florets after aerobic and anaerobic storage, *Acta. Hort. 343*:105 (1993).
18. Kjaer, A., Glucosinolates in the *Cruciferae*, in *The Biology and Chemistry of the Cruciferae* (J. G. Vaughn, A. J. MacLeod, and B. G. M. Jones, eds.), Academic Press, London, 1976.
19. Tookey, H. L., C. H. Van Etten, and M. E. Daxenbichler, Glucosinolates, in *Toxic Constituents of Plant Foods*, (2nd ed.) (I. E. Liener, ed.), Academic Press, New York, 1980.
20. Fenwick, G. R., R. W. Heaney, and W. J. Mullin, Glucosinolates and their breakdown products in food and food plants, *CRC Crit. Rev. Food Sci. Nutr. 18:123* (1983).
21. Nishie, K., and M. E. Daxenbichler, Toxicology of glucosinolates, related compounds (nitriles, R-goitrin, isothiocyanates), and vitamin U found in *Cruciferae*, *Food Chem. Toxicol. 18*:159 (1980).
22. Van Etten, C. H., and M. E. Daxenbichler, Glucosinolates and derived products in cruciferous vegetables. Gas liquid chromatographic determination of the aglucone derivatives from cabbage, *J. Assoc. Off. Anal. Chem. 60*:950 (1977).
23. Goodrich, R. M., J. L. Anderson, and G. S. Stoewsand, Glucosinolate changes in blanched broccoli and Brussels sprouts, *J. Food Proc. Preserv. 13*:275 (1989).
24. Lewis, J. A., G. R. Fenwick, and A. R. Gray, Glucosinolates in Brassica vegetables: Green-curded cauliflowers (*Brassica oleracea* L. Botrytis group) and purple-headed broccoli (*B. oleracea* L. Italica group), *Lebensm. Wiss. 24*:361 (1991).

25. Goodrich, R. M., R. S. Parker, J. L. Donald, and G. S. Stoewsand, Glucosinolate, carotene and cadmium content of *Brassica oleracea* grown on municipal sewage sludge, *Food Chem 27*:141 (1988).
26. Shelp, B. J., L. Liu, and D. McLellan, Glucosinolate composition of broccoli (*Brassica oleracea* var. *Italica*) grown under various boron treatments at three Ontario sites, *Can. J. Plant Sci. 73*:885 (1993).
27. National Research Council, Inhibitors of carcinogenesis, in *Diet, Nutrition, and Cancer*, National Academy Press, Washington, DC, 1982, p. 358.
28. Lindsay, A., D. J. Fink, and A. Lindsay, *The American Cancer Society Cookbook*, Hearst Books, New York, 1988, p. 269.
29. Graham, S., Results of case-controlled studies of diet and cancer in Buffalo, New York, *Cancer Res. 43*:2409s (1983).
30. Steowsand, G. S., J. B. Babish, and H. C. Wimberly, Inhibition of hepatic toxicities from polybrominated biphenyls and aflatoxin B1 in rats fed cauliflower, *J. Environ. Pathol. Toxicol. 2*:399 (1978).
31. Wattenberg, L. W., Inhibition of neoplasia by minor dietary constituents, *Cancer Res. 43*:2448s (1983).
32. Stoewsand, G. S., Bioactive organosulfur phytochemicals in *Brassica oleracea* vegetables—a review, *Food Chem. Toxicol. 33*:537 (1995).
33. Zhang, Y., T. W. Kensler, C. G. Cho, G. H. Posner, and P. Talalay, Anticarcinogenic activities of sulforaphane and structurally related synthetic norbornyl isothiocyanates, *Proc. Natl. Acad. Sci. USA 91*:3147 (1994).
34. Barcelo, S., J. M. Gardiner, A. Gescher, and J. K. Chipman, CYP2E1-mediated mechanism of anti-genotoxocityc of the broccoli constituent sulforaphane, *Carcinogenesis 17*:277 (1996).
35. Chen, M. F., Chen, L. T., and H. W. Boyce, Jr., Cruciferous vegetables and glutathione: Their effects on colon mucosal glutathione level and colon tumor development in rats induced by DMH, *Nutr. Cancer 23*(1):77 (1995).
36. Beecher,. C. W. W., Cancer preventive properties of varieties of *Brassica oleracea*: A review, *Am. J. Clin. Nutr. 59*:1166s (1994).
37. Martini, M. C., D. R. Campbell, M. D. Gross, G. A. Grandits, J. D. Potter, and J. L. Slavin, Plasma carotenoids as biomarkers of vegetable intake: The University of Minnesota Cancer Prevention Research Unit Feeding Studies, *Cancer Epidemoil. Biomarkers Prev. 4*:491 (1995).
38. Goodman, M. T., L. N. Kolonel, L. R. Wilkens, C. N. Yoshizawa, L. Le-Marchand, and J. H. Hankin, Dietary factors in lung cancer prognosis, *Eur. J. Cancer 28*:495 (1992).
39. Walaszek, Z., J. Szemraj, M. Hanusek, A. K. Adams, and U. Sherman, D-Glucaric acid content of various fruits and vegetables and cholesterol-lowering effects of dietary D-glucarate in the rat, *Nutr. Res. 16*:673 (1996).
40. Michaelsen, K. F., A. V. Astrup, L. Mosekilde, B. Richelsen, M. Schroll, and O. H. Sorensen, The importance of nutrition for the prevention of osteoporosis, *Ugeskr. Laeger. 156*:958 (1994).
41. Lust, K. D., J. E. Brown, and W. Thomas, Maternal intake of cruciferous vegetables and other foods and colic symptoms in exclusive breast-fed infants, *J. Am. Diet Assoc. 96*:46 (1996).
42. Rappaport, L., W. J. Lipton, and A. E. Watada, Differential deterioration of broccoli varieties, *Univ. Calif. Veg. Crops Rep. 16*:53 (1950).
43. Hardenburg, R. E., A. E. Watada, and C. Y. Wang, *The Commercial Storage of Fruits, Vegetables, and Flroist and Nursery Stocks*, U.S. Department of Agriculture, Handbook No. 66, 1986, p. 54.
44. Morris, L. L., A study of broccoli deterioration, *Ice Refrig. 113*(5):41 (1947).
45. Rygg, G. L., and W. W. McCoy, Broccoli good prepackaging prospect, *West Grower Shipper 23*(66):50 (1952).
46. Ballantyne, A. R., R. Stark, and J. D. Selman, Modified atmosphere packaging of broccoli florets, *Intl. J. Food Sci. Technol. 23*:353 (1988).
47. Wang, C. Y., Quality maintenance in polyethylene packaged broccoli, *USDA Marketing Res. Rep.* 1085 (1977).
48. King, G. A., and S. C. Morris, Physiological changes of broccoli during early postharvest senescence and through the preharvest-postharvest continuum, *J. Am. Soc. Hort. Sci. 119*(2):270 (1994).

49. Brennan, P. S., and R. L. Shewfelt, Effect of cooling delay at harvest on broccoli quality during postharvest storage, *J. Food Quality 12*:13 (1989).
50. Tian, M. S., C. G. Downs, R. E. Lill, and G. A. King, A role of ethylene in yellowing of broccoli after harvest, *J. Am. Soc. Hort. Sci. 119*:276 (1994).
51. Bath, N. M., A. K. Perry, S. J. Schmidt, and B. P. Klein, Misting affects market quality and enzyme activity of broccoli during retail storage, *J. Food Sci. 57*(4):954 (1992).
52. Kasmire, R. F., A. A. Kader, and J. A. Klaustermeyer, Influence of aeration rate and atmospheric composition during simulated transit on visual quality and off-odor production by broccoli, *HortScience 9*:228 (1974).
53. Damen, P. P. M, Good quality broccoli requires cooling and wrapping film, *Groenten Fruit 36*:61 (1980).
54. Lieberman, M., and R. E. Hardenberg, Effect of modified atmospheres on respiration and yellowing of broccoli at 75°F, *Proc. Am. Soc. Hort. Sci. 63*:409 (1954).
55. Lieberman, M., and A. R. Spurr, Oxygen tension in relation to volatile production in broccoli, *Proc. Am. Hort. Sci. 65*:381 (1955).
56. Leberman, K. W., A. I. Nelson, and M. P. Steinberg, Postharvest changes of broccoli stored in modified atmospheres. I. Respiration of shoots and color of flower heads, *Food Technol. 22*:143 (1968).
57. Rij, R. E., and S. R. Ross, Quality retention of fresh broccoli packaged in plastic films of defined CO_2, transmission rates, *Pack. Technol. 17*(3):22 (1987).
58. Liptom, W. J., and C. M. Harris, Controlled atmosphere effects on market quality of stored broccoli, *J. Am. Soc. Hort. Sci. 99*:200 (1974).
59. Smith, W. H., The storage of broccoli and cauliflower, *J. Pomol. Hort. Sci. 18*:287 (1940).
60. Makhlouf, J., F. Castaigne, J. Arul, C. Willemot, and A. Gosselin, Long-term storage of broccoli under controlled atmosphere, *HortScience 24*:637 (1989).
61. Aharoni, N., S. Philosoph-Hadas, and R. Barkai-Golan, *Modified Atmospheres to Delay Senescence and Decay of Broccoli*, 4th National Controlled Atmosphere Research Conference, Raleigh, NC, 1985, p. 169.
62. Forney, C. F., and R. E. Rij, Temperature of broccoli florets at time of packaging influence package atmosphere and quality, *HortScience 26*(10):1301 (1991).
63. Dedolph, R. R., S. H. Wittwer, and H. E. Larzelere, Consumer verification of quality maintenance induced by N-benzyladenine in the storage of celery and broccoli, *Food Technol. 17*:111 (1963).
64. Dostal, H. C., R. R. Dedolph, and V. Tuli, Changes in nonvolatile organic constituents in broccoli following postharvest N-benzyladenine treatment, *Proc. Am. Soc. Hort. Sci. 86*:387 (1965).
65. Gilber, D. A., and R. R. Dedolph, Phytokinin effects of transporation and photosynthesis in roses and broccoli, *Proc. Am. Soc. Hort. Sci. 86*:774 (1965).
66. Kader, A. A., Respiration and gas exchange of vegetables, in *Postharvest Physiology of Vegetables* (J. Weichmann, ed.), Marcel Dekker, New York, 1987, p. 25.
67. Shewfelt, R. L., K. M. Batal, and E. K. Heaton, Broccoli storage: Effect of N^6 benzyladenine, packaging and icing on color of fresh broccoli, *J. Food Sci. 48*:1594 (1983).
68. Rushing, J. W., Cytokinins affect respiration, ethylene production and chlorophyll retention of packaged broccoli florets, *HortScience 25*:88 (1990).
69. Watada, A. E., K. Abe, and N. Yamaguchi, Physiological activities of partially processed fruits and vegetables, *Food Technol. 5*:116 (1990).
70. Zhuang, H., M. M. Barth, and D. F. Hilderbrand, Packaging influenced total chlorophyll, soluble protein, fatty acid composition and lipoxygenase activity in broccoli floerts, *J. Food Sci. 59*:1171 (1994).
71. Barth, M. M., E. L. Kerbel, A. K. Perry, and S. J. Schmidt, Modified atmosphere packaging affects ascorbic acid, enzyme activity and market quality of broccoli, *J. Food Sci. 58*:140 (1993).
72. Mohd-Som, F., E. Kerbel, S. E. Martin, and S. J. Schmidt, Microflora changes in modified atmosphere packaged broccoli florets stored at refrigerated temperatures, *J. Food Qual. 17*(5):347 (1994).
73. Forney, C. F., R. E. Rij, and S. R. Ross, Measurement of broccoli respiration rate in film wrapped packages, *HortScience 24*(1) 111 (1989).

74. Shewfelt, R. L., E. K. Heaton, and K. M. Betat, Nondestructive color measurement of broccoli, *J. Food Sci. 49*:1612 (1984).
75. Klein, B. P., Fruits and vegetables, in *Food Theory and Applications* (J. Bowers, ed.), Macmillan Publishing Co., New York, 1992, p. 687.
76. Salunkhe, D. K., J. Y. Do, and C. Srisangnam, Exploratory research on the evaluation of nutritive value and the development of more acceptable products from selected cruciferous vegetables, A report submitted to Agricultural Research Centre, ARS, Athens, GA, 1976.
77. Drake, S. R., and D. M. Carmichael, Frozen vegetable quality as influenced by high temperature short time steam blanching, *J. Food Sci. 51*:1378 (1986).
78. Brewer, S. M., S. Begum, and A. Bozeman, Microwave and conventional blanching effects on chemical, sensory, and color characteristics of frozen broccoli, *J. Food Qual. 18*:479 (1995).
79. Petersen, M. A., Influence of sous vide processing, steaming and boiling on vitamin retention and sensory quality in broccoli floerts, *Z. Lebensm. Unters. Forsch. 197*:375 (1993).
80. Leichter, J., Folate content in the solid and liquid portions of canned vegetables. *Can. Inst. Food Sci. Technol. J. 13*:33 (1980).
81. DeSouza, S. C., and R. R. Eitenmiller, Effects of processing and storage on the folate content of spinach and broccoli, *J. Food Sci. 51*:626 (1986).
82. Martin, M. E., J. P. Sweeney, G. L. Gilpin, and V. J. Chapman, Factors affecting the ascorbic acid and carotene content of broccoli, *J. Agri. Food. Chem. 8*:387 (1960).
83. Park, Y. W., Effect of freezing, thawing, drying, and cooking on carotene retention in carrots, broccoli and spinach, *J. Food Sci. 52*:1022 (1987).
84. Hendrickson, A., and J. Hendrickson, *Broccoli and Company: Over 100 Healthy Recipes for Broccoli, Brussels Sprouts, Cabbage, Cauliflower, Collards, Kale, Kohlrabi, Mustard Greens, Rutabaga, and Turnip*, Storey Communications, VT, 1989.
85. Lindsay, A., D. J. Fink, and A. Lindsay, *The American Cancer Society Cookbook*, Heart Books, New York, 1988, p. 269.
86. Products ingredients, Gerber Products Co., Fremont, MI, 1994, p. 5,7.

15

Other Crucifers

S. S. KADAM AND K. G. SHINDE
Mahatma Phule Agricultural University, Rahuri, India

I. BRUSSELS SPROUTS

Brussels sprouts (*Brassica oleracea* var. *germifera*) are one of the many variations of cabbage; instead of a single head at the top of the stem, a large bud or miniature head is borne in the axil of each leaf so that little heads are distributed all along a tall stem, which is crowned with a cluster of loose leaves (1). The little solid heads, or "sprouts," 1–2 inches in diameter, are tender and delicious. They are prepared for the table in the same manner as cabbage and are a popular frozen product. This crop has been grown in the vicinity of Brussels, Belgium (hence its name), for hundreds of years and is an important vegetable in most European countries, particularly England. Outside of Europe the importanmce of Brussels sprouts is rather limited, though Australia and California grow the crop widely.

A. Botany

Brussels sprouts are mainly classified into two categories.

1. Dwarf Cultivars

These cultivars have short stems, mostly less than 50 cm long. Improved Long Island, Catskill, Early Morn Dwarf Improved, and Friither Zwerg Kvik are important cultivars belonging to this type. Jade Cross, a Japanese F1 hybrid cultivar, has become popular for its short stem, early high yield, and single-harvest quality. In dwarf cultivars, the sprouts are of medium size and the distance between sprouts is small.

2. Tall Cultivars

In England, tall cultivars are preferred for a longer growing season. These cultivars are classified according to their maturity and the size of the sprout. Evesham, Bedfordshire, Cambridge No. 1, 3, and 5, Irish Elegance, and Sherradian are some of the important cultivars grown in England and Ireland. Early cultivars such as Breda (Netherland), Weibulls Rapid (Sweden), and Wilhelmsburg (Germany) have light green sprouts. Late, hardier cultivars include Red Vein, Hild's

Idea, DeRoshy Amager, and Polarstjernen. Some F1 hybrids, including Olaf, Thor, Jade Cross, Doreman, Alkazar, Merlon, Poster, Rampart, Fortress, Predora, Rovoka, Sonora, and Ladosa, are also popular with growers.

B. Production

1. Cultural Practices

The seedlings of Brussels sprouts are raised in a seedbed. About 300–400 g of seeds are required per hectare. Usually, 1 g of seeds produces about 100 seedlings. Seedlings are transplanted to flat beds or ridges and furrows. For cultivars with large sprouts, spacing of 60 × 60 or 70 × 70 cm is followed. At wider spacing, sprouts will be large but somewhat loose. When harvesting is done only once, a closer spacing (70 × 30 cm or 70 × 40 cm) would be preferable.

Brussels sprouts are grown like other cole crops. It is well known that heavy application of nitrogen results in loose sprouts. The amount applied varies from 100 to 300 kg per hectare. Split doses are applied at the time of planting, a few weeks after planting, after the first picking, and after the second picking. Sprout greening is influenced by nitrogen application. In Western European countries, 50–150 kg of phosphorus and often more than 200 kg of K are applied.

2. Maturity, Harvesting, and Handling

Brussels sprouts are picked from three to five times in common cultivars. Harvesting generally begins 3–3½ months after setting the plants. Early sprouts are picked over several times, the lowest on the plant being taken each time, before they open out and become yellow (2). The first picking of Brussels sprouts should not be delayed after the lower leaves begin to turn yellow, as the sprouts become tough and lose their delicate flavor.

While harvesting, the leaf below the sprout is first broken off and the sprout is then picked by breaking it away from the stalk. As the lower leaves and sprouts are removed, the plant continues to bear new leaves and buds (sprouts) at the top. Baskets or quart berry boxes are used to pick and carry sprouts to the packing houses. The individual boxes may be overwrapped with a transparent film, which delays wilting, avoids physical injury, and thus prolongs shelf life. Thompson and Kelly (2) observed that in California, most of the crop is packed in plastic film bags holding 12 ounces of sprouts, 12 of which are packed in fiberboard boxes.

3. Grading

Sprout grading is becoming increasingly popular. Loose sprouts are sometimes marketed as "blowers." The freezing industry requires a sprout less than 3 cm in diameter. Sprouts should satisfy the following conditions for marketing.

1. Sprouts should be solid and have their top leaves folded closely.
2. Sprouts should be dark or at least light green, not yellow.
3. Sprouts with small loose leaves at the base are considered poor in quality.

4. Packaging

Film packaging or film crate liners are useful in preventing moisture loss because transpiration by Brussel sprouts is high even if the relative humidity is kept at recommended levels (3,4). The film should be perforated because accumulated volatiles other than CO_2 produced by the Brussels sprouts result in objectionable odor or flavor and because a build-up of CO_2 in the atmosphere can cause injury. Hence, sufficient air circulation and spacing between packages are desirable to allow good cooling and to prevent yellowing and decay (3,4). Brussels sprouts should not be stored with fruits because ethylene from fruits accelerates yellowing and can cause abscission of leaves.

C. Chemical Composition

Compared to other cruciferous vegetables, a relatively small portion of the total plant is consumed. The chemical composition of Brussels sprouts (5) and kale is given in Tables 1–3. Brussels sprouts are rich in vitamin A and ascorbic acid and contain appreciable amounts of riboflavin, niacin, calcium, and iron.

D. Storage

The optimum storage conditions for Brussels sprouts are 0°C and 95–98% relative humidity to keep them in marketable condition for about 3–5 weeks (2). Longer storage may result in blackening of the leaves, loss of fresh bright color, decay, wilting, and discoloration of cut stems. The rate of deterioration is twice as fast at 5°C as at 0°C. At 10°C and above, deterioration—yellowing of sprouts and discoloration of the stem end—is rapid; yellowing becomes evident within 1 week at 10°C (6).

Nieuwhof (7) observed that Brussels sprouts turn yellow fairly soon after they are harvested; however, at 0°C and 90% relative humidity, they keep for at least 3 weeks. At –2°C, sprouts can be stored for a longer period of time, but they must be thawed slowly. A controlled atmosphere of 2.5–5.0% O_2 and 5–7.5% CO_2 is helpful to the quality maintenance of Brussels sprouts held at 5 or 10°C but not at 0°C (8). Oxygen levels below 1% can cause internal discoloration (9).

E. Processing

Brussels sprouts can be preserved by freezing. Of the total quantity of Brussels sprouts produced in the United States, 25% are used fresh and 75% are frozen (7). The sprouts are passed through a pregrader and trimmed by water-driven mechanical cutters to remove the butt end and outer leaves. Trimmed sprouts are size-graded into small, medium, and large sizes. Average-size sprouts are blanched for 4 minutes in water or 5 minutes in steam. After cooling and sorting, the sprouts are mechanically deposited in cartons for freezing (7).

Brussels sprouts for canning are harvested, sorted, blanched, and chilled in cold water to

TABLE 1 Proximate Composition of Kale and Brussels Sprouts

Constituent (per 100 g)	Curly kale	Brussels sprouts
Edible portion	0.85	0.69
Water (g)	88.4	84.3
Total nitrogen (g)	0.55	0.56
Protein (g)	3.4	3.5
Fat (g)	1.6	1.4
Carbohydrate (g)	1.4	4.1
Energy value (kJ)	113	114
Starch (g)	0.1	0.1
Total sugars (g)	4.9	3.8
Dietary fiber (g)	2.1	3.1

Source: Ref. 5.

TABLE 2 Mineral Composition of Kale and Brussels Sprouts

Mineral (mg/100 g)	Curly kale	Brussels sprouts
Sodium	43	6
Potassium	450	450
Calcium	130	26
Magnesium	34	8
Phosphorus	61	77
Iron	1.7	0.7
Copper	0.03	0.02
Zinc	0.4	0.5
Sulfur	—	93
Chloride	68	38
Manganese	0.2	0.2

Source: Ref. 38.

prevent matting in the can. They are then packed in cans with dilute brine, exhausted, and processed at 115°C for 25 minutes. Brussels sprouts are more commonly preserved by freezing (7), and only a very small portion of the harvest goes to canning.

II. KOHLRABI

Kohlrabi (*Brassica oleracea* var. *gongylodes*) is often called turnip-rooted cabbage and is closely related to cabbage. The stem, which is the edible part, is greatly enlarged immediately aboveground. Although it is an excellent vegetable if used before it becomes woody, tough, and stringy, it is not popular in the United States and Europe. In India it is known as *knolkhol*. Kohlrabi is characterized by the formation of a tuber, which arises as thickening of the stem tissue above the cotyledon. The tuber develops entirely above the ground. It is this portion that is used as a vegetable, although the young leaves are also sometimes used.

TABLE 3 Vitamin Contents of Kale and Brussels Sprouts

Vitamin (per 100 g)	Curly kale	Brussels sprouts
Carotene (μg)	3145	215
Vitamin E (mg)	1.70	1.0
Thiamine (mg)	0.08	0.15
Riboflavin (mg)	0.09	0.11
Niacin (mg)	1.0	0.2
Vitamin B_6 (mg)	0.76	0.37
Vitamin B_{12} (μg)	0	0
Folate (μg)	120	135
Pantothenate (mg)	0.09	1.0
Biotin (μg)	0.5	0.4
Vitamin C (mg)	110	115

Source: Ref. 38.

A. Botany

The most popular varieties of kohlrabi in India are White Vienna, Green Vienna, Purple Vienna, and Earliest Earfurt. White Vienna is an early, dwarf type with globular, light green, smooth bulbs and creamy white flesh of delicate flavor (10).

In Europe, a large number of kohlrabi cultivars are grown. Among early cultivars, Wiesmoor Forcing White is resistant to bolting. Roggli's Outdoor White is more suitable to less favorable conditions due to its high resistance to bolting; its tubers do not quickly become fibrous. Goliath is a late cultivar maturing 70–100 days after planting.

B. Production

Kohlrabi can be grown on all types of soil and makes no high demands upon it. However, good conditions and fertility favor uniform growth. Kohlrabi tuber production is good at 10–20°C, although quality is better at lower temperatures.

Kohlrabi is usually sown in seedbeds for raising seedlings. The seedlings are then transplanted to the field in flat beds. Seed requirement per hectare is at least 50% more than for cauliflower or cabbage. Since this crop can be grown with much closer spacing, the seedlings are planted out generally at 25 × 25 cm, 25 × 30 cm, 25 × 40 cm, or 30 × 45 cm, according to intercultural facilities. Closer spacing will decrease tuber size but will also increase the yield per hectare. Kohlrabi requires steady growth, and any check in growth will cause the tubers to become more fibrous, while sudden rapid growth may cause the tubers to crack.

High temperatures and nitrogen may also favor the production of elongated tubers as well as check of growth due to vernalization (11). Kohlrabi is harvested when the swollen stem is about 5–8 cm in diameter and before it becomes tough and fibrous. The roots are cut off and the plants, along with the bulbs, are tied together in bunches like beets and sold in bulk in markets (2).

C. Chemical Composition

The chemical composition of kohlrabi is presented in Table 4. Compared to broccoli, it contains higher amounts of sugars but lower amounts of minerals.

D. Storage

Topped kohlrabi can be kept in good condition for 2–3 months if stored at 0°C and 98–100% relative humidity. Some space between containers for air circulation is desirable. High humidity is recommended to prevent shriveling and toughening. Packaging in perforated film can be used to reduce moisture loss. Kohlrabi with leaves has a storage life of only 2 weeks at 0°C. Major diseases are bacterial soft rot and black rot (11).

E. Processing

Kohlrabi tubers are freeze-stored after washing and peeling of fibrous portions. The tubers are generally cut into 1 cc cubes or 2- to 4-cm-thick slices. Kohlrabi tubers are peeled, and generally sliced, for canning with a 1% salt solution as covering liquid (7).

III. MINOR VEGETABLES

A. Mustard

Mustard (*Brassica juncea* or *B. alba*) is a member of the cabbage family. It is used both as a salad ingredient and as a green vegetable. Black mustard (*B. nigra*) is grown commercially for

TABLE 4 Nutritional Value and Chemical Composition of Green Broccoli (Calabrese) and Kohlrabi

Constituent (per 100 g)	Green broccoli (Calabrese)	Kohlrabi
Water (g)	88.2	91.7
Total nitrogen (g)	0.71	0.26
Protein (g)	4.4	1.6
Fat (g)	0.9	0.2
Carbohydrate (g)	1.8	3.7
Energy value (kJ)	138	95
Starch (g)	0.1	0.1
Total sugars (g)	1.5	3.6
Dietary fiber (g) (Englyst method)	2.6	2.2
Sodium (mg)	8	4
Potassium (mg)	370	340
Calcium (mg)	56	30
Magnesium (mg)	22	10
Phosphorus (mg)	87	35
Iron (mg)	1.7	0.3
Copper (mg)	0.02	Tr
Zinc (mg)	0.6	0.1
Sulfur (mg)	130	N
Chloride (mg)	100	34
Manganese (mg)	0.2	0.1
Carotene (mg)	575	0
Vitamin A (mg)	0.10	0.11
Riboflavin (mg)	0.06	Tr
Niacin (mg)	0.9	0.3
Vitamin B_6 (mg)	0.14	0.10
Vitamin B_{12} (mg)	0	0
Folate (mg)	90	82
Pantothenate (mg)	N	0.17
Vitamin C (mg)	87	48

N = Data not available; Tr = traces.
Source: Ref. 5.

its seed. Although it is principally a oilseed crop, it is commonly grown as a leafy vegetable in the northern parts of India, China, and other tropical and subtropical regions. In the United States, mustard is grown in home gardens in the South and to a limited extent for commercial purposes (12).

Ford Hook Fancy, Ostrich Plume, and Southern Giant Curled are the popular varieties of *B. japonica* grown in the United States. Chinese, a broad, leafy variety, is less popular for its foliage use. Cabbage Leaf Mustard (*B. campestris* var. *rugosa* Roxb) has radical leaves forming a loose, cabbagelike head. It is grown in cooler parts of Himachal Pradesh, Kumaon, Bhutan, and Nepal. It is cultivated principally as a vegetable crop, and its leaves are sun-dried and stored after picking (13).

In the plains the seeds of mustard are sown from September to November, while in the hills they are sown during May-August. The seeds are broadcast or sown in rows 30 cm apart. For vegetable purposes, 30 × 10–15 cm spacing can be followed. About 3.0 kg of seeds are required for one acre. Yawalkar (10) recommended the application of 10–15 cartloads of farmyard manure (FYM) as a basal dose, as well as 68 kg of ammonium sulfate given as a top dressing 2–3 weeks after sowing. Crops need adequate irrigation and interculture operations.

Leaves of various varieties grown for vegetable use are harvested 25–30 days after sowing. Not leaving enough leaves on the plants may lead the flowering shoot to become thin and fibrous sooner than expected. The seed stalks appear in 30–35 days and are collected before the flower buds open. Additional shoots can be collected subsequently while still in the tender stage. The chemical composition of mustard leaves and other minor crucifers is given in Tables 5, 6, and 7.

B. Kale

Kale (*Brassica oleracea* var. *acephala*) ranks high among the greens in nutritive value since it is high in vitamins (2). Kale is hardy to cold and does not thrive well in hot weather. In the United States it is seldom grown as a summer crop except where summers are cool. Kale plants do not form heads like cabbage or produce edible flowers like cauliflower and broccoli. Some plants are grown as ornamentals, being variously curled with beautiful colors.

The varieties of kale differ greatly in form, plant size, and foliage character. There are two main groups of kale, the curly-leaved types, referred to as Scotch kales or borecole, and the broader, smooth-leaved types. Rape or Siberian Kales are similar to curly-leaved kales but belong to the species *Brassica napus* (14). The most popular variety of kale in the United States is Dwarf Blue Scotch, characterized by very finely curled and deep bluish-green leaves of better quality than those of the Dwarf Siberian variety, although the latter is considered more winter hardy (12). Dwarf Green Scotch is similar to Dwarf Blue Scotch but is deep-green in color. Thompson and Kelly (2) discussed two groups of kale varieties known as Scotch and Siberian. The foliage of the former is grayish-green in color and curled but not quite so much as Scotch. Both dwarf and tall forms are grown, but the former is more popular.

For domestic use only, the tender inner leaves are picked from the plant, while for market, the entire plant is cut off near the ground. The discolored and injured leaves are discarded and packed dry in ventilated barrels, hampers, or round bushel baskets. Some of the commercial crop is prepackaged in bags of transparent plastic film for the retail trade. These films permit gas exchange but maintain high humidity and prevent wilting of the leaves (2,12).

TABLE 5 Proximate Composition of Some Cruciferous Vegetables

Vegetable	Water (%)	Proteins (%)	Fats (%)	Carbohydrates (%)	Energy (kcal/100 g)
Kale	82.7	6.0	0.8	9.0	53
Watercress	93.3	2.2	0.3	3.0	19
Mustard green	89.5	3.0	0.5	5.6	31
Turnip green	90.3	3.0	0.3	5.0	28
Chinese cabbage	95.0	1.2	0.1	3.0	14
Rutabaga	87.0	1.1	0.1	11.0	46
Collard	85.3	4.8	0.8	7.5	45
Kohlrabi	90.3	2.0	0.1	0.6	29

Source: Ref. 5.

TABLE 6 Mineral Contents (per 100 g edible portion) of Some Cruciferous Vegetables

Vegetable	Calcium (mg)	Phosphorus (mg)	Iron (mg)	Sodium (mg)	Potassium (mg)	Magnesium (mg)
Kale	249	93	2.7	75	238	37
Watercress	151	54	1.7	52	282	20
Mustard green	183	50	3.0	32	377	27
Turnip green	246	58	1.8	40	250	58
Chinese cabbage	43	40	0.6	23	253	14
Rutabaga	66	39	0.4	5	239	15
Collard	250	82	1.5	40	450	57
Kohlrabi	41	51	0.5	8	372	37

Source: Ref. 5.

The postharvest diseases of kale are similar to those of other leafy crucifers, and their incidence can be controlled by holding kale near 0°C with 95% or more relative humidity. Under these conditions, kale maintains its freshness for about 10–14 days (12).

C. Collard

Collard (*Brassica oleracea* var. *acephala*) is a form of kale resembling nonheading cabbage. The edible portion is a cluster of numerous tender leaves at the top of the stalk. It is a winter-hardy crop and withstands hot weather better than cabbage. Collard is an important green vegetable crop in the United States.

The mature head or rosette is cut for fresh market or plants may be cut when they are about one-eighth to one-quarter their normal size and tied into a bunch. The latter method is used most extensively by the market gardeners and truck growers along the Gulf Coast. Smaller shipping containers are used for the bunched product. Sometimes the lower leaves of collard plant are cut and tied into bunches of about 1½ to 2 pounds (2).

Collards are susceptible to *Alternaria* leaf spot (*A. brassicae*, *A. oleracea*), bacterial leaf

TABLE 7 Vitamin Contents (per 100 g edible portion) of Cruciferous Vegetables

Vegetable	Vitamin A (IU)	Thiamine (mg)	Riboflavin (mg)	Niacin (mg)	Vitamin C (mg)	Pyridoxine (mg)	Pantothenic acid (mg)
Kale	10,000	0.16	0.26	2.1	186	2.1	10.0
Watercress	4,900	0.08	0.16	0.9	79	1.9	4.5
Mustard green	7,000	0.11	0.22	0.8	97	3.0	10.0
Turnip green	7,600	0.21	0.39	0.8	139	1.5	1.65
Chinese cabbage	150	0.05	0.04	0.6	25	—	2.1
Rutabaga	580	0.07	0.07	1.1	43	1.0	1.6
Collard	9,300	0.16	0.31	1.7	152	2.6	3.8
Kohlrabi	20	0.06	0.04	0.3	66	1.3	3.1

Source: Ref. 5.

spot (*Pseudomonas maculicola*), bacterial soft rot (*Erwinia caratovora*), bacterial zonate spot (*P. cichorii*), black rot (*Xanthomonas campestris*), and downy mildew (*Peronospora parasitica*). Storage near 0°C prevents most of these diseases (12). The recommended storage conditions for collard are similar to those for spinach.

D. Chinese Cabbage

Chinese cabbage, probably native to China, is little known in the United States and Europe. It is an annual with few characteristics in common with common cabbage. Two species of Chinese cabbage are grown: pe-tsai (*Brassica pekinensis*) and pakchoi (*Brassica chinensis*). The pe-tsai resembles cos lettuce but produces a much larger head, which is elongated and compact. The pakchoi resembles Swiss chard in growth habit; it does not form a solid head (2). The most popular varieties of Chinese cabbage are Chihili and Wong Bok.

Chinese cabbage is harvested when the heads are fully developed by cutting them from the stalk in the same manner as cabbage or cauliflower (15). The harvested plants should not be exposed to direct sunlight because of excessive wilting. Harvesting should be done early in the morning and the produce put in the shade immediately.

Flat baskets, boxes, and crates of various sizes are used for packaging. Lettuce boxes are fairly satisfactory for the long-headed varieties. The vegetable can be stored at 0°C with 95% relative humidity for several weeks (2). Ryall and Lipton (12) recommended a temperature of 0°C with 98% relative humidity for keeping Chinese cabbage in marketable condition for about 4–6 months. Cold stores with a temperature of about 0°C and 90% relative humidity are ideal for storing most cole vegetables, including Chinese cabbage. In these conditions, cabbage may be stored as long as 8 months. Cabbage destined for storage is trimmed of its leaves. Grading heads into different sizes facilitates stacking (7).

Chinese cabbage can be stored for 2–3 months at 0°C with 95–100% relative humidity (16). Storage life is shorter at higher temperatures. A low concentration of oxygen was reported to be beneficial in extending the storage life of Chinese cabbage. Diseased or injured leaves should be removed before the heads are stored. The heads should be packed loosely and preferably upright in crates. Spacing in storage should allow for air circulation. Chinese cabbage is often eaten with frozen noodles. However, the texture of frozen Chinese cabbage is undesirable due to high moisture and thin cell wall (17).

E. Horseradish

Horseradish (*Armoracia rusticana*) is an extremely hardy perennial of the mustard family, which originated in Eastern Europe. It is an extremely important condiment. Roots are used as fresh vegetable. The white flesh is pungent and biting to the taste. In grated form, it is treated with vinegar and used mainly as a relish with oysters and meats (12). Light-textured, well-drained fertile soils are quite suitable for the growth of horseradish. Horseradish requires a cool climate. It is propagated from root cuttings made from the side roots, which are trimmed off in preparing the roots for market. The longer cuttings are best. They are planted in a deep furrow or dibbled by making holes. The spacing should be 25–35 cm in a row and 7.5–10.0 cm between plants. Horseradish is usually planted during early spring. Adequate amounts of moisture and a fertilizer dose of 20–40 kg of N and 35–50 kg of P per hectare have been recommended. Potassium may be applied if the soil is deficient. In addition to these fertilizers, 40 tons of farmyard manure as a basal dose is recommended.

Horseradish roots are hardy and may be left in the ground throughout the winter, but they are stored better after harvest (2). The roots are ploughed out, and the tops and side roots are removed, washed, and packed for market or shipping. The roots are stored in cool, moist cellars or storage house. Temperature of 0 to -1°C and an essentially water-saturated atmosphere slow deterioration most effectively (18). Ryall and Lipton (12) stated that storage of horseradish in containers lined with perforated film is advantageous because it provides high relative humidity around the roots while permitting adequate ventillation between containers to remove respiratory heat. These authors (12) did not recommend use of tightly sealed polyethylene bags or wraps for roots to be used fresh because off-flavors could develop. At 0°C and 95% relative humidity or above, horseradish can be stored for one year under good conditions.

Reports of experiments on use of CA storage of horseradish show that the crop is sensitive to CO_2 but to a much lower degree than other root vegetables (19). Horseradish (mixture of different types) was stored at 0–1°C and relative humidity exceeding 95% for 2, 4, and 6 months using various CO_2 concentrations (0.03, 2.5, 5.0, and 7.5%). Although the modified CO_2 contents did not cause significant difference in weight loss, higher CO_2 levels in CA storage resulted in higher dry matter contents (19). The total sugars of horseradish were higher when the crop was held under increased CO_2 conditions of CA storage.

F. Rutabaga

The rutabaga (*Brassica napobrassica*), also known as the Swedish turnip, is similar to the turnip. It is grown mainly in Canada where summers are cool. When produced in warm weather, both rutabaga and turnip may become woody and bitter very quickly (12). The harvesting of rutabaga is completed before severe freezing. The roots store well in moist conditions and temperatures close to freezing. Rutabagas are often waxed for market to improve their appearance and keeping quality.

G. Turnip

Turnip (*Brassica campestris*) is thought to have originated in the Mediterranean region and also in the Near East, Afghanistan, Iran, and West Pakistan. Turnip was derived from annuals grown for oilseeds. They are grown as annuals for the fleshy roots (swollen hypocotyl), which can vary in shape from flat, long to globe shaped. The flesh color may be white or yellow, but the skin color ranges from shades of white-yellow to red-purple to black (20). The plant is vernalized in the winter; flower color is correlated with the flesh color of the roots. White-fleshed rooted plants have bright yellow petals, and yellow-fleshed ones have pale orange-yellow petals.

The most popular varieties of turnip as discussed by Thompson and Kelly (2) are the Purple Top Globe, White Milan, White Flat Dutch, White Egg, Yellow Globe, Golden Ball, and Yellow Aberdeen. Shogrin or Japanese is a popular variety grown for turnip greens in the United States. In addition to the greens, the roots are quick-growing and of good quality. The Asiatic varieties of turnip grown in India are Variety 4 (White), Variety 4 (Red), Pusa Swati, and Pusa Kanchan. The latter two have been released by the Indian Agricultural Research Institute (IARI), New Delhi (21). The European or temperate varieties grown in India are Purple Top White Globe, Golden Ball, Snow Ball, and Early Millan. The IARI-released varieties of this group are Pusa Chandrima and Pusa Swarnima.

Turnip is sown directly in the field about 12 mm deep. In the Philippines, transplants are made from seedbeds to the field. Deep soil having a pH of 5.5–7.0 is recommended. Fertilization with 60–120 kg N, 60–80 kg of P, and 60–80 kg of K is recommended, with about one-half the dose applied at seeding and the other half short after thinning.

Turnip roots are harvested similarly to beets when used as greens; the plants are pulled or cut at the soil surface. The greens are eaten after cooking. Young turnip roots are sometimes bunched like beets, but very often they are topped and packed in transparent film bags. The commercially grown crops are harvested mechanically using best-harvester–type machines (2). Washing turnips after harvest improves appearance and reduced postharvest decay.

Packaging turnips in perforated plastic bags helps to keep the humidity high around the roots during marketing and reduces shriveling. Dipping turnips in hot melted paraffin wax gives a glossy appearance and is of some value in reducing moisture loss during handling. However, waxing is primarily to aid in marketing and is not recommended before long-term storage (22). Turnip greens are usually stored only for a short period.

Turnip roots are high in provitamin A. Turnip greens are also eaten, and are more nutritious than roots (Table 7).

Turnips (topped) can be kept in good condition for 4–5 months at 0°C and 95% relative humidity. At higher temperatures (≥ 5°C), decay will develop much more rapidly than at 0°C. Turnips should be stored in slatted cribs or bins, and good air circulation around containers should be provided.

Ryall and Lipton (12) recommended 0°C and 95% relative humidity to store turnips in good condition for 2–4 months; under the same conditions, turnip greens can be stored for only 10–14 days. Due to their susceptibility to shriveling, turnips are frequently held at relative humidity above 95% during storage. The effects of CA storage on the shelf-life of turnips have not been established fully. Ryall and Lipton (28), however, stated that CA storage of turnips may be worth testing, in the case that concentrations of CO_2 that do not damage would retard toughening of turnips.

Due to their susceptibility to shriveling at higher temperatures, turnips are often waxed to reduce moisture loss. Maintenance of high humidity (above 95%) and packaging in perforated film bags might eliminate the need for waxing. Haller (22) reported that application of wax emulsion to the roots neither improved appearance nor retarded shrinkage. The beneficial effects of parrafin were attributed to retardation in respiration of roots and reduction in occurrence of internal breakdown. Franklin (23) recommended that waxing be done just before marketing and not before long-term storage. Alexander and Francis (24) reported that turnips can be prepeeled and then packaged in perforated film. Spraying turnips with maleic hydrazide before harvest virtually eliminates sprouting during normal storage (12).

Turnips are susceptible to boron deficiency. The interior of the affected roots becomes water soaked and may develop brown spots and cracks. The leaves of affected plants are dwarfed and curled (12). Turnips are also susceptible to chilling injury indicated by surface pitting, shriveling, and decay. The most important diseases of turnips are *Alternaria* root rot (*A. hercules*), black rot (*X. campestris*), gray mold rot (*B. cinerea*), *Rhizoctonia* root rot (*R. solani*), and water soft rot (*S. sclerotiorum*). Ryall and Lipton (12) described these diseases as well as possible preventive measures.

H. Watercress

Watercress (*Nasturtium officinale* R.Br.) grows wild in cool regions of the world. It has been used as a medicinal plant for cure of scurvy. Presently it is grown throughout the world. The tender shoots are eaten raw in salads or cooked as vegetables in the Far East.

Watercress is grown in areas of abundance of water supply. There are two species—diploid and tetraploid—and a triploid that is probably a cross of the two, which is sterile, also exists. Propagation is either vegetative or by seed. Roots emerge readily from nodes of the stem under damp conditions.

Watercress is a cool season crop. It can be grown year round in the temperate zone and in the winter in the semi-tropics. Plants can be started from seed in beds kept moist and shaded or from cuttings from established plants. Cuttings about 15 cm long are planted 15 cm apart and about 10 cm deep. The top 15 cm of shoot is harvested. Bunched shoots are precooled with ice and water and shipped refrigerated (12).

The high perishability of watercress makes prompt cooling imperative. Watercress leaves wilt and become yellow and slimy when improperly handled. Watercress should be precooled either by hydrocooling or vacuum cooling and stored at 0°C with high (95–100%) relative humidity. It is bunched and usually packed in alternate layers with flake ice. Watercress stored at 0°C in waxed cartons with top ice holds up well for 2–3 weeks. A similar storage life is possible using perforated polyethylene crate liners and package icing to minimize wilting (25). Naked bunches of watercress are highly perishable even at 0°C and may keep only 3–4 days. In watercress, a 15% loss will cause only a trace of wilting and a 40% loss will cause moderate wilting (26).

REFERENCES

1. Watts, R. L., and G. S. Watts, *The Vegetable Growing Business*, Orange Judd Publishing Co., Inc., New York, 1954.
2. Thompson, A. K., and W. C. Kelly, *Vegetable Crops*, 5th ed., McGraw-Hill Book Co., Inc., New York, 1957, p. 611.
3. Rygg, G. L., and W. W. McCoy, Prepacking brussels sprouts, *West Grower Shipper 24*(1):46 (1952).
4. Stewart, J. K., and W. R. Barger, Effects of cooling method pre-packing and top icing on nutritive value of brussels sprouts, *Proc. Am. Soc. Hort. Sci. 83*:488 (1963).
5. Salunkhe, D. K., and B. B. Desai, Cruciferous vegetables, in *Postharvest Biotechnology of Vegetables*, Vol. I, CRC Press, Boca Raton, FL, 1984, p. 157.
6. Platenius, H., Effect of temperature on the rate of deterioration of fresh vegetables, *J. Agric. Res. 59*:41 (1939).
7. Nieuwhof, M., *Cold Crops: Botany, Cultivation and Utilization* (N. Polunin, eds.), Leonard Hill, London, 1969.
8. Eaves, C. A., and F. R. Forsyth, The influence of light, modified atmosphere and benzimidazole on Brussels sprouts, *J. Hort. Sci. 43*:317 (1968).
9. Lyons, J. M., and L. Rappaport, Effect of controlled atmosphere on storage quality of Brussels sprouts, *Proc. Am. Soc. Hort. Sci. 81*:324 (1962).
10. Yawalkar, K. S., *Vegetable Crops of India*, 2nd ed., Agril. Horticultural Publishing House, Nagpur, Inda, 1980, p. 370.
11. Ramsey, G. B., and M. A. Smith, *Market Diseases of Cabbage, Cauliflower, Turnips, Cucumbers, Melons and Related Crops*, U.S. Dept. Agriculture, Agriculture Handbook, 184, 1961, p. 49.
12. Ryall, A. L., and W. J. Lipton, *Handling, Transportation and Storage of Fruits and Vegetables*, Vol. I, *Vegetables and Melons*, AVI Pub. Co. Inc., Westport, CT, 1972.
13. Chauhan, D. V. S., *Vegetable Production in India*, 3rd ed., Ram Prasad and Sons, Agra, India, 1981.
14. Fordham, R., Cabbage and related vegetables, in *Encyclopaedia of Food Science, Food Technology & Nutrition* (R. McRae, R. K. Robinson, and M. J. Sandler, eds.), Academic Press, 1993, p. 4719.
15. Choudhury, B., *Vegetables*, 8th ed., National Book Trust, 1990.
16. Wang, C. Y., Postharvest responses of Chinese cabbage to high CO_2 treatment or low O_2 storage, *J. Am. Soc. Hort. Sci. 108*:125 (1983).
17. Fuchigami, M., N. Hyakumoto, and K. Miyazaki, Texture and pectic composition differences in a raw, cooked and frozen thawed Chinese cabbage due to leaf position, *J. Food Sci. 60*:153 (1995).
18. Hoyle, B. J., Storing horse-radish stecklings, *Calif. Agric. 10*:6 (1956).
19. Weichmann, J., Physiological response of root crops to CA, *Michigan State Univ. Hort. Report 28*:122 (1977).
20. Yamaguchi, M., *World Vegetables: Principles, Production and Nutritive Value*, AVI Pub. Co. Inc., Westport, CT, 1989, p. 219.

21. Chatterjee, S. S., Cole crops, in *Vegetable Crops in India* (T. K. Bose and M. G. Som, eds.), Naya Prokash, Calcutta, 1986, p. 165.
22. Haller, M. H., Effect of root trimming washing on the storage of turnips, *Proc. Am. Soc. Hort. Sci. 50*:325 (1947).
23. Franklin, E. W., The waxing of turnips for the retail market, *Canada Dept. Agric. Publ. No.* 1120, 1967.
24. Alexander, B., and F. J. Francis, Packaging and storage of prepeeled turnips, *Proc. Am. Soc. Hort. Sci, 84*:513 (1964).
25. Franklin, E. W., The waxing of turnips for the retail market, *Canada Dept. Agric. Pub.* 1120, 1967, p. 4.
26. Hruschka, H. W., and C. Y. Wang, Storage and shelf life of packaged watercress, parsley and mint, *U.S. Dept. Agric. Market Res. Report*, 1102, 1979, p. 19.

16

Onion

S. D. Warade and S. S. Kadam
Mahatma Phule Agricultural University, Rahuri, India

I. INTRODUCTION

Onion, one of the oldest vegetable crops known to man (1), is a popular vegetable worldwide. It is valued for its distinctive pungent flavor and is an essential ingredient of the cuisine of many regions (2). It is used in both immature and mature bulb stages as a vegetable and spice as well as food for cattle and poultry (3). Onions can be eaten raw or cooked; mild-flavored or colorful bulbs are often chosen for salads. Onions have many uses as folk remedies, and recent reports suggest that onions play a part in preventing heart disease and other ailments (4–6). The world production of onion is summarized in Table 1. China, India, CIS, the United States, Turkey, Japan, and Spain are the major producers of onions. The total world production of onions in 1993 was estimated to be 26.0 million tons (7). The major exporters of onion are India, Mexico, the United States, Turkey, Pakistan, Egypt, and Chile, whereas Malaysia, United Arab Republic, Canada, Japan, Lebanon, and Kuwait are the major importers.

II. BOTANY

Onion belongs to the genus *Allium*, which includes several cultivated crops such as onion, garlic, leek, shallot, and Chinese chive. This genus contains about 500 widely distributed species. The most important cultivated species are onion (*A. cepa* L.), leek (*A. ampeloprasum* L.), Japanese bunching onion (*A. fistulosum* L.), rakkyo (*A. chinese* G. Don), Chinese chives (*A. tuberosum* Rott. ex-spr.), chives (*A. schoenoprasum* L.), and shallots (A. cepa L.) (8). The term shallot refers to the vegetatively propagated forms of *A. cepa* var. *ascalonicum*, which were included in the *aggregatum* group of the species by Jones and Mann (9). Shallots of this type appear to have been derived by selection from naturally occurring variants within *A. cepa*. The old botanical name *A. ascalonium*, formerly used for shallots, was originally applied to a distinct wild species from the Near East (10). Several other cultivars mentioned in the U.S. literature as shallots are actually derived from crosses between *A. cepa* and *A. fistulosum* (cultivar Delta Giant); they should not be confused with the shallots of the *A. cepa aggregatum* group. *A. cepa* shallots are

TABLE 1 Dry Bulb Onion Production

Country	Production (× 10^3 M^T)
World	26,319
China	3,826
USSR	2,500
India	2,480
United States	2,168
Turkey	1,300
Japan	1,274
Spain	1,008

Source: Ref. 11.

distinguished from normal bulb onions by their habit of multiplying vegetatively by lateral bud growth—a single shallot bulb usually contains several initial shoots, and after the bulb is planted, several leafy shoots grow out from it. Each shoot them rapidly produces a small bulb so that a cluster forms, which remains attached to the original base plate (11). The bulbs can then be separated and the process repeated during the next growing season.

A. Cultivars

Onion (*Allium cepa*) grown for bulbs belongs to the family Alliaceae. A huge variety of different onion cultivars that are locally adapted but highly heterogeneous exists today. Thomas and Dabas (12) reported that 2000 accessions were collected for the Indian National Bureau for Plant Genetic Resources Gene Bank in New Delhi. Every year, new hybrids and varieties are released for commercial cultivation. These are higher yielding and more uniform in size than old varieties. Thompson and Kelly (13) mentioned two general types of onion grown in the United States: the "American" or pungent type and the "foreign" or mild type, three colors being recognized in each class—red, white, and yellow. In general, American onions produce smaller bulbs with denser texture, stronger flavor, and better keeping quality, varying in shape from oblate to globular. The important cultivars grown around the world, several of which are shown in Figures 1 and 2, are listed in Table 2.

B. Growth and Development

The life cycle of the onion has been described by many authors (2,9,11). Currah and Proctor (2) outlined the sequence of events in the growth and development of the onion. The onion seed contains embryo and endosperm. The embryo, under favorable conditions of germination, starts to grow actively and derives nutrients from the endosperm. In the germinating seed, first the radicle emerges and becomes anchored in the soil. Then the cotyledon emerges and, by elongation, it forms a looped structure, which breaks the surface of the ground while the seed is still below the ground level. The seed leaf continues to elongate at the base until eventually the rest of the seed is carried up above ground level, still attached to the tip of the cotyledon (2). The meristematic growing point of the seedlings remains below ground where the cotyledon joins the radicle. This area is crucial to the development of the onion plant throughout its life. From the point of growth, the seedling develops a succession of leaves, which grow from the fattened stem or base plant.

FIGURE 1 Basawant 780, improved cultivar of onion grown in Maharashtra, India.

The foliage leaf is made up of a hollow green phytosynthetic blade, a cylindrical sheath that connects the blade to the base plant (11). Cell division takes place near the base of the leaf. The oldest part of each leaf is the tip, and the youngest is the base of the leaf sheath. Each leaf is formed inside the encircling older leaf sheaths and grows up through them so that the neck of the pseudostem is formed from the concentric leaf sheaths (9). The leaf sheath remains hollow in the form of a tube that opens at the tip. Each new leaf blade emerges through a small hole or pore at the junction of the blade and the sheath of the previous leaf. The hollow tapering leaf blades are arranged in two rows opposite one another. Roots are also produced from the base plate. New roots form irregular rings above and around older ones and emerge through the corky outer tissue. The leafy plant eventually ceases to form leaf blades, and instead the apex begins to initiate a number of bladeless concentric thickened leaf sheaths, forming the bulb scales (2). Together with the swollen lower leaf sheaths of the older leaves, they make up the fleshy part

FIGURE 2 N-53, improved cultivar of onion grown in Maharashtra, India.

Table 2 Important Onion Cultivars

Country	Cultivars
India	Pusa Red, Pusa Ratnar, Nasik Red, Patna Red, N 404, N 207-1, White Patna, N-53, N-2-4-1, Bellary Red, Bellary Big Onion, B-780, Udaipur 101, Udaipur 103, Hissar-2, Hissar-11, Kalyanpur Red Round, Arka Pragati, Arka Niketan, Arka Kalyan, Bangalore Rose and Punjab 48, Udaipur 102, Pusa White Round, Pusa White Flat, Phule Safed, Phule Suwama.
Pakistan	Phulkara, Faisalabad, Early and Desi Red, Local White (Kasmir).
Bangladesh	Faridpur Vati, Taherpuri
Egypt	Giza 6, Beheri
Israel	Haemek, Moab
Spain	Babosa, Valenciana, Temprana
South Africa	De Wildt, Pyramid, Bon Accord, Hojein
Mexico	Cojumatlan
Brazil	Baia Periforme, Grano 502, Texas Early Grano
United States	Brigham Yellow Globe, Yellow Globe Danvers, Early Yellow Globe, Mountain Danvers, Ebenezer, Red Wethersfield, Southport Red Globe, California Early Red, Southport White Globe, White Cresole, White Portugal or Silverstein, Yellow Bermuda, Crystal wax, Early Grano, Yellow Sweet Spanish, White Balbosa, White Sweet Spanish, Southport White Globe, Barletta, Beltsville, Bunching, and White Lisbon

of the onion bulb. Most of the dry matter in green leaves is transferred down to the bulb at this stage, contributing to both the swollen leaf sheaths and to the bladeless fleshy scales. The thin outer bulb scales are formed from the expanded dried-out bases of older leaf sheaths.

Once no further blades are being produced to support it from inside, the neck of the onion becomes hollow and the top of the leaf blades fall down. The green blades gradually senesce and die, but during this period nutrients from the leaf blades are still being exported into the bulb, which continues to store them. When this process is complete, the onion ceases to grow and is ready for harvest.

III. PRODUCTION

A. Soil and Climate

Onion is grown in different kinds of soils, ranging from light-textured sandy loam to heavier clay loams. Proper drainage, looseness, absence of persistent weeds, ample organic matter, and a pH of 5.8–6.5 are the major requirements for the successful production of onion (14). Bulb production occurs earlier in light-textured soils than in heavier soils. The size and the quality of bulb depends on the soil types, fertility, and variety. In general, the quality of bulb produced in clay is very poor. The best quality bulbs are produced in lighter soils, such as sandy loam or loamy or alluvial soils that are rich in organic matter, deep, friable, and well drained (15). Soils that are too alkaline or acidic do not favor normal bulb growth.

Onion thrives under a wide range of climatic conditions (16), but it does not perform well under all types of weather conditions. The temperature requirement for this crop depends largely on the stage of development; for vegetative growth in the initial stage the temperature should be between 12.8 and 23.9°C, whereas bulb formation is favored by temperatures between 15.6 and 21.0°C (17). Young plants are more tolerant to freezing temperatures than the older ones. Yields are higher and better quality bulbs are produced under mild climates.

Day length (photoperiod) and light intensity are other important environmental parameters of onion production. Onions need long days to initiate bulb formation. High temperatures coupled with long days appear to shorten the time needed for bulb initiation. Onions also require ample sunshine and grow well even in cold weather if the light intensity is above the critical level. Onions can be grown up to elevations of 1500–2330 m, but at higher altitudes they can be grown only as spring or summer crops. The onions grown in tropical regions are often referred to as short-day cultivars (this does not mean that these onions are physiologically short-day plants). In the tropics, short-day onions may complete their life cycle during a photoperiod that in theory is sufficiently long to permit them to form bulbs. In such circumstances, environmental factors such as nutrition and spacing and internal factors such as stage of plant development control the onset and progress of bulb formation.

High rainfall (> 100 cm) is detrimental to the growth and bulbing in onion. The best quality bulbs and higher yields can be obtained when weather conditions during the prebulbing stage are optimum—mild temperatures, plenty of sunlight, optimum soil moisture levels, mild rains, and proper day length—followed by bright dry weather during ripening.

B. Propagation

Onion is propagated through seeds and bulbs. When seeds are used, they are either seeded directly or by raising seedlings in the nursery. Onion seeds are small, black or dark brown in color, and contain approximately 240–300 seeds per gram. The seeds remain viable for about 6 months to a year. The best temperature for germination is around 20°C, and under ideal conditions, germination takes about 6–8 days. Seedlings are raised on raised beds (18) in a nursery for transplanting. The seed beds are protected from hot sun, heavy winds, heavy rains, and frost. The seedlings will be ready for transplanting in about 6–8 weeks. The age of the seedlings has tremendous impact on the subsequent growth of plants and bulb yield. The highest bulb yields were reported using 12-week-old seedlings (14).

The land selected for nursery beds should be fertile, well drained, and free from weed seeds and soilborne diseases and pests. All dead wood or debris should be removed by deep digging or ploughing. The soil should be worked to a fine tilth by repeated ploughing, planking. harrowing, and leveling. Generally, nursery beds are one meter wide and of convenient length. They are usually raised 10–15 cm above ground level. The furrows in between nursery beds should be large enough for easy working (watering, weeding, lifting of seedlings, etc.). About a week to 10 days before sowing, well-decomposed farmyard manure (FYM) should be applied at the rate of a half ton per bed.

A seed bed of 500 m^2 provides enough seedlings for one hectare of land (i.e., 5.68–9.0 kg seeds per hectare). The seed beds need to be watered lightly with sprinklers or rosehead watering cans immediately after sowing, and subsequent irrigations are given as necessary. The seedlings will be ready in about 8–10 weeks. This commonly followed method produces a good yield.

When bulbs are used, the size or weight of the bulb is an important parameter determining subsequent growth and yield of plants. Usually medium-sized bulbs are used for planting. Large-sized bulbs have been shown to produce flowers early in the season. The bulbs need to rest at least one month before they can be planted in the field. About 825 kg of bulbs are needed for planting one hectare of land (14). This method is more commonly followed for seed production.

C. Cultural Practices

1. Planting

The land for growing onions is prepared to a fine tilth by repeated ploughings. These ploughings need not be deep, as the majority of onion roots will be in the top 5–10 cm of the soil. Any old debris, weeds, or remains of the previous crop should be completely removed before final preparation.

Soil solarization consists of heating damp soil over a period of weeks to temperatures that are lethal for weed seeds and pathogens such as *Fusarium* pink root rot disease. This process is carried out by covering the soil with sheets of clear plastic under which high temperatures are attained in dry sunny climate. The soil must be wet at the start of the treatment to work effectively. This technique has been used in Israel to improve seed bed hygeine with a consequent increase in survival and vigor of seedlings (19,20).

Onion crops can be grown using any of the following methods, depending upon the planting material available: seed sowing, transplanting, or bulb planting.

Seed Sowing

When seeds are sown directly in the field, the soil should be properly prepared to obtain a fine tilth. The presence of old debris, weeds, or other extraneous material in the field can hinder onion seed germination.

The seeds are drilled in rows 15–20 cm apart. The seeds should be sown very shallow (about 1.0–2.0 cm deep). Approximately, 16–22 kg of seeds are required per hectare of land. The field should be irrigated lightly immediately after sowing. When the seedlings are large enough to be easily handled (about 8 weeks old), they can be thinned to achieve the desired plant population. This method is not very popular, as it requires a large quantity of seeds.

Transplanting

In this method, seedlings are raised in nursery beds and transplanted when they are 6–8 weeks old. About 9–10 kg of seeds are required per hectare. The ideal seedlings are 20–25 cm tall and 0.8–0.9 cm thick. When uprooting seedlings, it is important to make sure that the root system remains intact. Hardening the seedlings by withholding frequent irrigation helps them withstand transplantation shock better. If allowed to remain in the nursery bed for a long period, seedlings become tall and will not experience rapid growth after transplanting. Some farmers practice topping of seedlings at the time of transplanting. However, there are no advantages to topping or pruning. On the contrary, the trimming of transplants results in reduced bulb yield (22).

It is best to transplant seedlings 2.5–4 cm deep so that disease incidence is kept to a minimum. Both raised beds and flat beds have been used for transplanting onion seedlings (Fig. 3). It has been shown that bulb yield is higher in flat beds than in raised beds. Spacing depends on the season, variety, and soil. In general, 15 × 20 cm (kharif) or 12.5 × 7.5 cm (rabi) is recommended. Light irrigation should be given immediately after planting.

Bulb Planting

Bulbs are used for planting are dibbled on the sides of ridges. The distance between ridges is 45 cm and between plants within the ridge is 15 cm. Light irrigation is necessary after planting (23).

2. Manuring and Fertilization

Onion is a heavy feeder, requiring ample supplies of nitrogen, phosphorus, and potassium (24). According to Yawalkar et al. (25), an onion crop yielding 30 tons of bulbs requires 73 kg of

FIGURE 3 Transplanting of onion seedlings practiced in India.

nitrogen, 36 kg of phosphorus, and 68 kg of potash. Kale and Kale (23) reported that an onion crop yielding 27.4 tons of bulbs required 71 kg of nitrogen, 40 kg of phosphorus, and 35.5 kg of potash. Suboptimal levels of these nutrients in the soil adversely affects the yield and quality of bulbs. Nitrogen is most essential during the initial stages of growth. A deficiency of this element at this stage causes stunted growth, general yellowing, and weak plants. On the other hand, excess nitrogen results in succulent growth and robust plants. Low nitrogen levels have been associated with early bulb formation. Bulbs from plants with higher tissue nitrogen levels are of poor quality and do not store well for longer periods. Zink (24) reported that adequate nitrogen was vital throughout growth, particularly during bulbing, and that phosphorus should only be applied if needed. Brown et al. (26) showed the benefits of putting slow-release nitrogen fertilizer, sulfur-coated urea in preplant bands below the root region in direct-sown crops grown with furrows irrigation in Idaho.

Phosphorus is required throughout the life of an onion crop in that it promotes bulb formation. Phosphorus deficiency adversely affects quality. An inadequate supply of potassium inhibits bulb formation, reduces bulb quality and thickness of scales, and increases the tendency of the plants to bolt and flower.

Fertilizer recommendations vary both in quantity and kind of fertilizer to be applied (27–42). It also depends on the variety, season, and soil type. Kale and Kale (23) suggested that onion crops require 40–50 tons of well-decomposed FYM, 50–150 kg nitrogen, 25–135 kg of phosphorus, and 50–110 kg of potash per hectare. FYM, along with *Azospirillum* was found beneficial for onion crops (Table 3) in that it reduced the nitrogen requirement 25% (34). Other recommendations include application of 50 kg of nitrogen, 50 kg of phosphate, and 100 kg of potash per hectare (35); 50 cartloads of FYM, 40 kg of nitrogen, 20 kg of phosphate, and 40 kg of potash per hectare (25); and 25–50 tons of FYM along with 100 kg of nitrogen (14).

The entire dose of FYM should be added during land preparation and mixed thoroughly. The full dose of phosphorus and potassium should be applied at the time of final land

TABLE 3 Effect of Fertilizer Source on Yield of Onion Bulbs cv. B-780

Treatment	Yield (t/ha)
FYM 20 t/ha + *Azospirillum*	20.7
FYM 40 t/ha + *Azospirillum*	21.9
FYM 60 t/ha + *Azospirillum*	25.2
FYM 40 t/ha + 100:50:50 NPK kg/ha	27.7
FYM 40 t/ha + 75:50:50 NPK kg/ha	25.9
FYM 40 t/ha + 50:50:50 NPK kg/ha	26.3
FYM 40 t/ha + 75:50:50 NPK kg/ha + *Azospirillum*	27.6
FYM 40 t/ha + 50:50:50 NPK kg/ha + *Azospirillum*	27.0
100:50:50: NPK kg/ha	24.0
FYM 20 t/ha	20.5
FYM 40 t/ha	21.5
FYM 60 t/ha	20.9
Control	16.8
S.E.	0.09
C.D.	0.22

Source: Ref. 34.

preparation, whereas nitrogen should be applied in two equal doses: the first one month and the second 2 months after transplanting. Apart from major nutrients, minor elements may also be applied, if necessary. Sulfur has been shown to be a principal constituent of allylpropyl disulfide—responsible for bulb pungency. Soil application of zinc sulfate and ferrous sulfate increased ascorbic acid, pyruvic acid, and sulfur contents of onion bulbs (43,44).

3. Irrigation

Regular irrigation is an absolute necessity for successful onion production. Onion is a unique crop in its water requirement, which changes with the stage of development. Young plants immediately after transplanting require less water, and this situation continues for some time. Relative water consumption increases with plant age, reaching a maximum just before maturity. Again it drops with the onset of ripening stage. Therefore, the irrigation frequency should be adjusted accordingly.

The number and frequency of actual irrigations needed depend largely on the soil type, the season, and the nature of planting material used. Rainy season crops do not usually require supplementary irrigation unless rainfall is not adequate or unevenly distributed. Irrigation in such cases is given only occasionally. However, both winter and summer crops need regular irrigation depending upon the crop stage. Water stress during bulb formation is very detrimental to bulb development. Bulbs have a tendency to crack in dry soil. Therefore, precautions should be taken not to miss irrigation during that period. Generally irrigation is stopped 2–3 weeks before harvesting of bulbs (23).

4. Intercultural Operations

Weed growth should be checked at all times during the growth of onion plants. For this purpose, regular weeding is necessary, particularly during early stages of growth (14). Fields should be hoed regularly to provide adequate aeration for the roots. Weedicides can also be used (23).

D. Disorders, Diseases, and Pests

1. Disorders

Bolting

Under certain conditions onion crops bolt or produce premature seed stalks, suddenly breaking the normal life cycle. Bulbs become light and fibrous and have very poor keeping qualities. Bolting may be due to generic factors, changes in temperature, poor seed quality, poor soil and cultural practices affecting the growth, relative length of day and night, spacing, and seedling size (45–47). Singh and Dhankar (36) reported that very high nitrogen doses reduced bolting. Bolting increased with an increase in set size (36).

Splitting and Doubling of Bulbs

Under some adverse conditions and imbalanced nutrition (47), splitting and doubling of bulbs occur and affect bulb quality. Sometimes, mechanical injury of plants during cultural operations causes splitting and doubling of bulbs (47). Water deficiency at the initial growth stage and irrigation after a long spell of drought may also lead to splitting and doubling of bulbs (48).

2. Diseases

The major diseases of onions in tropical countries include *Stemphylium* blight (*S. vesicarium*), which is damaging to the seed crop, onion smudge (*Colletotrichum circinans*), neck rot (*Botrytis allii*), black mold (*Aspergillus niger*) and *Alternaria* rot (*Alternaria* sp.), onion yellow dwarf virus and mycoplasma or viruslike aster yellows (18). Serious diseases reported from Guatemala include damping-off (*Rhizoctonia solani*, *Pythium* sp., and *Fusarium* sp.), downy mildew (*Peronospora destructor*), purple blotch (*Alternaria porri*), and bacterial soft rot (*Erwinia carotovora*). These attack the bulbs in the field. *Botrytis* leaf spot, pink root rot (*Pyrenochaeta terrestris*), and white rot (*Sclerotium cepivorum*) were reported from Venezuela (2).

3. Pests

Thrips tabaki is the most damaging pest of onions. Other important pests include the head borer caterpillar (*Helicoverpa armigera*), which is particularly troublesome in the seed crop, leaf miner (*Chromatomyia horticola*), and mite (*Rhizoglyphus* sp.), which attacks onions both in the field and in storage. Onion pests reported from Venezuela include thrips, slugs, cutworms, crickets (*Gryllus assimilis*), leaf miner (*Liriomyza trifolii* and L. *huidobrensis*), red spider mites (*Tetranychus uritical* and *T. cinnabarinus*), bulb mite (*Rhizoglyphus echinopus*), subterranean plant bug (*Crytomentus bergi*), leaf-eating caterpillars (*Trichoplusis* sp.), and nematodes (48).

E. Harvesting, Handling, and Maturity

Green bunching onions are harvested as soon as they reach edible size. The plants are pulled manually, the roots trimmed, and the outer skin removed, leaving the bulbs clean. The onions are then washed, tied in bunches, and packed in crates for marketing or storing. Maturity is characterized by a breakover of the pseudostem and inability to support the leaf blades.

Onions that are to be stored should be harvested after the top begins to break over and before the leaves dry down completely (13). Bulbs left in the ground until the tops are dead may develop roots, resulting in low-quality onions. The best time to harvest onions in India is when about 50–60% of the tops are broken over (14).

The harvest maturity of onion depends upon the purpose for which they are grown. Thompson et al. (49) observed that harvesting takes place 45–90 days after field setting for green onions and 90–150 for bulbs, depending upon the variety. The bulbs are considered mature when the neck tissues begin to soften and the tops are about to abscise (50) and decolorize (51). In India and other tropical areas, the development of red pigment and the characteristic pungency of the variety are also considered important indices of harvest maturity (52,53). The postharvest shelf life of onion depends upon the maturity of onion at harvest (54-60). Thompson et al. (49) observed that early onion harvest results in sprouting of bulbs and late harvest gives rise to formation of secondary roots during storage (50). Wall and Corgan (57) evaluated the effects of bulb maturity and storage duration under ambient conditions on fresh market quality of short-day onions. For all cultivars, average bulb weight increased and firmness decreased as harvest was delayed (Table 4). When onions were stored for 10–20 days under ambient shed conditions (15–34°C, 76–100% RH, respectively), average bulb weight and firmness decreased linearly with time in storage (Table 5).

Ali and Shabrawy (61) reported an incidence of neck rot disease in onion caused by *Botrytis allii* with high irrigation frequency during the growing season and that delayed harvest from 120–140 to 150 days after transplanting decreased the disease severity during storage. Onion bulbs left with 1- or 2-cm-long necks were less affected by the disease than were bulbs without necks (61).

Hand-pulled onions are placed in the windrows long enough to become dry, but care should be taken to prevent sunscald. Under natural conditions in California, the onion bulbs lose weight during the curing whether they are topped or not, but more weight is lost during the first few days when the tops are left intact (49). Bulbs with tops left intact are much higher in percent dry matter after curing than are bulbs with their tops cut off. After the tops are dried down, they

TABLE 4 Onion Cultivar and Maturity Effects on Average Bulb Weight, Firmness, and Disease Incidence at Harvest

Variable	Wt[a] (g)	Firmness (N)	Disease incidence (%)	
			1991	1992
Cultivar				
NuMex Starlite	263 a[b]	54.4 a	21.5 a	8.4 a
NuMex Sunlite	239 ab	53.7 a	4.5 bc	5.2 ab
NuMex BRI	235 ab	53.5 a	10.9 b	7.8 a
Buffalo	220 b	55.3 a	1.9 c	2.0 b
Harvest[c,d]				
1	182 d	56.1 a	0.3 c	4.5 a
2	231 c	56.1 a	4.0 c	6.3 a
3	244 bc	54.2 b	10.5 b	4.5 a
4	265 ab	52.6 bc	13.7 b	7.8 a
5	274 a	52.2 c	20.0 a	6.3 a

[a] Average bulb weight and bulb firmness data combined for 1991 and 1992. Bulb evaluations were after curing but before storage.
[b] Mean separation within columns by Walter-Duncan k ration t-test at $p \leq 0.05$.
[c] Harvests 1 and 2 were when 20% and 80% of the tops had matured, respectively. Harvests 3, 4, and 5 were 5, 10, or 15 days after harvest 2, respectively.
[d] Cultivar × harvest was nonsignificant at $p \leq 0.05$ for all criteria.
Source: Ref. 57.

TABLE 5 Onion Cultivar and Storage Duration Effects on Average Bulb Weight, Firmness, and Disease Incidence

Cultivar	Storage (days)	Wt[a] (g)	Firmness N	Disease incidence (%)	
				1991	1992
Starlite	0	208	54.4	21.5	8.4
	10	195	52.3	47.7	25.0
	20	183	49.6	68.7	50.4
		L	L	L	L
Sunlite	0	209	53.7	4.5	5.2
	10	200	52.8	25.9	14.2
	20	189	48.8	53.2	49.7
		L	L	L	L
BRI	0	186	53.5	10.9	7.8
	10	177	52.1	36.9	25.8
	20	171	49.8	65.1	50.6
		L	L	L	L
Buffalo	0	186	55.3	1.9	2.0
	10	180	54.9	13.1	21.8
	20	171	50.8	53.9	52.7
		L	L	L	L

[a]Average bulb weights for 1992. Bulb firmness data combined for 1991 and 1992.
L = Linear response significant at $p \leq 0.001$.
Source: Ref. 57.

are cut off by hand with a knife or by a topping machine, which also sorts the bulbs into different sizes (13).

Harvesting systems commonly used with mechanized production systems in northern Europe deliberately sacrifice optimum bulb yield in favor of improved quality in terms of skin color and retention. The onions are topped (i.e., have their foliage cut off) and are lifted before the outer scales have started to dry and are then dried artificially with rapidly circulating warm air in bulk stores to obtain a well-colored, complete outer skin and dry neck. Some potential yield is lost by early toppnig and harvesting, and some of the actual weight harvested disappears during bulb skin drying. Although in theory the bulbs may be left in the field until they are completely mature, that is, fully dormant, so as to obtain maximum recovery of assimilates from the leaves, in practice bulbs are usually lifted or loosened from the soil when the leaves of 50–70% of the crop have fallen over, which indicates that most of the crop is reaching bulb maturity.

IV. CHEMICAL COMPOSITION

The proximate composition of selected *Allium* species is presented in Table 6. Brahmachari and Augusti (62) reported that fresh onions contained 86.8% moisture, 11.6% carbohydrates (including 6–9% soluble sugars), 1.2% proteins, 0.1% fats, 0.2–0.5% calcium, 0.05% phosphorus, and traces of Fe, Al, Cu, Zn, Mn, carotene, thiamine, riboflavin, nicotinic acid, and ascorbic acid. Bajaj et al. (63) reported that cultivars with higher phenolics had a greater amount of coloring

TABLE 6 Composition of Cultivated Alliums (fresh weight basis)

Allium species	Moisture (%)	Protein (%)	Fat (%)	Carbohydrate (%)	Ash (%)	Energy (kJ)
Onion	89.1	1.5	0.1	8.7	0.6	160
Garlic	61.3	6.2	0.2	30.8	1.5	575
Leek	85.4	2.2	0.3	6.0	0.9	218
Chive	92.0	2.7	0.6	4.3	0.4	113
Chinese chive	92.2	1.4	0.6	3.4	0.9	109
Rakkyo	86.2	0.6	ND	12.9	0.2	218

ND = No data available.
Source: Ref. 1.

matter in the dehydrated onions. The compositiuon of onion varies according to variety and agronomic and environmental conditons of growth (4). In general, varieties with high (up to 20%) dry matter are selected for processing. Onions are a rich source of amino acids and γ-glutamyl peptide, anthocyanins, flavonols, and phenolics. Nonstructural carbohydrates consisting of free sugars, trisaccharides, and fructans contribute the major portion of the dry weight of onions (4). High–dry-matter onion cultivars have reduced glucose and fructose contents and much higher fructan levels than varieties with low dry matter contents.

Onions are rich in sulfur-containing compounds (64–74). The enzyme alliinase hydrolyzes the S-alk(en)yl cystein sulfoxides to produce pyruvate, ammonia, and many volatile sulfur compounds associated with the flavor and odor of onion (1). This occurs when cells are damaged or disrupted. Onions contain primarily the S-(1-propenyl), propyl, and, to a lesser degree, methyl alliins. Thus, the typical flavor of onion is due to the presence of propyl- and 1-propenyl—containing allicins and di- and trisulfides. More than 80 compounds have been identified in freshly cut onions (17).

Onion flavor is influenced by variety, stage of the bulb, storage period, and nutrition of the crop (72–74). Application of large doses of sulfur increases flavor, depending on the sulfur content of soil itself. Randle et al. (72) evaluated onions grown in greenhouses with various applications of sulfur and assayed for changes in thiosulfinates and related compounds. The concentration of methyl, *n*-propyl, and 1-propenyl thiosulfinate, the zwiebelanes, and onion lachrymatory factor (LF) increased with a linear trend as sulfur fertility increased, with the exception of methane sulfinothionic acid, S-*n*-propyl ester, and methane sulfinothionic acid S-methyl ester. The thiosulfinate plus zwiebelane:LF ratio increased with increased soil fertility, suggesting that the former is produced more quickly the higher the sulfur content. Variation in the absolute and relative concentration of these three classes of sulfur compounds with changing sulfur fertility could cause subtle flavor differences in onion. The variation due to maturity in volatile sulfur compounds has been reported (66). There was a steady increase in volatile sulfur content with maturity, reaching a maximum just before the tops begin to fall (66).

Onion oil is used to treat stomach ulcers, eye disorders, gastrointestinal disturbances, high blood pressure, malarial fevers, and intestinal worms (5). The onion is known to possess insecticidal, antibacterial, antifungal, antitumor, hypoglycemic, hypolipidemic, and antiatherosclerotic properties. This has been attributed to its sulfur-containing compounds (5).

V. STORAGE

Onion cultivars vary in their suitability for storage (Table 7). Genetically controlled factors that may influence storage performance include dry matter content, pungency, skin color, skin number and quality, and length of natural dormancy of the particular onion variety (78–85). The preharvest cultural factors include fertilizer and irrigation regime under which bulbs are raised and use of maleic hydrazide as a sprout suppresant (86-91) before harvest. Physical injury during and after harvest, greening of onion due to exposure to sunlight, sprouting, and injuries during storage due to ammonia, controlled-atmosphere storage, and freezing also cause postharvest losses.

Saxena et al. (92) evaluated the percentage loss in white, yellow, and red cultivars of dry onions stored for various periods in Guyana and reported that red cultivars had a higher storage potential than yellow and white cultivars (Table 7). The cultivar White Creole had exceptionally better storage characteristics. Rajapakse et al. (58) evaluated ten short-day onion cultivars for their storage potential and reported that Burgundy, Texas Grano (TG) 10254, and Selection 91438 had good storage potential and lost less than 35% in weight after 100 days of storage (Table 8). Water loss contributed most to the total weight loss in Burgundy, while in TG 1105 Y and TG 1015 Y disease contributed most. Both water loss and disease contributed equally in other cultivars (Table 9). Sprouting was also a significant factor in TG 502 and TG 1015 Y in long-term storage. It is suggested that development of onion cultivars with tightly closed neck ends could greatly reduce the storage losses in onion cultivars with high water loss and that postharvest handling should be done carefully to avoid damage to dry outer scales.

Postharvest diseases of onion cause significant losses in the quantity and quality of onions during storage (93,94). Currah and Procter (2) have listed various fungal and bacterial diseases (95–102) of onion (Table 10). The yeast *Kluveromyces marxianus* var. *marxianus* has been identified as a pathogen of onions in the United States (103,104). The symptoms, a soft rot of the bulbs, were easily confused with those of bacterial soft rot. In Australia, Cother and Dowling (105) found seven genera of bacteria associated with onions and suggested that when physiological changes take place in onion bulbs at high temperature, some of the existing microflora usually present in bulbs can become pathogenic. Biochemical changes caused by fungal development may also allow such opportunistic bacteria to multiply. Mites are thought to transmit black mold between onions in store. A *Rhizoglyphus* sp. mite in India attacks onions both in the field and in storage, reportedly leading to sprouting. Mites of this genus were also reported to be numerous on stored onions in Brazil (106), where they were suspected of

TABLE 7 Percent Loss of Dry Onion in Storage in Guyana

Cultivar (bulb color)	Cultivars tested	Mean bulb loss in storage (%)			
		21 days	44 days	55 days	82 days
White cultivars[a]	5	33	64	75	86
Yellow cultivars	5	16	58	72	89
Red cultivars	3	0	30	42	62
White Creole	1	11	36	46	57

[a]Excluding White Creole.
Source: Ref. 92.

TABLE 8 Percentage Total Weight Loss of Onion Cultivars During Storage at Ambient Temperature

Cultivar	Days in storage			
	28	56	84	100
Burgundy	6.5[d]	17.2[cde]	29.0[d]	34.0[cd]
EEWSS	9.0[bcd]	28.0[bc]	49.0[ab]	58.0[a]
TG 502	9.0[bcd]	26.0[bcd]	41.0[bcd]	52.0[ab]
TG 1015Y	11.2[bcd]	24.3[bcde]	43.0[bc]	60.0[a]
TG 1020Y	12.7[bc]	32.6[ab]	46.1[ab]	51.0[abc]
TG 1025Y	11.4[bcd]	19.5[cde]	26.0[d]	30.0[d]
TG 1030Y	6.0[d]	20.6[cde]	33.9[bcd]	39.6[bcd]
TG 1105Y	19.9[a]	44.3[a]	60.9[a]	64.1[a]
91438	7.0[cd]	15.6[c]	28.3[d]	32.6[d]
91554	14.3[ab]	33.0[ab]	46.2[ab]	49.0[abc]
LSD_{0-05}	5.8	12.4	16.6	17.5

Numbers with a common letter within columns are not significantly different at the 5% level of significance.
Source: Ref. 58.

transmitting fungal and bacterial diseases. Some species of beetle (*Anthrenus ocenicus*, *A. jordanicus*, and *Alphitobius laeviqatus*) attack stored onions in India, allowing rot-causing pathogens to enter the bulbs.

A. Curing

Onions are thoroughly cured before being stored in crates. Crates filled with topped bulbs are usually stacked in the field and covered with boards, roofing paper, or some other type of

TABLE 9 Contribution of Water Loss, Disease, and Sprouting to Total Weight Loss of 10 Onion Cultivars After 28 and 100 Days Storage[a]

Cultivar	28 days			100 days		
	Disease	Sprouting	Water loss	Disease	Sprouting	Water loss
Burgundy	0(0)	0(0)	6.5(100)	5.0(15)	1.3(4)	27.7(81)
EEWSS	5.0(55)	0(0)	4.0(45)	30.0(52)	3.0(5)	25.2(43)
TG 502	3.0(33)	0(0)	6.0(67)	22.0(42)	7.2(14)	22.7(44)
TG 1015Y	5.0(45)	0(0)	6.1(55)	32.0(53)	10.0(17)	18.1(30)
TG1020Y	6.2(49)	0(0)	6.5(51)	25.5(49)	1.0(2)	24.9(49)
TG1025Y	7.2(63)	0(0)	4.2(37)	15.0(50)	0(0)	15.0(50)
TG1030Y	1.0(17)	0(0)	5.0(83)	17.5(44)	0(0)	22.1(56)
TG 1105Y	12.1(61)	0(0)	7.8(39)	41.0(64)	0(0)	23.1(36)
91438	3.0(43)	0(0)	4.0(57)	13.6(42)	0(0)	19.0(58)
91554	7.8(55)	0(0)	6.5(45)	23.5(48)	1.4(3)	24.1(49)
$LSD_{0.05}$	4.9	0(0)	1.9	15.0	5.2	4.1

[a]Percentage contribution of each component indicated in parentheses.
Source: Ref. 58.

Table 10 Postharvest Diseases of Onions

Diseases	Causal organism	Ref.
Fungal;		
Black mold	*Aspergillus niger*	95,96
Neck rot	*Botrytis allii*	97
Basal rot	*Fusarium Oxysporum* f.sp. *cepae*	97
Smudge	*Colletotrichum circinans*	98
Pink root rot	*Pyrenochaeta terrestris*	9
White rot	*Sclerotium cepivorum*	99
Downy mildew	*Peronospora destructor*	9
Twister or seven curls disease	*Glomerella cingulata*	9
Blue and green molds	*Penicillium* spp.	9
Bacterial		
Bacterial soft rot	*Erwinia herbicola*	100,101
Soft rot	*Erwinia carotovora*	100,101
Brown rot	*Pseudomonas aeruginosa*	100,101
Slippery skin	*Pseudomonas gladioli* Pr. *alliicola*	102

covering to protect the onions from injury by sun and rain (13). The crates may be stacked in open curing sheds. The time required to cure onions may be 3–4 weeks or longer depending upon the weather conditions. For mechanically harvested onions, bulk storage instead of crate storage is adopted. The onions are harvested and topped in the field with a machine and hauled by wagon or truck to the storage facility and stored in bulk piles 8–10 feet deep. With good aeration, shrinkage of bulk-stored onions due to curing is no greater than that occurring in crates in the field (107). Hoyle (108) reported that both field curing (windrows) and artificial curing (16 hr at 46°C) reduced losses of onion in storage compared with that non-cured bulbs (Table 11). Artificial curing immediately after harvest greatly reduces losses of onions from neck rot (109). The cured onions are further cleaned, graded, and packed. The USDA has specified grades for northern grown domestic onions and for Bermuda onions.

B. Low-Temperature Storage

Fully matured and thoroughly cured onions are used for storage. Immature, soft, thick-necked bulbs cannot be stored for long and should be disposed of immediately after harvest (13).

Table 11 Effects of Curing on Storage Losses of Onions

	31 days		63 days		87 days	
Method	Weight loss	Rot (%)	Weight loss	Rot (%)	Weight loss	Rot (%)
No curing	11.8	76.7	a	a	a	a
Artificial curing	6.3	12.3	10.8	16.3	13.2	24.8
Field curing	6.1	20.0	11.2	23.3	14.5	36.5
LSD at $p = 0.05$	3.3	14.9	NS	NS	NS	NS

[a]All rotted after first month.
NS = Not significant.
Source: Ref. 108.

Thorough ventilation, uniform low temperature, dry atmosphere, and protection against actual freezing are essential for storing onions successfully. Wright et al. (110) recommended a temperature of 0°C and a relative humidity of about 64% under these conditions. Sprouting is influenced little by the humidity but increases with increaseing temperature. It was also reported (110) that rotting increases with increase in humidity and is ltitle influenced by temperature. Decay increases only slightly as both temperature and humidity increases.

Ryall and Lipton (93) recommended a temperature of 0°C and relative humidity of about 65–75% for storing onions for fresh usage. In Europe, storage of onions below the freezing point has been recommended. For this, onions are quickly cooled to 0.5–1.1°C, and when the room is filled the temperature is lowered to –1.0 to –2°C. These onions are thawed for about 1–2 weeks with the air at about 4.4°C before they are removed from storage, since rapid thawing damages onions. Subfreezing retards sprouting and is a useful alternative when sprout inhibitors cannot be used (93). Singh and Singh (111) stored onions at a temperature near 0°C for 150 days with no rotting or sprouting occurring during this period. However, weight loss amounted to 13%. Kessler (112) reported that storage below freezing is most suitable for onions high in soluble solids intended for processing.

Onions can also be stored at high temperatures of 29.2–34.7°C (113). The storage structure used in India is shown in Figure 4. The external color of onions stored at high temperatures is less attractive than that of cold-stored onions; they are, however, desirable for subsequent dehydration because the dry flakes from such bulbs have better color retention (113).

Pantastico et al. (50) recommended temperatures of 1 and 0°C and a relative humidity of 70–75% for storing white and red onions, respectively, for 16–20 and 20–24 weeks. The requirements for cold storage of onions for their efficient use in the United Kingdom (114,115), the Netherlands (94), and the United States (95) have been described.

FIGURE 4 Improved structure for storage of onions at ambient conditions.

C. Controlled-Atmosphere Storage

The usefulness of controlled-atmosphere (CA) storage for onions has not been investigated thoroughly. Chawan and Pflung (116) found that 3–5% O_2 plus 10% CO_2 is superior to storage in air at 4.4°C, but these conditions were not compared with onions stored in air at 0°C. Ogata and Inoue (117) reported that low O_2 (1%) without added CO_2 was superior to air at room temperature, but no comparison was made at temperatures recommended for onions. Ryall and Lipton (93) observed that there is little need for CA storage, since onions can be kept most of the year if stored properly.

D. Chemical Control of Postharvest Losses

Maleic hydrazide (MH) has been successfully used to inhibit sprouting of onions during storage (118). Spraying onion with 2500 ppm MH 2 weeks before harvest completely inhibits sprouting during storage (118). Spraying with 908 g of acid equivalent of MH when 50% of the tops are down gives excellent control of sprouting (89).

Several chemicals are used as dips or dusts to control postharvest diseases of onions during storage. Hughes (119) reported that Dicloran 4–5% used as dust at harvest reduced *Botrytis*, *Rhizopus*, *Fusarium*, *Penicillum*, *Rhizoctonia*, and *Sclerotium* infections. After a postharvest dip of 908 g of Dicloran per 100 gallons, *Sclerotium* infection was reduced from 100% (untreated) to 18% after 33 days storage and to 34% after 61 days. Soaking onion sets in 0.2% benomyl before planting reduces *B. allii* and *F. oxysporum* infections to less than 5% during 6–7 months of storage (120).

Misra and Panda (121) soaked onion seedlings (cultivar Bombay Red) in solutions of ethephon, CCC (Cycocel), or MH at 10–150 ppm for 10 minutes and also applied preharvest sprays of these chemicals at 0, 500, 1000, or 2000 ppm. The subsequently harvested and cured bulbs were held at room temperature. After one month, the TSS and reducing sugars were highest after a preharvest spray of MH at 1000 and 2000 ppm, respectively; the ascorbic acid content was highest with ethephon at 2000 ppm.

In the field, good control of *Botrytis allii* has been obtained with Benlate (benomyl) or Bavastin (Carbendazim) used as bulb dips (0.5%) before planting. The bulbs treated with these fungicides gave the highest seed yield (52).

E. Irradiation

A dose of 5–15 krad of gamma radiation short after harvest effectively inhibits sprouting of onions (122–124). Gonzales et al. (125) found that a dose of 5–15 krad was sufficient to inhibit sprouting of Philippine varieties of onions (Red Creole and Granex). Matsuyama (126) reported that sprouting can be effectively controlled when onions are irradiated with 3–7 krad during their dormant period.

VI. PROCESSING

Onions are generally dehydrated and pickled. The ratio of raw material to finished processed product depends on the solid content of raw material, maturity at harvest, size and shape of the bulbs, deterioration in storage, and processing methods (107). Onions with high solid content (dry matter) are preferred for dehydration. Varieties with 15–20% TSS are the most desirable.

Onions used for processing should have high pungency, since the dehydrated product is primarily used as a flavoring agent and some of the pungency is lost during the dehydration process. White bulbs are preferred to either yellow or red varieties. The pigment quercetin in yellow onion is a bitter principle with inferior flavor. For economy in field harvesting and plant preparation, large bulbs are desired for processing. The bulbs should be able to hold up in common storage for 2–3 months with a minimum of rot, shrinkage, or sprouting.

A. Fermented Products

In some parts of West Arica, fermented preparations are made from crushed or ground onion leaves and tops. These products are used to flavor food at times when onions are not available. Onion scales may be sun-dried for the same purpose.

B. Dehydrated Products

Commercially prepared onion products include dehydrated flakes and powder usually made from white cultivars with high dry matter contents and high pungency. The dehydrated products are extensively used in processing industries. Onions are cleaned, peeled, sliced or chopped, and dehydrated at 75–60°C in a series of stages, the temperature being reduced as moisture content falls. A final moisture content of 4% is achieved by circulating warm air. The problems of scorching and agglomeration encountered during preparation of dehydrated products may be overcome if fluidized bed techniques are used. The yield of dehydrated product is about 10%. Rondriguez et al. (127) observed that Southport White Globe and White Creole are suitable varieties for dehydration.

C. Oil

Onion oil is obtained by distillation of minced onions; the yield of oil ranges from 0.002 to 0.03%, depending upon the raw material and processing conditions. It has a flavor strength 500 times that of the dehydrated product.

D. Juice

Onion juice contains both flavor and aromatic compounds, their precursors and sugars. They may be blended with volatile oils to restore a rounder flavor profile (8). The juice, which is viscous and dark brown in color, can be mixed with a support (such as propylene glycol, lecithin, or glucose) to produce an oleoresin having a flavor intensity 10 times that of dehydrated onion powder.

E. Other Products

Canned and bottled onions are used in catering industries in North America (128). Onion rings are common products in fast food industries.

Pickled onions are popular in the United Kingdom. These are steeped in 10% brine for 24–96 hours, with lactic acid being added to control fermentation. They are washed, covered in vinegar, and pasteurized; bisulfite may be added to maintain the color. The undesirable yellow spots caused by flavonoids reduce the quality of bulbs for pickling, hence flavonoid-free varieties should be used.

REFERENCES

1. Salunkhe, D. K., and B. B. Desai, Onion and garlic, in *Postharvest Technology of Vegetables*, Part II, CRC Press, Boca Raton, FL, 1984, p. 23.
2. Currah, L., and F. J. Proctor, *Onions in Tropical Regions*, Bulletin No. 35, National Resources Institute, Kent, United Kingdom, 1990, p. 20.
3. Chauhan, D. V. S., *Vegetable Production in India*, Ram Prasad and Sons, Agra, India, 1986.
4. Hanley, A. B., and R. G. Fenwick, Cultivated alliums, *J. Plant Food* 6:211 (1985).
5. Augusti, K. T., Therapeutic and medical values of onion and garlic, in *Onions and Allied Crops*, Vol. III (J. L. Brewster and H. D. Rabinowitch, eds.), CRC Press, Boca Raton, FL, 1990, p. 93.
6. Augusti, K. T., Gas chromatographic analysis of onion principles and study on their hypoglycemic action, *Indian J. Exp. Biol. 14*:110 (1976).
7. FAO, *Production Yearbook*, Food and Agriculture Organization, Rome, 1990.
8. Fenwick, R. G., Onions and related crops, in *Encyclopedia of Food Science, Food Technology and Nutrition* (R. Macrae, R. K. Robinson, and M. J. Sadler, eds.), Academic Press, London, 1993.
9. Jones, H. A., and L. K. Mann, *Onions and their Allies*, London, Leonard Hill, 1990, p. 286.
10. Hanelt, P., Taxonomy, evolution and history, in *Onions and Allied Crops*, Vol. I (H. D. Rabinowitch and J. L. Brewster, eds.), CRC Press, Boca Raton, FL, 1990.
11. Hayward, H. E., *The Structure of Economic Plants*, New York, MacMillan, 1983, p. 674.
12. Thomas, T. A., and B. S. Dabas, Genetic resources of onion in India, *Proc. 2nd Intl. Allium Conference*, Strasbourg, July 1986, p. 5.
13. Thompson, H. C., and W. C. Kelly, *Vegetable Crops*, McGraw-Hill Book Co. Inc., New York, 1957, p. 611.
14. Yawalkar, K. S., Bulb vegetables, in *Vegetable Crops of India*, 2nd ed., Agri-Hort. Pub. House, Nagpur, India, 1989, p. 370.
15. Dhesi, N. S., K. S. Nandpuri, and J. C. Kumar, Onion cultivation in Punjab,. *Agril. Inform. Selection, Punjab Agric. Univ.*, Ludhiana, 1965.
16. Pandey, U. B., Onion (*Allium cepa* L.), *Indian Hort. 33*:58 (1989).
17. Panedy, U. C., and J. Singh, Agro-Techniques for onion and garlic, in *Advance in Horticulture*, Vol. 5, *Vegetable Crops* (K. L. Chadha and G. Kalloo, eds.), Malhotra Pub. House, New Delhi, 1993, p. 433.
18. Shinde, N. N., and M. B. Sontakke, Bulb crops, in *Vegetable Crops in India* (T. K. Bose aznd M. G. Som, eds.), Naya Prokash, Calcutta, 1986, p. 544.
19. Katan, J., Solar heating (solarization) of soil for control of soil-borne pests, *Ann. Rev. Phytopathol. 19*:211 (1981).
20. Katan, J., A. Grinstein, A. Greenberger, O. Yarden, and J. E. Devay, The first decade (1976-1986) of soil solariation (solar heating): A chronological bibliography, *Phytoparasitica 15*:229 (1987).
21. Rabinowitch, H. D., J. Katan, and I. Rotem, The response of onions to solar heating, agricultural practices and pink-root disease, *Sci. Hort. 15*:331 (1981).
22. Sabota, C. M., and J. D. Downes, Onion growth and yield in relation to transplant pruning, size spacing and depth of planting, *Hort. Sci. 16*:533 (1981).
23. Kale, P. N., and J. Kale, *Vegetable Production*, Continental Press, Pune, 1984.
24. Zink, F. W., Studies on the growth rate and nutrient absorption of onion, *Hilgardia 37*:203 (1966).
25. Yawalkar, K. S., P. N. Kakade, and M. M. P. Srivastava, *Commercial Fertilizers in India*, Agric. Hort. Publishing House, Nagpur, 1962.
26. Brown, B. D., A. J. Hornbacher, and D. V. Naylor, Sulphur coated urea as a slow-release nitrogen sources for onions, *J. Am. Soc. Hort. Sci. 113*:864 (1988).
27. Tremblay, F. T., Nitrogen requirements is all that you have been putting on necessary, *Onion World 4*(2):28 (1988).
28. Brewster, J. L., Effect of photo-period nitrogen nutrition and temperature on inflorescence initiation and development in onion (*Allium cea* L.), *Ann. Bot. 51*:429 (1983).
29. Brewster, J. L., Cultural system and agronomic practices in temperate climate, *Onions and Allied Crops*, Vol. II (H. D. Rabinowitch and J. L. Brewster, eds.), CRC Press, Boca Raton, FL, 1990, p. 130.

30. Hassan, M. S., Effects of frequency of irrigation and fertilizer nitrogen on yield and quality of onions (*Allium cepa*) in the arid tropics, *Acta Hort. 143*:341 (1984).
31. Hassan, M. S., and A. T. Ayoub, Effect of N.P.K. on yield of onion in the Sudan, *Gezira Exp. Agric. 14*:29 (1978).
32. Balasundaram, C. S., A. Ramakrishnan, C. Kailasam, and C. Paulraj, Optimising fertilizer rates for Bellary onion, *Indian J. Agric. Sci. 53*:1022 (1983).
33. Hegade, D. M., Effect of irrigation and N fertilization on water relations, canopy, temperature, yield, N uptake and water use of onion, *Indian J. Agric. Sci. 56*:858 (1986).
34. Warade, S. D., S. B. Desale, and K. G. Shinde, Effect of organic, inorganic and biofertilizers on yield of onion (*Allium cepa* Cv. B-780), *Mah. J. Agril. Univ. 20(3)*:467 (1995).
35. Purewal, S. S., and K. S. Dargan, Effect of fertilizers and spacing experiment with onion crop, *Indian J. Agron. 7*(1):46 (1962).
36. Singh, J., and B. S. Dhankar, Effect of nitrogen, potash and zinc on growth, yield and quality of onion, *Veg. Sci. 12*:136 (1989).
37. Saimbhi, M. S., B. S. Gill, and K. S. Sandhu, Fertilizer requirement of processing onion (*Allium cepa* L.), *Punjab Agric. Univ. J. Res. 24*:407 (1987).
38. Gill, H. S., A. S. Kataria, K. A. Balakrishnan, and B. Singh, Vegetable research highlights (Agronomical practices), *FDVR Tech. Bull. 2*:36 (1986).
39. Choubey, P. C., K. Sharma, B. R. Sharma, and K. N. Tambi, Nutritional studies in kharif onion (*Allium cepa* L.), Nutritional Symp. Onion and Garlic, June 2-3, Y.S. Parmar University Hort, Solan, H.P., 1990.
40. Shukla, V., and B. S. Prabhakar, Response of onion to spacing, nitrogen and phosphorus levels, *Indian J. Hort. 46*:379 (1989).
41. Singh, K., M. S. Saimbhi, and U. G. Pandey, Response of onion to the application of nitrogen, phosphorus and potassium on sandy loam soils of Hissar, *Indian J. Hort. 29*:190 (1972).
42. Singh, J. R., and N. K. Jain, Response of onion (*Allium cepa* L.) to different fertilizer application, *Indian J. Hort. 16*:31 (1956).
43. Jawaharlal, M., and D. Veeraragavathatham, Effect of zinc and iron on the quality of onion (*Allium cepa* var. Cepa) bulbs, *South Indian Hort. 36*:114 (1988).
44. Randle, W. M., and J. E. Lancaster, Sulfur fertility affects growth and flavor pathway in onions, *Proc. Natl. Onion Res. Conf.* 91-102 (1993).
45. Chaugule, B. A., and V. S. Khuspe, Why do onions bolt? *Indian Farming* 7:18 (1957).
46. Patil, J. A., D. R. Waghmare, and A. K. Patil, Effect of age of seedlings on bolting in onion as a rabi crop, *Pune Agric. College Mag. 49*:83 (1958).
47. Pandey, U. C., J. Singh, and V. Virender, Onion and garlic as cash crops: Problems and solutions, *Haryana Farming, 19*:8 (1990).
48. Choudhary, B., Bulb crops, in *Vegetable Crops of India*, National Book Trust, New Delhi, 1967, p. 95.
49. Thompson, A. K., M. B. Bhatti, and P. P. Rubio, Harvesting and handling: Harvesting, in *Postharvest Physiology, Handling and Utilization of Tropical and Sub-tropical Fruits and Vegetables* (E. B. Pantastico, ed.), AVI Pub. Co. Inc., Westport, CT, 1975.
50. Bautista, O. K., W. A. Dancel, and D. T. Eligio, *Onion and Garlic*, The Philippine Recom. Vegetables, 1972–73, 1972.
51. Ware, G. N., and J. P. McCollum, *Raising Vegetables*, Interstate Printers and Publishers Inc., Danville, 1959, p. 11.
52. Pantastico, E. B., B. H. Subramanyam, M. B. Bhatti, N. Ali, and E. K. Akamine, Postharvest physiology, harvest indicies, in *Postharvest Physiology, Handling and Utilization of Tropical and Sub-tropical Fruits and Vegetables* (E. B. Pantastico, ed.), AVI Pub. Co., Westport, CT, 1975, p. 56.
53. Thompson, A. K., R. H. Booth, and F. L. Proctor, Onion storage in the tropics, *Trop. Sci. 14*:19 (1972).

54. Corgan, J. N., and N. Kedar, Onion cultivation in subtropical climate, in *Onions and Allied Crops* (H. D. Rabinowitch and J. L. Brewster, eds.), CRC Press, Boca Raton, FL, 1990.
55. Smittle, D. A., and B. W. Maw, Effects of maturity and harvest methods on storage and quality of onions, *HortScience 23*:141 (1988).
56. Yoo, K. S., C. R. Andersen, and L. M. Pike, Determination of postharvest losses and storage life of Texas Grano 1015Y onion, *J. Rio Grande Hort. Sci. 42*:45 (1989).
57. Wall, M. M., and J. N. Corgan, *Postharvest losses from delayed harvest and during common storage of short-day onions, HortScience 29*(7):802 (1994).
58. Rajapakse, N. C., C. R. Andersen, and L. M. Pike, Storage potential of short-day onion cultivars: Contribution of water loss, diseases and sprouting, *Trop. Sci. 32*(33) 1991.
59. Ali, A. A., and T. Elyamani, Studies on the effect of some cultural practices on storage diseases of onion, *Agric. Res. Rev. 55*:123 (1977).
60. Ladeinde, F., and J. R. Hicks, Internal atmopsphere of onion bulbs stored at various oxygen concentration and temperature, *HortScience 23*:1035 (1988).
61. Ali, A. A., and A. M. El-Shabrawy, Effect of some cultural practices and some chemicals on the control of neck rot disease caused by *Botrytis allii* during storage and in the field for seed onion production in A.R.E., *Agric. Res. Rev. Plant Pathol. 57*:103 (1971).
62. Brahmachari, H. D., and K. T. Augusti, Effect of orally effective hypoglycemic agents from the plants on alloxan diabetes, *J. Pharm. Pharmacol. 14*:617 (1967).
63. Bajaj, K. L., K. Gurdeep, J. Singh, and S. P. S. Gill, Chemical evaluation of some important varieties of onion, *Qual. Plant Pl. Foods Hum. Nutr. 30*:117 (1980).
64. Thomas, D. J., and K. L. Parkin, Quantification of alk(en)yl cysteine sulfoxides and related amino acids in *Allium* by high performance liquid chromatography, *J. Agric. Food Chem. 42*:1632 (1994).
65. Lancaster, J. E., and K. E. Kelly, Quantitative analysis of the s-alk(en)yl-L-cysteine sulphoxides in onion (*Allium Cepa* L.), *J. Sci. Food Agric. 34*:1229 (1983).
66. Lancaster, J. E., P. F. Reay, J. D. Mann, W. D. Bennet, and J. R. Sedcole, Quality in New Zealand grown onion bulbs—a survey of chemical and physical characteristics, *N.Z. J. Exp. Agric. 16*:279 (1988).
67. Block, E., The organosulfur chemistry of genus Allium—implication for organic sulphur chemistry, *Angew. Chem. Int. Ed. Engl. 31*:1135 (1994).
68. Freeman, G. G., and N. Mossadeghi, Effect of sulphate nutrition on flavor components of onion (*Allium cepa*), *J. Sci. Food Agric. 21*:610 (1970).
69. Vavrina, C. S., and D. A. Smittle, Evaluating sweet onion cultivars for sugar concentration and pungency, *HortScience 28*(8):804 (1993).
70. Thomas, D. J., K. L. Parkin, and P. W. Simon, Development of a simple pungency indicator test for onions, *J. Sci. Food Agric. 60*(4):499 (1992).
71. Bajaj, K. L., G. Kaur, J. Singh, and S. P. S. Gill, Lachrymatory factor and other chemical constituents of some varieties of onion (*Allium cepa* L.), *J. Plant Foods* (India) *3*:199 (1979).
72. Randle, W. M., E. Block, M. H. Littlejohn, D. Putnam, and M. L. Bussard, Onion (*Allium cepa* L.) thiosulfinates respond to increasing sulfur fertility, *J. Agric. Food Chem. 42*:2085 (1994).
73. Randle, W. M., and M. L. Bussard, Pungency and sugars of short day onions as affected by sulfur nutrition, *J. Am. Soc. Hortic. Sci. 118*:766 (1993).
74. Randle, W. M., Onion germplasm interacts with sulfur fertility for plant sulfur utilization and bulb pungency, *Euphytica 59*:151 (1992).
75. Randle, W. M., and J. E. Lancaster, Sulfur fertility affects growth and the flavor pathway in onions, *Proc. Natl. Onion Res. Conf.* 91-102 (1993).
76. Purewal, S. S., *Vegetable Cultivation in North India*, ICAR Farm Bull. No. 26, 1957.
77. Tieseen, H., T. L. Nonnecke, and M. Vaik, *Onion: Culture, Harvesting and Storage*, Ontario Min. Agr. Food., 1981, p. 27.

78. Davis, H. R., R. B. Furry, and F. M. Isenberg, Storage recommendations for Northern-grown onions, *N.Y. (Cornell Univ.) Inform. Bull. 148*:14 (1979).
79. Dewey, D. H., Using temperature and humidity as guides to curing and storing onions, *Mich. State Univ. Agr. Ext. Bull. E-1409*:4 (1980).
80. Buffington, D. E., S. K. Sastry, J. C. Gustashaw, and D. S. Burgis, Artificial curing and storage of Florida onions, *Trans. Am. Soc. Agric. Eng. 24*:782 (1981).
81. Stow, J. R., Effects of humidity on losses of bulb onions (*Allium cepa*) stored at high temperature, *Exp. Agric. 11*(2):81 (1975).
82. Smith, M. A., L. P. McColloch, and B. A. Friedman, Market diseases of asparagus, onions, beans, peas, carrots, celery and related vegetables, *U.S. Dept. Agr. Agri. Handbook 303*:65 (1982).
83. Tucker, W. G., J. R. Stow, and C. M. Ward, The high temperature storage of onions in the United Kingdom *Acta Hort. 62:181 (1977).*
84. Foskett, R. L., and C. E. Peterson, Relation of dry matter content to storage quality in some onion varieties and hybrids, *Proc. Am. Soc. Hort. Sci. 55*:314 (1950).
85. Vaughan, E. K., M. G. Cropsey, and E. N. Hoffman, Effect of field curing practices, artificial drying and other factors in control of neck rot in stored onions, Tech. Bull. 77, Oregon Agric. Expt. Sta. 1964, p. 22.
86. Wayse, S. B., Effect of N.P.K. on yield and keeping quality of onion bulbs (*Allium cepa* L.), *Indian J. Agron. 12*:379 (1967).
87.. Wittwer, S. H., R. C. Sharma, L. E. Weller, and H. M. Sell, The effect of pre-harvest foliage sprays of certain plant growth regulators on sprout inhibition and storage quality of carrots and onion, *Plant Physiol. 25*:539 (1950).
88. Isenberg, F. M. R., P. M. Ludford, and T. H. Thomas, Hormonal alterations during the post-harvest period, in *Postharvest Physiology of Vegetables* (J. Weichmann, ed.), Marcel Dekker, New York, 1987, p. 45.
89. Isenberg, F. M., and J. K. Ang, Effect of maleic hydrazide field sprays on storage quality of onion bulbs, *Proc. Am. Soc. Hort. Sci. 84*:378 (1964).
90. Komochi, S., Bulb dormancy and storage physiology in *Onions and Allied Crops*, Vol. I (H. D. Rabinowitch and J. L. Brewster, eds.), CRC Press, Boca Raton, FL, 1990, p. 89.
91. Bhalekar, M. N., P. B. Kale, and L. V. Kulwal, Storage behavior of some onion varieties as influenced by nitrogen levels and pre-harvest spray of maleic hydrazide, *Punjabrao Krishi Vidyapeeth Res. J. 11*:38 (1987).
92. Saxena, G. K., L. H. Halsey, D. D. Gull, and N. Persuad, Evaluation of carrot and onion cultivars for commercial production in Guyana, *Sci. Hort.* 2:257 (1974).
93. Ryall, A. L., and W. J. Lipton, *Handling, Transportation and Storage of Fruits and Vegetables*, Vol. 1, *Vegetables and Melons*, AVI Pub. Co. Inc., Westport, CT, 1972, p. 479.
94. Eckert, J. W., P. P. Rubio, A. K. Mattoo, and A. K. Thompson, Postharvest pathology, Part III. Diseases of tropical crops and their control, in *Postharvest Physiology Handling and Utilization of Tropical and Sub-tropical Fruits and Vegetables* (E. B. Pantastico, ed.), AVI Pub. Co., Westport, CT, (1975), p. 415.
95. Maude, R. B., M. R. Shipway, A. H. Presly, and D. O'Connor, The effects of direct harvesting and drying systems on the incidence and control of neck rot (*Botrytis allii*) in onions, *Plant Pathol. 33*:263 (1984).
96. Maude, R. B., and Burchill, R. T., Black mould (*Aspergillus niger*) storage rots of bulb onions in UK, *Proc. Intl. Congress Plant Pathology*, Kyoto, Japan, August, 1988, p. 419.
97. Miller, M. E., and J. Amador, Leaf wetness hours over onions fields in the lower Rio Grande Valley of Texas, *Report No. MP 1482 of Texas Agril. Exp. Station*, Weslaco, TX, 1981, p. 12.
98. Walker, J. C., Onion smudge, *Phytopathol. 11*:685 (1921).
99. Aycock, R., and J. M. Jenkins, Methods of controllig certain diseases of shallots, *Plant Dis. Rep. 44*:934 (1960).

100. Taylor, J. D., and H. L. Munasinghe, Storage rots of onions, bacterial rots, *Ann. Rep. Natl. Veg. Res. Stat. Wellesbourne*, United Kingdom, 1983, p. 85.
101. Tanaka, T., and I. Saito, Spread mechanism of bacterial soft rot of onions in a field, *Bull. Hokkaido Prefect. Agril. Exp. Sta.* (53):61 (1985).
102. Pike, L. M., E. S. Yoo, and T. H. Camp, A comparison between bags and boxes for shipping Texas short-day onions, *Hot.Sci. 24*:631 (1989).
103. Johnson, D. A., K. M. Regner, and J D. Lunden, Yeast soft rot of onion in Walla valley of Washington and Oregon, *Plant Dis. 73*:686 (1989).
104. Johnson, D. A., J. D. Rogers, and K. M. Regner, A soft rot of onion caused by the yeast *Kluyveromyces marxianus* var. *marxianus*, *Plant Dis. 72*:359 (1988).
105. Cother, E. J., and V. Dowling, Bacteria assicated with internal breakdown of onion bulbs and their possible role in disease expression, *Plant. Pathol. 35*:329 (1986).
106. Campos, H. R., J. Soava, C. J. Rossetto, and H. C. Mendes, Approdrecimento de bulbos de cebola, *Hort. Brasil. 4*:45 (1986).
107. Boyd, J. S., and J. F. Davis, Mechanical handling and bulk storage of onions, *Quart. Bull. Michigan Agric. Exp. Sta. 35*:259 (1952).
108. Hoyle, B. J., Onion curing: A comparison of storage losses from artificial field and noncured onions, *Proc. Am. Soc. Hort. Sci. 52*:407 (1948).
109. Harrow, K. M., and S. Harris, Artificial curing of onions for control of neck rot (*B. allii*), *N.Z. J. Agric. Res. 12*:592 (1969).
111. Singh, K., and D. Singh, Effect of various methods of storage on keeping quality of onion bulbs, *Haryana J. Hort. Sci. 2*:116 (1973).
112. Kessler, H., Fruits and Vegetables 2. Cold storage of various vegetables, in *Handbuch der Katatechnik*, Vol. 10 (H. Engerth, ed.), 1960, p. 520.
113. Yamaguchi, M., H. K. Pratt, and L. L. Morris, Effect of storage temperature on keeping quality and composition of onion bulbs and on subsequent darkening of dehydrated flakes, *Proc. Am. Soc. Hort. Sci. 69*:421 (1957).
114. Kristiaan, G. J., and P. S. Hak, Snel drogen en langzaam afkcelen is goed voor uien, *Boerderij 74*(22):38 (1989).
115. Matson, W. E., N. S. Mansour, and D. G. Richardson, *Onion Storage—Guidelines for Commercial Growers*, Pacific Northwest Extension Publication No. 277, 1985, p. 15.
116. Chawan, T., and I. J. Pflug, Controlled atmosphere storage of onions, *Q. Bull. Mich. Agric. Exp. Sta. 50*:449 (1969).
117. Ogata, K., and T. Inoue, Studies on the storage of onions. 7. Physiological changes of onions bulbs during storage in closed vessels with special reference to the use of soda-lime, *J. Jpn. Hort. Assoc. 25*:237 (1957).
118. Wittwer, S. H., R. C. Sharma, L. R. Weller, and H. M. Sell, The effect of preharvest foliage sprays of certain growth regulators on sprout inhibition and storage quality of carrot and onions, *Plant Physiol 25*:539 (1950).
119. Hughes, I. K., Onion storage: Control of white rot of onions in storage, *Queensland J. Agric. Anim. Sci. 27*:391 (1970).
120. Tahvonen, R., Storage fungi of onion and their control, *J. Sci. Agric. Soc. Finland 53*:27 (1987).
121. Mishra, R. S., and S. C. Panda, Biochemical changes in onion bulbs during storage, *Punjab Hort. J. 19*:86 (1979).
122. Salunkhe, D. K., Gamma radiation effects on fruits and vegetables, *Econ. Bot. 15*(1):28 (1961).
123. Chachin, K., and K. Ogata, Effect of delay between harvest and irradiation and of storage temperatures on the sprout-inhibition of onions by gamma irradiation, *J. Food Sci. Technol.* (Japan) *18*:378 (1971).
124. Chachin, K., M. Matsuzuka, H. Honjo, and K. Ogata, Sprout inhibition of 'Sapporo-ki' onions and

the changes of nucleic acids in their buds by gamma radiation, *J. Food Sci. Technol.* (Japan) *19*:232 (1972).

125. Gonzales, O. N., L. B. Diamaunanan, L. W. Pilac, and V. O. Alabastro, Effect of gamma radiation on peanuts, onion and ginger, *Philipp. J. Sci. 98*:279
126. Matsuyama, A., Present status of food irradiation research in Japan with special reference to microbiological and entomological aspects, Paper presented at the Intl. Symp. on 'Radiation Preservation of Food' in Bombay, India, 1972.
127. Rodriguez, R., B. L. Raina, E. B. Pantastico, and M. B. Bhatti, Quality of raw materials for processing, in *Postharvest Physiology, Handling and Utilization of Tropical and Sub-tropical Fruits and Vegetables* (E. B. Pantastico, ed.), The AVI Pub. Co., Westport, CT, 1975, p. 467.
128. Pruthi, J. S., *Spices and Condiments*, National Book Trust, India, New Delhi, 1979, p. 125.

17

Garlic

S. D. Warade and K. G. Shinde
Mahatma Phule Agricultural University, Rahuri, India

I. INTRODUCTION

Garlic (*Allium sativum* L.) is a hardy perennial vegetable crop whose probable origin is middle Asia and the Mediterranean regions (1). Garlic grows under much the same set of soil and climatic conditions as does onion, but garlic prefers moderate summer and winter temperatures. According to the Food and Agriculture Organization (FAO), China is the highest producer of garlic (2), followed by South Korea, Spain, and India (Table 1). The world production of garlic in 1990 was estimated to be 3012 thousand metric tons (2). Garlic is largely used as a condiment and a flavoring agent in soups, stews, pickles, and salads. A good tasty pickle is also prepared from garlic cloves. Spray-dried garlic products are also available (3). Garlic is also used to disguise the smell and flavor of salted meat and fish. Dehydrated garlic in powdered or granulated form is sometimes used in place of fresh bulbs.

II. BOTANY

A truly wild species of garlic (*A. Sativum* L.) is not known, but *A. longicuspis* Rgl. is considered to be the species most closely related to the cultivated crop and is considered to be garlic's wild ancestor. Since *A. longicuspis* is native to Central Asia, it is believed that garlic originated there (4).

Taxonomists have recognized at least four botanical varieties within *A. sativum* L., namely, *A. sativum* L. var. *sativum*, *A. Sativum* L. var. *ophioscorodon* (Link) Doll, *A. sativum* L. var *pekinense* (Prokh) Maekawa, and *A. sativum* L. var. *nipponicum* Kitamura.

Distinguishing between garlic cultivars using the following criteria is relatively simply and may be of practical use:

1. *Morphological characteristics*: bolting type, number and size of (the primary) cloves, number of leaf axils forming (primary) cloves, number of secondary cloves formed in a lateral bud (or primary clove), bulb weight, color of the outer protective leaf of the

TABLE 1 Garlic Production Worldwide

Country	Production (× 10^3 m^T)
World	3012
China	647
South Korea	400
Spain	229
India	296
Egypt	200

Source: Ref. 2.

cloves, number of protective leaves, width and length of foliage leaves, plant height, and tenderness of the green leaves.,

2. *Physiological and ecological characteristics*: time of bulbing and maturity, low temperature and long-day requirements for bulb formation, winter hardiness, and bulb dormancy.

In addition to the the above characteristics, the taste of the cloves may also be used in practical classification. Also, adaptation to agroclimatic zones may be of help. This is a complex characteristic determined by the above-mentioned environmental requirements and is possibly the most important factor for cultural practice, since garlic cultivars adapted to one area are sometimes worthless as a crop in other regions.

Garlic is a bulbous erect herb 30–100 cm in height with narrow flat leaves and bearing small white flowers and bulbils (5). It has a superficial adventitious root system. The bulb is composed of disclike stem, thin dry scales, which are the bases of foliage leaves, and smaller bulbs (bulblets) or cloves (Fig. 1). These are formed from axillary buds of younger foliage leaves. The cloves are enclosed by the dry outer scales (6,7). Each clove consists of a protective cylindrical sheath, a single thickened storage leaf sheath, and small control bud. The leaf blade is linear, flat, solid, 2–5 cm wide, and folded lengthwise (8,9). Flowers are variable in number and sometimes absent, seldom open and may wither in the bud.

Two distinct cultivars—Fawari and Royalle Gaddi—are reported from India (3). In

FIGURE 1 Garlic bulbs and cloves.

addition, some local cultivars—Godawari, Sweta, Madrasi, Tabiti, Creole, Eknalia, TSS-4, and Jamnagar (3)—are grown in different parts of India. Odeskii-13 and Imeruli-23 are grown in the CIS (5). The principal varieties grown in the United States for dehydration are California Late (Pink or Italian) and California Early (White or Mexican and Creole).

III. PRODUCTION

A. Soil and Climate

A wide range of soils with good drainage can be employed for garlic cultivation. The soil depth should be at least 45–60 cm. According to Rao and Purewal (10), garlic requires medium black, well-drained loamy soils, rich in humus with fairly good potash content. Crops raised on sandy or loose soils have poor keeping quality and the bulbs produced are lighter in weight. Bulbs produced in heavy soils are deformed and during harvest many bulbs are broken and bruised. Bulbs become badly discolored ill-drained soil. Katyal (11) suggested that acidic soils are not good for clove development, however, a pH range between 5 and 7 had little effect on growth and yield. Mangal et al. (12) reported that garlic crop can tolerate salinity between 5.60 and 7.80 dS m^{-1} EC depending on cultivar.

Garlic grows under a wide range of climatic conditions. However, it cannot tolerate weather that is too hot or too cold (13). It prefers moderate temperature in summer as well as in winter. Extremely hot or long dry periods are not favorable to bulb formation. It is a frost-hardy plant requiring a cool and moist period during growth and a relatively dry period during bulb maturity. Bulbing takes place during longer days and at high temperatures. An average temperature of 25–30°C is most conducive for bulb initiation. Bulbing in garlic occurs with lengthening photoperiods in the spring, with the process being hastened by increasing tempeatures up to 25°C. Low temperatures during growth may induce sprouting of already formed cloves (14).

B. Propagation

Garlic is propagated by planting cloves that have been stored at 5°C for several months. The size of the seed clove is important—larger cloves give higher yield. The seed should be virus-free. Shinde and Sontakke (15) recommended that cloves for planting weigh 3.6–5.8 g. Bigger bulbs may be produced for consumption and small bulbs for planting purposes. The size of bulbs can be regulated by plant spacing. Close spacing produces smaller bulbs. Nath (16) recommended 400–500 kg of cloves per hectare for tropical regions. Aerial bulbils or topsets are also used as planting material. Om and Srivastava (17) obtained 96.10 q/ha yield at Chaubatia (Dist. Nainital, India) using aerial bulbils with closer spacing.

C. Cultural Practices

1. Planting

Garlic is extremely hardy and survives long periods at temperatures below 0°C. Consequently, in temperate regions, it may be planted in autumn or in spring. Studies on successional sowing dates in temperate regions consistently showed that autumn planting gives higher yields than midwinter or spring planting (18–23). In Kalyani, India, mid-November planting gave higher yields than late December planting (24). In Bangladesh, late October planting gave higher yields than November or mid-December planting (25).

Machinery for mechanically separating and grading cloves prior to planting has been developed (26). Machine planting using a pneumatic planter gave more even planting depths and interplant spacing than the widely used tulip planting machines (27). The pneumatic planter gave higher emergence percentages, yields, and quality. If cloves are damaged, as can happen with machine grading and planting, their subsequent multiplication is reduced (28).

Planting of garlic is done by different methods (29–35), which includes dibbling (29,32,34), furrow planting (31), broadcasting (32), and transplanting (33). Likhatskii (36) showed that soaking garlic cloves in 0.01, 0.05, or 0.1% solutions of $ZnSO_4$ or $MnSO_4$ induced more vigorous growth of the aerial parts than soaking in distilled water. $MnSO_4$ at 0.01% and $ZnSO_4$ at 0.1% produced the highest yields (95.3 and 93.3 hg/ha, respectively). The highest yield was produced from 5.0 g of cloves planted at 30 × 6 cm, and the best quality crop was obtained from 5.0 g of cloves planted at 30 × 12 cm (37). Blyshchik and Furman (38) reported that the highest yields (16.2 t/ha) of autumn-planted garlic were obtained by planting aerial bulbils of >5 mm diameter with the scapes removed; a yield of 12.8 t/ha was obtained from bulbils planted with the scapes attached. The bulb yield was highest (131.8 and 131.5 q/ha) from planting on September 20 and declined with delayed planting to 59.4 and 59.5 q/ha when planted on October 30 (39). Numerous studies show that both yield and the size of harvested bulbs increases with planted clove weight (40–42). A clove weight of 4–5 g has been recommended for planting in Chile and Poland (42).

2. Manuring and Fertilization

Garlic responds well to organic manure. A good yield was achieved with an application of 40–50 t of farmyard mantures (FYM) per hectare. Nitrogen plays an important role in early vegetative growth. Nitrogen application ranging from 50 to 150 kg/ha has produced positive effects on garlic yield (43–47). Phosphorus application with the nitrogen helps to increase the yield (48). Combined application of nitrogen, phosphorus, and potassium has been found beneficial for garlic yield as well as clove flavor (31,49). Pal and Pandey (49) obtained significantly higher yield by applying 150 kg of N, 250 kg of P, and 75 kg of K per hectare. The application of micronutrients such as boron (0.5 ppm) was found to increase garlic yield (50). Yield also increased significantly as N, P, and K supply increased; N had the greatest effect (51). Setty et al. (52) showed that the largest bulb diameter (3.67 cm) and the highest yield (7.91 t/ha) were observed with 100 kg of N, 50 kg of P_2O_5, and 50 kg of K_2O per hectare. Both organic manure and sulfur were very beneficial for plant growth, total yield, and quality as well as N, P, and K contents in the plant tissues (53). High levels of nitrogen fertilizer have been reported to increase the percentage of plants showing secondary growth (54,55). Such plants give rise to undesirable bulbs, with rough surfaces.

3. Irrigation

Garlic needs irrigation at 8-day intervals during vegetative growth and 10- to 15-day intervals during maturation (33). Pandey and Singh (56) recommended irrigation at 60 mm CPE (cumulative pan evaporation) for a good garlic crop. Irrigation frequncies are decreased towards crop maturity as more moisture causes regeneration and emergence of sprouts, resulting in nonuniform maturity and bulb degernation, i.e., lower yield. As crop reaches maturity, irrigation is discontinued to allow the field to dry for harvest.

In Egypt, the yields of crops receiving between 50 and 150 kg of nitrogen per hectare were compared. These were watered when the avaialble soil water had declined by 20, 40, 60, 80 or 100%. The maximum yields were attained at the highest rate of nitrogen along with irrigation at the least water deficit (57). Greenhouse studies in Argentina indicated that irrigation with solutions of electrical conductivity greater than 3.0 dS m^{-1} may significantly reduce yields (58).

4. Interculture and Weed Control

Garlic is a clsoely planted crop, hence, manual weeding is tedious, expensive, and often damages the plant (5). Mollejas and Mata (59) reported that treatment with simazine-Diuron (0.75–1.0 kg/ha) postemergence and simazine-Nitrofen (0.75–3.0 kg/ha) preemergence was 20% more profitable than weeding twice. The control of weeds using Linuron (2.25 kg/ha), Pendimethalin (1.5 kg/ha), and Tribunil (2.1 kg/ha) has been reported (4). Hoeing the crop just before the formation of bulbs loosens the soil and helps in setting of bigger and compact bulbs.

D. Diseases, Disorders, and Pests

White rot caused by *Sclerotinum cepivorum* and leaf blight caused by *Cladosporium delicatulum* and *Stemphylium botryosum* are the major diseases of garlic (5). A typical disorder, sprouting of bulbs in the field, has been noticed in Saurashtra, India, mostly towards the start of the maturity stage of bulbs, particularly when there have been winter rains or excessive soil moisture and supply of nitrogen. However, this disorder is not of a permanent nature. The loss due to sprouting in the field was not mroe than 0.5%. Sprouting is more commonly seen in white cultivars than in pink or purple cultivars (6). Varietal differences, spacing, and early planting were possible causes of sprouting (59). The occurrence of mites (*Rhizoglyphus echinopus*) and aphids in garlic has been reported (5). Since garlic is propagated vegetatively, nematode infestation has been reported (60). These can be controlled by applying nematicides like Temik® and Nemafos® (30 kg/ha) to soil before planting. A disorder known as waxy breakdown, in which the clove shrinks and becomes amber, translucent, and waxy or sticky to the touch, occurs in stored garlic and occasionally in the field. It may be associated with sunscald in the field but is usually associated with poor ventilation and lack of oxygen during storage and transport.

E. Harvesting and Handling

Generally, the garlic crop is ready for harvest 130–150 days after planting, depending on cultivar, soil, and season. At this stage, the tops become partly dry and bend to the ground. Early harvest results in poor-quality bulbs which cannot be stored for long periods. Delayed harvest results in splitting and resprouting of bulbs in some cultivars. A special cutter with a horizontal cutter bar passing under the rows is used to loosen the garlic bulbs from the soil. The whole plant is then easily pulled. In piling, the tops are laid over adjacent bulbs to prevent sunscald (61).

Garlic is harvested when the foliage collapses and starts to senesce. In a warm dry climate, the bulbs may be left to dry and cure in windrows in the field for about 3 weeks, using the foliage to protect the bulbs from direct sun. Where rain or dew may occur at harvest time, the tops and roots are removed and the bulbs are cured indoors. Potato and root harvesting machines can be modified and used for garlic (62,63). Before lifting, tops can be removed to about 13 cm above the bulb using a rotary knife. Trimmed bulbs can be stored in 2-m-deep bins or stacks provided with forced ventilation, as well as in shallow containers or sacks. Cold air (down to 0°C) of low relative humidity (RH) is best for ventilating the stacks (62).

Curing of bulbs is carried out for about a week in the field to dry the bulbs thoroughly (64). Bulbs are covered with their tops to avoid sun injury. After drying, plants are tied in small bundles and then stored hanging on bamboo sticks or rope. In some areas, after drying tops and roots are removed, leaving about 1 cm of root and 2 cm of tops. Normally, 4–10 tonnes of bulbs are obtained per hectare (29). Recovery of cloves in the bulbs was 86–96% (5).

IV. CHEMICAL COMPOSITION

Garlic is a rich source of carbohydrates, protein, and phosphorus (Table 2). Ascorbic acid content was reported to be very high in green garlic (65,66). Garlic contains an amino acid called alliin. On crushing the bulb, the enzyme alliinase (EC 4.4.1.4) converts alliin into allicin (Fig. 2). Diallyl thiosulfinate (allicin) is the major flavor component in fresh garlic (67–71). This was found to undergo nonenzymatic degradation leading to formation of mono-, di-, and trisulfides (72). Shankaracharya (73) reported that garlic contained about 0.1% volatile oil. The chief constituents of the oil are diallyl disulfide (60%), diallyl trisulfide (20%), ally-propyl disulfide (16%), a small quantity of disulfide, and probably diallyl polysulfide (73). The main flavor constituents of a water extract of garlic bulbs were sulfur-containing compounds, such as alliin-S-methyl-L-cysteine sulfoxide and γ-L-glutamyl-S-allyl-L-cysteine (74). Shobahalan and Arumugam (75) reported that during bulb development, N, P, K, Ca, and Mg contents decreased in the leaves and roots but increased in the bulbs. The bulbs contained a higher percentage of reducing and nonreducing sugars and proteins than did the other two plant parts.

Garlic is a gastric stimulant and helps in the digestion and absorption of food (6). Augusti (76) reported that allicin, which has a hypocholesterolemic action, is present in aqueous extract of garlic and reduces the cholesterol concentration in human blood. Mahanta et al. (77) observed significant reduction in serum cholesterol levels when 50 g of raw garlic was fed daily to 10 volunteers for one month. Several reports are available on lowering of cholesterol on feeding of garlic to experimental animals (78–82). A blood sugar–lowering effect was also reported (83,84). This was ascribed to allicin and related disulfide-containing compounds. However, excess consumption may also lead to anemia. Therefore, consumption of only 7–10 g of garlic is recommended per day.

Singh et al. (85) found that garlic leaf extract had fungicidal effects on young fungi. Crude extract of garlic was found to be quite effective against gram-negative as well as gram-positive bacteria. It inhibited the growth of some bacterial cultures that are resistant to commonly used

TABLE 2 Composition of Garlic and Its Products

Nutrients	Fresh peeled garlic cloves	Dry garlic clove	Dehydrated garlic powder
Moisture (%)	62.80	62.00	5.20
Protein (%)	6.30	6.30	17.50
Fat (%)	0.10	0.10	0.60
Mineral matter (%)	1.00	1.00	3.20
Fiber (%)	0.80	0.80	1.90
Carbohydrates (%)	29.00	29.00	71.40
Calcium (%)	0.03	0.03	0.10
Phosphorus (%)	0.31	0.31	0.42
Potassium (%)	—	—	1.10
Iron (%)	0.001	0.001	0.004
Naicin (%)	—	—	0.70
Sodium (%)	—	—	0.01
Vitamin A (IU)	—	0.40	175.00
Nicotinic acid (mg/100 g)	0.40	0.40	—
Vit. C (mg/100 g)	13.00	13.00	12.00
Vitamin B_2 (mg/100 g)	—	0.23	0.08

Source: Ref. 65.

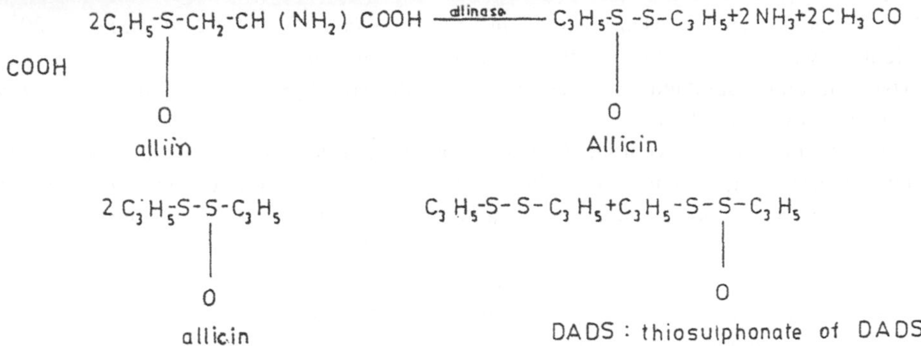

FIGURE 2 Enzymatic conversion of alliin into allicin. (From Ref. 67.)

antibiotics (86,87). About 10% dilution of aqueous extract of garlic completely inhibited the growth of these bacteria. In addition, reports showed that garlic extract exhibited promising antibacterial activity against several clinical strains (88). Garlic possesses insecticidal action and repellent properties. A formulation containing 1% garlic extract gave protection to persons against mosquitoes and black fly for 8 hours. Debkirtaniya et al. (89) reported garlic extracts possessing nematicidal and larvicidal properties, respectively.

V. STORAGE

Garlic cloves store best at 1°C. Garlic cloves sprout most rapidly at 4–18°C. Hence, prolonged storage at this temperature range should be avoided. Relative humidity should be lower for garlic than for most vegetables because high humidity causes rot and mold growth. It is essential that garlic destined for storage be well cured in the field and be provided with adequate air circulation through storage containers to remove transpired moisture (90,91). Incomplete curing results in excessive decay, particularly when bulbs are held above 0°C. Garlic can be held at 27–32°C satisfactorily for one month or less if humidity is as low as recommended. Storage life of garlic can be extended by treatment with maleic hydrazide before harvest or gamma irradiation after harvest. These treatments are effective in controlling sprout growth and weight loss and in decreasing external discoloration and diseases (90).

A. Dormancy and Storage of Bulbs

Immediately after harvest, garlic cloves are deeply dormant. The number of days to sprout emergence of cloves planted under optimal conditions for growth varies with the degree of maturity of the cloves and with storage temperatures (92). Cloves lifted when the foliage is still green require more time to sprout than mature cloves lifted when the foliage has senesced. It appears that the more immature the cloves are, the deeper their dormancy. However, if dormancy is estimated in terms of the growth rate of cloves while still attached to the mother plant, cloves from plants whose foliage has almost totally senesced can be considered the most dormant. Those from more immature plants where the foliage is still green are less dormant, since the storage leaf and sprout leaves are still growing.

In plants with near totally senescent foliage, the growth of storage leaves in the clove has almost ceased. Furthermore, the increase in number and length of the sprout leaves enclosed by the storage leaf ceases at this time (4,5). Growth of sprout leaves or of new roots does not occur for about 2 weeks when cloves are planted under optimal conditions for growth, i.e., at 20°C and with plentiful water. The growth inhibition of cloves is not caused by external conditions directly (e.g., rising air and ground temperatures) but by an internal (physiological) control, which slows down the growth of the cloves.

Cloves stored in well-ventilated sheds after total leaf senescence resumed sprout growth even later than those immediately transplanted to the field. The rooting of these cloves was much delayed. When such stored cloves were planted at approximately 2-week intervals in 20°C, cloves planted immediately after lifting and those stored for 2 weeks did not differ in the number of days required for sprout emergence. Cloves stored for 4 weeks sprouted considerably faster (Table 3). This indicates that the cloves were in total true dormancy for about 2 weeks after lifting, but that dormancy was in part imposed by the storage conditions in those stored for 4 weeks. Dormancy as measured by time to visible sprouting declined steadily and reached a minimum about 8 weeks after lifting (Fig. 3). The rate at which dormancy declines greatly depends on storage temperature and on cultivar.

B. Low-Temperature Storage

Garlic stores well under a wide range of temperatures but sprouts most quickly when stored at about 4.4°C (93). Low storage humidities are generally recommended to discourage mold development and formation of new rots. Ryall and Lipton (94) recommended temperatures of –0.6 to 0°C and 80% or less relative humidity to store garlic for at least 6–7 months. Higher temperatures of 26.4–32°C are satisfactory to store garlic for one month or less. Intermediate temperatures between 4.4 and 18.2°C are undesirable because they favor rapid sprouting, and high relative humidity results in rot and mold growth. Garlic may be kept in ventilated storage for 3–4 months (Fig. 4). The garlic bulbs must be thoroughly cured before they are stored since incomplete curing results in excessive decay, especially when storage temperatures are greater than 0°C. Patnastico et al. (61) recommended a temperature of 0°C with 65% relative humidity to store dry bulbs of garlic for 28–36 weeks.

TABLE 3 Sprouting and Root Growth of Cloves[a] of Garlic cv. Yamagata After Harvest

Observation date	Root length (mm) Field	Controlled temperature (20°C)	Ambient temperature
July 25	< 0.5	< 0.5	< 0.5
August 8	< 1.0	12.6 ± 3.4[b]	< 0.5
August 22	19.4 ± 6.6	54.2 ± 8.7	≤ 0.5
September 5	62.8 ± 6.3	136.8 ± 8.5	≤ 0.5
September 19	—	—	≤ 1.0

[a]Clove weight 5.0–7.0 g.
[b]± SE replication approximately 20.
Source: Ref. 92.

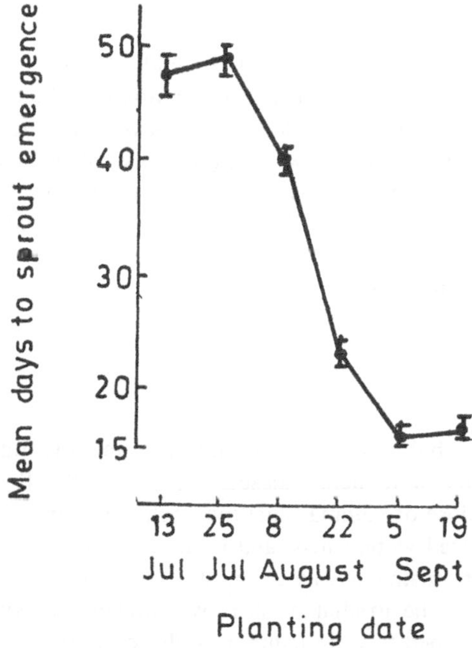

FIGURE 3 Effect of planting date on the time from planting to sprouting at 20°C in cloves of cv. Yamagata stored at natural temperatures after harvest. (From Ref. 92.)

FIGURE 4 Storage of garlic at ambient conditions.

C. Chemical Control of Postharvest Diseases

The important storage and market diseases of garlic include blue mold rot, bulb decay, *Aspergillus* rot, *Fuasrium* rot, dry rot, and gray mold rot. The most prevalent disease is blue mold rot, wherein lesions are produced. In advanced stages of decay the cloves become soft, spongy, and covered with powdery spore masses of *Penicullum* spp. (94). Dry rots are caused by *Fusarium solani*. Michail et al. (91) reported that cloves infected with dry rots do not sprout. Spraying with Bordeaux mixture (5,81), Ferbam, Zineb, and Nabam in the field controls dry rots effectively (95). Fumigation with methyl bromide at 32 g/m^3 for 2 hours at 21°C has been recommended for the control of the wheat curl mite (*Aceria tulipae*) in stored garlic (96).

D. Irradiation

Significant weight loss occurs in stored garlic bulbs due to sprouting. Several investigators have proved the efficacy of ionizing radiation in reducing these losses (97-100). Temkin-Gorodeiski et al. (101) reported that a 2 krad dose of gamma radiation applied within 8 weeks of harvest inhibited sprouting effectively, reduced weight loss, and prolonged the storage life of garlic for about one year. When irradiation treatment was given at a later date, sprouting was not inhibited (101). The respiration rate of the irradiated garlic was higher immediately after the treatment, but within a few days it dropped to values similar to those of nonirradiated garlic (101).

Croci and Curzio (100) reported that garlic bulbs treated with 50 Gy of ^{60}Cobalt 30 days postharvest can inhibit sprouting and reduce weight loss during storage. Joong and Hyung (102) evaluated the diallyl disulfide (DADS) content in garlic bulbs and reported that doses higher than 100 Gy caused an apparent reduction in the amount of DADS. Ceci et al. (103) reported that irradiation of garlic bulbs cv. Red did not affect the flavor. DNA was the cellular constituent sensitive to radiation in garlic cloves treated with 10 Gy of gamma radiation (104).

VI. PROCESSING

A number of garlic products including capsules, extracts, and tablets are now marketed, and odor-free garlic products have recently appeared (9).

A. Dehydrated Products

Garlic harvested for dehydration is brought to a dehydration plant in large bulb bins or open mesh bags. The bulk is broken into individual cloves by passing between rubber-covered rollers, which exert just enough pressure to crack the bulb without crushing the cloves. The loose paper shell is removed by screening and aspiration. The cloves are then washed in a flood washer, at which time the root stabs are floated off.

Garlic is sliced and dehydrated in a manner similar to that used for onions. After drying, that pink skin that adheres tightly to fresh cloves can be removed by screening and air aspiration (Fig. 5). Garlic is commercially dried to about 6.5% moisture. Dehydrated garlic is sold commercially as powder, as granules, or in sliced, chopped, or minced form. Packaging is similar to that used for onions. Dry garlic is widely used in the formulation of spice mixtures for luncheon meats, salad dressings, sauce or soup mixes, and pet foods (105).

Garlic cloves are dehydrated at 60–75°C in a series of stages (9), the temperature being

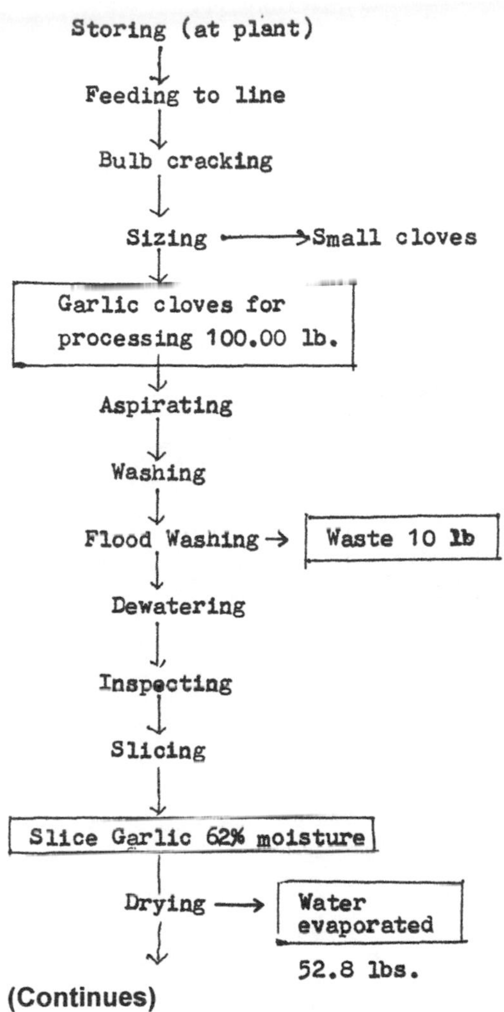

(Continues)

FIGURE 5 Flow sheet for garlic dehydration. (From Ref. 105.)

reduced as the moisture content falls. A final moisture content of 4% is achieved via warm air circulation. Yield is about 1 kg of dried product per 5 kg of fresh cloves.

B. Oil

Garlic oil is obtained by distillation of minced cloves. The oil comprises 0.1–0.25% of the fresh weight of the clove, and 1 g of oil is equivalent to 200 g of dried garlic (9). The pungency of garlic oil makes it difficult to use directly, and it is usually diluted in vegetable oil or encapsulated (9).

C. Garlic Salt

Specifications for the composition of garlic salt are shown in Table 4. Garlic salt has much wider culinary potential than garlic powder—one tablespoon is considered to be equivalent to a clove of fresh garlic.

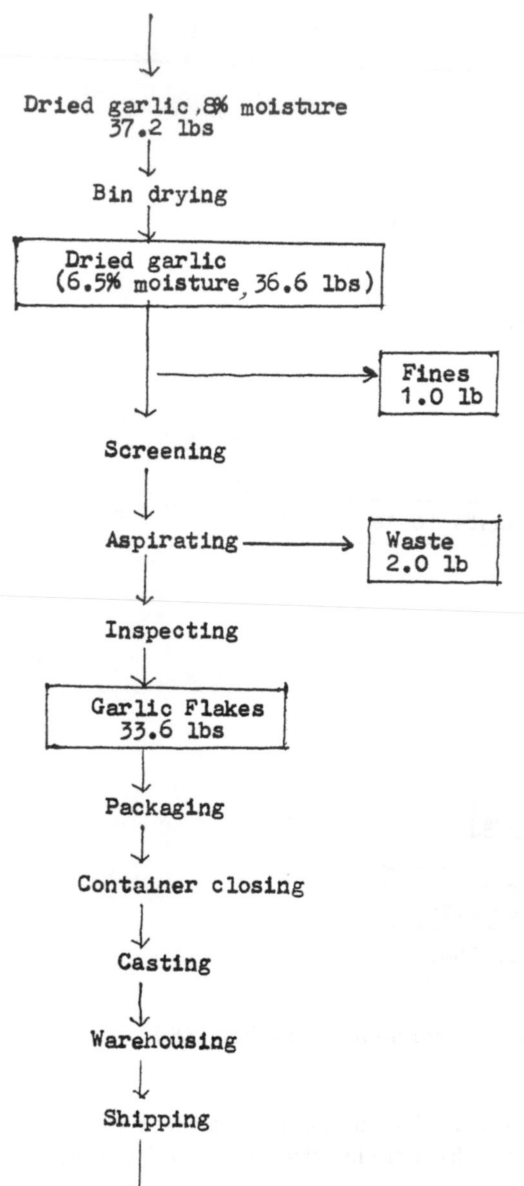

FIGURE 5 Continued.

TABLE 4 Approximate Composition (100 g) of Garlic Salt

Nutrient	Content
Water (g)	1.4
Enegy (kcal)	63
Protein (g)	3.2
Fat (g)	0.1
Carbohydrate (g)	13.8
Fiber (g)	0.4
Ash (g)	81.5
Ca (mg)	220
Fe (mg)	1
Mg (mg)	11
P (mg)	79
K (mg)	212
Na (mg)	31.4
Zn (mg)	1

Source: Ref. 106.

REFERENCES

1. Salunkhe, D. K., and B. B. Desai, Onion and garlic, in *Postharvest Technology of Vegetables*, Part II, CRC Press, Boca Raton, FL, 1984, p. 23.
2. FAO, Production Yearbook, Food and Agriculture Organization, Rome, 1990.
3. Yawalkar, K. S., Bulb vegetables, in *Vegetable Growing in India*, Agri. Hort. Pub. House, Nagpur, India, 1989, p. 370.
4. Etoh, T., Fertility of the garlic cloves collected in Soviet Central Asia, *J. Jpn. Soc. Hort. Sci. 55*:312 (1986).
5. Ishibasi, Y., T. Ogawa, and N. Matubara, Ecological and morphological classification of garlic cultivars (Japanese), *Bull. Nagasaki Agric. For. Exp. Stn.* (Sect. Agric.) *15*:95 (1987).
6. Pandey, U. G., and J. Singh, Agro-techniques for onion and garlic, in *Advances in Horticulture*, Vol. 5, *Vegetable Crops* (K. L. Chadha and G. Kalloo, eds.), Malhotra Pub. House, New Delhi, 1993, p. 433.
7. Hanley, A. B., and R. G. Fenwick, Cultivated alliums, *J. Plant Food 6:*211 (1985).
8. Choudhury, B., *Bulb Crop Vegetables*, N. B. Trust, New Delhi, 1967.
9. Aiyer, A.K.Y.N., *Field Crops of India*, 5th ed., The Bangalore Printing and Pub. Co. Ltd., Mysore, 1954.
10. Rao, H., and S. S. Purewal, Onion and garlic cultivation in India, *Farm Bull.* No. 3, Indian Council of Agricultural Research, New Delhi, 1957, p. 18.
11. Katyal, S. L., *Vegetable Growing in India*, Oxford and IBH Pub. Co., New Delhi, 1985, p. 123.
12. Mangal, J. L., R. K. Singh, A. C. Yadav, S. Lal, and U. C. Pandey, Evaluation of garlic cultivars for salinity tolerance, *J. Hort. Sci. 65*:657 (1990).
13. Singh, B. K., P. K. Ray, and K. R. Maurya, Optimum period of planting garlic in calcareous soil of subtropical North Bihar, *South Indian Hort. 32*:172 (1984).
14. Yamaguchi, M., *World Vegetables, Principles, Production and Nutritive Value*, AVI Pub. Co., Inc., Westport, CT, 1983.
15. Shinde, N. N., and M. B. Sontakke, Bulb crops, in *Vegetable Crops in India* (T. K. Bose and M. G. Som, eds.), Naya Prokash, Calcutta, 1986, p. 544.
16. Nath, P., *Vegetables for the Tropical Regions*, ICAR, New Delhi, 1976.

17. Om, H., and R. P. Srivastava, Influence of planting material and spacing on the growth and yield of garlic, *Indian J. Hort. 34*(2):152 (1977).
18. Rudge, T. G., Garlic: Time of planting, *N.Z. Commer. Grower 38*:28 (1983).
19. Ledesma, A., R. W. Racca, and M. I. Reale, Effect of storage conditions, and planting dates on garlic (*Allium sativum* L) cv. Rosado Paraguayo growth, *Phyton 43*:207 (1983).
20. Rudge, T., July planting of garlic desirable, *N.Z. Commer. Grower 39*:17 (1984).
21. Orlowski, M., and E. Kolita, Effect of some agrotechnical treatments on garlic yield. II. Effect of the date and density of planting on the yield, *Biul Warzywniczy 27*:165 (1984).
22. Scheffer, J., Red garlic-best yields from early plantings at Pukehohe, *N.Z. Commer. Grower 3*:25 (1985).
23. Buwalda, J. G., Nitrogen nutrition of garlic (*Allium sativum* L.) under irrigation; crop growth and development, *Sci.Hort. 29*:55 1986).
24. Das, A. K., M. K. Sadhu, M. G. Som, and T. K. Bose, Effect of time of planting on growth and yield of multiple clove garlic (*Allium sativum* L.), *Indian Agric. 29*:183 (1985).
25. Rahim, M.A., M.A. Siddique, and M. M. Hossain, Effect of time of planting, mother bulb size and plant density on the yield of garlic, *J. Agric. Res. 9*:112 (1984).
26. Bartos, J., and K. Holik, Mechanized preparation of garlic (*Allium sativum* L.) planting material, *Sb. UVTIZ Zahradnictvi 12*:131 (1985).
27. Bartos, J., and K. Holik, Intensification of garlic (*Allium sativum* L.) production by precision machine planting, *Sb. UVTIZ Zahradnictvi 12*:195 (1985).
28. Jarosova, J., and J. Bartos, Effect of grading garlic (*Allium sativum* L.) planting material by size on cultural and economical production indices, *Sb. UVTIZ Zahradnictvi 11*:60 (1984).
29. Purewal, S. S., and K. S. Dargan, Effect of fertilizers and spacing on the development and yield of garlic, *Indian J. Agron. 5*:262 (1961).
30. Singh, A. K., and J. P. Singh, Effect of spacing and level of nitrogen on growth, yield and quality of garlic, *Allahabad Farm. 58*(4):553 (1977).
31. Maurya, K. R., Effect of transplanting on the yield of garlic (*A. sativum*), *Prog. Hort. 19*:132 (1987).
32. Pal, R. K., and K. P. S. Phogat, Effect of different spacing on growth and yield of garlic (*Allium sativum* L.), *Prog. Hort. 16*:337 (1984).
33. Singh, L., K. P. S. Chauhan, and J. B. Singh, Effect of method of planting and size of cloves on yield and quality of garlic, *Natl. Symp. on Onion and Garlic*, June 2-3, 1990, Dr. Y. S. Parmar Univ. Hort., Solan, India, 1990.
34. Singh, J. R., and J. Tewari, Effect of source of organic manures and levels of nitrogen on growth characteristics of *Allium sativum*, *Indian J. Hort. 25*:191 (1968).
35. Maurya, A. N., and P. Bhuyan, Effect of nitrogen and plant density on growth and yield of garlic in acid soils of Assam, *Indian Cocoa Arecanut Spices J. 6*:10 (1982).
36. Likhatskii, V. I., The effect of treating garlic cloves with microelements before planting on the yield and market quality of the product, *Puti Intensif. Ovoshchevod Kiev, Ukr. SSR* 52 (1987).
37. Orlowski, M., and E. Rekowska, The effect of clove weight and planting density on garlic yield, *Biul. Warzywniczy* 85 (1989).
38. Blyshchik, B. I., and V. A. Furman, The effect of garlic productivity of grading aerial bulbils by size before planting, *Puti Intensif. Ovoshchevod. Kiev, SSR* 16 (1987).
39. Singh, R. V., and K. P. S. Phogat, Effect of different sowing times on the growth and yield of garlic, *Prog. Hort. 21*(1-2):145 (1989).
40. Lucero, J. C., C. Andreoli, M. Reyzabal, and V. Larregui, Influence of clove weight and planting density on garlic cv. Colorado, yields and quality, *An. Edafol. Agrobiol. 40*:1807 (1982).
41. Duranti, A., and L. Cuocolo, The effect of clove weight and distance between the rows on the yield of *Allium sativum* L. cv. Messidrome, *Riv. Ortofloro fruttic. Ital. 68*:25 (1984).
42. Orlowski, M., and E. Kolota, Effect of some agrotechnical treatments on garlic yield. I. Effect of clove size on garlic yield and quality, *Biul. Warzywniczy 27*:147 (1984).
43. Pandey, U. C., and J. Singh, Performance of new varieties of garlic (*A. sativum*), *Haryana Agric. Univ. J. Res. 19*(1):69 (1989).

44. Singh, J. R., and J. Tewari, Phosphorus nutrition of garlic, *Labdev* (Kanpur) *13*(2):119 (1969).
45. Augusti, K. T., *Allium cepa* Linn and *Allium sativum* L. in orally effective hypoglycemic principle from plant sources, Ph.D. thesis, University of Rajasthan, Jaipur, India, 1963, p. 52.
46. Lawande, K. E., R. D. Pawar, and P. N. Kale, Preliminary studies on sprouting in garlic, A note, *Haryana J. Hort. Sci. 12*:221 (1990).
47. Curzio, O. A., and C. A. Croci, Radio inhibition process in Argentina garlic and onion bulbs, *Radiation Phys. Chem. 31*:203 (1988).
48. Manjunath, B. L., *Allium cepa* Linn and *Allium sativum* Linn, in *The Wealth of India—Raw Material*, Vol. 1, Council of Scientific and Industrial Res. Govt. of India, New Delhi, 1948, p. 56.
49. Pal, R. K., and U. C. Pandey, Effect of different levels of nitrogen, phosphorus and potassium on the growth and yield of garlic, *Prog. Hort. 18*:256 (1986).
50. Singh, R. N., and J. R. Singh, Studies on the influence of boron nutrition on the growth characteristics of garlic (*Allium sativum*), *Indian J. Hort. 31*:255 (1974).
51. Wang, X. Y., J. W. Hou, X. M. Zhang, and W. C. Xu, Study on the pattern of fertilizer requirement for garlic bulbs, *North. Hort. 1*:10 (1992).
52. Setty, B. S., G. S. Sulikeri, and N. C. Hulamani, Effect of N, P and K on growth and yield of garlic *Allium sativum L., Karnataka Agril. Sci.* 2(3):160 (1989).
53. Khalaf, S. M., and E. M. Taha, Response of garlic plants grown on calcareous soil to organic manuring and sulphur application, *Ann. Agril. Sci.* (Cairo) *33*(2):1219 (1988).
54. Moon, W., and B. Y. Lee, Studies on factors affecting secondary growth in garlic (*Allium sativum* L.) I. Investigations on environmental factors and degree of secondary growth, *J. Korean Soc. Hort. Sci. 26*:103 (1985).
55. Buwalda, J. G., Nitrogen nutrition of garlic (*Allium sativum* L.) under irrigation. Component of yield and indices of crop nitrogen status, *Sci. Hort. 29*:69 (1986).
56. Pandey, U. C., and J. Singh, Effect of nitrogen and irrigation levels on garlic cv. HG-1, National Symposium on Onion and Garlic, June 2-3, 1990, Dr. Y. S. Parmar Univ. Hort., Solan (H.P.), India, 1990.
57. El-Beheidi, M. A., N. H. Kamel, and M. M. Abou El-Magd, Effect of water regime and nitrogen fertilizer on some mineral contents, yield and amino acid contents of garlic plant, *Ann. Agric. Sci. Moshtohor 19*:149 (1983).
58. Silenzi, J. C., A. M. Moreno, and J. C. Lucero, Effects of irrigation with saline water on sprouting of cloves of garlic (*Allium sativum* L.) cv. Colorado, *IDIA* (*Inf. Invest. Agric.*) *17*:433 (1985).
59. Mollejas, J. F., and R. H. Mata, Boil. Tec. Estac. Exp. Agri., Naya Prokash, Calcutta, 1986, p. 56.
60. Thompson, A. K., M. B. Bhatti, and P. P. Rubio, Harvesting and handling: Storage, in *Postharvest Physiology, Handling and Utilization of Tropical and Sub-tropical Fruits and Vegetables* (E. B. Pantastico, ed.), The AVI Pub. Co., Westport, CT, 1975, p. 236.
61. Pantastico, E. B., H. Subramanyam, M. B. Bhatti, N. Ali, and E. K. Akamine, Postharvest physiology 4. Harvest indices, in *Postharvest Physiology, Handling and Utilization of Tropical and Sub-tropical Fruits and Vegetables* (E. B. Pantastico, ed.), The AVI Pub. Co., Westport, CT, 1975, p. 56.
62. Peters, P., G. Banholzer, and H. Eysold, Preliminary results on the medium-term storage of garlic, *Gartenbau 31*:263 (1984).
63. Banholzer, G., G. Krumbein, and R. J. Maut, Mechanization of garlic production in the agricultural plant production collective "Progress" Krostitz, *Gartenbau 32*:266 (1985).
64. Whitaker, J. R., Development of flavor odor and pungency in onion and garlic, *Adv. Food Res. 22*:73 (1976).
65. Brondnitz, M. H., J. U. Pascale, and L. Van Derslice, Flavor components of garlic extract, *J. Agric. Food Chem. 19*:273 (1971).
66. Saghir, A. R., L. M. K. Mann, R. A. Bernhard, and J. V. Jacobsen, Determination of aliphatic mono- and disulfide in Allium by gas chromatography and their distribution in the common food species, *Proc. Am. Soc. Hort. Sci. 84*:386 (1964).
67. Block, E., The organosulfur chemistry of the genus *Allium*—implication for the organic chemistry of sulfur, *Angew. Chem. Int. Ed. Engl. 31*:1135 (1992).

68. Calvey, E. M., J. E. Matsik, K. D. White, J. M. Betz, E. Block, M. H. Littlejohn, S. Naganathan, and D. Putmani, Off-line supercritical fluid extraction of thiosulfinates from garlic and onion, *J. Agric. Food Chem. 42*:1335 (1994).
69. Block, E., Flavor artifacts, *J. Agric. Food Chem. 41*:692 (1993).
70. Lawson, L. D., S. G. Wood, and B. G. Hughes, Identification and HPLC quantitation of sulfides and dialk(en)yl thiosulfinates in commercial products, *Planta Med. 57*:363 (1991).
71. Saito, K., M. Horie, Y. Hoshino, N. Nose, E. Mochizuki, H. Nakazawa, and M. Fujita, Determination of allicin in garlic and commercial garlic products by gas chromotography with flame photometric detection, *Off. Anal. Chem. 72*:917 (1989).
72. Matusik, J. E., E. M. Calvey, and E. Block, Supercritical fluid extraction and LC/MS of thiosulfinate found in Allium species, presented at 41st ASMS Conference on Mass Spectrometry, May 30-June 4, San Francisco, CA, 1993.
73. Shanaracharya, N. B., Chemical composition of garlic, in *Proc. Symp. on Spice Industry in India*, CFTRI, Mysore, 1974, p. 24.
74. Ueda, Y., M. Sakaguchi, K. Hirayama, R. Miyasima, and A. Kimizuka, Characteristic flavor constituents in water extract of garlic, *Agril. Biol. Chem. 54*(1):163 (1990).
75. Halan, S. and R. Arumugam, Chemical changes in the quality parameters of garlic during growth and development, *South Indian Hort. 39*(2):93 (1991).
76. Augusti, K. T., Hypocholesterolmic effect of garlic (*Allium sativum* L.), *Indian J. Exp. Biol. 15*:489 (1977).
77. Mahanta, R. K., R. K. Goswami, D. Kumar, and P. Goswami, Effect of *Allium sativum* (garlic) on blood lipids. *Indian Med. Gaz.* (April):157 (1989).
78. Yu, T. S., C. M., Wu, and C. T. Ho, Meat-like flavor generated from thermal interaction of glucose and alliin or deoxyalliin, *J. Agric. Food Chem. 42*:1005 (1994).
79. Augusti, K. T., and P. T. Mathew, Lipid lowering effect of allicin (diallyl disulphide oxide) on long term feeding normal rats, *Experientia 30*:468 (1974).
80. Jain, R. C., Onion and garlic in experimental cholesterol induced atherosclerosis, *Indian J. Med. Res. 64*:1509 (1976).
81. Bordia, A. K., and S. C. Verma, Garlic on the reversibility of experimental atherosclerosis, *Indian Heart J. 30*:41 (1978).
82. Bordia, A., Effect of garlic on human platelet aggregation *in vitro*, *Atherosclerosis 30*:355 (1978).
83. Sainani, G. S., D. B. Desai, N. H. Gorhe, S. H. Natu, D. V. Pise, and P. G. Sainani, Effect of dietary garlic and onion on serum lipid profile in Jain community, *Indian J. Med. Res. 69*:776 (1970).
84. Brahmachari, H. D., and K. T. Augusti, Effect of orally effective hypoglycemic agents from plants on alloxan diabetes, *J. Pharm. Pharmacol. 14*:617 (1962).
85. Singh, U. P., K. K. Pathak, M. N. Khare, and P. B. Singh, Effect of leaf extract of garlic on *Fusarium oxysporum* sp. *ciceri*, *Sclerotinia lerotirum* on gram seeds, *Mycologia 71*:556 (1979).
86. Mathew, P. T., and K. T. Augusti, Studies on the effect of allicin (diallyl sulphide oxide) on alloxan diabetes, 1. Hypoglycemic action and enhancement of serum insulin effect and glycogen synthesis, *Indian J. Biochem. Biophys. 10*:209 (1973).
87. Sharma, V. D., M. S. Sethi, A. Kumar, and J. R. Ranotra, Antibacterial property of *Allium sativum in vivo* and *in vitro* studies, *Indian J. Exp. Biol. 15*:464 (1977).
88. Kumar, A., and V. D. Sharma, Inhibitory effect of garlic on enterotoxigenic *Escherichia coli*,. *Indian J. Med. Res. 76*:66 (1982).
89. Deb-Kirtaniya, S., M. R. Ghosh, A. N. Chaudhary, and A. Chatterjee, Extract of garlic as possible source of insecticides, *Indian J. Agric. Sci. 50*:507 (1980).
90. Singh, K. V., and N. P. Shukla, Activity on multiple bacteria of garlic (*Allium sativum* L.), *Fitoterapria 55*:313 (1984).
91. Michail, S. H., M. A. Abdel-Rehim, H. Elarosi, and E. A. Khairy, Dry rot of garlic cloves in U.A.R. (Egypt), *Photopath. Medit. 10*:202 (1971).
92. Takagi, H., Garlic, in *Onion and Allied Crops*, Vol. III (Brewster, J. L. and H. D. Rabinowitch, eds.) CRC Press, Inc., Boca Raton, FL, 1990, p. 109.

93. Hardenburg, R. E., *The Commercial Storage of Fruits, Vegetables and Florist and Nursery Stocks*, USDA, ARS, Agriculture Handbook No. 66,. 1986, p. 59.
94. Ryall, A. L., and W. J. Lipton, *Handling, Transportation and Storage of Fruits and Vegetables*, Vol. 1, *Vegetables and Melons*, AVI Pub. Co., Westport, CT, 1972, p. 479.
95. Eckert, J. W., P. P. Rubio, A. K. Mattoo, and A. K. Thompson, Postharvest pathology, Part-II, Diseases of tropical crops and their control, in *Postharvest Physiology, Handling and Utilization of Tropical and Sub-tropical Fruits and Vegetables* (E. B. Pantastico, ed.), The AVI Pub. Co., Westport, CT, 1975, p. 415.
96. Lammerink, J., Effect of fumigation, cold storage and fungicide treatment of planting cloves on yield and quality of garlic, *N.Z. J. Crop Hort. Sci. 18*(1):55 (1990).
97. Thompson, H. C., and W. C. Kelly, *Vegetable Crops*, McGraw-Hill Book Co., Inc., New York, 1957, p. 611.
98. Mathur, P. B., Extension of the storage life of garlic bulbs by gamma irradiation, *Int. J. Appl. Rad. Isot. 14*:625 (1963).
99. El-okshg, I. I., A. S. Abdel-Kader, Y. A. Wally, and A. E. El-Koholly, Comparative effects of gamma irradiation and maleic hydrazide on storage of garlic, *J. Am. Soc. Hort. Sci,, 96*:637 (1971).
100. Croci, C. A., and O. A. Curzio, The influence of gamma irradiation on storage life of red variety of garlic, *J. Food Proc. Preser. 7*:17 (1983).
101. Temkin-Gorodeiski, N., R. Pandova, and U. Zisman, Prolonging the storage quality of garlic by gamma radiation, in *Scientific Activities Inst. For Technol. and Storage of Agric. Products*, Pamphlet No. 184, 1978, p. 37.
102. Joong, H. K,. and S. Y. Hyung, Changes in flavour components of garlic resulting from gamma irradiation, *J. Food Sci. 50*:1193 (1985).
103. Ceci, L. N., O. A. Curzio, and A. B. Pomilio, Effects of irradiation and storage on the flavour of garlic bulbs, cv. Red, *J. Food Sci. 56*(1):44 (1991).
104. Croci, C. A., J. A. Arguello, and G. A. Orioli, Biochemical changes in garlic during storage following γ-irradiation, *Int. J. Radiation Bio. 65*(2):263 (1994).
105. Luh, B. S., L. P. Somogyi, and J. J. Meehan, Vegetable dehydration, in *Commercial Vegetable Processing* (B. G. Luh and J. G. Woodroof, eds.), AVI Pub. Co. Inc., Westport, CT, 1982, p. 361.
106. Farrell, K. T, *Spices, Condiments and Seasonings*, AVI, Westport, CT, 1985.

18

Other Alliums

S. D. Warade and K. G. Shinde
Mahatma Phule Agricultural University, Rahuri, India

I. LEEK

Leek is an important crop of the family Alliaceae which exhibits morphological differences with onions. It is larger than the onion. The leaf blades are flattened rather than radial. The leaf base of leek stores some reserves but does not thicken into a bulb. Leek has a milder and more delicate flavor than onion, though a coarser texture. When tender, it is eaten raw. It is also cooked with other vegetables or used as a flavoring in soups and stews. Leeks are mainly grown in northern Europe and less frequently in India, the United States, and Canada. Leeks are especially important in northern European countries such as Belgium, Denmark, and the Netherlands.

A. Botany

The leek (*Allium porrum* L.) is a tall, hardy, biennial with white, narrowly ovoid bulbs and broad leaves. It resembles the green onion but is larger. Leek, a cultivated form of *Allium ampeloprosum*, is a tetraploid ($2n = 32$). It differs mainly in its lesser tendency to form bulbs. Many cultivars selected for long, white, edible bases and green tops, winter hardiness, and resistance to bolting are avaialble for cultivation. These cultivars differ from one another mainly in length and diameter of the sheath part, leaf spacing, breadth and color of leaf blades, vigor and bolting, and resistance to cold. Resistance to cold is of special importance where leeks are to be harvested throughout the winter, while slowness to bolt permits a prolonged harvest period in the spring (1).

The cultivars listed by American seedmen include London Flag, American Flag, Elephant or Monstrous, Carentan, Giant Mussenburgh, and Lyon (2). Parijes (3) reported that cultivars like Acadia, Derrick, and Electra are suitable for autumn cultivation in Belgium, while Alberta, Brizzard, and Carina are better for winter. In Canada, promising cultivars mentioned by Maurer (4) are Longa, Odin, Kilima, Goliath, Cyberia, and Artico. Carina and Alberta cultivars gave the highest yield in the Netherlands (4). Menard (5) presented data on the characteristics of 30 leek cultivars; the highest yield was obtained from King Richard, Verina, Tivi, Kazan, Kilima, and

Albana, while Kanters (6) obtained highest yield from Alma during the summer season of the Netherlands.

B. Production

1. Soil and Climate

Leek can be grown on a variety of soils but grows luxuriantly on medium soils that are rich in plant nutrients and organic matter. Stone and Rawse (7) recommended thorough loosening of sandy clay loam subsoil, which helped in easy penetration of roots up to 70 cm and thereby increased the yield. The most favorable soil pH range for leek is 6–8 (3).

Leek withstands heat and cold better than onion. Although leek is a cool season crop, temperature is more important than the length of day in seed stalk development. In India and Sri Lanka, leek thrives well at higher altitudes.

2. Cultural Practices

Seedling Raising

Leek is propagated through seeds. Seed germination depends upon temperature, with 18–22°C being the optimum (Table 1). In India, seeds are sown during August to October in the nursery beds, and seedlings are ready to plant when they attain a height of 15 cm. About 5–7 kg of seeds are sufficient to raise seedlings for planting one hectare of land. A method of magnetic separation of leek seeds of low germination from commercial seed lots was described by Krishnan and Barlage (8).

Planting

Transplanting is done in trenches about 30–45 cm deep. Moneka (9) obtained the highest yield of leek transplanting at 80 days for February sowing and 70 days for March sowing. Perk (10) recommended planting dates for leeks grown in the mountains and lowlands of late June and mid-July, respectively. Kerba and Geriya (11) suggested planting in February, March, April, or October in Abkhazia (U.S.S.R.) for an uninterrupted supply of fresh leeks. Granges and Leger

TABLE 1 Effect of Temperature on Leek Seed Germination

	Number of days to 50% emergence at:			
Genotype	10°C	14°C	18°C	22°C
Elbeuf improved	27	19	15	13
Kurrat (Egypt)	19	19	12	9
Taree irani	15	12	9	9
Jambo garlic	19	12	9	9
Kajak	26	16	13	14
Alaska	19	13	12	12
Varna	19	13	12	12
Mean	21	15	12	11

Experiments were carried out in moist soil in the IVT phytotron in 1987.
Source: Ref. 24.

(12) obtained the highest yield planting between January 20 and February 5 with a spacing of 3 × 7 cm. Raising seedlings in "quick pots" with one plant per cell was better than sowing in earthen balls or clay pots. Kolota (13) suggested planting two or three plants per hole for increased yields compared with single planting, but the crop was less uniform. Synnevag and Stubhaug (14) indicated seedling raising in 260 plug trays with two seedlings per plug being the most satisfactory alternative to raising seedlings for bared root planting.

Transplanting is done in trenches about 30–45 cm deep. Kaniszewski (15) observed that planting in 20-cm-deep holes reduced yield, but, along with earthing up, increased the length and weight of the blanched part of the shaft. By increasing plant population from 196,000 to 256,000 plants per hectare, the yield was increased by 10–30% (16). However, Mayne (17) recommended 150,000 plants/ha for early leek planting and 180,000 plants/ha for autumn planting.

Manuring and Fertilization

Because the leek is larger than the onion, its requirements for manure and fertilizer are higher. A yield of 30 t/ha removes 100 kg of nitrogen, 65 kg of P_2O_5, and 130 kg of potash from the soil (1). The diameter and length of bulbs are increased by nitrogen fertilization (1). Kaniszewski (18) reported the highest yield of leeks with a preplanting application of 200 kg/ha nitrogen under irrigated and nonirrigated conditions in dry years. In wet years, split application of 600 kg of nitrogen recorded maximum yields. Weier and Scharpf (19) reported the highest yields of leeks when total nitrogen availability in the top 60–70 cm of soil was 220 kg/ha. In soil with good nitrogen and water-holding properties, the necessary fertilizers could be applied in one or more doses.

Irrigation

Leeks require a large amount of water during growth. Kolota (13) observed that sprinkler irrigatin significantly improved both the yield and the quality of leeks. Hartman et al. (20) concluded that 375–400 mm of water was required for high yields. Water requirements were low during the juvenile phase but subsequently increased rapidly. The diameter of the neck was the best indicator of water requirement (20).

Weed Control

Weed control at the proper stage plays an important role in the growth, yield, and quality of leeks. Pelletier (21) observed that herbicides were sufficiently selective for leeks. In trials in Breada, 1 kg of Bladex® (Cyamozine) + 1.5 kg Tribunil® (Methabenzthiazuron) showed potential for weed control (22). Zander and Zozmann (23) observed that in pots and small plot trials, Pendimethialin® + Propachlore® applied pre- and postemergence, prometryn + propachlor preemergence and at the cotyledon at the one- to two-leaf stage of the weeds, and chlorprophan or propachlor + atrazine + fenuron at the crook and whip stages of crop and the cotyledon at the three-leaf stage of the weeds provided good weed control in leeks with little crop damage.

Blanching

Blanching is done by covering the plants to a certain height with soil to improve the quality of the crop. For this purpose, plants are sunk up to their center leaves in trenches or pits that are heavily manured to earth up soil as they grow (24). Care should be taken not to earth up soil early when the plants are young.

3. Disorders, Diseases, and Pests

Weibe (26) reported that the most effective temperature for inducing bolting was 8–11°C. Photoperiod during the chilling period had little effect. In order to prevent bolting, cultivation at 18°C is recommended (26).

Diseases like downy mildew, purple blotch, pink rot, smut, smudge, *Heterosporium* blight, and white tip are the common diseases of leek. White tip of leek has been long known in the British Isles. The disease is caused by the fungus *Phytophthora pori* Foister. The tips of leaves become yellow, die, and turn white for a distance up to 15 cm. Most affected leaves turn backward. Water-soaked areas in the vicinity of the midrib may appear halfway down the leaf or near the base. The disease can be effectively controlled by spraying with Bordeaux mixture. Purple blotch (*Alternaria pori*) and powdery mildew (*Leveillula taurrca*) observed in leeks were effectively controlled by Prochlorz® and Fenpropimorp® chemicals (26). Leek moth (*Acrolepiopsis assectella*) is a serious pest in Italy. The moth had five overlapping generations in one year. The third generation, which occurred in July, was the most injurious (26).

4. Harvesting and Yield

Leek is harvested like green onions and marketed in bunches. Yield depends upon several factors, however, an average yield of 30 tons/ha can be obtained from a well managed crop (2).

C. Chemical Composition

The nutritional composition of leek is given in Table 2. Compared to onion, leek contains more proteins and minerals on a fresh weight as well as a dry weight basis. The energy value of 100 g of the edible portion of leek is also higher than that of onion.

TABLE 2 Chemical Composition of Leek (per 100 g fresh weight)

Constituent	Content
Water (g)	90
Protein (g)	2
Fat (g)	0.3
Carbohydrates (g)	5.0
Minerals (g)	1.5
Sodium (mg)	5
Potassium (mg)	250
Calcium (mg)	60
Iron (mg)	1
Phosphorus (mg)	30
Vitamins	
β-Carotene (mg)	600
Thiamine (B_1)	120
Nicotinic acid	500
Pyridoxine (B_6)	250
Ascorbic acid (vitamin C)	25

Source: Ref. 24.

D. Storage

After harvesting, leeks should be kept in cool conditions. The green crop can be stored for 1–3 months at a temperature of 0°C with a relative humidity of 85–90% (27). Goffings and Herregods (28) reported that freshly harvested unwashed leeks stored at 0°C and 94–95% relative humidity in atmosphere containing 2% oxygen, 2% carbon dioxide, and 96% N_2 had improved quality. *Botrytis porri* is the most important reason for quality loss of leek during storage (29,30).

II. SHALLOTS

Allium cepa (*aggregatum* group) is an onionlike plant that originated in western Asia. The crop is very popular in the southern United States and some European countries (31). In the tropics, shallots are often grown in areas where onion culture is difficult because the climate is humid and bulb onion is susceptible to leaf diseases that shallot can withstand. Shallot has a very short growing season of only 2 or 3 months, which allows it to be grown between other crops or during a short-day season. In the lowland tropics, lack of distinct cool period can prevent onion from flowering; under such conditions, growing of shallot is advantageous.

A. Botany

The term shallot refers to the vegetatively propagated forms of *Allium cepa* var. *ascalonicum*, which were included in *aggregatum* group of the species. Shallots of this type appear to have been derived by selection from a naturally occurring variant within *Allium cepa*. Several cultivars are actually derived from *A. cepa* × *Allium fistolosum* crosses (e.g., Delta Giant). These should not be confused with the shallots of the *Allium cepa aggregatum* group. *Allium cepa* shallots are distinguished from natural bulb onion by their habit of multiplying vegetatively by laterals and growth—a single shallot bulb usually contains several initial shoots. The bulb can be planted (32), and several leafy shoots will grow out from it. Each shoot then rapidly produces a small bulb, forming a cluster that remains attached to the original base plate (33). The bulbs can be separated and the process repeated in the next growing season.

Improved lines of multiplier onion have been bred at Tamil Nadu Agricultural University, Coimbatore (India), by crossing with bulb onion CO-1 to CO-4 series of cultivars developed there, and breeding work is still continuing (32). In cultivars where flowering can be induced, a cool period of 40 days at 14°C is required. Although seeds can be produced from some lines, vegetative multiplication is usually practiced. The CO cultivars are pink or red in color. The multiplier onion splits into several daughter bulbs, ranging in number from 4–8 to 8–10. Both multiplier onion and tropical shallots take only 60–75 days to multiply and die down again, and the bulbs can be stored for considerable periods (over 5 months in some trials). Individual clusters of multiplier onion in the CO series have an average weight range of 25–85 g (31). Dalbellary et al. (34) redommended the new shallot variety Milrac with higher yield potential. It is also tolerant to onion yellow dwarf virus. Other cultivars reported from Indonesia are Ampenan, Cloja, Bima, Bima Kuning, Bauji, Balijo, Suminep, Bawang Lambung, Betawi Cipanos, and Hajakuning.

B. Production

1. Soil and Climate

Shallots can be grown in all types of soil with a pH of more than 5.6, however, well-drained alluvial soils are preferred for better growth and development. Kusumo and Muhadjir (35) grew shallots traditionally after the harvest of the rice crop on raised beds of subsoil obtained from the deep furrows. However, seed origin and soil type had no significant effect on yields.

An average temperature of 25–32°C is optimum during the growing period of shallots. Jenkins (36) reported that high temperature favored bulbing. Plants grown at temperatures of 21°C and higher all formed bulbs, but larger bulbs were produced with a 15-hour photoperiod than with a 10-hour photoperiod. When the temperature was lower than 21°C, no bulbs were formed regardless of day length.

2. Cultural Practices

Planting

Three shallot crops are grown a year, the major seasons being April to August, January to March, and September to December (32). A few plantings are made in August, although the bulk of the crop is planted during October with little planting until January (36). Sinnadurai (37) described the growing system for shallots in the coastal area of Ghana. The shallots are planted on raised beds in sandy soils 7 cm apart. About 4 tons of bulbils are required for planting one hectare. In Indonesia, 900–1000 kg of planting material is needed for one hectare. Plants are spaced 15 × 15 cm to 20 × 20 cm according to the cultivar. At planting, the tops are cut if the bulbs are dormant. The crop is grown in beds 1.2 × 1.8 m wide with up to a half meter width and 50–60 cm depth between beds (32). Individual bulbs are planted in 5-cm-deep rows 30 cm apart (3). Alludim (38) reported that higher yields of shallots can be obtained with a density of 128 plants/m^2.

Manuring and Fertilization

The shallot crop is given a basal dressing of fertilizers or mixed fertilizers 10–15 days after planting and is then fertilized at 2-week intervals until 2 weeks before harvest, with 200 kg or more of urea per hectare on each occasion. Muhadjir and Kusumo (39) planted cultivars Ampenan and Medan with a basal dressing consisting of 100 kg each of N and P_2O_5 and 0, 50, and 100 kg of K_2O per hectare. The highest yields were obtained with 100 kg each of N and P_2O_5 and 50 kg K_2O. The quantity of nitrogen used had no influence on growth leaf color, susceptibility to bolting, or number of bulbs, however, increasing nitrogen had an adverse effect on the uniformity of leaf canopy at maturity, and 60 kg of nitrogen is recommended (40).

3. Diseases and Pests

A variety of diseases caused by organisms like *Alternaria porri* and *Colletotribum* species can be controlled with sprays of Maneb. To control *Spotoptera* species as well as thrips, twice-weekly sprays of monocrotophos or other insecticides are used.

4. Harvesting

Harvest takes place when 70–80% of the leaves have turned yellow, i.e., 65–70 days after planting in the lowlands and 80–100 days after planting in highland areas. The shallots are pulled by hand after they have obtained a diameter of at least 0.5–0.6 cm. The outer skin is peeled off and the roots are trimmed, after which they are washed and tied into 1-kg bunches. For dry bulb production, shallots are dried for 5–14 days in the field, covered by plastic if it rains.

C. Composition and Utilization

Shallots may contain more fat and soluble solids, including sugars, than bulb onions (32). The Indonesian shallot variety Sumenap is said to have a high fat content. In Ethiopia, small local red shallots grown in the highlands are highly valued in the traditional *wat* sauce to accompany *Injera* bread made from wheat flour (32). Shallots are also used in certain sauces.

III. MINOR ALLIUMS

A. Chives

Chive (*Allium schoenoprassum* L.) is a perennial plant of which only the leaves are eaten. Chives have never been of great importance, as their cultivation was confined to domestic gardens. More recently, they have been grown and processed commercially. Worldwide commercial chive production is approximately 1000 ha (41).

1. Botany

A. schoenoprassum L. belongs to the family Alliaceae. There are three types:

1. diploids, which include numerous biotypes of *A. schoenoprassum* L., e.g., var. *sibiricum* L., var. *ledebourianum* Roem et Schult, and var. *alpinum* Heg. These have a pollen diameter of 28–30 μm.
2. The autotetraploid gigas type, which includes *A. schoenoprassum* var. *sibiricum* Garke. The pollen diameter is 36–40 μm.
3. The allotetraploid types, which have a pollen diameter of 36–37 μm, form very distinct bulbs and are late flowering (41).

A. schoenoprassum L. is highly polymorphic (42). The bulbs are clustered on a very small rhizome and are covered by a membranous skin. Leaves are hollow, cylindrical, and semi-cylindrical and as long as the scape or even longer. The umbel is many-flowered; pedicels vary in length but are shorter than flowers. Sepals are 6–15 mm long. Stamens are one-third or one-half the length of the sepals and are joined to the perianth and to each other for one quarter or one third of their length (41).

2. Production

Seeds are sown at a rate of 7 kg per hectare in April-May. If sowing can be done in early August, seed is sown in double rows 5 cm apart. The distance between double rows is 55–65 cm (41). Irrigation is essential throughout the growing season, as water stress promotes chlorosis and necrosis (43). Perennial fields must be supplied with about 25 kg of P_2O_5 and 100 kg of K_2O per hectare. Nitrogen must be supplied throughout the growing season after every harvest. A weekly application of about 10–12 kg N per hectare is recommended, with a total of 220 kg/ha per year (41).

Puccinia allii Rud. causes serious damage, especially during August and September. The intensity of disease depends upon the relative humidity of the air (44). White rot (*Sclerotium cepivorum*), downy mildew (*Peropnospora destructor*), foot rot (*phythium* ssp.) are other important diseases of chives. Onion fly (*Delia antiqua*) can cause serious damage to young seedlings. Application of parathion between the rows gives good control (45). Thrips (*Thrips tabaci*) are not a serious problem when fields are irrigated (43). Stem and bulb nematodes (*Ditylenchus dipsaci*) rarely attack chives (43).

Dormant chives can be induced to grow following heat treatment. Chives can be produced year round utilizing greenhouses for winter production. Such winter production is termed "forcing." Immersion of plants in warm water (approximately 40°C) is now practiced commercially to break dormancy. Plants are transferred to a greenhouse at approximately 20°C, where leaf growth occurs rapidly. Such immersion for 3 hours breaks the dormancy effectively (41). Immersion at a temperature of 44°C for more than one hour damages the plants and reduces subsequent growth (46). Poulsen (41) described standard warm water treatment and warm air treatment for breaking dormancy.

Bulbs grown for forcing are sown in April-May at a depth of 5–10 mm when soil temperature reaches about 10°C (41). Leaf harvesting during the summer is not recommended, since this depletes the reserves used to form thc bulb, and, consequently, regrowth upon forcing is lessened. The growth can control bulb diameter to some degree by varying the seed rate. By using a high seed rate and consequent high plant density, nearly all the bulbs produced are suitably small. The production of bunches of fresh leaves is of limited economic importance. Such bunches are sold in fresh vegetable markets and to a lesser extent under contract to caterers (43).

After spring sowing, the field can be harvested three times in the first season. It is important to avoid overfrequent harvest in the first year, or productivity the following year will be reduced (41). From the second season onwards, the field can be first harvested in early May. The second harvest is taken in late June, but the quality is usually low because of chlorosis of older leaves (41). Maximum productivity occurs in July and August when about four harvests can be taken. In early September, the productivity declines and plants enter the rest phase (41). Unless the field is to be discontinued at the end of the season, only one harvest should be taken after this date. Otherwise, late harvesting will reduce early growth the next season. In September, the quality is also reduced due to chlorisis of older leaves. About six or seven harvests annually are typical in Denmark, where yield is about 7.5 t of fresh leaves per hectare (41).

3. Chemical Composition

The mineral composition of chives is presented in Table 3. The climatic vitamin C content varies from 60 to 140 mg/100 g depending upon the variety and climatic conditions (47,48). Thiamine content ranges from 71.5 to 100 μg/100 g fresh weight (48). The two most frequent alliins are propyl alliin and methyl alliin. The lacrimatory substance propenyl sulfenic acid is produced from propenyl cysteine sulfoxide when leaves are crushed (49). The volatiles 2-methyl-2-butenal, 2-methyl-2-pentenal, methyl propyldisulfide, and *cis,trans* forms of propenyl-propyldisulfide were identified (50,51), but allyl derivatives were not found (51). Two new sulfur compounds, methyl pentyl disulfide and pentyl hydrodisulfide, have been detected in chives (52,53). Three glucosides, kaempferol-3-glycoside, quercetine-3-glycoside, and

TABLE 3 Mineral Composition of Chives (% dry matter)

Mineral	Germany	Sweden	Italy
Nitrogen	—	3.6	—
Potassium	3.3	2.6	2.5
Calcium	2.1	1.2	0.8
Phosphorus	1.5	0.44	0.5
Magnesium	0.5	0.17	0.5
Sodium	6.4	0.03	0.06
Manganese (ppm)	—	83	—

Source: Ref. 41.

isorhamnetine-3-glycoside, were found in chives (54). Chives also contain malic acid (240 mg/100 g fresh weight) and citric acid (19 mg/100 g fresh weight) (55).

4. Processing

Frozen chive leaves have been marketed for many years. These retain their green color but lose their turgidity when thawed (41). Hot air drying is a common method of preservation of chive leaves. In this case, quality is reduced because of loss of aroma and some decrease in green color, as well as some loss of structure. The best method of preservation is freeze-drying. This retains most of the aroma, the structure, and the green color. This is practiced in the United States, Taiwan, Peru, and Denmark (41).

B. Chinese Chives

Chinese chives (*Allium tuberosum* Rottl.) are known as Kau tsai in China and Nira in Japan (56). It is perennial plant, and both the leaves and the inflorescences are eaten. It has also been used as an herbal medicine and is considered effective for treatment of fatigue.

1. Botany

Chinese chives spread via rhizomes. These bear long, flattish (about 15 mm wide), and slightly keeled leaves (56). Spread by rhizome results in the formation of dense clumps of leaves. The rhizomes may branch and are covered by a pale brown fibrous reticulum, which is formed from the remains of old folaige leaf bases. The leaves typically bend downward at the tip, are solid in transverse section, and do not have a ligule. The extent of bulbingvaries with genotype. The rhizome rather than the bulb is the main storage tissue (56).

2. Production

Chinese chives grow well in most soils, but rich well-fertilized fields and strains that are well adapted to climatic zone are essential for the production of commercial yields. The optimum temperature for Chinese chives is 20°C (56). Although chives can be cultured throughout the year, yields are low in winter, as growth is slowed by low temperature.

Chinese chives are grown as a perennial crop with a life ranging from 7 years to 20–30 years (57). Seeds are sown 1 cm deep in spring in 1-m-wide beds. During the first season, watering is minimal in order to encourage a deep root system, and leaves are not cut so as to maximize the built-up reserves. Early in the next season, decayed leaves are removed, the soil is pulled around the base of the plants, and a further 4–6 cm of sandy soil may be put into the furrows in which the plants are growing to blanch the lower parts of the leaves (57). Three cuts of leaves are made at 20- to 30-day intervals when leaves reach a height of approximately 20 cm. To allow root reserves to build up, leaves are not cut during the autumn. In later seasons, the beds must be raked to remove old and dead roots. The beds must also be top-dressed with 1–5 cm of sandy soil, since the roots have a tendency to work themselves up out of the soil. Fertilizers must be applied between the cuts. Chives have been raised in seedbeds and transplanted to beds in the second season. Clumps of 20 seedlings are transplanted 16 cm apart in rows 35 cm apart (57).

Bleached Chinese chives are produced in China (5). In the field, light may be excluded and blanched chives produced using "tents" of straw matting, roofing tiles, or dark paper (56). In the Guangzhou Province of southern China, alternate crops of blanched and green leaves are produced. Blanched chives can be produced in winter in colder areas by excluding light from plants growing in heated greenhouses.

3. Chemical Composition

The chemical composition of Chinese chives is presented in Table 4. Chinese chives are rich in minerals, especially calcium and iron. The leaves contain 3% sugar on a fresh weight basis. The main sugars are fructose, sucrose, and glucose. Chinese chives contain vitamin C and carotene as well as vitamins B_1 and B_2 (56).

4. Storage and Processing

Chinese chive does not store well for more than 2-4 days and it must be stored at 0-2° if quality is to be retained (56). Carotene content falls to less than half its original value during these few days of cold storage (Table 5). However, such a storage is much superior to ambient conditions for maintaining turgidity and carotene content (58). Chinese chives are used fresh and are not processed. They are fried together with other vegetables and meat or used as seasoning with meat. Vitamin C content is reduced to approximately one third of that in fresh leaf by boiling and to about 80% of that in fresh leaf by frying (56).

C. Japanese Bunching Onion

Japanese bunching onion (*Allium fistulosum*) is similar to leek (*Allium porrum* L.) and is cultivated throughout the world, ranging from Siberia to tropical Asia including China, Taiwan, Korea, Japan, Malaysia, the Philippines and Indonesia. Japan, Korea, China, and Taiwan grow most of the world production. In these countries it is a very important vegetable, ranking among the top 10 and supplied to the market year round (59).

1. Botany

A. fistulosum probably originated in China. The closely related wild species *A. altaicum* Pall. is still common in Mongolia and Siberia. There are many local and commercial cultivars with

TABLE 4 Chemical Composition and Nutritional Properties of Chinese Chives (per 100 g edible fresh weight)

Constituent	Fresh	Boiled
Water (g)	93.1	90.8
Proteins (g)	2.1	2.3
Fat (g)	0.1	0.0
Sugars (g)	2.8	5.2
Crude fiber (g)	0.1	1.1
Vitamins		
Vitamin A (IU)	1800	2200
Thiamine (μg)	60	40
Riboflavin (μg)	190	110
Niacin (μg)	600	300
Vitamin C (mg)	25	10
Minerals		
Calcium (mg)	50	46
Iron (mg)	10.6	0.6
Phosphorus (mg)	32	23
Energy (kcal)	19	28

Source: Ref. 56.

TABLE 5 Changes in Fresh Weight and Carotene Contents of Chinese Chives During Storage

Storage day	Storage	
	Cold room	Room temperature
Fresh weight (g)		
0	100	100
1	99.6	84.4
2	99.5	54.9
Carotene (mg/100 fr. wt.)		
0	13.9	13.9
1	10.6	10.2
2	10.0	5.4
3	6.6	6.7
4	5.2	—

Source: Ref. 56.

distinct morphological characters, which are adapted to a variety of climatic conditions. Kumazawa (61) classified Japanese bunching onion into four groups: Kaga, Senju, Kujyo, and Yagura negi.

A. fistulosum is a perennial herbaceous plant usually grown as an annual. The leaves are circular in cross section as compared to the flattened onion leaf. All leaves have blades, with the basal parts—the sheaths—forming a storage structure. Usually, the lateral buds in the leaf axils elongate and develop as tillers to form a vigorous clump. Tillering characteristics are more pronounced in cultivars grown for green leaves than in those grown for long, blanched pseudostems. Unlike the common onion, it forms only a poorly developed bulb with a diameter somewhat larger than that of the neck (pseudostem). The adventitious root system is rather shallow, as most of the roots penetrate less than 30 cm in length. The flowers open in consecutive order, starting at the top and continuing to the base of the umbel.

2. Production

Japanese bunching onion can be grown on a variety of soils. However, soils must be well drained and excessive moisture should be avoided. Well-drained loams or sandy loams containing high organic matter are optimal, especially for the production of blanched sheaths. A soil pH between 5.7 and 7.4 and soil rich in phosphate are suitable. The optimum temperature for seed germination and growth is between 15 and 25°C. High temperatures affect the ultrastructure of the epicuticular wax layers on leaves. Japanese bunching onions are mainly grown for their blanched pseudostems or green tops. The cultivars of the Kaga and Senju groups are mostly grown for the blanched pesudostem. In warmer regions of Japan, some cultivars of the Kujyo group are cultivated by this system, and the green tops as well as blanched pseudostem are harvested. Green tops are harvested year round (59). Since freshness is important, this crop is grown near population centers. The cultivars of the Kujyo group are grown because of the high quality of their tender juicy leaves and their tendency for fast tillering. The cultivar Iwatsuki, which belongs to the Kaga group, is also grown for green tops in central Japan (59).

Seeds are sown in early spring for a winter or spring crop and in the autumn for the following summer to winter harvests. The quantity of seed required for one hectare varies from 2 to 4 kg in nurseries and 8 to 16 kg for direct sowing in the field. The seeds are broadcast or drilled in nursery beds, in shallow trenches, or in rows. Seedlings are planted in furrows, and the roots and bases are lightly covered by soil. The depth of the furrow is about 15 cm for pseudostem production and 5 cm for green top production. The distances between rows and within rows are 55–85 and 3–15 cm, respectively, depending upon tillering tendency (61). In summer, some cultivars of the Kujyo group are transplanted immediately after a drying treatment of 1–2 weeks. This increases the tillering and resistance to heat and drought and results in high production. The recommended doses of fertilizers includes 200–300 kg of nitrogen, 100–200 kg of phosphorus, and 150–200 kg of potash per hectare. The fertilizer is split over three or four doses (59).

Long blanched pseudostems are produced by mounding soil around the lower leaf bases to a height of more than 30 cm. This is done in three to four stages, the first 50 days after transplanting and the last 20 days before harvesting for summer harvest (40 days for winter harvest), otherwise growth is retarded due to poor aeration.

Rust (*Puccinia allii*) causes serious damage to bunching onion leaves. Alternaria leaf spot (*Alternaria porri*), downy mildew (*Peronospora destructor*), *Phytophthora* blight (*Phytophthora nicotianae* var. *parasitica* (Dastur) Waterhouse), leaf spot (*Pleospora herbarum* (Persoon) Rabenhorst), black spotted leaf blight (*Septozia alliaceae* Cooke) Siroiro-eki-byo (*Phytophthora porri* Foister), botrytis leaf spot (*Botrytis squamosa* Walker, *Botrytis cinerea* Persoon), and fusarium wilt *(Fusarium oxysporum)* are other fungal diseases that may affect the crop. White rot (*Sclerotinum cepivorum* Berkeley) may damage the plants in the field under continuous cropping (59) and can be a serious problem because of the long persistance of this pathogen in soil. Onion yellow dwarf virus is the most important virus, causing mosaic-type symptoms including chlorotic mottle, chlorotic streaking, stunting, and distorted flattening of leaves. The Kujyo group tends to be rather resistant to this virus (59). Other viruses pathogenic to Japanese bunching onion are cycas (NSV) and tomato spotted wilt virus (TSWV). Onion thrips (*Thrips tabaci* Lind) is the most important pest of this crop. Cutworm (*Agrotis fucosa* Butler), mite (*Rhizoglyphus echinopus* Fumoze at Robin), stone leaf miner fly (*Dizygomyza cepae* Hering) and stone leek miner (*Acrolepia alliella* Semenov et Kuznestov) are important pests of this crop (59).

At harvest, plants are lifted by hand, trimmed, and packed into bundles, which may then be boxed. harvesting is very laborious, especially in blanched pseudostem production. Hence, mechanical lifters have been developed. Trimmers that peel the outer leaves using high-pressure water or air are also used.

3. Chemical Composition

The chemical composition of Japanese bunching onion is given in Table 6. Glucose, fructose, sucrose, maltose, and fructose-oligosaccharide are the major sugars present in the bunching onion (62). The leaf blades are rich in rhamnose, galactose, glucose, arabinose, and xylose (62). The leaves contain also small quantities of α-cellulose and lignin, but starch is not found. Mucilages increase in bolting plants (63). The leaf blades contain high levels of vitamins A, B_2, and C compared to other leafy vegetables, and they are low in calories (63). Like garlic, bunching onion contains alliin, a precursor of allicin, which plays an important role in the intake of thamine (B_1) and also has a strong antimicrobial effect (59). The odor of Japanese bunching onions is attributed to volatile allyl sulfides (64). It is thought that bunching onion improves eye-sight and internal organ function, enhances metabolism, and prolongs life (59). It is also effective in aiding digestion, perspiration, recovery from the common cold, and as treatment for head-aches, wounds, and sores.

TABLE 6 Chemical Composition and Nutritional Properties of Japanese Bunching Onion (per 100 g edible fresh weight)

Constituent	Plant tissue	
	Pseudostem	Green tops
Water (g)	91.6	92.0
Proteins (g)	1.1	1.7
Fat (g)	0.1	0.2
Digestible carbohydrates	6.7	5.4
Vitamins		
Vitamin A (IU)	85	480
Vitamin B_1 (μg)	40	60
Vitamin B_2 (μg)	60	100
Niacin (μg)	300	400
Vitamin C (mg)	14	33
Minerals		
Ca	47	80
Fe	0.6	1.0
Na	1.0	1.0
K	180	200
P	20	38
Energy (kcal)	27	25

Source: Ref. 59.

4. Storage and Processing

Precooling at 0°C is a common method of improving the shelf life of Japanese bunching onions. Japanese bunching onion was not proccssed until quite recently, when a dehydration industry started. The dehydrated product is mainly used as an additive to preprocessed food such as instant noodles (59).

D. Rakkyo

Rakkyo (*Allium chinese* G. Don, *Allium bakery* Rgl.) is a perennial herb belonging to the family Alliaceae. It is native to China, the wild species being found in an area extending from China to the Himalayan region of India (65). It is now widely cultivated in China and Japan.

1. Botany

Rakkyo has a chromosome number of $2n = 16$. Khurita (66) reported a chromosome number of $2n = 32$.) Katayama (67) confirmed it to be autotetraploid by their observations of cell division at root tips. However, most of the commercial cultivars are tetraploid and infertile. They are, therefore, multiplied vegetatively by offsets (68).

The Rakkyo bulb is elliptical in shape, 2–4 cm long, with a tapering top where the leaves are easily recognized (65). The external color is purplish or greyish white and the bulb is covered by a semitransparent dry membranous skin. The leaf is angular and hollow and 30–60 cm long. When the leaves wither, the bulbs are harvested and are either used as a food or kept as planting material. Rakkyo flowers in the fall. The scapes are 40–60 cm long and bear 6–30 flowers. The pedicels are about 3 cm long and are much longer than the flowers. The inflorescence is a spherical to bell-shaped umbel, and the flowers are purple in color tinged with red (65).

Several cultivars of rakkyo are grown in China and Japan. The recognized cultivars can be divided into two groups according to bulb size. Tama Rakkyo is a small-bulbed group. The number of side bulbs produced from one mother bulb is very high, ranging from 10 to 25, but they are small. The bulbs are white, soft, and almost odorless and have a firm neck, and a high proportion of bulbs suitable for processing (65). Rakuda is a large-bulbed group that is cultivated all over Japan. It has fewer (6–9) side bulbs. The bulbs are elliptical with a long neck. Yatsufusa is another group that produces more side bulbs with firmer necks than cv. Rakuda but of lighter weight. They are suitable for processing. Both yield and quality are low (65).

2. Production

Rakkyo requires well-drained soil. In fertile soil, the yield increases but bulbs become too large and therefore have a lower market value. Moreover, these bulbs are too soft and are much inferior in quality to those grown on sand dunes (65). Rakkyo is grown in temperate as well as semitropical areas. Air and soil temperature, relative humidity, and light intensity influence the yield of rakkyo.

Nitrogen, potassium, phosphorus, and magensium are required in substantial quantities for growth and development of the crop (69). Chlorosis in rakkyo is caused by a decrease in zinc supply in the soil resulting in zinc deficiency. This may be caused by increased pH and the antagonism between phosphate and zinc (70).

Rakkyo is resistant to drought and can be grown with no irrigation in Japan. However, it is almost impossible to obtain good growth in dry months from May to September in Japan, especially sand dune fields (71,72). Sato and Tanabe (71) reported that spring irrigation of sand dune fields results in an increase in both bulb weight and yield. Autumn irrigation promotes lateral bud formation and growth.

3. Composition

The chemical composition of rakkyo plants is given in Table 7. The content of true protein in the bulbs is low compared to the other parts. Soluble sugars are high in the bulbs and middle leaves. Most of the reducing sugars are glucose and fructose (65). Allyl sulfides are the source of the characteristic odor of rakkyo (65). The volatiles of rakkyo include thilanes, alcohols, ketone, and the oily compound 2,3-dihydro-2-hexyl-5-methylfuran-3-one (75).

TABLE 7 Chemical Composition of Different Parts of Rakkyo (% dry matter)

	Green leaves	Middle leaves	Bulb	Roots
Dry matter	37	22	35	6
Crude ash	11.67	9.22	4.30	18.95
Crude protein	17.57	15.38	18.38	16.29
Crude fat	5.09	4.75	3.20	3.08
Crude fiber	29.82	20.04	3.27	40.28
N-free extract	25.83	50.61	70.85	21.40
Pentosan	3.71	5.83	7.16	4.45
Moisture in fresh sample	91.67	87.35	76.70	77.35

Source: Ref. 65.

4. Processing

Small rakkyo bulbs are preferred for pickles. The bulbs are usually first steeped in brine for several days before being transferred to a vinegar, salt, and sugar solution to produce sweet pickle or a vinegar and salt solution to produce sour pickle. Soy sauce may be added to produce an amber-colored pickle. The pickles are bottled or canned and heat-sterilized (65).

REFERENCES

1. McCollum, G. D., *Evolution of Crop Plants* (N. W. Simond, ed.), Longman, London and New York, 1976.
2. Thompson, H. C., and W. C. Kelly, *Vegetable Crops*, McGraw-Hill Book Co. Inc., New York, 1957.
3. T. K. Bose and M. G. Som, *Vegetable Crops*, Naya Prokash, Calcutta, 1986, p. 602.
4. Rijbroek, V. Van, and P. Riepma, *Groeten Fruit 37*:56 (1982).
5. Maynard, D. W., Evaluation of leek cultivars in west-central Florida (U.S.), *Proc. Ann. Meeting Florida State Hort. Soc. 101*:385 (1989).
6. Kanters, F., Research on summer leeks: Only four cultivars remain upright, *Groenten Fruit Vollegrondsgroenten, 1*(12):29 (1991)
7. Stone, D. A., and H. R. Rowse, Proceeding of 9th Soil Tillage Research Organization, U.K., 1982.
8. Krishnan, P., and A. G. Barlage, Magnetic conditioning of seeds of leeks (*Allium porrum*) to increase seed lot germination percentage, *J. Seed Technol. 10*:79 (1986).
9. Maneka, S., The interaction between the time of transplanting and age of leek seedings, *Bull. Shkencave Bujgesore 28*(1):27 (1989).
10. Perk, J. E., Leeks for overwintering: Considerations on varicties, cultivation sites and planting depth, *Rev. Suisse Vitic. Arboric. Hort. 21*(3):159 (1989).
11. Karba, I. P., and R. I. Geriya, The effect of the sowing date on the yield and qualitative evaluation of leek cultivars, *Inst. Rastonieovodstva Imeno, N. J. Vavilova 178*:64 (1988).
12. Granges, A., and A. Leger, Spring leeks in protected cultivation, trial of varieties and cultural methods. *Rev. Suisse Vitic. Arboric. Hort. 22*(5):349 (1990).
13. Kolota, E., Irrigation of leeks planted and grown in different ways, *Biul. Warzywniczy 26*(1):65 (1986).
14. Synnevag, G., and E. Stubhaug, Raising methods and plant numbers for leek, *Oppalsmetoder Plantetall Purre Gartneryr Ket 80*(17):24 (1990).
15. Kaniszewski, S., J. Ruppel, and K. Elkner, Yield and quality of leek (*Allium porrum*) as affected by method of growing, *Acta Hort. 244*:229 (1989).
16. Maity, T. K., and P. Hazra, Leek, in *Vegetable Crops in India* (T. K. Bose and M. G. Som, eds.), Naya Prokash, Calcutta, 1986, p. 602.
17. Mayne, A., Methods of successful leek cultivation, *Wegezum Orfolgreichen Porree Anbau Gemüse 26*(2):74 (1990).
18. Kaniszewski, S., Effect of irrigation and fertilization on the yield and nutrient status of leek, *Biul. Warzywniczy 26*(1):95 (1986).
19. Weier, U., and H. C. Scharpf, Nitrogen fertilization of leek, *Stickst. Dungung Porree Gemüse 26*(2):84 (1990).
20. Hartmann, H. D., K. H. Zengerle, and E. Pfulb, Water requirement and irrigation of autumn leeks, *Wasserverbrauch Bewasserung Herbst Porree Gemüse 26*(2):87 (1990).
21. Pelletier, J., Selective weed control in leeks using chemicals, a point of clarification, *P. H. N. Rev. Hort.* No. 275 (1987).
22. Alofs, W. J., Weed control in leeks—before you spray first know which weed it is, *Groenten Fruit 43*(43):54 (1988).
23. Zander, M., and G. Zozmann, Studies of the application of herbicides before emergence and at early development stages of drilled leek, II. Results of yield pot trial and far scale experiments, II. *Arch. Gartenbau 37*(8):541 (1989).

24. Van der Meer, Q., and P. Hanelt, Leek (*Allium ampeloprasum*), in *Onions and Allied Crops*, Vol. III (J. L. Brewster and H. D. Rabinowitch, eds.), CRC Press, Boca Raton, FL, 1990, p. 170.
25. Katyal, S. L., *Vegetable Growing in India*, Oxford and IBH Publishing Co., New Delhi, 1977, p. 12.
26. Weibe, H. J., Temperature and photo-period influence bolting in leek (in German), *Gemüse 26*(2):62 (1990).
27. Ryall, A. L., and W. J. Lipton, *Handling, Transportation and Storage of Fruits and Vegetables*, Vol. 1, *Vegetables and Melons*, AVI Pub. Co. Inc., Westport, CT, 1972, p. 479.
28. Goffings, G., and M. Herregods, Storage of leeks under controlled atmosphere, *Acta Hort. 258*:481 (1989).
29. Hoflum, H., Storage of leeks, IV. Effect of temperature and atmospheric composition on growth and *Botrytis porri*, *Meld. Norges Landbrukshoegsk. 57*:38 (1973).
30. Tahvonen, R., *Botrytis porri* Bucw. on leek as an important storage fungus in Finland, *J. Sci. Agric. Soc. Finland 52*(4):331 (1980).
31. Splittstoesser, W. F., *Vegetable Growing Handbook* , AVI Pub. Co. Inc., Westport, CT, 1984, p. 262.
32. Currah, L., and F. J. Proctor, *Onions in Tropical Regions*, Natural Product Research Institute, Kent, 1990, p. 36.
33. Vadivelu, B., and I. R. Muthukrishna, CO-4 onion—a high yielding good storing hybrid onion, *South Indian Hort. 30*:142 (1982).
34. Dalbellay, C., A. Granges, and J. Perko, Milrac, a new shallot variety, *Rev. Suisse Vitic. Arboric. Hort. 80*(4):219 (1988).
35. Kusumo, S., and F. Muhadjir, Effect of seed origin and soil cultivation on yield of shallot. *Bull. Penelitian Hort. 15*(1):1 (1987).
36. Jenkins, J. M., Some effects of different day length and temperature upon bulb fermentation in shallots, *Am. Soc. Hort. Sci. Proc. 64*:311 (1959).
37. Sinna'durai, K., Shallot farming in Ghana, *Econ. Bot. 27*:2138 (1973).
38. Alludim, A., The effect of plant density and rice straw mulching on shallots (*Allium ascalonicum* L.), *Bull. Penelitian Hort. 16*(1):40 (1988).
39. Muhadjir, F., and S. Kusumo, Effect of K and seed size on yield and quality of shallot, *Bull. Penelitian Hort. 13*(4):31 (1986).
40. Vanparys, L., Sallots No. 3, nitrogen manuring and cultivar research with shallot, *Beitem Roeselare* No. 319:4 (1991).
41. Poulsen, N., Chives, in *Onions and Allied Crops*, Vol. III (J. L. Brewster and H. D. Rabinowitch, eds.), CRC Press, Boca Raton, FL, 1990, p. 231.
42. Levan, A., Zytologische Studien an *Allium schoenoprasum*, *Hereditas 22*:1 (1936).
43. Heinze, W., and H. Werner, Fruhtreiberei von Schnittlauch, *Gemüse 7*:245 (1971).
44. Dale, W. T., Aecidia of *Puccinia allii* Rud. in chives in Britain, *Plant Pathol. 19*:149 (1970).
45. Noodegaard, E., and K. E. Hansen, Forg med plante beky Helsemidler i landbrugs-og specialafgroder, 1970, *Tidsskr. Planteavl. 76*:63 (1972).
46. Folster, E., and H. Krug, Influence of the environment on growth and development of chives (*Allium schoenoprasum* L.), II. Breaking of rest period and forcing, *Sci. Hort. 7*:213 (1977).
47. Rosenfield, H. J., Ascorbic acid in vegetables grown at different temperature, *Acta Hort. 93*:425 (1979).
48. Franke, W., On the content of vitamin C and thiamine during the vegetative period in leaves of three spice plants (*Allium schoenoprasum* L.) *Melissa officinalis* L. and *Petroselinum crispum* (Mill), Nym. ssp. *Crispum*), *Acta Hort. 73*:205 (1978).
49. Whitaker, J. R., Development of flavor, odor and pungency in onion and garlic, *Adv. Food Res. 22*:73 (1976).
50. Shankarnarayan, M. L., B. Raghaven, K. O. Abraham, and C. P. Natarajan, Volatile sulfur compounds in food flavors, *CRC Crit. Rev. Food Technol. 4*:395 (1973).
51. Wahlroos, O., and A. I. Virtanen, Volatiles from chives (*Allium schoenoprasum*), *Acta Chem. Scand. 19*:1327 (1965).

52. Kameoka, H., and S. Hashimoto, Two sulfur constituents from *Allium schoenoprassum*, *Phytochemistry 22*:294 (1983).
53. Hashimoto, S., M. Miyazawa, and H. Kameoka, Volatile flavor components of chive (*Allium schoenoprasum* L.), *J. Food Sci. 48*:1858 (1983).
54. Starke, H., Über die Flaavonoleder Zweibel des Porrees, des Schnittlauchs und der Schwarzen Johannisbeeren, Diss. Tech. Univ. Hannover, 1975, p. 65.
55. Ruhl, I., and K. Herrmann, Organische Säuren der Gemüsearten I. Kohlarten Blatl und Zwiebelgemüse sowie Mohren und Sellerie, *Z. Lebensm. Unters. Forsch. 180*:215 (1980).
56. Saito, S., Chinese chives (*Allium tuberosum* Rottl.), in *Onions and Allied Crops*, Vol. III (J. L. Brewster and H. D. Rabinowitch, eds.), CRC Press, Boca Raton, FL, 1990, p. 219.
57. Larkcom, J., Chinese chives, *J. R. Hort. Soc. 112*:432 (1987).
58. Takama, F., and S. Saito, Studies on the storage of the vegetables and fruits, II. Total carotene contents of sweet pepper, leek and parsley during storage, *J. Agric. Sci. 19*:11 (1974).
59. Inden, H., and T. Asahira, Japanese bunching onion (*Allium fistulosum* L.), in *Onions and Their Allied Crops*, Vol. III (J. L. Brewster and H. D. Rabinowitch, eds.), CRC Press, Boca Raton, FL, 1990, p. 159.
60. Tachibana, Y., Meaning of drying treatment and spring sowing in Kujyo cultivars of Japanese bunching onion (in Japanese), *Breed Agric. 4*:257 (1949).
61. Kumazawa, S., *Vegetable Crops* (in Japanese), Yokendo, Tokyo, 1956, p. 325.
62. Mizuno, T., and T. Kinpyo, Studies on carbohydrate of *Allium* species, I. Kind of carbohydrates of *Allium fistulosum* L. (in Japanese with English summary), *Nippon Nogei Kagaku Kaishi 29*:665 (1955).
63. Mizuno, T., and T. Kinpyo, Studies on carbohydrates of *Allium* species II. On the mucilage of *Allium fistulosum* L. (1), *Nippon Nogei Kagaku Kaishi 31*:200 (1957).
64. Aoba, T., *Vegetables in Japan-Fruit Vegetables and Allium* (in Japanese), Yasakashobo, Tokyo, 1982.
65. Toyama, M., and I. Wakamiya, Rakkyo (*Allium chinense* G. Don), on *Onions and Allied Crops*, Vol. III (J. L. Brewster and H. D. Rabinowitch, eds.), CRC Press, Boca Raton, FL, 1990, p. 197.
66. Kurita, M., On the karyotypes of some *Allium* species, from Japan Mem. Eshime Univ. Sec. II *1*(3):179 (1952).
67. Katayama, Y., Chromosome studies in some alliums, *J. Coll. Agric. Tokyo Imp. Univ. 13*:431 (1936).
68. Jones, H. A., and L. K. Mann, *Onions Their Allies*, Interscience, New York, 1963.
69. Sato, I., Studies on growing Baker's garlic (*Allium bakery* Regal) in sand dune field III changes of mineral contents following the growth, *Bull. Sand Dune Res. Inst. Tottori Univ. 9*:9 (1970).
70. Yamagawa, T., and S. Fujii, Studies on the chlorisis of leaves in Baker's garlic grown on sandy soils IV. Foliar sprays for correcting deficiency in garlic plants, *J. Jpn. Soc. Hort. Sci. 41*(1):61 (1972).
71. Sato, I., and K. Tanabe, Studies on growing Baker's garlic in a sand dune field. Effect of irrigation in the autumn and in the spring on growth and yield, *Bull. Sand Dune Res. Inst. Tottori Univ. 11*:11 (1972).
72. Ueda, H., Irrigation and soil management in rakkyo fields, preparation of ground basis and introduction of irrigation to the fields in Fukuke Sand Dune, *Field Soil 13*(10):103 (1981).
73. Yamaguchi, F., K. Yoshida, and N. Takiguchi, *Guide Book on Vegetables*, Public Department Kagawa Nutrition College, Tokyo, 1982, p. 64.
74. Mizuno, T., M. Yokoyama, and T. Kinpyo, Studies on the carbohydrates of *Allium* species, VI. Free sugars and polysaccharides of *Allium bakery*, Regal, *Bull. Fac. Agric. Shizuoka Univ. 11*:117 (1961).
75. Kameoka, H., H. Ilda, S. Hashimoto, and M. Miyazawa, Sulfides and furanones from steam volatiles oils of *Allium fistulosum* and *Allium chinense*, *Phytochemistry 23*(1):155 (1984).

19

Garden Pea

S. S. Deshpande
IDEXX Laboratories, Inc., Sunnyvale, California
R. N. Adsule
Mahatma Phule Agricultural University, Rahuri, India

I. INTRODUCTION

The garden pea (*Pisum sativum*, also commonly known as the English pea or green pea; these terms are used interchangeably here) is one of the oldest cultivated vegetables in the world. Currently, it ranks among the top 10 vegetable crops. It is also one of the most popular vegetables grown for home use by home gardeners. The word is derived from the Greek *pison*, which in Middle English became *pease* and was later shortened to pea. Peas are widely cultivated in all but the tropical and subtropical countries of the world.

Peas of various genera, types, and varieties are closely related to beans. Similarly, there is no sharp demarcation in the use of the words pea and bean. In some countries, certain kinds of peas may be called beans and vice versa. The majority of plants and their pods and seeds that are classified as peas belong to the genus *Pisum*; those called beans are of the genus *Phaseolus*. However, this is not an exclusive differentiation. Even within the group "peas," the vegetable is known by several different names. Some of the common terms used to describe the green pea worldwide are as follows:

Baby garden pea: Pisum sativum var. *humile*. It is an early dwarf pea.

Dry pea: Pods and seeds are left to full maturity prior to harvesting.

Edible-podded pea: *P. sativum* var. *macrocarpum*. Pods are consumed as food and picked just as the seeds commence to form. This pea is also popularly known as Oriental or Chinese pea, French pea, snow pea, sugar pea, sugar snap pea, and Turkey pea.

English pea: Common garden pea available in numerous varieties. Also referred to as green pea, shelling pea, or Austrian winter pea in honor of the Austrian monk, Gregor Mendel, who carried out and established the principles of genetic heredity based on his work with garden pea cultivars.

Field pea: *P. arvense*. Seeds are smaller than *P. sativum*. In the past, it was used mainly as a cattle fodder in the developed countries. Ripe mature seeds are now commonly used as a grain legume in human nutrition, while the unripe young seeds can also be used as a vegetable.

Green pea: Pods are picked before the seeds are fully mature, as contrasted with the dry pea, which is harvested after the seeds have fully matured.

Petits pois pea: It is the French equivalent for pea, and is also used to describe a variety of pea known as the French Canner pea or the Turkey pea.

Smooth- and wrinkled-skinned pea: The seeds of several varieties of pea have perfectly smooth skins, as contrasted with the highly wrinkled skins of other varieties. The wrinkled-skinned varieties generally are regarded favorably for their flavor, although the Marrowfat variety with a smooth skin is also quite popular.

According to the current nomenclature, *P. sativum* is used for all edible forms of peas. The "garden" and "field" peas are treated as subspecies within *P. sativum.* Garden peas are recognized as cultivars of *P. sativum* ssp. *hortense* and varieties of field peas as *P. sativum* ssp. *arvense.*

Garden peas are always shelled before use in the green or the dry state, since their pods are fibrous and not particularly tasty. They are freshly picked, eaten as young and small as possible, and are the most popular summer vegetable. Shelled peas are savored as a steamed vegetable and also enjoyed in soups, casseroles, stews, and salads. Green peas are also cultivated commercially for canning, deep freezing, and dehydration. Dried peas and split peas are available year round and are primarily used for soups and purees.

Snow peas or the edible-podded peas are savored for their crisp, succulent pods, and not for their peas. In fact, if peas are allowed to develop in the pods, the edible quality of the pods declines. Snow pea pods lack the parchmentlike lining present in the pods of shelling peas. These peas are quite popular in Oriental stir-fry recipes. Sugar snap peas blend the best of both worlds. They are relished for both their juicy pods and their sweet-tasting peas. Here, the pods and peas are allowed to mature fully for the finest flavor and texture. Although sugar snaps can be lightly steamed and stir-fried, they are in their glory when popped fresh from vine to mouth.

In this chapter, various production, processing, and nutritional aspects of this important vegetable crop are described.

II. ORIGIN AND HISTORY

The garden pea is of very ancient origin, and its wild prototype has never been found. Furthermore, efforts to trace its history are complicated by the lack of definition of its ancient name. Hedrick (1) states that the ancients did not distinguish carefully between peas, beans, vetches, chickpeas, and lentils. It thus is difficult to determine from ancient writings what plants are being discussed.

Many kinds of peas were known to the ancient Greeks and Romans. Burned peas have been found in the ruins of lake dwellings in Switzerland and in the ancient city of Troy, which was first built during the Stone Age. Similarly, archaeological findings of carbonized pea seeds in western Asia and Europe have been dated to 7000 B.C. (2).

Peas were also known to and used by the Chinese in 2000 B.C. (3). Apparently the Chinese were also the first to use the green pods and seeds as food. Wild peas were first sold in England in the eleventh century. However, green peas did not appear on the European menus until the sixteenth century, when they were popularized by French royalty. About this time, agricultural writings also began to distinguish between field peas and garden peas. By 1597, the famous herbalist Gerard was able to list four varieties grown in British gardens: Rounceval peas (*Pisum majus*), which were cooked in the pod; garden and field peas (*P. minus*); tufted or Scottish peas (*P. umbellatum*); and peas without skins in the pods (*P. excorticatum*), which were cooked whole like Rounceval peas. In 1787, Thomas Andrew Knight experimented by crossing different strains of pea, and many of the modern American varieties orgiinated in England as a result of Knight's investigations.

Peas were supposedly brought to the West Indies by Christopher Columbus in 1493 and planted on Isabela Island. By 1614, they were being cultivated at Jamestown, Virginia. New England's first peas were planted by Captain Bartholomew Gosnold on the island of Cuttyhunk in 1602.

From the evidence available, it seems fairly certain that the pea is probably indigenous to southeastern Europe and western Asia. According to Vavilov (4), Ethiopia, the Mediterranean, and Central Asia with a secondary center in the west are the centers of origin of pea. The exact primary center is, however, still not known. Through the centuries, the crop spread westward and northward throughout Europe, southward into Africa, and eastward to India and China.

Peas of early times were small and were dried before cooking. They belonged to the variety now called field peas. Field peas (*P. sativum* ssp. *arvense*) are believed to have originated from the gray pea, which still grows wild in Greece and the Levant (5). Field peas are now grown primarily as a grain legume. Their flowers are colored and the seeds are green or yellow.

Garden peas (*P. sativum* spp. *hortense*), in contrast, are larger and sweeter than field peas and are of more recent origin (<1000 years). They have white flowers and round, smooth or wrinkled seeds, which are either white or yellow in color. Garden peas appear to have been derived from field peas by centuries of cultivation and selection for certain desired characteristics. This theory is lent credence by the fact that its wild cultivars were never found, whereas the field pea still grows wild in the Georgian Republic of the old Soviet Union (3).

Europeans found the garden pea to be much more appealing for use as a green vegetable than the field pea. Furthermore, peas appear to be the first crop scientifically bred to produce new varieties with more desirable characteristics (3,5). By the end of the nineteenth century, many cross-breeding trials had been made, the most notable of which were those conducted by the Austrian monk Gregor Mendel at Brunn (now Brno, Czechoslovakia), which provided the foundation for the science of genetics.

III. WORLD PRODUCTION

Almost 80% of the world's pea crop is utilized in the form of dry peas and only about 20% as green peas. Global green pea production amounted to approximately 4.6 million metric tons (unshelled weight basis) in 1993 (Table 1). North America, Europe, and Asia are the leading producers of vegetable green pea. The United States leads with approximately 28.5% of the annual global production, followed by China (10.3%), France (9.1%), India (5.8%), and the United Kingdom (5.7%). Italy, Hungary, Russia, the Belgium-Luxembourg area, and Australia are other leading producers of green pea. Together, all these countries account for approximately 80% of the world's total green pea production. The production of green pea is highly decentralized geographically, with nearly 50 countries reporting production quantities of a notable size when viewed regionally.

In the United States, Wisconsin leads in green pea production with nearly one third of the

TABLE 1 World Production of Garden Peas

Continent/Country	Production (1000 MT)
World	4602
Africa	229
Egypt	108
Tunisia	24
North-Central America	1410
United States	1310
Canada	68
South America	144
Chile	36
Peru	54
Asia	902
China	473
India	267
Japan	51
Europe	1570
France	418
Hungary	210
Italy	140
United Kingdom	261
Oceania	129
Australia	97

Source: Ref. 6.

total annual crop, followed by Minnesota, Washington, Oregon, California, New York, Idaho, Delaware, and Michigan. These states represent nearly 90% of the total U.S. production. Twelve other states represent green pea production of importance regionally. Not taking into account home gardens and unreported regional small truck farms for the local markets, less than 1% of the total production is intended for the fresh market. The amount of green peas for the frozen pack averages about 50% of the total, whereas approximately 40% and 9% are used for canning and dehydration purposes, respectively. During the past decade or so, there also seems to be a gradual shifting of preference from canned peas to frozen peas.

The decline in demand for fresh market peas in the developed countries has been primarily due to the high labor costs involved in hand-picking of the peas. The shelling of fresh peas can also be very time-consuming, and where crops have been left too long in the field or samples have been in shops for some time before pruchase, there is a noticeable drop in the quality of fresh peas. The reduction in demand can also be attributed to the growing popularity of frozen peas where quality comparable to freshly harvested peas can be obtained.

IV. BOTANY

A. Taxonomy

Peas are members of the subfamily Papilionoideae (or Faboideae; sometimes also treated as a family, Papilionaceae or Fabaceae) of the Fabaceae (or Leguminosae) family. They belong to the tribe *Vicieae* (or *Fabeae*). An early work, *Index Kewensis*, recognizes the following species of *Pisum* (5):

1. *P. arvense* Linn., *Pisum* Linn. System ed. 1935. Europe, northern Asia
2. *P. elatius* Bieb.-Fl. taur. Cane II Mediterranean countries, Orient
3. *P. formosum* Alef. orient, Caucasia, Persia
4. *P. fulvum* Sibth & Sm. Asia Minor, Syria
5. *P. humile* Boiss et Noe. Syria
6. *P. jomardi* Schrank. Egypt
7. *P. sativum* Linn. Europe, northern Asia

However, the taxonomic status of these species has been often questioned. For example according to White (7), there is a striking relationship between *P. arvense*, *P. elatus*, and *P. jomardi*, since crosses among these produce fertile hybrids. Several investigators also consider *P. sativum* and *P. arvense* to be one species. Govorov (8) first suggested the inclusion of all cultivated forms of pea in one species, *P. sativum* L., and subdivided it into two subspecies, *sativum* L and *arvense* L. Based on certain morphological characteristics and the prevalence of combinations of characters rare in the other subspecies of *P. sativum* L., he also suggested a third subspecies, *asiaticum*. Vinokur (9) reported that the carbonized remains among archaeological finds in Ukraine were identified as subspecies *asiaticum*.

Lamprecht (10) regarded *P. abyssinicum*, *P. asiaticum*, *P. elatius*, *P. puschiki*, and *P. tibetenicum* as geographical races of *P. sativum*. On the basis of comprehensive genetical, cytological, and taxonomic studies of the hybrid progeny of *P. sativum* and *P. abyssinicum*, von Rosen (11), however, concluded that the two species essentially differ in the polygene complement and are justifiably regarded as distinct taxonomic entities.

Lamprecht (12) suggested that *P. humile* was not a separate species but regarded it only as a race or variety of *P. sativum*. Menjkova (13) observed that *P. sativum* and *P. arvense* crossed readily and gave fertile progeny and normal segregates. He concluded that the two kinds pertain to the same species. Both these have the same number of chromosomes ($2n = 14$). Lamprecht (14) also observed high fertility, segregation ratios, and linkage relationship in crosses between *P. elatius* and *P. jomardi*, on one hand, and *P. arvense*, on the other, and suggested that the three specific names may be regarded as synonyms.

It is now customary to use *P. sativum* for all edible forms of peas and to differentiate between "garden" and "field" types as subspecies. Garden peas are recognized as cultivars of *P. sativum* ssp. *hortense* and varieties of field peas as *P. sativum* ssp. *arvense*. The principal types of garden or culinary peas are shown in Figure 1.

B. Morphology

Garden pea is an annual herbaceous plant, as small as 15 cm and as tall as 1.5 m or more. Its taproot is well developed with many slender, lateral branches. Being a member of the legume family, it is capable of fixing the atmospheric nitrogen.

The stem is angular or round, fistular, glabrous, nonpigmented and in "fasciated" types normal below and expanded above. Branching is extremely variable, with some varieties producing laterals freely, others only rarely. The leaves are pinnately compound, having one to three pinnae. The pinnae are oval to oblong, 25–50 mm long, and entire with mucronate tips. Leaves are normally green and glaucous; nonglaucous, yellow, and variegated forms are also known. The upper pinnae may be modified as tendrils; the terminal pinnae invariably modified into tendrils. Stipules are large and leafy with dentate margins.

Garden pea inflorescence is an axillary long-peduncled raceme with one or two flowers

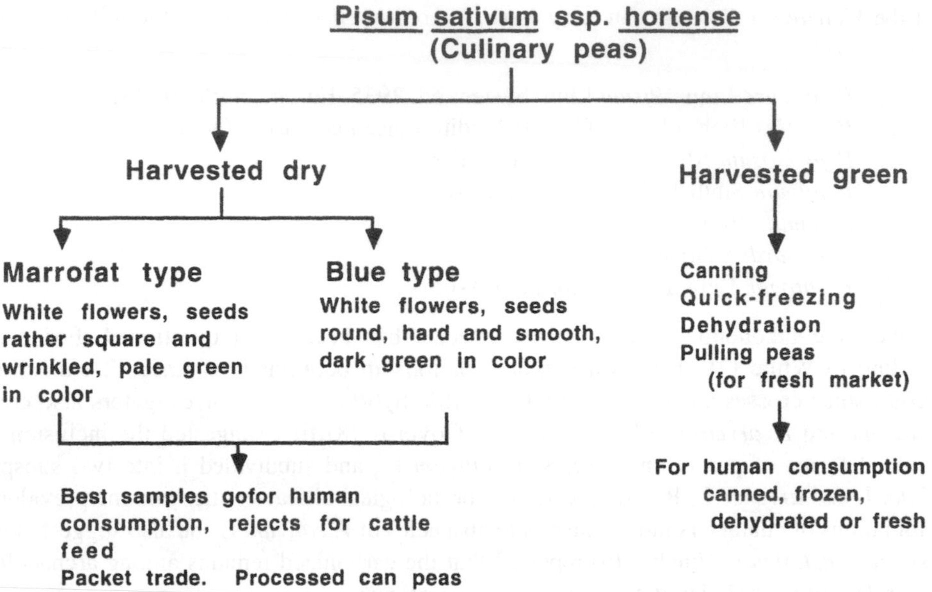

FIGURE 1 General classification of culinary peas.

having minute bracts and bracteole. Occasionally, three- to six-flowerd types are also found. Flowers are large, white in color; the calyx five-lobed, companulate; the corolla papilionaceous; standard large, orbicular, and clawed; the wings clawed with a prominent spur; the keels united and broader than the wings; stamens 10, diadelphous with uniform oblong anthers; and the ovary flattened with a sharply bent style and hairy on the concave face.

The fruit is a typical legume pod of variable length and breadth, curved or straight, yellowish-green to dark green, with either acute or blunt apices. In most varieties, it is lined by a thin parchmentlike sclerenchymatous membrane—the endocarp of the wall. This membrane at the time of maturity drying contracts and splits the pods. Varieites without parchment are known as "edible-podded" or "sugar peas."

The seeds are 4–10 in number and normally free. The testa is thin, either colorless (white) or green (the intensity of the greenness changes with the stage of maturity). The hilum is quite distinct and is colorless, brown, or yellow. The seed surface varies greatly with variety, being smooth, wrinkled, indented, or dimpled. Depending on the variety, individual seeds may weigh up to 250 mg.

Field pea cultivars (*P. satifum* ssp. *arvense*) resemble the garden pea in most of their characteristics, except as follows: leaves sometimes with greyish mottling; short peduncles; pigmented leaf axils; colored flowers with purple- or lavender-colored standards, purplish red wings, and greenish keels. The pods are small and sometimes purple in color. The seeds of field pea cultivars are distinctly smaller than thsoe of garden peas: angular and grey-green, grey-yellow, grey-brown, or grey-speckled with darker spots.

C. Varieties and Cultivars

Peas are grown for use as a fresh or processed vegetable and mainly wrinkled seeded varieties are grown. The other commercially important types include smooth peas or dimpled peas. These may be white, blue, green, or grey seeded. In the United States and several European countries, smooth peas are primarily grown for harvesting as dry peas, later used in the form of split peas.

In selecting the right variety of garden pea for a given situation, there are four major considerations: (1) timing, because there are early, medium, and late varieties, (2) size, whether dwarf, semi-dwarf, or tall, (3) smooth or wrinkled seeds, and (4) edible or nonedible pods.

The Britisher Thomas Andrew Knight (1758–1838) was a pioneer in producing garden pea varieties. Prior to his times, most early pea varieties appeared to be mainly round-seeded forms. Knight first selected two wrinkled-seeded varieties, Knight's Tall Green Marrow and Knight's Dwarf Green Marrow, which appear to be the forerunners of our modern wrinkled-seeded varieties (5). Subsequently, Thomas Laxton (1830–1893) produced varieties such as Alderman, Gradus, Supreme, and Thomas Laxton. Some of these are still popular today. Earlier this century, a number of British companeis introduced several early varieties of garden peas. The popular ones include Laxtonian, Progress, Superb, Pioneer, Foremost, Bountiful, Early Giant, Little Marvel, Telephone, Daisy, Giant Stride, Kelvedon Wonder, Pilot, and Freedom. Kelvedon Wonder is the foremost dwarf variety int he United Kingdom today for the market gardener. The English varieties now commonly cultivated in the United Kingdom include the round-seeded types British Lion, Clipper, Feltham Advance, Feltham First, Foremost, and Meteor. Wrinkled-seeded varieties include Exquisite, Kelvedon Wonder, Thomas Laxton, and Laxton Progress. Two varieties, Meteor and Clipper, are reportedly the most popular round-seeded varieties, while Kelvedon Wonder is the most popular wrinkled-seeded variety in Britain. All of these are of the first early group.

In the second early group, mainly wrinkled-seeded varieties such as Little Marvel and Kelvedon Triumph are widely grown. The round-seeded variety in this group is Laxton Superb.

In the United Kingdom, the main-season varieties are all wrinkled seeded, and popualr ones include Onward, Lincoln, Alderman, Kelvedon Monarch, Duplex, and Giant Stride. For crops of tall peas, Alderman is favored. There are also some late varieties, such as Gladstone. These, however, are not widely grown, since for late supplies of pods, early- or main-season varieties can also be sown late.

Several varieties grown as green peas for the market are also canned. Some early and main crop varieties, such as Gregory's Surprise, Witham Wonder, Canner's Perfection, Sharpes 99, and Canner, are grown primarily for canning purposes.

These English varieties of garden peas also proved to be quite successful as direct introductions in the United States. In the United States, most of the breeding work, especially breeding for disease resistance, was carried out at the Wisconsin Agricultural Experiment Station. The newly bred varieties include Wando (Laxton Progress × Perfection), Early Badger (involving Perfection Horsford and Notts Excelsior as parents), Delwiche Commando (Admiral × Pride), and Bonneville. New Era, which is a cross between Wisconsin Perfection and Delwiche Commando, is primarily used in the canning industry.

Canning varieties usually have a tough skin that holds its shape during the canning process. Most of the late-maturing canning varieties have larger seed than canners usually desire. The canning industry prefers the small peas because the consumer associates small size with high quality. The varieties grown for freezing, in contrast, must have a dark green color and tender skins. The size is not as important since large peas seem to be more acceptable as a frozen product. Thomas Laxtno is the most widely grown freezing variety.

Other wrinkled-seeded varieties include Emerald and Louisiana Purchase, from the Florida and Louisiana Agricultural Experiment Stations, respectively. The latter is a semi-viny type producing pods with 14–16 peas.

For edible-podded peas intended for home gardens and small, specialty truck operations, the Dwarf Grey Sugar variety is the earliest and smallest of the sugar peas. A larger and later variety, Mammoth Melting Sugar, is resistant to *Fusarium* wilt and requires support to climb upon. The characteristics of some of the cultivars of garden and edible-podded peas commonly grown in the United States are summarized in Table 2.

In the United States, the leading varieties of dry, edible field peas include Alaska, Blue

TABLE 2 Characteristics of Garden Pea Cultivars Commonly Grown in the United States

Cultivar	Average days to maturity	General description and comments
Garden pea cultivars		
Alderman (also Tall Telephone)	70–75	A tall variety; oval, curved pods are 4.5–5 in. long with 8–10 peas per pod; needs good support; retains color when cooked
Alaska	55	Earliest of all; the commonly grown garden pea; vines are short; smooth-seeded type recommended for green shell use for drying for split pea soup; resistant to *Fusarium* wilt
Freezonian	63	Popular freezing pea with short vines; pods are 3.5 in. long, holding 7–8 seeds per pod; vines are wilt-resistant
Little Marvel	63	Long a popular and dependable variety; vines are short; pods grow 3 in. long with 7–8 peas per pod; equally suited for fresh use and freezing
Thomas Laxton	62	An old standard; a heavy yielder and good freezer with short vines; pods grown 5 in. long with 8–10 peas per pod; Laxton's Progress, a successor variety, is very similar, with vines that are only 18 in. long
Wando (also Main Crop)	68	A heavy producer; withstands hot, dry spells; vines are short; pods grow 3.5 in. long with 6–8 peas per pod
Snow pea types		
Dwarf Grey Sugar	65	Vines reach 2–2.5 ft. long; pods grow 2.5–3 in. long; needs no staking
Mammoth Melting Sugar	75	Vigorous-growing vines reach 5 ft. long; pods are 4–4.5 in. long; yields heavily over a long harvest period; needs support
Sugar Snap	70	Pods and peas are edible at all stages of maturity; unlike standard snow peas, the peas are at their sweetest and tastiest when mature; vining type reaches 6 ft. or more and needs sturdy, tall support; dwarf types need no support; pods are 2.5–3 in. long

Bell (Blue Prussian), First and Best, Extra Early, and White Canada. In other parts of the world, the early English varieties also appear to be the forerunners of modern varieties. These have been successfully bred into new cultivars and varieties to suit the regional growing conditions.

V. PRODUCTION

A. Climate

Peas are adapted to a cool moist climate within a temperature range of 7–21°C. The optimum is in the region of 18–21°C. Green peas are cultivated in several temperate areas throughout the world. However, it will only produce satisfactorily in tropical countries at altitudes of 1500–1800 m (15). Excessive heat prevents blossoms from setting fruit and also retards the development of pods that have formed. This in turn reduces the size and weight of the seed. Light freezes or frosts do not hurt the young seedlings—only the blossoms are seriously damaged. The crop is most productive where rainfall is abundant, but it also yields satisfactorily in cool, semi-arid regions. Dry atmosphere is desirable at harvest time.

In the United States, green peas are grown in the South and in the warmer parts of California during the fall, winter, and early spring. The only regions where peas are grown successfully during midsummer and late summer are those having relatively low temperatures and good rainfall or where irrigation is practiced. The most important regions from which peas are marketed during the hottest part of the summer are the cool coastal sections of northern California, Washington, and Oregon; regions of high elevation in Colorado and other western states; and parts of New York and New England.

A few successful plantings may be made at 10-day intervals. Later plantings rarely yield as well as earlier ones. Planting may be resumed as the cool weather or fall approaches, but the yield is seldom as satisfactory as that from the spring planting.

Alaska and other smooth-seeded varieties are frequently used for planting in the early spring, since they germinate well in cold, wet soil. Thomas Laxton, Greater Progress, Little Marvel, Freezonia, and Giant Stride varieties are considered suitable early varieties with wrinkled seeds. Alderman and Lincoln develop approximately 2 weeks later than Greater Progress, but under favorable conditions they yield heavily. Alderman is a desirable variety for growing on brush or a trellis, since they are less susceptible to bird damage.

B. Soils

Peas are adapted to a wide range of soils but grow best on well-drained clay or sandy loams rich in or well supplied with lime. They do not yield well when planted in sandy or gravely soils. Heavy soils are also undesirable because of their high water-holding capacity and because the deep tilth required by this crop is often impossible to obtain on such soils. However, relatively heavier soils that are well drained, deep, friable, loose, and rich in organic matter with a high lime content and a good soil structure can also be used. Generally, the lighter soils produce earlier crops than the heavier soils. Thus, sandy loams are preferred for sowing of early crops, while the silt or clay loams can be used for late crops and better yields.

The optimum soil pH range for peas is 6.0–7.5. Peas do not thrive on highly acidic soils unless well limed.

C. Soil Preparation

Thorough preparation is necessary for peas as for other vegetables. It is especially important where the seed is broadcast or planted with a grain drill, since under these methods no further cultivation is given to the crop. Fall plowing is desirable for the early crop since planting often is greatly delayed when the land is plowed in the spring. The surface should be smooth and free from clods so that the drill plants all the seeds at the same depth. Uneven sprouting of the seed results in wide variations in maturity of peas at harvest time. This increases the difficulty of timing the harvest and may result in a decrease in grade and price. When the crop is grown for the cannery, an effort is made to leave the surface smooth by rolling either before or after planting or both, as a rough, uneven surface interferes with the use of the harvesting machinery.

D. Seed Treatment

Like other legumes, peas are capable of fixing atmospheric nitrogen. Therefore, when seeds are first planted in areas where they have not been previously grown, they should be inoculated with the appropriate species of nitrogen-fixing *Rhizobium* bacteria. In such cases, the soil may not contain sufficient amounts of these microorganisms to promote optimal growth. It is generally advisable to inoculate seed to be planted on highly acid soil even if peas were grown in the same area the previous year. In the developed countries, most growers do not inoculate their seed because of the labor costs involved and since it is simpler to provide the nitrogen for peas in the fertilizer rather than treat the seed.

Under moist soil conditions, the use of thiram- or captan-based seed dressings is also advised as an insurance against soilborne pathogens should temperatures be low and brairding slow. Such seed treatment is generally hgihly beneficial with exceptionally early sown peas.

E. Planting and Seed Rates

The seeds should not be planted too deeply, which results in slow sprouting and poor stands. A depth of 1 in. in moist heavy soil and about 1.5 in. (3.8 cm) on a dry heavy soil is sufficient. A covering of 1.5–2 in. (3.8–5 cm) might be given on a dry sandy loam, but a depth grater than 2 in. (5 cm) is seldom if ever justified in humid regions. The seeds germinate in about a week's time.

Depending on the growing habit of the variety, there are several ways to arrange the rows. Tall or short crops can be planted in single conventional rows with 30–36 in. (75–90 cm) between rows for tall crops and 24 in. (60 cm) for shorter ones. Tall varieties also do well in double rows, 8 in. (20 cm) apart, with a vertical support run between the rows. At least 30 in. (75 cm) should be allowed between the sets of double rows. In all row plantings, the seed should be spaced 1–2 in. (2.5–5.0 cm) apart within the row. Dwarf and semi-dwarf varieties can be broadcast in wide rows 18 in. (45 cm) across; both dwarf and tall varieties can be planted in raised beds. Generally, the spacing to be followed varies with the variety, the taller being given wider spacing; the early crop is sown more thickly to allow for losses caused by rather unfavorable conditions.

Garden peas in home gardens and some of the peas for fresh market are planted in rows 2–3 ft. (60–90 cm) apart. The tall-growing varieties are supported on trellises of wire or heavy string. Another method used in home gardens is to stick brush into the ground along the row. Most of the peas grown for processing are planted with a grain drill in the same manner used for small grain cereals.

The seed rate depends on the method of sowing. Generally, broadcasting requires more seed (50–75 kg/ha) than planting by dibbling or drill (20–25 kg/ha). Enough seed should be planted so that for single-stem varieties, such as Alaska and Surprise, there will be at least 18 plants per meter of row. Varieties that branch, such as Perfection, should be sown to attain a plant density of 14–16 plants per meter of row.

F. Fertilization, Irrigation, and Interculture

Because pea is a legume, only phosphatic and potassic fertilizers are generally recommended. These fertilizers are applied before sowing. Depending on the inherent soil fertility, a small dose of nitrogen is valuable in stimulating early growth. It is common to apply 20–30 kg of nitrogen, 70–110 kg of phosphorus, and 50–60 kg of potassium per hectare at the time of sowing. A small dose of nitrogen at the time of flowering and pod formation results in higher yields. Generally, the crop should not be fertilized heavily unless ample water is available.

The water requirement of pea crop is very low, and it can generally be grown without irrigation. Irrigation, however, is beneficial in dry areas and during prolonged hot spells. Overirrigation is undesirable because it causes excessive vegetative growth. Standing water can also cause sun-scalded plants, resulting in seed with reduced germination. In the United States, the first water application is usually at the time of blossoming and a second one before pods form.

It is beneficial to thin the seedlings to 3–4 in. (7–10 cm) apart in conventional and wide rows. Tall varieties also need support when they are 3 in. (7 cm) tall; without support, the vines will form dense mats, air circulation will be impeded, and production will be reduced greatly. Fences, trellises, or any vertical surface that is at least 5 ft (1.5 m) tall and to which the tendrils can cling should be used. Delay in stacking reduces the yields considerably. Stacking should also be done in the direction of the wind. Otherwise, wind blowing in a perpendicular direction can uproot the stakes and damage the plants. Regular hoeing and weeding are also necessary to check weed growth and to provide proper aeration to the roots.

G. Diseases and Pests

The important fungal and bacterial diseases of peas include the following: powdery mildew caused by *Eryshiphe polygoni* D.C.; *Fusarium* wilt or near wilt caused by *Fusarium oxysporum* f. *pisi* (Linford) Snyder and Hansen; root rots caused by several organisms including *Rhizoctonia*, *Fusarium*, *Pythium*, and other soilborne pathogens; bacterial blight caused by *Pseudomonas pisi* Packett; and gray mold rot caused by *Botrytis cinerea*. For controlling the *Fusarium* wilt, the best method is to use resistant varieties, and for controlling powdery mildew, dusting with sulfur is very effective.

Several viral diseases of pea have also been reported and are becoming serious in some parts of the world. Disease-free seed should be used for controlling bean mosaic, as it is transmitted through the seed.

The important pests include pea aphid (*Macrosiphum pisi*), pod borer (*Heliothes* spp.), pea weevil (*Bruchus pisorum*), and leaf miner (*Agromyza flaveola*). These can be controlled with insecticides such as malathion. Nematodes also affect pea yield in some infected areas (16).

H. Harvesting

The time to harvest is determined largely by the appearance of the pods. These should be well filled with tender young peas and changing in color from dark to light green. The harvest should be made when the peas are still in prime condition, but without sacrificing the yield. There is an inverse relationship between yield and quality after the peas reach a certain maturity. As the harvest is delayed, the proportion of small peas decrease steadily, thereby increasing the yield of the crop. However, the tenderomoter value and the starch content also incresae, which greatly reduces the quality and the total value of the harvested peas.

High quality of garden peas is generally associated with tenderness and high sugar content. Tenderometers or other similar instruments are used to measure the toughness of the seed coat and the firmness of the pulp. The tenderometer and texturemeter values, specific gravity, and percent alcohol-insoluble solids all increase with maturity and are negatively correlated with the organoleptic quality of harvested peas. The tenderometer is used by most canners to assess quality—a high value indicates low quality. The highest price is paid for peas with a low tenderomoter value. In some growing areas, the emphasis is on size, and the price paid to the growers is based on the sieve size of the peas.

Several processors use a system of accumulated degree-hours, or "heat units," to determine dates of planting and to schedule subsequent harvesting and processing of peas. The number of degree-hours above 4.4°C required to bring a certain variety to maturity is calculated each season. This accumulated information is used to determine the interval between plantings. The number of degree-hours between plantings is equivalent to the number that is expected to accumulate during the desired interval of harvest. By planting on the basis of degree-hours rather than actual calendar days, it is possible to handle a fairly uniform volume of peas over a long season (17).

In common practice, the appropriate maturity of peas for harvesting is determined by (1) use of physical instruments such as tenderomoter, maturometer, or shear press, (2) ratio of peas to pods, (3) content of alcohol-insoluble solids (AIS), (4) starch content, or (5) floating in 5% brine solution (18). Recommended ranges are 11–16% AIS for canning, less than 11% for fancy grade, 11–15% for extra standard grade, and more than 15% for standard grade canned peas. Using the tenderometer technique, vining peas used for canning fresh, quick-freezing, or artificial drying are ready when the crop is just starting to lose its green color and the peas are still soft. The reading for freezing peas is about 100, and for canning they can be slightly fimrer at 120. When ready, the crop must be cut as soon as possible and the shelled peas rushed to the processing plant for further processing. Alternatively, the seeds are graded by floating on a 5% brine solution; the sinkers are taken for canning while the floaters are used for freezing.

Peas grown for home use and for the fresh market are picked by hand. This is the most expensive operation involved in garden pea cultivation. Green peas must be picked at just the right stage of maturity becasue premature picking yields small-sized seeds whereas delayed pickings lower quality. Some growers make two or three pickings, while others make only one. In the latter case, the vines are pulled and all the pods are picked off. Although it costs less to harvest peas by this method, the quality of the product is better and the yield higher when two or more pickings are made.

In China and certain other countries, the pods are picked at a very immature stage so that they may be used as a green vegetable. Sugar snap peas are best when picked when pods are full-sized and peas are large and completely mature. Snow peas are harvested when the pods are flat and before the seeds have begun to form bumps. Both these peas should be utilized as quickly as possible.

Peas intended for commercial processing are harvested with machines of various types. The vines are cut with a mowing machine, windrowed, and loaded on trucks with a hay loader. Pea harvesters which mow the peas and load directly into a truck or wagon are commonly used in the main growing areas. The vines are hauled to a vining station, where the peas are removed from the vines and pods by machinery. In some areas, self-propelled combines are used for harvesting pea vines. The combines sweep over previously laid windrows of pea vines, engulf the vines, gently separate the tender vegetable from their pods, and convey the product into trucks while moving at about an acre an hour (19). Reid (20) built a mechanical harvester with harvesting efficiencies of 67–97% with commercial cultivars. An average pulling force of 2 kg was required to remove green pods from the vine.

To facilitate mechanical harvesting of pea vines, chemical defoliation was also tried to remove the leaves prior to harvesting. However, such treatment lowered the quality of both the fresh and processed peas (19). Mechanically harvested peas were entirely suitable for processing for a few hours, but when held for longer periods, the peas changed color and showed signs of increased enzymatic action.

The average yield of shelled green peas in the developed countries is about 5500 kg/ha (3).

I. Grading and Packing

When garden peas are intended for fresh market, it is necessary to remove the overmatured yellow pods, flat pods, the diseased and insect-injured pods, and the trash from the harvested peas before they are packed. Peas are commonly packed for shipment in baskets, hampers, and boxes of various types.

Peas for processing are graded into four grades based on the size of shelled peas. The smaller sizes are considered to have the best quality and fetch the highest prices from canners. The sieves used for separating the sizes have mesh ranging from 18/64 to 24/64 in. in diameter. Several processors also buy peas based on tenderometer readings rather than sieve size. Generally, such quality measures are much better than size alone for grading peas. For example, small peas are sometimes overmature and of poor quality, while large peas may be tender and not yet mature.

VI. STORAGE

The popularity of garden peas for fresh market has decreased markedly in the developed countries, primarily due to harvest expense and the ready availability of frozen peas throughout the year. However, they are still popular in the developing countries. Unless they are promptly cooled to near 0°C after picking, green peas tend to lose part of their sugar content, on which much of their flavor depends.

Hydrocooling is a preferred method of cooling freshly harvested peas. After precooling, peas should be packed with crushed ice to maintain their freshness and turgidity. The ideal holding temperature is 0°C. Even at this temperature, peas cannot be kept in salable condition for more than 1–2 weeks unless they are packed in crushed ice. Peas keep better unshelled than shelled (21).

Some of the common methods for extending the shelf life of fresh garden peas are as follows.

A. Low-Temperature Storage

Fresh garden peas can be stored for 2 weeks at 0°C. Peas that are to be shipped to distant markets should be kept cool if they are to reach the market in edible conditon. Peas precooled by immersing in water (hydrocooling) must be kept under refrigeration until they reach their destination. Precooling can also be done by placing crushed ice in the package or over the top of the lead or both. They should also be shipped in refrigerated trucks. A temperature of –1 to 0°C with 90% relative humidity is recommended for storage of garden peas in marketable conditions for up to 2 weeks (17,22,23).

B. Controlled-Atmosphere Storage

Ordinary refrigeration at 0°C can hold green peas for only 7–10 days (22). In contrast, controlled atmosphere storage with 5–10% O_2 and 5–7% CO_2 at 0°C can extend the storage life of garden peas for up to 8–10 weeks (17,21).

VII. PROCESSING

In the developed countries, almost 99% of the crop of fresh garden peas is processed in some form, since fresh peas spoil rapidly and there is little demand for them. In contrast, because of lack of the proper processing infrastructure, the reverse is often true in many developing countries. The varieties used for processing must be resistant to *Fusarium* wilt and downy mildew. Furthermore, their pods should be of uniform size and maturity and should shell out easily (17). Fresh green peas are processed in three ways: dehydration, canning, and freezing. These processes are briefly described below.

A. Dehydration

Sun-drying is the common method for drying peas in the tropics. In most developed countries, it is carried out with mechanical driers. Uniformity of texture is important for dried peas later intended for further processing. Most cultivars that make an acceptable frozen or canned pea also appear to be satisfactory for dehydration. In the United States, good-quality dried products have been made from Thomas Laxton and Dark Skinned Perfection cultivars.

In the past, drying green peas was often regarded as a difficult process. Becausing of the hardening of the outer tissues, peas dry very slowly toward the end of the process (24). The remaining moisture in the dried seed is deterimental to its quality. Conventional air-drying under a variety of conditions is therefore unsatisfactory. It does not produce dehydrated peas of satisfactory low moisture content without causing damage to the product. Low moisture levels are thus absolutely essential for storage stability of dehydrated peas.

To overcome the limitations of the conventional drying process, Moyer et al (25,26) first developed a mechanical technique for slitting the seed coat of the pea prior to dehydration. As a result of the slitting operation, the rate of drying was increased, lower final moistures were obtainable, the rehydration ratio and rehydration rate were increased, and the quality of the rehydration product was greatly improved. Shah and Sufi (27) also reported that pricking or slitting peas prior to dehydration can accelerate the drying process and also improve the rehydration characteristics of dried green peas.

To produce a good-quality dried product, peas should be harvested at full maturity. Too tender or overripe peas are not suitable for dehydration. Shah and Edwards (28) studied the influence of pea size and maturity on drying rates. They reported that smaller, less mature peas showed higher drying rates than larger and comparatively more mature peas. Smaller peas also rehydrated more completely. The seeds should be of uniform maturity, and have a sweet taste, good flavor, fine texture, and a bright dark green color, which should remain stable during the drying process.

A schematic diagram of the drying process is shown in Figure 2. Peas are harvested, vined, dry-cleaned, and size-graded. Sizes 3, 4, and 5, which will pass through a 13⁄32-in. (1.0-cm) mesh

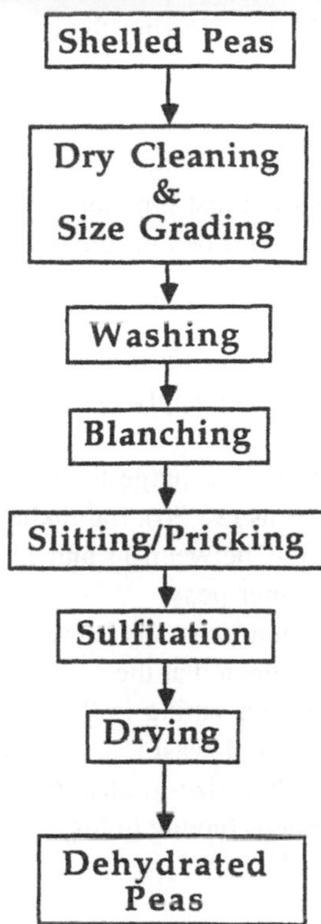

Figure 2 Flow chart for the preparation of dehydrated green peas.

screen but not through a 10⁄32-in. (0.8-cm) screen, are recommended for good-quality dried product. Smaller sizes usually do not rehydrate well, and larger sizes may be too high in starch content (29). The shelled peas are thoroughly washed in cold water and are sorted to eliminate pods, stems, off-color peas, and other defects.

Blanching is used as a pretreatment to inactivate quality-degrading enzymes (peroxidases, catalases, and lipoxygenases) and to improve the color and texture of processed green peas. Traditional methods such as hot water (99°C for 1.5–2 min) and steam blanching (1–2 min) are being increasingly replaced by chemical (dipping peas in a boiling 5% brine containing 0.15% sodium bicarbonate and 0.1% magnesium oxide) or microwave methods (18). Compared to steam blanching, hot water blanching generally destroys significant amoutns of ascorbic acid, nicotinic acid, pyridoxine, and other water-soluble nutrients. By increasing the severity of the blanching treatment, it is also possible to increase the drying rates of overmature peas (26).

After blanching, peas are run through a slitting machine. Commercial machines are available that handle about 6000 lb (2724 kg)/hr; they make about a ⅛-in (0.3-cm) slit in each pea (29). The slit peas are then dipped into, or sprayed with, a sulfite-bisulfite solution of sufficient concentraiton to p[roduce a dried product containing 300–500 ppm SO_2.

The peas are then dried in tray-tunnel driers, cabinet driers, continuous belt driers, or belt-trough driers. It is recommended to remove peas from any of these drying systems when they are dried to about 8% moisture content. The drying is then continued to 4% moisture in bin driers using desiccated air at 49°C. In two-stage tunnel driers, temperatures of 82°C for the initial stage and 71°C for the second stage are recommended (17,29). With thorough-circulation continuous belt driers, higher temperatures of up to 88–93°C can be used during the early part of the drying cycle.

B. Canning

Prior to the development of quick-freezing processes, canning was the most popular method of processing green peas. Even today, approximately 40% of the total green pea production in the United States is processed for canning.

Peas for canning may be classified into two types: the early smooth-seeded cultivars and the later sweet, wrinkled-seeded types. The latter have better flavor but do not produce as heavily as the smooth-seeded types. The words "smooth" and "wrinkled" primarily apply to the ripe, dry peas, since before drying they are all smooth.

The content of alcohol-insoluble solids (AIS) in peas is often used to judge the quality of peas for canning. An AIS range of 11–16% is ideal for canning of peas (18,30). Salunkhe et al. (18) have given the requirement of less than 11% AIS for fancy grade, between 11 and 15% for extra standard grade, and more than 15% AIS for standard grade canner peas.

The canning process is shown schematically in Figure 3. Green peas are harvested and vined or shelled at vining stations in the field. The shelled peas are collected at the viners in bins with 4.5–5 kg capacity. Normally no more than 4 hours elapse between vining in the field and canning of the peas. The shelled peas first pass through a cleaning and washing operation to remove dirt and other foreign material. The cleaned, shelled peas are then size-graded in a slowly revolving screen cylinder or through a series of nested or parallel screens having holes of varying

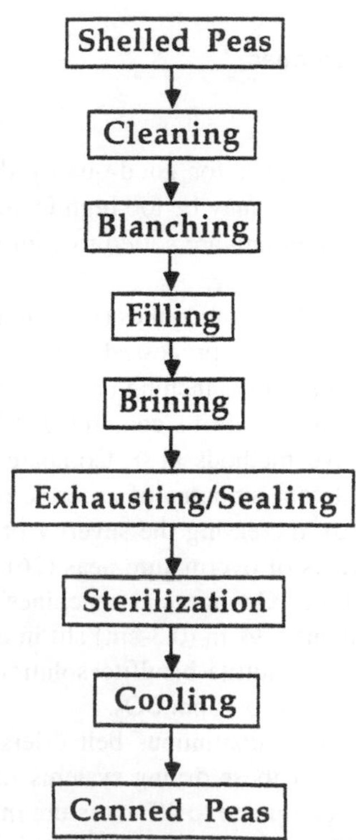

FIGURE 3 Schematic diagram of the process for canning of green peas in brine.

diameters. An important requirement for satisfactory size grading is that the grader be fed at a uniform rate. Frequently two or more sizes are combined into a single grade (31).

The peas are then subjected to hot water blanching in a pea blancher of special design for 3 minutes at 88–93°C. If the peas are fairly mature, the blanching time may be increased to 4–5 minutes; young tender peas are blanched for 1.5 minutes. Blanching removes the slight raw taste and odor of shelled peas and the air entrapped in the pea tissues. It also softens the peas so that a more uniform fill of the can is obtained. Water for blanching should be soft because the calcium in hard water will harden the peas. A continuous fluidized-bed steam blanching process for about 30 seconds can also be used. However, steam blanching does not have as great a cleansing and plumping action on peas as water blanching.

After the blanching treatment, peas are graded and sorted for quality. Generally, only the largest peas of any cultivar need to be quality-graded. After the blanched peas are cooled, the larger ones are fed into a brine of 38–40° Salometer (9.5–10% salt). The younger, tender Grade A peas, being of lower density, will float in brine of this concentration, whereas the more mature Grade B peas will sink. After separation, both grades are thoroughly washed immediately to remove the adhering brine. Only Grade B is used for canning purposes; Grade A floating peas are used for drying. The blanched and graded peas are then sorted on a slowly moving belt. Off-color peas, splits, and any undesirable trash are removed by the sorters.

The sorted peas are fed by gravity into cone-shaped hoppers on the can-filling machine. The peas fall into the cups of the automatic rotary can filler. Each cup delivers exactly the desired volume of peas into a can, which are then carried to the briner. Cans for peas are usually enameled with C-enamel to prevent blackening of the tin plate due to FeS formation (31).

The brine for canned peas usually contains sugar as well as salt. The composition of the brine differs considerably according to the preference of canners or distributors and with the maturity of the peas. Because less mature peas are sweeter, they need less sugar in the brine. It is not unusual to add aromatic herbs and spices, stock, or vegetable juice up to a maximum of 15% of the total drained vegetable ingredients (32). Butter or other edible animal or vegetable fats may also be used in amounts not less than 3% of the final product weight. Brine is added at 71°C by adjustable cups.

Filled cans are sealed under steam flow by the steam-vac procedure, or the lid may be clinched by the can sealer on the top of the can and the can prevacuumized and sealed in vacuum. The vacuum in the can after retorting and cooling should be at least 178–200 mm (31). The sealed cans are heat-processed in vertical or large horizontal still retorts. Continuous agitating retorts are also used for No. 303 cans in some canneries. The recommended processing times and temperatures for still retorts are shown in Table 3.

TABLE 3 Process Times and Temperatures for Canned Peas in Brine

Can name	Dimensions	Initial temperature (°C)	Time (min) at 115°C	Time (min) at 121°C
No. 2	303 × 409	21	34	17
and smaller	and smaller	60	31	15
No. 3 Cylinder	404 × 700	21	50	22
		60	45	20
No. 10	603 × 700	21	57	28
		60	48	21

Source: Ref. 33.

Canned peas should be cooled rapidly following sterilization to avoid excessive softening and the formation of a cloudy brine through gelatinization of starch.

A mixrure of diced carrots and green peas can also be processed for canning. In such instances, the vegetables must be well mixed before canning, and the peas should not be too heavily blanched before mixing. The ratio of carrots to peas is usually between 50:50 and 60:40 (31).

The FAO/WHO (32) have formualted quality standards for canned green peas. These allow certain additives in canned peas (Table 4) and make allowances for certain defects. For example, canned peas may contain a slight amount of sediment and be reasonably free of other defects. The permissible maximum limits for various defects (as percentage based on the weight of the drained peas) are as follows: blemished peas, 5%; seriously blemished peas, 1%; pea fragments, 10%; yellow peas, 2%; and extraneous plant material, 0.5%. The standard also calls for appropriate labels on the cans.

C. Freezing and Dehydrofreezing

Frozen green peas have virtually replaced fresh peas in the American diet. They rank third behind processed potato products and corn in their importance among frozen vegetables (34). Two cultivars of garden peas, Dark Skin Perfection and Thomas Laxton, are especially prized by freezers.

Since peas deteriorate in flavor soon after they have been picked, they need to be immediately processed for freezing. Hydrocooling or icing is generally advisable if there is any delay prior to processing. To obtain a high-quality frozen product, peas must be harvested before they become overmature. Once they arrive at the processing factory, peas are blanched in hot water for 1–2 minutes, cooled rapidly in a cold-water flume, and transferred to a specific gravity grader in which starchy, overmature peas sink, while Grade A peas float and are separated. The brine concentration depends upon the speed at which the peas are moving, the temperature of the water, and the final quality desired.

Peas are not always sized prior to freezing. However, when they are size-graded, sizing usually precedes blanching. If needed, special vibratory sorters, which do not bruise or damage the soft, blanched peas, can also be used to size peas after the blanching treatment.

Green peas are then frozen by air-freezing or air blast–freezing or in a fluidized bed freezer.

TABLE 4 Food Additives Permitted in Canned Green Peas

Additive	Maximum level of use
Monosodium glutamate	No limit
Mint flavor (mint oil)	No limit
Natural mint flavor	No limit
Coloring matters	100 mg/kg singly or in combination
Firming agents	350 mg/kg Ca in the final product
Modified starches	10 g/kg singly or in combination
Vegetable gums	10 g/kg singly or in combination
Alginates	10 g/kg singly or in combination
Propylene glycol alginate	10 g/kg singly or in combination
Tin (contaminants)	250 mg/kg, calculated as Sn

Source: Ref. 32.

The latter method is highly effective, the freezing rates being so rapid that the peas are brought down to an internal temperature of –18°C in about 4 minutes (34). Peas may be packed and frozen in the carton, or they may be individually quick-frozen for packaging or bulk storage. A process flow chart for the freezing of green peas is shown in Figure 4.

Frozen peas can be stored at –18°C for over a year without any significant loss in quality (18). Freezing retards the growth of microorganisms, metabolic activity of intact plant tissue, moisture loss, and deteriorative chemical changes, such as oxidative browning, lipid oxidation, and color degradation. Aerobic respiration, however, continues at a slow rate. Since losses in both nutrient and sensory quality are minimal, freezing is generally regarded as the best method of preservation.

For dehydrofreezing, green peas are air-dried to about 50% moisture prior to freezing in order to reduce packaging volume. Partial dehydration of fresh peas can also be achieved using saturated salt solutions (35). The salt treatment reduces the drying time by more than 30%. Salt-treated peas also retain chlorophyll better during storage.

Dehydrofrozen peas are equal in quality to fresh peas except for small losses of sugar and ascorbic acid during processing (18). The dehydrofrozen peas can also be subsequently freeze-dried to 2–3% moisture. Quick rehydration qualities and greater rehydrated volume of freeze-dried peas are achieved with this combination process at a significantly lower cost, because the initial air drying results in greater utilization of the freeze-drying equipment (29).

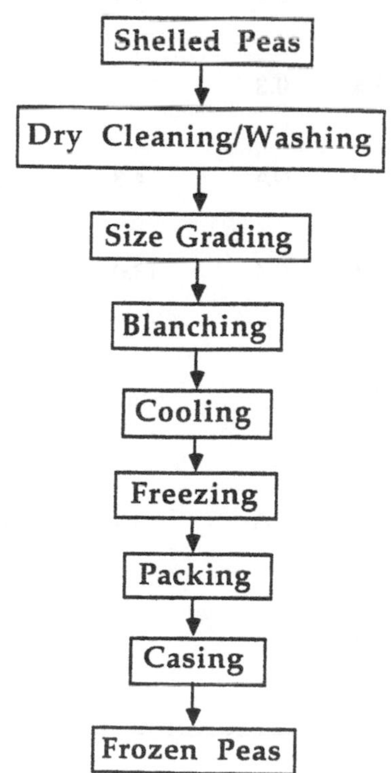

FIGURE 4 Process chart for quick-freezing of green peas.

VIII. NUTRITIONAL VALUE

Data on proximate composition and mineral and vitamin contents of fresh and processed green peas are summaried in Tables 5–7. Fresh green peas are one of the richest sources of proteins among commonly grown vegetables in a relatively low-cost food substance. About 1 cup of cooked green peas provides an amount of protein (8 g) equivalent to 1 oz (28 g) of cooked lean meat. However, peas provide about twice as many calories per gram of protein. Pea protein is moderately deficient in the sulfur-containing amino acids methionine and cysteine. However, it contains enough lysine to cover the deficiency of these amino acids in cereal products.

TABLE 5 Proximate Composition of Raw and Processed Green Peas

Type	Food energy (Cal)	Moisture (%)	Protein (%)	Fat (%)	Carbohydrate (%)	Fiber (%)
Edible podded, raw	53.0	83.3	3.4	0.2	12.0	1.2
Edible podded, boiled, drained	43.0	86.6	2.9	0.2	9.5	1.2
Green, immature, raw	84.0	78.0	6.3	0.4	14.4	2.0
Green, immature, sweet, boiled, drained	71.0	81.5	5.4	0.4	12.1	2.0
Green, immature, sweet, canned, drained solids	80.0	79.0	4.6	0.4	15.0	2.2
Green, immature, sweet, canned, solids and liquid	57.0	84.8	3.4	0.3	10.4	1.4
Green, immature, sweet, canned, drained liquid	22.0	93.3	1.3	NA	4.3	NA
Green, immature, sweet, canned, drained solids (low sodium)	72.0	81.8	4.4	0.4	13.0	2.0
Green, immature, sweet, canned, solids and liquid (low sodium)	47.0	87.8	3.3	0.3	8.2	1.3
Green, immature, sweet, canned, drained liquid (low sodium)	18.0	94.9	1.3	NA	3.4	NA
Green, immature, frozen, boiled, drained	68.0	82.1	5.1	0.3	11.8	1.9

NA = Data not available.
Source: Refs. 3, 17, 18, 36, 37.

TABLE 6 Mineral Composition (mg/100 g) of Raw and Processed Green Peas

Type	Ca	P	Na	Mg	K	Fe	Zn	Cu
Edible podded, raw	62.0	90.0	NA	6.0	170.0	0.70	NA	NA
Edible podded, boiled, drained	56.0	76.0	NA	NA	119.0	0.50	NA	NA
Green, immature, raw	26.0	116.0	2.0	35.0	316.0	1.90	0.90	0.05
Green, immature, sweet, boiled, drained	23.0	99.0	1.0	21.2	196.0	1.80	0.70	0.16
Green, immature, sweet, canned, drained solids	25.0	67.0	236.0	20.0	96.0	1.70	0.80	NA
Green, immature, sweet, canned, solids and liquids	19.0	58.0	236.0	20.0	96.0	1.50	NA	NA
Green, immature, sweet, canned, drained liquid	9.0	42.0	236.0	NA	96.0	1.10	0.60	NA
Green, immature, sweet, canned, drained solids (low sodium)	25.0	67.0	3.0	20.0	96.0	1.70	NA	NA
Green, immature, sweet, canned, solids and liquid (low sodium)	19.0	58.0	3.0	20.0	96.0	1.0	NA	NA
Green, immature, sweet, canned, drained liquid (low sodium)	9.0	42.0	3.0	NA	96.0	1.10	NA	NA
Green, immature, frozen, boiled, drained	19.0	86.0	115.0	24.0	135.0	1.90	0.69	0.11

NA = Data not available.
Source: Refs. 3, 17, 18, 36, 37.

Green peas also provide substantial amounts of vitamins A and C and are rich in calcium, potassium, phosphorus, and iron. Fresh green peas are also low in sodium. In contrast, peas canned in brine contain significant amounts of sodium. In recent years, canned green peas that are low in sodium are being marketed for populations who need to lower their dietary intake of sodium for health reasons.

Generally, processing of green peas results in significant losses of various water-soluble vitamins and minerals. For example, they may lose as much as 50–70% of their vitamin C during the freezing process (i.e., blanching, freezing, storage, and thawing).

Most homemakers use canned and/or frozen green peas. The latter is the leading frozen vegetable in the United States. Processed green peas are preferred primarily because dried peas (the most economical form of this vegetable) require longer cooking times and fresh green peas sold in pods require shelling. Peas go well with most other food items such as cereal products, cheeses and other dairy products, fish and seafood, meats, poultry, and other vegetables. They may be used in casseroles, salads, soups, stews, and vegetable side dishes. Snow peas are commonly used in Oriental stir-fry dishes.

Table 7 Vitamin Content (per 100 g basis) of Raw and Processed Green Peas

Type	Vit. A (IU)	Vit. E (mg)	Vit. C (mg)	Thiamine (mg)	Riboflavin (mg)	Niacin (mg)	Pantothenic acid (mg)	Pyridoxine (mg)	Folic acid (μg)	Biotin (μg)
Edible podded, raw	680.0	NA	21.0	0.28	0.12	NA	0.82	0.15	25.0	9.40
Edible podded, boiled, drained	610.0	NA	14.0	0.22	0.11	NA	NA	NA	NA	NA
Green, immature, raw	640.0	0.10	27.0	0.35	0.14	2.90	0.75	0.16	25.0	9.4
Green, immature, sweet, boiled, drained	540.0	0.55	20.0	0.28	0.11	2.30	0.33	0.11	NA	0.43
Green, immature, sweet, canned, drained solids	690.0	NA	8.0	0.11	0.06	1.0	NA	0.05	25.0	NA
Green, immature, sweet, canned, solids and liquid	450.0	0.02	9.0	0.11	0.06	1.0	0.15	0.05	29.0	2.10
Green, immature, sweet, canned, drained liquid	NA	NA	10.0	0.12	0.05	1.10	NA	NA	34.0	NA
Green, immature, sweet, canned, drained solids (low sodium)	690.0	NA	8.0	0.11	0.06	1.0	NA	0.05	25.0	NA
Green, immature, sweet, canned, solids and liquid (low sodium)	450.0	0.02	9.0	0.11	0.06	1.0	0.15	0.05	29.0	2.10
Green, immature, sweet, canned, drained liquid (low sodium)	NA	NA	10.0	0.12	0.05	1.10	NA	NA	34.0	NA
Green, immature, frozen, boiled, drained	600.0	0.25	13.0	0.27	0.09	1.70	0.28	0.09	84.0	0.43

NA = Data not available.
Source: Refs. 3, 17, 18, 36, 37.

REFERENCES

1. Hedrick, U. P., *Peas of New York, N.Y. State Stn. Rep.*, 1928.
2. Verma, V., *Economic Botany*, Emkay Publ., New Delhi, India, 1987.
3. Ensminger, A.H., M. E. Ensminger, J. E. Konlande, and J. R. K. Robson, *Foods and Nutirtion Encyclopedia*, 2nd ed., CRC Press, Boca Raton, FL, 1994, p. 1722.
4. Vavilov, N. I., Origin, variation, immunity, and breeding of cultivated plants, *Chron. Bot. 13*:1 (1949).
5. ICAR, *Pulse Crops of India*, Indian Council of Agricultural Research, New Delhi, India, 1970, p. 256.
6. FAO, *Production Yearbook*, Vol. 47, Food and Agriculture Organization, Rome, 1993.
7. White, O. E., Studies of inheritance in *Pisum*, *Proc. Am. Phil. Soc. 56*:487 (1917).
8. Govorov, L. I., The peas of Afghanistan. A contribution to the problem of the origin of cultivated peas, *Bull. Appl. Bot. Genet. Pl. Breed. 19*:497 (1928).
9. Vinokur, I. S., Peas from the Bronze Age, *Priroda (Nature) Leningrad 8*:102 (1960).
10. Lamprecht, H., The relation between *P. tibetanicum* and *P. sativum* in light of results of hybridization, *Svensk Bot. Tidskr. 38*:365 (1944).
11. Rosen, G. von, Interspecific crossings in the genus *Pisum* in particular between *P. sativum* L. and *P. abyssinicum* Braun., *Hereditas 30*:261 (1944).
12. Lamprecht, H., Further linkage studies on *P. sativum* especially in chromosome II (Ar.), *Agric. Hort. Genet. Landskrona 11*:51 (1952).
13. Menjkova, K. A., The results of crossing the garden pea (*P. sativum*) with the field pea (*P. arvense*) and rye (*S. cereale*) with bromus (*B. secalinus*), *Trud. Inst. Genet. 21*:179 (1954).
14. Lamprecht, H., On the specific status of *P. elatius* Stev. and *P. jomardii* Schrank. *Agric. Hort. Genet. Landskrona 14*:5 (1956).
15. FAO, *Agricultural and Horticultural Seeds*, Food and Agriculture Organization, Rome, 1961.
16. Pandita, M. L., and P. Singh, Pea, in *Vegetable Crops in India* (T. K. Bose and M. G. Som, eds.), Naya Prakash, Calcutta, 1986, p. 469.
17. Salunkhe, D. K., and B. B. Desai, *Postharvest Biotechnology of Vegetables*, Vol. I, CRC Press, Boca Raton, FL, 1984.
18. Salunkhe, D. K., S. S. Kadam, and S. S. Deshpande, Peas and lentils, in *Encyclopaedia of Food Science, Food Technology and Nutrition* (R. MaCrae, R. K. Robinson, and M. J. Sadler, eds.), Academic Press, London, 1993, p. 3482.
19. Woodroof, J G., Preparing vegetables for processing, in *Commercial Vegetable Processing*, 2nd ed. (B. S. Luh and J. G. Woodroof, eds.), Van Nostrand Reinhold, New York, 1988, p. 175.
20. Reid, J. T., A mechanical harvester for southern peas and lima beans, *Agric. Eng. 50*:412 (1969).
21. Wager, H. G., Physiological studies of the storage of green peas, *J. Sci. Food Agric. 15*:245 (1964).
22. Ryall, R. L., and W. I. Lipton, Handling, in *Transportation and Storage of Fruits and Vegetables*, Vol. 1, *Vegetables and Melons*, AVI Publ. Co., Westport, CT, 1972, p. 473.
23. Pantastico, E. B., T. K. Chattopadhyay, and H. Subramanyam, Storage and commercial storage operation, *Postharvest Physiology, Handling and Utilization of Tropical and Sub-tropical Fruits and Vegetables* (E. B. Pantastico, ed.), AVI Pub. Co., Westport, CT, 1975, p. 314.
24. Cruess, W. V., and G. MacKinney, *The Dehydration of Vegetables*, Calif. Univ. Agric. Exp. Stn. Bull. 680, Riverside, CA, 1943.
25. Moyer, J. C., *Factors in Raw Material and Processing Influencing the Reconstitution of Dehydrated Peas*, U.S. Quartmaster Food & Container Ins. Armed Forces, Contract Res. Proj. Rep. V-308, No. 20, New York State Agr. Exp. Stn. Geneva, NY, 1959.
26. Moyer, J. C., H. R. Pallesen, and R. S. Shallenberger, The interaction between blanching and drying rates of peas, *Food Technol. 13*:581 (1959).
27. Shah, W. H., and N. A. Sufi, Effect of maturity and pricking on dehydration and rehydration. Characteristics of various varieties of green peas, *Pak. J. Sci. Res. 31*:174 (1979).
28. Shah, W. H., and R. A. Edwards, Effect of pea size and storage conditions on the quality, drying rate and rehydration characteristics of dehydrated peas, *Pak. J. Sci. Ind. Res. 19*:154 (1976).
29. Somogyi, L. P., and Luh, B. S., Vegetable dehydration, in *Commercial Vegetable Processing*, 2nd ed, (B. S. Luh and J. G. Woodroof, eds.), Van Nostrand Reinhold, New York, 1988, p. 387.

30. Lynch, L. J., and R. S. Mithcell, *Physical Measurement of Quality in Canned Peas*, CSIRO, Australia, Bull. No. 354, 1950.
31. Luh, B. S., and C. E. Kean, Canning of vegetables, in *Comercial Vegetable Processing*, 2nd ed. (B. S. Luh and J. G. Woodroof, eds.), Van Nostrand Reinhold, New York, 1988, p. 195.
32. FAO/WHO, *Recommended International Standard for Canned Mature Processed Peas*, CAC/RS 81-1976, Joint FAO/WHO Food Standards Program, Rome, 1976.
33. National Canners Association, *Processes for Low-Acid Canned Foods in Metal Containers*, 11th ed., Natl. Canners Assoc. Res. Lab. Bull. 26L, 1976.
34. Luh, B. S., and M. C. Lorenzo, Freezing of vegetables, in *Commercial Vegetable Processing*, 2nd ed. (B. S. Luh and J. G. Woodroof, eds.), Van Nostrand Reinhold, New York, 1988, p. 343.
35. Shah, W. H., Studies on the effect of salt treatment and length of drying period on the quality and storage stability of dehydrated peas, *Pak. J. Sci. Res. 27*:159 (1972).
36. Chin, S. B., and J. A. Dudek, Composition and nutritive value of raw and processed vegetables, in *Commercial Vegetable Processing*, 2nd ed. (B. S. Luh and J. G. Woodroof, eds.), Van Nostrand Reinhold, New York, 1988, p. 647.
37. Adsule, R. N., K. M. Lawande, and S. S. Kadam, Pea, in *Handbook of World Food Legumes. Nutritional Chemistry, Processing Technology and Utilization*, Vol. II (D. K. Salunkhe and S. S. Kadam, eds.), CRC Press, Boca Raton, FL, 1989, p. 215.

20

French Bean

R. N. Adsule
Mahatma Phule Agricultural University, Rahuri, India

S. S. Deshpande
IDEXX Laboratories, Inc., Sunnyvale, California

S. K. Sathe
Florida State University, Tallahassee, Florida

I. INTRODUCTION

French bean (*Phaseolus vulgaris*) is known by several names, such as common bean, kidney bean, navy bean, pinto bean, field bean, haricot bean, China bean, marrow bean, frijole, snap or string bean, wax bean, black bean, or white bean. Over 14,000 cultivars of common bean are known to exist, and the International Center for Tropical Agriculture in Colombia is the main repository for the germplasm (59). The common bean (genus *Phaseolus*) includes 150–200 species of plants, many of which are cultivated as food or garden ornamentals. The common bean is therefore a highly polymorphic and perhaps the most widely cultivated of all beans in temperate and semi-tropical regions. Kaplan (60) suggests that the common bean is native to the New World, with origin(s) in Mexico, Central America, and northern areas of South America (60), and has been cultivated for at least 7000 years in Mexico. These beans are used as a green vegetable, green shelled, or dry as pulses according to the stage of harvest. In temperate regions, green mature pods are cooked and eaten as a vegetable (1). Immature pods are marketed as fresh, frozen, or canned, whole or French cut. Mature dry beans called navy beans, white beans, northern beans, or pea beans are widely consumed. The total world production of the French bean as a vegetable (green beans) in 1995 was 3,119,000 metric tons (MT), of which Asia accounted for 1,597,000 MT (51.2% of the total world production). China (560,000 MT), Turkey (440,000 MT), Spain (228,000 MT), Italy (158,000 MT), Indonesia (155,000 MT), the United States (122,000 MT), Egypt (108,000 MT), and France (103,000 MT) were the leading producers of French bean in 1995 (58). The name snap beans is applied to beans in the tender young stage when the pods are eaten. Snap beans are one of the most valuable vegetable crops grown by home and market gardeners and truck farmers for shipment and canning. Green shell beans are shelled and used just before the seeds become ripe and hard. Dry or field beans are produced in the fully matured and dried stage and usually classed as field crop. In addition to classification according to use, beans may be divided into dwarf or bush and pole or climbing groups, each including green pods and wax pods, which in turn may be flat, round, or intermediate in cross section (2).

II. BOTANY

French bean belongs to the family *Leguminosae*. Two general types of this species are cultivated: the bushy and the climbing or pole types. Both are annual plants with trifoliate leaves. In the bush types, inflorescences usually develop almost simultaneously, whereas on the pole types flowers appear one after another over a long period of time. The flowers are either white, yellowish, or pinkish-purple. The pods are 13 or more cm long and flat or rounded. The mature seeds may be white, black, or one of a wide range of colors.

French bean cultivars are of two types: dwarf or bush and pole or climbing (Fig. 1). Thompson and Kelly (15) classified French bean cultivars as follows:

1. snap beans for vegetable pods,
2. green shell beans, used in the green shelled conditions, and
3. dry shell beans, used as dry seeds.

Some of the important cultivars of snap beans are the flat types Bountiful, Plentiful, Green Ruler, and Golden Ruler; oval types Pusa Parvati, Contender, Spartan Arrow, Premier Tendergreen, and King Green; and pole types Blue Lake, Kentucky Wonder, Phenomenal Long, and Podded.

The Indian Agricultural Research Institute, New Delhi, has released the following varieties: Giant Stringless (early, bushy, green podded), Contender (an introduction from the United

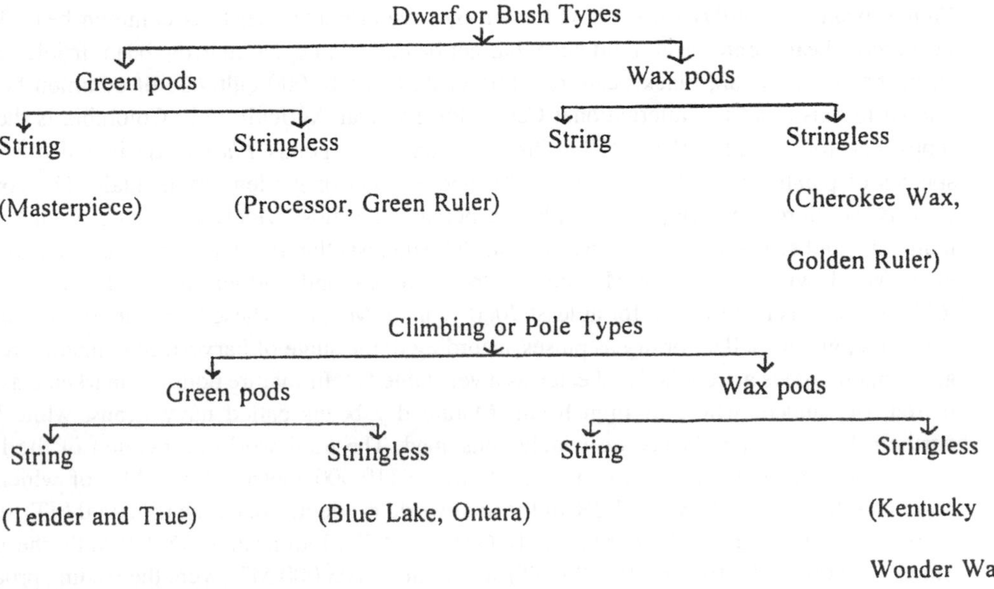

FIGURE 1 French bean cultivars.

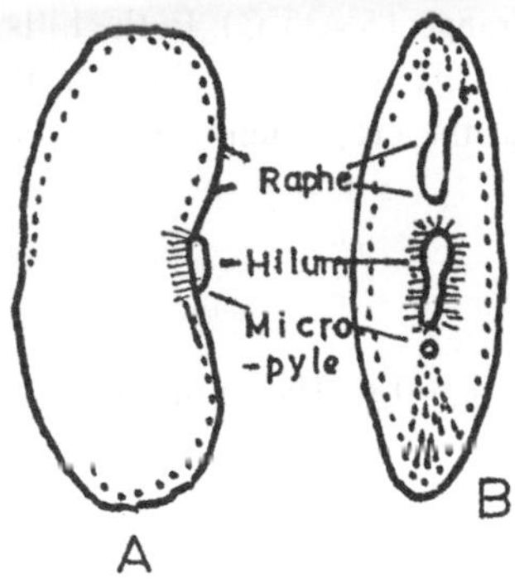

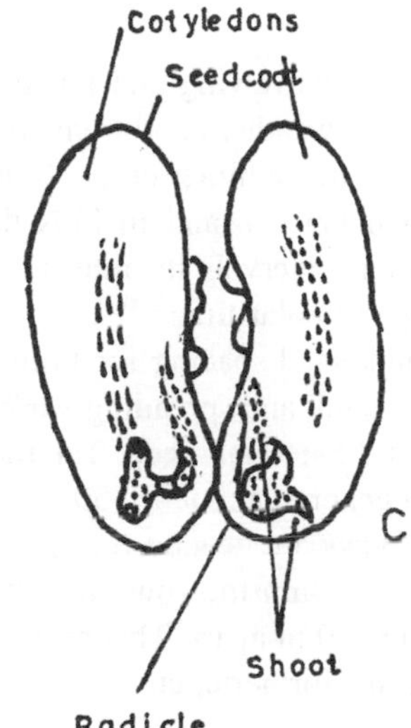

Figure 2 Anatomical parts of French bean seed.

States, early bushy, dark green stringless pods, resistant to mosaic and powdery mildew), Pusa Parvati (new variety developed through irradiation of the American Wax Pod, early bushy with attractive, round, neat light green pods), and Jampa (Mexican variety, nonstringy, with flat, smooth, pale green pods and black seeds).

Various anatomical parts of French bean seed are shown in Figure 2.

III. PRODUCTION

A. Soil and Climate

Like other vegetables, French bean is a crop that grows and yields best on loamy or sandy loamy soils. However, it can be easily grown on many other soil types both lighter or heavier than loamy soils. Soils that are very rich in organic matter promote more vegetative growth at the cost of pod production. The soil pH should never be lower than 5.5. The ideal pH range is 6.0–6.8.

French bean is a cool season crop but can tolerate higher temperatures better than peas. The young plants are more tolerant to low temperatures than the flowers and immature fruit (pods). This crop typically requires temperatures below 35°C (range 15°–35°C). In the hills, this crop is grown during spring and summer, whereas it can be easily grown in winter in plains. In India, French bean is grown at altitudes up to 2000 m, but the crop cannot tolerate heavy rains (>750 mm).

B. Propagation

The seeds of French bean are medium to large—about 200–300/100 g. They remain viable for about 2 years and require 8–15 days for germination.

C. Cultural Practices

1. Sowing

The land is prepared by ploughing 20–25 cm deep, followed by harrowing and leveling. The best time for sowing French bean depends upon the place and the weather conditions. In places that are mild throughout the year, they are grown as *kharif*, *rabi*, and summer crops. In general, in plains they are sown from August to October (autumn-winter) (3), and in hills they are typically sown from April to June. Seeds can be sown at 15-day intervals in order to provide fresh pods continuously to prevent a glut in the market at any particular time.

The seeds are broadcast, drilled, or dibbled. The recommended spacing for French bean (4) is 45–90 cm between rows by 7.5–10 cm within the row. The tall and spreading varieties are sown further apart. Two to three seeds are sown together about 2.5–5.0 cm deep. The total seed requirement varies with the variety, spacing, and seed size. In general, about 40–50 kg of seeds of small-seeded types or 60–90 kg of seeds of larger-seeded types are needed for one hectare. Some growers maintain a plant-to-plant distance of 20–30 cm within the row and 60–95 cm between rows. The trailing type of French beans are sown 1.0 × 1.0 m apart. This crop can also be grown as a mixed crop with field crops like cotton, sugarcane, turmeric, etc.

2. Manures and Fertilizers

According to Kale and Kale (5), French bean yielding about 7.2 quintal seeds and 6.3 quintal straw and leaves removes approximately 66 kg of nitrogen, 27 kg of phsophorus, and 55 kg of potash per hectare. Laske (6) reported that a French bean crop yielding 8 tons of pod and 7 tons of straw requires approximately 60 kg of nitrogen, 15 kg of phosphorus, 55 kg of potash, and 80 kg of calcium oxide, whereas Vishwanathan (7) reported 30–40 kg of nitrogen being required per hectare for French bean crop. This variation in fertilizer requirement for French beans is partly due to the soil type and its inherent fertility status (8–12). Kale and Kale (5) recommended 30–40 tons of farmyard manure, 20–60 kg of nitrogen, 40–60 of phosphorus, and 40–50 kg of potash per hectare. The entire dose of farmyard manure should be applied at the time of land preparation and thoroughly mixed in the soil. Half the total dose of nitrogen and the full dose of phosphate and potash fertilizers are applied at the time of sowing, placed 6–8 cm away and 5 cm below the seed. The remaining nitrogen should be given 3–4 weeks after sprouting of seeds.

In general, this crop responds well to calcium, phosphorus, and potassium nutrition, while excess nitrogen results in poor pod yield. Like other legume crops, it also fixes atmospheric nitrogen and improves the soil fertility. Therefore, this crop should also be allowed to remain in the soil until the plants are fully developed. The quality of French bean pods is significantly influenced by application of fertilizers.

3. Irrigation and Intercultural Operations

Watering is necessary immediately after sowing. Subsequent irrigations are given at 5- to 10-day intervals. Clean cultivation is necessary to check weed growth. Precautions to avoid root injuries should be taken at the time of hoeing and weeding. The pole type varieties need mechanical support for proper growth and fruiting. Earthing up may also be done to provide protection against winds. When the rains are heavy, additional drainage facilities may be provided to prevent waterlogging. Poor water drainage leads to yellowing of leaves and new growth. The level of irrigation influences the bean pod quality (Table 1).

D. Diseases and Pests

Anthracnose (*Colletotrichum lindemuthianum* Sacc. and Magn. Briosi & Cac), bean rust [*Vromyces appediculatus* Pers. (Fries)], leaf rot (*Cercospora cruenta* Sacc.), powdery mildew (*Ersiphe polygoni* DC), dry root rot and wilt [*Fusarium solani* f. sp. *Phaseoli* (Burk)], bacterial blight [*Xanthomonas phaseoli* (Smith) Dowson], and viral diseases caused by bean mosaic virus and yellow mosaic virus arc the important diseases of French bean. The use of sulfur fungicides like Thiram or Dithane Z-78 and the systemic fungicides like Benlate or Bavistin gives satisfactory control of anthracnose. Spraying with copper oxychloride or organic sulfur fungicides is recommended for control of leaf spot. Powdery mildew can be controlled by dusting with powdered sulfur or wettable sulfur preparations. Systemic fungicides like Benlate or Bavistin and Calixin also give effective control of powdery mildew. Seed treament with

TABLE 1 Effects of Irrigation and Row Spacing on Sensory and Objective Quality Measures of Bush Green Beans

	Soil moisture						
Cultivar	Code[a]	Water applied (cm)	Row space (cm)	Sensory texture score[b]	Fiber (%)	Work of compression and shear (cm-kg/100 g)[c]	Seed (%)
OR 1604	M1	15	15	5.0	0.090	81	6.3
	M2	19	15	5.5	0.049	71	5.6
	M3	23	15	5.8	0.019	36	5.9
	M1	15	91	5.9	0.090	52	6.3
	M2	19	91	5.9	0.024	62	4.1
	M3	23	91	5.9	0.014	46	6.1
Glamor	M1	15	15	5.7	0.017	64	6.4
	M2	19	15	6.0	0.022	70	7.6
	M3	23	15	6.5	0.016	59	6.5
	M1	15	91	6.3	0.017	56	6.5
	M2	19	91	6.1	0.010	64	6.5
	M3	23	91	5.5	0.018	64	7.1
	LSD at 5%			0.68	0.015	10.9	0.9
	Correlation coefficient[d]				−0.56	−0.45	0.65

[a]M1, Low; M2, intermediate; M3, high; during pod growth.
[b]Nine point preference scale: 9 = excellent; 5 = average; 1 = poor. Panel of 10 members.
[c]Method: Lee-Kramer Shearpress.
[d]Simple correlation with sensory texture.
Source: Ref. 13.

organomercurial fungicides and antibiotic preparations is recommended for control of bacterial diseases. The use of disease-free seeds for bean mosaic virus and spraying of crops with Dimethoate, Monocrotophos, or 1% mineral oil for vector control can give effective control of these diseases.

The important pests of French bean include bean aphid (*Aphis craccivora*), jassids (*Empoasca fabae*), Lygacid bug (*Chauliops fallex*), hairy caterpillars (*Ascotis imoarata* and *Spilosoma obliqua*), stem fly (*Ophiomyia phaseoli*), bean lady bird beetle (*Epilachna varivestis*), and root weevil (*Sitona lineatus*).

Bean aphids can be controlled by application of systemic organophosphatic insecticides—spraying with 0.05% phosphomidon or 0.025% methyl demeton. Hairy caterpillar can easily be controlled by systematic collection of larvae during early generation stages or by spraying with Endosulfan at 0.07%. Stem fly can be controlled by spraying crops with 0.03% Diazinor or 0.05% Quinolphos. Spraying with Malathion or Carbaryl, taking care that the lower surface of the leaves are well covered, can control bean lady bird beetle. Soil treatment with Dieldrin at 2.24–4.48 kg a.i./ha has been found effective in controlling the larvae of the weevil.

E. Maturity, Harvesting, and Handling

Snap beans are usually harvested before the pods are fully grown and while the seeds are small. Most of the varieties become tough and stringy if left on the plants until the seed develops to a considerable size (15). To fetch a better market price, crops of all varieties should be harvested before the seeds become large enough to cause the pod to bulge around the seeds. Snap beans are usually ready for harvest 12–14 days after the first blossoms open, but the time required varies with weather conditions. The time of harvest of snap beans is a compromise between yield and quality and should be carried out so that the grower will obtain the maximum yield of high-quality beans.

Picking is usually done by hand. In large-scale commercial plantings of the bush type snap bean, one or two pickings are done. Thompson and Kelly (15) observed that most of the growers make the first picking by hand and the second by machine. Pole beans blossom over a long period of time, and harvests are made at 3- to 5-day intervals depending upon the quality desired and weather conditions. Sometimes beans are sprayed with various chemicals to delay seed development and provide a higher-quality product without requiring numerous pickings. Mitchell and Marth (16) reported that beans sprayed 4 days before harvest with 4-chlorophenoxyacetic acid were delayed in maturity. The chemical treatment reduced water loss from the pods and slowed down undesirable color changes of the pods during storage. The development of seed fiber was retarded without affecting the final yield when the beans were sprayed with *p*-chlorophenoxyacetic acid (17), however, the quality of processed bean products (color, flavor, and shape of pods) was adversely affected.

Kish and Ogle (18) determined the accuracy of the heat unit system in predicting the maturity date of snap beans. The formula that gave the smallest coefficient of variation was one using the daily heat unit multiplied by a ratio of the available soil moisture to a constant soil moisture value.

IV. CHEMICAL COMPOSITION

Like other legume seeds, green beans are a good source of proteins, minerals, and vitamins (19). The chemical composition of French bean green pods is presented in Table 2. Many legume seeds are known to contain undesirable factors such as protease inhibitors, lectins, phytates,

TABLE 2 Chemical Composition of French Bean[a]

	French bean[b]	French bean, green—frozen[c]	Snap cut green can[c]	Snap beans green, raw, boil and snap beans, wax, raw, boil[c]	Snap beans yellow wax-can[c]
kcal	32	26	20	35	20
Protein (g)	1.9	1.4	1.2	1.9	1.2
Carbohydrates (g)	7.1	6.1	4.5	7.9	4.5
Dietary fiber (g)		1.6	1.3	1.8	1.3
Fat (g)	0.2	0.14	0.1	0.28	0.1
Saturated fatty acids (g)		0.03	0.02	0.06	0.02
Monounsaturated fatty acids (g)		0.005	0.004	0.011	0.004
Polyunsaturated fatty acids (g)		0.069	0.051	0.015	0.051
Vitamin A (retinol equivalents)	600[d]	53	35	67	35
Vitamin E (mg)		0.24	0.05	0.11	—
Vitamin C (mg)	19	8.22	4.80	9.68	4.80
Thiamine (mg)	0.08	0.048	0.015	0.074	0.015
Riboflavin (mg)	0.11	0.074	0.056	0.097	0.056
Niacin (mg)	0.50	0.42	0.20	0.61	0.20
Vitamin B_6 (mg)		0.06	0.04	0.06	0.04
Folic acid (μg)		8.2	31.8	33.3	31.8
Sodium (mg)		13	251	3	251
Potassium (mg)		112	109	300	109
Magnesium (mg)		21	13	25	13
Calcium (mg)	56	45	26	46	26
Phosphorus (mg)	44	24	19	21	19
Iron (mg)	0.8	0.8	0.9	1.3	0.9
Zinc (mg)		0.6	0.3	0.4	0.3
Selenium (μg)		0.74	0.74	0.74	—

[a]Data are expressed on a 100 g weight (as is basis). Data were recalculated and rounded to appropriate integer when required.
[b]From Ref. 19.
[c]From Ref. 66.
[d]Data expressed as IU.

polyphenols, and oligosaccharides of the raffinose family. Such information for dry seeds of French bean is documented (1).

The chemical composition and quality of French bean pods and grains are influenced by the fertility status of soil (20–23), spacing (24–27), irrigation (28–32), maturity (33–34), and variety (35–38). Volatile compounds of cooked snap beans after boiling for 30 minutes were identified as hydrogen sulfide, acetaldehyde methanethiol, propionaldehyde acetone, dimethyl sulfide, *n*-propanethiol, and 2-methyl butanal (39). Stevens et al. (40) studied the volatile compounds of canned snap beans. Forty of the compounds present in the thermally processed snap beans were characterized. They believed that *cis*-hex-3-en-1-ol, oct-1-en-ol, linalool, α-terpineol, pyridine, and furfural were of primary importance in canned snap bean aroma. Furfural was probably formed during processing by dehydration of pentose sugars. The presence of 3-alkyl-2-methoxypyrazines in the volatiles of raw beans was reported by Murray and Whitfield (41,42). A possible biogenesis of methoxypyrazines involving amino acids and glyoxal as precursors was suggested by the same investigators. Hemple and Böhm (61) recently reported the presence of glycosylated flavonoids in the French bean. These investigators found that quercetin-3-*O*-glycosides ranged from 19.1 to 183.5 μg/g and kaempferol-3-*O*-glycosides from 5.6 to 14.8 μg/g of fresh beans in green and yellow varieties. These flavonoids occurred as 3-*O*-glucuronides and 3-*O*-rutinosides.

V. STORAGE

Green beans are highly perishable and should be cooled rapidly after harvest, preferably to 4–5°C. They can be vacuum-cooled or forced-air-cooled, but hydrocooling is preferable not only because cold water cools beans rapidly but also because the free moisture helps to prevent wilting or shriveling (43). Green beans lose moisture rapidly if not properly protected by packaging or a relative humidity of 95% or above. At this humidity, temperatures above 7°C must be avoided or decay is likely to be a serious problem within a few days (44).

Green beans should be stored for a short period (1 or 2 weeks) at temperatures of 4–7°C. These temperatures cause some chilling but are best for short storage (45). Beans exhibit chilling injury if stored at 3°C or below for several days. This condition is characterized by surface pitting and rosseting a day or two after removal to warm temperature for marketing. Tender green beans can be held for about 2 days at 0.5°C, 4 days at 2.5°C, or 12 days at 5°C before chilling injury is induced.

Anthracnose, watery soft rot, and gray mold rot are the most common postharvest diseases affecting beans (46). Anthracnose caused by *Colletotrichum lindemuthianum* is characterized by deep black spots with reddish-brown borders. The pathogen of watery soft rot is *Sclerotinia sclerotiorum*, and *Botritis cinerea* causes gray mold of beans.

Ryall and Lipton (47) have described the following market diseases of snap beans: Cottony leak (*Pythium butleri*), anthracnose (*C. lindemuthianum*), soil rot (*Rhizoctonia solani*), and watery soft rot (*Sclerotinia* spp.). High temperatures and humidities initiate cottony leak and anthracnose; soil rot and watery soft rot are caused by soil fungi and can cause serious losses in snap beans in seasons when cool, moist weather precedes harvest.

Containers of beans should be stacked to allow abundant air circulation. If containers are packed close together, the heat of respiration may cause the temperature to rise and beans will deteriorate rapidly. When beans are stored in larger bins or pallet boxes, provision should be made for rapid cooling.

Low temperature (0.6–2.8°C) can cause chilling injury (rosseting) and symptoms such as surface pitting, diagonal brown streaks, surface dullness and higher susceptibility to decay. Physical injury can also be inflicted by pods rubbing against each other during machine harvest or rough transport. Snap beans lose their marketable quality through solar injury and wilting (15).

Stanwood and Ross (48) stored seeds of 14 vegetable species placed in paper envelopes in liquid nitrogen (–196°C) for periods up to 180 days with no adverse effect of percentage germination. In tests with *Phaseolus vulgaris*, lettuce, and peas, exposure of seed to –190°C did not indicate any adverse effect on seed germination or vigor. Stanwood and Ross (48) concluded that liquid nitrogen offers promise as a medium for long-term storage of seed germplasm.

Snap beans can be stored for several days at 4.4°C and about 90% relative humidity. Platenius et al. (49) reported that beans kept better at 4.4°C than at either 0° or 10°C. Beans intended to be transported to long-distance markets during warm weather must be shipped in refrigerated cars (15). Hydrocooling the beans immediately after harvest is beneficial provided the beans are kept refrigerated after the operation. Temperatures of 3.3–5.5°C with 88% relative humidity have been recommended to prolong the shelf life of snap beans for 2 or 3 weeks (50).

VI. PROCESSING

Green (snap) beans or French beans are generally canned and frozen. The desirable characteristics for processing are (1) resistance to mosaic disease, (2) long, straight, tender pods of medium size with thick walls and small seeds, (3) lack of fiber and strings, (4) bright green color after blanching, and (5) firm skins that do not split or slough off during cooking. Tendergreen bush beans and Blue Lake pole beans are generally used for drying, canning, and freezing. The leucoanthocyanins of beans are converted to anthocyanins during processing, resulting in an undesirable grayish-brown discoloration. The cultivar, as well as the preparation methods, significantly affects the sensory quality and therefore the acceptance of processed green beans (64).

A. Canning

The harvested beans are subjected to size grading (51) and shipping. If beans are smaller in size, they are canned whole. Large beans are cut crosswise by machine into 1- to 1.5-in. lengths for regular cut beans. Some smaller beans are French-cut (e.g., 0.5 in.) after blanching. The cross-cut beans are cut before blanching. The cut beans as well as small whole beans are blanched in hot water at 82°C rather than in steam for about 1.5–2.0 minutes to give better-filled cans. Some of the blanched whole beans are then returned to the French style cutters and cut into shoestring-shaped pieces. The blanch time is influenced by variety (52), pretreatment with $CaCl_2$ (53), and type of blanching medium (54).

Stanley et al. (62) reported that blanching of fresh green beans (cultivar Acclaim) in acid or $CaCl_2$ resulted in a firmer canned product compared with the corresponding control. The cut beans are filled by volume, mechanically. The cans are filled with 2% salt brine. The cans are heated in a steam exhaust box for 5 minutes and sealed hot or steam-flow sealed. The process time for canned green beans is presented in Table 3. Textural defects are the major problems in canned green beans. Sloughing is one of the major problems which causes the pods to separate into layers and gives it an undesirable texture.

Van Buren and coworkers (65) found that softening of the snap bean texture occurred when the pH was raised from 5.2 to 6.2. They also noted that the presence of cations ($Li^+ > Na^+ = K > NH_4+$ and $Ca^{2+} > Mg^{2+}$) as well as anions ($SO_42- > CH_3COO^- > Cl^- > NO_3-$) in the cooking medium typically decreased bean firmness. They attributed this softening to β-elimination of pectin. High-pressure processing (600 MPa for 15 min at 70°C) followed by immediate cooling in cold tap water followed by either drying or freezing has been reported to retain texture and color properties similar to that of raw beans and superior to that of water-blanched green beans (63).

B. Freezing

Green beans are harvested and transported to the plant and dumped onto a belt and shaker screen with blower to remove field dirt and extraneous material. The beans are then subjected to size grading. The size-graded beans are fed into shippers, which remove the stems and most blossom ends from the pods. The beans are then moved to mechanical cutters, where they are cross-cut into 1- to 1.5 in. lengths. After the cutting operation, the beans are blanched in steam or in water at 99°C for 2–3 minutes (56). The large beans used for French cut are first blanched and then

TABLE 3 Process Time for Canned Green Beans

Can name	Dimensions	Initial temperature (°F)	Time (min) at 240°F	Time (min) at 250°F
Whole or cut				
No. 2 and smaller	307 × 409	70	21	12
	and smaller	120	20	11
No. 2½	401 × 411	70	26	15
		120	25	14
No. 3 cylinder	404 × 700	70	32	18
		120	30	16
Sliced lengthwise or French style				
8 oz. Tall	211 × 304	70	26	17
		120	25	16
No. 303	303 × 406	70	26	17
		120	25	16
No. 2	307 × 409	70	26	16
		120	25	17
No. 1	603 × 700	70	50	35
		120	45	30

Source: Ref. 55.

cut. Smaller beans can also be cut after blanching, a technique that reduces the flavor loss encountered when cut beans are blanched. This practice frequently leads to sanitation problems. The product is then quickly cooled after blanching, then sorted, packed, and frozen. Cross-cut beans are frequently individually quick-frozen on belts or immersion-frozen in a refrigerant and bulk-stored (56).

C. Dehydration

Beans are graded for size, sorted for defects, and mechanically shipped to remove pod ends. The beans are washed either before or after these operations. Sizing of beans is done by an automatic grader. The cut beans are steam-blanched for about 4 minutes and promptly cooled with water sprays. A sulfite-bisulfite solution is then sprayed over the blanched material to produce a SO_2 content of about 500 ppm in the product (57). The blanched, sulfited, cut beans are frozen for later dehydration. The product (cross-cut dehydrated beans) has good rehydration and cooking properties. Precooked dehydrated green beans prepared by extending the blanching operation until the beans have been completely cooked, frozen, and dehydrated have been successfully manufactured. This product is ready to serve after the addition of boiling water and 15 minutes holding time (57).

REFERENCES

1. Salunkhe, D. K., S. K. Sathe, and S. S. Deshpande, French bean, in *Handbook of World Food Legumes: Nutritional Chemicstry, Processing Technology and Utilization*, Vol. II (D. K. Salunkhe and S. S. Kadam, eds.), CRC Press, Inc., Boca Raton, FL, 1989, p. 23.
2. Watts, R. L., and G. S. Watts, *The Vegetable Growing Business*, War Dept. Education Manual, Orange Judd Pub. Co. Inc. Washington, DC, 1940, p. 520.

3. Singh, J. P., and R. K. Sharma, Effect of different sowing dates on yield in French bean, *Prog. Hort.* 2:39 (1970).
4. Prasad, R. D., K. N. Singh, and K. V. Bondal, Effect of inter and intra row spacing and number of seeds per hill on yield of French bean in Kodai hills, *Ghana J. Sci 16*:25 (1978).
5. Kale, P. N., and S. P. Kale, *Bhajipala Utpadan*, Continental Press, Pune, India, 1984.
6. Laske, P., *Manuring of Vegetables in Field Cultivation*, Pub. Vee. Taqesellschaft. fur Ackerpou, Hannover, 1962.
7. Viswanath, M. L., French bean from America, *Indian Agric. 4*:5 (1963).
8. Srivastava, B. K., and B. P. Singh, Improved way to French bean, *Indian Farm Digest 12*(1):27 (1979).
9. Singh, K. N., R. D. Prasad, and V. P. S. Tomar, Response of French bean to different levels of nitrogen and phosphorus in Nilgiri hills under rainfed conditions, *Indian J. Agron. 26*:101 (1981).
10. Chandra, R., C. B. S. Rajpur, K. P. Singh, and S. J. P. Singh, A note on the effect of nitrogen, phosphorus, and rhizobium culture on growth and yield of French bean Cv. Contender, *Haryana J. Hort. Sci. 16*:145 (1987).
11. Jana, B. K., and J. Kabir, Influence of micronutrients on growth and yield of French bean cv. Contender under poly house conditions, *Veg. Sci 14*:124 (1987).
12. Pandey, G. K., J. N. Seth, K. P. S. Phogat, and S. N. Tiwari, Effect of spacing and P_2O_5 on the green pod yield of dwarf French bean variety Black Prince, *Prog. Hort 6*:41 (1974).
13. Mack, H. J., and G. M. Verseveld, Response of bush snap beans (*Phaseolus vulgaris* L.) to irrigation and plant density, *J. Am. Soc. Hort. Sci. 107*:286 (1982).
14. Verseveld, G. M., and H. J. Mack, French beans: effect of modified cultural practices and varietal improvement on sensory quality, in *Evaluation of Quality of Fruits and Vegetables* (H. E. Pattee, ed.), AVI Pub. Co., Inc. Westport, CT, 1973.
15. Thompson, H. C., and W. C. Kelly, *Vegetable Crops*, 5th ed., Agri. Horticultural Pub. House, Nagpur, India, 1980.
16. Mitchell, J. W., and P. C. Marth, Effect of growth regulating substances on the water-retaining capacities of bean plants, *Bot. Gaz. 112*:273 (1950).
17. Guyer, R. B., and A. Kramer, Objective measurements of quality of raw and processed snap beans as affected by maleic hydrazide and p-chlorophenoxy acetic acid, *Proc. Am. Soc. Hort. Sci. 58*:263 (1951).
18. Kish, A. J., and W. L. Ogle, Improving the heat unit system in predicting maturity date of snap beans, *Hort. Sci. 15*:140 (1980).
19. Roy, S. K., and A. K. Chakrabarti, Vegetables of tropical climate—commercial and dietary importance, in *Encyclopaedia of Food Science, Food Technology and Nutrition* (R. MaCrae, R. K. Robinson, and M. J. Sandler, eds.), Academic Press, London, 1993, p. 4743.
20. Shugard, J. P., The influence of soil levels of nitrogen, phosphorus and potassium and method of application on the growth and quality and nutritional composition of snap beans (*Phaseolus vulgaris* L.), *Diss. Abstract. Int.* (B) *30*:4656 (1970).
21. Lipton, W. J., Interpretation of quality evaluation of horticultural crops, *HortScience 15*:64 (1980).
22. Lee, J. M., P. E. Read, and D. W. Davis, Effect of irrigation on interlocular cavitation and yield in snap beans, *J. Am. Soc. Hort. Sci. 102*:276 (1977).
23. Guyer, R. B., and A. Kramer, Studies of factors affecting the quality of green and wax beans, *Md. Agric. Exp. Stn. Bull. A. 68* (1952).
24. Mack, H. J., and D. L. Hatch, Effect of plant arrangement and population density on yield of snap beans, *J. Am. Soc. Hort. Sci. 92*:418 (1968).
25. Mack, H. J., and G. M. Verseveld, Response of bush snap beans (*Phaseolus vulgaris* L.) to irrigation and plant density, *J. Am. Soc. Hort. Sci. 107*:286 (1982).
26. Drake, S. R., and M. J. Silbernagel, The influence of irrigation and row spacing on the quality of processed snap beans, *J. Am. Soc. Hort. Sci. 107*:239 (1982).
27. Smittle, D. A., Response of snap beans to irrigation, nitrogen fertilization and plant population, *J. Am. Soc. Hort. Sci. 101*:37 (1976).

28. Baggett, J. R., and C. W. Verseveld, Green beans tested for yield and quality, *Oreg. Veg. Dig. 31*(2):1 (1982).
29. Kattan, A. A., and J. W. Fleming, Effect of irrigation at specific stages of development on yield, quality and growth of snap beans, *Proc. Am. Soc. Hort. Sci. 68*:329 (1956).
30. Mack, H. J., L. L. Boersma, J. L. Wolfe, W. A. Sistrunk, and D. D. Evans, Effect of soil moisture and nitrogen fertilization on pole beans, *Oreg. Agric. Exp. Stn. Tech. Bull. 97*:1 (1966).
31. Gabelman, W. H., and D. D. F. Williams, Development studies with irrigated snap beans, *Wis. Agric. Exp. Stn. Bull. 227* (1960).
32. Doss, B. D., C. E. Evans, and J. L. Turner, Irrigation and applied nitrogen effects on snap beans and pickling cucumbers, *J. Am. Soc. Hort. Sci. 102*:654 (1977).
33. Robinson, W. B., D. E. Wilson, J. C. Moyer, J. D. Atkin, and D. B. Hand, Quality versus yield of snap beans for processing, *Proc. Am. Soc. Hort. Sci. 84*:339 (1964).
34. Weckel, K. G., and P. R. Freund, Effects of variety, age, and size at harvest on preference for canned vegetables, *Res. Rep. Univ. Wis. Coll. Agric. Life Sci. 56* (1970).
35. Polesello, A., G. Criveili, and A. Locatelli, Preliminary studies on quality characteristics of green beans for deep freezing, *Ind. Agrar. 10*:127 (1972).
36. Sistrunk, W. A., K. B. Reddy, and A. R. Gonzalez, Relationship of cultivar, and maturity to protein, fiber and seed of canned green beans, *Arkansas Farm Res.* (May-June):6 (1982).
37. Zaehringer, M. V., K. R. Davis, and L. L. Dean, Persistent green color snap beans (*Phaseolus vulgaris* L.). Color-related constituents and quality of cooked fresh beans, *J. Am. Soc. Hort. Sci. 99*:89 (1974).
38. Kaldy, M. S., Fiber content of green snap beans as influenced by variety and environment, *Proc. Am. Soc. Hort. Sci. 89*:361 (1966).
39. Self, R., J. C. Casey, and T. Swain, The low boiling volatiles of cooked foods. *Chem Ind.* (London) *21*:863 (1963).
40. Stevens, M. A., R. C. Lindsay, L. M. Libbey, and W. A. Frazier, Volatile components of canned snap beans (*Phaseolus vulgaris* L.), *Proc. Am. Soc. Hort. Sci. 91*:833 (1967).
41. Salunkhe, D. K., and J. Y. Do, Biogenesis of aroma constieutns of fruits and vegetables, *CRC Crit. Rev. Food Sci. Nutr. 161* (1976).
42. Murray, K. E., and F. B. Whitfield, The occurrence of 3-alkyl-2-methoxypyrazine in raw vegetables, *J. Sci. Food Agric. 2*:973 (1975).
43. Gorini, F., G. Borinelli, and T. Maggiore, Studies on precooling and storage of some varieties of snap beans, *Acta Hort. 38*:507 (1974).
44. Thompson, B. D., Pre-packaging pole beans at grower level, *Proc. Fla. State Hort. Soc. 77*:131 (1964).
45. Watada, A. E., and L. L. Morris, Effect of chilling and non chilling temperature on snap bean fruits, *Proc. Am. Soc. Hort. Sci. 89*:368 (1966).
46. Ryall, R. L., and W. I. Lipton, *Handling, Transportation and Storage of Fruits and Vegetables*, Vol. I, *Vegetables and Melons*, AVI Pub. Co. Inc., Westport, CT, 1972, p. 473.
47. Pantastico, E. B., T. K. Chattopadhyay, and H. Subramanyam, Storage and commercial storage operation, in *Postharvest Physiology, Handling and Utilization of Tropical and Sub-tropical Fruits and Vegetables* (E. B. Pantastico, ed.), AVI Pub. Co., Westport, CT, 1975, p. 314.
48. Standwood, P. C., and E. E. Ross, Seed storage of several horticultural species in liquid nitrogen, *Hort. Sci. 14*:628 (1979).
49. Platenius, H., F. S. Jamison, and H. C. Thompson, Studies on cold storage of vegetables, *Cornell Expt. Sta. Bull.* No. 602 (1934).
50. Anandaswamy, B., and N. V. R. Iyengar, Pre-packaging of fresh snap beans (*P. vulgaris*), *Food Sci.* (Mysore) *10*:279 (1961).
51. Lopez, A., *A Complete Course in Canning*, 9th ed., The Canning Trade, Baltimore, 1969.
52. Mundt, J. O., and McCarthy, I. E., Factors affecting the blanching of green beans, *Food Technol. 14*:309 (1960).
53. Van Buren, J. P., Adding calcium to snap beans at different stages in processing: Calcium uptake and texture of canned product, *Food Technol. 22*(6):132 (1968).

54. Freemman, D., and W. A. Sistrunk, Comparison of steam and water in the blanching of snap beans, *Arkansas Farm Res. 22*(5):11 (1973).
55. National Canners Association Process for low-acid canned foods in metal containers, *Natl. Canners Assoc. Lab. Bull. 26-L*:29 (1966).
56. Luh, B. S., B. Feinberg, and J. J. Meehan, Freezing preservation of vegetables, in *Commercial Vegetable Processing* (B. S. Luh and J. G. Woodroof, eds.), AVI Pub. Co., Inc., Westport, CT, 1982, p. 317.
57. Luh, B. S., L. P. Somogyi, and J. J. Meehan, Vegetable dehydration, in *Commercial Vegetable Processing* (B. S. Luh and J. G. Woodroof, eds.), AVI Pub. Co. Inc., Westport, CT, 1982, p. 361.
58. Food and Agriculture Organization, *Production Yearbook*, Vol. 49, 1995, Rome, Italy, p. 142.
59. Nwokolo, E., Common bean (*Phaseolus vulgaris*L.), in *Legumes and Oilseeds in Nutrition* (Nwokolo, E., and J. Smartt, eds.), Chapman & Hall, London, 1996, p. 159.
60. Kaplan, L., Archeology and domestication in American *Phaseolus* (beans), *Econ. Bot. 19*:358 (1996).
61. Hempel, J., and H. Böhm, Quality and quantity of prevailing flavonoid glycosides of yellow and green French beans (*Phaseolus vulgaris* L.), *Agric. Food Chem. 44*:2114 (1996).
62. Stanley, D. W., M. C. Bourne, A. P. Stone, and W. V. Wismer, Low temperature blanching effects on chemistry, firmness and structure of canned green beans and carrots, *J. Food Sci. 60*:327 (1995).
63. Eshtiaghi, M. N., R. Stute, and D. Knorr, High pressure and freezing pretreatment effects on drying, rehydration, texture and color of green beans, carrots and potatoes, *J. Food Sci. 59*:1168 (1994).
64. Baron, R. F., and M. P. Penfield, Panelist texture preferences affect sensory evaluation of green bean cultivars (*Phaseolus vulgaris* L.), *J. Food. Sci. 58*:138 (1993).
65. Van Buren, J. P., W. P. Kean, B. K. Gavitt, and T. Sajjaanantukul, Effects of salts and pH on heating-related softening of snap beans, *J. Food Sci. 55*:1312 (1990).
66. Wardlaw, G. M., and P. M. Insel, *Perspectives in Nutrition*, 2nd ed., Mosby, St. Louis, 1993, p. A-31.

21

Other Legumes

S. S. KADAM AND J. K. CHAVAN
Mahatma Phule Agricultural University, Rahuri, India

I. LIMA BEAN

Green lima beans (*Phaseolus lunatus* L.) are grown for home use, for fresh market, and for canning and freezing in the United States, Canada, Latin America, and elsewhere. The origin of the lima bean is tropical America (1). Both small-seeded and large-seeded types of lima beans have climbing bush or dwarf forms. Lima bean is a tender plant that cannot withstand frost, and seed germination requires temperatures well above 15.5°C. It also requires a longer growing season than the common varieties of snap beans. Unlike snap beans, it is grown for seed. Pole varieties require a longer season than do bush varieties.

A. Botany

Lima Bean (*Phaseolus lunatus* L. Syn. *Phaseolus limensis* Macf. and *Phaseolus inamoenus* L.) is one of the important legume vegetables grown in South and Central America, Africa, and many parts of Asia (1). There are two species of lima bean: the large-seeded types, *Phaseolus limensis* and the smaller baby limas, *Phaseolus lunatus*. However, many reports in the literature have grouped both types under a single species, *Phaseolus lunatus*.

Lima bean is a perennial or annual herb that shows considerable variation in vines, pods, and seeds (2). The pole types are generally perennial, tall, with an enlarged taproot. The bush types are normally annual with a dwarf habit. The leaves are trifoliate and often hairy on the lower surface. The pods are oblong, generally curved with a sharp beak, somewhat pubescent, and contining two to six seeds. The seeds are variable in size, shape, and color but are usually divided into small or baby limas and larger limas.

All lima beans grown for processing and most of those for market are dwarf or bush types. Henderson is a small, thin-seeded baby lima, and Fordhook is a large, thick-seeded potato-type lima bean. Henderson Bush Lima and Fordhook account for most of the bush lima bean acreage of the United States (3). Cangreen and Thorogreen are similar to Henderson in all respects except that they have green cotyledon rather than white. Fordhook has been largely replaced by Fordhook 242 and Concentrated Fordhook, which are more vigorous and resistant to hot weather.

Evergreen, Triumph, and Peerless have small thick seeds and are classed as baby potato types. Challenger is a Fordhook-type pole lima bean. King of the Garden is a popular climbing variety in home gardens, having large but thin seeds. The most important climbing variety is the Carolina or Sieva, the butter bean found in the southern United States, which is quite resistant to drought and hot weather. Ventura, Wilbur, and Weston are dry lima bean varieties grown in California (3).

B. Production

Well-drained and fertile soils with light texture are most suitable for the cultivation of lima bean. Loamy and silt soils are also preferred. Lima beans require a slightly warmer climate for growth than snap beans. Mean monthly temperatures of 15–24°C and a frost-free period favor lima bean growth (3). The crop is propagated by seeding. Seed germination takes place at 15–30°C, with an optimum soil temperature of 27°C. Spacing of 10–15 cm between plants and 60–75 cm between rows has been recommended. The culture of lima beans with respect to water and fertilizer is similar to that of French bean (3).

Downy mildew, bacterial spot, pod spot, and leaf blight are the major diseases of lima beans causing serious losses in recent years. Downy mildew is caused by *Phytophthora phaseoli* Thaxt, which is primarily injurious to the pods and attacks them in all stages of growth from flower to maturity. Bacterial spot, caused by *P. syringae*, results in young diseased pods dropping off the plant, resulting in considerable loss in yield. Pod spot is caused by *Diaporthe phaseolorum*, producing brown patches on leaves and on older pods (3).

Green lima beans for fresh market are picked by hand when the seeds are nearly fully grown but before the pods begin to turn yellow (3). Usually several pickings are needed, and the high picking cost limits the acreage of green lima beans planted for fresh market in the United States. Lima Beans that are grown for processing are harvested by mowing machines; the vines are hauled to the vining station or to the cannery where the beans are shelled mechanically with machines. In some places, self-propelled viners are employed so that only the shelled beans are hauled to the cannery or freezing plant. As in the case of other beans and peas, the choice of harvesting process for lima beans is also a compromise between yield and quality. The highest yields of top-quality beans are obtained when the beans show 3–5% white seeds. Thompson and Kelly (3) mentioned that seeds of lima beans are not considered satisfactory for canning or freezing after they turn white. As the seeds mature, the starch content increases and the sugar content decreases and the color of the seed changes from dark to pale green and then to white.

Salunkhe et al. (4–6) evaluated the relationship of harvest time of lima bean with yield and processing quality. The yield of shelled beans increased as the harvest time was delayed. The percentage of shriveled and dry beans also increased with later harvests, although not consistently. High shear-press values were obtained when dry and shriveled beans were included (Table 1), the values being related to the size of beans. Similarly, when dry and shriveled beans were included in the samples, the percent of total and alcohol-insoluble solids increased with later harvest.

C. Chemical Composition

The chemical composition of lima beans is presented in Table 2. Compared to the cotyledons and the embryo, the seed coat contains a higher proportion of crude fiber and calcium. Globulins represent a major class of proteins in lima beans. The globulin proteins of lima beans have been characterized as α-globulin and β-globulin. Lima beans contain a small quantity of albumin.

TABLE 1 Effects of Harvest and Variety on Average Shear-Press Values[a] of Lima Beans

Harvest[b] (1956)	Clark's Bush	Early Thorogreen	Concentrated Fordhook	Lima-green	LSD (0.05%)
1. With	1175	1285	880	1036	330
Without	925	975	695	695	349
2. With	1130	1755	800	1210	330
Without	1065	1155	650	865	349
3. With	1575	1840	830	1845	330
Without	1095	1065	800	1020	349
4. With	1430	1900	1765	1920	330
Without	1085	1135	865	1085	349
5. With	2300	2275	2205	2490	330
Without	1120	1155	965	1205	349

[a]Numbers represent pounds of pressure.
[b]First harvest was started when most plants showed indications of prime maturity, and five successive harvests were made at intervals of 2–5 days. With: With dry and shriveled beans; Without: without dry and shriveled beans.
Source: Ref. 6.

Lima beans, like other legumes, are a good source of protein. Lima beans were found to be a good source of most amino acids except methionine (1).

D. Storage

Lima beans are highly perishable and also sensitive to chilling injury, so they are precooled (prefarably by hydrocooling) immediately after harvest and kept at a low temperature (7). The pods are more sensitive to chilling injury than the beans. Hence, unshelled beans should be kept

TABLE 2 Chemical Composition of Lima Bean

Constituent (per 100 g edible portion)	Content
Energy (kcal)	123
Moisture (g)	67.5
Protein (g)	8.4
Fat (g)	0.5
Carbohydrate (g)	22.0
Vitamin A (IU)	290
Thiamine (mg)	0.24
Riboflavin (mg)	0.12
Niacin (mg)	1.4
Ascorbic acid (mg)	29.0
Calcium (mg)	52
Phosphorus (mg)	142
Iron (mg)	2.87

Source: Ref. 2.

at 5–6°C. Shelled beans can be kept at 3–4°C. Unshelled lima beans can be kept for about 5 days and shelled beans for about a week at the respective suggested temperatures (7).

Pantastico et al. (8) recommended temperatures of 4.4–7.2°C with 90–95% relative humidity for lima beans in the pod and at 4.4°C with 95% relative humidity for shelled lima beans. Under these conditions, lima beans can be stored for 1.5–2 weeks. Ryall and Lipton (9), however, stated that 2.8°C and 95% relative humidity for shelled lima beans will extend its marketable life for about one week; lima beans in pods stored at 4.4–7.2°C and 95% relative humidity have a storage life of only 3–5 days. it appears that the stage of maturity of lima bean is an important factor determining the shelf life of the pods.

Anandaswamy and Iyengar (10) observed that a combination of low O_2 (2–3%) and high CO_2 (5–10%) retards yellowing of lima beans at 7.2°C; higher CO_2 may result in development of off-flavor.

E. Processing

1. Canning

Fresh lima beans may be canned and preserved for a considerable time, thus they are an important product of the U.S. vegetable-processing industry (1). The beans are first washed carefully and stored. They are size-graded and then blanched in hot water at 88–95°C—the amount of time depends upon the maturity of the beans, varying from 2 to 3 minutes for young, tender beans to 7 or 8 minutes for more mature, tougher beans. For canned products, the beans are then filled hot into cans, 2% hot brine is added, and the cans are processed at 115°C. The suggested processing time for smaller cans is 35 minutes at 115°C if the initial temperature is 60°C or more. For larger cans, the processing time at 115°C is 50–55 minutes. It is not necessary to exhaust the smaller size cans if the brine added is already boiling, but many canners pass large cans through an exhaust box with steam full on to prevent buckling. Immeidately after processing, the cans are watercooled to between 35 and 65°C.

2. Other Methods

Quick-frozen lima beans are processed by packing the blanched beans in cartons and freezing them in plant freezers. Dehydrated lima beans are often processed by treating the blanched beans with a 1.5% sulfite solution at pH 7.2 to preserve their color and dehydrating them in a through-flow atmospheric drier at 50°C for approximately 12 hours. The dehydrated beans are then vacuum-packed in plain cans.

Mature lima bean pods may be harvested, sun-dried, and the beans then separated from the pods. The beans may be stored at ambient temperature for several months.

Salunkhe et al. (6) processed lima beans by canning and freezing as rapidly as possible to preserve maximum color, flavor, and nutritional value. As the concentration of brine increased, the amounts of total and alcohol-insoluble solids, specific gravity, and drained weight increased and the riboflavin and thiamine contents generally decreased. Although the total yield increased with harvest, the yield of high-quality beans (floater in 18% brine) increased from the first to second harvest and then decreased rapidly as the days increased in subsequent harvests (Fig. 1).

II. CHICKPEA

The green chickpea (*Cicer arietinum* L.) is another staple food crop in many tropical and subtropical countries. In India, it is mostly consumed as cooked, dehusked *dhal* or as whole seeds. Several traditional food products are prepared from dhal and flour. Chickpeas hold good

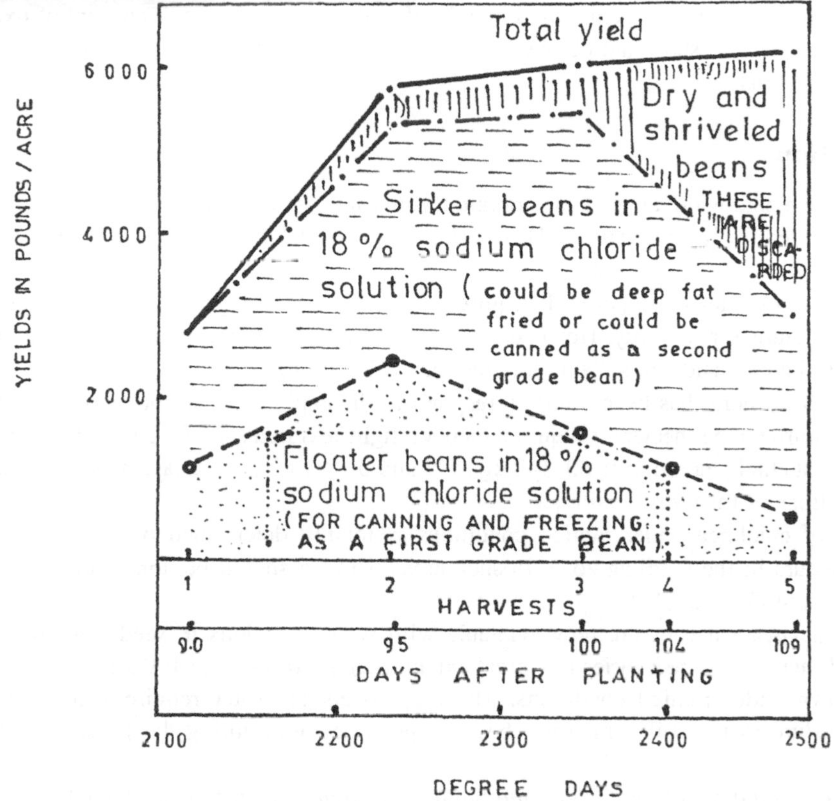

FIGURE 1 Effect of harvest time on the yield of shelled lima beans of different grades of Limagreen variety.(From Ref. 6.)

promise in the preparation of fermented food products, such as temphe, miso, sausages, bakery products, and weaning foods (11).

A. Botany

The chickpea, also known by several other names, namely, Bengal gram, boot, chana, chola, chhole, garbanzo bean, gram, hommes, and pois chiche (12,13), has been classified in the family *Fabaceae*. The genus *Cicer* comprises 39 known species. *Cicer reticulatum*, the progenitor of *C. arietinum*, is closely related in terms of crossability and morphological similarity. The cultivated *C. arietinum* is closely related to *C. pinnatifidum*, *C. echinospermam*, and *C. bijugam*. Based on seed color and geographic distribution, the chickpea is grouped ito two types: *desi* (Indian origin) and *kabuli* (Mediterranean and Middle Eastern origin). Kabuli cultivars are white to cream colored and used almost exclusively as cooked whole seeds as a vegetable. The seeds of *desi* cultivars are wrinkled at the beak with brown, light brown, fawn, yellow, orange, black, or green color. These cultivars are normally dehusked to obtain *dhal*, which is directly cooked or milled to flour. The bold-seeded cultivars of the *desi* type are often used for roasting. The seed coat contributes 14.5–16.4% of the seed weight. The cotyledons contribute 82.9–84% and the germ 1.2–1.5% of the seed weight.

Chickpea is a bushy upright plant, 25–50 cm in height and covered with glandular hairs.

These hairs exude an acrid liquid composed mainly of malic acid and a small amount of oxalic acid. The mature pod are short and oblong, 2–3 cm long and 1–2 cm wide, containing one to two angular seeds (14).

B. Production

Chickpea is grown on a wide range of soils ranging from sandy loam to clay loam. Medium type soils are most suited for its cultivation. The soil should be free from excessive salts and neutral in reacton. It should also be well drained (15).

Chickpea is a winter season crop. It requires cool climate for its growth and development and high temperature for maturity. However, severe cold and frosts are injurious. Areas with moderate rainfall are considered suitable for chickpea cultivation.

Chickpea in general has been found to respond to the application of phosphorus. A dose of 10–20 kg of nitrogen per hectare is required on poor, light-textured soil. Saxena and Sheldrake (16) reported that chickpea removes 60–143 kg of N and 5–10 of P per hectare depending upon growing conditions of the crop.

Sowing of chickpea under rainfed conditions is mainly determined by availability of conserved moisture in the seedling zone. In such areas, sowing should be done earlier, during the first 2 weeks of October (17).

Many chickpea varieties are now available with wide variations in seed size. Accordingly, the seed rate of chickpea varies to a great extent, ranging from 50 to 100 kg/ha. Chickpea is mainly grown under rainfed conditions. The crop meets its water requirement from the conserved moisture in the soil, which has been properly restored during the preceding monsoon season.

Wilt (*Fusarium oxysporum*), *Sclerotinia* blight (*Sclerotinia sclerotorum*), *Ascochyta* blight (*Ascochyta* sp.), and gray mold (*Botrytis cinera*) are important diseases of chickpea. The major pests reported from India include cutworms and pod borer.

Crops should be harvested when the beans become fully matured. The pods are shelled and green beans are used as vegetables.

C. Chemical Composition

The chemical composition of the green seed of chickpea is presented in Table 3. It is a good source of calcium, phosphorus, and iron (14). The ascorbic acid content is also very high. However, it is a poor source of β-carotene (14).

D. Processing

Canning and dehydration of immature green chickpeas are becoming popular in many countries including India and the United States (18). The green seeds are a rich source of proteins, total sugars, and minerals and vitamins. Blanching of seeds followed by canning results in loss of the original green color (19). However, preliminary soaking of seeds in a sulfite solution helped in uniform absorption of green color added to the brine. Similarly, soaking in sodium sulfite and addition of 300 ppm EDTA to the canning brine improved color retention in canned chickpea without affecting chemical composition (20). A method for dehydration of green chickpeas that does not affect chemical composition has been standarized (20). The seeds floating in 10–15% NaCl were found to have the best color and reconstitution properties.

TABLE 3 Nutritional Composition of Green Chickpea Seeds

Composition	Water	Floaters in NaCl (%) 10	15	25	Sinkers in 25% NaCl Green	Yellow
Seeds (%)	10.0	11.5	11.0	23.5	44.0	32.5
Moisture (%)	85.3	79.8	77.3	69.9	61.6	50.5
Protein (%)	43.8	37.9	36.2	27.5	25.7	19.8
Ash (%)	4.5	5.2	5.8	6.6	5.5	4.1
Crude fiber (%)	13.2	17.2	15.1	14.6	13.0	11.2
Total sugars (mg%)	58.0	37.2	21.5	17.1	14.3	8.2
Acidity (mg%) (citric acid)	1646	1225	1114	862	743	569
Calcium (mg%)	261.9	246.0	267.9	215.1	179.9	146.1
Phosphorus (mg%)	448.1	325.8	301.6	276.1	224.8	209.0
Iron (mg%)	67.8	66.7	61.6	48.9	48.5	41.3
Ascorbic acid (mg%)	371.6	118.9	93.8	80.4	50.7	11.6
β-Carotene (mg%)	2.0	1.1	1.0	0.9	0.7	0.3

Source: Ref. 14.

III. CLUSTER BEAN

Cluster bean (*Cyamopsis tetragonoloba* L. Taub) or guar is a valuable drought-tolerant legume grown in Asia and the United States. India, Pakistan, and the United States are the major producers of cluster beans. This crop can be grown on soils of low fertility in the arid and semiarid areas of the tropics and subtropics. The green immature pods of the cluster bean have been used for centuries in Asia as a nutritious vegetable. The young pods are sweet and are often cooked like *Phaseolus* beans. In certain countries like the United States, the cluster bean is grown as a source of vegetable gum, which is widely used in the food, paper, and textile industries (21).

A. Botany

The cluster bean is a robust annual with a long taproot and a well-developed lateral root system. There is considerable variability; some types are erect and others have numerous strong branches arising from basal nodules; some have small hairs on all parts of the plant, whereas others are completely glabrous. The leaves are alternate trifoliate and are borne on long petioles. The flowers are small, typically papilionaceous, with standard petal and keel being white and wings being pinkish purple. The pods are oblong, 5–12 cm in length, normally containing 5–12 oval or cube-shaped seeds of variable size and color (22).

There are numerous cultivars of cluster bean. In general, branched types are considered more suitable for seed production, and erect single-stemmed types are more suitable for pod production. In some parts of India, three main types of cluster bean are recognized (23–25). The Indian Agricultural Research Institute, New Delhi, has released three varieties: Pusa Sadabahar, Pusa Mausami, and Pusa Navbahar. The last one is produced by crossing the first two varieties with pod quality of Pusa Mausami (smooth, bright green) and growing habit of Pusa Sadabahar (single stem and high yield). Pusa Navbahar gives an excellent crop in both summer and the rainy season.

B. Production

Cluster bean can be grown on almost all types of soils, but a well-drained light soil such as sandy loam soil is preferred. A soil pH up to 7.5–8.0 is tolerable for this crop, but waterlogging adversely affects both growth and productivity. Unlike some other legume crops, cluster bean is a warm season crop. It grows well both during *kharif* and summer seasons. It is very resistant to drought, so this crop can be grown in an area of low rainfall and high temperatures.

The seeds are directly sown in the field by broadcasting or dibbling (26). The land preparation for cluster bean is similar to that for other vegetables. It is grown both for pods and fodder; when grown for pods the field should be well prepared. The seeds can be sown almost throughout the year if there are no heavy rains. The spacing depends upon the soil type and weather. The common spacing recommended for this crop is 45–60 cm (row to row) and 20–30 cm (plant to plant in a row). At this spacing, it takes about 14–24 kg of seeds to cover one hectare by the broadcast method and even less by the dibbling method. More seeds are required when beans are grown for fodder. Both flat beds and ridges and furrow methods are followed. Cluster bean is often grown as a mixed crop with other crops like cucurbits, cotton, and sugarcane.

Because it is a legume, cluster bean does not need too much fertilizer, as it can fix atmospheric nitrogen. However, when it is grown for pods, application of additional fertilizer is beneficial for increasing the yield. The amount of fertilizer to be added depends on the soil type, availability of irrigation facilities, and weather. The application of 10–14 cartloads of farmyard manure, 12–24 kg of nitrogen, and 60–74 kg each of phosphorus and potash will provide the required levels of nutrients for this crop (27,28). The entire quantity of farmyard manure is added before sowing and mixed with the soil while preparing it, while the full dose of nitrogen, phosphorus, and potassium is given at the time of sowing.

Rainy season crops do not need any additional irrigation if rains are well distributed, but summer crops may require additional water. Weeding may also be necessary to keep weeds under control.

Tender green pods are harvested by cutting or twisting. Pod formation starts from the base, and new flower buds and pods continue to form as the plant grows, requiring a prolonged harvest period (29). Three to four pods in the axil of a single leaf are ready at the same time. The pods should be picked with care to avoid injury to the plant.

C. Chemical Composition

The chemical composition of cluster bean pods is presented in Table 4. Cluster bean seeds contain galactomannan polysaccharide. The endosperm contains about 68–70% galactomannan gum (30), also known as guar gum. Das and Arora (31) reviewed the chemistry of cluster bean gum. The gum is composed of D-galactopyranose and D-mannopyranose units. Excellent monographs are available on the chemistry of cluster bean gum (32–35). The biodegradation of galactomannan requires the presence of at least three enzymes in germinating seeds (36): α-D-galactosidase, β-D-mannanase, and β-D-mannosidase. These enzymes have been reported in guar seeds (37–39). Guar is a good source of vitamin A and iron (40).

D. Processing

Tender cluster bean pods are used as a vegetable, and in southern parts of India they are dehydrated and stored for future use. Cluster bean seeds are a good source of gum, a galactomannan used in foods and cosmetics. The residual meal after separation of the gum fraction is a potentially valuable protein source (45% protein) for animal feed but contains about

TABLE 4 Nutritional Composition of Cluster Bean

Constituent (per 100 g edible portion)	Content
Energy (kcal)	16
Moisture (g)	81
Protein (g)	3.2
Fat (g)	1.4
Carbohydrate (g)	10.8
Vitamin A (IU)	65.3
Thiamine (mg)	0.09
Riboflavin (mg)	0.03
Niacin (mg)	0.6
Ascorbic acid (mg)	49.0
Calcium (mg)	130
Phosphorus (mg)	57
Iron (mg)	4.5

Source: Ref. 2.

10% saponin, limiting its use to low levels in poultry diets because of its adverse effect on chick growth and egg production (41).

IV. COWPEA

Cowpea (*Vigna unquiculata* L. Walp) is mainly cultivated for its pods and seeds in the tropics and subtropics (42). The botanical classification of cowpea and its subspecies has been reported by several investigators. In African countries, cowpea is consumed as a boiled vegetable using fresh or rehydrated seeds, as an ingredient in soups, and as a paste in steamed and fried dishes (42). In Nigeria, cowpea is commonly consumed in the form of fermented products (43). In India, it is mostly consumed as cooked green immature pods or cooked whole seeds in the form of curry with rice or other cereals. The young shoots and leaves are often consumed as spinach in fresh or dry form. In the United States, immature seeds are used for canning (42).

A. Botany

Cowpea is native to Central America. There are distinct types (species) of cowpea, characterized by growth habit and pod character. The immature pods are used as vegetables. The plant is erect or semi-erect, and pods are pendant and 10–30 cm long. The seed has a hilum, which may be black, dark purple, brown, or maroon (13). Different cultivars respond differently to photoperiod. Pusa Phalguni, Pusa Barsati, Pusa Dofasli, and Sel-1552 are some of the cultivars of vegetable cowpea recommended by the Indian Agricultural Research Institute, New Delhi, India.

B. Production

Cowpea is not particular in its soil requriements. It can be grown in all types of soils ranging from very light to heavy clays as long as the soil is well drained and deep, but it is very susceptible to waterlogging. Acid soils are not suitable for its growth.

Being a warm season crop, cowpea can tolerate a hot dry climate, but it is susceptible to low temperatures, particularly below 20°C. Flowering and fruiting are inhibited by low temperatures. If frost occurs late in the season, the flowers and immature fruits will drop. This crop grows well when temperatures are high (40°C) and water is not a constraint. The varieties differ in their response to temperatures and day length. For example, when grown in summer the rainy season varieties put on only vegetative growth. Therefore, it is imperative that proper varieties be selected for each season.

Cowpea crop can be grown both in *kharif* (June-July sowing) and summer (February-March sowing). In places where the winter is not too severe it can be grown throughout the year. Selecting the proper variety is absolutely necessary. For the spring season, varieties like Pusa Phalguni and Pusa Dofasli are suitable, whereas other varieties like Pusa Barsati and Pusa Dofasli are good for the rainy season.

Fertilizer doses are determined by the variety, growing season, soil type, and soil fertility status (44–48). Being a legume crop, cowpea does not need much fertilizer. Fertilizer recommendations given by different investigators vary. Mehta (45) recommended the application of 20–40 kg of nitrogen per hectare while according to the Indian Council of Agricultural Research, New Delhi, a fertilizer dose consisting of 10–20 kg of nitrogen and 50–70 kg each of phosphorus and potash per hectare should be applied. It should be noted that the nitrogenous fertilizers delay flowering and fruiting.

Yawalkar (29) recommended 25–30 tons of farmyard manure and 40 kg of nitrogen per hectare. Kale and Kale (49) summarized different recommendations, which include 25–50 tons of well-decomposed farmyard manure, 12–50 kg of nitrogen, and 60-74 kg each of phosphorus and potash per hectare. The entire quantity of farmyard manure should be added at the time of land preparation and should be thoroughly mixed in the soil, while full doses of nitrogen, phosphorus, and potash are applied just before sowing.

The first irrigation is given immediately after sowing and subsequent irrigations are given according to the need. Under normal conditions, seeds take 3–4 days for germination. Weeding and hoeing help to check weed growth and provide aeration to the roots. Pod yield can be significantly improved by spraying maleic hydrazide (15 ppm) just before flowering.

Anthracnose (*Colletotrichum lindemuthianum* (Sacc. & Magn.) Briosi and Cac), dieback (*Collectotrichum capsici* (Syd), Butl. & Bis.), ashy stem blight (*Macrophomina phaseolina* (Tassi), Goid), and powdery mildew (*Erysiphe polygoni* DC) are important fungal diseases of cowpea. Bacterial blight caused by *Xanthomonas vignicola* Busk occurs commonly in subtropical and tropical climates. The major pests of cowpea include Galerucid beetle (*Madurasia obscurella*), aphid (*Aphis* sp.), jassid (*Amrasca kerri*), pod borer (*Heliothis armigera*) and bean weevil (*Bruchus* sp.).

The plants start flowering in about 6–8 weeks depending upon the variety, climate, and soil conditions. Pods will be ready for harvesting 2–3 weeks after fruit set. The pods are harvested when they are still tender or immature. When the crop is grown for dry seed, the seeds are harvested when the pod is fully mature. The yield depends upon the variety, soil fertility, irrigation facilities available, and climatic conditions. In general, approximately 5–7.5 t of green pods or 1–1.5 t of dry seeds can be harvested from one hectare.

C. Chemical Composition

The chemical composition of green cowpea seeds is presented in Table 5. The large variations in the nutrient contents may be due to the genetic background as well as to climate, fertilization, season, and agronomic practices. Cowpea seeds are a good source of proteins, minerals, and

TABLE 5 Nutritional Composition of Cowpea

Constituent (per 100 g edible portion)	Content
Energy (kcal)	48
Moisture (g)	85.8
Protein (g)	3.5
Fat (g)	0.2
Carbohydrate (g)	8.1
Vitamin A (IU)	1861
Thiamine (mg)	0.07
Riboflavin (mg)	0.09
Niacin (mg)	0.90
Ascorbic acid (mg)	14.0
Calcium (mg)	72
Phosphorus (mg)	59
Iron (mg)	2.5

Source: Ref. 2.

energy. However, the presence of polyphenols, phytate, oligosaccharides, and protease inhibitors has been reported in dry seeds of cowpea (46).

D. Storage

Cowpeas can be stored at low temperature in either the shelled or the pod form. Smittle and Hayes (50) studied the effects of short-term storage on cowpea quality (southern pea). Mechanically shelled peas cv. Purple Hull Pinkeye were stored at temperatures of 5, 25, and 45°C for 3, 6, and 12 hours. A response curve relating the rate of loss of green seed color to storage temperature was found to assist in the coordination of harvesting, transport, and processing operations for maintenance of a high-quality product. The percentage of green seed was not influenced by storage at 5°C for up to 12 hours or storage at 25°C for up to 6 hours. A decrease in green seed, however, occurred after 12 hours when produce was stored at 25°C. The hemicellulose and cellulose concentrations were not affected by storage treatment.

E. Processing

Cowpea seeds are generally preserved by either freezing or drying. The dried seeds are soaked to produce quick-cooking cowpeas. Cowpeas are canned to a limited extent as whole beans or mixed beans (48).

V. HYACINTH BEAN

A. Botany

The hyacinth bean (kidney bean) is an important food legume in many parts of Asia, notably in southern India. It is also known as Bonavist(a) bean, Dolichos bean, Egyptian bean, Indian (butter) bean, and lablab bean (51). The hyacinth bean (*Lablab purpureus* L. Sweet syn. *Dolichos lablab* L., *Lablab niger* medik) supplies a considerable portion of the protein in the diet of the rural population in India. The pods of the hyacinth bean may be inflated, white, green, or purplish in color, and can vary in length from approximately 5 to 20 cm and in width from 1 to 5 cm. They may be crescent-shaped to more or less straight and oblong or sometimes dorsally straight and ventrally deeply curving. Cultivars grown for vegetables have pods with thick, fleshy skins with practically no fiber. The pods may be septate or nonseptate; in the former, each seed occupies a separate compartment in the pod, while in the latter the pods have a bloated appearance. Each pod normally contains three to six seeds. These are generally less than 1.25 cm in length and may be rounded or oval or rather flattened white, cream, red, brown, or black. The hilum is white, prominent, and oblong, usually covering one third of the seed. There are numerous cultivars of hyacinth bean adapted to local conditions and different purposes (52–54). Pusa early Prolific, JDL-37, and Kalianpur T-1 are some of the cultivars most widely grown in India.

B. Production

Hyacinth bean can be grown in a wide range of soils of average fertility (54). It grows very well in tropical and subtropical regions. It is a relatively cool season crop. Like other beans, hyacinth bean requires a good seed bed for sowing. About 20–30 kg of seed is required for sowing one hectare. The crop is fertilized with 90 kg of ammonium sulfate and 40 kg of super phosphate per hectare after the first weeding. Intercultural operations are performed to control weeds until vines spread between rows.

The most important diseases include leaf spot caused by *Cercospora dolichii*, ashy stem blight, powdery mildew caused by *Laveillula taurica* var. *macrospora*, and yellow mosaic caused by a virus. Major pests attacking hyacinth bean are those described for French bean (52).

Pods are harvested when they are green and succulent and have not become fibrous. It produces on average 5–8 t of green pods per hectare (52).

C. Chemical Composition

Immature pods of the hyacinth bean are commonly used as a vegetable. The chemical composition of pods and dry seeds is presented in Table 6. Like other legumes, hyacinth beans are a good source of protein, minerals, and vitamins. Studies on the nutritive value of proteins have indicated that the amino acid composition of hyacinth beans is comparable to that of other legume proteins (54). Methionine is the most limiting amino acid in hyacinth beans. Like other legumes, hyacinth bean seeds contain trypsin inhibitor, lectin, phytic acid, and polyphenols. The lectin from hyacinth beans has been isolated and purified (55), and its toxicity has been studied by feeding rats with different levels of lectins (56). The toxicity of lectin can be eliminated by autoclaving the meal.

D. Processing and Utilization

Hyacinth beans are a nutritious food source. The culinary types are popular in Asia, where they are widely consumed as a vegetable or used in curries. Sometimes the immature green seeds are extracted from the pods and eaten as a vegetable either boiled or roasted. In Egypt, hyacinth beans are sometimes used as a substitute for broad beans in the preparation of fried bean cake (*tanniah*). In Asia, the mature seeds are utilized as a pulse, often as *dhal*. The seeds are sometimes soaked in water overnight, and when germination starts they are sun-dried and stored for future use. Pod husks are often used as cattle feed (57).

TABLE 6 Nutritional Composition of Hyacinth Bean

Constituent (per 100 g edible portion)	Content
Energy (kcal)	48
Moisture (g)	86.1
Protein	3.8
Fat (g)	0.7
Carbohydrate (g)	6.7
Vitamin A (IU)	617
Thiamine (mg)	0.1
Riboflavin (mg)	0.06
Niacin (mg)	0.7
Ascorbic acid (mg)	9
Calcium (mg)	210
Phosphorus (mg)	68
Iron (mg)	1.7

Source: Ref. 2.

VI. PIGEON PEA

A. Botany

Pigeon pea is one of the oldest food crops and ranks fifth in importance among edible legumes. The pigeon pea has been classified under the family Fabaceae with one polymeric species—*Cajanus cajan* (58). The genus *Atylosia* is closely related to *Cajanus*. It is a commercially important crop in India and many countries of East Africa, the Caribbean, and Latin America. The green seeds are consumed as a vegetable. Pigeon pea is a short-lived woody perennial but is grown as an annual in some countries.

B. Production

Pigeon pea is mainly grown in tropical and subtropical climates. it is highly susceptible even to light frost. It can tolerate heavy rains provided waterlogging does not take place.

Pigeon pea requires light-textured, well-drained soil, although it is grown on a wide range of soils ranging from sandy to clay. Soils should be neutral in reaction and well drained. Soils should have a pH range ploughing of 6.5–7.5. Pigeon pea, being a deep-rooted crop, responds well to proper tilth. A deep plouging followed by two or three discings and harrowings before planting is essential.

Several varieties of pigeon pea have been developed. At present emphasis is being laid on the development of short-duration, high-yielding varieties permitting a second crop or wheat or other *rabi* crop in the winter. These varieties mature in about 130–180 days (57).

Pigeon pea requires heavy doses of nutrients, hence care should be taken that the crop does not lack nutrients. Pigeon pea invariably responds to application of phosphate fertilizers. A starter dose of 15–20 kg N/ha may be given to ensure quick, early growth of the crop. On the basis of available data, it is recommended that on phosphorous-deficient soils 50 kg P_2O_5/ha and on soils medium in P status 25–30 kg P_2O_5/ha be applied.

Pigeon pea is a traditionally *kharif* crop sown in June-July with the onset of monsoon in various agroclimatic zones of India. Most of the traditional pigeon pea areas in India are planted with late-maturity varieties where the crop is adversely affected by frost.

Long-duration varieties of pigeon pea are tall, spreading, and occupy the field for about 250–270 days. These varieties are planted with wider row spacing of 90–120 cm and about 30 cm between plants particularly under rainfed conditions. Under irrigated conditions, early-maturing varieties are planted with a row spacing of 50–75 cm and plant-to-plant spacing of 15–20 cm. Pigeon pea is generally broadcast. Broadcasting results in an uneven plant population, which ultimately results in low yield. In such areas where temporary waterlogging occurs, planting on ridges has been found superior.

Pigeon pea is attacked by a number of diseases. Among these, wilt, stem rot, canker, leaf spot, and sterility mosaic are most improtant. Wilt is caused by the fungus *Fusarium oxysporum* and is the most important disease of pigeon pea. It is damaged by various insects from the seedling stage to maturity. Pod borer, pod fly, and pod bug are the major insect pests attacking pigeon pea.

Pigeon pea is an indeterminate type and its growth continues into the reproductive phase. Pigeon pea is harvested when beans are fully matured and green in color. The pods are shelled and the beans separated and used as a vegetable.

C. Chemical Composition

The chemical composition of green seeds and mature seeds is presented in Table 7. Generally green seeds used for vegetables contain more moisture than mature seeds. Green seeds also contain fewer proteins, carbohydrates, fats, fiber, and mineral matter than mature seeds on a fresh weight basis (58). The proteins of immature seeds are richer in cysteine and methionine than mature seeds. The cotyledons contribute about 90% of the protein, 95% fat, 86% of carbohydrate, 83% of minerals, and most of the phosphorus of the whole seed (59,60). Although the germ is rich in protein (48.1%), fat (13.5%), and minerals, its contribution is negligible on a total seed weight basis.

D. Processing

Canning of green seeds of pigeon pea is a common practice in African countries as well as the West Indies and the Dominican Republic (61,62). In the Dominican Republic, over 80% of seeds produced is canned in brine and exported to the United States and other Latin American

TABLE 7 Proximate Composition of Immature Pods and Dry Seeds of Pigeon Pea

Constituent (per 100 g)	Immature pods	Dry seeds
Moisture (g)	82.4	9.6
Protein (g)	4.5	24.6
Fat (g)	0.1	0.8
Fiber (g)	2.0	1.4
Carbohydrate (g)	10.0	60.1
Ash (g)	1.0	3.2
Calcium (mg)	50.0	60.0
Phosphorus (mg)	60.0	450.0
Iron (mg)	10.0	2.0
Nicotinic acid (mg)	0.8	1.8

Source: Ref. 56.

countries. The cultivars *Kaki*, *Pinto*, *Villalba*, USSD, and *Cajunus indicus* var. *Semper florens* are commonly used for canning (61). Sanchez-Nieva (62) reported that a satisfactory product can only be obtained from fully mature but still green pigeon peas. Mansfield (61) has described the technology for pigeon pea canning in the Dominican Republic and suggested possible improvements in the existing techniques (Fig. 2). The natural green color of pigeon pea can be stabilized by treatment with compounds yielding SO_2 at a pH of 6.0 (63–65). Frozen green seeds are usually consumed after cooking separately or mixed with other vegetables. The freezing process is outlined in Figure 3. Green pigeon peas are nutritionally superior to mature seeds of *dhal* since they are harvested at a stage when the concentration of desirable nutrients such as sulfur amino acids are higher (66). Hence, utilization of pigeon pea in the form of processed green peas needs to be popularized among all segments of the populations where it is widely grown.

VII. WINGED BEAN

The winged bean, a high-protein crop of the tropics, has been cited as an underexploited legume (67). The winged bean has an exceptionally high nitrogen-fixing capacity. This appears to be the

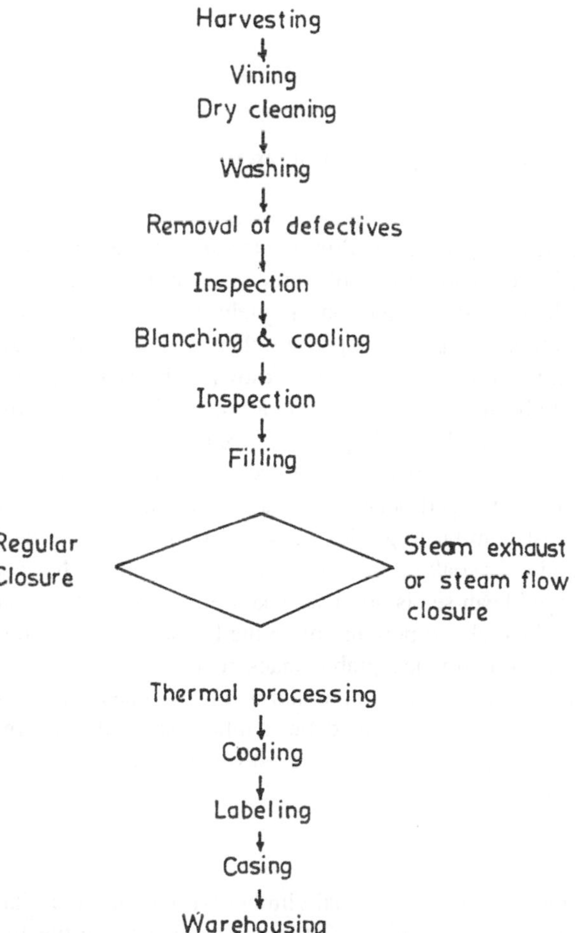

FIGURE 2 Flowchart for canning of pigeon peas. (From Ref. 61.)

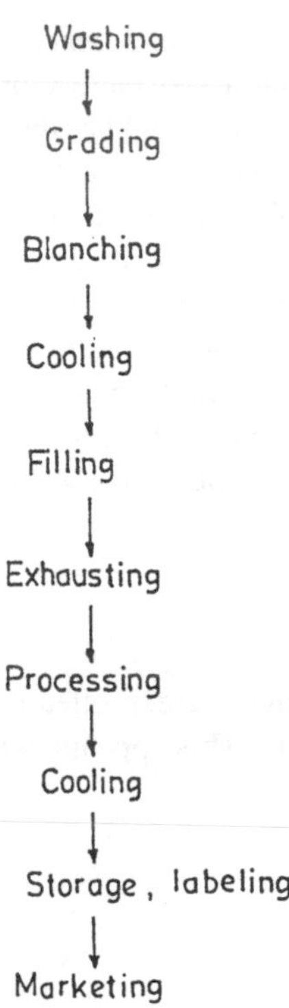

FIGURE 3 Flowchart for freezing of pigeon peas. (From Ref. 48.)

basis for two attributes which give this crop its agricultural interest: (1) the unusually high protein content of the seeds, pods, leaves, and roots (all parts to be utilized by humans or animals), and (2) the reportedly excellent residual effect on the yield of subsequent crops (68). This crop may be capable of yielding more usable protein per unit than any other cultivated crop. Like the soybean, the seeds of the winged bean have long been known to be high in protein and oil (69). The tender pods of the winged bean are the most popular part of this crop and the only part commonly eaten in most of Asia. Young pods are chopped and are used either raw or cooked. Winged bean pods can be stir-fried, deep-fried, or pressure-cooked in earthen pots, as is done in Papua New Guinea (70). The taste of the winged bean is similar to that of most green beans. Immature winged bean seeds have a taste similar to garden peas. Fully ripe seeds are occasionally eaten. Seeds are steamed or boiled after soaking in water overnight to loosen the seed coat and to shorten the cooking time. Winged bean seeds are also baked either inside or outside the dry pod and are stir-fried in fat or oil. The high temperature of frying breaks open the tough seed coat, and the peeled cooked seeds make a highly acceptable snack food.

Young winged bean leaves, shoots, and flowers are popular meals in some areas, providing color and flavor as well as nutrients. The use of winged bean tubers as food appears to be restricted to Burma and Papua New Guinea (70). These are most often eaten after cooking. In fact, the entire plant has nutritional properties.

A. Botany

Winged bean is a herbaceous perennial grown as an annual climber with a wiry stem having a tendency for twining (71). Depending upon the cultivar, stems are green to purple in color.

Leaves are trifoliate and range in shape from oval to ovate-lanceolate with an entire margin. The leaves are alternately borne subtended by a stipule. The plant requires a day length shorter than 12 hours with a temperature of 27°/22°C for reproduction (71). The inflorescence is a raceme bearing many flowers. Flowers are basically blue or purple, ranging from almost white to deep reddish-purple in some lines. The pods have four sides with wings protruding from angles. Pods vary in length and shape.

Winged bean cultivars vary in seed color, seed size, ring around the hilum, seed hardness, vigor of vegetative growth, height of the plant, response to photoperiod, tuberous root production, and many other morphological and physiological characteristics. Several local cultivars are identified in Venezuela, Ghana, Sri Lanka, Hawaii, Papua New Guinea, and India (71).

B. Production

Winged bean can be grown on any soil with good drainage. The crop is tropical in nature and grows well in the subtropics. It is sensitive to photoperiod. The plant may be planted any time that day length falls between 11 and 13 hours (71).

The seeds, presoaked for 2 days before planting, are sown in flat beds or on ridges at a distance of 60 cm. The distance between rows varies from 60 to 120 cm. The spacing varies according to the cultivar and location. Seedlings show signs of twining a few weeks after germination. Hence, support to the vine is essential. This is given with local wood material such as bamboo or *subabul* (71). Crops of winged beans grown for tubers are traditionally pruned. Flower removal and pruning increase the tuber yield. The information on nutritional requirements is meager, but dry matter yield increased with phosphatic fertilizer. The winged bean is free from major diseases and pests. False rust caused by *Synchytrium psophocarpi* and leaf spot caused by *Cercospora psophocarpi*, root-knot nematode (*Meloidogyne* spp.), and cowpea aphid (*Aphis craccivora*) are some of the diseases and pests of winged bean.

Green pods to be used as a vegetable are ready for harvest 10 weeks after sowing and for a few months thereafter. Each vine yields about 25 pods every week. The yield of pods varies from 5 to 20 t per hectare.

C. Chemical Composition

Winged bean tubers have a pleasant and slightly sweet taste, ivory flesh, and a firm texture. The protein content of winged bean tubers is at least five times higher than that of yams and 10 times higher than that of cassava root (72). In some cultivars, as much as 15% protein was found in roots when calculated on a fresh weight basis (72). Not only is the protein content high, but the tuberous roots are also rich in carbohydrates. The values for calcium, phosphorus, and iron compare well with those found in most tropical tubers (Table 8). Compared to other tubers, winged bean tubers have less sulfur-containing amino acids. The chemical score calculated according to the FAO provisional amino acid scoring pattern is significantly lower than that of other tubers. Thus, the amino acid composition of winged bean tuber protein cannot be regarded as satisfactory (72).

The leaves of the winged bean are also consumed in some areas where it is commonly grown. They are a good source of protein and rich in vitamin A precursors, other vitamins, and minerals. The crude protein compares well with that of the cassava and many other dark green tropical leaves. Platt (73) gave the average value for protein content in this type of leaf as 5%. According to the National Academy of Sciences, 15% protein could be recovered from the leaves of the winged bean cultivar. The vitamin A value as determined in fresh and relatively young leaves amounted to more than 20,000 IU. Winged bean leaves have fairly high levels of lysine,

TABLE 8 Chemical Composition of Root Foods (in 100 g edible portion)

Constituent	Winged bean	Cassava	Sweet potato	Yam
Moisture (g)	56.5	65.5	70.7	71.8
Calories	150	135	115	108
Fat (g)	0.4	0.2	0.3	0.1
Crude protein (g)	10.9	1.0	1.2	2.0
Carbohydrate (g)	30.5	32.4	27.1	25.1
Fiber (g)	1.6	1.0	0.8	0.5
Ash (g)	1.7	0.9	0.7	1.0
Calcium (mg)	25.0	2.6	36.0	22.0
Phosphorus (mg)	30.0	32.0	56.0	39.0
Iron (mg)	0.5	0.9	0.9	1.0
Manganese (mg)	10	—	387	—
Zinc (mg)	1.3	—	2.0	1.1

Source: Refs. 70, 72.

leucine, aspartic acid, and glutamic acid, whereas cysteine and methionine appear to be the limiting amino acids (74).

The young green pods are mostly consumed as vegetables. They can be eaten raw or cooked. In Thailand, the Philippines, Vietnam, Malaysia, and other countries, the young pods are most often eaten after cooking with chilies. In Bangladesh, the pods are fried and eaten with fish or meat. The composition of green pods compares well with cowpeas or kidney beans (Table 9). The calcium, iron, and vitamin A contents are fairly higher than for other legumes.

TABLE 9 Composition of Immature Pods of the Winged Bean (in 100 g edible portion)

Composition	Winged bean	Cowpea	Kidney bean
Moisture (%)	76–93	88.3	90.6
Fat (g)	0.1–3.4	0.2	0.2
Crude protein (g)	1.9–3.0	3.0	2.1
Carbohydrates (g)	1.1–7.9	7.9	6.4
Fiber (g)	0.9–2.6	1.6	1.3
Ash (g)	0.4–1.9	0.6	0.7
Calcium (mg)	53–236	44	50
Phosphorus (mg)	26.60	45	48
Sodium (mg)	3	6	6
Potassium (mg)	205	233	250
Iron (mg)	0.2–12.0	0.7	0.7
Vitamin A (IU)	340–595	225	110
Vitamin B_1 (mg)	0.06–0.24	0.12	0.07
Vitamin B_2 (mg)	0.08–0.12	0.11	0.08
Niacin (mg)	0.5–1.2	1.0	18
Ascorbic acid (mg)	21–37	22	16

Source: Ref. 72.

Immature winged bean seeds are eaten either raw or cooked. In some places they are consumed in preference to whole pods. The chemical composition of immature seeds depends upon the stage of maturity (75).

D. Processing and Utilization

The entire winged bean plant is consumed because each part of the plant has nutritional and medicinal properties (70). The mature seeds are either cooked or roasted, but never eaten raw. The seeds from immature pods are consumed as vegetables or added to other dishes without any adverse effect. The flowers and leaves of the winged bean are eaten raw or cooked and added to salad, fish, and prawn soup. They can also be used as green fodder and as a crop for animals (70). Like sweet potatoes, the tubers are consumed after cooking in places like Burma and Papua New Guinea (70).

REFERENCES

1. Salunkhe, D. K., N. R. Reddy, and S. S. Kadam, Lima bean in *Handbook of World Food Legumes: Nutritional Chemistry, Processing Technology and Utilization*, Vol. II (D. K. Salunkhe and S. S. Kadam, eds.), CRC Press, Boca Raton, FL, 1989, p. 153.
2. Roy, S. K., and A. K. Chakrabarti, Vegetables of temperate climate, in *Encyclopaedia of Food Science, Food Technology and Nutrition* (R. MaCrae, R. K. Robinson, and M. J. Sadler, eds.), Academic Press, London, 1993, p. 4715.
3. Thompson, H. C., and W. C. Kelly, *Vegetable Crops*, 5th ed., McGraw-Hill Book Co. Inc., New York, 1957.
4. Salunkhe, D. K., and L. H. Pollard, Further studies in microscopic examination of starch grains in relation to maturity of lima beans, *Food Technol. 9*:521 (1955).
5. Salunkhe, D. K., and L. H. Pollard, A simple method to determine maturity and quality in lima beans, *Utah State Agric. Expt. Stn. Farm Home Sci. Bull. 15*:42 (1954).
6. Salunkhe, D. K., L. H. Pollard, E. B. Wilcox, and H. K. Burr, Evaluation of yield and quality in relation to harvest time of lima beans grown for processing in Utah, *Utah State Agric. Expt. Stn. Bull. 407* (1959).
7. Brooks, C., and L. P. McColloch, Stickiness and spotting of green lima beans, *USDA Tech. Bull.* No. 625 (1938).
8. Pantastico, E. B., T. K. Chattopadhyay, and H. Subramanyam, Storage and commercial storage operations, in *Postharvest Physiology, Handling and Utilization of Tropical and Sub-tropical Fruits and Vegetables* (E. B. Pantastico, ed.), AVI Pub. Co., Westport, CT, 1975, p. 314.
9. Ryall, A. L., and W. J. Lipton, *Handling, Transportation and Storage of Fruits and Vegetables*, Vol. 1, *Vegetables and Melons*, AVI Pub. Co., Westport, CT, 1972.
10. Anandaswamy, B., and N. V. R. Iyengar, Prepackaging studies on fresh produce, *Food Sci. 10*:279 (1961).
11. Duke, J. A., *Handbook of Legumes of Economic Importance*, Plenum Press, New York, 1981.
12. Janoria, M. P., V. K. Gaur, and C. B. Singh, *Perspectives in Grain Legumes*, J. N. Krishi Vishwavidyalaya, Jabalpur (M.P.), India, 1984.
13. Yamaguchi, M., *World Vegetables: Principles, Production and Nutritive Value*, AVI Pub. Co. Inc., Westport, CT, 1983, p. 27.
14. Ramanatha, L. A., and B. S. Bhatia, dehydrated green Bengal gram (*Cicer arietinum* L.), *J. Food Sci. Technol. 7*:208 (1970).
15. Yadav, D. S., *Pulse Crops—Production Technology*, Kalyani Publishers, New Delhi, 1991, p. 90.

16. Saxena, M. C., and S. C. Sheldrake, Ann. Rep. 1976–1977, Part II, *Chickpea Physiology*, ICRISAT, Hyderabad, 179, 1977.
17. Kanwar, K. S., A. S. Faroda, D. S. Malik, I. P. S. Ahlawat, K. K. Dhingra, M. R. Rao, R. C. Singh, D. R. Datriya, and R. P. Roy, *Quarter Century of Agronomic Research in India* (1955-80), ISA, New Delhi, 1981.
18. Chavan, J. K., S. S. Kadam, and D. K. Salunkhe, Chickpea, in *Handbook of World Food Legumes*, Vol. I, *Nutritional Chemistry, Processing Technology and Utilization* (D. K. Salunkhe and S. S. Kadam, eds.), CRC Press, Boca Raton, FL, 1989, p. 247.
19. Siddappa, G. S., Canning of dried bengal gram (*Cicer arietinum*), *Indian J. Hort. 16*:170 (1959).
20. Daoud, H. N., B. S. Luh, and M. W. Miller, Effect of blanching, EDTA and $NaHCO_3$ on colour and vitamin B_6 retention in canned garbanzo beans, *J. Food Sci. 42*:375 (1977).
21. Kay, D. E., *Food Legumes, Crop and Product Digest* No. 3, Tropical Products Institute, London, 1979, p. 72.
22. Poats, F. J., Guar a summer row crop for Southwest, *Econ. Bot. 14*:241 (1961).
23. Sastri, B. N., *Cyamopsis*, in *The Wealth of India, Raw Material*, Vol. 2, Council of Scientific and Industrial Research, New Delhi, 1950, p. 427.
24. Singh, H. B., and S. D. Mittal, Cluster bean *Cyamopsis tetragonoloba* (l.) Taub., in *Pulse Crops of India* (P. Kachroo and M. Arif, eds.), Indian Council of Agricultural Research, New Delhi, 1970, 334.
25. Hymowitz, T., The trans-domestication concept as applied to guar, *Econ., Bot. 26*:49 (1972).
26. Bains, D. S., and A. S. Dhillon, Influence of sowing dates and row spacing of patterns on the performance of two varieties of cluster bean, *Punjab Agric. Univ. J. Res. 14*:157 (1977).
27. Singh, H. G., Present status of pulses research in Rajastan, in *Pulse Production* (H. C. Srivastava, S. Bhaskaran, K. K. G. Menon, S. Ramanujam, and M. V. Rao, eds.), Oxford and IBH, Pub. Co., New Delhi, 1984, p. 89.
28. Singh, S. J. P., B. S. Rajput, and K. P. Singh, Effect of various levels of nitrogen, phosphorus and cycocel on yield and yield contributing attributes of cluster bean green pods under rainfed conditions, *Gujrat Agric. Univ. Res. J. 13*:1 (1987).
29. Yawalkar, K. S., *Vegetable Crops of India*, 2nd ed., Agri Horticultural Pub. House, Nagpur, India, 1980.
30. Das, B., S. K. Arora, and Y. P. Luthra, Variability in structural carbohydrates and *in vitro* digestibility of forage and guar (*Cyamopsis tetragonoloba*), *J. Dairy Sci. 58*:1347 (1975).
31. Das, B., and S. K. Arora, Guar seed: Its chemistry and industrial utilization of gum, in *Guar: Its Improvement and Management* (R. S. Paroda and S. K. Arora, eds.), Indian Society of Forage Research, Hissar, 1978, p. 80.
32. Glicksman, M., *Gum Technology in the Food Industry*, Academic Press, New York, 1969, p. 120.
33. Rafique, C. M., and F. Smith, The constitution of guar gum, *J. Am. Chem. Soc. 72*:4632 (1950).
34. Ahmed, Z. F., and R. L. Whitler, The structure of guran, *J. Am. Chem. Soc. 72*:2524 (1950).
35. Whistler, R. L., and C. L. Smart, *Polysaccharide Chemistry*, Academic Press, New York, 1953.
36. Rheese, E. T., and Y. B. Shibata, Mannanases of fungi, *Can. J. Microbiol. 11*:167 (1965).
37. Courtois, J. E., and P. Le Dizet, Action of some enzyme preparations on galactomannans of clover and gleditsia, *Bull. Soc. Chim. Biol. 45*:731 (1963).
38. Sehgal, K., H. S. Nainawatee, and B. M. Lal, Galactomannan degrading enzyme from germinating *Cyamopsis tetragonoloba* seed, *Biochem. Physiol. Pflanz. 164*:423 (1973).
39. Lee, S. R., Purification and properties of enzymes which attack guar gum, *Diss. Abstr. 27*:2626 (1967).
40. Gunjal, B. B., and S. S. Kadam, Cluster bean, in *Handbook of World Food Legumes*, Vol. 2, *Nutritional Chemistry, Processing Technology and Utilization* (D. K. Salunkhe and S. S. Kadam, eds.), CRC Press, Boca Raton, FL, 1989, p. 301.
41. Coxon, D. T., and J. W. Wells, 3-epi-ketononic acid from guar meal (*Cyamopsis tetragonoloba*), *Phytochemistry 19*:881 (1980).
42. Bliss, F. A., Cowpea in Nigeria, in *Proc. Symp. Nutritional Improvement of Food Legumes by Breeding* (M. Milner, ed.), PAG/FAO, Rome, 1972, p. 151.
43. Stanton, W. R., *Grain Legumes in Africa*, Food and Agriculture Organization, Rome, 1966, p. 20.

44. Purseglove, J. W., *Tropical Crops: Dicotyledons*, I. Longmans, London, 1968, p. 321.
45. Mehta, Y. R., *Vegetable Growing in Uttar Pradesh, Bureau of Agric. Inf.*, UP, Lucknow, 1959.
46. Jayram, R., and S. Ramiah, Response of cowpea to phosphorus and growth regulators, *Madras Agric. J. 5*:102 (1980).
47. Mohan Kumar, B., P. Balakrishnan Pillai, and P. V. Prabhakaran, Effect of levels of nitrogen, phosphorus and uptake nutrients and grain yield in cowpea, *Agric. Res. J. Kerala 17*:289 (1979).
48. Chavan, J. K., S. S. Kadam, and D. K. Salunkhe, Cowpea, in *Handbook of World Food Legumes, Volume II, Nutritional Chemistry, Processing Technology and Utilization* (D. K. Salunkhe and S. S. Kadam, eds.), CRC Press, Boca Raton, FL, 1989, p. 1.
49. Kale, P. N., and J. Kale. *Bhajipala Lagwad* (Marathi), Continental Press, Pune, India, 1984.
50. Smittle, D. A., and M. J. Hayes, Influence of short term storage conditions on quality of shelled Southern pea, *J. Am. Soc. Hort. Sci. 104*(6):783 (1979).
51. Gohl, B., Dolichos lablab (*Lablab vulgaris* Medic L. niger Medic), in *Tropical Foods: Feeds Information Summeries and Nutritive Value*, Food and Agriculture Organization, Rome, 1975, p. 186.
52. Sastri B. N., ed., *Dolichos lablab* Linn. *The Wealth of India: Raw Materials*, Council of Scientific and Industrial Research, New Delhi, 1952, p. 236.
53. Westphal, E., The proposed retypification of Dolichos: A review, *Taxonomy 24*:189 (1975).
54. Kay, D. E., Hyacinth bean, in *Food Legumes Crop Products*, Digest No. 3, Tropical Products Institute, London, 1979.
55. Salgarkar, S., and K. Sohonie, Hemagglutinins of field bean *Dolichos lablab* isolation, purification and properties of hemagglutinins, *Indian J. Biochem. Biophys. 2*:193 (1965).
56. Salgarkar, S., and K. Sohonie, Effect of feeding field bean hemagglutinin A on rat growth, *J. Food Sci. Technol. 3*:79 (1965).
57. Kadam, S. S., N. R. Reddy, and G. D. Patil, Other legumes in *Handbook of World Food Legumes*, Vol. III, *Nutritional Chemistry, Processing Technology and Utilization*, (D. K. Salunkhe and S. S. Kadam, eds.), CRC Press, Boca Raton, FL, 1989, p. 67.
58. Morten, J. F., The pigeonpea (*Cajanus cajan* L. Mill sp.) a high-protein tropical bush legume, *HortScience 11*:11 (1976).
59. Jambunathan, R., and U. Singh, Grain quality of pigeonpea, in *Proc. Symp. Intl. Workshop on Pigeonpea*, Vol. I, December 15-19, 1980, ICRISAT, Patancheru, India, 1981, 351.
60. Singh, S., H. D. Singh, and K. C. Sikka, Distribution of nutrients in anatomical parts of common Indian pulses, *Cereal Chem. 45*:13 (1958).
61. Mainsfield, R. G., Processing and marketing of pigeonpea (*Cajanus cajan*): The case of Dominican Republic in Proc. Intl. Workshop on Pigeonpea, December 15-19, 1980, ICRISAT, Patancheru, India, 1981, p. 344.
62. Sanchez-Nieva, F., The influence of degree of maturity on the quality of canned pigeonpea, *J. Agric. Univ. P.R. 45*:217 (1961).
63. Barlow, J. W., Color stabilization of peas, U.S. Patent 3,583873 (1971).
64. Sanciez-Nieva, F., M. A. Gonzalez, and J. R. Benero, The effect of some processing variables on quality of canned pigeonpea, *J. Agric. Univ. P.R. 45*:217 (1961).
65. Sanhez-Nieva, R., Processing characteristics of pigeonpea of kaki and Saragateado selections, *J. Agric. Univ. P.R. 46*:23 (1962).
66. Sanchez-Nieva, F., Application of the shear press to determine the degree of maturity of pigeonpea, *J. Agric. Univ. P.R. 47*:212 (1964).
67. NAS, *The Winged bean—a High Protein Crop for the Tropics*, 2nd ed., National Academy of Sciences, 1978.
68. Masfield, G. B., *Psophoscarpus tetragenolobus* a crop with future, *Field Crop Abstr. 26*:157 (1973).
69. Pospisil, F., S. K. Karikari, and K. Boamah-Mensah, Investigation on winged bean in Ghana, *World Crops 23*:260 (1971).
70. Kadam, S. S., and D. K. Salunkhe, Winged bean in human nutrition, *CRC Crit. Rev. Food Sci. Nutr. 21*:1 (1984).

71. Parthasarathy, V. A., Winged bean, in *Vegetable Crops in India* (T. K. Bose and M. G. Som, eds.), Naya Prokash, Calcutta, 1986, p. 535.
72. Cerny, K., Comparative nutritional and clinical aspects of the winged bean, in *Proc. Intl. Symp. on Developing Potential of the Winged Bean*, Manila, 1978, p. 281.
73. Platt, B. S., *Tables of Representative Value of Foods Commonly Used in Tropical Countries*, 2nd ed., London, 1962.
74. Okezie, B. O., and F. W. Martin, Chemical composition of dry seeds and fresh leaves of winged bean varieties grown in the U.S. and Puerto Rico, *J. Food Sci. 45*:1045 (1980).
75. Kadam, S. S., L. S. Kute, K. M. Lawande, and D. K. Salunkhe, Changes in chemical composition of winged bean, (*Psophocarpus tetragonolobus*) during seed development, *J. Food Sci. 47*:2051 (1982).

22

Lettuce

S. S. Deshpande
IDEXX Laboratories, Inc., Sunnyvale, California

D. K. Salunkhe
Utah State University, Logan, Utah

I. INTRODUCTION

Lettuce is most popular of the salad crops and is of great commercial importance; its commercial value is exceeded only by potatoes in the United States, where it is found in the market throughout the year. Lettuce occupies the largest area of the salad crops worldwide (1). It is a pleasure food with low nutrient density. It has a crisp texture, a large surface to volume ratio, and serves as a source of bulk for diet-conscious consumers.

II. BOTANY

According to Ryder and Whitaker (2), all four species of *Lactuca* (*L. sativa*, *L. serriola*, *L. viross*, and *L. saligna*) originated in the Mediterranean region. This is a group with 9 pairs of chromosomes, which cross with varying degrees of success with each other. Lettuce (*L. sativa*) is botanically related to wild lettuce (*L. serriola*), a common weed. Lettuce is an annual belonging to the family Compositae. According to Thompson and Kelly (3), there are four distinct types of subspecies or botanical varieties: (a) Head, var *capitata* (b) Cutting or leaf, var. *crispa*, (c) Cos or Romaine, var. *Longifolia*, and (d) asparagus or stem lettuce, var. *asparagina*. There are two classes of head lettuce known as crisphead and butterhead.

A. Morphology

Lettuce produces a taproot that grows in the surface layer of the soil. A large number of leaves are produced; the outer leaves are usually green to dark green and the inner leaves are light green to whitish and very tender and crisp (4). The inflorescence is a head or capitulum consisting of a large number of flowers. Numerous flowers form closely together. The floret has a single ligulate yellow petal, five stamens, and two carpels. There is a high degree of self-pollination.

Both pollen and ovule are usually highly fertile. Anthesis starts at sunrise and flowers open fully by late morning. The fruit is a typical achene and the seed exalbuminous. Lipton and Ryder (5) grouped modern lettuce cultivars into six morphological types: crisphead, butterhead, cos or romaine, leaf, Latin, and stem. The varieties under each category differ greatly in texture, color, and other characteristics.

1. Crisphead

Crisphead is the most important commercial lettuce type grown in the United States. It has a firm head formed of tightly overlapping leaves and crisp or brittle foliage. Great Lakes, Imperial New York, and Pennlake are important crisphead cultivars (Table 1).

2. Butterhead

Butterhead lettuce is mainly grown in northern Europe. Compared to crisphead, it has small soft heads with loosely overlapping leaves. It has a tendency to bolt faster than crisphead lettuce in warm weather. Butterhead varieties have a cabbage heading habit with comparatively smooth, finely veined leaves with a soft texture and buttery flavor. The butterhead varieties are of relatively less commercial importance because they do not ship and handle as easily as crisphead varieties (5).

3. Cos

This type of lettuce is divided into two types, open and self-closing, the latter of which is found in the United States. Cos leaves are longer than they are broad and overlap only at the sides. They are dark green on the outside and creamy yellow inside the head. They have a coarse but crisp and tender texture. Cos lettuce is usually grown early in the spring or during winter, as it goes to seed quickly in hot weather (6).

4. Leaf

Loose leaf lettuce varieties have a spreading leaf habit, clustered or bunched leaves with little or no overlapping of inner leaves. They can be grown outdoors as an early spring crop (7). Blackseeded Simpson has tender leaves, tolerates hot weather, and is suitable for home gardens in all seasons.

5. Others

Two other lettuce types are rarely grown in the United States. Grasse or Latin lettuce is grown in Western Europe and in South America. The leaves are shorter than Cos lettuce leaves but less

TABLE 1 Important Lettuce Cultivars

Type	Cultivars
Crisphead	New York, Imperial, Imperial 44, Imperial 459, New York 515, New York 12, Great Lakes, Pennlake, Calmar, Montemar, Merit, Climax, Salinas, E-I-Toro, Sea Grean, Amaral 400, Winterhaven, Vanguard 75, Red Coach 74, Domingos 43, Moranguard
Butterhead	White Boston, Big Boston, May Kings, Salamander Wayahead, Buttercrunch, Bibb
Cos or romaine	Parris Island, Dark Green, Valmaina
Leaf	Grand Rapids, Blackseeded Simpson, Early Curled Simpson, Slobalt, Prizehead, Deep Red, Some Ruby, Salad Bowl, Waldmann's Green

tender than butterhead leaves and form rudimentary or no heads. Stem lettuce is found in China, where the stems are used as a cooked vegetable, and in Egypt, where the stems are eaten raw in the same manner as celery stalks.

B. Cultivars

Thompson and Kelly (3) divided the important commercial varieties of lettuce into five general classes: crisphead, butterhead, cos or romaine, leaf or bunching, and stem. The varieties differ greatly in texture, color, and other characteristics. The important cultivars are listed in Table 1. The Indian Agricultural Research Institute, New Delhi, has recommended Great Lakes, Chinese Yellow, Imperial 859, and Slow Bolt as improved varieties of lettuce (9) for cultivation in India.

III. PRODUCTION

A. Soil and Climate

A well-drained fertile soil with a pH of about 6.0 is most desirable for the growth of lettuce. It is fairly salt tolerant. Lettuce is a cool season crop requiring a mean air temperature of 10–20°C. Cool nights are essential for good-quality lettuce (10). Lettuce is grown mainly in areas having cool summers and mild winters. In the United States, it is mainly confined to California, Arizona, New York, and New Jersey.

B. Cultural Practices

1. Planting

Lettuce is propagated by direct seeding 0.5–1 cm deep on 1-m beds (two rows per bed, 36 cm between rows). Seed germination is favored at 7–24°C. Plants should be thinned to 30–40 cm between plants. Lettuce can also be raised by transplanting seedlings from nursery beds. Seedlings should be hardened before transplanting. Hardening is done by withholding water for about 6–8 days. Seedlings are transplanted in flat beds at a spacing of 45×45 cm or 45×30 cm.

2. Manuring and Fertilization

The growth and yield of lettuce is influenced by application of fertilizers (11–17). For growing lettuce, 10–45 t of farmyard manure supplemented with 25 kg of both N and K and 90 kg of P per hectare has been recommended (4). On sandy and sandy loam soils without manure, 40–50 kg of N and 75–100 kg each of P and K per hectare should be applied. For good growth on silt loam and clay loam soils, 25 kg each of N and K and 50–75 kg of P per hectare may be used (6). However, fertilization would depend upon the nutrient availability and the status of soil fertility with respect to essential nutrients. About one third to one half of the nitrogen should be applied at the time of seeding and the rest after thinning operation.

3. Irrigation

There are various systems of irrigation, which include furrow, surface, trickle, and sprinkler. Frequent and light irrigations have been found to be more effective in achieving high yield and quality production (18). After planting, crops should be irrigated at 8- to 10-day intervals. Trickle irrigation increased yield about 30% over furrow irrigation (4).

C. Disorders, Diseases, and Pests and Their Control

1. Disorders

Tipburn

Tipburn is a necrosis of portions of the inner leaf margins of lettuce. Calcium deficiency associated with soil and environmental interaction is one of the causes of tipburn (19–26). The chelation of calcium by organic acid under high temperature may lead to a local deficiency of calcium (27–29). This disorder can be decreased by increasing the nitrogen supply (Table 2).

Russet Spotting

This disorder, caused by ethylene, is characterized by the occurrence of reddish, tan, olive, and/or brown elongated pitlike spots on the midribs of leaves. An ehtylene concentration as low as 0.5 ppm may lead to spotting in susceptible lettuce. This also occurs when lettuce is held between 3 and 10°C or during prolonged storage. This disorder can be minimized or even prevented by keeping lettuce at 0–2.5°C or away from ethylene.

Rib Discoloration

This disorder is induced when lettuce matures during hot weather. It is characterized by yellow to black lesions mostly discoloring the cap leaf and the next few inner leaves. Good refrigeration practices will reduce decay development in the injured tissue.

Pink Rib

This is characterized by a discoloration of the large midribs to a light or dark pink. In severe cases, small veins are also discolored.

Other Disorders

Carbon dioxide injury is seen when the CO_2 concentration reaches 1–2%. A brown stain may develop within a week at temperatures usually encountered during marketing of crisphead

TABLE 2 Incidence of Dry Tipburn Harvested at Three Planting in 1990 as Affected by Cultivar and Total Nitrogen Supply

	Tipburn incidence (%)		
	Early	Midseason	Late
Cultivar			
Marius	32a	15a	48a
Saladin	24a	5b	63b
Nitrogen (kg/ha)			
50	30a	12a	72a
100	29a	12a	60ab
150	30a	7a	50bc
200	22a	8a	41c

Mean separation among cultivars, nitrogen supply, and plant age within columns by LSD at $p = 0.05$.
Source: Ref. 39.

lettuce. This type of injury is certain when the CO_2 level is more than 4%. A deficiency of O_2 (below 0.5%) leads to development of a reddish-brown discoloration of the heart leaves. Freezing injury is characterized by blistering and subsequent drying and darkening of the epidermis of outer leaves when lettuce freezes in the field or during storage. Since lettuce freezes at -6 ± 0.2°C (18), care must be taken that lettuce is never exposed to low temperatures long enough to freeze the leaves.

2. Diseases

Damping-off and root rot caused by *Rhizoctonia solani* Kuehn and *Phythium ultimum* Trow, rot caused by *Phythium polymastum*, downy mildew caused by *Bremia lactucae* Reg., lettuce drop caused by *Sclerotinia sclerotiorum* (Lib.) de Bary and *S. minor* Jagg., botrytis rot caused by *Botrytis cinerea* Persex Fr., and lettuce mosaic and big vein caused by viruses are some of the important diseases of lettuce. Lettuce leaves are susceptible to infection by *Botrytis cinerea*, *Bremia lectucae*, *Sclerotinia* spp., *Stemphylum botryosum*, and *Alternaria tenuis* (30). Ryall and Lipton (6) described bacterial soft rot, big vein (*Olpidum brassicae*), downy mildew (*Bromia lactucae*), gray mold rot (*B. conerea*), and watery soft rot (*S. sclerotiorum* or *S. minor*). Such infections occur in the form of discoloration or water-soaked spots, turning the leaves into a soft, slimy wet mass (31). Mold growth or sclerotial bodies are found on the surface of infected tissues.

3. Pests

The most important pests of lettuce include (1) cabbage looper (*Trichoplusia ni* Hubber), (2), corn earworm (*Heliothis zea* Boddie), (3), beet army worm (*Spodoptera exigue* Hubner) (4), and tobacco bud worm (*Heliothis virescens* I). Aphids and white flies are important as sucking insects and as transmitters of disease organisms. Aphid types include the following: green peach aphid (*Myzus persicae* Sulz), lettuce seed stem aphid (*Macrosiphum barri* Essig), lettuce aphid (*Nasonovia ribis-negri* Mosley), and potato aphid (*Macrosiphum euphorbiae* Thomas). The lettuce root aphid (*Pemphigus bursarius* L.) feeds on lettuce roots and can be quite destructive, causing severe wilting, retarded growth, and even death. The white flies important on lettuce are the green house whitefly (*Trialeurodes vaporariorium* Westov) and the sweet potato whitefly (*Bemisia tabaci* Genn.). Both are important as transmitters of yellowing virus disease. The principal means of control of insect populations is the application of insecticides of various types and integrated chemical control with various forms of biological control, such as parasitism, predation, and genetic resistance.

D. Maturity and Harvesting

Harvesting of lettuce begins as soon as plants reach acceptable size and firmness (7) and should be completed before the leaves become tough and bitter and before seed stalks start to bold (32). The stage of harvest maturity depends on the variety of lettuce and the purpose for which it is grown (6). Head lettuce for market is allowed to grow to full size and to develop a solid head, but for home use it is often harvested before the head is well formed. Leaf lettuce plants may be thinned at various times, removing the largest leaves for use and leaving smaller ones to develop.

Lettuce is usually cut with a long-handled sharp knife. Since the development of vacuum cooling, marked changes have taken place in California and Arizona in the harvesting and handling methods for lettuce. Shed packing of two to six dozen heads of lettuce in wooden crates with ice has yielded to dry packing of two dozen heads in paperboard cartons (6) or fiberboard containers.

Sevila and Gratnaud (33) described a self-propelled lettuce harvester with metal fingers in front, inclined at an angle of 35°. As the machine moves forward, the lettuces are uprooted together with the plastic mulch through which they are generally grown. A bandsaw cuts the roots and part of the stump without damaging the conveying plastic film. The lettuces are then manually trimmed and packed.

Polyethylene-lined packaging reduces lettuce leaf rot by 30–40%, compared with open polyethylene bags (34). Tumbling lettuce into bins is the most efficient system of handling mechanically harvested lettuce (35). However, lettuce in bins must be settled either by vibration or by vacuum-cooling. It was further reported (35) that lettuce is packed into cartons more efficiently by the harvester than at a central location. About 320 dry-pack paperboard cartons are mechanically placed in a large steel precooling tube at one time during the vacuum-cooling (27). With this method, the center of the lettuce head is cooled to about 10°C in less than 30 minutes. The cartons are then transferred immediately to refrigeration cars.

E. Grading

According to U.S. standards, there are several grades of crisp lettuce, which include U.S. Fancy, U.S. No. 1, U.S. commercial, and U.S. No. 2 (36), based on degree and timeliness of precooling, tolerance to defects and diseases, maturity, and presence of seed stalks and insects. The higher the grade, the lower the allowed tolerance of various defects and diseases (19).

IV. CHEMICAL COMPOSITION

Lettuce is a rich source of vitamin A and minerals such as calcium and iron (Table 3). The chemical composition of lettuce varies according to the type of lettuce, variety, cultural practices, and environmental conditions under which it is grown (38–47). Application of nitrogen signifi-

TABLE 3 Nutritional Composition of Lettuce Leaves

Constituent	Butterhead	Cos (romaine)	Crisphead (Great Lakes)
Energy (kcal)	11	16	11
Water (g)	96	94	95
Protein (g)	1.2	1.6	0.8
Fat (g)	0.2	0.2	0.1
Carbohydrates (g)	1.2	2.1	2.3
Vitamin A (IU)	1200	2600	300
Thiamine (mg)	0.07	0.10	0.03
Niacin (mg)	0.4	0.5	0.3
Vitamin C (mg)	9	24	5
Minerals			
Calcium (mg)	40	36	13
Iron (mg)	1.1	1.1	1.5
Magnesium (mg)	16	6	7
Phosphorus (mg)	31	45	25

Source: Ref. 47.

TABLE 4 Effect of Total Nitrogen Supply on Nutritive Element Content of Crisphead Lettuce (fresh weight basis)[a]

Constituent	Total nitrogen supply (kg N/ha)				LSD 0.05
	50	100	150	200	
Dry matter (g/100 g)	4.49	4.22	4.09	4.05	0.11
Vitamin C (mg/kg)	68.4	55.5	51.9	51.5	5.2
Nitrate (mg/kg)	330	562	719	774	63
Fructose (g/100 g)	1.44	1.38	1.33	1.31	0.07
Glucose (g/100 g)	1.16	1.10	1.08	1.05	0.04
Phosphorus (mg/kg)	213	190	183	179	NS
Calcium (mg/kg)	154	148	143	194	NS
Iron (mg/kg)	3.0	2.66	2.69	2.69	NS
Magnesium (mg/kg)	54.4	56.4	59.0	60.8	2.0
Zinc (mg/kg)	1.40	1.32	1.39	1.43	NS

[a]Mean of cultivars, plantings, and plant age at harvest.
NS = Not significant.
Source: Ref. 39.

cantly influences nitrate content in lettuce leaves (Table 4). Lettuce also contains proteins, carbohydrates, and vitamins. Vitamin C is essentially lost if lettuce is cooked. Lettuce is also known to act as a sedative, diuretic, and expectorant.

V. STORAGE

A. Postharvest Losses

Mechanical injury is one of the major causes of loss of quality and waste in unwrapped lettuce (48). Much of this type of damage is inflicted during harvest, packaging, and closing of boxes if too much lettuce is packed into the box (48). Packaging of romaine lettuce in unsealed, polyethylene-lined (0.04-mm-thick) packaging caused a marked decrease in the rate of decay during storage in spite of high relative humidity (48). The beneficial effect of packaging was attributed to the high CO_2 and low O_2 levels in the package as a result of lettuce respiration. The method is found effective in preventing rots caused by *S. sclerotiorum*, *Stemphylium botryosum*, and *Bremia lettucae* (48). When lettuce reaches the market, wrapper leaves are discarded. Removal of the outer leaves before shipping the lettuce to prevent decay is particularly advantageous during long transit storage as in shipment overseas.

Removing the wrapper leaves at harvest and wrapping the heads with plastic film reduces the amount of injury sustained by lettuce during packing. If the lettuce is wrapped in plastic film, the material must be vented sufficiently to permit air and moisture to escape during vacuum-cooling.

B. Precooling

Lettuce should be precooled to −1°C soon after harvest and stored at 0°C and 98–100% relative humidity (29). The rate of general deterioration at 15°C is about 2.5 times greater than that at 0°C (47). Precooling is commonly accomplished by vacuum-cooling. Since most head lettuce is field-packed in corrugated cartons, vacuum-cooling is more suitable (49). For this purpose, film wraps are perforated or readily permeable to water vapor. To aid vacuum-cooling, clean water is sprinkled on lettuce heads prior to carton closure if they are dry and warmer than 25°C. The lettuce should be thoroughly precooled because mechanically refrigerated rail cars or trucks cannot cool warm lettuce during transit (48).

C. Low-Temperature Storage

Lettuce is highly perishable and deteriorates rapidly with increasing temperature. The respiration rate increases greatly and storage life decreases concomitantly as storage temperature increases over the temperature range from 0 to 25°C (48). Leaf lettuce respires at about twice the rate of head lettuce (50). Lettuce is generally not stored for long periods, but it can be stored for 3–4 weeks at about 0°C if it arrives at the storage facility in good condition (51). Ryall and Lipton (6) recommended that lettuce be precooled to 1 ± 0.1°C in the leafy main portion of the head before it is loaded onto refrigeration cars equipped for maintaining the desired temperature (0°C) and with a relative humidity of $\geq 95\%$ (52). Lettuce is easily damaged by freezing, so all parts of the storage room must be kept above the highest freezing point of lettuce (–0.2°C).

D. Controlled-Atmosphere Storage

Controlled atmosphere (CA) is of limited benefit in storing lettuce (53,54). About 3–5% O_2 and 1.5% CO_2 have been recommended for CA storage (6). This atmosphere prevents russet spotting and butt discoloration, although the latter effect persists as long as the product is under controlled atmosphere. Lettuce is sensitive to high CO_2 and low O_2. Oxygen below 1% and CO_2 above 2.5% are injurious to lettuce (55). The symptoms of high CO_2 injury are brown stain, death of leaflets at the growing point, and yellowish to reddish-brown discoloration of midribs and adjacent tissue (55). Ryall and Lipton (6), however, stated that neither low O_2 nor high CO_2 inhibits pink rib, and that low O_2 may even increase its severity at high temperatues. Lettuce is extremely sensitive to injury from ethylene and russet spotting (57); low O_2 and high CO_2 reduce the adverse effects of ethylene, but the former is not completely effective and the latter can injure the lettuce.

Singh et al. (57) studied the effects of CA on the composition and quality of lettuce. At 1.6°C, lettuce heads were field fresh and light green with no apparent sign of quality deterioration in any of the three combinations of gases up to 3 weeks. When heads were examined after 40 days of CA storage, 2.5% O_2 and 2.5% CO_2 proved to be the best combination for prolonging the shelf life of lettuce (Table 5). The effects of CA (2.5% O_2 plus 2.5% CO_2) in combination with Captan® (*N*-trichloromethylthio-4-cyclohexanil, 2-dicarbonimide, 100 ppm), Phaltan® (N-trichloromethylthio-phthalimide, 100 ppm), Mycostatin® (3-emino-3,6-dideoxyl-*D*-aldohexose, 400 ppm), and N^6-benzyladenine (200 ppm) are shown in Table 6. After 40 days of storage, the lettuce heads treated with Phaltan and Phaltan + N^6-benzyladenine were of higher quality, and, in general, the chemicals had detrimental effects on the quality of lettuce under CA at 1.6°C. The effects of CA on physiological changes such as release of CO_2, uptake of O_2 and quality rating are given in Table 7. Compared to the control sample (normal air and 1.6°C), the CA-stored lettuce had a significantly slower rate of respiration on the 15th, 30th, 45th, and 60th days of storage. The lettuce headds treated with Phaltan alone or in combination with polyethylene packing before

Table 5 Effects of Combinations of CO_2 and O_2 Storage on Lettuce Quality

Days[a]	Conc. of gases % O_2	% CO_2	Rating	Descriptive rating	Visual observation
20	21	1	9	Excellent	Generally lettuce stored in these atmospheres had a fresh, green appearance and no defects except lettuce held in 1:1 O_2 to CO_2 mixture, which had slight yellowing on the tip of outer leaves
	5	5	8[b]	Excellent	
	5	0	9 (ns)	Excellent	
	2.5	2.5	9 (ns)	Excellent	
	1	1	8[b]	Excellent	
40	21	1	7	Good	Minor defects only, red butt
	5	5	5[c]	Fair	Outer leaves having brown spotting, red butt
	5	0	7 (ns)	Good	Minor defects
	2.5	2.5	8[b]	Excellent	Field fresh, bright green
	1	1	5[c]	Fair	Red heart leaves, reddish midrib, soft in texture

ns, Not significant at 0.05.
[a]After storage at 0°C and 95% relative humidity.
[b]Significant at 0.05.
[c]Significant at 0.01.
Source: Ref. 57.

Table 6 Effects of Chemical Treatments on Storage Life of Lettuce

Treatment: Chemical	Concentration (ppm)	Rating	Descriptive rating	Visual observation
Control		8.5	Good	Minor defects, red butt
Captan	1000	5.0[a]	Fair	Slight yellowing on outer leaves
Phaltan	1000	7.0[b]	Good	Minor defects
Mycostatin	400	6.5[a]	Fair	Little red spotting and midrib
N^6-BA[c]	20	5.0[a]	Fair	Slight yellowing, red midrib on outer leaves
Captan + N^6-BA	1000 20	5.0[a]	Fair	Slight red rib and yellowing
Phaltan + N^6-BA	1000 20	7.0[a]	Good	Minor defects
Mycostatin + N^6-BA	400 20	5.5[a]	Fair	Slight yellowing, red rib on the outer leaves

Lettuce was stored in 2.5% O_2 and 2.5% CO_2 at 1.6°C and 95% relative humidity for 40 days.
[a]Significant at 0.01.
[b]Significant at 0.05.
[c]N^6-benzyladenine.
Source: Ref. 57.

TABLE 7 Physiological Changes in Lettuce During CA Storage with Chemical Treatments

Days of storage	Treatment	Dry wt.[a] (%)	CO_2 release[b] (mg/hr/kg)	O_2 uptake[b] (mg/hr/kg)	Quality rating[c]
0	Control	4.9	15.1	176.1	9
15	Control	5.0	25.5	170.0	9
	CA	4.7	21.5[d]	152.4[d]	9 (ns)
	CA + Ph[e]	4.7	29.3	238.1	9 (ns)
	CA + Ph + Pkg[f]	4.3	28.0[d]	260.7[d]	9 (ns)
	CA + Pkg	4.6	21.3[d]	135.3[d]	9 (ns)
30	Control	4.9	19.8	135.5	9
	CA	4.7	15.0[d]	104.4[d]	9 (ns)
	CA + Ph	4.7	23.8[d]	202.6[d]	7[d]
	CA + Ph + Pkg	4.5	22.6[d]	186.5[d]	7[d]
	CA + Pkg	4.5	17.0	118.7[d]	9 (ns)
45	Control	4.9	34.96	171.4	8
	CA	4.7	30.19[d]	152.0[d]	9[d]
	CA + Ph	4.7	45.45[d]	275.1[d]	5[g]
	CA + Ph + Pkg	4.7	40.20[d]	243.9[d]	5[g]
	CA + Pkg	4.8	32.16 (ns)	167.4 (ns)	7[d]
60	Control	4.4	31.69	210.7	7
	CA	3.7	26.95[d]	183.0[d]	8[d]
	CA + Pkg	3.3	27.25[d]	188.2	5[g]
75	Control	4.2	10.65	188.2	5
	CA	3.4	10.17 (ns)	175.3 (ns)	7[g]

Lettuce in controlled-atmosphere storage (2.5% O_2, 2.5% CO_2) at 1.6°C.
ns, Not significant at 0.5.
[a]Results expressed on fresh wet basis.
[b]Results expressed on dry weight basis.
[c]8,9 = excellent; 6,7 = good; 4,5 = fair; 1,2,3 = unsalable.
[d]Significant at 0.05.
[e]Ph, Phaltan.
[f]Pkg, Polyehylene packaging.
[g]Significant at 0.01.
Source: Ref. 57.

storage in CA had a higher rate of respiration and poorer quality than the CA- or conventionally stored lettuce (Table 7).

Singh et al. (58) reported the effects of CA (2.5% CO_2 plus 2.5% O_2) and CA in combination with Phaltan and polyethylene packaging on pH, titratable acidity, and organic acids (Table 8) reducing sugars, total sugars, and total starch (Table 9), amino acids, soluble proteins, and total proteins (Table 10) and chlorophyll and total carotene contents of lettuce (Table 11). Compared to conventional refrigeration storage, the total organic acids in CA and CA in combination with packaging were lower on the 15th, 30th, and 45th days. After 60 days, however, lettuce in CA with packaging showed 5% higher and lettuce in CA alone showed 6% higher total organic acid content than the conventional low-temperature storage. Starch and total sugars of all the treatments decreased during storage. CA and CA in combination with packaging showed a higher retention of starch and total sugars throughout the storage period than the conventional refrigeration. Total protein (total N × 6.25) did not change significantly with CA in

TABLE 8 Effects of CA, Phaltan, and Polyethylene Packaging on pH, Titratable Acidity, and Organic Acids of Lettuce

Days of storage	Treatment	pH	Titratable acidity (mEq/g)	Organic acids (mg/g)
0	CR	6.41	0.1678	30.61
15	CR	6.35	0.1695	33.67
	CA	6.42 (ns)	0.1610[b]	31.69 (ns)
	CPh	6.36 (ns)	0.1680 (ns)	27.69[b]
	CPP	6.43 (ns)	0.1513[b]	25.00[b]
	CPK	6.42 (ns)	0.1581[b]	32.54 (ns)
30	CR	6.40	0.1712	38.88
	CA	6.41 (ns)	0.1617[b]	34.73 (ns)
	CPh	6.37 (ns)	0.1832[b]	39.74 (ns)
	CPK	6.42 (ns)	0.1746[c]	37.91 (ns)
45	CR	6.47	0.1723	35.49
	CA	0.49 (ns)	0.1699 (ns)	34.26 (ns)
	CPh	6.30 (ns)	0.2015[b]	39.14[c]
	CPP	6.15[d]	0.2033[b]	41.21[c]
	CPK	6.44 (ns)	0.1775[b]	36.20 (ns)
60	CR	6.45	0.1725	34.78
	CA	6.42 (ns)	0.1776[b]	35.88 (ns)
	CPK	6.47 (ns)	0.1802[b]	36.93 (ns)
75	CR	6.40	0.1764	36.32
	CA	6.41 (ns)	0.1979[c]	38.50 (ns)

ns, Not significant at 0.05; CR, conventional refrigeration; CA, controlled atmosphere (2.5% O_2 and 2.5% CO_2); CPh, CA plus Phaltan (1000 ppm); CPP, CA plus Phaltan (1000 ppm) plus packaging in polyethylene bags; CPK, CA plus packaging in polyethylene bags.
[a]Results expressed on dry weight basis.
[b]Significant at 0.01.
[c]Significant at 0.05.
Source: Ref. 58.

any of the treatments. However, the soluble proteins of lettuce were significantly lower with CA and CA combined with packaging throughout the storage period. The lettuce treated with Phaltan had a higher content of soluble proteins by the 45th day of storage, when it started to decay. Compared to the controls, chlorophyll retention was significantly higher in lettuce under CA and CA combined with packaging (Table 11), but it was lower in lettuce treated with Phaltan. Although none of the treatments significantly affected the concentration of total carotene, the carotene content was 10% higher in the CA and CA combined with packaging than with conventinal refrigeration on the 60th and 75th days of storage (Table 11). This was ascribed to the probable lower rate of carotene destruction by lipoxidase in the lettuce stored under these conditions.

Lettuce should be held at high relative humidity—98–100%. Film liners or individual polyethylene head wraps are desirable for attaining high relative humidity. However, they should be perforated or permeable to maintain a noninjurious atmosphere and to avoid 100% relative humidity on removal from storage.

TABLE 9 Effects of CA, Phaltan, and Polyethylene Packaging on Sugars and Starch Contents of Lettuce[a]

Days of storage	Treatment	Reducing sugars (mg/g)	Total sugars (mg/g)	Total starch (mg/g)
0	CR	49.36	132.70	5.38
15	CR	51.90	100.72	4.81
	CA	48.41 (ns)	103.24 (ns)	5.09 (ns)
	CPh	39.67[b]	83.08[c]	4.04 (ns)
	CPP	39.55[b]	90.04 (ns)	4.55 (ns)
	CPK	46.44 (ns)	102.61 (ns)	5.25 (ns)
30	CR	57.15	103.34	4.05
	CA	46.29 (ns)	96.18 (ns)	4.07 (ns)
	CPh	42.63[c]	35.88[c]	4.08 (ns)
	CPP	44.42[c]	99.14 (ns)	4.15 (ns)
	CPK	52.60 (ns)	109.79 (ns)	4.40 (ns)
45	CR	53.95	69.16	2.67
	CA	46.48 (ns)	68.73 (ns)	2.61 (ns)
	CPh	38.99[b]	68.81 (ns)	2.67 (ns)
	CPP	40.96[b]	65.77 (ns)	2.95 (ns)
	CPK	50.24 (ns)	79.56	2.72 (ns)
60	CR	55.67	85.02	2.81
	CA	48.99[b]	87.43 (ns)	2.90 (ns)
	CPK	46.80[b]	86.40 (ns)	3.33c
75	CR	50.66	77.99	2.64
	CA	45.56[c]	83.47 (ns)	2.64 (ns)

ns, Not significant at 0.05; CR, conventional refrigeration; CA, controlled atmosphere (2.5% O_2 and 2.5% CO_2); CPh, CA plus Phaltan (1000 ppm); CPP, CA plus Phaltan (100 ppm) plus packaging in polyethylene; CPK, CA plus packaging in polyethylene bags.
[a]Results expressed on dry weight basis.
[b]Significant at 0.01.
[c]Significant at 0.05.
Source: Ref. 58.

E. Chemical Control of Postharvest Losses

The effects of exogenous application of growth substance on changes in the levels of endogenous hormones and on the rate of senescence in detached lettuce leaves have been reported by Aharoni (59). Application of kinetin delayed chlorophyll loss and slowed down the increase in abscissic acid (AB) content. A combined application of 10 ppm of GA3 and 0.1 ppm of the cytokinin isopentenyladenine (IPA) 2 days before harvest delayed lettuce senescence most effectively. Bacterial soft rot, the most serious disease of lettuce, often starts on bruised leaves, but it is much less serious at 0°C than at higher temperatures (60). Tipburn is also a major market disease of lettuce. It is of field origin but occasionally increases in severity after harvest (61). Fumigation with acetaldehyde killed 100% of the aphids (*Myzus persicae*) on head lettuce without injuring the lettuce (62).

VI. PROCESSING

Lettuce is one of the few horticultural food crops used exclusively as a fresh raw product (63–65). It is never processed (canned, dried, or frozen) and rarely cooked. The quality traits

TABLE 10 Effects of CA, Phaltan, and Polyethylene Packaging on Amino Acids and Protein of Lettuce[a]

Days of storage	Treatment	Amino acids (mg/g)	Soluble proteins (mg/g)	Total proteins
0	CR	48.85	62.77	19.90
15	CR	49.97	63.54	21.17
	CA	44.97 (ns)	54.20[b]	20.32 (ns)
	CPh	43.56[c]	62.86[b]	20.13 (ns)
	CPP	40.71 (ns)	49.09	20.92 (ns)
	CPK	47.96 (ns)	54.05	21.06 (ns)
30	CR	60.85	71.65	19.09
	CA	52.98[c]	68.86 (ns)	19.74 (ns)
	CPh	59.07 (ns)	69.27 (ns)	17.76 (ns)
	CPP	57.37 (ns)	70.47 (ns)	18.61 (ns)
	CPK	57.81 (ns)	65.66 (ns)	18.85 (ns)
45	CR	73.62	64.14	21.93
	CA	96.25 (ns)	55.82[b]	21.16 (ns)
	CPh	76.98 (ns)	84.34[b]	19.78 (ns)
	CPK	75.15(ns)	58.48 (ns)	21.84 (ns)
60	CR	77.48	64.29	19.34
	CA	79.04[b]	53.79[c]	19.57 (ns)
	CPK	82.13[b]	55.22[c]	19.01 (ns)
75	CR	77.84	66.58	21.81
	CA	80.96 (ns)	53.56[c]	22.31 (ns)

ns, Not significant at 0.05; CR, conventional refrigeration; CA, controlled atmosphere (2.5% O_2 and 2.5% CO_2); CPh, CA plus Phaltan (1000 ppm); CPP, CA plus Phaltan (1000 ppm) plus packaging in polyethylene bags; CPK, CA plus packaging in polyethylene bags.
[a]Results expressed on dry weight basis.
[b]Significant at 0.05.
[c]Significant at 0.01.
Source: Ref. 58.

important in lettuce, therefore, are simple in the sense of its relatively restricted, through widespread use. They are complex, however, in that many requirements are placed upon lettuce to ensure its arrival in the store, restaurant, or home in the best possible condition.

The popµlarity of salad bars and fast food outlets has led to a substantial market for shredded lettuce (65). Even though shredded lettuce is simply cut-up lettuce, it requires special care because maintenance of its quality is a special problem and in some respects very different from that for a whole product. The major potential problems are wilting, discoloration of cut surfaces, and contamination by organisms of public health concern. The processing of lettuce into shredded lettuce (66–70) involves (a) slicing of lettuce with a sharp blade, (b) maintaining lettuce at 0°C, (c) packaging lettuce in sealed plastic bags that are minimally permeable to CO_2 and O_2 so that modified atmosphere can be achieved, and (d) rinsing in iced water for 5 minutes just before use to restore freshness. Self-generated modification of the atmosphere within the packages was distinctly advantageous when O_2 was between 2 and 5% and CO_2 at about 10%. Even though modified-atmosphere storage is desirable, storage under partial vacuum (low-pressure or hypobaric storage) was highly detrimental, causing breakdown of the lettuce (65). It should be noted that shredded crisphead lettuce, unlike whole heads, is not injured by CO_2 levels up to at least 10%.

TABLE 11 Effects of CA, Phaltan, and Polyethylene Packaging on Chlorophyll and Carotene Contents of Lettuce[a]

Days of storage	Treatment	Chlorophyll a	Chlorophyll b	Chlorophyll Total	Total carotene (ppm)
0	CR	35.5	18.9	54.6	33.39
15	CR				
	CA	25.5 (ns)	12.9 (ns)	33.5 (ns)	25.52 (ns)
	CPh	19.6 (ns)	12.4 (ns)	31.9 (ns)	25.96 (ns)
	CPP	20.0 (ns)	12.1 (ns)	32.1 (ns)	25.28 (ns)
	CPk	20.9 (ns)	12.9 (ns)	33.9 (ns)	25.91 (ns)
30	CR	26.2	14.6	40.8	24.79
	CA	17.6[b]	15.2 (ns)	42.8 (ns)	27.00 (ns)
	CPh	26.7[b]	12.7 (ns)	39.4 (ns)	29.95 (ns)
	CPP	24.1 (ns)	13.1 (ns)	37.2[b]	30.78[b]
	CPk	27.3 (ns)	15.1	42.4 (ns)	27.06 (ns)
45	CR	28.3	16.9	45.2	18.49
	CA	30.4 (ns)	17.7 (ns)	48.1 (ns)	20.51 (ns)
	CPh	23.7	14.7[b]	38.4[b]	18.50 (ns)
	CPP	24.1[b]	14.3[b]	38.4[b]	18.17 (ns)
	CPk	29.7 (ns)	17.5 (ns)	47.2 (ns)	19.57 (ns)
60	CR	31.9	17.7	49.7	18.41
	CA	34.5 (ns)	18.8 (ns)	53.3[b]	20.46 (ns)
	CPk	34.6 (ns)	19.3 (ns)	53.9[b]	20.34 (ns)
75	CR	31.2	20.3	51.5	26.34
	CA	33.9 (ns)	22.2 (ns)	56.0 (ns)	29.00 (ns)

ns, Not significant at 0.05; CR, conventional refrigeration; CA, controlled atmosphere (2.5% O_2 and 2.5% CO_2); CPh, CA plus Phaltan (1000 ppm); CPP, CA plus Phaltan (1000 ppm) plus packaging in polyethylene bags; CPK, CA plus packaging in polyethylene bags.
[a]Results expressed on dry weight basis.
[b]Significant at 0.05.
Source: Ref. 58.

REFERENCES

1. Ryder, E. J., Sea green lettuce, *Hort. Sci. 16*(4):571 (1981).
2. Ryder, E. J., and T. W. Whitaker, The lettuce industry in California: A quarter century changes 1954-79. *Hort. Rev. 2*:164 (1980).
3. Thompson, H. C., and W. C. Kelly, *Vegetable Crops*, 5th ed., McGraw-Hill Book Co. Inc., New York, 1957.
4. Kalloo, G., Salad vegetable, in *Vegetable Crops in India* (T. K. Bose and M. G. Som, eds.), Naya Prokash, Calcutta, 1986, p. 690.
5. Lipton, W. J., and E. J. Ryder, Lettuce, in *Quality and Preservation of Vegetables* (N. A. M. Eskins, ed.), CRC Press, Boca Raton, FL, 1989, p. 217.
6. Ryall, A. L., and W. J. Lipton, *Handling, Transportation and Storage of Fruits and Vegetables*, Vol. I, *Vegetables and Melons*, The AVI Pub. Co., Westport, CT, 1972.
7. Watts, R. L., and G. S. Watts, *The Vegetable Growing Business*, Orange Judd Publishing Co. Inc. New York, 1940, p. 520.

8. Singh, H. B., M. R. Thakur, and P. M. Bhagchandani, Vegetable seed production in Kulu Valley, IV. Seed production of peas, lettuce, celery, parsley, onion, leek and spinach, *Indian J. Hort. 20*:148 (1962).
9. Yawalkar, K. S., *Vegetable Crops of India*, 2nd ed., Agri-Horticultural Publishing House, Nagpur, India, 1980, p. 370.
10. Yamaguchi, M., *World Vegetables: Principles, Production and Nutritive Values*, The AVI Pub. Co., Westport, CT, 1983.
11. Everett, P. H., Influence of fertilizer rates and plastic mulch on the production of two cultivars of crisphead lettuce, *Proc. Fla. State Hort. Soc. 93*:243 (1980).
12. Walworth, J. L., D. E. Carling, and G. J. Michaelson, Nitrogen sources and rates for direct seeded and transplanted head lettuce, *HortScience 27*:228 (1992).
13. Blom-Zandstra, M., and J. E. M. Lampe, The role of nitrate in the osmoregulation of lettuce grown at different light intensities, *J. Exp. Bot. 36*:1043 (1985).
14. Bakker, M. J., J. H. G. Slangen, and W. Glas, Comparative investigation into the effect of fertilization and broadcast fertilization on the yield and nitrate content of lettuce, *Neth. J. Agric. Sci. 32*:330 (1984).
15. Greenwood, D. J., and J. Hunt, Effect of nitrogen fertilizer on nitrate content of field vegetables grown in Britain, *J. Sci. Food Agric. 37*:373 (1986).
16. Hansen, H., The content of nitrate and protein in lettuce grown under different conditions, *Qual. Plant. Plant. Foods Human Nutr. 28*:11 (1978).
17. Shinohara, Y., and Y. Suzuki, Effect of light and nutritional conditions on the ascorbic acid contents of lettuce, *J. Jpn. Soc. Hort. Sci. 50*:239 (1981).
18. Ellison, J. H., G. Vest, and R. W. Langlois, 'Jersey Centennial' asparagus, *HortScience 16*(3):349 (1981).
19. Burton, W. G., *Postharvest Physiology of Food Crops*, Longman, London, 1982.
20. Kalloo, G., Lettuce, in *Vegetable Crops in India* (T. K. Bose and M. G. Som, eds.), Naya Prokash, Calcutta, 1986, p. 690.
21. Colliev, G. F., and T. W. Tibbits, Tipburn of lettuce, *Hort. Rev. 4*:49 (1982).
22. Barla, D. J., and T. W. Tibbitts, Effect of artificial enclosure of young lettuce leaves on tipburn incidence and leaf calcium concentration, *J. Am. Soc. Hort. Sci. 111*:413 (1986).
23. Thibodeau, P. O., and D. L. Minotti, The influence of calcium on the development of lettuce tipburn, *J. Am. Soc. Hort. Sci. 94*:372 (1969).
24. Ashkar, S. A., and S. K. Ries, Lettuce tipburn as related to nutrient imbalance and nitrogen composition, *J. Am. Soc. Hort. 96*:448 (1971).
25. Fermohlen, G. P., and A. P. Vander, Hoeven tipburn symptoms in lettuce, *Acta Hort. 4*:105 (1966).
26. Peck, N. H., M. H. Dickson, and G. E. MacDonald, Tipburn susceptibility in semi-isogenic inbred lines of cabbage as influenced by nitogen, *HortScience 18*:726 (1983).
27. Welch, J. E., and T. W. Whitaker, Recent developments in California's lettuce industry, *Am. Veg. Grower 1*:12 (1953).
28. Misaghi, I. J., and R. G. Grogan, Effect of tempeature on tipburn development in head lettuce, *Phytopathology 68*:1738 (1978).
29. Misaghi, I. J., and R. G. Grogan, Physiological bases for tipburn development in head lettuce, *Phytopathology 68*:1744 (1978).
30. Ramsey, G. B., B. A. Friedman, and M. A. Smith, Market diseases of beets, chicory, endive, escarole, globe artichoke, lettuce, rhubarb, spinach and sweet potato, *U.S. Dept. Agric. Handbook*, 155, 1959, Washington, DC, p. 42.
31. Eckert, T. W., P. P. Rubio, A. K. Matioo, and A. K. Thompson, Postharvest pathology, Part 2, diseases of tropical crops and their control, in *Postharvest Physiology, Handling and Utilization of Tropical and Sub-tropical Fruits and Vegetables* (E. B. Pantastico, ed.), The AVI Pub. Co., Westport, CT, 1975, p. 415.
32. Pantastico, E. B., T. K. Chattopadhyay, and H. Subramanyam, Storage and commercial storage

operations, in *Postharvest Physiology, Handling and Utilization of Tropical and Sub-tropical Fruits and Vegetables* (E. B. Pantastico, ed.), The AVI Pub. Co., Westport, CT, 1975, p. 415.
33. Sevila, F., and J. Gratnaud, Mechanizing lettuce, harvesting *Am. Veg. Grower 28*:53 (1980).
34. Aharoni, N. R., Barkai Golan, and H. Aviram, Reducing rot in stored lettuce with polyethylene lined packaging, in *Scientific Activities 1974-1977, Pamphlet No. 184, Institute for Technology and Storage of Agriculture, Product*, Bet Dagan, Israel, 1978, p. 57.
35. Lenker, D. H., M. Zahara, and P. A. Adrian, Three systems for handling mechanically harvested lettuce. *Transactions of the A.S.A.E. 24*:1114 (1981).
36. Beraha, L., and W. F. Kwoled, Prevalence and extent of eight market disorders of western-grown head lettuce during in 1973 and 1974 in the greater Chicago Illinois area, *Plant Dis. Rep. 59*:1001 (1975).
37. U.S. Department of Agriculture, *United States Standards and Grades of Lettuce*, Consumer and Marketing Service, U.S. Department of Agriculture, Washington, DC, 1970.
38. Feigin, A., E. Pressman, P. Imas, and O. Miltau, Combined effects of KNO_3 and salinity on yield and chemical composition of lettuce and Chinese cabbage, *Irrig. Sci. 12*:223 (1991).
39. Sorensen, J. N., A. S. Johansen and N. Poulsen, Influence of growth conditions of the value of crisphead lettuce. 1. Marketable and nutritional quality as affected by nitrogen supply, cultivar and plant age, *Plant Foods Human Nutr. 46*:1 (1994).
40. Poulsen, N., J. N. Sorensen, and A. S. Johansen, Influence of growth conditions on the value of crisphead lettuce, 2. Weight losses during storage as affected by nitrogen, plant age and cooling system, *Plant Foods Human Nutr. 46*:13 (1994).
41. Brunsgaard, G., U. Kidmose, J. N. Sorensen, K. Kaack, and B. O. Eggum, Influence of growth conditions on the value of crisphead lettuce. 3. Protein quality and energy density as determined in balance experiments with rats, *Plant Foods Human Nutr. 46*:255 (1994).
42. Hansen, H., The influence of nitrogen fertilization on the chemical composition of vegetables, *Qual. Plant. Plant Foods Human Nutr. 28*:45 (1978).
43. Poulsen, N., A. S. Johansen, and J. N. Sorensen, Influence of growth conditions on the value of crisphead lettuce. 4. Quality changes during storage, *Plant Foods Human Nutr. 47*:157 (1995).
44. Forney, C. F., and R. K. Austin, Time of day at harvest influences carbohydrate concentration in crisphead lettuce and its sensitivity to high CO_2 levels after harvest, *J. Am. Soc. Hort. Sci. 113*:581 (1988).
45. Alberecht, J. A., Ascorbic acid content and retention in lettuce, *J. Food Qual. 16*:311 (1993).
46. Stevens, M. A., Varietal influence on nutritional value, in *Nutritional Qualities of Fresh Fruits and Vegetables* (P. L. White and N. Selvey, eds.), Futura, Mount Kisco, NY, 1974, p. 87.
47. Senti, F. R., and R. L. Rizek, Nutrient levels in horticultural crops, *HortScience 10*:243 (1975).
48. Hardenburg, R. E., A. E., Watada, and C. Y. Wang, *The Commercial Storage of Fruits, Vegetables and Florist and Nursery Stocks*, U.S. Department of Agriculture, Agriculture Handbook No. 66, Washington, DC, 1986.
49. Barger, W. R., Vacuum-cooling precooling a comparison of the cooling of different vegetables, *U.S. Dept. Agric. Market Res. Rep. 600*, Washington, DC, 1973, p. 12.
50. Scholz, E. W., H. B. Johnson, and W. R. Buford, Heat evolution rate of some Texas grown fruits and vegetables, *J. Rio Grande Valley Hort. Soc. 17*:170 (1963).
51. Morris, L. L., H. K. Pratt, and C. L. Tucker, Lettuce handling and quality West, *Grower Shipper 26*(5):14 (1955).
52. Lipton, W. J., Market quality and rate of respiration of head lettuce held in low oxygen atmosphere, *U.S. Dept. Agr. Market Res. Rep.*, Washington, DC, 1967, p. 777.
53. Lipton, W. J., Controlled atmosphere effects on lettuce quality in simulated export shipments, *U.S. Dept. Agric. ARS*, 51-45, Washington, DC, 1971, p. 14.
54. Brecht, P. E., A. A. Kader, and L. L. Morris, The effect of composition of atmosphere and duration of exposure on brown stain of lettuce, *J. Am. Soc. Hort. Sci. 98*:536 (1973).
55. Kader, A. A., D. E. Brecht, R. Woodruff, and L. L. Morris, Influence of carbon monoxide, carbon dioxide and oxygen level on brown stain respiration rate and visual quality of lettuce, *J. Am. Soc. Hort. Sci. 98*:485 (1973).

56. Hatton, T. T., Jr., E. B. Pantastico, and E. K. Akamine, Controlled atmosphere storage: Part 3, Individual commodity requirements, in *Postharvest Physiology, Handling and Utilization of Tropical and Sub-tropical Fruits and Vegetables* (E. B. Pantastico, ed.), The AVI Pub. Co., Westport, CT, 1975, p. 201.
57. Singh, B., C. C. Young, D. K. Salunkhe, and A. R. Rahman, Controlled atmosphere storage of lettuce: Effects on quality and rate of respiration of lettuce heads, *J. Food Sci. 37*:48 (1972).
58. Singh, B., D. J. Wang, D. K. Salunkhe, and A. R. Rahman, Controlled atmosphere storage of lettuce: Effect on biochemical composition of leaves, *J. Food Sci. 37*(1):52 (1972).
59. Aharoni, A., Hormonal regulation during senescence and water stress of detached lettuce leaves, in *Scientific Activities, 1974-1977*, Pamphlet No. 184, Institute for Technology and Storage of Agricultural Products, Bet. Dagan, Israel, 1978, p. 66.
60. Ramsey, G. B., B. A. Friedman, and M. A. Smith, Market diseases of beets, chicory, endive, escarole, globe artichoke, lettuce, rhubarb, spinach and sweet potato, *U.S. Dept. Agr. Agric. Handbook*, 155, Washington, DC, 1959, p. 49.
61. Lipton, W. J., Postharvest changes in amount of tipburn of head lettuce and the effect of tipburn on incidence of decay, *Plant Dis. Rep. 47*:875 (1963).
62. Stewart, J. K., Y. Aharoni, P. L. Hartsell, and D. K. Young, Acetaldehyde fumigation at reduced pressures to control the green peach aphid on wrapped and packed head lettuce, *J. Econ. Entomol. 73*:149 (1980).
63. Gupta, K., G. R. Singh, J. L. Mangal, and D. S. Wagale, Salad Crops, Dietary Importance, in *Encyclopaedia of Food Science, Food Technology and Nutrition* (R. MaCrae, R. K. Robinson, and M. J. Sandlers, eds.), Academic Press, London, 1993, p. 3963.
64. Couture, R., M. I. Cantwell, and M. E. Sattveik, Physiological attributes related to quality attributes and storage life of minimally processed lettuce, *HortScience 28*(7):723 (1993).
65. Thompson, A. K., M. B. Bhatti, and P. P. Rubio, Harvesting and handling: Harvesting, in *Postharvest Physiology, Handling and Utilization of Tropical and Sub-tropical Fruits and Vegetables* (E. B. Pantastico, ed.), The AVI Pub. Co., Westport, CT, 1975, p. 236.
66. Bolin, H. R., A. E. Stafford, A. D. King Jr., and C. C. Huxsoll, Factors affecting the storage stability of shredded lettuce, *J. Food Sci, 42*:1319 (1977).
67. Herner, R. H., and T. R. Krahn, Copped lettuce should be kept dry and cold, in *Produce Market Association 1973 Year Book*, Produce Market Association, Newark, DE, 1973, p. 130.
68. Krahn, T. R., Improving the keeping quality of cut head lettuce, *Acta Hort. 62*:79 (1977).
69. Maxcy, R. B., Lettuce salad as a carrier of micro-organisms of public health significance, *J. Food Prod. 41*:435 (1978).
70. Priepke, P. E., L. S. Wei, and A. F. Nelson, Refrigerated storage of pre-packaged salad vegetables, *J. Food Sci. 41*:379 (1976).

23

Asparagus

P. M. Kotecha and S. S. Kadam
Mahatma Phule Agricultural University, Rahuri, India

I. INTRODUCTION

Asparagus is an important crop that is grown in temperate as well as tropical regions. It has 150 species which are perennials, and many species are grown for ornamental purposes. Asparagus is commercially grown in many parts of Europe and Asia, Australia, New Zealand, and the United States. In the United States, production is concentrated in California, Michigan, and Washington state. In Canada, Ontario and British Columbia are the major producers.

II. BOTANY

Asparagus (*Asparagus officinalis* L.) is a delicate, nutritious, and appetizing vegetable belonging to the genus *Asparagus* of the family Lily. The genus has more than 150 species native to Europe, Asia, and Africa, and it includes herbaceous, woody, and erect as well as climbing forms of asparagus (2,3). The commonly grown asparagus (var. *altilis*, L.) is a dioecious perennial herb, about 4–10 feet tall, having male (yellowish-green) and female (less conspicuous) flowers on separate plants. Asparagus is indigenous to Europe and Asia, and its medicinal properties were known to the ancient Greeks and Romans. The roots of cultivated asparagus are numerous and fleshy. They occur horizontally in the soil and serve as a storage organ. Spear production occurs primarily at the expense of the sugars stored in the fleshy roots. The crown is made of roots and rhizomes, which are underground stems. The harvested spear is a stem that arises from a bud on the rhizome. The crown of the roots grows closer to the surface of the soil each year, and older fibrous roots die and rot away (1,2).

Very few varieties of asparagus have been well established. Thompson and Kelly (3) stated that Mary Washington is the main variety grown in the United States on a large scale; others grown on a small scale include Martha Washington, Reading Giant, Palmetto, and Aggenteuil. The Washington varieties are more resistant to rust in addition to having good market quality. Other commonly grown cultivars are MSU-1, Cal.66, Cal.72, Cal.309, N.J. Approved, RH-201, and Cal.711. Jersey Centennial asparagus is a recently released cultivar that is more vigorous, rust resistant, and productive than Mary Washington (2).

New selections, such as UC 157 and UC 72, exhibit tolerance to *Fusarium* root rot caused by *Fusarium oxysporium* (sp. *asparagi*). UC 157 lacks the characteristic purple pigmentation of the spears. Several other newer varieties include Lucullus varieties from Germany and Jersey Centennial and Jersey Giant from the United States. In Canada, a series of Mary Washington progenies, identified as Viking (V 35), Viking Select, and Viking KB 3, have been important varieties. In Europe there are new selections, e.g., Limbros 10 and 22, developed for culture or forcing under glass or plastic. There are also all-male lines, such as Franklin F (also known as Limbros 126), used for either white or green asparagus production. The impetus to produce staminate rather than pistillate plants lies in their higher, more stable yields, although pistillate spears are generally heavier. The rapid and efficient production of thousands of unique, high-yielding staminate asparagus clones has been achieved by tissue culture and, in particular, meristem culture. There is no botanical distinction between green and white asparagus. Production of the two types is based on differing cultivation practices—white (blanched asparagus) is devoid of pigment since it is grown in dark (4).

III. PRODUCTION

A. Soil and Climate

Well-drained, light-textured soils with high contents of organic matter are most suitable for asparagus cultivation. The spears should emerge from the soil without damage. A cool season with a mean air temperature of 16–24°C favors crop growth. The crop should be protected from freezing.

B. Propagation

The asparagus crop is either direct-seeded or one-year-old crowns are transplanted in the field. Seeds are drilled 2.5–5 cm deep in furrows 15–20 cm below ground level for direct seeding. Crowns are transplanted 20–25 cm below ground level and 30 cm apart. The distance between rows is 1.8 m for green asparagus and 2–2.5 m for white asparagus.

C. Cultural Practices

Seedlings are raised from good-quality seeds and used for planting. Young plants seeded the previous year, called crowns, are removed in the spring before growth start and planted in the field within a few hours. Potassium and phosphorus are added to the soil before planting the crowns. Nitrogen is used for top-dressing. A fertilizer dose of 65–90 kg of N, 35 kg of P, and 35 kg of K per hectare has been recommended (4). In young asparagus plants, there is usually a 1:1 sex ratio. Male plants generally produce more spears than female plants (3).

Asparagus is one of the few perennial vegetables. The spring production of spears is at the expense of food stored in the roots from the previous year. The cutting season should be limited in duration so the plants have ample time to manufacture and store food for the succeeding year's crop. (5–7).

D. Diseases and Pests

The symptoms of and possible preventive measures for bacterial soft rot (*Erwinia carotovora*) and rots caused by various species of *Fusarium* and *Phytophthora* have been described by Ryall and Lipton (5). Asparagus rust (*Puccinia asparagi*), *Fusarium* wilt, asparagus beetles (*Crioceris asparagi* L. and *C. duodecimpunctata* L.), and garden centipedes (*Scutigerella immaculata*) are the important pests of asparagus, causing serious losses (3).

E. Maturity, Harvesting, and Handling

The edible portions of asparagus are the spears or stems (cladophylls) that develop from the crown. Harvesting (first cutting) of asparagus usually starts with the third season, or after the completion of two full growing seasons, although in some areas having longer seasons, it is harvested during the second season (3). Green asparagus, which is the greater portion of the crop, is cut 1 or 2 inches below the surface of the soil. White asparagus is produced by blanching treatment. Ryall and Lipton (5) advocated selective harvesting of asparagus, since new shoots appear daily over a period of 6–10 weeks. Mechanical harvests of asparagus have been reported to decrease yield and quality.

Since deterioration sets in and asparagus loses freshness and quality quickly after harvest, it is desirable to market or process it as soon as possible. Instant cooling of the vegetable helps to maintain quality and improves marketability considerably.

The harvested asparagus shoots are washed, graded, bunched, and packed. Straight spears, green-tinged with purple on the bracts and with tightly closed tips, are preferred. Spears should snap before reaching at 90° angle without bending into a U-shape. The spears also should be harvested when they are rounded and before they become ribby, which indicates old age (5). Asparagus is packaged either in plastic film bags or in boxes or crates without bunching. The butts are cut off and spears are laid in boxes lined with oily paper with a layer of wet moss on the bottom when the loose-pack method is employed.

Harvest maturity of asparagus is usually reached when spears reach about 12.5–20 cm above the ground, and they should be cut before they are too long and before the heads or tips begin to spread. Mack et al. (9), however, reported that the greatest yield of asparagus can be realized when the spears are allowed to grow up to 20–25 cm in length.

Pinkau (10) tabulated the effects of stem length on the yield, quality, and market price. Daily growth of 0.5–1.6 cm was temperature dependent, with average temperatures being more than 18°C. Pinkau (10) recommended daily harvesting and stated that harvesting every second day is more profitable at lower temperatures. Kaufmann and Starz (11) emphasized the importance of frequent cutting of asparagus, with one day's delay causing the spears to deteriorate from grade A to grade B (except for the cultivar Start, which can be left for an extra day). These workers (11) further stated that grades A and B should be cut 2–3 cm below soil level and that grade C spears should be cut at ground level.

F. Packaging

The shelf life of asparagus can be extended by packing only spears with tightly closed tips. Spears with open (feathered) tips decay more rapidly, toughen, and are unattractive. Carolus et al. (12) reported that packaging of asparagus in film bags maintained the market quality by reducing moisture loss, the edible quality by preventing fiber formation, and the nutritional value by decreasing the loss of ascorbic acid. However, impermeable plastic bags can produce off-flavors due to anaerobic decomposition (13). Ryall and Lipton (5) stated that such packaging would lead to uncontrolled accumulation of CO_2 and depletion of O_2 with possible injury to the produce. They (5) further stated that if asparagus is packed in film bags or overwrapped, the unit must be sufficiently ventilated to permit exchange of CO_2 and O_2 and to dissipate the respiratory heat. About half a dozen 0.63 cm holes are adequate to pack one pound of spears 17.5 cm long. The number of holes should be doubled for shorter spears, because of their higher rate of respiration. A schematic diagram of postharvest handling operations for stem vegetables, including asparagus, is shown in Figure 1.

IV. CHEMICAL COMPOSITION

Asparagus is a good source of vitamin A. The vitamin A content of green asparagus is higher than that of white or blanched asparagus. Green asparagus (100 g edible portion) contains 93 g of water, 2.2 g of protein, 21 mg of calcium, 700 I.U. of vitamin A, 30 mg of ascorbic acid, 0.20 mg of thiamine, 0.16 mg of riboflavin, and 1.0 mg of niacin (1). The nutritional composition of asparagus reported by Gupta et al. (2) is given in Table 1.

V. STORAGE

Fresh asparagus is highly perishable and deteriorates rapidly at temperatures above 5°C. Significant changes in respiration (14,15), nitrogen metabolism (16), and other metabolic changes (17) occur within 3 hours of harvest. The spears should be cooled immediately after cutting, preferably by hydrocooling, and placed at a low temperature. In addition to general deterioration, growth, loss of tenderness, loss of flavor, loss of vitamin C, and development of decay take place at moderately high temperatures. Asparagus can be kept successfully for about 3 weeks at 20°C (Fig. 2).

The biochemical changes in harvested asparagus depend upon the temperature and storage atmosphere (18–20). The most significant changes include an increase in spear length and weight and in sugars and fiber (5). Bisson et al. (21) demonstrated that spears grow in length if butts have sufficient moisture. They also noted that the growth rate was least at 0.6°C and increased with a rise in temperature to 5.0, 13.3, 25, and 35°C. The increase in weight was attributed to absorption of water. The losses in reducing and total sugars were pronounced at higher temperatures, a maximum rate of loss occurring during the first 24 hours of storage. The crude fiber content, however, increased at all temperatures. Crude fiber and lignification were least at the lowest temperature.

Asparagus can be held for about 10 days at 0°C, but it is subject to chilling injury when held longer at this temperature (22). High humidity is essential to prevent dessication, particularly at the butt ends. The desired relative humidity is obtained by placing the butts of asparagus on wet pads. A high relative humidity can also be obtained by prepacking spears in perforated film. Asparagus with white butts is less perishable than all-green asparagus. Bacterial soft rot, which can occur at either the tip or butt of the asparagus, is the principal decay (23).

A. Low-Temperature Storage

Owing to the highly perishable nature of asparagus, the quickest possible cooling to temperatures below 4.4°C is very advantageous. The field heat should be removed by employing hydrocooling or vacuum-cooling (5). Although vacuum-cooling is feasible, it is less desirable due to incomplete cooling (24). Hydrocooling reduces the temperature from 21.1 to 2.2°C within about 15 minues; chlorine is generally added to the water used for cooling to reduce the bacterial population.

The storage life of asparagus decreases rapidly as the temperature rises above 4.4°C. Storage conditions of 0°C with 95% relative humidity are generally recommended to extend the

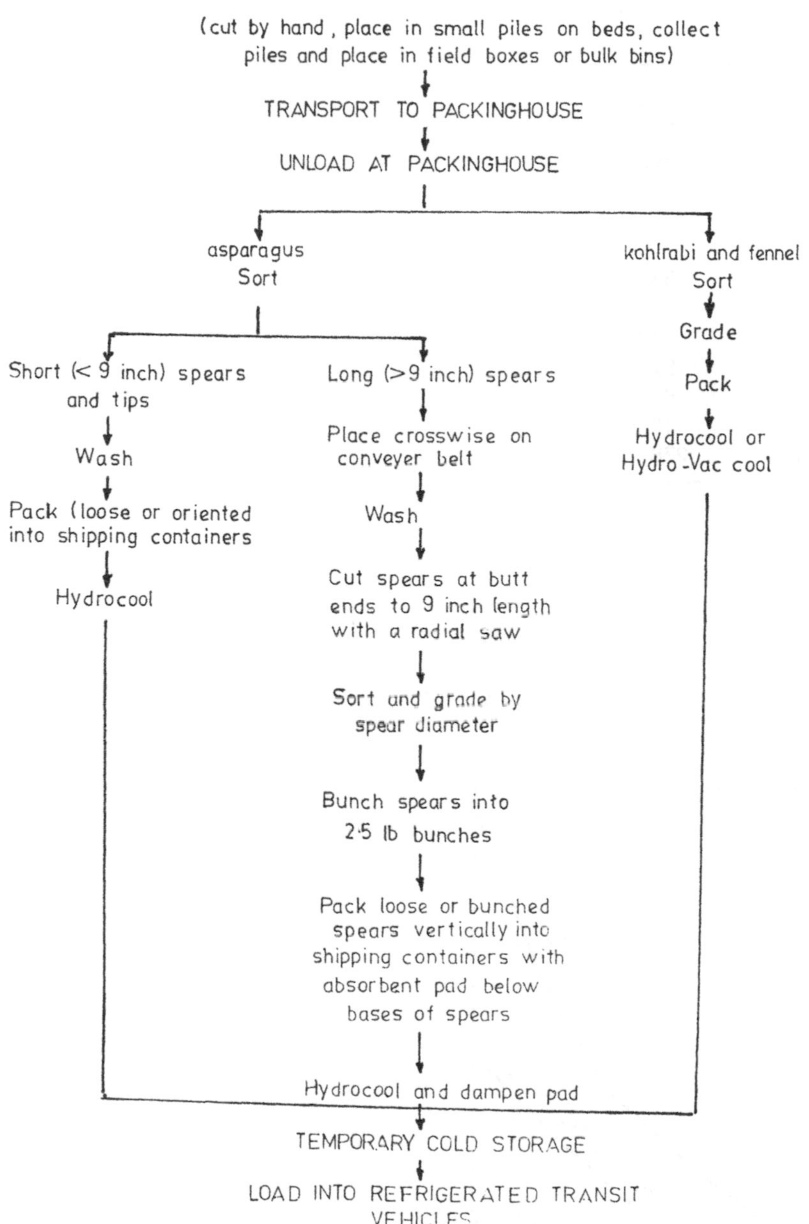

FIGURE 1 Harvest and postharvest operations for asparagus, kohlrabi, and fennel. (From Ref. 5.)

marketable life of asparagus for about 3–4 weeks (25). Rapid sorting, packing, and cold storage at 0–6°C and 90% relative humidity maintain high quality (11). Adamicki (26) cooled harvested (after 2 or 8 hours) spears in water at 0.5–1.0°C for 9–12 minutes and then stored them at 2°C for 1–5 days followed by 2 days at 20°C (simulated shelf life). This significantly increased the fiber content (26).

Table 1 Nutritional Constituents of Asparagus (per 100 g edible portion)

Constituent	Content
Water (g)	91.7
Energy (kcal)	26.0
Fat (g)	0.2
Carbohydrate (g)	5.0
Vitamin A (IU)	900
Vitamin C (mg)	33
Thiamine (mg)	0.18
Niacin (mg)	1.5
Calcium (mg)	22
Phosphorus (mg)	62
Iron (mg)	10.2
Potassium (mg)	278

Source: Ref. 2.

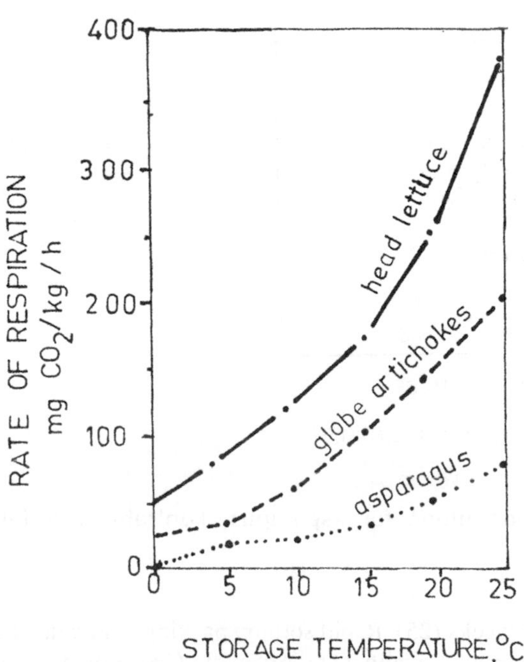

Figure 2 Effect of storage temperature on rate of respiration of lettuce, globe artichoke, and asparagus. (From Ref. 5.)

B. Controlled Atmosphere Storage

Controlled-atmosphere (CA) storage of asparagus is known to reduce postharvest deterioration. CA storage prevented the development of off-flavor (27), water loss (28), toughening, and chlorophyll loss (30). CA effects on vitamin C content depend upon the atmosphere composition. Exposure of asparagus to an atmosphere of 7 ± 2% CO_2 and 2% O_2 for 24 hours retards bacterial rots. Lipton (27–29) demonstrated that, while 5% CO_2 levels are ineffective in controlling the growth of *Phytophthora*, 10% or higher concentrations of CO_2 caused pits at 6.1°C. No chilling injury was noticeable at 1.7°C and 15% CO_2 concentration (7). Lipton (29) reported that atmosphere lacking O_2 or containing 20 or 30% CO_2 caused severe injury at 3 or 6°C, while lower CO_2 concentration caused injury only at 6°C (Table 2). CA storage of asparagus, however, delayed the rate of degradation of chlorophyll (30) and prevented the development of *Phyptophthora*, toughening, and fibrousness (31). Ryall and Lipton (5) concluded that CA storage of asparagus is a useful adjunct to good refrigeration, even for relatively brief periods. Asparagus particularly benefits from CO_2 concentrations of 7 ± 2%, if postharvest decay is to be reduced and if temperature control is uncertain. About 12 ± 2% CO_2 is permissible at 0–2.8°C if both decay and toughening are to be controlled. Lipton (29) concluded that CA storage of asparagus has three advantages: reduction of decay, retardation or even reversal of toughening, and a more

TABLE 2 Percentage of Physiological Injury Caused by Holding in Various Concentrations of CO_2 and O_2

Concentration		Holding time and temperature			
CO_2 (%)	O_2 (%)	7 days 3°C	7 days 3°C + 2 days 15°C	7 days 6°C	7 days 6°C + 2 days 15°C
0.1	21	0	0	0	0
	0	81	100	100	100
0.1	1	2	2	5	10
	5	14	27	9	15
	10	5	14	8	19
5	5	0	0	5	5
	10	0	0	—	—
	20	0	0	5	5
10	5	0	0	12	18
	10	0	5	11	12
15	5	0	0	24	26
	10	0	0	—	—
20	5	8	2	3	9
	10	15	11	20	22
	20	42	14	12	18
30	5	25	48	47	69
	10	28	49	—	—
	20	50	57	19	17

All asparagus spears were held 7 days in atmospheres as indicated in columns 1 and 2, all 2-day periods in air. Values are percentage of physiological injury.
Source: Ref. 29.

TABLE 3 Radiation and Storage Effects on Taste Preference for Asparagus

Rate ($\times 10^6$ rad/hr)	Dose ($\times 10^5$ rad)	Taste preference score 24–48 hr	10 days
Control	Control	6.8	7.0
	0.93	5.6	5.8
	2.79	5.4	7.0
	5.58	5.5	3.4
	Mean	5.5	5.4
0.465	0.93	5.6	7.1
	2.79	6.0	5.7
	5.58	5.7	6.0
	Mean	5.8	6.3
0.93	0.93	6.3	6.1
	2.79	6.1	6.6
	5.58	6.1	6.1
	Mean	6.2	5.9
Dose mean	Control	6.8	7.0
	0.93	5.8	6.3
	2.79	5.8	6.4
	5.58	5.8	4.8
$LSD^{0.05}$	Individual	1.4–1.2	1.4–1.2
	Dose means	0.7	0.7
	Rate means	0.6	0.8–0.7

Source: Ref. 34.

desirable color. It provides a useful technique not only to retain a high quality in fresh asparagus but also to improve the quality of the processed product.

Baxter and Walter (20) studied the effect of CA storage on the quality of asparagus stored for 28 days at 2°C in air, a flow-through CA system, or 14 days in polymeric film consumer packages. CA-stored spears retained more sugars, organic acids, and soluble proteins than spears stored in air. Spears stored in vented consumer packages had a useful life of 14 days, whereas those in nonvented packages started to break down after 8 days. Spears from

TABLE 4 Radiation Effects on Average Shear Press Values for Asparagus

Rate ($\times 10^6$ rad/hr)	Dose ($\times 10^5$ rad) Control	0.93	2.79	5.58
0.093	—	850	695	665
0.465	900	800	685	660
0.930	900	795	665	615

Source: Ref. 34.

vented packages lost more weight but retained more sugars and organic acids than those from nonvented packages.

Andre et al. (32) reported that the vacuum prerefrigeration of asparagus spears in a controlled atmosphere (in polyethylene bags with silicon elastomer windows) within 6–8 hours of harvest followed by cold storage at 1°C was the best long-term method of storage (20–30 days). Prerefrigeration by water was possible for short-term storage. Fungal development during storage could be prevented by flushing with CO_2 (3.0% for 24 hours) or by maintaining a CO_2 concentration of 5–10%. Shelf life of several leafy vegetables including asparagus is reported to be doubled by immersing the cut ends or dipping the vegetable in solutions of CCC (33).

C. Irradiation

Salunkhe et al. (34,35) reported that acceptability of irradiated asparagus decreased with an increase in the dose of radiation when evaluated 24–48 hours and 10 days after irradiation (Table 3). The tenderness of asparagus spears measured in terms of shear-press values increased as the dose and rate of radiation increased (Table 4).

VI. PROCESSING

Most asparagus produced in the United States is canned, with only part of the harvest being frozen (1). From the canning belt, spears are sorted for size and color and placed blossom-end up in cans. A dilute brine of about 2–2.5% salt is added and the cans are exhausted in live steam. It is common to add salt from an automatic dispenser and fill with hot water or alternatively fill with hot water and then add salt by dispenser. In either case, the cans are sealed at once in an atmosphere of steam. The filled cans may be exhausted and sealed in plain double steamers. The sealed cans are processed at 115.5–120°C for 25–35 minutes depending upon size and grade of cans (1).

Green asparagus is preferred to white for frozen storage because of its richer flavor and the popular preference for green fresh asparagus. Field-fresh asparagus should be frozen 5–6 hours after cutting because it rapidly toughens, becomes stringy, and acquires a bitter taste on standing at room temperature. A thorough blanching is necessary before the produce is frozen. The collapsing of asparagus on freezing and thawing can be minimized if it is quick-frozen (1). The quick-frozen product is not markedly different in appearance from cooked fresh asparagus.

REFERENCES

1. Salunkhe, D. K., and B. B. Desai, *Postharvest Biotechnology of Vegetables*, Vol. 1, CRC Press, Boca Raton, FL, 1984.
2. Gupta, K., G. R. Singh, J. L. Mangal, and D. S. Wagle, Salad crops—dietary importance, in *Encycloppaedia of Food Science, Food Technology and Nutrition* (R. MaCrae, R. K.Robinson, and M. J. Sandler, eds.), Academic Press, London, 1993, p. 3963.
3. Thompson, H. C., and W. C. Kelly, *Vegetable Crops*, 5th ed., McGraw-Hill Book Co., Inc. New York, 1957.
4. Yamaguchi, M., *World Vegetables, Principles, Production and Nutritive Values*, AVI Pub. Co., Westport, CT, 1983.
5. Ryall, A. L., and W. J. Lipton, *Handling, Transportation, and Storage of Fruits and Vegetables*, Vol. 1: *Vegetables and Melons*, The AVI Pub. Co., Westport, CT, 1972.
6. Lipton, W. J., Physiological changes in harvested asparagus as related to temperature of holding, Ph.D. thesis, University of California, Davis, 1957.

7. Lipton, W. J., Postharvest repsonses of asparagus spears to high carbon dioxide and low oxygen atmospheres, *Proc. Am. Soc. Hort. Sci. 86*:347 (1965).
8. Pantastico, E. B., H. Subramanyam, M. B. Bhatti, N. Ali, and E. K. Akamine, Postharvest physiology: harvest indices, in *Postharvest Physiology, Handling and Utilization of Tropical and Sub-tropical Fruits and Vegetables* (E. B. Pantastico, ed.), The AVI Pub. Co., Westport, CT, 1975, p. 56.
9. Mack, W. B., R. E. Larson, D. G. White, and H. O. Sampson, *Vegetable and Fruit Growing*, Lippincott Co., New York, 1956.
10. Pinkau, H., Reduction of green asparagus crop losses by correct harvesting, *Gartenbau 28*(5):138 (1982).
11. Kaufmann, F., and K. Starz, Assuring high quality yields with green asparagus, *Gartenbau 27*(4):109 (1980).
12. Carolus, R. L., W. J. Lipton, and S. B. Apple, Effect of packaging on quality and market acceptability, *Bull. Mich. Agric. Exp. Sta. 35*:330 (1953).
13. Herregods, M., The storage of asparagus, *Tuinbouwberichten 25*:48 (1961).
14. King, G. A., D. C. Woollard, D. E. Irving, and W. M. Borst, Physiological changes in asparagus spear tips after harvest, *Physiol. Plant. 80*:393 (1990).
15. Lill, R. E., G. A. King, and E. M. Donoghue, Physiological changes in asparagus spears immediately after harvest, *Sci. Hort. 44*:191 (1990).
16. Hurst, P. L., G. A. King, D. C. Woollard, and W. M. Borst, Nitrogen metabolism in harvest asparagus: No differences between light and dark storage at 20°C, *Food Chem. 47*:329 (1993).
17. Davies, K. M., and G. A. King, Isolation and characterization of C-DNA clone for harvest induced asparagine synthetase from *Asparagus officinalis L, Plant Physiol. 102*:1337 (1993).
18. Lill, R. E., G. A. King, and E. M. Donoghue, Physiological changes in asparagus spears immediately after harvest, *Sci. Hort. 44*(3-4):191 (1990).
19. Lipton, W. J., Postharvest biology of fresh asparagus, *Hort. Rev. 12*:69 (1990).
20. Baxter, L., and L. Walter, Jr., Quality changes in asparagus spears stored in flow through CA system or in consumer packages, *HortScience 26*(4):399 (1991).
21. Bisson, C. W., H. A. Jones, and W. W. Robbins, Factors influencing the quality of fresh asparagus after it is harvested, *Bull. of California Agric. Exp. Sta. 410*(1926).
22. Lipton, W. J., Effect of temperature on asparagus quality, in *Proc. Conf. Transport. Perishables*, University of Claifornia, Davis, 1958, p. 147.
23. Smith, M. A., L. P. McCulloch, and B. A. Friedman, Market diseases of asparagus, onions, beans, peas, carrots, celery and related vegetables, *U.S. Dept. of Agric. Handbook* 303, 1982, p. 65.
24. Stewart, J. K., and W. R. Barger, Effect of cooling methods on quality of asparagus and cauliflower, *Proc. Am. Soc. Hort. Sci. 78*:295 (1961).
25. Pantastico, E. B., T. K. Chattopadhyay, and H. Subrmanyam, Storage and commercial storage operations, in *Scientific Activities, 1974-1977, Pamphlet No. 184, Institute for Technology and Storage of Agriculture Product*, Bet Dagan, Israel, 1978, p. 57.
26. Adamicki, F., Effect of water-cooling on the storability and commercial quality of asparagus, *Biul. Warzywniezy 23*:419 (1979).
27. Lipton, W. J., Market quality of asparagus: Effect of maturity at harvest, and of high carbon dioxide atmospheres during stimulated transit, U.S.D.A. Marketing Res. Rep. Mo., 1968.
28. Lipton, W. J., Effect of atmospheric composition on quality in California asparagus: Effect of transit environments on market quality, U.S.D.A. Marketing Res. Rept. No. 428, 17 (1960).
29. Lipton, W. J., Controlled atmospheres for fresh vegetables and fruits—why and when, in *Postharvest Biology and Handling of Fruits and Vegetables* (N. F. Haard and D. K. Salunkhe, eds.), The AVI Pub. Co., Westport, CT, 1975, p. 130.
30. Wang, S. S., N. F. Haard, and G. R. Dimarco, Chlorophyll degradation during controlled atmosphere storage of asparagus, *J. Food Sci. 36*:657 (1971).
31. Hatton, T. T., Jr., E. B. Pantastico, and E. K. Akamine, Controlled atmosphere storage, Part 3. Individual commodity requirements, in *Postharvest Physiology, Handling and Utilization of Tropi-*

cal and Sub-tropical Fruits and Vegetables (E. B. Pantastico, ed.), The AVI Pub. Co., Westport, CT, 1975, p. 201.

32. Andre, P., M. Buret, Y. Chambroy, P. Dauple, C. Flanzy, and C. Pelisse, Conservation trials asparagus spears by means of vacuum pre-refrigeration, associated with controlled atmospheres and cold storage, *Rev. Hort.* (205):19 (1980).
33. Halevy, A. H., and S. H. Wittwer, Chemical regulation of leaflet senescence, *Qt. Bull. Mich. Agric. Extp. Sta. 48*:9 (1965).
34. Salunkhe, D. K., L. H. Pollard, and R. K. Gerber, Effects of gamma radiation dose, rate and temperature on the taste preference and storage life of certain fruits, vegetables and their products, *Proc. Am. Soc. Hort. Sci. 74*:414 (1959).
35. Salunkhe, D. K., R. K. Gerber, and L. H. Pollard, Physiological and chemical effects of gamma radiation on certain fruits, vegetables and their products, *Proc. Am. Soc. Hort. Sci. 74*:423 (1959).

24

Celery and Other Salad Vegetables

S. S. KADAM
Mahatma Phule Agricultural University, Rahuri, India
D. K. SALUNKHE
Utah State University, Logan, Utah

I. INTRODUCTION

The salad crops are normally consumed uncooked. They are attractive in appearance and are valued for their appetizing succulence, bulk, and vitamin and mineral contents. With improved packaging and transportation, demand for salad vegetables, especially celery and lettuce, is increasing. Most salad crops thrive best during cooler, moist weather and do not tolerate high temperatures. They are grown on a large scale and are found in the market throughout the year. Because salad vegetables are consumed fresh, they can be classified on the basis of edible portion consumed (Table 1). This chapter presents information on celery and other salad vegetables such as globe artichoke, parsley, endive, and chicory.

II. CELERY

Celery is native to the Mediterranean region, its habitat extended from Sweden south to Algeria, Egypt, Abyssinia and in Asia to the Caucasus, Baluchistan, and the mountains of India. It is one of the most important vegetable crops of the United States, being in large part consumed in its raw state, although considerable quantities are used in vegetable juices, soups, stews, and as a cooked vegetable (1). Some celery is grown for processing in soups or pickled vegetable combinations (2).

A. Botany

1. Taxonomy and Morphology

Celery (*Apium graveolens* L.) is a biennial plant but is grown as an annual crop. It belongs to the family Apiaceae. The most common form bears 5–12 thick petioles in a tight bunch or head. A less popular plant is celery root, also known as celeriac (var. *rapeceum*). Celery is a cool

TABLE 1 Classification of Salad Crops

Edible part	Botanical name	Family
Underground portion		
Enlarged taproot		
Beet	*Beta vulgaris*	Chenopodiaceae
Carrot	*Daucus carota* var. *sativus*	Umbelliferae (Apiaceae)
Celery	*Apium gracolens* var. *rapeceum*	Apiaceae
Radish	*Raphanus sativus*	Cruciferae (Brassicaceae)
Turnip	*Brassica compestris* var. *rapifera*	Cruciferae
Parsnip	*Pastinaca sativa*	Umbelliferae
Tuber		
Jerusalem artichoke	*Helianthus tuberosus*	Compositae (Asteraceae)
Bulb		
Florence	*Foeniculum vulgare* var. *azoricum*	Umbelliferae
Onion	*Allium cepa*	Alliaceae
Leek	*Allium ampeloprasum*	Alliaceae
Aboveground portion		
Stem		
Asparagus	*Asparagus officinalis*	Liliaceae
Kohlrabi	*Brassica oleracea* var. *gongylodes*	Cruciferae
Petiole		
Rhubarb	*Rheum rhababarum*	Polygonaceae
Celery	*Apium graveolens* var. *dulce*	Umbelliferae
Leaf		
Cabbage	*Brassica oleracea* var. capitata	Cruciferae
Chicory	*Cichorium intybus*	Compositae
Cress	*Lepidium sativum* L.	Cruciferae
Dandelion	*Taraxicum officinale*	Compositae
Garlic	*Allium sativum*	Alliaceae
Lettuce	*Lactuca sativa*	Compositae
Mustard	*Brassica junceae*	Cruciferae
Onion	*Allium cepa*	Alliaceae
Endive	*Cichorium endiva*	Compositae
Parsley	*Petroselinum crispum* var. *neopolitanum*	Umbelliferae
Watercress	*Rorippa nasturtinum* var. *aquaticum*	Cruciferae
Flowers		
Cauliflower	*Brassica oleracea* var. *botrytis*	Cruciferae
Artichoke (globe)	*Cynara scolymus*	Compositae
Immature fruits		
Cucumber	*Cucumis sativus* L.	Cucurbitaceae
Mature fruits		
Tomato	*Lycopersicum esculentum*	Solanaceae

Source: Ref. 2.

season crop, but it is sensitive to prolonged cold temperature. Growth through the first year normally produces a tight cluster of petioles and leaves attached to a very compressed stem. Celery leaves are divided, pinnatified, and smooth with almost triangular, toothed leaflets that are dark green in color. The leaf stalk is rather broad, furrowed, and concave on the inside. The stem, which appears in the second year, is furrowed and branched. Early vegetative growth is spreading, but leaves arising from the apex of the short stem form compact heads. The root system of celery is spreading and fibrous with many feeder roots close to the soil surface. The plant produces compound clusters, or umbels, of small, white perfect flowers with five petals and five stamens. The seeds (actually fruits formed from two compressed carpels enclosing the actual seed) are small (3).

2. Cultivars

The two general classes of celery varieties are the green varieties, with dark green foliage, and the yellow or "self-blanching" varieties, with yellowish-green foliage. Thompson and Kelly (1) report that since 1920, green varieties have increased in importance as compared to the yellow varieties. The green varieties can be further divided into two groups: those that are early and easy to blanch and others that are late and slower to blanch. The most important green varieties are Utah, Pascal, Emerson Pascal, Summer Pascal, and Ford Hook. Strains of the same variety differ in resistance to diseases and bolting (premature seeding) and in other properties. All green varieties are resistant to *Fusarium* wilt. Emerson Pascal is resistant to early and late blights and to *Fusarium* wilt but has tendency to bolt. Early blanching Sanford Superb or Newark Market is thought to have originated from Golden Self Blanching, the most extensively grown yellow variety. Other important yellow or self-blanching varieties are Michigan Improved Golden, Cornell 19, Supreme Golden, Golden Plume, Wonderful, Golden No. 15, and Cornell 619. Varieties belonging to the Cornell group (Nos. 19 and 619) have thick petioles, whereas Pascal varieties are resistant to *Fusarium* yellowing but have a bolting tendency. White Cornell 19 is susceptible to brown spot, while others such as Michigan Golden, Michigan Improved Golden, and Supreme Golden are resistant to *Fusarium* yellowing (4). The Indian Agricultural Research Institute, New Delhi, recommended Standard Beares and Wright Gieve Grant as improved varieties of celery in India (5).

B. Production

1. Soil and Climate

Well-drained loam, clay loam, and soils with high organic matter content are most suitable for growing celery. The crop requires a mean air temperature of about 16–21°C; lower temperatures cause bolting and shorten the time required for vernalization. Transplanted celery plants should be grown at temperatures above 16°C.

2. Cultural Practices

Celery is propagated by seeding in beds 60–100 cm apart in the field. It can also be transplanted when the plants are about 10–15 cm high. Since the seeds are very fine, depth of sowing should not exceed 1 or 2 cm. Thinning is carried out to maintain a 15–20 cm distance between plants.

Adequate moisture is necessary thoughout germination and growth, and a fertilizer dose of 330 kg of N, 80 kg of P, and 70 kg of K has been recommended per hectare (5). Depending upon nutrient availability and the fertility status of the soil, doses up to 220–450 kg of N, 120 kg of P, and 180 kg of K per hectare may be used. About half of the nitrogen and all of the phosphorus and potassium are applied at the time of planting and the remainder used as side dressing. Deficiencies of boron, magnesium, and calcium have been known to occur in celery (6–8). A heavy dose of nitrogen and nitrogen and potassium fertilization appear to favor development of boron deficiency. Application of 10–20 kg of borax per hectare has been recommended (9). Plants sprayed thoroughly several times during growth with 1.5% boric acid solution are able to absorb enough boron to reduce the incidence of cracked stem (10). A deficiency of magnesium results in chlorosis of the leaves, usually those on the older petioles.

Soil application of magnesium sulfate or spraying with magnesium sulfate can overcome this deficiency (11). Calcium deficiency in celery causes a condition known as blackheart. Spraying celery leaves with calcium nitrate is recommended to avoid this disorder.

The small shallow roots of celery make the frequent replenishment of soil moisture important. Water stress during growth causes serious losses in yield and quality. Gibberellin has been used to increase the length of petioles of celery in tropic (12). Sprays of 50–100 ppm gibberellic acid applied one month before harvest increased the length of petioles of some varieties by 5–7.5 cm over those of the controls (12).

3. Diseases and Pests

There are several diseases of celery. Damping-off of seedlings caused by *Phythium debaryanum* Hesse, *Rhizoctonia solani* Kuhn, and *Sclerotinum rolfail* Sacc. is a major problem. This can be controlled by seed treatment with Captan, Chloronil, or Thiram dust (11). Rhizoctonia stalk rot caused by *Rhizoctonia solani* Kuhn is another major disease of celery (11). Celery is susceptible to bacterial soft rot caused by *Erwinia carotovora*, brown spot by *Caphalosporium apii*, gray mold rot by *Botrytis cinerea*, late blight by *Septoria apicola*, and watery soft rot or pink rot by *Sclerotinia* spp. (13). A few insects such as cutworms, aphids, and mites attack celery.

4. Harvesting

Celery harvested before plants reach full size can be of better quality, and prices received for early-harvested crops may more than compensate for lower yields (14). The plants are either pulled off or cut below the soil surface. The petiole should be attached to the base (15). Since all sizes of celery are marketable (16), once-over harvesting machines are well adapted to this crop. Such machines cut the roots at the proper place, top the stalks, and load them for transport to a central packing plant. The harvested plants should be moved quickly from the field since the plants wilt easily (8). The fact that stalk weight correlates with diameter (17) has been used for machine-sizing celery. Preparations for market include removal of suckers, small lateral branches, and damaged leaves; washing; grading; packaging; and precooling. All operations except the last may be done in the field or the packing plant (1).

C. Chemical Composition

Celery is quite popular in western countries. One hundreds grams of celery contains 6.3 g of protein, 2.1 g of minerals, 1.4 g of fiber, and 1.6 g of carbohydrates (Table 2). Celery contains calcium, phosphorus, iron, carotene, thiamine, vitamin C, niacin, riboflavin, and vitamin A. The composition varies at different stages of maturity.

D. Storage

1. Changes During Storage

Celery stored at 0°C is active, and during early parts of the storage period, normal ripening processes are still going on. Later there is a breaking down of the cells followed by decay (1). Marked changes in the pectic substances of the middle lamella and changes in sugars and nitrogen have been reported (18). The more resistant pectic compounds change to less resistant ones, these changes being correlated with the other changes in celery.

Reducing and total sugars decline markedly in the leaf blades from harvest time to the end of storage. This decrease is accompanied by an increase in the petioles until the end of storage, when there is a marked decline. Soluble nitrogen increases in both the leaf blades and petioles

TABLE 2 Nutritional Constituents[a] of Celery

Constituent	Self-blanching (petiole)	Green (petiole)
Energy (kcal)	29	34
Water (g)	96	95
Protein (g)	0.7	0.9
Fat (g)	0.1	0.1
Carbohydrate (g)	1.2	1.2
Vitamin A (IU)	90	120
Thiamine (mg)	0.03	0.03
Riboflavin (mg)	0.02	0.04
Niacin (mg)	0.3	0.3
Vitamin C (mg)	7	10
Ca (mg)	25	70
Fe (mg)	0.3	0.5
Mg (mg)	10	14
P (mg)	27	34

[a]Per 100 g edible portion.
Source: Ref. 2.

until the end of storage, when there is a marked decrease in the blades and a slight decrease in the petioles. Insoluble nitrogen is generally high in the blades at harvest time, which decreases by the end of storage. Low temperature slows down these changes (1). Celery in storage absorbs foreign flavors. Hence, it should be kept away from the odors of other products.

2. Low-Temperature Storage

Celery is generally hydrocooled or vacuum-cooled. If ice is used in crates or waxed cartons, it is advisable to store celery near 0°C to avoid decay (13). To overcome wilting of celery, packaging in ventilated wraps is advantageous. Parsons (19) recommended that box or crate liners should have about one hole per square foot; bags for individual stalks should have four to six holes. Pantastico et al. (20) recommended temperatures of 0–0.6°C and 92–95% RH to store celery for about 8 weeks. Trenching in the field or storing celery in pits is practiced in the fall when there is a danger of freezing before the crop is harvested, but usually for only a short period of time (1).

3. Controlled-Atmosphere Storage

Parsons (19) found that O_2 levels of < 5% retard yellowing of celery only slightly, and CO_2 > 2.5% may be damaging; all wraps should, therefore, be perforated to avoid potentially dangerous accumulation of CO_2 or reductions in O_2 below 1%. Ryall and Lipton (13) stated that there is little advantage to holding celery under controlled-atmosphere conditions during transit, but for a storage period of one month, CO_2 levels up to 9% may be adopted without harm (13).

4. Chemical Control

Aharoni et al. (21) treated export-grade celery with 250–550 ppm thiabendazole (TBZ). The higher concentrations increased the disinfection efficiency and also increased the level of residues up to 7–9 ppm. Dipping the upper part of petioles in 500 ppm TBZ resulted in good fungal control without increasing TBZ residues more than slightly. Spraying the upper part of petioles with a 200 ppm solution or drenching the whole celery with a 500 ppm solution markedly reduced storage rots caused by *S. sclerotiorum* and *B. cinerea* while the residues level did not exceed 5 ppm. Aharoni et al. (22) reported that a combined disinfection with chlorine (100 ppm) and TBZ (500 ppm) immediately after cutting of petiole ends reduced storage decay in mechanically harvested celery.

Tasca (23) tested nine fungicides during storage of celery over 140 days at 2–4°C. Dusting with Tecto-2 (thiabendazole) at 3.5 kg/t celery or dips of TBZ-20 (tiabendazole) 0.1% or Benlate (benomyl) 0.1% before storage gave the best control of *B. cinerea*, *S. sclerotiorum*, and *Phoma spiicola.* Barkai-Golan and Aharoni (24) reported that dipping celery in a TBZ + chlorine solution inhibited the postharvest development of fungi (*S. sclerotiorum* and *B. cinerea*) and bacteria (mainly *Erwinia carotovora*) during storage. Dipping the upper part of the celery head disinfected the petioles without exceeding the permitted level of TBZ residue. Application of GA_3 to celery delayed yellowing of both petioles and leaflets and further reduced the incidence of storage rots (27).

III. OTHER SALAD VEGETABLES

A. Chicory

Chicory (*Cichorium intybus*) is consumed as a salad vegetable in Europe and the United States. It is also known as French endive, witloof, witloof-chicory, and succory. The first mention of chicory culture was made in 1616 in Germany. Cultivation began in England in 1886 and in France in 1926 (25). It is debatable whether chicory should be classified as a root or a leafy crop. Radichetta and Rosso di Verona are the principal chicory cultivars in the United States and Italy, respectively. Penninck, Christaens, Mueninck, and Van Espen are Belgian cultivars, while Normato, Manila Bubbal Blank, and Slusia Meilof are Dutch cultivars. Flambor, Bergera, and Zoom are chicory hybrids.

Green chicory is harvested when the heads are compact, bright-colored, and well sized; the plants are cut at the base, and damaged or diseased leaves are trimmed. The trimmed heads are washed and cooled by vacuum-cooling or hydrocooling to 0°C. The heads are then packed in wire-bound crates or cartons or baskets for transit. Chicory is rich in vitamin A (Table 3), containing 1400 IU of vitamin A per 100 g and 100 mg of vitamin C (26). Chicory can be stored at 0°C and 90–95% relative humidity for 2-3 weeks. It can be stored for a longer period surrounded by crushed ice.

Chicory is an important salad vegetable especially in France, Belgium, and Holland. Chicory roots can be dried, roasted, and ground and used as a coffee substitute or blended with coffee.

B. Endive

Endive (*Cichorium endiva*) is a popular salad vegetable in Europe and increasingly in the United States. It is believed to be a native of Egypt or India (27). Endive is an important market garden crop in Florida. The endive can be classified into two general groups: the curled- or fringed-leaved cultivars and the broad-leaved cultivars. The green curled Ruffic Deep Heart Fringed, Green Curled Pancalier, and White Curled are curled- or fringed-leaf types (28). The cultivars representing the broad-leaved class are Broad-Leaved Batavian, Full-Heart Batavian or Escarole, and Florida Deep Heart.

Endive grows a loose head of leaves, which are usually ruffled and serrated. The outside leaves are green and bitter, but the inner leaves are light green to whitish. It is a plant with

Table 3 Nutritional Composition of Parsley, Endive, and Chicory Leaves

Constituent	Parsley	Endive	Chicory leaves
Energy (kcal)	16	11	13
Moisture (g)	90	95	92
Protein (g)	2.2	1.3	1.7
Fat (g)	0.3	0.2	0.3
Carbohydrate (g)	1.3	1.2	1.1
Vitamin A (IU)	5200	3500	4000
Thiamine (mg)	0.08	0.07	0.06
Riboflavin (mg)	0.11	0.08	0.10
Niacin (mg)	0.7	0.4	0.5
Ascorbic acid (mg)	90	8	24
Calcium (mg)	125	42	100
Phosphorus (mg)	40	30	47
Iron (mg)	2.0	2.0	0.9

numerous smooth-lobed radical leaves, more or less deeply cut and spreading into a rosette. The stem is hollow, 50 cm to > 1 m tall, ribbed, and branching.

Curly endive provides 81 mg of calcium, 3300 IU of vitamin A, and 10 mg of vitamin C per 100 g as well as moderate amounts of phosphorus, potassium, and iron (Table 3). All endive cultivars contain moderate amounts of minerals.

Endive and escarole are leafy salad greens not adapted to long storage. Even at 0°C, they cannot be expected to keep satisfactorily for more than 2–3 weeks. Vacuum-cooling or hydrocooling can help maintain their fresh appearance. They should keep somewhat longer if stored with cracked ice in or around the packages. The relative humidity in rooms where endive or escarole is kept should be above 95% to prevent wilting.

Fringed-leaved endive is more ornamental and more popular as a salad vegetable than the broad-leaved type. Broad-leaved types are used mainly unblanched in stews and soups.

C. Globe Artichoke

The artichoke or globe artichoke (*Cynara scolymus*) is a thistlelike perennial herb grown for its flower head or bud and belonging to Compositae family. It is a native of the Mediterranean region and appears in its wild form in parts of southern Europe and North Africa.

The shoots and crowns of artichoke senesce after a year's growth and die but are renewed by offshoots from the rootstocks. Both the main stem and lateral shoots bear flowers germinally. Bracts, which are fleshy at the lower end, form the edible portion (1). Mild winters and cool summers favor the growth of artichoke, but the crop can withstand temperatures several degrees below freezing if this does not occur during active growth period. The Green Clode cultivar is grown commercially in the coastal area of California (13).

Globe artichoke is a cool season crop grown in regions having a mean air temperature of 13–18°C. Flower buds become tough in hot and dry climates, and petals tend to spread, shortening the yield (3). Globe artichokes are usually propagated by suckers emerging from the crown. Plants propagated from seeds are variable and of inferior quality.

The harvesting of artichokes begins as the buds mature in the fall and continues through the spring. The stem is cut about 2.5–3.75 cm below the base of the bud by hand. The harvested buds are put in suitable containers to be emptied into bulk bins for transport to the packing shed, where they are graded according to their size and quality. They are then packed in different layers in paper-lined boxes of different sizes; the large and half-size boxes are used for transporting artichokes to distant marketplaces under refrigeration.

Globe artichoke is a good source of calcium, phosphorus, and iron. It contains moderate amounts of thaimine, riboflavin, and ascorbic acid. Ryall and Lipton (13) described various market disorders of artichokes: physical injury (scuffed and split bracts), freezing injury (detached epidermis), and wilting (soft and pliable rather than crisp). Kasmire et al. (29) reported that rubbing of flower buds against each other and against the surface of containers during harvesting and packing is the main cause of scuffing. This may remove the epidermis, causing the underlying tissues to turn brown or black.

Artichokes must be precooled as quickly as possible to retain their quality after harvest. This can be accomplished by hydrocooling or room-cooling to 4.4°C within 24 hours. For prolonged storage, temperatures as close to 0°C as possible are recommended, although 1.7–2.8°C is satisfactory for normal marketing periods (13). The development of gray mold rot (*Botrytis cinerea*) can be effectively prevented by holding artichokes below 4.4°C. Gentle handling aids in maintaining high quality, because physical injury can accelerate decay as much as rise in temperature of 10°C (30). Rappaport and Watada (31) showed that length of storage life of artichokes is related to their initial quality; injured and decaying buds at harvest deteriorate 1½ times more rapidly than sound buds, especially when they are held at 4.4°C or above. Lipton and Steward (27) reported that wilting of artichokes can be prevented by hydrocooling or by washing the buds in cool water followed by immediate cold storage.

Poma-Treccani and Anoni (26) reported that CA (3% CO_2 + 3% O_2) extended the storage life of artichoke for about one month, mainly by reducing the browning of bracts. Controlled and low-oxygen atmospheres preserve the shelf life of several vegetable crops, including artichokes and asparagus, for a considerable time (32).

D. Parsley

Parsley (*Caruro petroselinum*) is a native of the Mediterranean region and Europe. It was popular in early times among the Greeks and Romans. Its value was reputed to be in aiding digestion and in supressing odors of onion and wine. Parsley is produced commercially in Texas, California, New Jersey, Florida, and New York (28).

Parsley is a rosette of divided leaves on a short stem. The most common is the curled-leaf type, which contains three subtypes: doubled-curled (moss-curled), evergreen, and triple-curled. These subtypes are distinguished by the degree of leaf curling, leaf coarseness, and plant growth habit. Plain leaf parsley (Dark Green Italian) has deeply cut leaves but no curling or fringing. Moss-curled (double-curled) has a 30-cm-tall stem and virogous, compact, and very dark green leaves, which are finely cut, deeply curled, and frost resistant. Evergreen leaves are coarsely cut, while extra triple-curled leaves are finely cut and very closely curled; the triple-curled variety has slightly shorter leaves, which are closely curled. The leaves of Paramount are tall, very uniform, triple-curled, and very dark green. Plain-leaved types include plain (singles) and have flat leaves. The leaves of the Dark Green Italian cultivar are heavy and glossy green (28).

Five types of parsley cultivars are available: plain-leaved, celery-leaved or Neapolitan, curled, fern-leaved, and turnip-rooted. The curled type is common in the United States. Plain and turnip-rooted parsley are also grown in the United States. The best known varieties are Moss, Extra Triple Curled, Fern-Leaved, Curled Dwarf, Evergreen, Extra Triple Curled, Plain, and Dark Green Italian. Morgo was developed in Sweden, and Hamburg is the only turnip-rooted cultivar.

Parsley can be harvested over a long period of time, cutting the outer and larger leaves only and tying in bunches for market. The leaves are packed loose or bunched and washed thoroughly before packing in crates or bushel baskets. Crushed ice is usually placed in the package, and shipment is made under refrigeration to distant markets. Much of the crop goes to the institutional trade (restaurants and hotels) and for dehydration. Under storage at 0°C and high humidity, parsley can be held for up to 2 months.

Parsley is a rich source of vitamin A, vitamin C, and calcium (Table 3). It is also a good source of potassium and a moderate source of iron, sodium, and phosphorus.

Parsley is popular as a garnish for salads, sandwiches, and cooked dishes, as a flavoring for soups and pasta, and as a salad ingredient in small amounts because of its pungent flavor. The Hamburg type is used mainly in stews and soups. The swollen root of turnip-rooted parsley is eaten as a cooked vegetable. Parsley is one of the best vegetable sources of vitamin A and C, but its contribution to the diet is negligible.

REFERENCES

1. Thompson, H. C., and W. C. Kelly, *Vegetable Crops,* 5th ed., McGraw-Hill Book Co. Inc., New York, 1957.
2. Gupta, K., J. L. Mangal, G. R. Singh, and D. S. Wagle, Salad crops, in *Encyclopaedia of Food Science, and Food Technology and Nutrition* (R. McRae, R. K. Robinson, and S. J. Sandler, eds.), Academic Press, New York, 1993, p. 3974.
3. Ryder, E. J., *Leafy Salad Vegetables*, The AVI Pub. Co., Westport, CT, 1979.
4. Newhall, A. G., Blights and other ills of celery, *U.S. Dept. Agric. Year Book,* 1953, p. 408.
5. Choudhury, R., *Vegetables*, National Book Trust of India, New Delhi, 4th ed., 1967, p. 214.
6. Yamaguchi, M., and P. A. Minges, Brown checking of celery, a symptom of boron deficiency, I. Field observation variety, susceptibility and chemical analyses, *Proc. Am. Soc. Hort. Sci. 68*:318 (1956).
7. Pope, D. T., and H. N. Munger, Heredity and nutrition in relation to magnesium deficiency chlorosis in celery, *Proc. Am. Soc. Hort. Sci. 61*:472 (1953).
8. Geraldson, C. M., The control of blackheart of celery, *Proc. Am. Soc. Hort. Sci. 63*:353 (1954).
9. *Annual Work Progress Report*, Directorate of National Agriculture, Republic of Vietnam, 1963.
10. Yamaguchi, M., F. W. Zink, and A. R. Spurr, Cracked stem of celery, *Calif. Agric. 7*(5):12 (1953).
11. Knott, J. E., and J. R. Deanon, Jr., *Vegetable Production in Southeast Asia*, University of Philippines, College of Agriculture, College Los Banos, Laguna, Philippines, 1967.
12. Takatori, F. H., O. A. Lorenz, and F. W. Zink, Gibberellin sprays on celery, *Calif. Agric. 13*(T):3 (1959).
13. Ryall, A. L., and W. J. Lipton, *Handling, Transportation and Storage of Fruits and Vegetables*, Vol. 1, *Vegetables and Melons*, The AVI Pub. Co., Westport, CT, 1972.
14. Aharoni, N. R., Barkai-Golan, R., Ramaraz, and H. Aviram, Mechanical packaging of celery for export, in *Research Activities of Fruit Section of the Institute of Crop Science*, Technische University Obstabau, Berlin, 1979.
15. Thompson, A. K., M. B. Bhatti, and P. P. Rubio, Harvesting and handling: Harvesting, in *Postharvest Physiology of Fruits and Vegetables* (E. B. Pantastico, ed.), The AVI Pub. Co., Westport, CT, 1975, p. 236.
16. Showalter, R. K., Mechanizing the harvesting and postharvest handling of snap beans, celery and sweet corn, *Proc. Fla. Sta. Hort. Soc. 80*:203 (1967).
17. Gull, D. D., Homogeneity of sized celery in Florida, *Proc. Fla. Sta. Hort. Soc. 77*:137 (1964).
18. Corbett, L. W., and H. C. Thompson, Physical and chemical changes in celery during storage, *Proc. Am. Soc. Hort. Sci. 30*:346 (1925).
19. Parsons, C. S., Effects of temperature, packaging and sprinkling on the quality of stored celery, *Proc. Am. Soc. Hort. Sci. 75*:463 (1960).
20. Pantastico, E. B., H. Subramanyam, M. B. Bhatti, N. Ali, and E. K. Akamine, Postharvest physiology:

Harvest indices, in *Postharvest Physiology, Handling and Utilization of Tropical and Sub-tropical Fruits and Vegetables* (E. B. Pantastico, ed.), The AVI Pub. Co., Westport, CT, 1975, p. 56.

21. Aharoni, N., R. Barkai-Golan, and H. Aviram, Reducing rot in stored lettuce with polyethylene lined packaging, in *Scientific activities 1974-1977*, Pamphlet No. 184, Institute for Technology and Storage of Agriculture Product, Bet Dagan, Israel, 1978, p. 57.
22. Aharoni, N., R. Barkai-Golan, R. Ramaraz, and H. Aviram, Mechanical packaging of celery for export, in *Research Activities of Fruit Section of the Institute of Crop Science*, Technische Universität Obstabau, Berlin, 1979.
23. Tasca, O., Studies on the control of the principal fungi causing celery rot during storage. Lucrari, *Stiinifice Inst. de Cercetari si Proiectari Pentru Valorifires ai Industrializerea Legume 1 or si Fructelor 11*:29 (1982).
24. Barkai-Golan, R., and N. Aharoni, Combined disinfectant and growth substance treatment to delay deterioration in stored celery, *Hassadeb 61*:1484 (1981).
25. Gupta, K., J. L. Mangal, G. R. Singh, and D. S. Wagle, Salad crops—leaf types, in *Encyclopaedia of Food Science, Food Technology and Nutrition* (R. MaCrae, R. K. Robinson, and M. J. Sandler, eds.), Academic Press, London, 1993, p. 3968.
26. Poma-Treccani, C., and A. Anoni, Controlled atmosphere packaging in polyethylene and defoliation of the stalks in the cold storage of artichokes, *Riv. Ortoflorofruttic. Ital. 53*:203 (1969).
27. Lipton, W. J., and J. K. Stewart, Effects of precooling on the market quality of globe artichokes, *U.S.D.A. Marketing Res. Rept. No. 633* (1963).
28. Gupta, K., J. L. Mangal, G. R. Singh, and D. S. Wagle, Salad crops—leaf stem crops, in *Encyclopaedia of Food Science Food Technology and Nutrition* (R. MaCrae, R. K. Robinson, and M. J. Sandler, eds.), Academic Press, London, 1993, p. 3974.
29. Kasmire, R. F., A. S. Greathead, and M. J. Snyder, A demonstration of the effect of careful handing and rapid pre-cooling on the market quality of California globe artichokes, Unnumbered Mimeo of the University of California, 1958.
30. Lipton, W. J., and J. M. Harvey, Decay of artichoke bracts inoculated with spores of *Botrytis cinerea* Fr. at various constant temperatures, *Plant Dis. Reptr. 44*:837 (1960).
31. Rappaport, L., and A. E. Watada, Effect of temperature on artichoke quality, *Proc. Conf. Transport Perishables*, University of California, Davis, 1958, p. 142.
32. Dewey, D. H., R. C. Herner, and D. R. Dilley, Controlled atmospheres for the storage and transport of horticultural crops, *Proc. Natl. Controlled Atmosphere Res. Conf.*, Michigan State University, 1968.

25

Leafy Vegetables

Rekha S. Singhal and Pushpa R. Kulkarni
University of Bombay, Bombay, India

I. INTRODUCTION

The importance of dark green leafy vegetables in the nutrition of populations in developing countries has been recognized. However, edible green leafy vegetables appear to be underutilized throughout the world and may in some areas even be diminishing in use. They are inexpensive, high yielding, already a part of the local diet, and often easily available. The importance of edible leaves in human diets seems to be inadequately stressed. They supply the roughage required in the daily diet. Dieticians recommend a daily consumption of at least 116 g of vegetables for a balanced diet. The consumption of dark green leafy vegetables that contain β-carotene or provitamin A is highly recommended in nutrition education.

Most cultivated greens are grown mainly in the cooler regions. Chad and New Zealand spinach are relatively tender, cold sensitive, and generally grown during the summer in the north (1). Spinach is the only leafy vegetable that is grown on a commercial scale for shipment to distant markets and for canning and freezing. Turnips, beets, and sea kale are also used as leafy vegetables. In addition to these, many wild plants such as dandelion, several species of dock, lamb's quarters, wild cress, pokeweed, mountain spinach or orach (*Chakwat*) (*Atriplex hortensis* L.), *Amaranthus* spp., purslane (*Portulaca oleracea* L.), colocasia (*C. esculenta* L.), kale (*Brassica oleracea* var. *acephala*), and mustard (*B. juncea*) are grown as leafy vegetables (2).

II. SPINACH

Spinach is probably a native of southwest Asia and is grown extensively for spring and fall in the northern United States, for late fall, winter, and early spring shipment in the south, and for canning in the Pacific Coastal states and other areas (3). Spinach thrives best during the cooler and moist seasons and exhibits a marked tendency towards seeding when grown during warm periods, especially when the days are long.

Spinach was first cultivated by the Arabs. The Moors took it to Spain, after which it spread to other parts of the world. The word spinach comes from the Spanish word *Hispania*. It belongs to the family *Chenopodiaceae*, genus *Spinacia*, and species *oleracea*.

A. Botany

Spinach (*Spinacia oleracea*) develops a rosette of leaves soon after germination, and the short stem begins to elongate as the plant grows. Secondary/Lateral branches arise from the leaf axis of both the central and lateral stems, and the flower clusters are borne axillary on both the larger stems and the smaller branches (1).

Flowering normally begins on the middle portion of the larger stems and proceeds towards the base and the top. Four types of spinach plants are known with reference to sex expression: extreme males, vegetative males, monoecious, and females. Spinach are usually dioecious, with occasional monoecious plants (3).

There are two distinct types of spinach. Yawalkar (4) called *Spinacia oleracea* Vilayati Palak and *Beta vulgaris* Common Palak or spinach beet. The latter has long leaves with entire margins, and its flowers have both male and female sexual parts (dioecious). Spinach beet grows well in hot wealther, whereas *Spinacia oleracea* is a cold season plant. The Indian Agricultural Research Institute, New Delhi, has released Virginia Savoy and Early Smooth Leaf as important varieties of *Spinacia oleracea*. Depending upon leaf types, Juliana (blue-green), Long Standing Bloomsdale (dark green), Virginia Savoy, and Dark Green Bloomsdale have plants that are very early seeding and medium-early seeding. While Old Domination has medium-savoyed leaves, Hollandia and Vitro Flay leaves are not savoyed or slightly savoyed in cool weather. Amsterdam Giant plants are not early seeding and have prickly seeds. King of Denmark and Nobel can be distinguished by their blue-green and medium green leaves, respectively.

B. Production

1. Soil and Climate

Spinach grows best in sandy loam or alluvial soils. However, it can be grown in any good soil with pH between 7.0 and 10.5 (5). The freezing tolerance of spinach is attributed to the presence of cold-regulated proteins in the leaves (6).

2. Propagation

Spinach is propagated by seed. Spinach seeds are sown in August-September in the hills and between October and December in the plains. The crop sown in spring-summer usually has a greater tendency to bolt. The seeds are either broadcast or sown in rows, with a row-to-row spacing of 20–22 cm to encourage optimum plant growth. The recommended sowing depth is about 3.5–4.0 cm for both kinds of seeds. The seed rate varies with the method of sowing and species. In spinach about 11.5 kg of seeds are required for one acre. Due to its dioeceous nature and because male plants grow poorly, spinach requires a higher seed rate than spinach beet. Proper land preparation and adequate soil moisture are needed to ensure satisfactory germination.

3. Irrigation and Manuring

The first irrigation, which is usually a light one, is given immediately after sowing, while subsequent irrigations should be given according to the need. As a general rule, one irrigation every 4–6 days in summer and every 10–12 days in winter can provide soil moisture levels adequate for optimum growth of plants.

Singh and Joshi (7) recommended the application of 10 cartloads of farmyard manure as a basal dose followed by three top dressings each of 45.5 kg of ammonium sulfate per acre. Based on investigations carried out at various centers, Kale and Kale (5) summarized manurial recommendations consisting of 25 cartloads of farmyard manure, 50–190 kg of nitrogen, and 88 kg each of phosophorus and potassium per hectare. However,t he actual amount of fertilizer to be applied should be based on soil analysis. Farmyard manure is applied before sowing and mixed into the soil properly. One third of the nitrogen and the full dose of phosphorus and potassium is given before sowing, with the remaining two-thirds of the nitrogen being applied in two equal split doses after the first and second cutting. In addition, urea (1.5%) can be applied as a foliar spray 15 days after germination and repeated later after every cutting (5).

4. Diseases and Pests

Bacterial soft rot, caused by *Erwinia carotovora* and other species of bacteria, and downy mildew, caused by *Peronospora effusa* (*farinosa*), cause severe losses of spinach (8,9). In bacterial soft rot, the infected leaves are gray-green, appear water-soaked, and feel mushy. The bacteria generally enter through wounds or via fungal lesions. The decay progresses at a faster rate at high humidites and temperatures above 7°C. The first signs of downy mildew include pale-yellow, irregularly shaped areas without distinct margins. Whitish-gray mycelium grows on the leaf surface at high humidities; the mycelium is generally absent under low humidity. The infection of downy mildew occurs primarily in the field, but it becomes prominent after harvest, especially at temperatures of 14–24°C and above 85% relative humidity.

5. Maturity, Harvesting, and Handling

Spinach is usually harvested from the time plants have five to six leaves until just before the seed stalks develop. Thompson and Kelly (1) stated that a larger yield is obtained when the plants are allowed to develop to full size. Spinach for fresh market is generally harvested by cutting the taproot just below the lowest leaves; for processing, the plants are cut off about an inch above the solid surface. Separate machines have been developed for harvesting spinach for fresh market (to cut the whole plants) and processing (to cut the leaves). The spinach used for processing is grown during short days. More than one harvest can be made before seed stalks develop. "Clip-topped" spinach is preferred, since less labor is required for packaging it than the crown-cut product and the former yields more than the latter (10).

Thompson and Kelly (1) do not recommend cutting plants immediately after rain or heavy dew because the leaves are crisp and break easily when wet. Yellowed and diseased leaves should be removed and the product handled carefully to prevent bruising or damaging of leaves and stems. Water spraying of harvested spinach on conveyor belts is preferred to washing the produce in tanks to avoid injury. Ryall and Lipton (9) recommended that spinach be harvested only during the cooler parts of the day to minimize wilting, a major disorder of spinach. Spinach loses water more rapidly when warm than when cool, even at high humidity.

Round bushel baskets, hampers, and crates are used to pack spinach for shipment, and crushed ice is generally placed in the containers when refrigerator cars or trucks are used for long hauls. Spinach sold in retail stores is prepackaged in transparent plastic film bags, which allow gas exchange but maintain high humidity by reducing water evaporation (1).

C. Chemical Composition

The chemical composition of spinach is given in Table 1. Spinach is rich in iron and calcium (Table 2), but the calcium is probably unavailable, as it forms a complex with oxalic acid, leading to the formation of calcium oxalate. Kim and Zemel (14) found calcium from spinach to be less

TABLE 1 Percent Chemical Composition (on DM basis) of *Spinacia oleracea*

Constituent	Content (%)
Dry matter	8.50
Total ash	5.90
Silica	1.60
Crude protein	22.75
NDF	31.06
ADF	7.02
Hemicellulose	24.04
Cellulose	5.02

Source: Ref. 11.

soluble than that from skim milk. Poor availability of calcium from spinach has been confirmed in humans (15) and rats (16). Boiling spinach in water for 20 minutes had no effect on the availability of calcium (17). Similarly, only about 2.8% of the total iron is available in vitro (18). Foliar application of s-triazines has been shown to increase the protein content as well as the iron, magnesium, phosphorus, and potassium contents in spinach leaves (19). Spinach leaf protein has been shown to have hypocholesterolemic properties (20). The amino acid composition of the protein in spinach leaves is shown in Table 3.

Spinach leaves are known to contain about 0.6% fat. The fatty acid composition is shown in Table 4. A notable feature is the presence of high amounts of 18:3 fatty acid (22).

Plant materials, including leafy vegetables, contain many antinutritional components, of which nitrate occurs in abundance. The nitrite content of vegetables is generally low, but plants that accumulate high levels of nitrate are known to contain significant amounts of nitrite. Postharvest storage of vegetables also leads to accumulation of nitrite arising from the reduction of nitrate caused by plant enzymes and/or microbial activity. *Spinacia oleracea* leaves are reported to contain 1075–2300 ppm nitrate (mean 1703 ppm) (24–26) and 4.0–5.3 ppm nitrite (mean 4.7 ppm). Variations in nitrate levels may appear due to cultivar (27), levels of nitrogen

TABLE 2 Mineral Composition of Spinach

Mineral	Content (mg/100 g raw material)
Sodium	140
Potassium	500
Calcium	170
Magnesium	54
Phosphorus	45
Iron	2.1
Copper	0.04
Zinc	0.7
Sulfur	20
Chloride	98
Manganese	0.6

Source: Refs. 12, 13.

TABLE 3 Amino Acid Composition of Spinach Leaf Protein

Amino acid	Content (g/g N)
Arginine	0.35
Histidine	0.14
Lysine	0.40
Tryptophan	0.10
Phenylalanine	0.33
Tyrosine	0.31
Methionine	0.11
Cystine	0.08
Threonine	0.29
Leucine	0.53
Isoleucine	0.30
Valine	0.35

Source: Ref. 21.

fertilizer used (26,27), growth temperatures (28,29), or nitrate reductase activity in the leaf blades (30). Frequent consumption of large quantities of leafy vegetables containing nitrate may prove hazardous, particularly to infants. Nitrate reduced to nitrite, either in the saliva on chewing or in the upper gastrointestinal tract, not only causes methemoglobinemia in infants but also increases the risk of formation of carcinogenic nitrosamines. Exposure to light before harvesting decreases the nitrate levels in spinach, and hence harvesting the leaves should be done at an appropriate time. Storage of spinach at 8 and 22°C for 24 hours is reported to increase the nitrate by 54% and 27%, respectively, suggesting that the storage time for spinach should be minimized

TABLE 4 Fatty Acid Composition of *Spinacia oleracea* in the Lipids of Raw, Frozen, and Canned Samples

	Fatty acid (%)		
Fatty acid	Raw	Frozen	Canned
16:0	17.3	16.1	13.6
16:1	1.4	1.8	1.1
16:2	0.3	1.1	1.8
16:3 + 18:0	6.7	6.6	9.0
Total C16	25.7	25.6	25.5
18:1	7.8	6.7	6.0
18:2	13.5	13.0	15.4
18:3	53.0	54.7	53.1
Total C18	74.3	74.4	74.5
U/S ratio	4.8	5.2	6.4
L/O ratio	6.8	8.2	8.9

U/S = Unsaturated fatty acids/Saturated fatty acids; L/O = linoleic/oleic acid.
Source: Ref. 23.

(31). High nitrate content also leads to processing problems due to can corrosion, and hence developing cultivars that accumulate less nitrate would be advantageous to the consumers (32). Secondary amines are important since biogenic formation of N-nitroso compounds is possible. Spinach leaves contain about 275 μg/g dry weight of secondary amines (33). However, specific and highly sensitive tests on raw whole and various processed forms of spinach such as frozen, frozen and thawed, and cooked spinach failed to show the presence of any nitrosamines (34), even in samples that were considered inedible and had very high nitrite levels of 300–500 ppm.

Oxalate levels of 658–1760 mg/100 g have been reported in spinach. The oxalate content increases with growth (35), an increase in air temperature and light intensity (36), but is less influenced by the type of fertilizer used (37). It is thought that *L*-ascorbic acid might be metabolized to oxalic acid in oxalate-accumulating plants, including spinach (38). Glycolate is also perceived to be an efficient precursor of oxalic acid in spinach leaves (39).

Spinach contains appreciable quantities of ascorbic acid and riboflavin (Table 5) and small quantities of thiamine (1). It is rich in vitamins, particularly carotene, a precursor of vitamin A (12). Addition of spinach to wheat flour dough increases the carotene content (41). Other carotenoids identified by Khacik et al. (42) are *trans*-lutein, *trans*-plus 15, 15'-*cis*-β-carotene, *cis*-neoxanthin, and violaxanthin, which comprise about 39.6, 18.4, 13.7, and 11.4%, respectively, of the total carotenoid content (346 μg/g) in spinach. An increase in the illumnination intensity and air temperature increases the productivity of the biomass and the contents of ascorbic acid and carotene (43). Ascorbic acid content also increases with plant age (44). Application of nitrogen fertilizers has a positive effect on contents of vitamins B_1 and B_2, although nitrogen and potassium fertilizers have a negatvie effect on vitamin E content (45). Vitamin K_1 or phylloquinone at a level of 299.5 μg/100 g is reported for spinach (46). The phylloquinone content of spinach varies according to geographical location, which could be due to differences in the soil, climate, and other growth conditions. The level increases with maturity (47). Folate is present at 123 μg/100 g edible portion, of which 51.0 μg is in the free form. About 58.5% of the folate exists in the conjugated form (48). Histamine has also been reported in *Spinacia oleracea* leaves (49).

Modori is a term used to describe the loss of elasticity of fish gel during the heating process in Kamaboko manufacture. This is undesirable. Many plant extracts have been evaluated for their potential to inhibit *modori*, of which spinach leaf extract showed the greatest activity. This inhibitor has been characterized by gel filtration to reveal its molecular weight to be less than

TABLE 5 Vitamin Contents in Spinach

Vitamin	Content (per 100 g raw material)
Carotene (μg)	35.35
Vitamin E (mg)	1.71
Vitamin K (mg)	25.00
Thiamine (mg)	0.07
Riboflavin (mg)	0.09
Niacin (mg)	1.2
Vitamin B_6 (mg)	0.27
Folate (μg)	150
Pantothenate (mg)	0.27
Biotin (μg)	0.1
Vitamin C (mg)	26

Source: Refs. 12, 13, 40.

10,000 daltons (50). Proteasome, a multicatalytic protease complex containing at least three types of activities—chymotrypsinlike, *Staphylococcus aureus* V8 protease–like, and trypsinlike—has been isolated from spinach leaves (51).

D. Storage

Low humidity and rapid air movement often result in rapid wilting of leafy vegetables, making the product less attractive. Wilting is known to hasten the loss of ascorbic acid in spinach, but is of much less importance than unfavorable temperatures (52).

1. Low-Temperature Storage

Continuous refrigeration near 0°C from packaging to final retail sale is generally recommended. Ryall and Lipton (9) reported that spinach can be package-iced, hydrocooled, or vacuum-cooled equally well. If hydrocooled, it should be centrifuged for about 3 minutes to remove excess water to retard decay under higher than ideal temperatures. Spinach that is held above 24°C should be wetted before it is vacuum-cooled. The use of chlorinated water (100 ppm) for washing or cooling spinach avoids build-up of bacteria in the water but has little effect on its decay. Plantenius (53) reported that spinach becomes unsalable after 24 days at 1.6°C, 7 days at 10°C, or 2–2½ days at 18.3°C. Ryall and Lipton (9) commented that these periods probably should be about halved to meet the present day standards of quality. Yamauchi et al. (54) reported that chlorophyll and ascorbic acid contents in spinach stored at 20°C for 6 days decreased greatly and that yellowing increased. By the third day at 20°C, about 83% of ascorbic acid was lost. In contrast, thiamine is reported to increase sharply during storage and riboflavin is unaffected (55). Peroxidase degraded chlorophyll in the presence of hydrogen peroxide, but degradation was inhibited by the addition of L-ascorbic acid. The degradation was slower at 0°C than at 20° C (55).

Vitamin A content decreased by 21.6% and 23.7% on storage at room temperature (29–32°C) in a closed vessel and in a refrigerator (5°C) for 2 days, respectively (56).

2. Controlled-Atmosphere Storage

Murata and Ueda (57) found that CO_2-enriched atmosphere (10% CO_2) retarded yellowing of spinach and maintained the product in good quality for 3 weeks at 5°C. The controlled atmosphere (CA) thus prolongs the shelf life of spinach by about one week at 5°C. These workers (57) noticed no injuries in spinach held 3 weeks in 4% CO_2, but the higher level of CO_2 (above 10%) had no beneficial effect.

E. Processing

Sistrunk et al. (58) investigated the relationship between processing methodology and quality attributes and nutritional value of canned spinach. They found that the volume of water utilized for washing did not affect either the quality attributes or the nutritional value of the canned spinach. Dipping spinach in a detergent solution prior to washing decreased greenness, optical density of liquor, nitrates, and ascorbic acid (Table 6). Color was rated slightly higher for spinach that was blanched in steam. Water-blanching leached out more nitrate (Table 7), riboflavin, and ascorbic acid than steam-blanching, but the carotene content was not affected. Defrosting frozen spinach in a refrigerator after steaming or cooking in a microwave oven results in higher residual ascorbic acid than other methods of thawing (59). Steam blanching is known to convert 20% of the calcium into insoluble and nonabsorbable calcium oxalate (60). Magnesium, in contrast, is less sensitive, and iron is quite resistant to cooking (61). Dipping spinach in a magnesium

Table 6 Characteristics of Canned Spinach

Treatment	Drained wt (g)	Shearpress (kg/150 g)	Panel rating		Color attributes			NO_3N (ppm)
			Color	Texture	L	–a	b	
Blanch method								
Rotary steam	275a	54a	7.4ab	7.3a	20.9a	0.23a	10.9b	1707a
Still steam	260b	54a	7.5b	7.0b	70.7b	0.05b	11.0a	1225b
Water	256c	55a	7.3a	7.2ab	20.9a	0.20a	10.9b	1013a
Hydrocooling								
Yes	251b	54b	7.4a	7.1a	20.7b	0.16a	10.9b	1223b
No	277a	56a	7.4a	7.2a	20.9a	0.16a	11.0a	1407a
Detergent dip								
Yes	265a	55a	7.4a	7.2a	20.8a	0.10b	11.0a	1258b
No	263a	55a	7.4a	7.4a	20.8a	0.21a	10.9b	1372a

Mean separation in columns within variable by Duncan's multiple range test, 5% level.
Source: Ref. 59.

carbonate solution maintains adequate quantities of chlorophyll and carotenoids in the freezing and drying treatments after storage for 3 months (62). Cleanliness and convenience (e.g., spinach ready for cooking or precut and premixed salad vegetables) are the principal selling points of these vegetables (63).

Canning and freezing at –18°C also affects ions—both cations and anions (Table 8). These processes can lead to cell rupture and hence ion loss. In particular, nitrate reduction as a result of freezing (64) and canning (65) has been reported. Lead and cadmium contents in canned spinach of 0.13–0.95 and 0.03–0.18 ppm, respectively, have been reported (66). Cadmium is believed to be originally present in spinach, but lead may arise from contamination during the canning process or be due to the release of lead from the can.

Cooking of spinach in distilled water for 60 minutes decreases vitamin A, as shown in Table 9. The effect of different processing techniques such as microwaving and steaming on the carotenoid profile of spinach is shown in Table 10. Epoxycarotenoids are somewhat sensitive to

Table 7 Contents of Canned Spinach

Treatment	Ascorbic acid (mg/100 g)	Riboflavin (ppm)	Carotene (ppm)	OD liquor (475 nm)	Grit and sand (mg/200 g)
Blanch method					
Rotary steam	17a	24.0a	183a	0.266a	7a
Still steam	15a	23.5a	183a	0.258b	8a
Water	9c	21.0b	179a	0.213c	5a
Hydrocooling					
Yes	13b	21.5b	182a	0.227b	6a
No	14a	24.1a	181a	0.266a	8a
Detergent dip					
Yes	13b	22.6a	182a	0.236b	8a
No	14a	22.6a	183a	0.237a	6a

Mean separation in columns within variable by Duncan's multiple range test, 5% level.
Source: Ref. 59.

TABLE 8 Effect of Processing on Cations and Anions in Spinach[a]

	Raw	Frozen	Canned
Anions			
Chloride	2769	1017	3483
Nitrate	1062	590	490
Phosphate	Not detected	Not detected	Not detected
Sulphate	921	111	214
Oxalate	2592	1376	1200
Cations			
Sodium	1387	960	1001
Ammonium	543	323	343
Potassium	10200	5400	4100

[a]Values are in ppm on dry weight basis.
Source: Ref. 67.

heat treatments, but lutein and hydrocarbon carotenoids such as neurosporene, α- and β-carotene, lycopene,ζ-carotene, phytofluene, and phytoene survived heat treatments (42).

Vitamin C is present in spinach but is destroyed by even simple processing techniques, such as cutting and washing (Table 11). Folate is present at about 251.4 μg/100 g in spinach but it lost due to degradation or into the processing effluent (70). Free and total folate contents in spinach at different stages of processing are presented in Table 12. While water-blanching could cause a 33% loss in folate, microwave-blanching resulted only in 14% loss (71). About 35% of folate was lost during canning (72). Leichter *et al.* (73) found 22% retention of folate in cooked spinach, while 65% of the original folate leached into the cooking water. Klein *et al.* (74) reported 77% retention of folate in cooked spinach. Irradiation at 10 kGy improves the bioavailability of folate in spinach leaves (75). Blanching also decreases nitrate and nitrite in spinach (76).

Delay in the canning process may lead to contamination of raw spinach by nitrate-reducing organisms, which generate nitric oxide, carbon dioxide, and hydrogen sulfide gas. These cause bloating of the cans (77).

Treatment of composite effluent from spinach canning with 40 mg/liter of chitosan reduced suspended solid concentration from 298 to 0 mg/liter (78).

Spinach leaves are sun-dried before storage in many developing countries. This is, however, a very destructive procedure for β-carotene. The effect of drying on the β-carotene content of spinach is presented in Table 13. Solar drying is reported to be better than oven-drying with respect to retention of ascorbic acid (80). Processing of spinach to powders using β-cyclodextrin gives a superior product with less tendency to fade (81). Dried spinach flour in

TABLE 9 Vitamin A Activity in *Spinacea oleracea*

	Vitamin A activity (μg/g)
Uncooked	33.3
Cooked	25.0
Loss of vitamin A activity (%)	25
Residue in cooking water	3.1

Source: Ref. 68.

TABLE 10 Effect of Processing on Carotenoid Profile of Spinach[a]

Carotenaid	Raw	Steamed	Microwaved
Neoxanthin	2.36	2.47	2.26
Violaxanthin	7.40	4.90	4.80
Lutein 5,6-epoxide	0.50	nd	nd
trans-Lutein	8.06	9.26	7.46
cis-Lutein	1.40	1.36	1.33
Total lutein	9.50	10.60	8.73
β-Carotene	8.90	9.86	9.10

[a]mg/100 g edible leaves.
Source: Ref. 42.

pasta increases the iron content (82). It has been reported that ultraviolet light is quite destructive, and even after a short period of radiation, 65% of the carotene may be lost (79). The effect of various processing and preservation methods on the phylloquinone or vitamin K_1 content of spinach is as shown in Table 14. Cooking does not significantly alter the phylloquinone content. This is mainly due to its poor solubility in water.

III. SPINACH BEET

Spinach beet (*Palak*) is one of the most common leafy vegetables of tropical and subtropical regions. The succulent leaves and stem are used in the preparation of various dishes. It is a biennial cultivated as an annual for its edible leaves. It has a number of synonyms, including silver chard, seakale beet, silver beet, and perpetual spinach. The leafy beets are closely related to the beet root (*B. vulgaris* ssp. *rubra*) and probably developed from an ancestral form native to the Mediterranean region. It is grown to a limited extent in temperate regions (12).

A. Botany

Spinach beet (*Beta vulgaris* var. *bengalensis* Hort.) is closely related to Swiss chard. It belongs to the family *Chenopodiaceae*. It has chromosome number $2n = 18$.

TABLE 11 Effect of Processing Techniques on Vitamin C Content of *Spinacea oleracea*

	Vitamin C content (mg/100 g)
Control, no treatment	34.4
Chopping with stainless steel knife	22.4
Washing in tap water	30.9
Washing in tap water after cutting	15.9
Washing in tap water before cutting	24.8
Cooking with lid for 2 minutes	17.1
Cooking without lid for 2 minutes	11.2

Source: Ref. 69.

TABLE 12 Free and Total Folate in Spinach at Different Stages of Processing[a]

Product	Free folate (μg/100 g)	Total folate (μg/100 g)	% Retention	% Total folate in processing media
Spinach				
Raw	160.9	251.4		
Water-blanched	39.1	42.5	17	
Steam-blanched	34.9	146.7	58	
Canned	7.1	21.2	50 (12)	
Canned/stored (3 months)	1.4	17.0	40 (10)	
Frozen (3 months)	6.0	30.8	72 (12)	
Processing media				
Water-blanch effluent	2.0	14.5		40
Steam-blanch condensate	0.8	1.1		1
Canning medium	7.0	34.0		14
Canning medium (3 months)	12.3	44.6		18

[a]Retention based on total folate in the product at each processing effect. Numbers in parenthesis indicate retention values relative to raw sample.
Source: Ref. 70.

1. Morphology

Spinach beet is a herbaceous annual for edible leaf production (84), while it is biennial for seed production. The early stages of plant growth are characterized by production of succulent tender edible leaves on a small thick stem. During the reproduction phase, the stem elongates after about 75 days, which results in bolting. The plant has a taproot system along with numerous feeder roots. Leaves are long with entire margin and long petioles.

2. Cultivars

The pigmentation of midrib and leaf veins is used to classify varieties into two groups: reddish midribs and leaf veins, and green midribs and leaf veins.

Improved spinach cultivars include All Green, Pusa Palak, Pusa Jyoti, Pusa Harit, Jobner Green HS 23, Palak No. 56-6, and Banaresee Giant, released in India.

TABLE 13 β-Carotene Content of Spinach

	β-Carotene (mg /kg fresh material)
Fresh spinach leaves	29
Dried leaf powder	7
Dried leaf powder boiled with potash	8
Dried whole leaves, boiled	11
Dried whole leaves, rehydrated by soaking	15
Mean value for dried spinach	10

Source: Ref. 79.

Table 14 Phylloquinone Content of Spinach Before and After Various Manipulations for Preparation or Preservation

Treatment	Phylloquinone (μg/100 g)
Raw	385
Cooked	562
Deep-frozen	561
Canned or potted	609
Irradiated	
2.1 kGy	393
10.5 kGy	388

Source: Ref. 83.

B. Production

1. Soil and Climate

Spinach beet can be grown in any type of soil having sufficient fertility and proper drainage, but it does best in sandy loam soil. It can tolerate slightly alkaline soil. Spinach beet is a highly salt-tolerant vegetable and can be grown successfully in saline soils. This crop can withstand frost and tolerate warm weather, but high temperature leads to early bolting. During hot weather, leaves pass the edible stage quickly.

2. Propagation

After preparing soil thoroughly, seedbeds of convenient size and irrigation channels are prepared. Seeds are soaked in water overnight to hasten germination. The sowing is done either by broadcasting or in rows 3–4 cm deep. If optimum soil moisture and temperature are present, seeds will germinate about 10 days after sowing. In case of insufficient soil moisture, a light irrigation should follow the sowing to ensure good germination.

3. Cultural Practices

Spinach beet is a leafy vegetable requiring a great deal of nitrogen for crown (leaves) growth (84). According to Choudhury (85), a basal dressing of 34–40 tons of farmyard manure per hectare should be added to soil at the time of field preparation and a top dressing of 20 kg of nitrogen per hectare after every cutting for quick growth of tender succulent leaves.

Crop irrigation depends upon the soil condition and season. Irrigation is needed more frequently in light than in heavy soils. The spring-summer crop needs frequent irrigation at 6- to 7-day intervals, whereas autumn-winter crops require irrigtion at about 10- to 15-day intervals. The rainy season crop requires little irrigation.

4. Diseases and Pests

The most common diseases of spinach beet are damping-off caused by *Phythium* sp. and leaf spot caused by *Cercospora beticola* Sacc. Aphids and leaf-eating caterpillar (*Laphyqma exiqua*) can also cause damage to this crop (86).

5. Harvesting

The crop becomes ready for cutting about 4 weeks after sowing. Cutting is done by hand with the help of a sharp knife. Winter crops yield more than spring-summer crops—about 4–6 cuttings per season. Spinach beet harvesting is not done early in the morning because the leaves are crisp

TABLE 15 Nutrient Composition of Spinach Beet

Constituent	Content/ per 100g
Energy (kcal)	46.0
Moisture (g)	86.4
Protein (g)	3.4
Fat (g)	0.8
Carbohydrate (g)	6.5
Vitamin A (IU)	5862
Thiamine (mg)	0.26
Riboflavin (mg)	0.56
Niacin (mg)	3.30
Ascorbic acid (mg)	70
Calcium (mg)	380
Phosphorus (mg)	30
Iron (mg)	16.20

Source: Refs. 12, 13.

and break easily when wet. The crop gives an average yield of about 50–60 q/ha (87). Spinach beet generally bolts 75 days after sowing, and seeds mature in 150–180 days (87).

C. Chemical Composition

The chemical composition of spinach beet is presented in Table 15. It is a rich source of vitamin A, calcium, and iron (87,88). Swiss chard, a closely related species, is a good source of folacin (89–92).

D. Storage

Spinach beet leaves cannot be stored for long periods and hence should be sent to market immediately after harvesting. Leaves cannot be stored at room temperature, but at low temperature (0°C) and high relative humidity (90–95%) they can be stored for about 10–14 days (87).

E. Processing

Spinach beet is a popular vegetable in Southeast Asia. It is cooked along with garlic and spices and consumed along with flatbread, or it is fried in oil along with chopped garlic. Cooking spinach beet leaves for 35 minutes did not appreciably alter the carotenoid profile. The major carotenoids were lutein (1503 and 1960 μg/100 g edible portion in raw and cooked leaves, respectively) and β-carotene (1095 and 1360 μg/100 g edible portion for raw and cooked leaves, respectively) (88).

IV. OTHER VEGETABLES

A. Amaranth

Amaranths are hardy, wild, uncultivated, fast-growing cereal-like plants with a seed yield of about 3 tons per hectare in 3–4 months and a vegetable yield of 4–5 tons of dry matter per hectare after 4 weeks (93). About 60 members of the genus *Amaranthus* are widely distributed throughout the world in tropical, subtropical, and temperate regions as grain crops, pot herbs, ornamentals, and dye plant. Their growth habits vary from horizontal to erect and branched to unbranched; their leaves and stems are many shades of green and seeds vary from black to white. Amaranths photosynthesize via the C_4 pathway and utilize carbon dioxide more efficiently to produce sugars than plants possessing the classical C_3 pathway (94,95). The distribution, ecology, physiology, altitude, temperature, rainfall, soil requirements, pests, and diseases of *Amaranthus* have all be described in detail (96–102).

1. Botany

Several species of *Amaranthus* (*A.* spp., family *Amaranthaceae*) are used as a leafy vegetable. The important ones include *A. tricolor*, *A. magnostanus*, *A. blitum*, *A. lividus*, and *A. caudatus* L. These species differ from each other in height, size and color of leaves, growth habits, response to cutting, and flowering timing (4). Varieties such as *Chhoti Chaulai*, *Bari Chaulai*, CO-1, CO-2, and *Pusa Chaulai* are grown along with several local types (4).

Amaranth is primarily a warm season crop and as such grows well in the rainy season. As most Amaranth varieties are short-day plants, they tend to flower when grown in winter. Although it is not very specific in soil requirements, loamy soils and moderately fertile, well-drained soils are best suited for its cultivation.

The soil should be brought to a fine tilth by repeated ploughing and working and beds of desirable size (3 m × 2 m) are made, keeping in mind the irrigation facilities. The best time for sowing in the plains is March to June, but in the hills is May to July. In areas with milder climates, particularly south India, it can be sown throughout the year. The seeds are sown by broadcasting or in rows 20–30 cm apart to get 75 or 45 plants per m^2, respectively. In south India, Amaranth is grown as a transplanted crop. About 1.5–2.0 kg of seeds are required for one hectare of land.

Amaranth is usually grown as a second crop after a previous crop has been harvested without any further fertilizer. However, in some places growers apply 20–30 cartloads of farmyard manure (FYM) as a basal dose followed by 20 kg of nitrogen per hectare. Yawalkar (4) recommended a manurial schedule consisting of 25 t of FYM, 50 kg of nitrogen, 100 kg of phosphorus, and 50 kg of potassium per hectare. The entire quantity of FYM, half of the nitrogen, and the full dose of phosphorus and potassium are applied at sowing, while the remaining quantity of nitrogen is applied 20 days after sowing.

Amaranths need ample soil moisture and timely hoeing and weeding. Harvesting commences about 3 weeks after sowing in the case of *Chhoti chaulai* and 4 weeks in the case of *Bari chaulai*. These varieties are usually harvested when they attain a height of 25–30 cm. Delay in cutting makes the stem fibrous but increases iron content (4). The subsequent cuttings are taken at weekly intervals. In general, six to eight cuttings can be taken from a good amaranth crop. Yield may vary from 7.5 to 10 t of greens per hectare.

2. Chemical Composition

Amaranth is one of the few double-duty plants that can supply tasty leafy vegetables as well as grains of high nutritional quality. The dry seeds of Amaranth are used in the preparation of snack foods (13). The leaves of all 50–60 species of Amaranth are edible and are mostly eaten in the humid tropics. Their spinach-like flavor, high yield, and nutritional advantages over other leafy greens (95) have made them popular vegetable crops in many tropical islands of India, southeast China, and the South Pacific, where they have been cultivated for more than 2000 years; they also form a significant part of the diet in Africa. The composition of some vegetable amaranth

TABLE 16 Compositional Analysis (dry weight basis) of Some *Amaranth* Species

Constituent (%)	*A. edulis*	*A. retroflexus*	*A. tricolor*	*A. spinosus*
Ash	22.0	20.9	—	22.1
Protein (N × 6.25)	18.0	23.9	32.7	28.4–31.0
Crude lipids	1.60	2.72	3.5–10.6	1.8–4.5
Crude fiber	12.4	14.3	7.0	9.4
Calcium	6.2	3.0	3.5	1.1–1.8
Phosphorus	0.3	0.42	0.7	0.4
Iron	0.02	0.043	—	0.01
Oxalic acid	—	5.4	—	—
Nitrate	1.7	0.76	—	—

Source: Ref. 109.

species is given in Table 16. Amaranth greens have been reported to have a supplementary value in poor Indian rice diets (103). The high nutritive potential of these leafy greens (104–108) justify their wide consumption.

The harvested amaranth plant is 50–80% edible. The composition of vegetable amaranths is influenced by nitrogen fertilizer and maturity (110) and by application of soil insecticides (111). Nutrient content varies with seasonal variations (112) and cultivar (113).

The vegetable amaranths have a protein content of 17.4–33.5% on a dry weight (dw) basis (114), a protein digestibility of 80%, and a biological value of 75–78 (115). The leaves contain 5% lysine per 100 g protein and many other essential amino acids (116,117). The total lipids in vegetable amranths vary from 1.0 to 10.6% dw (118) and have been analysed in *A. gangeticus*. Palmitic, linoleic, and linolenic acids are the major components, with significant amounts of oleic and stearic acids. Hydrocarbons, alcohols (120), and sterols from amaranth leaves have been isolated and characterized (121,122).

Vegetable amaranths are also an important source of vitamins, especially provitamin A, β-carotene (123–126). About 220 ppm of vitamin K is also found in amaranth leaves (40). The carotenoid concentrations in vegetable amaranths is given in Table 17.

Trace quantities of a vitamin B_{12}–like substance have been reported in amaranth leaves (128). Vitamins C and B_6 are present in nutritionally significant levels, averaging 420 mg/100 g dw and 0.15 mg/100 g edible portion (129); niacin, thiamine, and riboflavin are also present (114). Choline at levels of 31 μg/100 g is also reported (21).

TABLE 17 Carotenoid Concentrations of *Amaranthus viridis*

Type	Concentration (μg/g)
β-Carotene	110
α-Cryptoxanthin	1.3
Lutein + violaxanthin	237
Zeaxanthin	8.2
Neoxanthin	43
Total	400

Source: Ref. 127.

Folic acid in foods occurs either in the free form, which is readily absorbed from the gut, and in conjugated forms, from which free folate may be released by enzymic treatments. Free and total folate contents in leafy vegetables of *Amaranthus* species are reported to be 41.0–50.1 and 61.0–149 μg/100 g, respectively (48,130). Conjugated folate makes up about 72% of the total folate. However, folate may be partially or wholly destroyed during processing or storage, and, in particular, large losses may occur during preparation and cooking, especially by leaching of this water-soluble vitamin in the cooking water. Ascorbic acid is also lost on processing in boiling water, the loss ranging from 46 to 97%, depending on the water used for boiling and the amaranth species (131).

Amaranths are also good sources of calcium and iron (132–134), although they are high in oxalates (Table 16). The absorption of calcium from amaranth leaves is, however, poorer than from milk (135). The percentage of available iron is only about 15.2–53.6% of the total iron and depends on the species of vegetable amaranthus (136). Magnesium, phosphorus, zinc, and potassium levels are similar to those for other leafy vegetables. *A. blitorides* contains the trace elements cobalt, molybdenum, manganese, and copper and also appreciable amounts of glycosides, alkaloids, saponins, and organic acids (137). A fluoride content of 27 μg/g in *A. hybridus* makes it an important source for this element, since the drinking water in certain areas is not fluorinated and these leaves are consumed regularly in the diet (138). Selenium at 3.0 μg/100 g and chromium at 14.6 μg/100 g are also reported in amaranth leaves (139). This trace element selenium acts as a defense mechanism against diseases and other environmental harm, is involved in immune response, and is necessary for the fucntioning of the hepatic cytochrome *p*-450 system. Chromium, on the other hand, plays an important role in normal carbohydrate and lipid metabolism, serves as a cofactor for several enzymes, and is required for insulin receptor interaction.

Many amaranthus species are prolific weeds, including *A. spinosus*, *A. hybridus*, and *A. retroflexus*. Nevertheless, these are important and should be studied because of the presence of desirable traits, such as rapid maturation, disease resistance, and adaptability, which might be genetically manipulated into the cultivated grain forms (95).

Wild uncultivated amaranth plants are similar to oats in palatibility and nutritive value to sheep (140). In *A. spinosus*, the crude protein, acid detergent lignin, and in vitro digestibility is higher than in cultivated forages. For nonruminants, *Amaranthus* forages are poorly utilized as protein sources and are unpalatable. Leaf protein concentrates offer a possible solution for nonruminants. Table 18 compares rat growth and feed intake for diets containing different *Amaranthus* leaf protein concentrates with those containing alfalfa. The poor growth rate of rats fed on amaranth leaf protein concentrates could be attributed to phenolic content, saponins, oxalates, and high ash content. A low in vitro digestibility of 28–48% of leaf protein concentrates from several *Amaranthus* species has been reported (142). Table 19 compares the essential amino acid composition of some *Amaranthus* species with other vegetables. Factors responsible for poor animal acceptance have to be overcome before *Amaranthus*, or leaf protein concentrates prepared from it, can be successfully utilized as feedstuff.

The high oxalate content of 1–2% (143) in vegetable amaranths can be reduced by boiling for 15 minutes (144,145). Oxalate content can also be reduced by tissue culture (114) and breeding (115). Amaranths require and absorb large amounts of nitrate; levels of 1.8–8.8 g/kg dry matter have been reported for leaves of various amaranthus species. Leaf position and age influence the nitrate content (146). It is concentrated most in the stems (145), which are generally discarded. Boiling the leaves also removes most of the nitrate (95). Amaranths can be a potential source of nitrite (147), which is not decreased greatly on cooking. In an in vivo test, Met-Hb formation paralelled the content of nitrite in vegetables; therefore, caution must be exercised in

Table 18 Diet Composition Containing Various Amaranth Leaf Protein Concentrates (LPC), Rat Growth, and Feed Intakes

	Diet composition (%)			Rat performance			
Source of LPC	LPC	Corn	Soybean	Days on test	Average daily gain	Gain as % of control	Average daily feed intake (g)
Corn-soy control	—	69	22	6	8.3 ± 0.3	100	—
				9	8.2 ± 0.4	100	—
				16	8.5 ± 0.3	100	22.0 ± 0.6
A. anclancalius	44.6	41.4	5	13	5.9 ± 0.3	69	21.6 ± 1.5
A. flavus	54.3	29.7	7	6	4.8 ± 1.0	58	18.1 ± 1.1
A. gangeticus	46.2	39.8	5	15	4.8 ± 0.3	56	19.5 ± 1.8
A. cruentus HH2	55.2	27.8	8	8	5.4 ± 0.9	66	18.5 ± 1.1
A. hypochondriacus	48.5	36.5	6	13	4.9 ± 0.9	58	18.9 ± 1.3
Mixed amaranths (unwashed)	68.5	11.5	11	16	2.2 ± 0.4	26	15.9 ± 1.0
High-saponin alfalfa	20.6	70.4	—	16	5.9 ± 0.7	69	17.7 ± 1.2
Low-saponin alfalfa	19.3	71.7	—	14	6.9 ± 0.4	81	19.3 ± 1.0

Source: Ref. 141.

Table 19 Essential Amino Acid Composition (g/100 g protein) of Some Species of Amaranth Leaf Protein Concentrates

	Source		
Amino acid	*A. retroflexus*	*A. hybridus*	*A. cruentus*
Lysine	6.26	6.46–6.83	7.36
Phenylalanine	3.78	4.45–4.82	4.91
Methionine	1.76	2.02–2.42	2.16
Threonine	3.06	3.55–2.94	4.50
Isoleucine	4.85	5.46–6.11	4.69
Leucine	7.54	8.58–9.65	8.54
Valine	4.80	5.60–6.14	5.95
Tryptophan	—	—	—

Source: Ref. 109.

the use of some of these vegetables in infant foods. Aliphatic and aromatic secondary amines at 279 and 33 μg/g dry weight, respectively, in *Amaranthus viridis* Linn has been reported (33).

The well-characterized dye from amaranth leaves (148,149) is used to color maize dough in Mexico and the southwestern United States and to dye foods and beverages in Ecuador. The effect of some processing treatments on pigment retention in amaranths has been reported (150).

Some other antinutritional factors and flatus factors in vegetable amaranth are given in Table 20. Amaranth leaves have high antitryptic activity and in general a low concentraiton of flatus factors.

B. Basella

Basella is a popular summer leafy vegetable grown in tropical Asia and Africa. The plant has a fleshy stem and leaves and has a trailing habit. The fresh tender leaves and stems are consumed as leafy vegetables after cooking. The coloring matter present in the red cultivar is reported to have been used in China as a dye. The ripe fruit also contain a deep violent coloring matter, which is also used for coloring food.

Basella belongs to the family *Basellaceae* and genus *Basella*. There is only one species,

Table 20 Antinutritional and Flatus Factors in Vegetable Amaranth

Chemical constituent	Value
Total sugar (%)	5.6
Sucrose (S) (%)	0.8
Raffinose (R) (%)	0.2
Stachyose (St) (%)	0.2
S + R + St	1.2
Saponin (%)	1.9
Antitryptic activity (TIU/mg protein)	8.1
Phenol (mg tannic acid equivalent/g dry weight)	11.0
Phytate (mg/g)	32.9

Source: Ref. 151.

Basella rubra ($2n = 24$), which is extremely variable. *B. alba*—a white flowered form, *B. caninifolia*, *B. cordifolia* (wheat-shaped leaves), *B. crasifolia*, *B. japonica*, *B. lucida*, *B. ramosa*, and *B. volubilis* are different forms of *B. rubra*. Basella is a fleshy annual or biennial twining, much-branched herb with alternate and ovate and pointed leaves at the apex. Flowers are white or pink, small sessile, in clusters or elongated thickened peduncles in an open-branched inflorescence. Fruit is enclosed in a fleshy perianth. There are two distinct types—one with reddish petioles and stems and another with green leaves, petioles, and stems. Both types are consumed as a vegetable, but the green type is commercially cultivated and belongs to var. *alba*.

Basella grows well in fertile sandy loam soil provided it is well drained and aerated. It is grown during warm and moist seasons. Frost is injurious to the growth of this crop. Basella is propagated through seeds, but stem and root cuttings are also used. Seeds may be sown in situ or seedlings transplanted when they can be easily handled. A seed rate of 12.5–16 kg/ha is optimum. Seedlings are spaced in beds 45 cm apart, and plants are allowed to sprawl over the ground.

Rich soil is essential for a good crop, and nitrogen fertilizer application has been found to be beneficial. In general, basella requires five to six irrigations when grown in summer, and the frequency of irrigation depends on the soil type.

Basella is subject to attack by a number of diseases. These include damping-off *Pythium aphanidermatum*, leaf spots due to *Acrothesium basellae* Alvarez-Garcia, *Fusarium moniliforme* Sheldon, and *Cercospora* sp. The crop is almost impervious to insect attack, although a minor incidence of omnivorous wooly bear caterpillar of ermine moth is reported (12).

Basella crops raised from seed provide edible leaves and stems 8–10 weeks after sowing. Plants raised from root or stem cuttings are ready for harvest about 6 weeks after planting. The yield varies from 140 to 185 q/ha. A higher yield is achievable from a well-managed crop (86).

The chemical composition of *Basella rubra* is given in Table 21. Its mineral content is presented in Table 22. The composition of desmethylsterols in *Basella rubra* lipid is presented in Table 23. A glycoprotein containing 20% carbohydrate and 80% protein has been identified as an inhibitor of tobacco mosaic virus from the leaves of *Basella alba* (153). The nutritive value of the young shoots and leaves is very high in terms of minerals and vitamins. The nitrate content is about 764 ppm on a dry weight basis (13). There is no appreciable loss of nitrate even after 48 hours of cold storage (13). *Basella alba* contains about 16 μg/g fluoride. This is especially important in countries like Nigeria, where the drinking water is not fluorinated and where the consumption of these leafy vegetables is high in the form of soup. The levels of fluoride in

TABLE 21 Chemical Composition (on DM basis) of *Basella rubra*

Constituent (%)	Content
Total ash	3.70
Silica	1.20
Crude protein	34.12
NDF	33.04
ADF	15.01
Hemicellulose	18.03
Cellulose	11.02

Source: Ref. 11.

TABLE 22 Mineral Content of *Basella alba*

	Mineral content (mg/g dry weight)
Total ash	25.0
Sodium	0.52
Potassium	0.79
Calcium	0.58
Phosphorus	0.24
Magnesium	0.70
Iron	0.37
Zinc	0.081
Manganese	0.118

Source: Ref. 107.

Nigerian populations are below 1 mg/kg—the amount recommended for the prevention of dental caries (138).

Basella alba is a good source of β-carotene (58,154) and is quite stable during processing (Table 24). Ascorbic acid is present to the extent of 76.6 mg/100 g of fresh material. However, processing basella leaves in excess water degrades this vitamins (loss of 74.2%). Processing in small amounts of water causes only about a 34% loss (131). The leaves of *Basella alba* lose relatively little ascorbic acid during wilting compared to other leafy greens (155). Vitamin K is present at about 200 ppm (40).

C. Chard

Chard or Swiss chard (*Beta vulgaris* var. *Cicla*) is a foliage beet developed for its large, fleshy leaf stalks and broad, crisp leaf blades. It is commonly grown in home gardens and is of little commercial importance. The leaf blades are prepared for table use like spinach; the stalks and midribs are cooked and served like asparagus. Chard is one of the best pot herbs of the summer, and it withstands hot weather better than most greens.

TABLE 23 Desmethylsterol Composition of *Basella rubra* Leaves

Sterol	% of total desmethylsterol
Cholesterol	—
24-Methylenecholesterol	—
Campesterol	3
Stigmasterol	18
Sitosterol	11
Isofucosterol	—
7-Ergostenol	2
Spinasterol	24
7-Stigmastenol	40
7,25-Stigmastadienol	—
7,(Z)-24(28)0-Stigmastadienol	3

Source: Ref. 152.

TABLE 24 β-Carotene Content of *Basella alba,* Raw and After Processing

Processing treatment	β-Carotene (mg/kg fresh material)
Raw (whole or ground)	19
Boiled (15–120 min)	19
Boiled, with potash	19
Stewed, 25–30 ml oil, 100 ml water	20
Frying, 100–150 ml oil	—

Source: Ref. 79.

The popular varieties of chard are Fordhook, Giant Perpetual, Lucullus, and Silver Leaf. Lucullus has very large crumpled, dark-green leaves with greenish-white leaf stems. Giant Perpetual has broad, light-green leaves. Other varieties mentioned by Thompson and Kelly (1) are Lyon (broad stem and midrib), Large Ribbed White (broad, white stalks and white midrib) and Rhubarb (dark-green crumpled leaf blades and bright crimson petioles or leaf stalks).

For domestic use, the tender outer leaves are removed 1 or 2 inches from the ground. Care must be taken to avoid injury to the bud. For market, leaves above the growing point in the center are cut off. Subsequent harvestings are made in the same way when fresh leaves achieve marketable size. The harvested leaves are often bunched and tied for market. Several pickings may be made during the season.

Chard and Swiss chard are closely related to beet and spinach (tops) and are subject to the same diseases. Ryall and Lipton (9) mentioned bacterial soft rot caused by *Erwinia carotovora* and *Cercospora* leaf spot by *Cercospora beticola* as important diseases of these crops. Cold storage near 0°C retards these diseases. Chard must be handled carefully and should not be overpacked, otherwise physical injury in the form of abrasion of petioles and crushing of leaves may take place. Prompt cooling and shipping at 0°C is necessary to retard discoloration and wilting (156).

The chemical and mineral composition of Swiss chard and its vitamin profile are given in Tables 25 and 26. Vitamin K_1 or phylloquinone content in Swiss chard varies according to the geographic location; maturation also increases Vitamin K_1 content in the leaves (47).

A folacin content for Swiss chard of about 297 μg/100 g fresh weight has been reported (89), although lower values of about 11 μg/100 g were reported earlier (90). A study of folacin degradation during storage (91) has revealed it to be more stable when stored in plastic bags under moist conditions and least stable in open air at 21°C. Degradation is temperature dependent and follows first-order kinetics. The data on the effect of storage conditions on first-order rate constant and half-life for folacin degradation in Swiss chard are presented in Table 27. The relative stability of folacin under refrigerated and room storage conditions is attributed to the presence of about 50–95 mg of ascorbic acid per 100 g of chard (92), at which level it exerts an antioxidative effect.

D. Fenugreek (Methi)

Fenugreek is considered to be a native of eastern Europe and Ethiopia. There are two types of methi. Common methi or fenugreek is known botanically as *Trigonella foenumgracum*, and the other Champa or Kasturi methi is *Trigonella corniculata.* It belongs to the family *Leguminosae.*

TABLE 25 Proximate and Mineral Composition of Swiss Chard

Constituent (per 100 g raw material)	Content
Water (g)	92.70
Total nitrogen (g)	0.29
Protein (g)	1.80
Fat (g)	0.20
Carbohydrate (g)	2.90
Starch (g)	2.30
Total sugar (g)	0.60
Mineral composition	
Sodium (mg)	210
Potassium (mg)	380
Calcium (mg)	51
Magnesium (mg)	81
Phosphorus (mg)	46
Iron (mg)	1.8

Source: Refs. 12, 13.

Fenugreek is cultivated widely all over India and several other tropical regions. Common methi is quick growing and produces upright shoots, whereas Kasturi methi is initially slow growing and remains in rosette form during its vegetable period (88). The crop is used as a green vegetable as well as pulse (86).

Pusa Early Bunching and Pusa Kasuri are the two improved varieties of methi released by the Indian Agricultural Research Institute, New Delhi (4). Pusa Early Bunching is quick growing with upright shoots, bold seed, high yield, and is suitable for cutting. Pusa Kasuri is a late-flowering rosette type, a heavy yielder having leaves with special fragrance. A broad-leafed, succulent, and vitamin C-rich variety, Methi No. 47, has been released by the Department of Agriculture, Maharashtra, India.

Methi, particularly the *Kasturi methi*, is a cool season crop. Its growth is affected under warm climate. Therefore, in the plains, it is grown during winter. Successful methi production needs ample water supply and rich medium heavy soils.

For a continuous supply of leaves, crops can be sown at weekly intervals. The seeds are

TABLE 26 Vitamin Profile in Swiss Chard

Vitamin (per 100 g raw material)	Content
Carotene (μg)	4625
Thiamine (mg)	0.04
Riboflavin (mg)	0.09
Niacin (mg)	0.40
Vitamin B_6 (mg)	0.17
Folate (μg)	165
Pantothenote (mg)	0.17
Vitamin C (mg)	20
Vitamin K_1 (μg)	7.43–9.17

Source: Refs. 12, 13, 47.

TABLE 27 Storage Conditions on First-Order Rate Constant and Half-Life for Folacin Degradation in Swiss Chard

Storage condition	Correlation coefficient	Rate constant, k (hr^{-1})	Half-life, $t_{1/2}$ (hr)
In plastic bag at 21°C	0.998	0.00810	108
Under moist conditions at 21°C	0.996	0.00896	77
In open air at			
4°C	0.925	0.00086	806
21°C	0.988	0.0123	56
35°C	0.992	0.0625	11
40°C	0.980	0.0987	7

Source: Ref. 89.

sown in beds by broadcasting or in rows. The common bed size for methi is 3 × 2 m and the recommended row-to-row spacing is abuot 20–25 cm. The seed rate for Kasturi methi is 11.3–14.00 kg/acre and for common methi about 18.0 kg for the same area. The soil moisture level should be adequate to ensure satisfactory germination.

Yawalkar (4) stated that application of 10 carloads of FYM as a basal dose followed by three top dressings each of 45 kg of ammonium sulfate per acre is needed for securing optimum plant growth. The first nitrogen dose is given 4 weeks after sowing, while the other two are given after every two cuttings. Timely irrigation and the interculture operations are needed to obtain good-quality produce.

Common methi is harvested either by clipping the young plants from the base or by pulling them out. The clipped plants are allowed to grow further, and their tops are nipped periodically until flowering. Kasturi methi plants are pulled out entirely or the leafy growth is cut several times during the season (86).

Common methi is generally ready for harvesting 20 days after sowing, while Kasturi or Champa methi is ready 25–30 days after sowing. Subsequently cuttings may be performed every 12–15 days. If methi is harvested late, its leaves develop a bitter taste.

The proximiate principles and mineral composition of fenugreek leaves is given in Table 28. Aykroyd (157) reported that 100 g of fresh methi leaves contains 3900 IU of vitamin A and 140 mg of vitamin C. The protein and ash contents decrease on maturity, but fiber constituents such as cellulose, lignin, hemicellulose, acid detergent fiber, and neutral detergent fiber increase (158). The in vitro available iron in fenugreek leaves is only about 4.6% of the total iron (18). Addition of fenugreek leaves significantly improves the total iron in a cereal meal (159). The amino acid composition of fenugreek leaf protein is given in Table 29. A general deficiency in sulfur amino acids, typical of all leafy greens, is present here too. Neutral lipids, glycolipids, and phospholipids at 0.98, 0.38, and 0.14 g/100 g dry weight, respectively, are reported in fenugreek leaves. The fatty acid composition of these lipid fractions is presented in Table 30. Very high amounts of unsaturated fatty acids are present in all these fractions, with 18:2 predominating in the neutral fraction and 18:3 in the glycolipid and phospholipid fractions.

Vitamin B_6 is reported to be present at 0.18 mg/100 g edible portion (129), and vitamin K is present at 240 ppm (40). α- and β-Tocopherols are reportedly present at levels of 0.87 and 0.37 mg/g dry weight, respetively (160). Vitamin C is present in methi leaves but is destroyed by cutting and washing (Table 31). The green color, attributed to chlorophyll pigments, is stable on steam cooking compared to pressure cooking. Processing in an alkaline media is known to

TABLE 28 Proximate Constituents and Mineral Composition of Fenugreek Leaves (on dry weight basis)

Constituent	Content
Proximate principles	
Total ash (%)	9.8
Crude protein (%)	28.5
Ether extractives (%)	4.0
Neutral detergent fiber (%)	21.2
Acid detergent fiber (%)	19.9
Hemicellulose (%)	1.3
Mineral composition	
Calcium (%)	0.9
Phosphorus (%)	1.0
Magnesium (%)	0.3
Sodium (mg/100 g)	1.03
Potassium (mg/100 g)	1.2
Copper (ppm)	25.0
Iron (ppm)	312.5
Zinc (ppm)	27.5
Manganese (ppm)	37.5

Source: Ref. 151.

TABLE 29 Amino Acid Composition of Fenugreek Leaf Protein

Amino acid	g/16 g N
Aspartic acid + Asparagine	21.9
Threonine	2.66
Serine	5.31
Glutamic acid + Glutamine	10.7
Proline	5.41
Glycine	3.28
Alanine	4.72
Valine	5.52
Cystine	—
Methionine	0.09
Isoleucine	8.86
Leucine	6.76
Tyrosine	—
Phenylalanine	0.302
Lysine	4.22
Histidine	1.00
Arginine	3.08
Tryptophan	0.142

Source: Ref. 151.

Table 30 Fatty Acid Composition[a] of the Neutral Lipids, Glycolipids, and Phospholipids of Edible Leaves of *Trigonella foenumgraecum*

Fatty acid	Neutral	Glycolipid	Phospholipid
12:0	1.3	0.2	—
14:0	2.2	0.3	0.6
16:0	7.6	5.0	11.1
16:1	—	0.4	1.6
18:0	4.3	—	2.8
18:1	7.1	0.4	0.8
18:2	44.6	0.3	20.2
18:3	32.9	93.4	62.9

[a]Fatty acid composition as weight percent of total fatty acids.
Source: Ref. 160.

enhance the green color, which has been confirmed from computation using color parameters in the CIElab color system (161).

Methi leaves contain several flatus and antinutritional factors (Table 32). In addition, it also contains about 296 and 59 μg/g dry weight of aliphatic and aromatic secondary amines, respectively (33).

Traditionally, methi leaves are sun-dried and used in the off season. Dried methi leaves can be stored for one year. Kasturi methi is available (packed) in packets of different sizes (86). During dehydration, chlorophyll is oxidized and ascorbic acid is also lost. Hence, blanching by boiling for 3–6 minutes and treating with 0.2% sodium meta-bisulfite are practiced. The effect of these treatments on ascorbic acid, β-carotene, and chlorophyll retention in dehydrated methi leaves is shown in Table 33. The retention of chlorophyll is considerably improved by blanching in sulfited water (163).

The palatability of fenugreek leaves improves with steaming and with addition of seasonings rather than with boiling or frying (164).

Incorporation of methi leaves in a sorghum-based diet improves the PER from 0.92 to 1.54. The biological value also improves from 79 for sorghum to 89 for methi-incorporated sorghum. Methi protein by itself has a biological value of 83 (165). Methi leaves contain 77.25 μg/100 g

Table 31 Effect of Different Processing Techniques on the Vitamin C Content of Fenugreek Leaves

Processing treatment	Vitamin C content (mg/100 g)
Control, no treatment	91.5
Cutting into pieces using stainless steel knife	84.4
Washing in tap water	87.7
Washing in tap water after cutting	48.2
Washing in tap water before cutting	82.1
Cooking with lid for 2 minutes	36.1
Cooking without lid for 2 minutes	27.9

Source: Ref. 69.

TABLE 32 Antinutritional and Flatus Factors in Fenugreek Leaves

Chemical constituent	Content
Total sugar (%)	8.1
Sucrose (S) (%)	0.5
Raffinose (R) (%)	0.2
Stachyose (St) (%)	0.2
S + R + St	1.0
Nitrate (mmol/100 g)	1.6
Oxalate (%)	4.6
Saponin (%)	1.7
Antitryptic activity (TIU/mg protein)	6.0
Phenol (mg tannic acid equivalent/g dry weight)	53.4
Phytate (mg/g)	32.1

Source: Ref. 151.

of selenium and about 15.2 μg/100 g of chromium (139). Fenugreek leaves are reported to contain 200–500 (mean 310) ppm nitrate and 1.1–5.3 (mean 4.1) ppm nitrite (32).

Bioflavonoids is a term used to denote all flavonoids exhibiting some pharmacological activity. Although their pharmacological activity has not been proved conclusively, flavonoids have been found to exert beneficial effects in more than 50 diseases. In particular, flavonoids with free hydroxyl groups in the 3′ and 4′ positions, such as quercetin, gossypetin, quercetagetin, and luteolin, exert beneficial effects on the capillaries by chelating metals and thus sparing ascorbate from oxidation, prolonging epinephrin action by inhibiting *O*-methyl transferase, and stimulating the pituitary axis. The flavonoids with multiple methoxy groups such as 7,4′ -diOMe

TABLE 33 Effect of Processing on Retention of Ascorbic Acid and Chlorophyll in Dehydrated Methi

	% Retention of chlorophyll		% Retention of ascorbic acid		% Retention of β-carotene	
Treatment	Sun-dried	Cabinet-dried	Sun-dried	Cabinet-dried	Sun-dried	Cabinet-dried
W_0	49.90	37.34	79.65	57.21	42.54	34.18
W_1C_0	34.31	23.70	60.97	42.48	29.26	20.62
W_1C_1	28.84	20.64	59.64	44.12	25.34	18.78
W_1C_2	24.02					
W_2C_0	29.99					
W_2C_1	26.05					
W_2C_2	14.93					
W_3C_0	26.15					
W_3C_1	22.84					
W_3C_2	10.00					

W_0 = No blanching; W_1 = blanching at 80°C; W_2 = blanching at 90°C; W_3 = blanching at boiling temperature; C_0 = no chemical; C_1 = 2% NaCl + 0.2% KMS; C_2 = 0.5% KMS + 0.1% MgO + 0.1% $NaHCO_3$.
Source: Ref. 162.

kaempferol, 3′, 4′ -diOMe quercetin, etc., play an important role in the circulatory system by reducing aggregation of erythrocytes (during illness or injury) by site-specific membrane surface effects and improving microcirculation within the body. The presence of flavonoids that are regularly consumed is noteworthy. The flavonoids identified in the leaves of *Trigonella foenum-graecum* L. are kaempferol and 3′4′-diOMe quercetin (166). Diosgenin has been identified as the major steroid sapogenin in the leaves of *Trigonella foenum-graecum*, along with smaller amounts of tigogenin and gitogenin (167).

E. New Zealand Spinach

New Zealand spinach (*Tetragonia expnase*) is a member of the *Aizoceae* family. It is not a true spinach but has similar leaves, and the product is used in the same way (1). New Zealand spinach has thick, dark green leaves that are somewhat triangular in shape. The crop thrives in hot weather but is not as hardy as spinach and hence is generally planted in the spring.

The edible portion is the tender tips of the many shoots, which are cut to lengths of 3–4 inches. If they are cut too long, the quality of the product suffers from inclusion of tough, woody stems (3). The quality also depends upon liberal fertilization and an abundance of moisture to produce rapid and succulent growth. When grown for market, a more common practice is to cut the entire plant 2 or 3 inches above ground by grasping a bunch in one hand and cutting it off with a large sharp knife or sickle held in the other hand. Two, three, or more cuttings are made in this way. New Zealand spinach is a substitute for spinach and is handled and processed in the same way; its main harvest period is summer, when spinach is not available (1). Watts and Watts (3) stated that New Zealand spinach is quite resistant to insects and diseases.

F. Purslane

Purslane (*Portulaca oleracea*) belongs to the *Portulacaceae* or Purslane family. It is a common weed (168,169) in many parts of the world. Many weedicides have been evaluated for its eradication (170,171).

Since ancient times, purslane has been used as a vegetable, eaten raw in salads or cooked as greens. The entire plant is edible. It is a prized garden vegetable over much of Europe and Asia, and several different varieties have been developed. Purslane is widely distributed in the tropics and subtropics, including many parts of United States. Feeds continuing *Portulaca oleracea* leaves are useful for immunostimulation and prevention of diarrhea of domestic animals and fowl (172). The chemical composition of *Portulaca oleracea* leaves is given in Table 34.

Iron is present at a level of 3.6 mg%, of which only 30.6% is available iron (136). Purslane contains about 1679 mg of oxalic acid per 100 g of edible material (21,173) and accumulates nitrate. The crop is cultivated commercially in Egypt and Sudan.

Curvulin and *O*-methylcurvulinic acid have been identified as phytotoxic metabolites of the fungus *Drechslera indica*, which causes necrosis on purslane leaves (174).

Protein in *Portulaca* leaves is around 29% on a dry weight basis, in addition to free amino acids such as phenylalanine, valine, alanine, tyrosine, and aspartate and a predominance of glutamate (175). The amino acid composition of purslane leaves is given in Table 35.

Whole purslane plants contain 3.5% lipids on a dry weight basis, of which 25% are free fatty acids. This vegetable plant has been used throughout history as a medicine, a fact attributed to the presence of large quantities of omega-3 polyunsaturated fatty acids (176,177). Purslane is in fact the richest source of omega-3 fatty acids examined to date (178) at levels of 4 mg/g wet weight (179). It has been recommended as a cholesterol-lowering agent and as a therapy for

Table 34 Chemical Composition of the Leaves of *Portulaca oleracea*

Constituent	Content
Moisture (%)	90.5
Protein (%)	2.4
Fat (%)	0.6
Minerals (%)	2.3
Fiber (%)	1.3
Carbohydrates (%)	2.9
Minerals	
Calcium (mg/100 g)	111
Phosphorus (mg/100 g)	45
Iron (mg/100 g)	14.8
Magnesium (mg/100 g)	120
Sodium (mg/100 g)	67.2
Potassium (mg/100 g)	716
Copper (mg/100 g)	0.9
Sulfur (mg/100 g)	63
Chlorine (mg/100 g)	73
Vitamins	
Carotene (μg/100 g)	2292
Thiamine (mg/100 g)	0.10
Riboflavin (mg/100 g)	0.22
Niacin (mg/100 g)	0.70
Vitamin C (mg/100 g)	29

Source: Ref. 21.

Table 35 Amino Acid Composition of *Portulaca* Leaf Protein

Amino acid	Content (g/g N)
Arginine	—
Histidine	—
Lysine	0.45
Tryptophan	0.03
Phenylalanine	0.09
Tyrosine	—
Methionine	0.02
Cystine	—
Threonine	0.12
Leucine	0.29
Isoleucine	0.35
Valine	0.28

Source: Ref. 21.

arteriosclerosis. Suggestions of purslane as an alternative to fish oils with respect to omega-3 fatty acids have been made. However, this has been disputed (180). Health benefits associated with a modest increase in dietary linoleic acid include blood-clotting tendency and reduced blood pressure (181). Hens feeding on purslane plants produce eggs with high levels of n-3 fatty acids (17.66 vs. 1.73 mg/g egg yolk) (182). Fatty acids in the whole plant, stems, and leaves of different ages are shown in Table 36. The occurrence of 18:2 omega6 and 18:1 omega9 in high levels compared to other vegetable crops further stresses the potential benefits of purslane in human, animal, and fish nutrition. The ratio of omega-3 fatty acid to other major fatty acid families is a critical indicator of essential fatty acid status and has been found to be low in stems, lower in whole plant, and lowest in the leaves (176).

The sterols, accounting for 19% of the total lipids, include sitosterol (72%), campesterol (14%), and stigmasterol (14%) (182). Besides triterpene alcohols, including β-amyrin, butyrospermol, parkeol, cycloartenol, 24-methylene-24-dihydroparkeol, and 24-methylenecycloartenols, other C_{28} to C_{30} linear alcohols have been identified in purslane leaves. These comprise about 23% of the total unsaponifiables. Hydrocarbons represent 18% of unsaponifiables (183). Phenolic constituents, namely scopoletin, bergapten, isopimpinellin, lonchocarpic acid, lonchocarpenin, robustin, and genistein, having antimicrobial activity have been isolated from *Portulaca oleracea* (184).

Carotenoids are present to the extent of 89 μg/g. β-Carotene is found in substantial amounts but is lost to the extent of 43% due to faulty processing techniques (185). Table 37 gives the carotenoid concentrations of *Portulaca oleracea* leaves. α-Tocopherol levels are seven times higher in purslane leaves than in spinach (186). Phylloquonine or vitamin K_1 is present at 381 μg/100 g and is quite stable to cooking (83). Glutathione at 14.8 mg/100 g fresh purslane leaves is also reported (186).

A polysaccharide complex in the form of a clear and viscous mucilage having physiochemical properties appropriate for industrial uses as food extenders and viscosifiers has been extracted from purslane leaves. It is tentatively identified to be a neutral arabinogalactan and a polydisperse pectinlike polysaccharide (187).

Oxalate content in dry forages of *Portulaca* from Mexico is reported to be between 21.48 and 41.59% and is mainly responsible for the interaction with calcium and the resulting hypocalcemia suffered by animals that have ingested it (188).

Ascorbic acid is present at levels of 46.8 mg/100 g fresh material; however, it is very unstable and is almost totally destroyed during processing (131). The presence of the bioflavonoid liquiritin in purslane leaves has been suggested (40).

Water-soluble yellow food dyes have been prepared from the culture medium of *Portulaca grandiflora* yellow flowers (189). The color is attributed to the betaxanthin humilixanthin (190).

V. MINOR VEGETABLES

A. Chenopodium

Chenopodium quinoa is native to the Andean regions of South America (191). Leafy vegetables and grains are abundantly consumed in Andean region of South America, Africa, and some parts of Asia and Europe. The plant is a native of Peru, and the highly nutritious vegetables are used whole in soups or ground into flour, which is made into bread and cakes (192). In contrast to maize, potatoes, and *Phaseolus* beans, all of which are staple crops originating from the Andes,

TABLE 36 Composition of Selected Fatty Acids in Various Parts of Purslane Harvested at Various Ages After Planting

Fatty acid (mg/kg wet weight)	30th day[a]			49th day[a]			59th day[a]		
	Leaf	Stem	Whole plant	Leaf	Stem	Whole plant	Leaf	Stem	Whole plant
16:0	66.89	27.50	38.36	24.46	12.38	12.14	52.08	43.12	73.48
18:0	4.64	0.03	3.65	2.75	0.02	1.94	10.42	0.05	8.26
18:1omega9	5.96	0.03	3.65	3.98	0.02	4.13	10.42	0.04	13.91
18:2omega6	54.97	4.42	17.35	14.68	5.24	8.98	32.64	4.13	18.45
18:3omega3	290.73	2.50	60.73	97.25	3.02	18.45	120.83	1.83	72.83
20:5omega3	7.28	7.88	16.89	1.22	6.51	13.83	36.80	27.06	13.04
22:5omega3	5.96	Tr	1.37	4.89	Tr	1.94	9.72	Tr	2.61
22:6omega3	1.32	Tr	0.91	7.95	Tr	3.61	18.75	Tr	2.61

[a]Number of days after planting.
TR = < 0.02 mg/kg wet weight of purslane.
Source: Ref. 176.

TABLE 37 Carotenoid Concentrations of *Portulaca oleracea* Leafy Vegetable

	Concentration (μg/g)
β-Carotene	30
α-Cryptoxanthin	0.6
Lutein + violaxanthin	48
Zeaxanthin	0.7
Neoxanthin	9
Total	89

Source: Ref. 127.

C. quinoa has not attained global importance. One reason for this may be the need for removal of bitter and antinutritional saponins (193) from the seed prior to cooking or processing (194). The seeds are also used as poultry feed, in medicine, and in making beer.

C. quinoa is one of the lesser-known food crops, possessing an exceptionally attractive amino acid balance for human nutrition.

1. Botany

Chenopodium quinoa is a dicotyledonous plant. The family *Chenopodiaceae* consists of herbs and shrubs, and rarely small trees, usually occurring in alkaline situations. These plants are usually scruffy because of external cells that dry into white flakes. The leaves are simple, sometimes more or less succulent or reduced to small scales, usually alternate, but rarely opposite. There are no stipules. Flowers are bisexual or sometimes unisexual (195).

The family is seen worldwide, but it is centered in alkaline areas. Some species are restricted to wet, strongly salty or alkaline soil, such as that of coastal salt marshes or alkaline plains and deserts. On the whole, the family is made up of weedy plants, some of the more important of which are (1) *Chenopodium* or goosefoot or pigweed or lamb's quarter, (2) *Kochia* or red sage, and (3) *Salsola.* The genus *Chenopodium* has a worldwide distribution and consists of about 260 species (194). About eight species occur in India (197). Some of the different species of *Chenopodium* and the country in which they are found are listed in the Table 38.

Chenopodium seeds have been considered to be weeds (198), and many efforts have been directed towards its eradiation (199).

Interest in *C. quinoa* as a valuable food source has been renewed in recent years due to its versatility and its ability to grow under conditions normally inhospitable to other grains. These include low rainfall, high altitude, thin cold air, hot sun, and subfreezing temperatures.

The average yield of the fruit is 840–3000 kg/ha (12), while Weber (200) reports yields for *C. quinoa* from as little as 450 kg/ha to as much as 2000 k/ha, a record yield of 5000 kg/ha, and an average yield of 800–1000 kg/ha. *C. quinoa* is extensively grown in Peru and Bolivia. By virtue of its resistance to frost and drought, *C. quinoa* is very suitable for cultivation in highlands and temperate regions. Production of *C. quinoa* in Ecuador has shifted from backyards to extensive cultivation. During 1987, around 431 ha were harvested, producing 720 tons. Highland farmers often cultivate *C. quinoa* in rotation with other crops because they believe it can prevent disease among other crops (200).

2. Chemical Composition

The leaves of *Chenopodium* are widely used as a vegetable in human diets and livestock feeding (200) and constitute an inexpensive source of vitamins and minerals. Generally the younger

Table 38 Some *Chenopodium* Species and Their Occurrence

Species	Country
C. quinoa	South America, Peru, Guatemala, Bolivia, Chile, Argentina
C. pallidicaule	South America
C. berlandieri	Mexico
C. album	India (Valleys of Himalayas)
C. ambrosoides	India
C. amaranticolor	India
C. murale	India
C. striatum	Czechoslovakia
C. opulifolium	Czechoslovakia
Hybrid	
C. foliosum-polyspermum	Finland

leaves of chenopods are consumed as human food. Correlations of nutrient content with leaf age (position) is an important factor in choosing leaves for harvesting. A comparative composition of some vegetable *Chenopodium* species with some other selected vegetables is presented in Table 39. Chenopod leaves have more proteins and minerals than commonly consumed spinach and cabbage, but less than *amaranth* leaves. The mineral composition of *Chenopodium album* is given in Table 40. Iron is present at 6.6 mg%, of which only 10.6% is available (136). Vitamin K is present at 250 ppm in chenopod leaves (40). The leaves of chenopodioum species are also known to accumulate nitrate in the range of 3.1–5.0% on a dry weight basis (196). Comparative values of nitrate on a dry weight basis for vegetable amaranths are in the range of 0.76–1.70%. The corresponding values for spinach and cabbage are 2.40 and 0.84% (109). It appears that the nitrate content in vegetable chenopods is higher than in commonly consumed vegetables. However, most of the nitrate is concentrated in the stem portion, which is generally discarded, and is therefore not of much interest (196). Besides nitrate, nitrite in the range of 3.4–5.3 (mean 4.2) ppm (32) and oxalates ranging from 0.9 to 3.9 g/100 g fresh weight in vegetable chenopods is reported. This is also mainly concentrated in the stems (196).

Flavonoids have been identified in five species of chenopodium. The flavonol quercetin

Table 39 Comparison of % Composition of Some Vegetable *Chenopodium* Species with Some Selected Vegetables

Leafy vegetables	Moisture	Carbohydrate	Protein	Fat	Minerals	Crude fiber
A. gangetieus	85.7	6.1	4.0	0.5	1.0	1.0
A. gangetieus	78.6	8.6	5.9	1.0	3.8	2.1
Cabbage	91.9	4.6	1.8	0.1	0.6	1.0
Cauliflower	80.0	7.6	5.9	1.3	3.2	2.0
Coriander leaves	86.3	4.1	3.3	0.6	2.3	1.2
Radish	90.8	2.4	3.8	0.4	1.6	1.0

Source: Refs. 21, 203.

TABLE 40 Mineral Composition in Leaves of *Chenopodium album*

Mineral	mg/kg dry matter
Iron	45.0
Zinc	20.1
Copper	12.0
Manganese	90.0
Calcium	1452.0
Phosphorus	860.0

Source: Ref. 204.

was found in all species, kaempferol in four, and isorhamenlin in one (201). The biological function of these flavonoids could be to resist viruses (202).

The amino acid composition of *C. quinoa* leaves as compared to the other leafy vegetables is given in Table 41. Higher content of lysine and lower content of methionine, as compared to other leafy greens, is the most distinguishing feature.

Chenopodium leaves can also be eaten in salads. In certain regions where vegetables are scarce, the leaves are important. The leaves and stalks are also fed to ruminants, and the chaff and gleanings from threshing arc generally fed to pigs.

B. Colocasia

The genus *Colocasia* belongs to the family *Araceae*. It is a small genus of about 13 species of perennial herbs distributed in the tropical parts of southeastern Asia and Polynesia. It occurs in the world over the greater part of tropical India and is also cultivated throughout India on account of its corms, which are boiled and eaten.

The nomenclature of the cultivated species is somewhat confused. *C. esculenta* and *C. antiquorum* are regarded as distinct species by some, whereas others regard the former as a variety of *C. antiquorum*. This is because of the variation in chromosome numbers in the same species.

The origin of *Colocasia* is believed to be southern Asia, from where it is thought to have travelled east to the islands of Oceania, the Philippines, China, and Japan. It is a tropical crop and prefers an abundant supply of water. It can be grown in heavy soils and can withstand waterlogging. The optimum soil pH is 5.5–7.0. It cannot withstand cooler climates and frost. The propagation of *Colocasia* is accomplished by vegetative means.

Colocasia is infected by a number of leaf diseases. *Phytophthora* leaf blight and *Pythium* rot or soft rot are the most destructive. The former occurs under wetland conditions and is favored by temperatures of 20–22°C and high relative humidity. The disease manifests itself as purple to brownish circular water-soaked lesions 1–2 cm in diameter, usually at the tip, base, and margin of the leaves. Losses ranging from 25 to 50% have been recorded due to this disease. Protective sprays with copper oxychloride can prevent this disease. Other diseases of minor importance are leaf spots caused by *Cladosporium colocasiae* Sawada and *Phyllosticta colocasiophilia.* Taro leaf hopper (*Tarophagus proserpina*) causes severe losses in most growing areas (205).

Colocasia leaves can also be eaten like spinach, but like the root they must be well cooked in order to destroy the acridity peculiar to aroids. A pinch of baking soda can be added during

TABLE 41 Comparison of Amino Acid Composition (g/g N) of Vegetable *C. quinoa* with Some Selected Leafy Vegetables

Leafy vegetables	Total N (g/100 g)	Arg	His	Lys	Trp	Phe	Tyr	Met	Cys	Thr	Leu	Isc
Amaranth	0.64	0.24	0.13	0.25	0.07	0.18	0.19	0.07	0.04	0.14	0.37	0.29
Cabbage	0.29	0.45	0.13	0.24	0.07	0.20	0.12	0.06	0.07	0.22	0.34	0.23
Drumstick	1.07	0.38	0.14	0.32	0.10	0.29	—	0.11	0.13	0.25	0.46	0.28

Source: Ref. 21.

cooking to remove the acridity, which increases with age. Leaves that are unopened or just about to open are preferable to older leaves. In Ashanti and southern Ghana, the most popular vegetable is *Colocasia esculenta* or cocoyam leaves. In these countries, the consumption of leaves has been increasing in view of the shortage of all other commodities.

The proximate and mineral composition of *Colocasia* leaves dried at 60°C and leaf protein derived therefrom is shown in Table 42. Only about 3.6% of the total iron is available in vitro (18). The amino acid composition of cocoyam leaf protein is given in Table 43. Comparison of the essential amino acids of *Colocasia* protein with the provisional FAO pattern (Table 44) shows that all of the essential amino acids provide over 100% of the pattern. Only the total sulfur-containing amino acids was found to be 54.8% of the pattern (206).

Comparison of dried *Colocasia* leaves and *Colocasia* leaf extract with casein with respect to protein quality is shown in Table 45. The PER value of *Colocasia* leaf extract is higher than that of dried cocoyam leaves. The NPU and BV obtained for the leaves is lower than that for casein, attributed mainly to the low content of sulfur-containing amino acids. These leaves are eaten in Ghana incorporated in stews or soups, which may or may not contain meat or fish. It has been suggested that protein extracts from *Colocasia* leaf could be incorporated in the diets of infants and preschool children. it can be easily extracted at the village level. The question of acceptability does not arise, since leaves in general and *Colocasia* leaves in particular form a major component of diet. It should be noted that palaver sauce made from colocasia leaves is considered a national dish in Ghana (206).

TABLE 42 Proximate and Mineral Composition of *Colocasia esculenta* Leaves

Constituent	Content
Proximate constituents	
Moisture (%)	7.55
Ash (%)	10.6–11.20
Protein (%)	18.0–23.4
Ether extract (%)	4.00–6.30
Fiber (%)	17.50
Neutral detergent fiber	37.0
Acid detergent fiber	21.7
Hemicellulose	15.3
Mineral profile	
Phosphorus(%)	0.35–0.60
Potassium (%)	1.70–3.86
Calcium (%)	1.13–1.30
Magnesium (%)	0.30–0.32
Manganese (ppm)	43.7–115
Copper (ppm)	8.3–27
Zinc (ppm)	17.5–58
Iron (ppm)	100–245
Sodium (mg/100 g)	0.04

Source: Ref. 151, 206.

Table 43 Amino Acid Composition of Colocasia Leaf Protein

Amino acid	g/16 g N
Aspartic acid	5.54
Threonine	4.44
Serine	4.42
Glutamic acid	4.42
Glycine	5.36
Alanine	6.46
Valine	5.91
Cystine	0.72
Methionine	1.66
Total sulfur amino acids	2.38
Isoleucine	4.84
Leucine	5.83
Tyrosine	4.16
Phenylalanine	6.26
Lysine	6.06
Histidine	1.97
Arginine	6.06
Ammonia	0.98

Source: Ref. 206.

Colocasia esculenta leaves contain about 9.7% lipids on a dry weight basis. The various lipid classes present in *Colocasia* leaves are given in Table 46. It can be seen that nonpolar lipids are present to a significant extent, with pigments being the major type. Among the glycolipids, monogalactosyl diglyceride is the major component, and among the phospholipids, phosphatidylcholine is the main component. The fatty acid composition of the lipid classes in *Colocasia esculenta* leaves is given in Table 47. It can be seen that the major fatty acids of leaves are linolenic (18:3), palmitic (16:0), and linoleic (18:1) acids. Sterol content is reported to be at

Table 44 Comparison of Some Essential Amino Acids of *Colocasia* Leaf Protein with the FAO Provisional Pattern

Amino acid	*Colocasia* leaf protein	FAO pattern	% of pattern
Threonine	277	180	153
Valine	364	270	136
Total sulfur amino acids	148	270	54.8
Isoleucine	302	270	111
Leusine	533	306	174
Total aromatic amino acids	657	360	184
Lysine	378	270	140

Source: Ref. 206.

TABLE 45 Protein Quality of Dried *Colocasia* Leaves and Leaf Extract as Compared to Casein

	Total digestibility	PER	NPU (%)	NPR	BV (%)
Dried colocasia leaves	87.9	0.89	38.9	2.86	44
Colocasia protein extract	89.3	1.37	40.3	2.99	45
Casein	96.2	3.58	68.7	4.60	71

Source: Ref. 206.

35 mg per 100 g fresh leaves (208). These sterols do not have any effect on serum cholesterol, but do lower liver cholesterol.

Colocasia leaves are known to have about (209). The mineral composition in *C. esculenta* leaves vegetable is as given in Table 42. Colocasia leaves contain about 10,657 IU of vitamin A per 100 g fresh material, of which 6.7% and 9.2% are lost after 2 days of storage at 29–32°C and 5°C, respectively (56). Recent reports indicate the presence of about 9000 μg of β-carotene/100 g fresh *Colocasia* leaves (210).

TABLE 46 Lipid Composition (wt%) of *C. esculenta* Leaves

Lipid class	wt %
Nonpolar lipids	39.6
Pigments	9.9
Hydrocarbons	5.0
Sterol esters	4.1
Ester waxes	2.3
Triglycerides	3.3
Fatty acids	2.9
Alcohols	0.0
Sterols	2.2
Diglycerides	3.0
Monoglycerides	2.9
Unidentified	4.3
Glycolipids	33.2
Esterified steryl glycosides	1.8
Monogalactosyl diglycerides	16.5
Cerebrosides	2.0
Digalactosyl diglycerides	7.5
Sulfoquinovosyl diglycerides	5.4
Phospholipids	27.1
Cardiolipin	1.8
Phosphatidylglycerol	5.4
Phosphatidylethanolamine	2.9
Phosphatidylinositol	4.4
Phosphatidylcholine	12.7

Source: Ref. 207.

Table 47 Fatty Acid Composition of Lipids in *C. esculenta* Leaves

Lipid class	Fatty acid (wt%)									
	12:0	14:0	14:1	16:0	16:1	18:0	18:1	18:2	18:3	20:0
Nonpolar lipids										
Sterol esters	2.5	2.1	tr	12.8	0.4	18.0	16.5	19.6	25.8	2.3
Ester waxes	4.3	1.8	0.2	22.3	0.7	13.6	21.2	20.9	10.8	4.8
TAG	0.4	1.4	tr	23.8	tr	10.6	15.8	16.1	31.3	0.6
Fatty acids	1.1	1.4	0.1	26.4	tr	11.8	12.6	21.4	24.8	0.4
DAG	3.4	3.4	nd	23.2	nd	8.5	9.4	22.1	24.5	3.5
MAG	3.6	3.3	0.4	22.0	tr	5.8	17.6	19.1	28.2	nd
Glycolipids										
ESG	9.4	4.6	0.3	28.1	0.3	9.9	13.7	16.8	16.9	nd
MGDG	nd	0.2	nd	4.0	nd	5.6	5.7	6.1	78.4	nd
Cerebrosides	6.0	4.1	nd	18.6	nd	11.1	12.8	9.8	33.3	4.3
DGDG	0.6	1.1	nd	6.8	nd	8.9	9.4	10.6	62.6	nd
SQDG	2.2	1.3	nd	14.8	nd	10.3	14.6	20.9	35.9	1.1
Phospholipids										
Cardiolipin	3.3	4.6	tr	20.1	0.6	6.8	11.8	19.1	33.4	0.3
PG	1.3	0.9	nd	21.9	6.2	5.1	8.6	11.8	44.2	0.1
PE	2.0	1.1	nd	25.8	0.6	7.8	14.9	20.5	27.3	nd
PI	0.8	0.3	nd	33.6	0.5	10.1	17.8	21.1	15.6	0.2
PC	0.7	0.4	tr	28.2	1.0	6.8	9.8	17.6	35.1	0.4

Source: Ref. 207.

TABLE 48 Antinutritional and Flatus Factors in *Colocasia* Leaves

Chemical constituent	Value
Total sugar (%)	8.6
Sucrose (S) (%)	1.4
Raffinose (R) (%)	0.2
Stachyose (St) (%)	0.5
S + R + St	2.2
Nitrate (mmol/100 g)	0.3
Oxalate (%)	5.1
Saponin (%)	0.9
Antitryptic activity (TIU/mg protein)	3.3
Phenol (mg tannic acid equivalent/g dry weight)	36.0
Phytate (mg/g)	32.5

Source: Ref. 151.

The antinutritional and flatus factors in *Colocasia* leaves are given in Table 48. The concentration of flatus factors is quite low, but the concentration of phytate is quite high (151). Aliphatic and aromatic secondary amines to the tune of 549 and 7 μg/g dry weight, respectively, are reported to be present in *Colocasia* leaves (33).

Calcium oxalate is present in *Colocasia* leaves at 0.008% (211). Sunell and Healy (212) refer to the irritant properties of calcium oxalate crystals in the corms and leaves of *Colocasia esculenta*, the major crystal type being raphides and druses. The crystal types of calcium oxalate in *C. esculenta* leaves are shown in Table 49. An investigation into the crystal type is generally made due to a possible relationship between the presence of raphides and aroid toxicity.

The flavonoids identified in *C. esculenta* leaves are apigenin, 3′,4′-diOMe luteolin, and anthocyanins (166). Soluble fibers, which are abundantly available in fruits and vegetables, regulate the cholesterol level by sequestering the bile acids needed for metabolism. To determine the effect of fibers from *Colocasia* leaves on cholesterol-fed hypercholesterolemic rats, a study was undertaken that revealed *Colocasia* leaves to aggravate the serum and tissue levels of cholesterol in cholesterol-fed rats. This is suggested to be due to some component in *Colocasia* leaves that interferes with lipid metabolism in general and especially with cholesterol metabolism (214).

TABLE 49 Occurrence of Various Crystal Types in *C. esculenta* Leaves

Crystal type	AdE	PM	SM	AbE
Druses	–	+	+	–
Raphides	–	+	(+)	–
Crystal sand	–	–	–	–
Prismatic crystal	nd	–	–	nd

– = crystal type absent; + = crystal type present; (+) = idioblastic cell (specialized spindle-shaped cell); AdE = adaxial epidermis; PM = palisade mesophyll; SM = spongy mesophyll; AbE = abaxial epidermis.
Source: Ref. 213.

TABLE 50 β-Carotene Content of *Colocasia esculenta*, Raw and After Processing

Processing treatment	β-Carotene (mg/kg fresh material)
Raw (whole or ground)	78
Boiled (15–120 min)	77
Boiled, with potash	79
Stewed, 25–30 ml oil, 100 ml water	78
Frying, 100–150 ml oil	—

Source: Ref. 79.

β-Carotene is quite stable to different processing methods, as shown in Table 50.

C. Cassia

This plant is a gregarious annual shrub 1–2 feet in height found everywhere in Bengal and widely spread and abundant throughout the tropical parts of India. The plant, like several others of the same genus, owes its medicinal activity to the presence of chrysaphanic acid, sometimes called rhein (215).

The sap is used locally to treat ringworm and other skin diseases. The flowers may be eaten. The leaves are made into an infusion, which is taken internally for gonorrhea in its subacute stage. It is anthelmentic, used externally for syphillis. Various anthraquinone derivatives have been reported to be present in cassia leaves (216–219). The antifungal activity of *Cassia alata* has been documented, a fact attributed to the presence of 4,5-dihydroxy-1-hydroxymethylanthrone and 4,5-dihydroxy-2-hydroxymethylanthrone (220). Antiviral activity attributed to emodin has also been reported for leaves of *Cassia augustifolia* (221). Antibacterial activity of *Cassia obovata* extracts against *Salmonella*, *Bacillus*, and *Corynebacterium* species are also reported (222). Fungal infection in *Cassia tora* is manifested as decreased catalase and increased peroxidase activity (223).

Cassia leaves are used for human consumption and as a feed. Replacement of 5% of the standard mash with *Cassia tora* leaf meal is reported to increase feed consumption and also increase the intake of magensium, calcium, and phosphorus by poultry (224). About 9.7 mg% of iron is present in *Cassia tora* leaves, of which only 15.5% is available (136). The disagreeable smell is eliminated by boiling. Seeds of this shrub are used to produce a blue dye. About 15–20 tons of seeds per year are collected in south India. Green dyes are made from the barks of rhamnus and *Cassia* in China. Naphthalene glycosides have been isolated from the leaves of *Cassia senna* and *C. augustifolia* (225). Sugars identified in the leaves of *C. augustifolia* are glucose, fructose, sucrose, and pinitol (226). Other constituents identified in the leaves of *Cassia tomentosa* are quercetin-7,3'-dimethyl ether-3-*O*-galactoside, hyperin, anthrone-*O*-glucoside, and kaempferol-3-*O*-galactorhamnoside in the ethyl acetate fraction and long-chain saturated alcohols such as 24-ethyl-δΔ^{24}-cholesten-3β-ol, α-amyrin, and β-sitosterol (227) in the hexane fraction (228).

Fresh leaves of *Cassia obtusifolia* are fermented to make *kawal*, a food product indigenous to Sudan, which is used as a meat substitute. The conventional process consists of pounding *Cassia obtusifolia* leaves into a paste, which is then packed firmly into an earthenware vessel. This is then buried in the ground in a cool or shaded location. The surface of the leaf paste is covered with a layer of green sorghum leaves, and the earthenware vessel, called a *zeer*, is fitted with a lid sealed with mud. The vessel is unsealed at intervals of 3 days, the now-dried sorghum leaves are removed, and the paste mixed thoroughly by hand. Fresh sorghum leaves are placed on the repacked paste and the *zeer* resealed. After a total fermentation of 11–15 days, the *kawal* is removed, molded into small balls, and sun-dried for 5 days. The dried *kawal* is consumed in the form of a stew or broth with okra. It can also be stored indefinitely in the dry form until required. The dominant microflora indicated in *kawal* fermentation are *Bacillus subtilis*, *Propionobacterium* spp., *Lactobacillus plantarum*, *Staphylococcus sciuri*, and two yeasts—*Candida krusei* and *Saccharomyces* spp. (229). The chemical composition of *Cassia obtusifolia* leaves, its mineral profile, and its amino acid composition are given in Table 51. The leaves of *Cassia obtusifolia* contain threonine, valine, isoleucine, leucine, and lysine in amounts greater than the

TABLE 51 Proximate Composition, Mineral Profile, and Amino Acid Composition of *Cassia obtusifolia* Leaves

Constituent	Content
Proximate constituents	
Ash (%)	12.6
Crude protein (%)	24.3
Oil (%)	2.5
Crude fiber (%)	13.5
Minerals	
Calcium (%)	3.85
Phosphorus (%)	0.26
Magnesium (%)	0.30
Manganese (mg/kg)	75.00
Iron (mg/kg)	534.00
Zinc (mg/kg)	32.00
Amino acids (g/16 g N)[a]	
Aspartic acid	12.1
Threonine	6.2 (4.0)
Serine	4.6
Glutamic acid	13.6
Proline	7.7
Glycine	6.7
Alanine	7.5
Valine	7.5 (5.0)
Cystine	1.4
Methionine	2.1 [Cys + Met = 3.5]
Isoleucine	6.0 (4.0)
Leucine	10.4 (7.0)
Tyrosine	5.3 [Tyr + Phe = 6.0]
Phenylalanine	6.8
Histidine	3.3
Lysine	7.7 (5.5)
Arginine	7.2

[a]Values in parentheses indicate FAO values for reference protein.
Source: Ref. 229.

FAO reference protein. Protein concentrates prepared from two species of *Cassia—C italica* and *C. holospericea*—have been analyzed with a view to using them as a raw material for supplementing animal dietary proteins. Phenylalanine was found to be absent in both these species. Qualitative amino acid compositions were identical, except for the absence of glutamic acid in *C. italica* and of aspartic acid in *C. holospericea* (230).

D. Peucedanum

Peucedanum graveolens is synonymous with *Anethum graveolens* and is indigenous to southern Europe and is cultivated in England, Germany, Rumania, and the Mediterranean region. Its chemical composition is shown in Table 52. The leaves contain about 109 mg/kg iron, of which only 3.7% is available in vitro (18). The lipid composition of *P. graveolens* leaves is presented in Table 53. The fatty acid composition of the lipid in *Peucedanum graveolens* leaves is given in Table 54. The major fatty acids in leaves are linolenic acid (18:3), palmitic (16:0) and linoleic (18:1).

E. Rumex

Rumex is a large genus of annual, biennial herbs, rarely shrubs, widely distributed in the temperate regions of the world. Many of the species belonging to this genus are eaten as pot herbs. The main species are *Rumex acetosa* or garden sorrel, *Rumex acetosella* or sheep sorrel, *Rumex crispus* (also called yellow dock or curled dock), *Rumex dentatus*, *Rumex hastatus*, *Rumex maritimus*, *Rumex nepalensis*, *Rumex patientia*, *Rumex scutatus* or French sorrel, and *Rumex vesicarius* or bladder dock. The leaves of all of these species are edible and are eaten as salad vegetables or cooked like spinach. The chemical composition of some of these species is

TABLE 52 Chemical Composition of the Leaves of *Peucedanum graveolens*

Constituent	Content
Moisture (%)	88.0
Protein (%)	3.0
Fat (%)	0.5
Minerals (%)	2.2
Fiber (%)	1.1
Carbohydrates (%)	5.2
Minerals	
Calcium (mg/100 g)	190
Phosphorus (mg/100 g)	42
Iron (mg/100 g)	17.4
Vitamins	
Carotene (μg/100 g)	7182
Thiamine (mg/100 g)	0.03
Riboflavin (mg/100 g)	0.13
Niacin (mg/100 g)	0.20
Vitamin C (mg/100 g)	25

Source: Ref. 21.

TABLE 53 Lipid Composition (wt%) of *P. graveolens* Leaves

Lipid class	wt%
Nonpolar lipids	32.2
Pigments	5.9
Hydrocarbons	4.6
Sterol esters	4.0
Ester waxes	0.0
Triglycerides	2.3
Fatty acids	4.2
Alcohols	0.0
Sterols	4.4
Diglycerides	3.7
Monoglycerides	0.0
Unidentified	3.1
Glycolipids	44.7
Esterified steryl glycosides	0.0
Monogalactosyl diglycerides	21.9
Cerebrosides	0.0
Digalactosyl diglycerides	22.5
Sulfoquinovosyl diglycerides	0.3
Phospholipids	23.1
Cardiolipin	0.7
Phosphatidylglycerol	12.8
Phosphatidylethanolamine	0.3
Phosphatidylinositol	0.2
Phosphatidylcholine	9.1

Source: Ref. 207.

given in Table 55. Glycolipids, neutral lipids, and phospholipids in *Rumex vesicarius* at levels of 0.17, 0.17, and 0.13 g/100 g dry weight, respectively, are reported (160). The fatty acid composition in these lipid classes is given in Table 56. It can be seen that all these lipid classes contained 18:3 in maximum amounts. The phospholipids and the glycolipids also contained a sufficient amount of 16:0.

Carotene levels in the leaves of *Rumex tianschanicus* are known to vary from 25.7 mg% in the first year to 31.7 and 36.4 mg% in the second and third years, respectively (231). α- and β-Tocopherol are reportedly present at 6.19 and 3.37 mg/100 g dry weight, respectively (106). The leaves of *Rumex vesicarius* are reported to contain 125 μg of folic acid per 100 g edible portion, of which 40 μg are in the free form. Folate existing as conjugated folate accounts for about 68% of the total folate (48). Pantothenic acid and pyridoxine at 0.160 and 0.018–0.214 mg fresh weight % has also been reported (232). These leaves contain about 4938 IU of vitamin A per 100 g of fresh material, of which about 20.5% is lost during refrigerator storage (5°C) for 2 days, 74.1% on frying in oil, and about 35% by boiling in water (56). The effect of different processing techniques on vitamin C content in *Rumex vesicarius* leaves is as given in Table 57, which indicates a substantial loss in ascorbic acid during cooking followed by the combined process of washing and cutting.

TABLE 54 Fatty Acid Composition of Lipid Classes of *P. graveolens* Leaves

Lipid class	Fatty acid (wt%)											
	12:0	14:0	14:1	16:0	16:1	18:0	18:1	18:2	18:3	20:0	22:0	Other
Nonpolar lipids												
Sterol esters	5.3	9.0	0.6	23.5	0.7	6.0	18.8	14.8	20.6	0.5	0.2	nd
TAG	1.0	3.2	0.6	17.8	tr	9.5	15.1	36.6	20.2	nd	1.0	nd
Fatty acids	nd	nd	nd	4.1	nd	14.9	14.8	21.6	44.5	nd	nd	nd
DAG	3.8	9.5	1.1	27.7	tr	9.5	17.2	11.9	5.3	8.7	6.2	nd
Glycolipids												
MGDG	0.1	0.1	nd	1.1	tr	0.1	46.5	1.4	49.9	0.2	nd	0.5
DGDG	nd	0.1	nd	0.5	tr	0.1	47.6	0.9	50.1	0.1	nd	0.5
Phospholipids												
Cardiolipin	5.3	4.7	0.9	27.5	2.9	7.1	13.9	9.0	28.3	0.2	nd	nd
PG	0.3	0.5	0.1	22.3	5.4	1.6	5.2	13.6	50.4	0.2	nd	0.2
PE	1.6	0.9	nd	27.7	3.8	4.4	6.1	12.0	43.4	nd	nd	nd
PI	7.0	4.2	0.8	42.8	nd	17.6	14.9	6.3	5.9	0.5	nd	nd
PC	0.1	0.2	nd	23.0	5.4	1.0	4.0	19.3	46.5	0.2	0.2	nd

Source: Ref. 207.

TABLE 55 Chemical Composition of Some *Rumex* Species Consumed as Leafy Vegetables

Constituent	*R. acetosa*	*R. crispus*	*R. dentatus*	*R. vesicarius*
Moisture (%)	90.0	92.6	89.3	95.2
Protein (%)	2.6	1.5	3.5	1.6
Fat (%)	0.5	0.3	0.4	0.3
Carbohydrates (%)	2.6	4.1	4.1	1.4
Fiber (%)	nr	0.9	0.7	0.6
Potassium (mg/100 g)	400	nr	nr	nr
Magnesium (mg/100 g)	46	nr	nr	nr
Phosphorus (mg/100 g)	44	56.0	53.8	17.0
Calcium (mg/100 g)	43	74.0	611.5	63.0
Manganese (mg/100 g)	0.59	nr	nr	nr
Copper (mg/100 g)	0.30	nr	nr	nr
Zinc (mg/100 g)	0.22	nr	nr	nr
Iron (mg/100 g)	nr	5.6	53.8	8.7
Iodine (mg/100 g)	0.007	nr	nr	nr
Ascorbic acid (mg/100 g)	124.0	30.0	115.0	12.0
β-Carotene (mg/100 g)	11.0	nr	nr	nr
Thiamine (mg/100 g)	0.08	0.06	nr	0.03
Riboflavin (mg/100 g)	nr	0.08	nr	0.06
Nicotinic acid (mg/100 g)	nr	0.40	nr	0.20
Vitamin A (mg/100 g)	nr	1.38	11,700 IU	6,100 IU
Vitamin B_6 (mg/100 g)	nr	nr	nr	0.15

nr = Not reported.
Source: Refs. 129, 197.

TABLE 56 Fatty Acid Composition[a] of the Neutral Lipids, Glycolipids, and Phospholipids of Edible Leaves of *Rumex vesicarius*

Fatty acid	Neutral	Glycolipid	Phospholipid
12:0	2.2	6.9	1.2
14:0	2.9	5.7	1.4
14:1	0.2	—	—
16:0	17.2	38.2	34.3
16:1	0.6	—	3.4
18:0	3.5	1.7	—
18:1	4.5	1.0	3.6
18:2	19.8	6.3	16.7
18:3	48.8	40.2	39.4
20:4	0.2	—	—
22:0	0.1	—	—

[a]Fatty acid composition as weight percent of total fatty acids.
Source: Ref. 160.

TABLE 57 Effect of Different Processing Techniques on the Vitamin C Content of *Rumex vesicarius* Leaves

Processing treatment	Vitamin C content (mg/100 g)
Control, no treatment	28.8
Cutting into pieces using stainless steel knife	21.4
Washing in tap water	24.5
Washing in tap water after cutting	10.9
Washing in tap water before cutting	17.4
Cooking with lid for 2 minutes	10.6
Cooking without lid for 2 minutes	4.3

Source: Ref. 69.

The oxalate content of *Rumex* species on a dry weight basis is reported to be between 1.69 and 44.67, causing interaction with calcium and hypoglycemia (188). Acute oxalate poisoning attributable to ingestion of *Rumex crispus* in sheep has been reported (233). Fertilizer applications of nitrogen are known to increase the content of both oxalate and nitrate in *Rumex* leaves (234).

Leaves of *R. acetosa* contain the flavone glycosides rutin (1.28%), hyperin (37 mg/100 g), vitexin, and traces of oxymethylanthraquinone; *R. dentatus* contains about 7.8% flavonoids on a dry weight basis and includes rutin, avicularin, quercitrin, and quercitin in addition to about 0.04% anthraquinone derivatives (235,236). Antifungal principles have been isolated from nine species of *Rumex* and have been identified as nepodin, emodin, physcion, and chrysophanol (237,238). These have been traditionally used in Chinese medicine. An antitumor agent, believed to be a polysaccharide, has been reported in the leaves of *Rumex acetosa* (239). Extracts of *Rumex japonicus* have been demonstrated to have antioxidant properties and, in combination with ascorbic acid or its derivatives, surpassed BHA and BHT (240).

REFERENCES

1. Thompson, H. G., and W. C. Kelly, *Vegetable Crops*, 5th ed., McGraw-Hill Book Co., New York, 1957, p. 611.
2. Salunkhe, D. K., and B. B. Desai, *Postharvest Biotechnology of Vegetables*, Vol. I, CRC Press, Boca Raton, FL, 1984, p. 181.
3. Watts, R. L., and G. S. Watts, *The Vegetable Growing Business*, Orange Judd Pub. Co., Inc., New York, 1940, p. 520.
4. Yawalkar, K. S., *Vegetable Crops of India*, 2nd ed., AgriHorticultural Pub. House, Nagpur, 1980, p. 370.
5. Kale, P. N., and S. Kale, *Bajipala Utpadan* (Marathi), Continental Press, Pune, 1984.
6. Kazuoka, T., and Oeda, K., Heat stable COR (cold-regulated) proteins associated with freezing tolerance in spinach, *Plant Cell Physiol. 33*(8):1107 (1992).
7. Singh, H. B., and B. S. Joshi, *Our Leafy Vegetables, Farm Bull. No. 56*, Farm Information Unit, Director of Extension, Ministry of Food & Agriculture, New Delhi, 1960.
8. Hardenburg, R. E., A. E. Watada, and C. Y. Wang, *The Commercial Storage of Fruits, Vegetables and Florist and Nursery Stocks*, U.S. Department of Agriculture, Handbook 66, 1986, p. 69.
9. Ryall, A. L., and W. J. Lipton, *Handling, Transportation and Storage of Fruits and Vegetables*, Vol. 1, *Vegetables and Melons*, AVI Pub. Co. Inc., Westport, CT, 1972, p. 473.

10. Friedman, B. A., M. Lieberman, and J. Kauffman, A comparison of crown-cut & clip-topped spinach prepacked at terminal market, *Proc. Am. Soc. Hort. Sci. 57*:285 (1951).
11. Chakraborti, N., I. Mandal, and G. C. Bannerjee, Chemical composition of some common herbs and shrubs, aquatic plants and kitchen greens, *Indian J. Dairy Sci. 42*(2):375 (1989).
12. Fordham, R., Vegetables of temperate climate—leafy vegetables, in *Encyclopaedia, Food Science, Food Technology and Nutrition* (R. MaCrae, R. K. Robinson, and M. J. Sandler, eds., Academic Press, London, 1993, p. 3964.
13. Holland, B., I. D. Unwin, and D. H. Buss, *Vegetables, Herbs and spices: Fifth supplements to McCance & Widdowson's. The Composition of Foods*, HM SO, London, 1991.
14. Kim, H., and M. B. Zemel, *In vitro* estimation of the potential bioavailability of calcium from sea mustard (*Undaria pinnatifida*), milk and spinach under simulated normal and reduced gastric acid conditions, *J. Food Sci. 51*:957 (1986).
15. Heaney, R. P., C. M. Weaver, and R. R. Recker, Calcium absorbability from spinach, *Am. J. Clin. Nutr. 47*:707 (1988).
16. Kikunaga, S., M. Arimori, and M. Takahashi, The bioavailability of calcium in spinach and calcium oxalate to calcium deficient rats, *J. Nutr. Sci. Vitaminol, 34*:195 (1988).
17. Reykdal, O., and K. Lee, Soluble, dialyzable and ionic calcium in raw and processed skim milk, whole milk and spinach, *J. Food Sci. 56*(3):864 (1991).
18. Chawla, S., A. Saxena, and S. Seshadri, In-vitro availability of iron in various green leafy vegetables, *J. Sci. Food Agric. 46*:125 (1988).
19. Singh, B., O. P. Vadhwa, M. T. Wu, and D. K. Salunkhe, Effects of foliar application of s-triazines on protein, amino acids, carbohydrates and mineral composition of pea and sweet corn seeds, bush bean pods and spinach leaves, *J. Agric. Food Chem. 20*:1256 (1972).
20. Satoh, T., M. Goto, and K. Igarashi, Effects of protein isolates from raddish and spinach leaves, *J. Nutr. Sci. Vitaminol. 39*(6):627 (1993).
21. Gopalan, C., B. V. Ramasastri, and S. C. Balasubramaniam, Nutitive value of Indian Foods. Hyderabad, India, National Institute of Nutrition, ICMR, 1985.
22. Browse, J., G. Roughan, and R. Slack, Light control of fatty acid synthesis and diurnal fluctuations of fatty acid composition in leaves, *Biochem. J. 196*:347 (1981).
23. Murcia, M. A., A. Vera, and F. Garcia-Carmona, Effect of processing methods on spinach: proximate composition in fatty acids and soluble protein, *J. Sci. Food Agric. 59*:473 (1992).
24. Hertz, J., and U. Baltensperger, Determination of nitrate and other inorganic anions (NO_2, PO_4-, Cl-, SO_4-) in salad and vegetables by ion chromatography, *Anal. Chem, 318*:121 (1984).
25. Lara, W. H., and M. Y. Takahashi, Nitrate levels in vegetables, *Revista Instituto Adolfo Lutz 42*:53 (1982).
26. Takruri, H. R., and M. A. Humeid, Nitrate levels in edible wild herbs and vegetables common in Jordan, *Nutr. Health 6*:89 (1988).
27. Baker, A. V., D. N. Maynard, and H. A. Mills, Variations in nitrate accumulation among spinach cultivars, *J. Am. Soc. Hort. Sci. 99*(2):132 (1974).
28. Cantliffe, D. J., Nitrate accumulation in spinach grown under different light intensities, *J. Am. Soc. Hort. Sci. 97*:152 (1972).
29. Cantliffe, D. J., Nitrate accumulation in spinach grown at different temperatures, *J. Am. Soc. Hort. Sci. 97*:674 (1972).
30. Olday, F. C., A. V. Baker, and D. N. Mainard, A physiological basis for different patterns of nitrate accumulation in two spinach cultivars, *J. Am. Soc. Hort. Sci. 101*:217 (1976).
31. Machackova, I., Z. Zmrhal, and M. Trckova, Nitrate content in lettuce and spinach leaves in relaton to the day harvested and to conditions and duration of storage, *Rostl. Vyroba 31*(11):1151 (1985).
32. Teotia, M. S., S. K. Berry, S. G. Kulkarni, and S. Kaur, Nitrate and nitrite content in vegetables, *J. Food Sci. Technol. 25*(5):272 (1988).
33. Shah, A. S., and S. V. Bhide, Content of nitrate and secondary amines in vegetables, legumes and marine foods, *J. Food Sci. Technol. 22*(4):266 (1985).
34. Heisler, E. G., J. Siciliano, S. Krulick, J. Feinberg, and J. H. Schwartz, Changes in nitrate and nitrite

content, and search for nitrosamines in storage-abused spinach and beets, *J. Agric. Food Chem. 22*(6):1029 (1974).
35. Yamanaka, H., M. Kuno, K. Shiomi, and T. Kikuchi, Determination of oxalate in foods by enzymic analysis, *Shokuhin Eiseigaku Zasshi 24*(5):454 (1983).
36. Limongelli, J. C. H., C. A. M. Mundt, E. Bernardi, and M. Berreiro, Oxalic acid in spinach, *Cienc. Invest. 30*(1-2):22 (1974).
37. Van Maercke, D., Spinach oxalate content as influenced by nitrogen fertilization, *Meded Fac. Landbouwentensch. Rijksuniv. Gent. 38*(1):173 (1973).
38. Yang, J. C., and F. A. Loewcis, Metabolic conversion of L-ascorbic acid to oxalic acid in oxalate accumulating plants, *Plant Physiol. 56*(2):283 (1975).
39. Fujii, N., M. Watanabe, Y. Watanabe, and N. Shimada, Rate of oxalate biosynthesis from glycolate and ascorbic acid in spinach leaves, *Soil Sci., Plant Nutr. 39*(4):624 (1993).
40. Ganju, K., and B. Puri, Bioflavonoids from Indian vegetables and fruits, *Indian J. Med. Res. 47*:563 (1959).
41. Jurkovic, N., and M. Lackovic, Determination of total carotenoids in dough made from *Triticum vulgare* wheat flour, *Kem. Ind. 20*(8):379 (1971).
42. Khacik, F., G. R. Beecher, and N. F. Whittaker, Separation, identification and quantification of the major carotenoid and chlorophyll constituents in extracts of several green vegetables by liquid chromatography, *J. Agric. Food Chem 34*:603 (1986).
43. Dmitrieva, L. V., N. M. Gluntsov, and T. A. Lomakova, Effect of various illumination intensities and temperatures on the yield and quality of spinach and Peking cabbage, *Dokl. Vses. Akad. S-Kh. Nauk 11*:21 (1974).
44. Stino, K. R., M. Abdelfattah, and N. H. Abdelaziz, Vitamin C and oxalic acid concentrations in spinach, *Agric. Res. Rev. 51*(5):109 (1973).
45. Kochubei, I. V., Effect of the properties of sod-podzolic soil and fertilizer on the yield and level of vitamins in spinach leaves, *Vestr. Mosk. Univ. Biol. Pochvoved 30*(6):88 (1975).
46. Booth, S. L., K. W. Davidson, and J. A. Sadowski, Evaluation of an HPLC method for the determination of phylloquinone (vitamin K_1) in various food matrices, *J. Agric. Food Chem. 42*:295 (1994).
47. Ferland, G., and J. A. Sadowski, Vitamin K (phylloquinone) content of green vegetables: Effects of plant maturation and geographical growth location, *J. Agric. Food Chem, 40*:1874 (1992).
48. Lakshmaiah, N., and B. V. Ramasastri, Folic acid conjugase from plasma. III. Use of the enzyme in the estimation of folate activity in foods, *Int. J. Vit. Nutr. Res. 45*(3):262 (1975).
49. Matsumoto, T., and T. Kaya, Distribution of amines in plants, *Nippon Nogei Kagaku Kaishi 56*(3):209 (1982).
50. Toyohara, H., M. Kinoshita, K. Sasaki, S. Yamaguchi, Y. Shimuzu, and M. Sakaguchi, Occurrence of a modori inhibitor in spinach leaves, *Nippon Suisan Gakkaishi 58*(9):1705 (1992).
51. Ozaki, M., K. Fujinami, K. Tanaka, Y. Amemiya, T. Sato, N. Ogura, and H. Nakagawa, Purification and initial characterization of the proteosome from the higher plant Spinacia oleracea, *J. Biol. Chem. 267*(30):21678 (1992).
52. Ezell, B. D., and M. S. Wilcox, Loss of vitamin C in fresh vegetables as related to wilting and temperature, *J. Agric. Food Chem. 7*(7):507 (1959).
53. Plantenius, H., Effect of temperature on the rate of deterioration of fresh vegetables, *J. Agric. Res. 59*:41 (1939).
54. Yamaguchi, N., S. Hamaguchi, and K. Ogata, Physiological and chemical studies on ascorbic acid of fruits and vegetables VIII. Mechanism of chlorophyll degradation and action of ascorbic acid in the inhibition of yellowing in harvested parsley leaves, *J. Jpn. Soc. Hort. Sci. 49*:414 (1980).
55. Watada, A. E., S. D. Kim, K. S. Kim, and T. C. Harris, Quality of green beans, bell peppers and spinach stored in polyethylene bags, *J. Food Sci. 52*(6):1637 (1987).
56. Rao, C. N., True vitamin A value of some vegetables, *J. Nutr. Dietet. 4*(1):10 (1967).
57. Murata, T., and Y. Ueda, Studies on the CA storage of spinach, *J. Jpn. Soc. Hort. Sci. 36*:449 (1967).

58. Sistrunk, W. A., M. K. Mohan, and D. W. Freeman, Relationship of processing methodology to quality attributes and nutritional value of spinach, *Hort. Sci. 12*:59 (1977).
59. Hosotani, K., and N.Kuratomi, Effect of blanching and thawing process on the content of β-carotene and L-ascorbic acid in raw and frozen spinach, *Wakayama Daigaku Kyoikugakubu Kiyo, Shizen Kagaku 41*:95 (1992).
60. Richter, E., and S. Hondke, Influence of blanching and preservation by air drying at different temperatures, deep freezing and freeze drying on oxalic acid of spinach, *Z. Lebensm. Unters. Forsch. 153*(1):31 (1973).
61. Astier-Dumas, M., Changes in the content of nitrates, vitamin C, Mg and Fe during the cooking of spinach, *Ann. Nutr. Aliment. 29*(3):239 (1975).
62. El-Din, A. E. M. N., and M. S. El-Feky, Chemical changes during the preservation of some vegetables. I. Egyptian mallow, spinach and peas, *Egypt. J. Food Sci. 19*(1-2):115 (1991).
63. Hanson, L. P., *Commercial Processing of Vegetables*, Noyes Data Corporation, London, 1975, p. 305.
64. Phillips, W. E. J., Changes in the nitrate and nitrite contents of fresh and processed spinach during storage, *J. Agric. Food Chem. 16*:88 (1968).
65. Lee, C. Y., R. S. Shallenberger, D. L. Sowning, G. S. Stroewsand, and N. M. Peck, Nitrate and nitrite nitrogen in fresh, stored and processed table beets and spinach from different levels of field nitrogen fertilization, *J. Sci. Food Agric, 27*:109 (1971).
66. Thomas, B., J. A. Roughan, and E. D. Watters, Lead and cadmium content of some canned fruit and vegetables, *J. Sci. Food Agric. 24*:447 (1973).
67. Vera, A., M. A. Murcia, and F. Garcia-Carmona, Ion levels of fresh and processed spinach using ion chromatography, *J. Food Qual. 18*:19 (1995).
68. Nagra, S. A., and S. Khan, Vitamin A (β-carotene) losses in Pakistani cooking, *J. Sci. Food Agric. 46*:249 (1989).
69. Tapadia, S. B., A. B. Arya, and P. R. Devi, Vitamin C contents of processed vegetables, *J. Food Sci. Technol. 32*:513 (1995).
70. DeSouza, S. C., and R. R. Eitenmiller, Effects of processing on the folate content of spinach and broccoli, *J. Food Sci. 51*(3):626 (1986).
71. Chen, T. S., Y. O. Song, and A. J. Kirsch, Effects of blanching, freezing and storage on folacin contents of spinach, *Nutr. Rep. Int. 28*:317 (1983).
72. Schroeder, H. A., Losses of vitamins and trace minerals resulting from processing and preservation of foods, *Am. J. Clin. Nutr. 24*:562 (1971).
73. Leichter, J., V. P. Switzer, and A. F. Landymore, Effect of cooking on folate content of vegetables, *Nutr. Rep. Int. 18*:475 (1978).
74. Klein, B. P., C. H. Y. Kuo, and G. Boyd, Folacin and ascorbic acid retention in fresh raw, microwave and conventionally cooked spinach, *J. Food Sci. 46*:640 (1981).
75. Pfeffer, C., J. F. Diehl, and W. Schwack, Effect of irradiation on bioavailability of food folates, *Ber. Bundesforschungs, Ernaehr.* (1993).
76. Janicki, J., J. Miklaszewska-Gawecka, and M. Urbanowicz, Effect of blanching on changes in the contents of nitrate and nitrite in different varieties of spinach, *Rocz. Wyzsz. Szk. Roln. Poznanice 47*:89 (1970).
77. Hawat, H., and M. L. Achtzehn, Formation and determination of NO_2- in sterile spinach preserve, *Nahrung 15*(8):869 (1971).
78. Bough, W. A., Reduction of suspended solids in vegetable canning waste effluents by coagulation with chitosan, *J. Food Sci. 40*:297 (1975).
79. Renqvist, U.H., A. C. De Vreeze, and B. Evenhuis, The effect of traditional cooking methods on carotene content in tropical leafy vegetables, *Indian J. Nutr. Dietet. 15*:154 (1978).
80. Joshi, H. C., L. Kusum, N. Joshi, P. C. Pant, N. C. Joshi, B. P. Gupta, and M. C. Joshi, Quality comparison of vegetables dehydrated in solar drier and electric oven, *Def. Sci. J. 41*(1):87 (1991).

81. Sato, M., M. Watanabe, and Y. Yagi, Green vegetable powders and their manufacture using cyclodextrin, *Jpn. Kokkai Tokkyo Koho* JP 04, 271, 745 (1992).
82. Cirill, G., and C. A. Ridolf, Noodles and macaroni made from eggs and spinach flour, *Tec. Molitoria 26*(17):98 (1975).
83. Langenberg, J. P., U. R. Tjaden, E. M. De Vogel, and D. I. Langerak, Determination of phylloquinone (vitamin K_1) in raw and processed vegetables using reversed phase HPLC with electrofluorometric detection, *Acta Aliment. 15*(3):187 (1986).
84. Pandita, M. L., and S. Lal, Leafy vegetables, in *Vegetable Crops in India* (T. K. Bose and M. G. Som, eds.), Naya Prokash, Calcutta, 1986, p. 775.
85. Choudhury, B., Vegetables, 41st ed., National Book of India, New Delhi, 1967.
86. Chauhan, D. V. S., *Vegetable Production in India*, Ram Prasad & Sons, Agra, India, 1981, p. 532.
87. Muthukrishnan, C. R., and I. Irulappan, Leay vegetables, in *Vegetable Crops in India* (T. K. Bose and M. G. Som, eds.), Naya Prokash, Calcutta, 1986.
88. Granado, F., B. Olmedilla, I. Blanco, and E. Rojas-Hidalgo, Carotenoid composition in raw and cooked Spanish vegetables, *J. Agric. Food Chem. 40*:2135 (1992).
89. Gami, D B, and T. S. Chen, Kinetics of folacin destruction in Swiss chard during storage. *J. Food Sci. 50*:447 (1985).
90. Olsen, O. E., R. H. Burris, and C. A. Elvehjem, A preliminary report of the folic acid content of certain foods, *J. Am. Dietet. Assoc. 23*:200 (1947).
91. Mullin, W. J., D. F. Wood, and S. G. Howsam, Some factors affecting folacin content of spinach, Swiss chard, broccoli and Brussels sprouts, *Nutr. Rep. Int. 26*:7 (1982).
92. Zepplin, M., and C. A. Elvehjem, Effect of refrigeration on retention of ascorbic acid in vegetables, *Food Res. 9*:100 (1944).
93. Grubben, G. J. H., and D. H. Van Slotten, *Genetic Resources of Amaranths*, International Board for plant genetic resources, FAO, Rome, 1981.
94. Olusegun, O. L., in *Handbook of Tropical Foods,* (H. T. Chan, Jr., ed.), Marcel Dekker, New York, 1983, p. 1.
95. National Research Council, *Amaranth: Modern Prospects for an Ancient Crop*, National Academy Press, Washington, DC, 1984, p. 1.
96. Schurmann, P. and H. Matsubara, Ferredoxins from fern and amaranth—two diverse plants with similar ferredoxins, *Biochem. Biophys. Acta 223*:450 (1970).
97. Hendricks, S. B., and R. B. Taylorson, Promotion of seed germination by nitrates and cyanides, *Nature 237*:169 (1972).
98. Hendricks, S. B., and R. B. Taylorson, Variation in the germination and amino acid leakage of seeds with temperature related to phase change, *Plant Physiol. 58*:7 (1976).
99. Ikenaga, T., M. Matuyo, and H. Ohashi, Studies on physiology and ecology of *Amaranthus viridis.* II. The effect of nutritional conditions on growth and chlorophyll content in *Amaranthus viridis. Zasso Kenkyu 20*:156 (1975).
100. Schinninger, R., Influence of single and combined pollutants on the dessication resistance and transpiration of *Amaranthus chlorostachys* wild, *Flora 171*:187 (1981).
101. Yunus, M., and K. Srivastava, Response of some plants to sulfur dioxide, *Kalikasan 10*:115 (1981).
102. Toru, M., O. Daisaku, and T. Euchi, Effect of sodium application on growth of *Amaranthus tricolor* L., *Plant Cell Physiol. 27*:187 (1986).
103. Devadas, R. P., M. V. K. Devadoss, and R. Rowlands, The supplementary value of wild greens (a mixture of leafy vegetables) to poor Indian wild rice diet, *J. Nutr. Dietet. 1*:34 (1964).
104. Santos, O. J., and M. F. De Carvalho, Nutritional value of some edible leaves used in Mozambique, *Econ. Bot. 29*:255 (1975).
105. Ifon, E. T., and O. Bassir, The nutritive value of some Nigerian leafy green vegetables. 2. The distribution of protein carbohydrates (including ethanol soluble simple sugars), crude fat, fibre and ash, *Food Chem. 5*:231 (1980).

106. Ifon, E. T., and O. Bassir, The nutritive value of some Nigerian leafy green vegetables. 1. Vitamin and mineral contents, *Food. Chem. 4*:263 (1979).
107. Faboya, O. O. P., The mineral content of some green leafy vegetables commonly found in western part of Nigeria, *Food Chem. 12*:213 (1983).
108. Ezeala, D. O., Nutrients, carotenoids and mineral compositions of the leaf vegetables *Amaranthus viridis* and *A. caudatus* L., *Trop Agric. 62*:95 (1985).
109. Singhal, R. S., and P. R. Kulkarni, Amaranths: an underutilized resource, *Int. J. Food Sci. Technol. 23*:125 (1988).
110. Fetuga, B. L., V. A. Oyonuga, and O. A. Taylor, Accumulation of mineral elements in five tropical leafy vegetables as influenced by nitrogen fertilisation and age, *Sci. Hort. 18*:313 (1983).
111. Srimathi, M. S., and N. G. K. Karanath, Insecticide induced shifts in the nutritive quality of vegetables grown in BHC treated soil, *Indian J. Nutr. Dietet. 20*:216 (1983).
112. Devadas, R. P., V. Anuradha, and U. Chandrashekar, Seasonal variation in the nutrient content of *Amaranthus flavus*, *J. Nutr. Dietet 6*:305 (1960).
113. Olufolaji, A. O., and Tayo, Growth, development and mineral contents of three cultivars of amaranth, *Sci. Hort. 13*:181 (1980).
114. Teutonico, R. A., and D. Knorr, Amaranth: Composition, properties and applications of a rediscovered food crop, *Food Technol. 39*:49 (1985).
115. Kidwai, I. M., and B. K. Zain, Nutritive value of some common plants, *Pak. J. Biochem. 2*:28 (1969).
116. Behari, M., and C. K. Andhiwal, Chemical examination of *Amaranthus spinosus* L., *Curr. Sci. 45*:481 (1976).
117. Vasi, I. G., and V. P. Kalintha, Amino acid composition of some leafy vegetables, *J. Inst. Chem. (India) 52*:13 (1980).
118. Gollamudi, L., A. J. Pantulu, and K. S. Rao, Lipid class and fatty acid composition of young *Amaranthus gangeticus* L leaves, *J. Agric. Food Chem. 32*:1361 (1984).
119. Bajwa, S. S., and P. S. Sastry, Positional distribution of fatty acids in leaf lecithins, *Indian J. Biochem. Biophys. 9*:133 (1972).
120. Rehari, M., and R. K. Sharma, Isolation and characterisation of hydrocarbons, alcohols and sterols from *Amaranthus tricolor*, *Acta Ciencia Indica (Chemistry) 10*:42 (1984).
121. Behari, M., V. Shri, T. Akihisa, and T. Matsumoto, 24-Alkyl-7 sterols of the herb *Amaranthus viridis*, *Filoterapia 57*:276 (1986).
122. Tudor, F., and B. George, Fatty acids and sterols of *Amaranthus tricolor* L., *Food Chem. 15*:233 (1984).
123. Lala, V. R., and V. Reddy, Absorption of beta-carotene from green leafy vegetables in undernourished children, *Am. J. Clin. Nutri. 23*:110 (1970).
124. Rao, C. N., and B. S. N. Rao, Absorption of dietary carotenes in human subjects, *Am. J. Clin. Nutr. 23*:105 (1970).
125. Devadas, R. P., and N. K. Murthy, Biological utilization of β-carotene from papaya fruit and amaranth in preschool children, *Indian J. Nutr. and Dietet. 17*:41 (1978).
126. Srinkantia, S. G., Prevention of vitamin A deficiency, *World Rev. Nutr. Dietet. 31*:95 (1978).
127. Mercadenate, A. Z., and D. B. Rodriguez-Amaya, Carotenoid composition and vitamin A value of some native Brazilian green leafy vegetables, *Int. J. Food Sci. Technol. 25*:213 (1990).
128. Jathar, V. S., P. R. Kulkarni, and D. V. Rege, Vitamin B_{12}-like activity in leafy vegetables, *Indian J. Biochem Biophys. 11*:71 (1974).
129. Bapu Rao, S., and P. G. Tulpule, Vitamin B_6 content of some Indian foods and regional diets and effect of cooking on the vitamin content, *Indian J. Nutr. Dietet. 18*(1):9 (1981).
130. Keshinro, O. O., The free and total folate activity in some commonly available tropical foodstuffs, *Food Chem. 11*(2):87 (1983).
131. Sreeramulu, N., G. D. Ndossi, and K. Mtotomwema, Effect of cooking on the nutritive value of

common food plants of Tanzania: Part 1—Vitamin C in some of the wild green leafy vegetables, *Food Chem. 10*(3):205 (1983).

132. Schmidt, D. R., H. A. Macdonald, and W. C. Kelly, Solubility of iron, calcium, magnesium in amaranth and collard leaves, *Qual. Plant. Plant Foods Human Nutr. 23*:373 (1974).
133. Bassir, O., The efficiency of utilizing iron in leafy green vegetables from haemoglobin systhesis by anemic rats, *Nutr. Rep. Int. 18*:481 (1978).
134. Murthy, N., S. Annapurni, R. Premjothi, J. Rajah, and K. Shubha, Bioavailability of iron by in vitro method. 1. From selected foods and effect of fortification, promoters and inhibitors, *Indian J. Nutr. Dietet. 22*:68 (1985).
135. Pankaja, N., and J. Prakash, Availability of Ca from kilkeerai (*Amaranthus tricolor*) and drumstick (*Moringa oleifera*) greens in weanling rats, *Nahrung 38* (2):199 (1994).
136. Reddy, N. S., and K. Kulkarni, Availability of iron from some uncommon edible green leafy vegetables determined by *in vitro* method, *Nutr. Rep. Int. 34*:(5):859 (1986).
137. Saina, Z. I., Phytochemical study of spreading amaranth (*amaranthus blitorides*) and *Anaphalis racemiberae, Inst. Fiziolog Akad. Nauk. Kazakh SSR 18*:44 (1973).
138. Sanni, S. B., The fluoride contents of common Nigerian vegetables, *J. Sci. Food Agric. 33*:686 (1982).
139. Giri, J., K. Usha, and T. Sunitha, Evaluation of selenium and chromium content of plant foods, *Plant Foods Human Nutr. 40*(1):49 (1990).
140. Marten, G. C., and R. N. Anderson, Nutritive value and palatability of 12 common annual weeds, *Crop Sci. 15*:821 (1975).
141. Cheeke, P. R., R. Carlsson, and G. O. Kohler, Nutritive value of leaf protein concentrates prepared from *Amaranthus* species, *Can. J. Animal Sci. 61*:199 (1981).
142. Ol'Shanskaya, L. E., Chemical composition and feed value of protein concentrates from green plant Sap, *Visnik. Sil 'S' kogospodars'koi Nauki 1*:68 (1980).
143. Shrivastava, S. K., and P. S. Krishnan, Oxalic acid in higher plants, *J. Sci. Indust. Res. (India) 18C*:220 (1959).
144. Gomez, R. G., M. H. Bertoni, and G. Covas, The leaf general chemical composition of American amaranth species *Amaranthus* spp.: Total oxalate and nitrate and residual levels after cooking, *Anal. Assoc. Quim. Argentina 74*:333 (1986).
145. Hill, R. M., and D. R. Prabhu, Evaluation of food potential, some toxicological aspects and preparation of protein isolate from the aerial part of amaranth (pigweed), *J. Agric. Food Chem. 30*:465 (1982).
146. Prakash, D., and M. Pal. Nutritional and antinutritional composition of vegetable and grain amaranth leaves, *J. Sci. Food Agric. 57*:573 (1991).
147. Okiei, I., and I. Adamson, Nitrate, nitrite, vitamin C and in-vitro methemoglobin formation from some vegetables, *Nutr. Rep. Int. 19*:241 (1979).
148. Tai, Hyeun, Y., L. Sang Jik, and K. Kwang Soo, Studies on the utilisation of plant pigments. 1. Isloation and identification of anthocyanin pigments in Ganges amaranth, *Hanguk Sik'Pum Kwahakhoe 10*:194 (1978).
149. Shen, H. G., and L. Sun Hwang, Red pigment of amaranth. II. Effect of some preharvest treatments on pigment retention, *Shih P'ink'o 1 Hsueh 12*:12 (1985).
150. Huang, A. S., and J. H. Von Elbe, Stability comparison of two betacyanin pigments—amaranthine and betanine, *J. Food Sci. 51*:670 (1986).
151. Gupta, K., G. K. Barat, D. S. Wagle, and H. K. L. Chawla, Nutrient contents and antinutritional factors in conventional and non-conventional leafy vegetables, *Food Chem. 31*:(2):105 (1989).
152. Xu, S., and G. W. Patterson, Sterol composition of the Phytolaccaceae and closely related families, *Lipids 25*:(1990).
153. Ushasri, V., K. Nagarajan, and T. S. N. Reddy, Isolation and characterization of inhibitor of tobacco mosaic virus from *Basella alba* L., *Tobacco Res. 8*(1):39 (1982).

154. Bhattacharjee, L. I., S. R. Mudambi, B. Bhushan, and M. V. Patankar, Provitamin A content of delected Indian foods, *J. Food Sci. Technol. 31*:249 (1994).
155. Kailasapathy, K., and T. Koneshan, Effect of wilting on the ascorbate content of selected fresh green leafy vegetables consumed in Sri Lanka, *J. Agric. Food Chem. 34*(2):259 (1986).
156. Nath, P., *Vegetables for Tropical Regions*, ICAR, New Delhi, 1976, p. 109.
157. Aykroyd, A., Health Bulletin No. 23, Nutrition Res. Lab., Koonoor, India, 1941.
158. Gupta, P. C., and K. Pradhan, Effect of stage of maturity on chemical composition and *in vitro* nutrient digestibility of legume forages, *Indian J. Animal Sci. 45*(9):614 (1975).
159. Jonnalagadda, S. S., and S. Seshadri, In vitro availability of iron from cereal meal with the addition of protein isolates and fenugreek leaves (*Trigonella foenum-graecum*), *Plant Foods Human Nutr. 45*(2):119 (1994).
160. Sridhar, R., and G. Lakshminarayana, Lipid classes, fatty acids and tocopherols of leaves of six edible plant species, *J. Agric. Food Chem. 41*:61 (1993).
161. Archana, G. N., J. Prakash, M. R. Asha, and N. Chand, Effects of processing on pigments of certain selected vegetables, *J. Food Qual. 18*:91 (1995).
162. Patil, V. R., D. N. Kulkarni, K. Kulkarni, and U. N. Ingle, Effect of blanching factors on quality and durability of sun dried and dehydrated fenugreek (*metho*), *Indian Food Packer 32*(1):43 (1978).
163. Chaudhary, A. T., and B. Y. Rao, Retention of chlorophyll and ascorbic acid in dried fenugreek (*Trigonella foenumgraecum*), *Indian Food Packer 33*:35 (1979).
164. Rowlands, R., U. S. Arondekar, and R. P. Devdas, Effect of different methods of cooking on the ascorbic acid and palatability of fenugreek leaves (*Trigonella foenumgraecum*), *J. Nutr. Dietet. 3*(1):15 (1966).
165. Talwalkar, R. T., and S. M. Patel, Biological evaluation of proteins of ambadi (*Hibiscus cannbinus*) and methi (*Trigonella foenumgraecum*) and their supplementary effect on jowar (*Sorghum vulgare*), *Indian J. Nutr. Dietet. 7*(1):13 (1970).
166. Daniel, M., Polyphenols of some Indian vegetables, *Curr. Sci. 58*(23):1332 (1989).
167. Varshney, I. P., and A. R. Sood, Sapogenins from *Trigonella foenum-graecum* stem and leaves and *T. corniculata* leaves and flowers, *Indian J. Appl. Chem. 34*(5):208 (1971).
168. Ricotta, J., and J. B. Masiunas, The effects of black plastic mulch and weed control strategies on herb yield, *HortScience 26*(5):539 (1991).
169. Bedmar, F., Evaluation of different pre-emergence herbicides in sunflower, *Tests Agrochem. Cultiv. 11*:62 (1990).
170. Beste, C. E., Terbacil selectivity for watermelon, *Brighton Crop Prot. Conf. Weeds. 3*:1045 (1989).
171. Tiwari, J. P., C. R. Bisen, and K. K. Trivedi, Herbicides to control weeds in paddy nursery, *Pesticides 21*(11):21 (1987).
172. Kato, S., and T. Kawaguchi, *Portulaca oleracea* in feeds for diarrhea prevention, *Jpn. Kokai Tokkyo Koho* JP 06,141,784 (1994).
173. Bharadwaj, K., Screening of weeds for oxalic acid, *Res. Ind. 33*(3):249 (1988).
174. Kenfield, D., Y. Hallock, J. Clardy, and G. Strobel, Curvulin and O-methylcurvulinic acid: Phytotoxic metabolites of *Drechslera indica* which causes necrosis on purslane and spiny amaranth, *Plant Sci. 60*(1):123 (1989).
175. Mirajkar, P. B., B. G. Gujarathji, and T. M. Patil, Studies on leaf protein of Portulaca species and other leafy vegetables, *Curr. Trends Life Sci. 11*:95 (1984).
176. Omara-Alwala, T. R., T. Mebrhatu, D. E. Prior, and M. O. Ezekwe, Omega-three fatty acids in purslane (*Portulaca oleracea*) tissues, *J. Am. Oil Chem. Soc. 63*(3):198 (1991).
177. Koch, H. P., Purslane. Omega-3 fatty acids in an old medicinal plant, *Dtsch. Apoth. Ztg. 128*(47):93 (1988).
178. Simopoulos, A. P., Terrestrial sources of omega-3 fatty acids: Purslane, *Epitheor. Klin. Farmakol. Farmakokinet. 2*:89 (1987).

179. Simopoulos, A. P., and N. Salem, Purslane: A terrestrial source of omega-3 fatty acids, *New Engl. J. Med. 315*(13):833 (1986).
180. Schneider, K., and W. Kubelka, Omega-3-fatty acids from Portulaca—an alternative for fish oil? *OAZ, Oesterr. Apoth. Ztg. 44*(15):287 (1990).
181. Hunter, J. E., n-3 Fatty acids from vegetable oils, *Amer. J. Clin. Nutr. 51*(5):809 (1990).
182. Simopoulos, A. P., and N. Salem, Jr., n-3 Fatty acids in eggs from range-fed Greek chickens, *New Engl. J. Med. 321*(20:1412 (1989).
183. Boschelle, O., S. Sblattero, C. Da Porto, N. Frega, and G. Lercker, Lipid composition of *Portulaca oleracea, Riv. Ital. Sostanze Grasse 68*(6):287 (1991).
184. Awad, N. E., Lipid content and antimicrobial activity of phenolic constituents of cultivated *Portulaca oleracea* L., *Bull. Fac. Pharm. 32*(1):137 (1994).
185. Rahman, M. M., M. A. Wahed, and M. Akbar Ali, β-Carotene losses during different methods of cooking green leafy vegetables in Bangladesh, *J. Food Comp. Anal. 3*(1):47 (1990).
186. Simopoulos, A. P., H. A. Norman, J. E. Gillaspy, and J. A. Duke, Common purslane: a source of omega-3-fatty acids and antioxidants, *J. Am. Coll. Nutr. 11*(4):374 (1992).
187. Wenzel, G. E., J. D. Fontana, and J. B. C. Correa, The viscous mucilage from the weed *Portulaca oleracea* L., *Appl. Biotechnol. Biotechnol. 24-25*:341 (1990).
188. Rodriguez, M. M., M. V. Rivas, and M. R. Rosiles, Oxalate levels in wild forage from the states of Hidalgo, Guanajuato, Mexico, Tlazcala and the federal district, *Veterinaria 16*(1):21 (1985).
189. Boehm, H., L. Boehm, H. Nixdorf, and E. Rink, Manufacture of yellow betalains with plant cell cultures, *Ger. (East) DD 264*:012 (1989).
190. Strack, D., D. Schmitt, H. Reznik, W. Boland, L. Grotjahn, and V. Wary, Humilixanthin, a new betaxanthin from *Rivina humilis*, *Phytochemistry 26*(8):2285 (1987).
191. Sanchez-Marroquin, A., Dos Cultivos Olvidados de Importanoia Agroindustrial. El Amaranto Y la Quinoa, *Arch. Latinoamerican. Nutr. 33*(1):13 (1983).
192. Chauhan, G. S., R. R. Zillman, and N. A. Michael Eskin, Dough mixing and bread making properties of quinoa–wheat flour blends, *Int. J. Food Sci. Technol. 27*:701 (1992).
193. Morales, P. P., and C. Curl, A physicochemical method for total saponin determination in quinoa samples, *Re. Boliv. Quim. 6*(1):13 (1986).
194. Risi, J. C., and N. W. Galwey, the *Chenopodium* grains of the Andes. Inca crops for modern agriculture, *Adv. Appl. Biol. 10*:142 (1984).
195. Trease, G. E., and W. C. Evans, The pharmacological action of plant drugs, in Pharmacognosy, 12th ed., Bailliere Tindall, Eastbourne, UK, 1983, p. 154.
196. Prakash, D., P. Nath, and M. Pal, Composition variation of nutritional contents of leaves, seed protein, fat and fatty acid profile in *Chenopodium* species, *J. Sci. Food Agric. 62*:203 (1993).
197. *Wealth of India. Raw Materials*, Vol. IX, Publications and Information Directorate, CSIR, New Delhi, 1972.
198. Khurana, S. C., Y. S. Malik, and M. L. Pandita, Herbicidal control of weeds in potato c.v. kfri badshah, *Pesticides 20*(11):55 (1986).
199. Sarmah, S. C., M. Borgohaill, and A. K. Pathak, A note on herbicidal control of weeds in summer soyabean, *Pesticides 20*(11):56 (1986).
200. Weber, E. J., The Inca's ancient answer to food shortage, *Nature 272*:486 (1978).
201. Bahrman, N., M. Jay, and R. Govenflot, Contribution to the chemosystematic knowledge of some species of the genus *Chenopodium*, *Lett. Bot. 2*:107 (1985).
202. French, C. J., and G. H. L. Towers, Inhibition of infectivity of potato virus x by flavonoids, *Phytochemistry 31*(9):3017 (1992).
203. Chauhan, G. S., N. A. M. Eskin, and R. Tkachuk, Nutrients and antinturients in quinoa seed, *Cereal Chem. 69*(1):85 (1992).
204. Pushpanjali and S. Khokhar, The composition of Indian foods, *J. Sci. Food Agric. 67*:267 (1995).

205. Parthasarthy, V. A., Other root crops, in *Vegetable Crops in India* (T. K. Bose and M. G. Som, eds.), Naya Prokash, Calcutta, 1986, p. 720.
206. Dako, D. Y., Potential of dehydrated leaves and cocoyam leaf protein in the Ghanian diet, *Nutr. Rep. Int. 23*(1):181 (1981).
207. Rao, K. S., and G. Lakshminarayana, Lipid class and fatty acid compositions of edible tissues of *Peucedanum graveolens, Mentha arvensis*, and *Colocasia esculenta* plants, *J. Agric. Food Chem. 36*:475 (1988).
208. Sayeed, S., and K. Ahmed, Hypocholesterolemic effect of sterol of colocasia leaves, *Bangladesh J. Biol. Sci. 8*(1):17 (1979).
209. Osada, Y., K., Kawana, T. Nakaoka, and K. Ito, Determination of phosphate in herb vegetables, *Kangawa-ken Eisei Kenkyosho Kenkyu Hokoku 18*:46 (1988).
210. Dikshit, S. N., S. A. Udipi, A. Rao, and V. Manohar (1988). Separation of carotenoids and estimation of beta-carotene content of selected Indian food and food preparations by HPLC, *J. Food Sci. Technol. 25*:39 (1988).
211. Nakanishi, Y., E. Maruyama, T. Kajita, and C. Hasegawa, Acrid components of vegetable foods. Acrid componens of taro, *Kaseigaku Kenkyu 29*(1):75 (1982).
212. Sunell, L. A., and P. L. Healy, Distribution of calcium oxalate crystal idioblasts in corms of taro (*Colocasia esculenta*), *Am. J. Bot. 66*:1029 (1979).
213. Genua, J. M., and C. J. Hillson, The occurrence, type and location of calcium oxalate crystals in the leaves of fourteen species of Araceae, *Ann. Bot. 56*:351 (1985).
214. Mani, U. V., M. Sharma, K. Waghray, U. Iyer, and I. Mani, Effect of colocasia leaves (*Colocasia antiquorum*) on serum and tissue lipids in cholesterol-fed rats, *Plant Foods Human Nutr. 39*(3):245 (1989).
215. Duquenois, P., and R. Anton, Chemical study of *Cassia sieberiana* leaves, *Planta Med. 16*(2):184 (1968).
216. Richter, G., and H. Hauenstein, Chemical quantitative determination of senna drugs, extracts, and preparations by separate quantitative analysis of anthraquinone glycosides and aglycons, *Deutsch Apoth. Ztg. 197*(48):1751 (1967).
217. Friedrich, H., and S. Baier, Constuents of senna leaves, *Planta Med. 23*:74 (1973).
218. Saber, A. H., S. I. Balbaa, and A. T. Awaa, Anthracene derivatives of the leaves and pods of *Cassia acutifolia* cultivated in Egypt, their nature and determination, *Bull. Fac. Pharm. 1*(1):7 (1961).
219. Saber, A. H., S. I. Balbaa, and A. J. Awad, Identification of anthracene derivatives of the leaves and pods of Cassia obovata grown in Egypt, *Lloydia 25*:238 (1962).
220. Fuzellier, M. C., F. Mortier, and P. Lectard, Antifungal activity of *Cassia alata* L., *Ann. Phar. Fr. 40*(4):357 (1982).
221. Sydiskis, R. J., D. G. Owen, J. L. Lohr, K. H. A. Rosler, and R. N. Blomster, Inactivation of enveloped viruses by anthraquinones extracted from plants, *Antimicrob. Agents Chemother. 35*(12):2463 (1991).
222. Patel, R. D., and K. C. Patel, Antibacterial activity of *Cassia tora* and *Cassia obovata*, *Indian J. Pharm. 19*:70 (1957).
223. Vora, A. B., and P. K. Gopalkrishnan, Effect of fungi infections on catralase and peroxidase activities of leaves of some monsoon weeds, *Sci. Culture 45*(4):167 (1979).
224. Murthy, V. N., *Cassia tora* leaf as a component in poultry rations *Poultry Sci. 41*:1026 (1962).
225. Lemli, J., S. Toppet, J. Curevle, and G. Janssen, Napthalene glycosides in *Cassia senna* and *Cassia augustifolia*. Studies in the field of drugs containing anthracene derivatives. XXXII, *Planta Med. 43*(1):11 (1981).
226. Lemli, J., and J. Cuvecle, Sugars from the leaves of *Cassia augustifolia* (Sene), *Planta Med. Phytother. 10*(3):175 (1976).
227. Varshney, S. C., S. A. I. Rizvi, and P. C. Gupta, Chemical and spectral studies of novel ket alcohols from leaves of *Cassia auriculata*, *Planta Med. 23*(4):363 (1973).

228. Sayed, H. M., On the chemistry of *Cassia tomentosa* (Linn). Part I. Constituents of the leaves, *Bull. Fac. Sci Assiut. Univ. 20*(1):115 (1991).
229. Dirar, H. A., D. B. Harper, and M. A. Collins, Biochemical and microbiological studies on kawal, a meat substitute derived by fermentation of *Cassia obtusifolia* leaves, *J. Sci. Food Agric. 36*:881 (1985).
230. Kaniz, F., H. Raisul, and I. Mohammad, Amino acid analysis of leaf protein concentrates prepared from 2 species of Cassia, *Karachi Univ. J. Sci. 11*(1):876 (1983).
231. Pipynus, J., and D. Smaliukas, Biological and biochemical characteristics of *Rumex tianschanicus*. 2. Dynamics and distribution of pigments and ascorbic acid in the organs above ground, *Liet. TSR Mokslu Akad. Darb. Ser. C 3*:43 (1974).
232. Slapkauskaite, G., and R. Varhaite, Content of pantothenic acid and pyridoxin in representatives of *Daucus, Allium, Cyperus, Arachis, Capsicum, Rumex, Capsella, Urtica, Primula, Lepidium, Brassica, Lactuca genera, Liet. TSR Mokslu Akad. Darb. Ser. C 4*:25 (1988).
233. Panciera, R. J., T. Martin, G. E. Burrows, D. S. Taylor, and L. E. Rice, Acute oxalate poisoning attributable to ingestion of curly dock (*Rumex crispus*) in sheep, *J. Am. Vet. Med. Assoc. 196*(12):1981 (1990).
234. Tendaj, M., Use of sorrel as a raw material for processing, *Przem. Spozyw. 32*(11):424 (1978).
235. Grznar, K., and K. Rada, Isolation and identification of anthraquinone derivatives from the leaves of some *Rumex* species, *Far. Obz. 47*(5):195 (1978).
236. Midiwo, J. O., and G. M. Rukunga, Distribution of antraquinone pigments in Rumex species of Kenya, *Phytochemistry 24*(6):1390 (1985).
237. He, L. Y., B. Z. Chen, and P. G. Xiao, Survey, identification and constituent analysis of Chinese herbal medicines from the genus *Rumex*, *Yao Hsueh Tung Pao 16*(4):289 (1981).
238. Yangti, H., K. Pei, L. I. Ho, P. C. Chen, and H. C. Kuo, Botanical and chemical studies of Chinese herbal medicine, *Yao Hsueh Tung Pao 15*(7):48 (1980).
239. Ito, H., and H. Hidaka, Antitumor agents from *Rumex acetosa*, *Jpn. Kokai Tokkyo Koho* 80,157,516 (1980).
240 Nishina, A., and H. Hashimoto, Ascorbic acid and sorrel extracts as antioxidants for foods, cosmetics and pharmaceuticals, *Jpn. Kokai Tokkyo Koho* JP 06,220,450 (1994).

26

Okra

N. D. JAMBHALE AND Y. S. NERKAR
Mahatma Phule Agricultural University, Rahuri, India

I. INTRODUCTION

Okra has been a popular vegetable crop in the tropics because of its easy cultivation, dependable yield, adaptability to varying moisture conditions, and resistance to diseases and pests (1). Okra fruit is rich in vitamin C and calcium and has medicinal value in curing ulcers and relief from hemorrhoids (2). This crop is suitable for cultivation as a garden crop as well as on large commercial farms. The cultivated species, *Abelmoschus esculentus*, has been known by different names in different languages: okra or lady's finger in English, *gombo* in French, *guino-gombo* in Spanish, *guibeiro* in Portoguese, *bhindi* in Hindi, and *bamiah* in Arabic. Yet another species, *A. manihot* (ssp. *manihot* 'Guinean' type), is cultivated for its green fruits in the humid tropics in West Africa (3). The species *A. moschatus* (Syn. *Hibiscus abelmoschus*), known as musk mallow, is famous for the musk-scented perfume made from its seeds. The seeds contain 60% essential oils and are also used as a condiment. Leaves of this species are also used as a vegetable, which is cultivated in Africa, Asia, and America (4).

II. BOTANY

Okra belongs to the family Malvaceae. Cultivated okra and related wild species were originally grouped in the genus *Hibuscus*, section *Abelmoschus*. Hochreutiner (5) classified *Abelmoschus* as a separate genus having a deciduous (caducous) calyx. In this genus, the calyx, corolla, and stamens are fused together at the base and fall off as one piece after anthesis. About 40 species have been described by taxonomists under the genus *Abelmoschus*. The taxonomical revision undertaken by Van Borssum Waalker (6) and its continuation by Bates (7) constitute the most fully documented studies of the genus *Abelmoschus*. Van Borssum Waalkers (6) distinguished only six species on the basis of the epicalyx and fruit characteristics: *A. moschatus*, *A. manihot*, *A. esculentus*, *A. ficulneus*, *A. crinitus*, and *A. angulosus*. Many of the older species have been united under the same binomial, in certain cases forming subspecies and varieties (Table 1). Bates (7) maintained the species *A. rugosus* in place of the species *A. moschatus* ssp. *tuberosus*

TABLE 1 Different Species of *Abelmoschus*

Species	Synonyms (Hibiscus)	Wild/Cultivated
A. moschatus Medikus	*H. abelmoschus* L.	Wild and cultivated
A. manihot (L) Medikus	*H. manihot* L.	Wild and cultivated
A. esculentus (L) Moench	*H. esculentus* L.	Wild and cultivated
A. ficulneus (L) Wight and Arnott ex. Wight	*H. ficulneus* L.	Wild
A. crinitus Wallich		Wild
A. angulosus Wallich ex. Wight and Arnott		Wild

Source: Ref. 6.

and further considered the species *A. manihot* as a complex including continuous variation, whether cultivated or wild.

A. Origin and Distribution

Okra is mentioned in the ancient sacred books of India and the documents of ancient Egypt. It is thought to be of African or Asiatic origin, perhaps both. According to de Candolle (8), the domestication of okra probably took place in Africa. The presence of wild varieties in Ethiopia and of primitive perennial varieties in West Africa suggest an African origin. The perennial forms are seldom seen in other parts of the world. The geographical distribution of cultivated okra and the related wild species is overlapping in Southeast Asia. Van Borssum Waalkers (6) considered this region the center of diversification of the genus *Abelmoschus*. This view, however, ignores the diversity present in the Indian and West African regions.

The cultivated species of okra, *A. esculentus* L., is grown in the tropical and subtropical low-altitude regions of Asia, Africa, and America, with an extension to the temperate regions of the Mediterranean basin. It is an important crop in northeastern Brazil, the Indian subcontinent, West Africa, and the southern United States (Georgia, Florida, Texas, Alabama, and Louisiana). In West Africa, it is preferred in the Sudano-Sahelian zone. Guinean okra (*A. manihot* spp. *manihot*) is found in the forest regions of Guinea, Liberia, Ivory Coast, Ghana, and Nigeria (3).

B. Morphology

Okra is generally an annual plant, although perennial varieties with large treelike trunks have been found in West Africa. Its stem is robust, erect, variable in branching, and varying from 0.5 to 4.0 m in height. The stem is semiwoody and sometimes pigmented. The root is a deep taproot. Leaves are alternate and usually palmately five-lobed (Fig. 1). The degree of leaf incision increases with the age of the plant. Leaves are subtended by a pair of narrow stipules. The flower is axillary and solitary, borne on a peduncle 2.0–2.5 cm long. It is hermaphroditic, regular, and perfect. The flower diameter varies from 3.5 to 5.5 cm. The flower has an epicalyx formed by 8–10 narrow hairy bracteoles (1–5 cm long), a calyx consisting of three sepals, and a corolla consisting of five free petals. The calyx, which is completely fused to form a protective case for

FIGURE 1 Okra plant cv. Parbhani Kranti.

the floral bud, splits into lobes when the bud opens (Fig. 2). The petals vary in size from 3.5 × 2.5 to 5.0 × 4.5 cm and are yellowish to bright yellow in color, having large deep reddish-purple spots at the base (1). The petals are large, round, and thin towards the apex, and narrow and thickened at the base. Numerous stamens are united through their filaments to form a central staminal tube around the style. The staminal column is fused to the petals at the base. The pistil consists of five to nine lobed, deep purple stigma, a slender style, and five to nine carpeled ovaries with axial, central placenta. The number of ovules ranges from 40 to 80 (1).

The fruit is normally yellowish-green to green, but is sometimes purple or white. The fruit grows rapidly into a long (10–30 cm) and narrow (1–4 cm) pod with a tip that is either pointed like a beak or blunt (Fig. 3). It is pyramidal in shape due to longitudinal ridges (5 to 9 in number), or sometimes it is cylindrical. The fruit may be glabrous or spiny (hirsute). It matures into either a dehiscent, longitudinally splitting capsule or an indehiscent capsule, sometimes curved like a horn. The seeds are spherical, smooth, striated, and dark green to dark brown in color. The seeds average about 15–20 per gram.

C. Anatomy

Metcalfe and Chalk (9) studied the anatomy of some of the genera of Malvaceae. In general, it possesses multicellular capitate glandular hair comprising a few cells of epidermis. The epidermis has straight or undulating anticlinal walls, some of which are specialized for the secretion

FIGURE 2 Floral bud development and flowering in okra.

of mucilage. The stomata are of the ranunculaceous type and are present on the lower side. The palisade tissue is well developed on the upper side or on both the sides; vascular bundle of the large vein is accompanied by sclerenchyma. The xylem forms close rings in the petiole. The cortex of the young stem is generally collenchymatous and parenchymatus; the pericycle always contains fibers with simple pits. Mucilage cavities are present in the cortex and pith (9).

Idioplasts, the specialized mucilage-secreting cells, have been described (9). These are present in the cortex of the stem. Kundu and Biswas (11) studied the anatomical features of *Hibiscus* and *Abelmoschus*. These two genera have distinguishing anatomical features—nature and distribution of epidermal trichomes, structure and distribution of crystals of leaf lamina, petiolar vascular number, and arrangement of mucilage cells and canals and nature of wood rays.

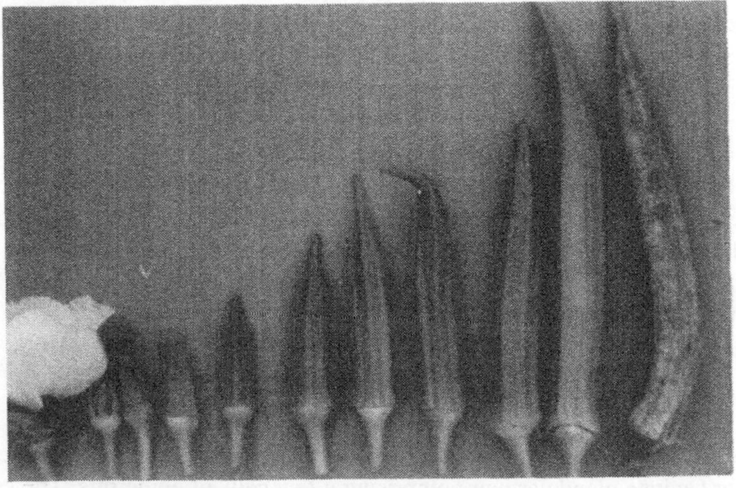

FIGURE 3 Okra fruits at various stages of development and maturity.

The seed coat of *Hibiscus* has a thin epidermal layer followed by a single subepidermal layer, while in *Abelmoschus* spp., the epidermal layer is thick without a subepidermal layer. Colchiploids of the interspecific hybrids of *Abelmoschus—A. esculentus* × *A. tetraphyllus*, *A. esculentus* × *A. manihot*, and *A. esculentus* × *A. manihot* spp. *manihot*—exhibited a significant increase in stomatal length and frequency of stomatal chloroplasts and a significant decrease in frequency of stomata per unit area compared to their respective F_1 hybrids (12).

D. Cytology

There is considerable variation in the chromosome numbers of cultivated okra. This is also true of the chromosome numbers observed in related species. There is a possibility of the existence of chromosome polymorphism in these species. The genus *Abelmoschus* is a polyspecies complex consisting of species with apparently three levels of ploidy: the diploids (2X), including *A. coccineus* ($2n = 38$), *A. angulosus* ($2n = 38$), *A. tuberculatus* ($2n = 58$), *A. manihot* ($2n = 60$–68), *A. moschatus* ($2n = 72$), and *A. ficulneus* ($2n = 72$); the tetraploids (4X), including *A. esculentus* ($2n = 120$–140), *A. tetraphyllus* ($2n = 130$–138), and *A. pungens* ($2n = 138$); and the hexaploids (6X), including *A. manihot*, Guinean type ($2n = 185$–198). However, the variation in chromosome number could also be attributed partly to erroneous taxonomical grouping of the species and to the difficulty of counting the chromosomes precisely. The mucilagenous nature of the material and the small size of the chromosomes make the chromosome count difficult.

Based on cytological studies of a large number of *A. esculentus* strains, Joshi and Hardas (13) arrived at the conclusion that the chromosome number of cultivated okra is $2n = 130$. Two haploid plants were detected by Joshi et al. (14) in advanced generations of intervarietal crosses. Meiotic metaphase plates showed 65 univalents in these haploids, thus confirming $n = 65$ as the chromosome number of *A. esculentus*. Some research workers have reported the diploid ($2n$ = 66 or 72) chromosome number in *A. esculentus* (15–18). However, such lines of *A. esculentus* could not be traced subsequently (4). Variation in the chromosome number, from $2n = 72$–130, was observed in the different stable okra lines derived from an interspecific cross (19). Such lines were obtained by back-crossing the F_1 hybrid *A. esculentus* ($2n = 130$) × *A. manihot* ($2n$ = 66) to the former species.

An amphidiploid was synthesized by Kuwada (20) by colchicine treatment of the F_1 hybrid *A. esculentus* × *A. manihot*. There were similarities between the sterility patterns observed by Siemonsma (3) and Kuwada (20) by crossing the induced amphidiploid to either of the parental species and those found by Siemonsma (3) in the crosses of *A. manihot* ssp. *manihot* Guinean form to *A. esculentus* and *A. manihot*. The morphological characteristics of the Guinean form were intermediate between *A. esculentus* and *A. manihot*. Based on these observations, Siemonsma (3) proposed a hypothesis that the Guinean form is an amphidiploid composed of the genomes of *A. esculentus* and *A. manihot*. However, neither cytological data nor sufficient morphological data have been provided in support of this hypothesis.

Jambhale and Nerkar (21) also obtained an induced amphidiploid from the cross of *A. esculentus* ($n = 65$) and *A. manihot* ($n = 33$). Comparison of the morphological and cytological behavior of this induced amphidiploid with the Guinea form showed that there were wide differences in their behavior (22), thus rendering no evidence in support of the hypothesis proposed by Siemonsma (3). Based on the similarities in chromosome association observed in the cross *A. esculentus* ($n = 65$) × *A. manihot* ($n = 33$) and *A. esculentus* ($n = 65$) × *A. manihot* ssp. *manihot* from Ghana ($n = 97$) (23), it may be concluded that *A. manihot* might have contributed its genome to *A. manihot* ssp. *manihot* Guinean form. There seems to be little affinity between the other two genomes of the Guinean form and the two genomes of *A. esculentus* (23).

E. Genetics

The genetics of different qualitative and quantitative characteristics in okra has been studied by various workers (24–49). Monogenic inheritance has been reported for stem, petiole, petal base, petal variation, and pod color (24). Di- and tetragenic inheritance for stem color and digenic inheritance for pod color (25) have also been reported. Chlorina, a mutant with yellowish-green foliage, was monogenic recessive (26). Simple inheritance was found for leaf lobation (27,28). Palmatisect, a mutant, was found to be monogenic recessive (29). High heritability has been reported for plant height, internode length (30), days to flowering, seeds per pod, seed weight and yield (31), fruit length (32), fruit diameter, fruit weight, vitamin C content, and fiber content of fruit (32). Pods per plant is under the control of additive gene action (33) or dominance (34).

Inheritance of yellow vein mosaic resistance derived from wild species has been found to be due to two complementary genes (35) or a single dominant gene (36). However, yellow vein mosaic susceptibility in the intervarietal crosses was found to be due to two complementary genes (37). Resistance to powdery mildew was found to be under the control of a incompletely dominant gene (38). Resistance to cotton jassid was due to additive and dominant × dominant type interaction (39). Pleiotropy was found between calyx color and petal vein color (27). Kolhe and D'Cruz (28) observed pigmentation of petal base, petal veins, and fruit, petiole, and leaf vein (40).

Heterosis has been reported for yield (41–46). Heterosis for earliness was reported by Kulkarni and Virupakshappa (47). Inbreeding depression for number of fruits per plant, days to flowering, and plant growth was observed (47). Plant height and number of pods per plant are important yield-contributing characteristics (48,49).

F. Floral Biology and Growth and Development of Fruit

Flower bud initiation, flowering, anthesis, and stigma receptivity are influenced by genotype and climatic factors like temperature and humidity (50). From studies made on six okra varieties, Sulikeri and Swamy Rao (51) concluded that flower buds are initiated at 22–26 days and the first flower opened 41–48 days after sowing. Once initiated, flowering continues for 40–60 days. Varieties like Emerald that bloom late continue to flower over a longer period. Anthesis was observed between 6 and 10 a.m. Anthers dehisce before flower opening, and hence self-pollination may occur at anthesis. Pollen stored for 24 hours at room temperature (27°C) and 88% relative humidity was not viable. The stigma was most receptive on the day of flowering (90–100%). Stigma receptivity was also observed the day before flowering (50–70%) and the day after (1–15%). Flowers open only once in the morning and close after pollination on the same day. The following morning the corolla withers.

Pods develop rapidly for 11 days following the opening of flowers, after which their development slows down. The fruits are ready for harvest 4–10 days after flowering. Because of the rapid growth of okra fruits, plants should be harvested at least every other day. In climates where growth is especially vigorous, it may be necessary to harvest every day.

Okra plants continue to flower and to fruit for an indefinite time, depending upon the variety, the season, and soil moisture, and fertility. Regular harvesting stimulates continued fruiting. Plants age rapidly when pods are not removed, exhibiting a reduction in leaf production and an increase in the depth of lobing. Okra pod yields range from 5 to 18 t/ha.

Mitidieri and Venocovsky (52) estimated that cross-pollination was as high 42% in Brazil, while others have reported cross-pollination rates of 4–32% (53). Under field conditions, cross-pollination in okra is high due to the activity of several kinds of bees. Thus, okra could be classified as an often cross-pollinated crop.

Self-pollination is done by tying the closed flower bud with a thread the day before anthesis. Parthsarathy and Sambandan (54) studied three methods of selfing, namely, using a paper bag, thread, and clay smear. Among the three methods, the paper bag method gave the highest fruit set (88%), seed number, and seed weight. Controlled pollination is done by emasculating buds in the afternoon before anthesis is expected. Pollination is done the following morning with the desired pollen. Pollinated flowers must be covered to prevent contamination by insect borne pollen. Giriraj and Swamy Rao (55) evolved a simple emasculation technique that involved removal of the fused calyx cap along with the corolla by making a circular cut in the base (without injuring the gynoecium), removal of the anthers using fine forceps, and replacement of the fused calyx cap at its original position. The next morning the cap is removed, the desired pollen is applied, and the cap is replaced. Fruit set by this method was observed to be normal. Artificial pollination by three methods resulted into 62–94%, 44–70%, and 40–76% pod set, respectively, compared with 93–100% pod set obtained after open pollination (56).

G. Cultivars

Okra has been bred more systematically in the southern United States, India, and Brazil. Many okra varieties developed in the United States are cultivated in the tropics. Okra varieties differ with respect to physiological characteristics such as response to season, adaptability, habit stature, fruit quality, and resistance to biotic and abiotic stress. Fruit color may be dark green, light green, reddish, purple, yellowish, or almost white. The pods may be glabrous or spiny. In addition to the improved varieties of okra, unnamed old local varieties are raised in some tropical areas.

Systematic breeding aimed at developing okra varieties resistant to yellow vein mosaic in India led to the development of resistant varieties (Table 2). Pusa Sawani was developed from an intervarietal cross between IC 1542 and Pusa Makhmali (57). Punjab Padmini was developed from a cross between *A. esculentus* and *A. manihot* ssp. *manihot* in 1982 at Punjab Agricultural University, Ludhiana. Jambhale and Nerkar (58) developed Parbhani Kranti by transferring a resistant gene from *A. manihot* into *A. esculentus* (cv. Pusa Sawani). Anamika (Sel. 10) was developed at the Indian Institute of Horticultural Research, Bangalore, from an interspecies cross (59).

H. Seed Production

Seed production is undertaken during a season that is most favorable for crop growth and when there is the least incidence of diseases and pests. Seed is sown where no okra crop was grown in the preceding year. Natural cross-pollination may occur involving other varieties of the cultivated species or some related *Abelmoschus* species. Therefore, utmost care must be taken during seed production to maintain a safe distance from such contaminants.

Breeder's and foundation seed plots are isolated by a distance of 400 m and certified seed plots by 200 m from (1) fields of other varieties, (2) fields of the same variety not conforming to the varietal purity requirements for certification, and (3) other *Abelmoschus* species (54). Diseased plants, especially yellow vein mosaic–infected ones, are uprooted and destroyed as

TABLE 2 Popular Okra Varieties Cultivated in the Tropics

Variety	Salient features
Anamika (Sel. 10)	Resistant to yellow vein mosaic
Campinas (1 and 2)	Early, productive, drought resistant
Chifre-de-Veado (several varieties)	Large, much branched
Clemson Spineless	High yields
Dwarf Green	Traditional variety
Dwarf Green Long Pod	Suitable for fall, compact
Dwarf Prolific	Small stature
Emerald	Dark green, smooth, spineless pods; high yields
Gold Coast	Dwarf, heat-tolerant, long-bearing, short pods
Green Velvet	Early, high yields, smooth pods
Long Horn	Long, sutureless pods
Louisiana Market	Short pods
Native Brown	Purple stems and leaf margins
Parbhani Kranti	Resistant to yellow vein mosaic
Perkin's Mammoth	Long pods
Pusa Sawani	Resistant to yellow vein mosaic
Punjab Padmini	Resistant to yellow vein mosaic
Red Wonder	Red fruits
Sabour Selection	Branching tendency
St. John Bush	Enormous size, long life
Vaishali Vadhu	6–7 ridged fruits
White Velvet	White fruits

soon as they are noticed. Off-type plants, distinguished from the seed variety, are rouged out prior to flowering.

Fruits are harvested 30-35 days after anthesis. At this stage, the fruits become light brown in color and show cracks through their sutures. Dry fruits should be harvested quickly to avoid shattering and deterioration of seed quality due to rain. Fruits are picked by hand and dried in the sun. Twisting the dried pods in both directions usually breaks loose the locules of the capsule and releases the seed. Threshed and winnowed produce is sun-dried in thin layers on the threshing floor. If facilities are available, drying can be done by forced air-drying. To avoid seed deterioration, the moisture content of seed should be less than 13%.

For seed purposes, the best time to harvest pods is 30–35 days after anthesis when the pods are completely dried and the seeds are fully viable (60). Seeds can be stored under cool dry conditions for about 2 years. The seed yield of okra is 2–2.5 t/ha.

III. PRODUCTION

A. Soil and Climate

Okra does not have rigid soil requirements. It can be grown on any soil ranging from sandy to clay as long as it is rich in organic matter and well drained. Like other vegetable crops, it grows best and produces an excellent crop in well-drained, rich loamy or sandy loam soil. If the soil does not have adequate drainage, supplementary drainage facilities should be provided using drainage tiles. A neutral soil pH is ideal for growth and productivity. Soils that are too acidic or alkaline should be adjusted before planting.

Okra is a warm-season crop. It grows well in hot, dry and in hot, humid tropical areas. Cool climates are undesirable for the growth and productivity of okra. Okra seeds do not germiante below 17°C. High temperatures are correlated with large plant size, high production of flower and fruit, and large fruit size (61). Flowers drop when day temperature exceeds 42°C (62).

Okra growth is influenced by day length, with short day length stimulating early flowering with reduced vegetative growth (61). All 265 varieties tested in the dry season in Puerto Rico began to flower when day lengths were about 11 hours and achieved normal growth (63).

Okra requires a warm growing season free of frost. This crop remains in the field for a relatively long time. Therefore, when planting okra it is absolutely necessary to make sure that the crop is well protected from low temperatures and frosts. Generally, the immature fruits and young shoots are more susceptible to frost. The ideal temperature for growing okra is around 25–30°C—when the growing temperature is higher than 42°C, excessive flower drop takes place.

B. Propagation

Okra is mainly propagated by seeds. Seeds are round or globular, medium-sized, and bluish-green or blackish in color. The seed remains viable for about 2 years if stored under the proper conditions. It is always advisable to use fresh seeds. Seeds require 4–6 days for germination. Seed requirements are 18–22 kg seed/ha during summer.

C. Cultural Practices

1. Sowing

The land for growing okra should be prepared well to provide a suitable substrate for seeds. After the land is prepared, beds of suitable size, depending upon the irrigaton facility and convenience, are prepared. Okra can also be grown in ridges and as a mixed or border crop with other crops such as sugarcane, cotton, chilies, turmeric, and bananas. In such cases, land preparation is the same as for the main crop.

Time of sowing depends on the location. In the plains, okra can be grown twice a year, the early crop being sown in January-March and the second crop in July. In places where frost is common in the month of January, sowing of the early crop can be delayed until February-March. In the hills, okra is sown between April and July. Staggered plantings at 15-day intervals is practiced in order to supply fruits continuously over a long period.

Seeds are sown at 30 × 30 cm or 45 × 45 cm spacing depending on the season, with rainy season crops requiring more space for optimum growth and productivity. The tall varieties should be spaced further apart than the dwarf or medium types. The seeds are generally soaked in water for 24 hours or in acetone or alcohol for 30 minutes before sowing. This treatment causes faster and better germination (64). Two seeds are dibbled in a furrow at the desired spacing in a flat bed or about one third from the bottom of the furrow. When grown with other crops like colocasia or other leafy vegetables, okra is sown on the top of the hill while the intercrop is planted in the furrow.

2. Manuring and Fertilization

The importance of organic manures and fertilizers in okra production has been reported (65). The most common recommendation consists of 20 cartloads of farmyard manure (FYM), 45 kg of nitrogen, and 22.5 kg each of phosphorus and potash per hectare for production of 2725 kg of green fruits. Thakur and Arora (66) have compiled data on fertilizer requirements of okra in different types of soils in India. The entire dose of FYM is applied at the time of land preparation, whereas phosphatic and potassic fertilizers are applied in shallow furrows before sowing. Nitrogen is given in two equal split doses applied 1 and 2 months after sowing, respectively. The deficiency symptoms of phosphorus, potassium, and nitrogen in okra have been described (67). Several other deficiency disorders—for zinc, vanadium, and boron—have been reported (68–71).

3. Irrigation

The first irrigation, which is given immediately after sowing, is followed by regular irrigations depending on the requirements. Generally, irrigating fields every third or fourth day in summer and once every 8–10 days during winter provides adequate soil moisture for optimum growth and productivity . Sivanappan et al. (72) recommended drip irrigation system in place of the conventional furrow system for economy of water utilization (84.7%) without any loss of yield.

D. Diseases and Pests

Yellow vein mosaic virus is major disease of okra in the tropics. Under these conditions, growing of resistant varieties such as Pusa Sawani, Parbhani Kranti (58) in India, and St. John in the Caribbean has been recommended (1). Other important diseases include damping-off caused by *Pythium* sp., *Rhizoctonia* sp., and *Fusarium* sp.; powdery mildew caused by *Erysiphe* sp.; and blight caused by *Cercospora* and *Alternaria*. The important pests of okra are jassids (*Amrasca biguttula bigutulla*), spotted bollworm (*Earias* sp.), white fly (*Bemisia tabaci*), spider mite (*Tetranychus cinnabarinus* and *Tetranychus necoaledonicus*) and root knot nematode (*Meloidogyne incognila* and *Meloidogyne javania*).

E. Harvesting and Maturity

The tender young okra pods are gathered when they are about one-half grown. Picking may be necessary every day or 2 when development is rapid. Unless pods are harvested promptly, they become woody and the plant matures too rapidly and ceases bearing. The harvest starts when the first pods have developed and continues until finished if pods are kept picked. Perkins et al. (73) stated that harvesting the pods when they are 3 or 4 days old results in a continuous harvest over the entire growing season, whereas if pods are allowed to mature, plants will stop fruiting. Delay in harvesting of fruits results in a decrease in fruit quality (74). According to Culpepper and Moon (75) the eating quality of okra is rather high at the 4-day stage from blooming, increases to the 6-day stage, and then slowly declines to the 10- to 12-day stage, after which the pods become so fibrous as to be inedible. Singh et al. (76) reported that okra fruits should be harvested 6–9 days after fruitset for maximum crop and nutritional value. Okra that is bright green, firm, free of blemishes, and no longer than 7 cm will be enjoyable to eat, but very young pods tend to taste grassy (77). Pods up to 12.5 cm can be satisfactory, but they may be more fibrous than desirable. Pods that are dull, flaccid, and yellowish are inferior, particularly because of their high fiber content (78). The tender pods are broken from the stalks and must be handled carefully because they bruise and discolor. Okra pods are graded in various sizes and packaged in various ways, usually in a cardboard tray covered with plastic film (64).

IV. CHEMICAL COMPOSITION

A. Carbohydrates

The chemical composition of okra fruit is presented in Table 3. Sucrose is present in the developing and dry seeds of okra at all stages of development (75). The raffinose family of oligosaccharides are present in mature and dry seeds, while free glucose and fructose were detected at most stages. Dry seeds contained raffinose, stachyose, and verbacose sugars of oligosaccharide family. However, monosaccharides were not detected in any of the dry seeds. Mucilage present in okra fruits is acidic polysaccharide with associated proteins and minerals (78). Hydrolysis of the mucilage gave polysaccharide composed of galacturonic and glucuronic acids and minor contents of galactose, rhamnose, glucose, and arabinose. The sugar components of okra mucilage resembled many pectic substances. The chemical composition of the fruit is influenced by variety (79) and the environment under which crop is grown (Table 4).

B. Proteins

Total protein content in okra seeds and pods at various developmental stages has been determined (80). The maximum protein content was 2.08% in pods and 2.09% in seeds on the ninth day of pollination. The seed meal of okra contains albumins, globulins, and glutelins. Mature dry okra seeds contain 20.58% protein (80). The amino acid composition resembled that of

TABLE 3 Nutritional Composition of Okra

Constituent	Content (per 100 g edible portion)
Moisture (g)	89.6
Carbohydrates (g)	6.4
Proteins (g)	1.9
Fats (g)	0.2
Fiber (g)	1.2
Minerals (g)	0.7
Calcium (mg)	66
Magnesium (mg)	43
Oxalic acid (mg)	8
Phosphorus (mg)	56
Iron (mg)	1.5
Sodium (mg)	6.9
Potassium (mg)	103
Copper (mg)	0.19
Sulfur (mg)	30
Vitamin A (IU)	88
Thiamine (mg)	0.07
Riboflavin (mg)	0.10
Nicotinic acid (mg)	0.60
Vitamin C (mg)	13

Source: Refs. 79, 80.

TABLE 4 Effect of Organic vs. Inorganic Nitrogen on Fruit Quality of Okra (cv. Pusa Sawani)

	Total carbohydrate (%)		Crude fiber (%)		Protein (%)		Ascorbic acid (mg/100 g)	
Treatments	5th harvest	Last harvest	5th harvest	Last harvest	5th harvest	Last harvest	5th harvest	Last harvest
40 kg (FYM)	33.66	30.27	14.76	14.98	18.80	19.52	14.72	13.48
40 kg N (PM)	35.31	33.47	13.87	13.60	19.62	19.80	16.34	15.89
40 kg N (HM)	33.35	33.25	14.21	14.52	18.97	18.37	15.25	15.00
20 kg N (AS) + 20 kg N (FYM)	33.45	34.06	13.32	13.60	20.93	20.60	15.60	14.47
20 kg N (AS) + 20 kg N (PM)	35.95	33.30	13.13	13.55	22.76	22.28	17.63	16.87
20 kg N (AS) + 20 kg N (HM)	35.71	34.35	14.00	13.86	21.62	21.66	15.80	15.12
30 kg N (AS) + 10 kg N (FYM)	33.71	32.17	14.94	14.91	21.22	20.84	14.08	13.55
30 kg N (AS) + 10 kg N (PM)	34.95	34.28	13.42	13.71	22.38	21.84	16.50	16.12
10 kg N (AS) + 10 kg N (HM)	34.35	32.47	13.77	14.10	21.46	20.92	15.63	15.00
10 kg N (AS) + 10 kg N (FYM)	35.46	36.13	14.25	14.40	22.39	22.96	15.89	15.52
10 kg N (AS) + 30 kg N (PM)	35.65	37.22	12.38	13.25	24.62	24.21	17.86	17.35
10 kg N (AS) + 30 kg N (HM)	35.37	36.37	13.40	13.61	24.34	24.13	16.10	17.10
Control (No nitrogen)	30.02	25.07	15.70	16.24	17.02	16.07	13.31	12.25
CD ($p = 0.05$)	1.27	1.25	0.80	0.69	0.46	1.02	2.27	3.95

AS = Ammonium sulfate; PM = poultry manure; FYM = farmyard manure; HM = horse manure.
Source: Ref. 79.

soybeans, but the protein-energy ratio was higher in the case of okra seeds than that with soybean seeds.

C. Lipids

Okra seeds contain 14–19% oil having a good proportion of linoleic acid. The percentage of individual fatty acids revealed myristic (0.2%), palmitic (30.2%), stearic (4.0%), palmitoleic (22.4%), oleic (24.4%), and linoleic acids (40.8%) (80). Gopalkrishnan et al. (81) reported lipid content of stems (1–3%), cotyledons (3.7–9%), and seeds (2.2–20.2%). Lipid content fluctuates considerably during seed development, being low in the early stages and increasing gradually during development (82). A marked increase was noticed betwen days 12 and 20.

D. Vitamins and Minerals

Okra fruits and leaves contain high amounts of calcium (83,84). Phosphorus, sodium, sulfur, and nitrogen contents in the developing seeds, embryo, seedcoat, and fruit wall were analyzed (80). Developing embryos had a maximum amount of total nitrogen at day 14. Embryos were consistently richer in phosphorus and sulfur. Contents in fruit wall was almost constant during all stages of development. Sulfur content increased sharply after day 14 (80). Sodium content in the developing okra fruits was maximum in the developing wall followed by seedcoats (80).

V. STORAGE

Okra deterioates rapidly and is normally stored only briefly to hold for marketing or processing (85,86). Large quantities are canned, frozen, or brined. It has a very high respiration rate at warm temperatures and should, therefore, be promptly cooled to retard heating and subsequent deterioration.

A. Low-Temperature Storage

Okra in good condition can be stored satisfactorily for 7–10 days at 7–10°C. At higher temperatures, toughening, yellowing, and decay are rapid. A relative humidity of 85–90% is desirable to prevent shriveling. At temperatures below 7°C, okra is subjected to chilling injury, which is manifested by surface discoloration, pitting, and decay. Holding okra for 3 days at 0°C may cause severe pitting (85).

French okra bruises easily, and bruises blacken within a few hours. A bleaching type of injury may also develop when okra is held in hampers for more than 24 hours without refrigeration (85). Storage containers should permit ventilation.

Okra should be stored below 10°C with high humidity to prevent wilting (86–89). The high rate of deterioration and respiration demand rapid cooling of harvested okra. Ryall and Lipton (87) state that unless okra is cooled below 15.4°C soon after being packed, the heat of respiration will cause the temperature in the package to rise quickly and result in rapid deterioration. In spite of this situation, hydrocooling is generally not recommended, because water may cause pitting, as does prolonged contact with ice or with ice water (90).

Ryall and Lipton (87) also state that vacuum-cooling of okra may be feasible, although prewetting of pods probably would be necessary to avoid excessive water loss. Once okra is cooled, it should be held at 7.9–10°C; at higher temperatures toughening, yellowing, and decay are rapid, whereas pitting, induced by chilling injury, can develop below 4.4°C. Okra must be held at about 95% relative humidity if wilting is to be avoided. Pantastico et al.

(88) recommended a temperature of 8.8°C and 90% relative humidity to store okra for about 2 weeks in marketable condition.

B. Controlled-Atmosphere Storage

Anadaswamy et al. (86) suggested that 5–10% CO_2 extends the shelf life of okra stored at 11.1–12.7°C by about a week. Higher concentrations of CO_2 caused off-flavors that persisted even after cooking. Hatton et al. (90) reported that okra held in 10–21% CO_2 at 7.7°C maintained a fresh appearance and prevented peel discoloration.

C. Packaging

Prepackaging okra in perforated film is a great aid not only to avoid wilting but also to avoid physical injury to this tender vegetable. Singh et al. (91) studied the effect of prepackaging materials on the storage life of fresh okra fruits. The fruits of cultivar Pusa Sawani were packed in 100–400 gauge polyethylene and stored for up to 9 days at room temperature (32 ± 2°C). Fruits packed in 400 gauge polyethylene had the longest shelf life (9 days). The control fruits kept for only 2–3 days. Fruit retention of chlorophyll *a* and *b* was greater in packed than in control fruits (92).

D. Chemical Control

Singh et al. (93) investigated the effect of growth retardants and wax emulsion on the shelf life of okra. Fruits were dipped in CCC (chlormequat) or B9 (daminozide) at 100–300 ppm for 10 minutes followed by dipping in wax emulsion (12%) for one minute and stored in paper bags at room temperature or at 10°C. At room temperature weight loss was reduced by all treatments compared with water-dipped controls. Fruits dipped in CCC at 100 ppm and stored at room temperature and at 10°C remained in good condition even after the third and twelfth days, respectively, whereas fruits treated with B9 developed superficial black spots (92).

When okra fruits were dipped in 0–500 ppm morphactin (chloroflurenol–methyl ester chloroflurecol) or in 2.5–10 ppm morphactin + a wax emulsion, losses in weight and soluble sugars were reduced most by 2.5 ppm morphactin applied either pre- or postharvest. All treatments with more than 100 ppm morphactin induced severe pod blackening. All morphactin treatments reduced chlorophyll degradation and increased acidity, but amino acids increased with preharvest and decreased with postharvest application (92). Singh and Dhankhar (92) dipped freshly harvested, washed, and dried okra fruits (cv. Pusa Sawani) in solution containing GA, ascorbic acid, or CCC each at 100 ppm or 250 ppm for 10 minutes. The fruits were then packed in polyethylene bags and stored at room temperature (32 + 2°C) for 12 days. At room temperature, weight loss was least in fruits treated with ascorbic acid at 250 ppm, but at low temperature, all treatments gave similar results. At both temperatures, retention of chlorophyll *a* and *b* was best after treatment with ascorbic acid at 250 ppm.

VI. PROCESSING AND UTILIZATION

The young fruits of okra are used mostly as a fresh vegetable. The green fruits are picked in the young (3–6 cm long) stage, before they become fibrous and before the seeds are fully developed. In India, the sliced fruits are fried with condiments and salt; fruits and dry seed powder are used in soups and curries for thickening due to their mucous nature. In Africa, the fruits are consumed alone or in salad, after cooking in salty water, and are used in the preparation of certain sauces.

The fruits are also preserved in brine after boiling, dried (in the sun or oven) in the form of slices (Africa, India, and Turkey) or frozen and sterilized (United States) (94).

Okra seed oil is used for cooking or in the production of margarine as in Greece. Recently, okra seeds have been utilized as a source of protein and vegetable oil in the USA.

Fibers extracted from the stems of okra are sometimes used to make string and nets, in Mali (1). An extract of okra stems and of a wild species, *A. manihot* ssp. *tetraphyllus*, is used as a clearing agent in the manufacture of jaggery from sugarcane juice in rural India. Okra leaves are eaten like spinach in Africa. The species *A. manihot* (ssp. *manihot*) is also cultivated for consumption as a leaf vegetable in the Far East.

Okra is commercially preserved by salting for use in the manufacture of certain canned products, such as vegetable soup mixtures. The salted okra, which contains about 20% salt, is not desalted before use but rather is added to the product in such quantity that the need for salt seasoning is met. Soluble nutrient losses from salted okra are minimal as a result of this manner of use of the preserved product (1).

A. Dehydration

Green okra pods can be sun-dried to preserve them for the off-season. Young pods are first sliced to facilitate drying. Special varieties are sometimes used for this purpose, and in Turkey, okra is grown especially for drying (84). Pusa Sawani was found to be best for dehydration, followed by Sel.2 for color, texture, taste, flavor, and dehydration and rehydration ratio (84). For the dehydration of okra, less fiber, less mucilaginous substances, high protein, high dry matter, and high mineral content are important attributes.

B. Canning and Freezing

After being cut and blanched, the pods can be frozen for long storage (1 year) (80). Small tender pods should be taken for canning. For canning and freezing, high chlorophyll, high crude fiber and mucilage substance, low dry matter, and high amounts of protein, vitamin, and minerals are required. Pusa Sawani and Dwarf Green Smooth were found to be suitable for canning. Parbhani Tillu, a new variety suitable for processing, has been developed at Marathwada Agricultural University, Parbhani, India (95). Its fruits are shorter and do not break during processing. It also exhibits acceptable texture after freezing (95).

REFERENCES

1. Martin, F. W., and R. Ruberte, *Vegetables for the Hot Humid Tropics, Part 2, Okra*, Science and Education Administration, U.S. Department of Agriculture, New Orleans, 1978, p. 22.
2. Adams, C. F., *Nutritive Value of American Foods in Common Units*, U.S. Department of Agriculture, Agric. Handbook, 425, 1975, p. 29.
3. Siemonsma, J. S., West African Okra-morphological and cytological indications for the existence of natural amphidiploid of *Abelmoschus esculentus* (L) Moench and *A. manihot* (L.) Medikus, *Euphytica 31*:241 (1982).
4. Charrier, A., *Genetic resources of genus* Abelmoschus *Med. (Okra)*, International Board for Plant Genetic Resources, 1984, p. 61.
5. Hochreutiner, B. P. G., Centres nouveax et. disculees de la Famille des Malvacees, *Candolla* 2:79 (1924).
6. Van Borssum-Waalker, J. Van, Malesian malvaceae, revised, *Blumea 14*(1):1 (1966).
7. Bates, D. M., Notes on the cultivated Malvaceae 2, *Abelmoschus, Baileya, 16*:99 (1968).
8. de Candolle, A. P., *Origin of Cultivated Plants*, Noble Offer Printing Inc., New York,1983, p. 150.

9. Metcalfe, C. R., and L. Chalk, Okra, in *Anatomy of the Dicotyledons*, Clarendon Press, Oxford, 1957, p. 151.
10. Scott, F. M., and B. G. Bystrom, New Research in Plant Anatomy, *Ann. Bot. 63*:15 (1970).
11. Kuncu, B. C., and C. Biswas, Section of Botany, *Proc. 60th Ind. Sci. Part III, abstract*, 1974, p. 248.
12. Jambhale, N. D., and Y. S. Nerkar, Indirect selection criteria for isolation of induced polyploids in the *Abelmoschus* species hybrids, *Cytologia 47*:603 (1982).
13. Joshi, A. B., and M. W. Hardas, Alloploid nature of okra *Abelmoschus esculentus* (l). Moench, *Nature 178*:1190 (1956).
14. Joshi, A. B., V. R. Gadwal, and M. W. Hardas, Okra, in *Evolutionary Studies in World Crops, Diversity and Change in Indian Subcontinents* (J. B. Hutchinson, ed.), Cambridge University Press, Bombay, 1974, p. 99.
15. Teshima, T., Genetical and cytological studies in an interspecific hybrid of *Hibiscus esculentus* and *H. manihot*, *J. Fac. Agric. Hokk. Idoh Univ. 34*:1 (1933).
16. Ford, C. E., A contribution to a cytogenetical survey of Malvaceae, *Genetica 20*:431 (1938).
17. Ugale, S. D., R. C. Patil, and S. S. Khuspe, Cytogenetic studies in the cross between *Abelmoschus esculentus* and *A. tetraphyllus*, *J. Maharashtra Agric. Univ. 1*:106 (1976).
18. Kamalova, G. V., Cytological studies of same species of Malvaceae, *Uzb. Biol. Zurnali 3*:66 (1977).
19. Tekale, P. D., N. D. Jambhale, and Y. S. Nerkar, Cytomorphological studies in interspecific hybrid derivative lines of okra, *Indian J. Genet. 45*(27):224 (1985).
20. Kuwada, H., Studies on the interspecific crossing between *Abelmoschus esculentus* Moench and *A. manihot* Medic and the various hybrids and polyploids derived from the above two species, *Mem. Fal. Agric. Kagawa Univ.* No. 8, 1961, p. 1.
21. Jambhale, N. D., and Y. S. Nerkar, Induced amphidiploidy in the crosses *Abelmoschus esculentus* (l.) × *A. manihot* (L.) Medik spp. *manihot*, *Genet. Agr. 36*:19 (1982).
22. Jambhale, N. D., and Y. S. Nerkar, Evolution of Guinean okra cultivars from West Africa, *J. Maharashtra Agric. Univ. 14*(2):215 (1989).
23. Mamidwar, R. B., Y. S. Nerkar, and N. D. Jambhale, Cytogenetics of interspecific hybrids in genus *Abelmoschus*, *Indian J. Herd. 11*:35 (1979).
24. Erikson, N. T., and F. A. A. Couto, Inheritance of four plant floral characters in okra *Hibiscus esculentus* (L.). *Proc. Am. Hort. Sci. 83*:605 (1963).
25. Nath, P., and O. P. Putta, Inheritability of fruit hairiness, fruit colour and leaf lobing in okra *Abelmoschus esculentus*, *Can. J. Genet. Cytol. 12*(3):589 (1970).
26. Jambhale, N. D., and Y. S. Nerkar, Inheritance of chlorina, an induced chlorophyll mutant in okra (*A. esculentus* (L.) Moench, *J. Mah. Agric. Univ. 4*(3):316 (1979).
27. Kalia, H. R., and D. S. Padda, Inheritance of leaf and flower characters in okra, *Indian J. Genet. Plant Breed.* 22:282 (1962).
28. Kolhe, A. K., and R. D'Cruz, Inheritance of pigmentation in okra, *Indian J. Genet. 23*:112 (1966).
29. Jambhale, N. D., and Y. S. Nerkar, Inheritance of Palmatisect, an induced leaf mutation in okra, *Indian J. Genet. Plant Breeding 40*(3):600 (1980).
30. Ngah, A. W., and K. M. Graham, Heritability of four economic characters in okra (*Hibiscus esculentus*), *Malaysian Agric. Res. 2*(1):15 (1973).
31. Padda, D. S., M. S. Saimbhi, and D. Singh, Genetic evaluation and correlation studies in okra, *Indian J. Hort. 27*:39 (1970).
32. Singh, K., Y. S. Malik, G. Kalloo, and N. Mehrotra, Genetic variability and correlations studies in *bhindi* (*Abelmoschus esculentus* (L.) Moench, *Veg. Sci. 1*:47 (1974).
33. Swamy Rao, T., and G. P. Satyavati, Influence of environment on combing ability and genetic components in *bhindi* (*Abelmoschus esculentus*), *Genet. Pol. 18*(2):141 (1977).
34. Kulkarni, R. S., Biometrical investigations in *bhindi* (*Abelmoschus esculentus* (L.) Moench, *Mysore J. Agric. Sci. 10*:332 (1976).

35. Thakur, M. R., Inheritance of resistance to yellow mosaic in a cross of okra species *Abelmoschus esculentus* × *A. manihot* subsp. *manihot, SABRAQ J. 8*:69 (1976).
36. Jambhale, N. D., and Y. S. Nerkar, Inheritance of resistance to okra yellow vein mosaic in interspecific crosses of *Abelmoschus, Theor. Appl. Genet. 60*:313 (1981).
37. Singh, H. B., B. S. Joshi, P. P. Khanna, and P. S. Gupta. Brewing for field resistance to yellow vein mosaic in bhindi, *Indian J. Genet. Plant Breeding 22*(2):137 (1962).
38. Jambhale, N. D., and Y. S. Nerkar, Inheritance of resistance to powdery mildew in okra, *Proceedings of National Seminar on Breeding Crop Plants for Resistance to Pests and Diseases* (S. R. Rangaswamy, ed.), Tamilnadu Agricultural University, School of Genetics, Coimbatore, 1983.
39. Sharma, B. R., and B. S. Gill, Genetics of resistance to cotton jassids *Amrasca biguttula biguttula* (Ishida) in okra, *Euphytica 33*:215 (1984).
40. Mehetre, S. S. Genetics of pigmentation in okra (*Abelmoschus esculentus* (L) Moench, *J. Maharashtra Agric. Univ. 5*(1):19 (1980).
41. Akran, M., and M. Shafi, Superior quality F_1 hybrids in okra (*Hibiscus esculentus* L.), *J. Agric. Res. Pak. 9*:114 (1971).
42. Jalani, B. S., and K. M. Graham, A study of heterosis in crosses among local and American varieties in okra (*Hibiscus esculentus* (L.) Moench, *Malays Agric. Res. 2*(1):7 (1975).
43. Singh, S. P., J. P. Srivastava, and H. N. Singh, Heterosis in bhindi (*Abelmoschus esculentus* (L.) Moench, *Prog. Hort. 7*(2):5 (1975).
44. Singh, S. P., J. P. Srivastava, H. N. Singh, and N. P. Singh, Genetics divergence and nature of heterosis in okra, *Indian J. Agric. Sci. 47*:546 (1977).
45. Singh, S. P., and H. N. Singh, Hybrid vigour for yield its components in okra, *Indian J. Agric. Sci. 49*:596 (1979).
46. Elangovan, M., C. R. Muthukrishnan, and I. Irulappan, Hybrid vigour in bhindi (*Abelmoschus esculentus* (L.) Moench for some economic characters, *South Indian Hortic. 29*:4 (1981).
47. Kulkarni, R. S., and K. Virupakshappa, Heterosis and inbreeding depression in okra, *Indian J. Agric. Sci. 47*:552 (1977).
48. Swamy Rao, T., and P. M. Ramus, A study of correlation and regression coefficients in bhindi (*Abelmoschus esculentus* (L.), *Current Res.* (India) *4*:135 (1975).
49. Swamy Rao, T., and G. P. Sathyavathi, Genetic and environmental variability in okra, *Indian J. Agric. Sci. 47*:80 (1976).
50. Venkatramani, K. S., A preliminary study on some inter varietal crosses and hybrid vigour in *Hibiscus esculentus* (L.), *J. Madras Univ. 22*(2):183 (1952).
51. Sulikiri, G. S., and T. Swamy Rao, Studies on floral biology and fruit formation in okra (*Abelmoschus esculentus* (L.) Moench varieties, *Prog. Hort. 4*:71 (1972).
52. Mitidieri, J., and R. Vencovsky, Polinizaca'O crusado do quiabeiro em condicoe's de campo, *Rev. Agric.* (Piracicaba Braz) *49*(1):3 (1974).
53. Purewal, S. S., and G. S. Randhava, Studies in *Hibiscus esculentus* (Lady's finger) 1. Chromosome and pollination studies, *Indian J. Agric. Sci. 17*(3):129 (1947).
54. Parthasarathy, V. A., and C. N. Sambandam, Studies on self pollination techniques in bhindi (*Abelmoschus esculentus* (L.) Moench, *AVARA 6*:76 (1976).
55. Giriraj, K., and T. Swamy Rao, Note on a simple crossing technique in okra, *Indian J. Agric. Sci. 43*:1089 (1973).
56. Parthasarathy, V. A., and C. N. Sambandam, Studies on cross-pollination in bhindi, *Abelmoschus esculentus* (L.) Moench, *AVARA 6*:83 (1976).
57. Joshi, A. B., H. B. Singh, and B. S. Joshi, Is yellow vein mosaic disease a nuisance in your *bhindi*, then why not grow Pusa Sawani, *Indian Farming 10*:6 (1960).
58. Jambhale, N. D., and Y. S. Nerkar, Parbhani Kranti-a yellow vein mosaic resistant okra, *HortScience 21*(6):1470 (1986).

59. *Indian Minimum Seed Certification Standards*, Central Committee, Dept. Agriculture, Ministry of Food, Agriculture Community Development and Co-operation, New Delhi, 1971.
60. Chauhan, M. S., and Y. M. Bhandari, Pod development and germination studies in okra, (*Abelmoschus esculentus* (L.) Moench), *Indian J. Agric. Sci. 41*(10):852 (1971).
61. Arulrajah, T., and D. P. Ormrold, Responses of okra (*Hibiscus esculentus* (L.) to photoperiod and temperature, *Ann. Bot.* (London) *37*:331 (1973).
62. Chauhan, D. V. S., *Vegetable Production in India*, 3rd ed., Ram Prasad and Sons (Agra), 1972.
63. Pereira, A. A., F. A. A. Couto, and M. Moestri, Influencia do foto periodo na floracao do quiabo (*Hibiscus esculentus* L.), *Rev. Ceres 18*(96):131 (1971).
64. Thompson, H. C., and W. C. Kelly, *Vegetable Crops*, 5th ed., McGraw-Hill Book Co. Inc., New York, 1957.
65. Yawalkar, K. S., J. P. Agarwal, and S. Bokade, *Manures and Fertilizers*, Agri-Horticultural Pub. House, Nagpur, India.
66. Thakur, M. R., and S. K. Arora, Okra, in *Vegetable Crops in India* (T. K. Bose and M. G. Som, eds.), Naya Prokash, Calcutta, 1986.
67. Chhonkar, V. S., and S. N. Singh, Studies on inorganic nutrition of bhindi (*Abelmoschus esculentus* (L.) Moench in sand culture, *Indian J. Hort. 20*:51 (1963).
68. Hipp, B. W., and W. R. Cowley, Importance of the phosphorus-zinc interaction in okra production, *HortScience 6*:211 (1971).
69. Costa, M. C. B., H. P. Haag, and J. R. Sarnge, Anais da Escola Superior de Agricultura, *Leuiz Pueiroz 29*:109 (1972).
70. Singh, B. B. C., Effect of vanadium on the growth and yield of okra (*Abelmoschus esculentus* (L.) Moench, *Haryana J. Hort. Sci. 3*:177 (1974).
71. Al-Budrawy, R., and W. Bussier, *Z. Pflanzenernaeh. Bodenk. 140*:50 (1977).
72. Sivanappan, R. K., C. R. Muthukrishnan, P. Natarajan, and S. Ramadas, The response of bhindi (*Abelmoschus esculentus* (L.)) Moench to the drip system of irrigation, *South Indian Hort. 22*:98 (1974).
73. Perkins, D. Y., J. C. Miller, and S. L. Dallyn, Influence of pod maturity on vegetable and reproductive behaviour of okra, *Proc. Am. Soc. Hort. Sci. 60*:311 (1952).
74. Kanwar, J. S., and M. S. Saimbhi, Pod maturity and seed quality in okra, *Punjab Hort. J. 2*:234 (1987).
75. Culpepper, C. W., and H. H. Moon, the growth and composition of fruit of okra in relation to its eating quality, U.S. Department of Agriculture Circular 545 (1941).
76. Singh, S., A. B. Mandal, and T. Ram, Physico-chemical changes in developing fruits of okra, *Indian J. Plant Physiol. 33*(3):266 (1990).
77. Woodroof, J. G., and E. Shelor, Okra for processing, *Georgia Agric. Expt. Sta. Bull.* NS 56 (1958).
78. Ryall, A. L., and W. J. Lipton, *Handling, Transportation and Storage of Fruits and Vegetables*, Vol. 1, *Vegetables and Melons*, The Avi Pub. Co. Inc., Westport, CT, 1972.
79. Abusaleha, and K. G. Shanmugavelu, Studies on the effect of organic vs. inorganic source of nitrogen on growth, yield and quality of okra, *Indian J. Hort. 45*:312 (1988).
80. Malik, C. P., P. L. Pohalla, M. B. Singh, and H. Singh, The biology of okra, A review, in *Widening Horizons of Plant Sciences* (C. P. Malik, ed.), Cosmo Publ., New Delhi, 1985, p. 355.
81. Gopalkrishnan, N., T. N. B. Kainal, and G. Lakshminarayan, Fatty acid changes in *Hibiscus esculentus* tissue during growth, *Phytochemistry 21*(3):565 (1982).
82. Ahluwalia, K. J., and C. P. Malik, Lipids in okjra, in *3rd Annual Botanical Conference of the Society for Advancement of Botany*, New Delhi, 1980.
83. Aykrod, W. R., The nutritive value of Indian foods and the planning of satisfactory diets, *ICMR Special Report Series* No. 42 (1963).
84. Saimbhi, M. S., Agrotechnique for okra, in *Advances in Horticulture*, Vol. 5, *Vegetable Crops* (K. L. Chadha and G. Kalloo, eds.), Malhotra Publishing House, New Delhi, 1993, p. 529.

85. Scholz, F. W., H. B. Johnson, and W. R. Buford, heat evolution rates of some Texas grown fruits and vegetables, *J. Rio Grande Valley Hort. Soc. 17*:170 (1963).
86. Anandaswamy, B., Pre-packaging studies on fresh produce IV. Okra (*Hibiscus esculentus*), *Food Sci.* (Mysore) *12*:332 (1963).
87. Ryall, A. L., and W. J. Lipton, *Handling, Transportation* and *Storage of Fruits and Vegetables*, Vol. I, *Vegetables and Melons*, The AVI Pub. Co. Inc., Westport, CT, 1972, p. 45.
88. Pantastico, E. B., T. K. Chattopadhyay, and H. Subramanyam, Storage and commercial storage operations, in *Postharvest Physiology, Handling and Utilization of Tropical and Sub-tropical Fruits and Vegetables* (E. B. Pantastico, ed.), The AVI Pub. Co., Westport, CT, 1975, p. 314.
89. Hardenburg, R. E., A. E. Watada, and C. Y. Wang, *The Comercial Storage of Fruits, Vegetables and Florist and Nursery Stock*, U.S. Department of Agriculture Handbook No. 66, 1986.
90. Hatton, T. T. Jr., E. B. Pantastico, and E. K. Akamine, Controlled atmosphere storage, Part III, Individual commodity requirement, in *Postharvest Physiology, Handling and Utilization of Tropical and Sub-tropical Fruits and Vegetables* (E. B. Pantastico, ed.), AVI Pub. Co., Westport, CT, 1975.
91. Singh, B. P., B. S. Dhankhar, M. L. Pandita, Effect of pre-packaging materials on storage of fresh okra (*Abelmoschus esculentus* (L.) Moench) fruits, *Haryana J. Hort. Sci. 9*:175 (1980).
92. Singh, B. P., and B. S. Dhankhar, Effect of growth regulators and pre-packaging on the storage life of okra (*Abelmoschus esculentus* (L.) Moench) fruits, *Haryana Agric. Univ. J. Res. 10*:398 (1980).
93. Singh, B. P., D. K. Bhatnagar, and O. P. Gupta, Effect of growth retardants and wax emulsion on the shelf-life of okra (*Abelmoschus esculentus* (L.) Moench), *Haryana J. Agric. Sci. 8*:97 (1979).
94. Jones, I. D., Effects of processing by fermentation on nutrients, in *Nutritional Evaluation of Food Processing* (R. S. Harris and F. Karmas, eds.), AVI Pub. Co. Inc., Westport, CT, 1975, p. 324.
95. Kulkarni, U. G., and Y. S. Nerkar, Parbhani Tillu—an induced *bhendi* mutant suitable for processing. *J. Mah. Agric. Univ. 17*(3):496 (1992).

27

Sweet Corn

V. M. GHORPADE AND M. A. HANNA
University of Nebraska, Lincoln, Nebraska

S. J. JADHAV
Alberta Agriculture, Leduc, Alberta, Canada

I. INTRODUCTION

Corn (*Zea mays*) has been a very important food crop in the Western Hemisphere, as well as in the world's agricultural economy. Corn was probably gathered from the wild millennia before being domesticated some 7000 years ago. Early Americans spread this food from southern Mexico to as far south as Argentina and northward to Canada. Later on, Columbus took corn from the New World back to Spain. Within a short period it had spread all around the world (1).

Sweet corn (*Zea mays* saccharata) is of relatively recent origin. The earliest records date back to the eighth century A.D., when it was grown by Indians in the region that is now western Guatemala. A writer stated in the Aug. 3, 1822 *New England Farmer* that sweet corn was not known in the New England region until a gentleman from Plymouth, Massachusetts—a member of General Sullivan's expedition against the Indians in 1779—obtained sweet corn seeds from India. A similar story was reported by another writer in September of 1822, who stated that sweet corn was brought back by Lieutenant Richard Bagnal from General Sullivan's expedition against the Six Nations in 1779. It was described as having a shriveled grain and being red colored when mature and ripe, and was called *papoon corn*. In 1828, sweet corn was listed by seed growers as "a very sweet vegetable known as sweet corn" (2), and Galinat (3) reported Danting Early as the first named sweet corn cultivar.

The major difference between sweet corn and field corn is its genetic makeup rather than systematic or taxonomic characterization. Sweet corn has the gene *su* at the sugary locus on chromosome 4. The early literature on corn origin, development, and breeding is applicable to sweet corn. However, sweet corn mutant types differ in seed quality, storage, and end use; thus the evaluation procedures and breeding objectives differ greatly from those of field corn.

Sweet corn is grown in all parts of the United States. Sweet corn–producing areas are

TABLE 1 Sweet Corn Production in the United States[a] by Area

Region	1988	1990	1992
East[b]	18.3	16.4	18.6
Midwest[b]	74.9	119.4	122.5
West[b]	41.0	47.3	48.9
Other states[c]	2.9	4.9	7.4

[a]Sweet corn–harvested area (in 1000 ha; 1 ha = 2.47 acres).
[b]Area totals do not include quantities in other states.
[c]Michigan is included in Other states (1992, 2915 hectares).
Source: Ref. 4.

grouped into East, Midwest, West, and Other States regions (Table 1). The total acreage for processing is shown in Table 2. The Midwest region (Illinois, Wisconsin, and Minnesota) has the largest acreage under cultivation, combining to make up over 40% of total U.S. production. Per capita consumption of sweet corn in the United States during the 10-year period (1974–83) remained relatively constant at about 12.7 kg. Fresh consumption was about 3.2 kg, canned corn consumption declined from 6.7 to 5.8 kg, and frozen corn consumption increased from 2.6 to 3.7 kg (4).

In the United Kingdom, the canned sweet corn market has remained stable at around 36,000 MT over the last 5 years. Most canned sweet corn is imported from other countries. The United States holds 24% of the market share. Naturally sweet varieties became more popular as consumers moved away from products with added salt and sugar, primarily due to dietary concerns. Distribution of canned sweet corn is divided between the food service industry (30% of market) and retail outlets (70%). Consumers in the United Kingdom prefer corn from the United States, Canada, and France, even at a premium price, because of its high quality. This high-quality canned sweet corn is required by the 1990 Food Safety Act, which governs all trading of foods in the United Kingdom (5).

II. BOTANY

A. Morphology

Sweet corn is an herbaceous, monocotyledonous, annual seed-producing plant (Fig. 1). The plant is monoecious, having separate male (tassel) as well as female organs located on the same plant. Upon germination, a corn seed produces a single shoot from which a primary root grows down from the first node. An adventitious root system is developed from the second to the seventh nodes. The root system spreads up to 1 m in all directions from the plant, with maximum concentration within 0.5 m. The stem is made up of a series of pith-filled internodes separated by nodes that are larger in diameter than the internodes. Each internode bears a leaflet at the top and a bud or bud tissue at the lower end, which may or may not produce subsidiary branches. Branches produced at the lower-end internodes are referred to as suckrs or tillers. The leaf of the corn plant is made up of the sheath and the blade. The sheath covers the stem past most of the internode above the node from which it grows, strengthening the stem and protecting the

TABLE 2 Production of Sweet Corn for Processing in the U.S. by Area[a]

Processing	1988	1990	1992
Canning	107.3	117.0	118.9
Freezing	67.2	71.0	78.1
Total processing	174.5	198.4	197.1
Fresh market	78.7	76.2	83.3
Total U.S.	253.1	274.6	280.3

[a]Sweet corn–harvested area (in 1000 ha; 1 ha = 2.47 acres).
Source: Ref. 4.

developing region of the next higher internode. The leaf is flat and long with a strong midrib and smaller vein parallel to it. The blade tapers to a point at the tip and eventually has an auricle at each side, where it links to the stem. The tassel has branches that are arranged spirally and usually has secondary branching. The ear is at the central point of the spike with the purified spikelets borne in several longitudinal rows. Ear husks are the leaf sheath that generally have reduced leaf blades with ligules borne on short internodes below the pistillate spike (6).

B. Cultivars

Sweet corn is commonly classified into three groups based on the length of time required to produce edible ears: early, medium, and late. In the early 1820s seed industries listed sweet corn cultivars such as Eight-Rowed Early, Dwarf Early, and Sweet or Sugar. The cultivar Darling's Early is the first sweet corn cultivar having a detailed accounting of its breeding. Sweet corn also has been classified solely on color as white, yellow, and bicolor. The preferred sweet corn cultivar prior to 1902 was white, when the Golden Bantam cultivar was introduced. This yellow cultivar had desirable characteristics such as earliness and improved flavor. By 1919 most of the sweet corn grown in the United States was yellow. Bicolor hybrids are regionally popular, and breeding efforts have produced a number of such cultivars since 1960.

Today, many well-known top cross hybrids of sweet corn such as Mar Cross and Span Cross are available. Golden Cross Bantam (or Golden Cross) is the most widely grown type of hybrid sweet corn cultivated in the United States. The most popular open pollinated cultivars are Stowell Evergreen, County Gentleman, and Shoe Peg. Southern Shipper and Hickory King are also grown in some parts of the United States. In 1948, the yellow hybrid cultivars losweet, Jubilee, NK-199, and Golden Cup were developed and released. The most significant white hybrid, Silver Queen, was developed and released in 1960. This was the most popular variety for fresh market production, and its popularity resulted in development of a large number of white hybrids by the seed industries. A classification of edible sweet corn based on genetic makeup of the endosperm (7) is presented in Table 3. Wiley et al. (6) reported the following genetic types of sweet' corn in order of lowest to highest sugar content.

1. Sugary Hybrids *(su)*

These varieties contain a sugary gene, which blocks conversion of sugars (mostly sucrose) to starch after moving from leaves to the kernel. All sweet varieties disrupt, to a greater or lesser degree, the synthesis of sugar from starch. The wrinkled appearance of sweet corn is attributed to the comparatively smaller size of the sugar molecule as opposed to the starch molecule and

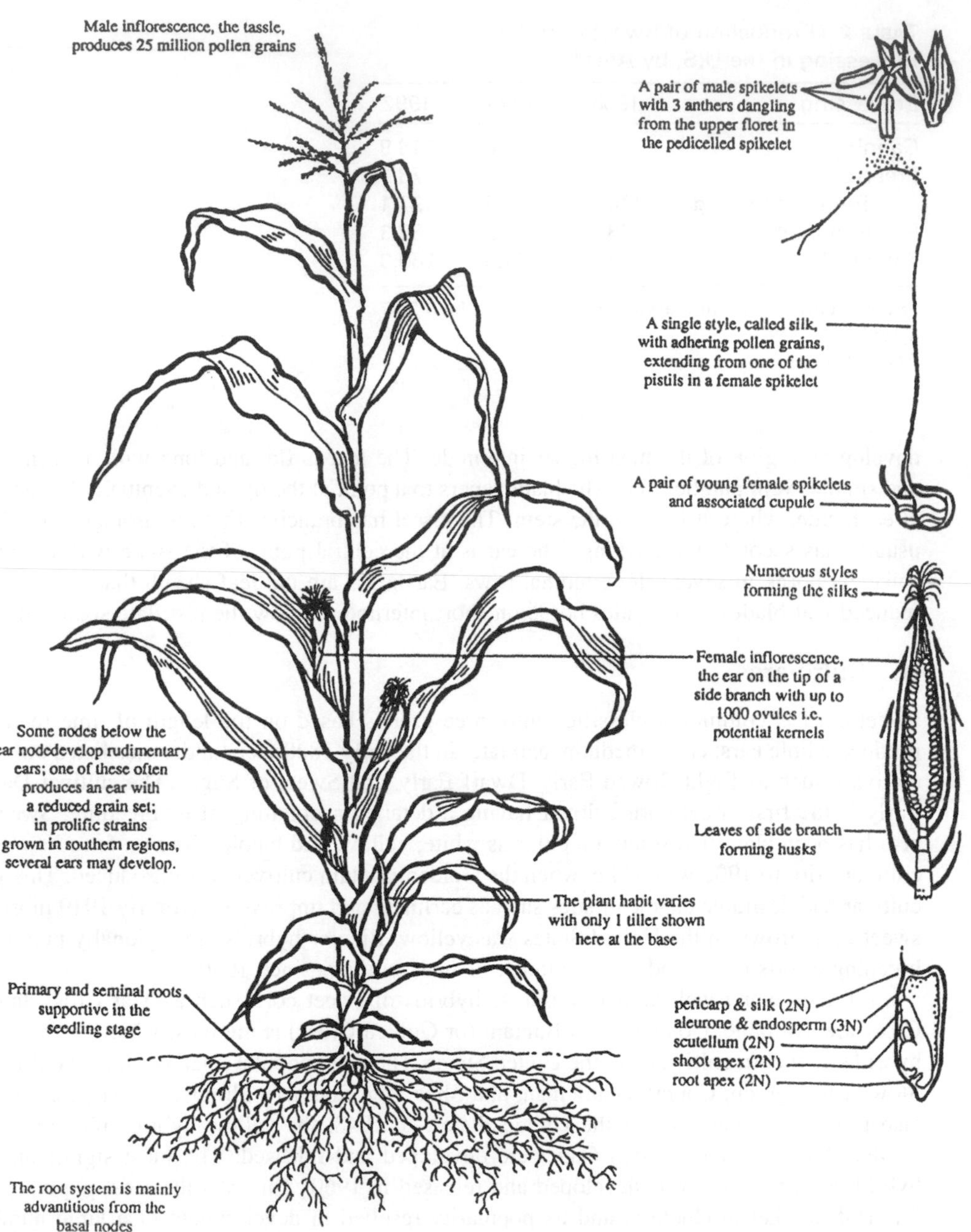
Male inflorescence, the tassle, produces 25 million pollen grains
A pair of male spikelets with 3 anthers dangling from the upper floret in the pedicelled spikelet
A single style, called silk, with adhering pollen grains, extending from one of the pistils in a female spikelet
A pair of young female spikelets and associated cupule
Numerous styles forming the silks
Female inflorescence, the ear on the tip of a side branch with up to 1000 ovules i.e. potential kernels
Some nodes below the ear nodedevelop rudimentary ears; one of these often produces an ear with a reduced grain set; in prolific strains grown in southern regions, several ears may develop.
Leaves of side branch forming husks
The plant habit varies with only 1 tiller shown here at the base
Primary and seminal roots supportive in the seedling stage
pericarp & silk (2N)
aleurone & endosperm (3N)
scutellum (2N)
shoot apex (2N)
root apex (2N)
The root system is mainly advantitious from the basal nodes

FIGURE 1 Morphology of sweet corn plant. (From Ref. 3.)

to the fact that the sugar molecules pack more tightly when dried. Varieties of this type are Silver Queen, Jubilee, and Gold Cup.

2. *Se* Hybrids

Dr. A. M. Rhode of the University of Illinois discovered the *se* gene in 1967. This specific gene enhanced the sweetness of the corn and extended the time sweet corn stayed tender and edible.

TABLE 3 Classification of Edible Sweet Corn Based on Genetic Model of Endosperm

I. Sugary (*su*) mutant
  A. "Standard" sweet corn
  B. Augmented sugary kernels; the sugars are increased by the action of other genes
    1. Partial modification
      a. Major effect genes; *bt*, bt_2, *sh*,sh_2, *se*
      b. Sugary enhancer (*se*) combined with one of the major effect genes
    2. 100% modification
      a. Major effect gene sugary enhancer (*se*)
      b. Minor effect genes; dull (*du*), floury (*fl*), floury 2 (fl_2), opaque (*o*), opaque-2 (o_2), sugary 2 (su_2), and waxy (*wx*)
II. Other mutants
  A. Single genes
    1. Shrunken types: *sh*, sh_2
    2. Brittle types: *bt*, bt_2
  B. Multiple genes
    1. Amylose-extender (*ae*), + dull (*du*) + waxy (*wx*)
III. Starchy corn selections, eaten immature

Source: Ref. 7.

This gene also incresaed the maltose content of the kernel, resulting in enhancement of taste. Varieties in this group include Tastee Treat, Tendertreat EH, and Snowbelle.

3. Sweet Gene Hybrids

This breeding process was developed by P. Bonucci of Sunseeds (Farmington, MN) and patented in 1986 (8). This process improved the sweetness of corn by 50% over normal sugary hybrids. The sh_2 gene resulted in much greater accumulation of sugar than the *su* gene. Therefore, the overall effect was the taste sensation of a 50% sweeter corn than normal sugary varieties. The high sugar content of the sh_2 kernels combined with water-soluble polysaccharide in the *su* kernels balanced flavor, texture, and sweetness in a desirable combination. Varieties of this type include Honey Comb and Sugarloaf.

4. Supersweet Hybrids

These hybrid varieties mainly depend on the sh_2 gene rather than the *su* gene for the sweetness. Kernels are sweeter at harvest and somewhat more watery due to the absence of water-soluble polysaccharide (WSP), and some appear to have a tighter pericarp than other types. They should be planted in warm, moist soil for best germination and seedling development. Examples of varieties within this category are Florida Staysweet, Wisconsin Natural Sweet, Sucro, How Sweet It Is, and Summer Sweet Series.

5. Improved Super Sweet Hybrids

The genetic combination of *su* and sh_2 genes resulted in a sweet corn that was 20% sweeter than the super sweets and almost 125% sweeter than the normal sugary (*su*) hybrids. The sweetie variety falls within this category.

Some of the most popular varieties of early-maturing types are Early Market, Surprise, Span Cross, and North Star. Carmel Cross and Whipple Early are medium-maturing types, and Golden Cross, Golden Security, Ioana, Iochef, Calumet, Narrow Grain Evergreen, and Long Island Beauty are late-maturing varieties of sweet corn grown in the United States (9). Varieties such as Tendergold, Golden Cross, Victory Golden, and Tendermost are most suitable for processing (10). In a field trial of 36 commercial hybrids of sweet corn, Dale et al. (11) found that Honey Comb and Azetec gave highest yields of marketable ears.

III. PRODUCTION

A. Soil and Climate

Warmth, moisture, and nutrient-rich soil are important factors for good sweet corn yields. It is not advisable to plant sweet corn until the danger of hard frost is over. However, the advantage of high prices for early sweet corn can be maximized by early planting and targeting of local markets. Temperature and soil moisture are the principal climatic factors for corn growth. In general, the higher the temperature (between 5 and 33°C), the greater the rate of growth and the shorter the time necessary for the plant to attain a particular stage of maturity. Sweet corn requires 70–110 frost-free days from planting to harvest. The crop is grown from 50°N to 40°S latitude from sea level to 3000 m in elevation. Required soil temperature for germination should be above 13°C (sweet corn will not germinate below 10°C). The optimum range of soil temperature for seed germination is 21–27°C (12).

1. Planting

Sweet corn is planted in rows, usually 90–100 cm apart and 2.5–5 cm deep. Spacing between the rows is maintained at 15–30 cm. Seeding rates vary from 11 to 14 kg/ha, depending on the seed size and plant population desired. Thinning is done when plants are 10–15 cm tall. Seed treatment with approved fungicides and insecticides may be required to protect seeds and seedlings from seedborne diseases (6).

Scheduling sweet corn planting for commercial processing and fresh market to ensure a continuous supply of high-quality sweet corn is very important. Wiley et al. (6) reported scheduling of sweet corn planting based on heat units required for optimum growth. Cultivars such as NK 199, Commander, Silver Queen, and Shoe Peg require about 1700, 1800, 1900, and 2000 heat units from planting to optimum harvest, respectively. Heat units are calculated as follows:

$$\text{Heat units} = \frac{\text{Maximum temperature} + \text{minimum temperature}}{2} - \text{BT}$$

where BT is baseline temperature. The baseline temperature of sweet corn is 10°C (50°F). At this temperature, the growth rate is zero. A linear growth is observed at elevated temperature up to 30°C (86°F). For example, if maximum temperature was 86°F and minimum temperature was 62°F, then the day's heat unit value would be 86°F + 62°F = 148°F ÷ 2 = 74°F – 50°F = 24°F, or 24 degrees/day. If the average temperature is below 10°C (50°F), zero is used instead of a negative number in calculating cumulative heat units (6).

2. Fertilization

Sweet corn can be grown in sandy loam to clay loam soil or in peat and muck soils. Soil pH should be in the range of 6.0–7.0, but the crop can be grown in soils with pH varying from 5.0 to 8.0. Sweet corn is moderately tolerant to salt and alkali. Depending upon soil fertility, the fertilization requirement will change. The typical fertilization requirement for corn is 100–112 kg of nitrogen per hectare for heavy soils and late planting. A heavier application of nitrogen

(115 kg/ha) is recommended for light soils and for early planting. Phosphorous requirement is 50–112 kg/ha, whereas 60 kg/ha of potash is recommended for other types of soils. Additionally, 27.4 MT/ha or 9–10 MT/ha (4–5 tons of poultry litter) of manure is recommended. Sweet corn response to nitrogen depends on the nitrogen source, time of application(s), and amount of rainfall. In regions of heavy precipitation (and if applied in nitrate form), most of the nitrogen is lost by leaching. Losses due to leaching may be reduced by application of nitrogen in ammonium form. Regardless of the source, it is better to apply part of the nitrogen fertilizer at the time of planting or preplanting, with the remainder as one or more applications during the growing season. During planting, fertilizer is applied in split bands 5 cm below and along both sides of the seed. Injury to seed or germinating plants can occur if the fertilizer is too close. Nitrogen and potassium are more likely to cause injury than phosphorus.

3. Irrigation

If rainfall is inadequate, irrigation is necessary for good yields. Sweet corn requires 15–25 acre inches of water throughout the cropping season. Water is critical to plant development and yield at the tasseling and silking stages. Either furrow or sprinkler irrigation with thorough watering is recomended (12).

B. Pests and Diseases

1. Pests

Many different kinds of insects can attack sweet corn under different conditions and cause serious injuries. Table 4 lists common sweet corn insects and their botanical names. Figure 2 shows representative sweet corn insects and damage caused by them (13).

Armyworm (*Pseudaletia unipuncta*) is a serious pest of corn in the early growth statges (< 20 cm tall), attacking and feeding on the leaves. In larger corn, however, armyworm will pass over the leaves but will destroy the center of the young stalk. Figure 2a shows a dark green armyworm 5 cm in length with white stripes on the sides and down the middle of the back. The armyworm can be found hiding under clods and stones or in the center of the leaves of the plant during the day. An effective way of controlling an outbreak or armyworm is to poison them by scattering a poison bran mixture in the field where they are feeding or across the line of march of the worms when they are leaving fields where food is scarce. Other suggested treatments for controlling armyworms are use of carbaryl (1.5 lb/acre), ethyl parathion (0.5 lb/acre), malathion 57 EC (1.25 lb/acre), and methyl parathion (0.5–0.75 lb/acre). Rates are based on active ingredients per acre.

Corn root worm larvae (*Diabrotica undecimpunctata*)–infested corn fields start growth in a normal way. Plants begin to show the effects of infestation when they are 20–50 cm tall. The plants grow very poorly or not at all and frequently die. Larger plants lodge after heavy rains. Close examination of the plant will reveal that the roots have been tunneled and eaten off by larvae (Fig. 2b). These larvae are about 1–2 cm in length with yellowish-white and somewhat wrinkled bodies and a brownish-white head (13). It is extremely difficult to control damage caused by the insects as the eggs are frequently laid in the fields after corn is well grown. The most effective cultural method is late planting on land that is cultivated frequently to minimize vegetation growth. Control should be applied when there are 18,000 beetles per acre (0.75 beetles per plant, based on a plant population of 24,000 per acre) and 10% of the females have mature eggs. Recommended levels of carbaryl (1.0 lb/acre), carbofuran (0.125–0.50 lb/acre), and methyl

TABLE 4 Insects of Corn

1. Insects chewing leaves or stalks aboveground	
Grasshopper	*Melanoplus* spp.
Corn flea beetle	*Chaetocnema pulicaraia Melsheimer*
Billbugs	*Sphenophorus* spp.
White-fringed beetle	*Graphognathus* spp.
Armyworm	*Pseudaletia unipuncta* (Haworth)
Fall armyworm	*Spodoptera frugiperda* (J. E. Smith)
Cutworms (black)	*Agrotis ipsilon* (Hufnagle)
Webworm	
2. Small insects sucking sap from leaves or stems not leaving visible holes, but causing wilting, spotting, discoloration, or death of plant	
Corn leaf aphid	*Rhopalosphum maidis* (Fitch)
Chinch bug	*Blissus leucopterus leucopterus* (Say)
3. Insects boring and tunneling in the stalks or stems	
Maize billbug or curlew bug	*Spenophorus maidis* Chittenden
European corn borer	*Ostrinia nubilalis* (Hubner)
Stalk borer	*Papaipema nebris* (Guenee)
Southern cornstalk borer	*Diatraea crambidoides* (Grote)
Southwestern corn borer	*Diatraea grandiosella* (Dyar)
Lesser cornstalk borer	*Elasmopalpus lignoselius* (Zeller)
4. Insects attacking the ear	
Japanese beetle	*Popillia japonica newman*
Corn ear worm	*Heliothis zea* (Boddie)
European corn borer	*Ostrinia nubilalis* (Hubner)
5. Insects attacking roots or underground stems	
Corn root aphid	*Anuraphis maidiradiois* (Forbes)
Grubs	*Phyllophaga* spp.
Wireworms	*Elateridae* spp.
Northern corn rootworm	*Diabrotica longicornis barberi* (Smith and Lawrence)
Southern corn rootworm	*Diabrotica undecimpunctata howardi* (Barber)
White-fringed beetle	*Graphognathurs* spp.
Cutworms (black)	*Agrotis ipsilon* (Hufnagle)
Cornfield ant	*Lasius alienus* (Foerster)
Grape colaspis	*Colaspis brunnea* (Fabricius)
6. Insects attacking planted seed, eating into or devouring the plantlet as it germinates, so that plants often fail to come up	
Pale-striped flea beetle	*Systena blanda* Melsheimer
Wireworms	*Elateridae* spp.
Seedcorn beetle	*Stenolophus lecontei* (Chaudoir)
Thief ant	*Solenopsis molesta* (Say)
Fire ant	*Solenopsis germinata* (Fabricius)
Seedcorn maggot	*Della platura* (Meigen)

Source: Ref. 13.

parathion (0.25–0.50 lb/acre) and parathion (ethyl or methyl; 0.25 lb/acre) as an active ingredient can be applied.

Cutworms are serious pests of sweet corn and have several different species. Often their injury to the plant is devastating and results in replanting of the corn. Injury can destroy 5–50% of the total crop. Cutworms cause injury in four different ways (13). Black cutworm (*Agrotis*

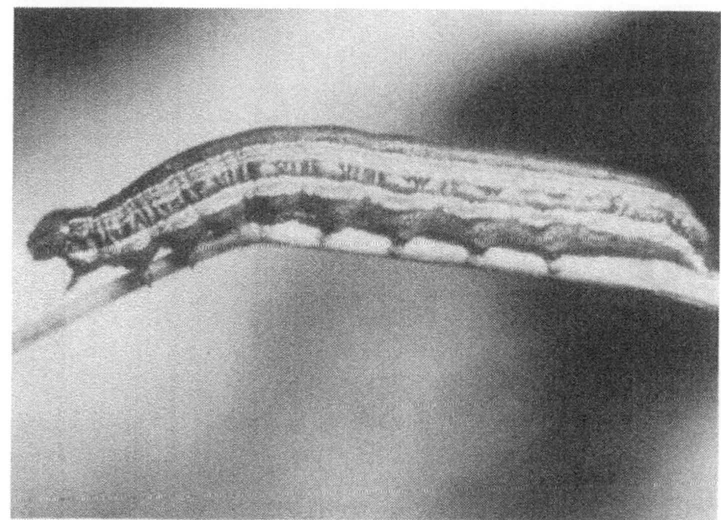

(a)

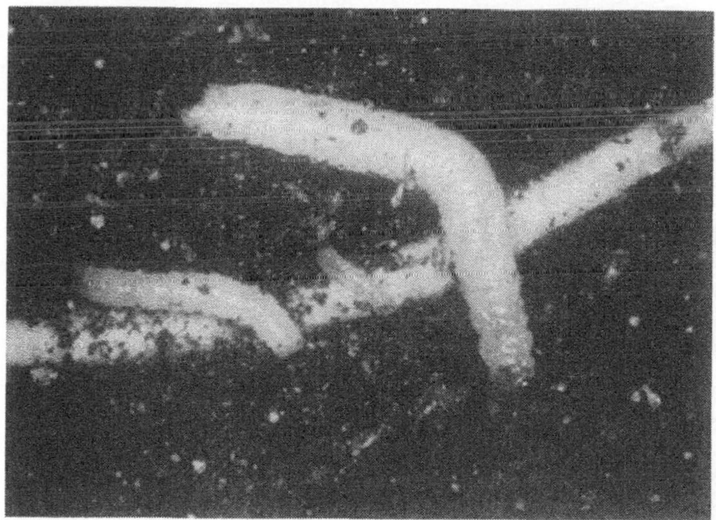

(b)

FIGURE 2 Representative insects of sweet corn: (a) armyworm; (b) corn rootworm larvae; (c) cutworm; and (d) European corn borer. (Courtesy of Department of Entymology, University of Nebraska, Lincoln.)

ipsilon) is a cosmopolitan species with a restless, pernicious habit of cutting off many plants while satisfying its appetite (Fig. 2c). Black cutworms lay eggs singly or a few together on the leaves or stems of plants frequently located in low or flooded land. The winter is spent in the larval and pupal stages. Because of the growing seasons, there are two generations per year in Canada and four in Tennessee. The larvae color is greasy gray to brown with faint larger stripes. The skin is strongly convexed and rounded, with isolated granules of large and small size (13).

(c)

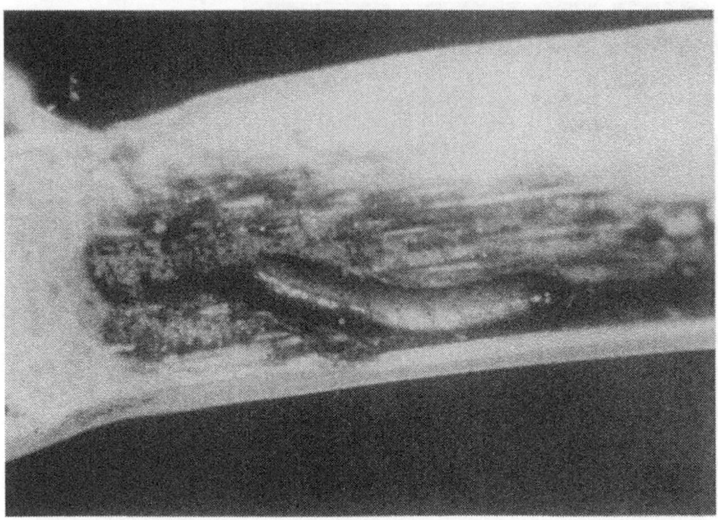

(d)
FIGURE 2 Continued

Cutworms are best controlled by rescue treatments applied after plants are up and early damage signs are detected. Preventive treatment applied at or before planting have generally given erratic control, especially where cutworm numbers have been high. For controlling soil cutworms in corn chlorpyrifos (1.0–1.5 lb/acre), esfenvalerate (0.025–0.05 lb/acre), fenvalerate (0.1–0.2 lb/acre), and permethrin (0.1–0.2 lb/acre) as active ingredient are recommended.

European corn borer infestation is evidenced by broken tassels or bent-over plants. Other indicators of an infestation are small areas of surface feeding on the leaf blades with fine sawdustlike casting on the upper side of leaves and stalks and small holes in the stalks. The larvae of the European corn borer are greatly enlarged. On the back of each segment it bears small spots. The distance between spots is usually less than the width of one spot. A faint stripe can usually be seen on the mid-dorsal line of the European corn borer (Fig. 2d). Corn borer can be effectively controlled by utilizing all crop residue in which corn borer can overwinter. Borers that have left the whorl and entered the stalk cannot be controlled. If most have left the whorl, it is too late to attempt control. Suggested treatment for controlling first-generation European corn borer in sweet corn is Pounce 1.5G at the rate of 6.7–13.3 lb/acre. Do not apply more than 0.4 lb active ingredient after the brown silk stage.

2. Diseases

Sweet corn is infected by corn smut caused by *Ustilage zeae*. The disease is apparent as black puffy masses on ears, tassels, and other parts (14). Early removal and destruction of the diseased parts checks further dissemination of the spores of the pathogen. Early, open-pollinated yellow varieties of sweet corn are particularly susceptible to the disease.

Sweet corn is also affected by a disease called Stewart's Wilt, caused by *Erwinia stewarti*. The disease causes severe losses of sweet corn in the warmer regions, resulting in complete failure of very susceptible varieties. Root rot of sweet corn is caused by a number of organisms. Table 5 shows diseases of sweet corn and their scientific names (15).

Bacterial leaf stripe (*Pseudomonas andropogonis*) usually affects leaves below the ear. Lesions are narrow, linear, and buff colored and extend up the leaf blade between the heavier veins (Fig. 3a). The disease is not economically important and has been severe on only a few susceptible inbred lines and hybrids. Use of resistant hybrids and varieties is recommended.

Crazy top (*Sclerophthora macrospora*) infection results in excessive tillering, rolling, and twisting of upper leaves and is sometimes coupled with proliferation of the tassel and/or ears into a mass of leafy structures. No pollen is produced since normal flower parts in the tassel are completely deformed (Fig. 3b). Control measures such as providing adequate soil drainage and control of grassy weeds are helpful.

Corn lethal necrosis results from the synergistic infections by two viruses: maize chloritic mottle virus (MCMV) and maize dwarf mosaic virus (MDMV). Infected upper leaves exhibit bright yellow mottling or mosaic patterns, later turning dull yellow or buff colored (Fig. 3c). Death of leaves occurs from the top of the plant downward. Ears are reduced in size, with small or aborted kernels. The disease is predominantly distributed in Kansas and Nebraska. It can be controlled by growing a nonhost crop for one year, such as soybean, sorghum, or a small-grain crops, in fields where corn lethal necrosis has occurred. Some varieties and hybrids are resistant to this virus.

Sting nematodes (*Belonolaimus nortoni*) damage young corn plants. Severely colonized seedlings may die, leaving gaps in the row. Surviving seedlings are chloritic and stunted, resulting in less thrifty plants that yield poorly (Fig. 3d). Below ground symptoms are deep necrotic lesions on roots. Root tips are frequently destroyed, resulting in thick, stubby roots.

C. Maturity, Harvesting, and Handling

Harvesting of sweet corn is based on its peak maturity. It is helpful to make a distinction between physiological and horticultural maturity in the case of sweet corn. Physiological maturity is the development stage during which a plant part will still continue ontogeny if detached. On the other hand, horticultural maturity relates to the development stage during which the plant product meets the quality attributes required for use by consumers. Obviously, optimum maturity of sweet corn expressed by horticultural maturity and a suitable maturity index is based mainly on physicochemical changes that occur in the kernels.

TABLE 5 Sweet Corn Diseases

1. Bacterial diseases	
Bacterial leaf blight	*Pseudomonas alboprecipitans*
Bacterial stripe	*Pseudomonas andropogonis* (E. F. Smith)
Bacterial stalk rot	*Erwinia chrysanthemi*
Chocolate spot	*Pseudomonas syringae* (Elliot)
Goss's bacterial wilt and blight	*Clavibacter michiganense*
Holcus spot	*Pseudomonas syringae* (Kendrick)
Stewart's wilt	*Erwinia stewartii* (Smith)
2. Fungal diseases	
Alternaria leaf blight	*Altemaria alternata*
Anthracnose leaf blight/stalk rot	*Collectotrichum graminicola* (ces)
Ascochyta leaf and sheath spot	*Ascochyta zeae*
Aspergillus ear rot	*Aspergillus* spp.
Banded leaf and sheath spot	*Rhizoctonia microsclerotia*
Northern leaf blight	*Helminthosporium turcicum*
Southern leaf blight	*Helminthosporium maydis* (Nisikado and Miyake)
Rust	*Puccinia sorghi (*Schwienitz*)*
Seedling diseases	*Pythium, Rhizoctonia* spp.
Black kernel rot	*Botryodiplodia theobromae*
Brown stripe downy mildew	*Sclerophthora rayssiae*
Charcoal rot	*Macrophomina phaseolina*
Common smut	*Ustilago maydis*
Crazy top	*Sclerophthora macrospora (Sacc.)*
Ergot	*Claviceps gigantea*
False smut	*Ustilaginoidea virens*
Fusarium stalk rots	*Fusarium* spp.
3. Nematodes	
String nematodes	*Belonolaimus* spp.
Stubby root nematodes	*Trichodorus* spp.
Root knot nematodes	*Meliodogyne* spp.
Lance nematodes	*Hoplolaimus* spp.
Spiral nematodes	*Heliocotylenchus* spp.
Root lesion or meadow nematodes	*Pratylenchua* spp.
Awl nematodes	*Dolichodorus* spp.
Com cyst nematodes	*Heterodera zeae*
Ring nematodes	*Criconemoides* spp.
4.Viruses	
American wheat striate mosaic	American wheat striate mosaic virus
Barley stripe mosaic	Barley stripe mosaic virus
Brome mosaic	Brome mosic virus
Maize chlorotic mottle	Maize chlorotic mottle virus
High plains disease	High plains virus
Cucumber mosaic	Cucumber mosaic virus
Maize chlorotic swarf	Maize chlorotic dwarf virus
Maize dwarf mosaic	Maize dwarf mosaic virus

Source: Ref. 15.

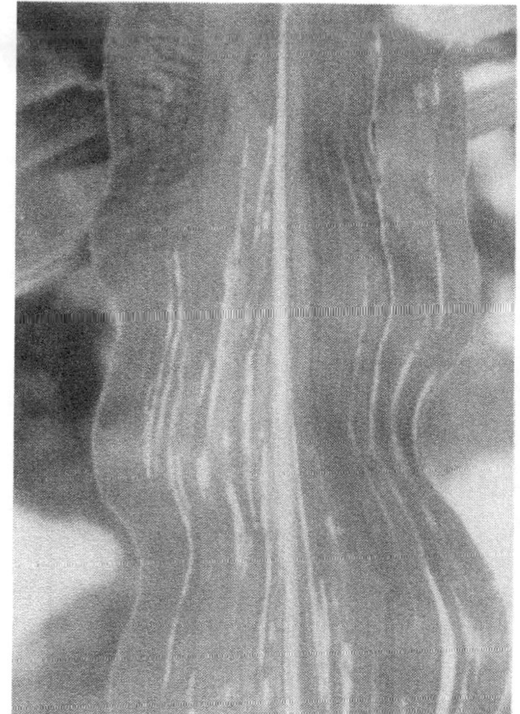

(a)

(b)

Figure 3 Representative diseases of sweet corn: (a) bacterial stripe; (b) crazy top; (c) corn lethal necrosis: (d) sting nematodes. (Courtesy of Department of Plant Pathology, University of Nebraska, Lincoln.)

(c)

(d)
FIGURE 3 Continued

The harvestable stage of kernels can be determined on the basis of external appearance of silk, ears, and husks. The kernels at a harvestable stage yield a milky juice when squeezed and are generally larger and sweeter compared to the premilk stage. After the milk stage, the kernel quality declines due to a rapid conversion of sugar to starch (16).

1. Maturity

Sweet corn should be harvested at the milk stage, considered to be the best edible stage. It is difficult for an inexperienced person to determine sweet corn maturity without pulling down the

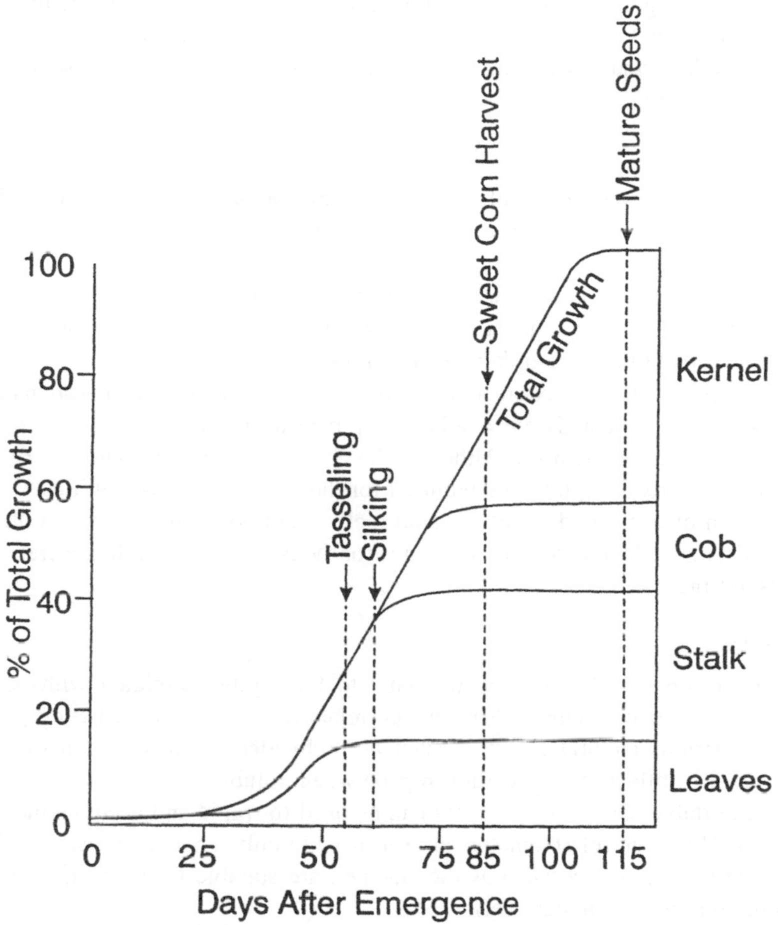

FIGURE 4 Stages of growth in sweet corn. (From Ref. 12.)

husk and examining the kernel. Figure 4 shows stages of corn growth maturity (12). Ware and McCollum (17) reported that premilk, milk, and dough stages comprise the three distinct phases in the maturation of corn kernels. The premilk stage is characterized by a sweet taste; lean, immature, and small kernels; and clear and watery juice. In the milk stage, the taste also is sweet but the kernels are bigger in size and the juice is milky in appearance. The highest-quality ears are harvested in the milk stage with filled tips (18). The dough stage is marked by rapid conversion of sugars into starch. The quality of the ear at this stage is poor and consequently the market value is low. The sugar:starch ratio is employed to determine the extent of conversion of sugars into starch—a critical quality measure. Browning of the silk also indicates that the ears are ready for harvest (17,19). Work and Carew (20) also have stressed timely harvest as a major element in delivering high-quality sweet corn to processors or consumers. Huelsen and Michaels (21) studied many factors that govern harvest dates in order to resolve the conflict between growers' desire for heavy yield and canners' desire for quality.

2. Maturity Tests

Objective tests of sweet corn maturity and quality have been of great importance to the processing industry. A number of objective tests have been developed to determine appropriate stage of maturity for predicting optimum sweet corn maturity. The selection of reliable objective tests has been problematic due to the influence of season and cultivar variables. However, U.S. sweet corn packers, breeders, and researchers have been successful in using selected objective tests, which are discussed here.

Percent Moisture

The increase in dry matter of sweet corn with growth and development can be determined and expressed in terms of percent moisture. The moisture content of kernels is related to juiciness. Therefore, moisture content is considered to be the most appropriate objective test of sweet corn maturity and quality. Also, it serves as an index of maturity for the determination of harvestable stage of sweet corn for processing. It is generally believed that the quality of a canned product depends largely on the percent moisture of kernels at harvest.

Also, there is a decline in flavor, texture, color, and endosperm quality with a drop in kernel moisture below 75%. Several methods are available for measuring kernel moisture content. Geise et al. (22) compared three methods. Although the Steinlite instrument and the Brown-Duvall methods had lower precision, the time required for these methods was significantly less than that for the vacuum oven method. Another rapid method employed by Becwar et al. (23) involved microwave drying and required only 3 minutes for the determinations. In general, rapid and simple methods are preferred.

Percent Soluble Solids

Objective measurements on soluble solids were found to be highly correlated with percent moisture determined by the oven method (24). This correlation is useful in predicting percent moisture. Apparently various cultural practices such as herbicides, fertilizers, cultivars, and irrigation methods do not influence the relationship between soluble solids and moisture of kernels. Inexpensive, portable hand refractometers can be used for rapid and accurate measurement of soluble solids. However, refractometers are sensitive to only a narrow range of soluble solids. Expanded scale refractometers, such as the Gaetner, are suitable for measuring soluble solids from very immature to very mature corn (25).

Percent Alcohol-Insoluble Solids (AIS)

The components of the kernels that are insoluble in hot 80% alcohol are measured by this test. Kernel components include starch, hemicellulose, pectin, and cellulose. The AIS test has been applied quite extensively to raw and processed sweet corn. It is intended for corn high in starch content or corn that transforms soluble carbohydrates into less soluble compounds during maturation and/or storage. The test is relatively rapid because the alcohol-insoluble residue is dried under infrared heat. The AIS test is not satisfactory for high-sugar genotypes because of lower levels of AIS (26). The change in AIS is 2.25% for every 5% change in moisture (27).

Percent Water-Soluble Polysaccharide (WSP)

The extract of kernels with 80% alcohol contains soluble sugars. The remaining traces of sugars are removed from the alcohol-insoluble residue by washing twice with 80% alcohol. The resultant residue contains WSP, which can be solubilized by refluxing in water for 30 minutes and filtering. This fraction contains both starch and phytoglycogen. WSP ranges between 12 and 18%, corresponding to a moisture content range of 74–70% (28). Percent WSP, like percent AIS, has a drawback in evaluating kernels that are high in sugar and low in polysaccharide.

Pericarp

The pericarp test involves the following steps (29). One hundred grams of kernels are blended with about 200 ml of water for 5 minutes, the macerate is washed through a preweighed 30-mesh screen, dried for 2 hours at 100°C, and weighed. This test is simple and relatively fast, but it lacks quality consideration of the endosperm. It gives a good estimate of the chewiness of the pericarp wherever it is applicable.

Trimetric Test

This test was developed in 1952 and appears to be the most significant objective test for sweet corn quality (30). The grade of raw sweet corn was predicted from moisture, pericarp, and kernel size measurements. The USDA canned corn grade was based on a series of multiple regression equations. The percent pericarp, percent AIS, and kernel diameter of 20 kernels was useful for predicting the grade of canned and frozen whole kernel corn (29). The monographic results correlated very well with the USDA grades. However, this approach to evaluation lacks commercial acceptance because of the length of time required to complete the tests. In short, this technique appears to be superior to any single test for determining sweet corn maturity and/or quality.

3. Harvesting

Sweet corn for processing is picked at different stages of maturity depending on the way it is to be processed. Corn for freezing is picked at about the same stage as corn for fresh market, while corn for whole kernel pack is picked at a slightly later stage of maturity (9). Work and Carew (20) stated that the thumbnail test, time of silking, and the degree-day system are used as maturity indices to judge the correct harvesting maturity of sweet corn for fresh market and processing. Ratios of sugars to starch and alcohol-soluble to alcohol-insoluble material are useful chemical indices of maturity. Derbyshire et al. (20) established a relationship between nonstructural carbohydrate content and the maturity of the sweet corn for processing. During the period when sweet corn ears are suitable for harvesting and grain moisture content decreases, water-soluble polysaccharide concentration increases rapidly but soluble sugar concentration declines only slightly. They suggested that decreasing sweetness as the grain matures is due to increasing water-soluble polysaccharide concentration making the sweetness attributable to a relatively constant sugar concentration. The rapid increase in water-soluble polysaccharides makes their measurement a sensitive indicator of small changes in maturity. Sweet corn often becomes heated after it is detached from the plant, especially when loaded in trucks for transportation to the cannery. The load may stand the whole night, and under such conditions the quality deteriorates rapidly. Thompson and Kelly (9) recommended prompt handling and cooling of sweet corn to maintain its high quality.

Hydrocooling of the ears (placing the ears in circulating cold water immediately after picking) is very advantageous. Hydrocooling sweet corn for 13 minutes at 5.5°C reduces the temperature at the center of the cob from 18°C to 11°C (31). On the other hand, corn stored in wire baskets in a 4°C refrigerator requires 5 hours to reach a temperature of 11°C in the center of the cob. Alben and Scott (32) reported that placing crushed ice in field containers at harvest maintained a high level of sugar content in the sweet corn. After the iced-down ears were placed in a 4°C refrigerator for 24 hours, the corn iced in the field was found to contain nearly twice as much sugar as corn that was not iced. In the United States, corn is marketed in various types of crates and bags. The two most popular containers are flat wire-bound crates holding 5 dozen ears and wet-strength paper bags, which hold 50 ears of sweet corn and 5–10 kg of crushed ice. Sweet corn ears may be husked and packed in trays overwrapped with transparent films or in transparent film bags (9). The husks of sweet corn make an excellent protective package for the kernels, but this package keeps the corn in good condition only at temperatures below 4°C and when ample water adheres to the husks (33).

Sweet corn grown for processing is harvested by machines, which pull the ear from the plant and convey it to a truck or wagon. The sweet corn is then transported to the processors. Wiley et al. (6) described the operation of sweet corn harvesters for fresh market and processing industries as follows.

Fresh Market Harvesters

Principal components of the machine include gathering unit, picker head, duel cleaning fans, rear elevator with optional hinged discharge section, and an optional hydraulic trailer hitch. The functions are hydraulically driven by two pumps. A slip clutch protects the head from damage in the event it is jammed by a foreign object. A hydraulically lifted picker head and an independently operated hydraulic rear discharge elevator permit a minimum turnaround time within a 6-m radius. The elevator discharge is to the rear of the machine, making it a one person operation with a trailing wagon. The discharge elevator will be furnished with a normal clearance height of 2 m. The elevator can be extended if a higher discharge is required. An optional 1-m extension is available for the discharge elevator, providing for more even corn distribution on longer wagons. Two gathering units with moving gathering chains gently guide the stalks with attached ears into the picker head. The corn is then grasped between stop bars, which support the ears, while two, sharp four-blade knives provide for adjusting shank below the butt of the ear. A two-positioned cutting height on the knives allows for adjusting the shank length as detailed by growers. Adjustable moving stop bars compensate for large and small ear varieties. Most stalks are discharged from the bottom of the head when the ears are severed. Weak or brittle stalks, which break off above the knives, are fed to the stalk ejector by the gathering chain above the stop bars, and the stop bars convey the ear back to the discharge elevator. As the ears drop into the discharge elevator, they fall through an air stream, which removes any trash.

Sweet Corn Harvester for Processing

Figure 5 shows a commercial sweet corn harvester. The functional elements of the machine include gathering units, gathering chains, knives, flighted first elevator conveyor, stalk ejectors, split-flow cleaning fan, and discharge elevator. The gathering units are sloped gently upward to lift and guide the stalk into the gathering chain. There are two gathering chains per picker head, which grasp the stalk and pull it rearward into the knives. The rotating knives have four sharp replaceable edges and are supported by a cantilever shaft. As the stalks are pulled downward by the knives, they pass between two stripper bars, which strip or husk the ear from the stalk. The ears are then delivered by the gathering chain lugs into the flighted first elevator conveyor. This conveyor moves the product upward and rearward to the discharge elevator hopper. The cleaning system, which consists of stalk ejectors and a cleaning fan, is located at the point where the product transfers from the first elevator to the discharge elevator. The stalk ejector rollers grasp the loose stalks and eject them out the rear of the machine. The cleaning fan blows loose leaves and trash from the product. From the upper end of the first elevator, the corn falls into a flared hopper, which diverts the corn toward the center of the machine and into the discharge elevator. The corn is conveyed up and into the trailer, which is pulled behind the corn harvester.

FIGURE 5 Sweet corn harvester.

Sweet corn processors are exploring a mobile processor developed by Byron Enterprises Inc., of Byron, New York.The field cutter, a truck-mounted sweet corn harvester, was developed as a result of joint effort between Byron Enterprises and Del Monte Foods of San Francisco, California. The field cutter picks the ears of corn from the stalk, husks them, orients them, and cuts the kernels from the cob. The cut corn is transferred to special stainless steel dump carts. Each dump cart services three to four field cutters. When the dump cars is full, the corn is dumped into conventional "dump" trucks equipped with a special food-grade molded box liner for transport to the plant. All remaining processing steps are performed at the plant. Waste materials are returned to the fields, where they are recycled as mulch.

4. Grading

Raw sweet corn for processing is classified by U.S. standard grades as U.S. No. 1, U.S. No. 2, and culls. These grades include ears of sweet corn with similar color characteristics and freedom from various types of damage. The maturity indices of each ear in the respective grades fall between the blister stage and the hard stage of maturity. The maturity groups A-1, A-2, A-3, A-B, BC, and C relate to blister, milk, cream, dough, and the hard stages of development, respectively. Obviously, the most mature grades are not acceptable for processing. These maturity grades are highly subjective and require considerable skill and experience to make the proper judgment. U.S. No. 1 and 2 grades also specify the amount of cob surface that must be undamaged. There must be at least 10 cm and 7.6 cm, respectively, of undamaged cob surface for U.S. No. 1 and 2 grades. Other factors that may be evaluated from a sensory viewpoint are appearance of the ear, ear diameter, kernel size, and color.

The positive and negative sensory quality factors of sweet corn include corn flavor (aroma) strength, sweetness, cobby flavor, and off-flavor. Proper kernel depth, tenderness, and juiciness are requirements for the product for fresh market or processing. The most important quality attributes of both market and processed sweet corn are listed in Table 6. The market quality calls for a large well-filled ear with plump kernels wrapped in a tight, dark green husk. It is expected

TABLE 6 Sensory and Related Factors to Be Considered in the Quality and/or Maturity of Sweet Corn

Texture-related factors		Flavor-related factors		Appearance-related factors	
Market	Processed	Market	Processed	Market	Processed
2.6	2.1 Tenderness	2.2	3.0 Sweetness	4.8	3.6 Kernel size
3.5	3.1 Pericarp	3.9	3.3 Flavor	4.9	3.8 Color of kernal
3.6	3.8 Texture	5.1	5.4 Mouthfeel	5.2	Color of husk
			5.7 Aroma	5.3	Size of ear
				6.1	Insect damage
				6.2	Shank length
				6.0	Size and color flag leaves

Rank Factor 1 is highest.
Source: Ref. 6.

that the kernel quality should last for a long period, impart a sweet taste, have tender pericarp, and have a cornlike aroma. Processed corn requires corn with a sweet fresh flavor (not sugarlike), bright, deep yellow color, tender but not starchy pericarp, and a good kernel shape with no off-flavor. As shown in Table 6, texture and sweetness have almost equal weight in market corn. On the other hand, sweetness is the most important factor in processed sweet corn (34).

IV. STRUCTURE AND COMPOSITION

Sweet corn includes edible kernels, husk, and cob. The former is the most important to the consumers and processors. The husks are used as a by-product for animals and also is used in some Mexican entrees for wrapping foods. The cob is important from a sensory point of view, and it must also be reliable for machine handling and processing. Chemical composition of the husk and the cob are not important from either processing or fresh market standpoints. However, the complex polysaccharide contents in the cobs are used in value-added processing for production of chemicals.

A. Kernel Structure

A schematic diagram of a longitudinal section of a sweet corn kernel is shown in Figure 6. Wolf et al. (35–38) extensively described the structure of a kernel. It has a flattened, wedge shape and is broader at the apex end than at the point of attachment to the cob. The structure is greatly affected by gene modification and maturity. Figure 7 shows the carbohydrate accumulation in sweet corn after silking (12). The immature structure of sweet corn is composed of pedicel, pericarp (epicarp and mesocarp), germ (coleoptile, plumule, and primary root), and endosperm (Fig. 6). The pedicel is a remnant of the tissue connecting the kernel to the cob and is not an important consideration in fresh market and processed sweet corn unless some of the tough cobby tissue stays connected to the kernel during cutting or is encountered during mastication by the teeth. The pericarp is formed of one-cell epicarp layers of pitted mesocarp cells, one or two layers of cross cells, and one or more layers of tube cells adjacent to the seed coat (39). As the kernel matures, the pericarp becomes more and more resistance to puncture (39,40). The germ comprises only 5% of the weight of the whole kernel (41). The coleoptile and primary root

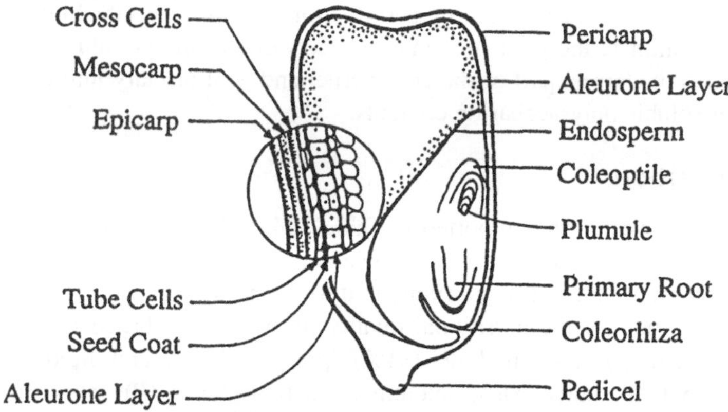

FIGURE 6 Schematic diagram of longitudinal section of a corn kernel. (From Ref. 29.)

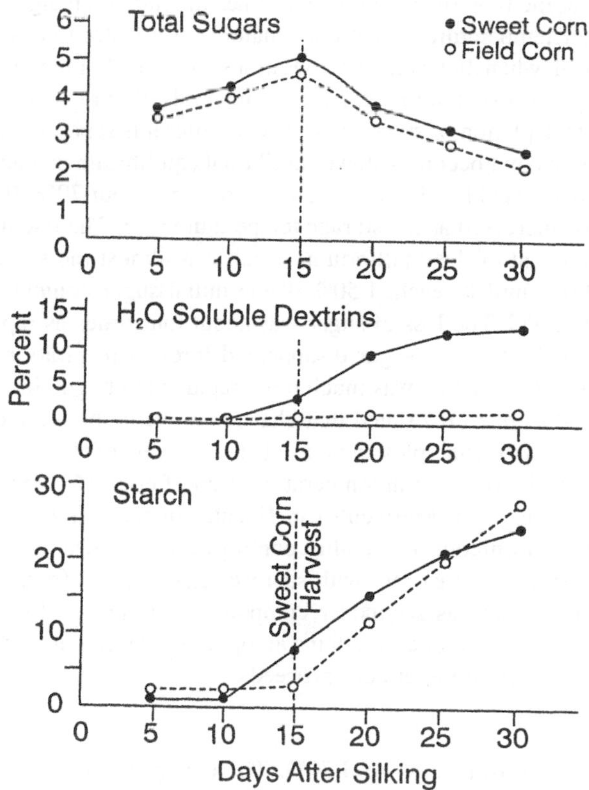

FIGURE 7 Carbohydrate accumulation in sweet corn after silking. (From Ref. 12.)

are important parts of the germ and are active during germination. The germ has the highest lipid content. Processors try to eliminate the germ in some products due to the possibility of lipid oxidation. The endosperm is the largest part of the corn kernel and is of primary interest due to sugar, starches, and water-soluble polysaccharide contents.

B. Chemical Composition

The chemical composition of sweet corn is reported in Table 7 (42). All reported values are on a fresh weight basis since most corn products are consumed in a fully hydrated condition. Composition is affected by environmental conditions, maturity, and storage (16). Data in the *USDA Handbook 1984* (43) indicate the nutritional quality of sweet corn based on moisture (72.7%) and total solids (22.3%). Carbohydrate (81%), protein (13%), and lipids (3.5%) comprise major portions of total solids. Also, calcium (Ca), phosphorus (P), iron (Fe), and potassium (K) contents are reported on fresh weight bases. In the following sections, composition is described in more detail.

1. Carbohydrates

Starch is the predominant carbohydrate component of corn. As sweet corn matures, there is a decline in the starch content (44). Sweet corn is at its highest edible quality in the milk stage. Sugar content decreases and the starch content increases as sweet corn passes from this stage to the dough stage (9). The temperature at the time of ripening determines the rate of change from the milk to the dough stage. At higher temperature, sweet corn matures at a faster rate. Close attention to harvest maturity is critical when hot weather conditions prevail. Appleman and Arthur (45) analyzed Stowell Evergreen sweet corn at 24-hour intervals during storage at different temperatures. They found that depletion of sugar in sweet corn after it is separated from the stalk does not proceed at a uniform rate but becomes slower until final equilibrium is reached. This took place when the total sugar decreased by about 62% and sucrose by about 70%. It was demonstrated that the rate of sugar loss increased as the storage temperature rose. The rise in the temperature simply hastened the attainment of the equilibrium, which was almost the same for all temperatures. The rate of sugar loss, until it reached 50% of the initial sugar content, was doubled for every 10°C increase up to 30°C. The loss of sugar was about four times as rapid at 10°C as at 0°C. At 0°C, more than 20% of initial sugar disappeared through respiration and conversion to starch in 4 days. The rate of sugar loss was much more rapid at the beginning than at the later stages of storage. Work and Carew (20) stated that changes in the carbohydrates of sweet corn during ripening follow vant Hoff's principles, which state that the rate of the chemical reactions is approximately double for each 10°C rise in temperature. Rumpf et al. (46) reported gas chromatic analysis of soluble substances of the Merit cultivar. The quantitative determination of soluble substances provide specific information for quality control of sweet corn. Figure 8 shows a chromatographic analysis of sweet corn extract with sucrose, glucose, and fructose as major components and sorbitol and malic acid as secondary components. It appears from the figure that the sucrose content increases until sweet corn attains an optimum degree of maturity and reducing sugars decrease as the maturity of sweet corn proceeds.

2. Proteins

The protein content of sweet corn ranges from 2.86 to 3.70%. The concentration of protein decreases from the outer region toward the center of the kernel (47). The sweet corn proteins are deficient in lysine and tryptophan (Table 7). Patterns of change in free amino acids, total proteins, and protein fractions during maturation of five sweet corn cultivars (Jubilee, NK 51036, Bonanza, Triumphant-2, and Yukon) were reported by Pukrushpan (48). The pattern of grain

TABLE 7 Composition of Sweet Corn

Composition		Protein	Fat	Carbohydrates
Energy value (average)	kJ	62.44	47.86	337.40
per 100 g edible portion	kcal	14.92	11.44	80.64
Amount of digestible constituents per 100 g	g	1.96	1.10	18.43
Energy value (average)	kJ	37.47	43.07	323.90
of the digestible fraction per 100 g edible portion	kcal	8.95	10.30	77.41

Constituents	Units	Average value/ 100 g edible portion
Water	g	74.700
Protein	g	3.280
Fat	g	1.230
Carbohydrates	g	19.200
Crude fiber	g	0.800
Minerals	g	0.800
Sodium	mg	0.300
Potassium	mg	300.000
Magnesium	mg	48.000
Calcium	mg	5.800
Manganese	mg	0.200
Iron	mg	0.550
Copper	mg	0.060
Phosphorus	mg	114.000
Iodine	mg	3.300
Boron	mg	0.070
Vitamin E	mg	0.640
Vitamin B_1	mg	0.150
Vitamin B_2	mg	0.120
Nicotinamide	mg	1.700
Pantothenic acid	mg	0.890
Vitamin B_6	mg	0.220
Folic acid	mg	0.043
Vitamin C	mg	12.000
Arginine	mg	160.00
Histidine	mg	85.00
Isoleuecine	mg	130.00
Leucine	mg	350.00
Lysine	mg	130.00
Methionine	mg	56.00
Phenylalanine	mg	200.00
Theronine	mg	130.00
Tryptophan	mg	16.00
Valine	mg	220.00
Malic acid	mg	29.000
Citric acid	mg	21.000
Quinic acid	mg	3.500
Succinic acid	mg	7.700
Pyrrolidone carb. acid	mg	6.400

TABLE 7 Continued

Constituents	Unit	Average value/ 100 g edible portion
Glucose	g	0.340
Fructose	g	0.310
Sucrose	g	3.030
Total Dietary Fiber	g	3.700

Source: Ref. 42.

composition was similar for all cultivars. The total protein, free amino acids, and water-soluble and -insoluble carbohydrates decreased, alcohol-soluble prolamine (zein) continuously increased, and alkali-soluble protein increased up to the second stage of maturity (moisture content 75–79%) and then gradually decreased. Available lysine remained fairly constant throughout the four stages of maturity and development as designated by the moisture content ranging from more than 80 to 64%. The quantities of each component varied among cultivars at the various stages of maturity (48).

3. Lipids

Compared to other cereals, sweet corn is relatively high in oil content. Corn oil is predominantly rich in linolenic (50%) and oleic (30%) fatty acids. It contains a higher proportion of sitosterol

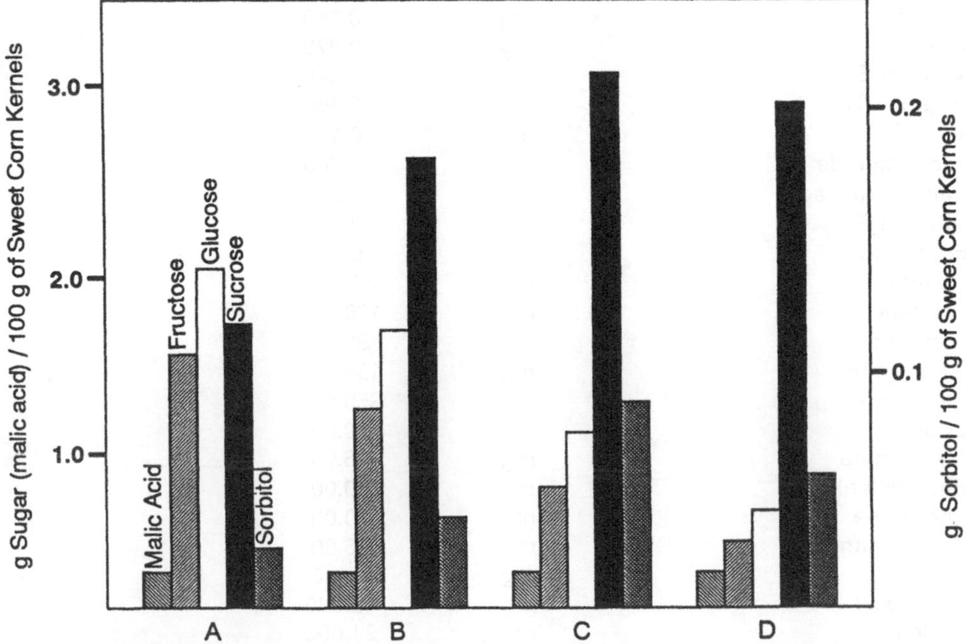

FIGURE 8 Maturity changes in sugars in the Merit cultivar of sweet corn over various maturity levels. (From Ref. 46.)

(47). Rancidity is a major problem in processed sweet corn. Due to high oil content, flavors and colors develop in processed corn. Corn on the cob also is susceptible to lipid rancidity. Pascual and Wiley (49) reported that corn oil is a good source of polyunsaturated fatty acids (PUFA). Oil content ranged from 5.2 to 18.4% among the 12 mutant genotypes they investigated (50). The major fatty acids in seven mutant maize genotypes with sweet corn backgrounds and commercial corn cultivars were palmitic (16:0), stearic (18:0), oleic (18:1), linoleic (18:2), and linolenic (18:3). The kernels contained 18.4, 1.4, 21.5, 57.1, and 2.1%, respectively, of these fatty acids. Small differences exist between mutant genotypes and commercial samples (51).

4. Vitamins and Minerals

With the current emphasis on nutritional labeling, research has focused on the adequacy of available data on the nutrient content of the foods. Sweet corn contains thiamine, riboflavin, niacin, vitamin B6, and vitamin A as major vitamins. Farrow et al. (52) reported the nutrient content of whole kernel corn. Lee et al. (53) reported carotenoid composition in four cultivars and six lots of commercially processed sweet corn. Seven carotenes and two monohydroxy carotenoids were separated (Table 8). The major pigments were zeinoxanthin and β-zeacarotene. Unlike many other vegetables, sweet corn contained monohydroxy carotenoids as a major fraction, followed by hydrocarbon and polyhydroxy carotenoids. Hydrocarbon carotenoids mainly consisted of α-carotene, β-carotene, γ-carotene, β-zeacarotene, and small amounts of ζ-carotenes, neurosporene, and lycopene. β-Zeacarotene contents were greatest in losweet, 70-2599, and Stylepak. Monohydroxy carotenoid mainly consisted of zeinoxanthin and β-cryptoxanthin. Petzoid and Quackenbush (54) reported a large amount of zeinoxantin in seed corn. It has long been established that niacin deficiency causes pellagra, associated with corn consumption (55). Unlike mature field corn, which contains 27 μg of total naicin per gram of dry matter, immature sweet corn has greater amounts of both total (51 μg/g) and bioavaialble (40 μg/g) niacin as determined by rat growth assay (55). Walls et al. (56) found that the total nicotinic acid content of sweet corn after calcium hydroxide treatment exceeded that of field corn by 13–16% at each of the immature stages of development (milky, dough, and denting) and was 44% greater than that of field corn at the mature 49-day stage (56).

Wolnik et al. (57) analyzed sweet corn from major sweet corn–growing regions for Ca, Cu, Fe, K, Mg, Mo, Na, Ni, P, Se, and Zn. They reported significant amounts of K (2900 μg/g wet weight), P (830 μg/g wet weight), and Mg (270 μg/g wet weight). One should keep in mind

TABLE 8 Major Carotenoids in Four Cultivars of Sweet Corn (μg/100 g)

Carotenoids	Jubilee	losweet	70-2499	Stylpak
Hydrocarbon	89.0	41.4	45.0	45.6
Monohydroxy	98.5	92.7	101.6	89.5
Polyhydroxy	58.5	19.6	35.0	51.5
α-Carotene	24.6	5.0	8.4	8.1
β-Carotene	9.8	5.3	145.0	8.4
β-Zeacarotene	15.7	16.8	14.5	17.0
γ-Carotene	15.1	8.9	7.5	10.6
Zeinoxanthin	79.2	66.0	77.2	68.9
β-Cryptoxanthin	18.9	18.3	15.1	18.4

Source: Ref. 53.

that most of the minerals and vitamins are in the germ, which is removed in some processed sweet corn products.

5. Enzymes

Quality deteriorations that result in off-flavor and off-odor formation after harvesting and during frozen storage are major concerns of the processing industries (58–61). Off-flavor formation in frozen, stored raw, or underblanched vegetables including sweet corn is caused by enzymatic action (60,61). Lipoxygenase and peroxidase are directly associated with off-flavor and other quality deteriorations (62,63). Valasco et al. (61) and Theerakulkait and Burrett (64) separated peroxidase and lipoxygenase and catalase from sweet corn kernels and evaluated off-odor formation by enzymes. They could not clarify which enzyme(s) were responsible. Theerakulkait et al. (65) reported that addition of purified lipoxygenase extract to blanched corn increased "painty" and "stale/oxidized" off-flavor descriptors and decreased "sweet" and "corn" descriptors (Table 9). They also suggested that sweet corn germ peroxidase was not responsible for off-odor formation. Table 9 shows odor descriptors, definitions, reference standards, their preparation, and amount of serving for descriptive sensory evaluations of sweet corn homogenate used in this study. The effects of blanching treatments and packaging materials on the lipoxygenase and peroxidase activity and fatty acid stability of two cultivars of sweet corn (Jubilee and GH-2684) were evaluated during 9 months of storage at –20°C by Rodriquez-Saona et al. (66). They reported that lipoxygenase and peroxidase were inactivated completely by 9 and 15 minutes of steam-blanching, respectively. Comparison of fatty acid content revealed no change in fatty acid composition during storage. Degradation of PUFA did not depend on oxygen permeability of the packaging materials.

6. Aroma Constituents

Raw sweet corn characteristically has little or no odor. However, Wade (67) reported that cut kernels have three characteristic aromas reminiscent of fresh corn. Two of these were similar to a fresh green vegetable-like aroma and the third aroma was identified as methanol. Bills and Keenan (68) reported acetaldehyde, acetone, and ethanol at levels of 2.5, 0.7, and 322 ppm, respectively, as they investigated raw blends of kernels. Off-odor formation due to enzymatic action is described in an earlier section. Some cornlike aromas are developed during processing. Wiley (7) reported that dimethyl sulfide (DMS) and hydrogen sulfide (H_2S) contributed the most odor to the cooked corn aroma at levels in the areas of low-boiling volatiles. DMS is found in many other sulfur-containing vegetables as well as in milk and meats. Dignan and Wiley (69) investigated shelf-life stability of DMS in frozen sweet corn and showed that the normal paperboard box with a wax overlap (commercial packaging) was not a satisfactory package for frozen corn. Both a polyester, flexible pouch boiled-in-bag type and a 303 C-enamel can closed under vacuum had approximately twice as much DMS as compared with the paperboard box (Table 10).

Evans (70) isolated and identified two high-boiling flavor compounds from cooked corn. The two compounds, vanillin and 4-ethyl guaiacol, mixed in the proper ratio produced a cornlike aroma. Lee (71) used a modified Likens-Nickerson distillation apparatus in which he cooked the corn for 5–6 hours and collected the volatiles in hexane. He found that dimethyl pyrazine and dimethylethylpyrazine had sweet corn–like aromas. Other aromas may be developed as a result of chemical reactions. Yoshikawa et al. (72) found that aroma compounds from Streaker-degraded proline and ornithine were 1-pyrroline. Free α-aminobutanol also yielded the same aroma compounds via formation of an internal Schiff base. These compounds have more or less the same aroma as cooked sweet corn. Wade (67) also found many other compounds from steam distillate of sweet corn. Acetal was very similar to cooked sweet corn aroma.

TABLE 9 Odor Descriptors, Definitions,Reference Standards, and Their Preparation and Amount Used for Serving[a] for Descriptive Sensory Evaluation of Sweet Corn Homogenate Samples

Descriptors	Reference standards and preparations	Definitions
Overall odor		The overall odor impact (intensity) of all compounds perceived by nose
Painty	Linseed oil: used 15 ml linseed oil (Grumbacher Artists Oil Medium)	Odor quality associated with the deterioration of the oil fraction—it may be described as linseed oil, paint thinner, shoe polish.
Stale/Oxidized	Wet masa harina; prepared by mixing 1 cup corn tortilla mix (Quaker Masa Harina de Maiz) with 1/2 cup of hot water	Cardboard, old corn flour, or the dusty/musty odor that does not include painty
Cooked cabbage	Sliced cooked cabbage; prepared by cooking 250 g sliced cabbage with 500 ml of spring water on gas stove at high (10) for 4 min and at low (2) for 30 min; used 10 ml liquid portion and 15 g cooked cabbage	All characteristic notes associated with odor of cooked cabbage; e.g., sour, cabbage, fermented
Straw Hay	Chopped straw and hay; prepared by chopping the dried straw and hay in a length about 1 to 2 cm; used 3 g chopped straw and 3 g chopped hay	All characteristic notes associated with straw and hay
Corn	Cooked fresh cut sweet corn; prepared by cooking 75 g fresh cut sweet corn with 5 ml spring water using microwave at full power for 1.5 min; used 30 g cooked cut corn	The characteristic note of "corn" associated with cooked sweet corn
Sweet	Liquid of canned whole kernel corn; used 30 ml of liquid portion of canned whole corn kernels (Del Monte brand Golden Sweet, Family Style)	The characteristic note of "sweet" associated with canned sweet corn
Cobby/Husky	Diced fresh corn cob and fresh corn husk; prepared by dicing fresh corn husk (thickness about 0.5 cm) and fresh corn cob (thickness about 0.1 cm); used 15 g for diced cob and 8 g for diced husk	The characteristic note associated with diced fresh corn cob and husk

[a]Served in 250-ml clear wine glasses covered with tight-fitting aluminum.
Source: Ref. 65.

TABLE 10 Effects of Packaging on Mean Dimethyl Sulfide (DMS) Levels in Aroma of Frozen Sweet Corn after Cooking

Cultivar	Paperboard box DMS (μg/ml)[b]	Plastic pouch DMS (μg/ml)[b]	303 Vacuum can DMS (μ/ml)[b]
Commander	0.9	1.8	1.8
NK-199	1.0	1.7	1.8
Buttersweet	1.1	2.0	1.3
Stylepak	1.1	2.5	1.9
Mean	1.0	2.0	1.7

[a]Only the paperboard box was significantly different ($p < 0.01$).
[b]36 determinations per mean.
Source: Ref. 69.

V. POSTHARVEST TECHNOLOGY

A. Storage

Storage of sweet corn for more than one week results in serious deterioration and loss of tenderness and sweetness. The changes in sucrose at different temperatures are presented in Figure 9. The loss of sucrose is about four times as rapid at 10°C as at 0°C (45). At 30°C, about 60% of sugars may be converted into starch in a single day as compared with only 6% at 0°C. However, sweet corn loses sweetness or desirable flavor fairly rapidly when stored at 0°C (73,74). Denting is an indication of loss of quality. A loss of 2% moisture from sweet corn may result in objectionable kernel denting.

Postharvest cooling of corn to near 0°C within an hour of harvest and handling at 0°C during marketing results in maximum quality retention. Crated corn can be vacuum-cooled from about 30° to 5°C in a half-hour. Hydrocooling by spraying, showering or immersion in water at

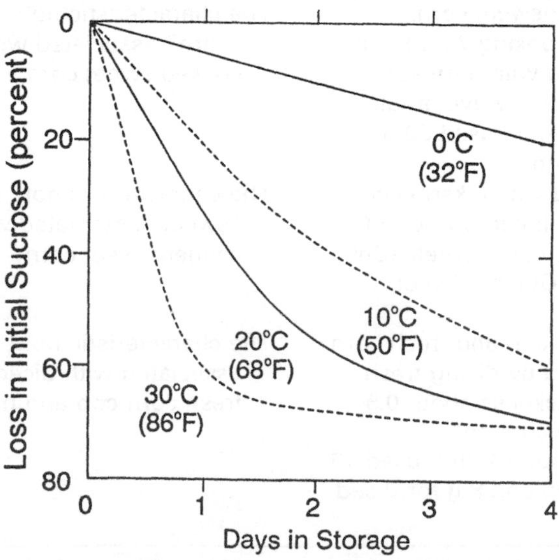

FIGURE 9 Sucrose depletion in sweet corn at four temperatures. (From Ref. 36.)

0–3°C is effective, although it takes longer than vacuum-cooling for the same temperature reduction if corn is packed before it is cooled. Crated corn takes about an hour to cool to 5°C in a hydrocooler. After hydrocooling, caping with ice is desirable during transport or holding to hasten continued cooling, remove the heat of respiration, and maintain husk freshness.

Sweet corn should not be handled in bulk unless iced because it tends to heat throughout the pile. Corn cannot be kept in marketable condition even in cold storage at 0°C for more than 5–8 days. Storage life at 5°C is about 3–5 days and at 10°C is about 2 days (51). Sweet corn may be packaged in moisture-retention film, with the husk removed after precooling. The film should be perforated to prevent development of off-flavor. This product is perishable and must be continuously refrigerated.

1. Low-Temperature Storage

Sweet corn is not stored to any great extent. However, it can be stored at 0°C and 90–98% relative humidity for a week or more, if precooled immediately after harvesting (9). Ryall and Lipton (33) also stated that to maintain the best quality, sweet corn must be cooled to as near 0°C as soon after harvest as possible. Time is of critical importance because sucrose rapidly changes to starch after harvesting. Sweet corn must also be kept as close to 0°C as feasible at wholesale and retail markets, otherwise all previous efforts may be quickly negated. The storage life of sweet corn is very limited because sweetness and tenderness are lost so rapidly. Under the best handling conditions, such as prompt cooling to a cob temperature of 4°C and continued refrigeration near 0°C, corn will have satisfactory culinary quality for a maximum of 6–8 days at 0°C, 3–4 days at 5°C, and 2 days at 10°C (24). Pantastico et al. (75) recommended temperatures of 0°C and 0–2°C for storage of green corn and sweet corn for about 1–1½ weeks at 90–95% relative humidity. Wills et al. (76) stated that sweet corn stored well for 1–2 weeks at –1 to –4°C.

2. Controlled-Atmosphere Storage

Corn derives little benefit from controlled-atmosphere storage, although 5–10% CO_2 slightly retards sugar loss (33). CO_2 levels greater than 10% caused injury to corn. Sweet corn (cv. Isobelle) ears stored for 3 weeks at 1.7°C and 90–100% relative humidity in controlled atmospheres of 2 or 21% O_2 and 6, 15, or 25% CO_2 or at 50 mmHg pressure did not differ significantly in appearnce or flavor (77). Sucrose content of stored ears remained higher in 2% O_2 at atmospheric or reduced pressure (76 mmHg) than in other atmospheres. Ethanol content increased during storage except in 21% O_2 without added CO_2, and was highest in ears stored in atmosphere containing 25% CO_2. The high sucrose content of Florida Sweet cultivar after 3 weeks of storage suggested that for maintenance of high market quality, breeding cultivars that retain quality in combination with prompt precooling offers more chance of success than use of a modified atmosphere (77).

3. Irradiation

Salunkhe (78) reported effects of gamma radiation on the shelf life of four cultivars of yellow sweet corn: F. M. Cross, Golden cross Bantam, Seneca Chief, and Victory Golden. The ears were harvested when the crop was of prime maturity and quality. Ears with husks were then irradiated to 0 (control), 2, 3, and 5×10^5 rad. Perforated cans containing corn were stored at 4.4°C for 16 days before evaluation. Quality evaluation indicated that the change in the preference score with the increase in the radiation dose at the 6th day of the storage period was not significant, but there was a decrease in the taste preference scores with an increase in the dose at the 16th day of the storage. F. M. Cross and Golden Cross Bantam cultivars maintained better quality than did Seneca Chief and Victory Golden cultivars with higher dosage at the 16-day interval. Slight growth of mold was observed on the husk in control and samples irradiated with 2×10^5 rad. The sweet corn from these samples could be cooked and served after dehusking. No mold was present on husks of the corn irradiated at 3 and 5×10^5 rad.

B. Processing

Sweet corn is the most important canned vegetable in the United States. In 1992, as shown in Table 2, 197,100 hectares of sweet corn were harvested for processing and 83,300 hectares for fresh market. Corn cultivars suitable for canning are Country Gentleman, Crosob, Golden Bantam, Mores's Golden Cream, Potter's Excelsior, Stowells Evergreens, and Yellow Evergreen. Golden Bantam was more popular before hybrids came into general use. Sweet corn qualities such as tender endosperm and uniformity of pericarp maturity and deep kernels are desirable for processing (10). Development in breeding and hybridizing technology has greatly improved cultivars for processing. Nearly all cultivars of sweet corn for processing are now hybrids. Hybrids give greatly increased yield, better uniformity, and excellent flavor, color, and canning qualities.

Sweet corn is processed in a variety of ways—as frozen corn on the cob, cut kernels, and cream style. Corn products are preserved by either canning or freezing. Figure 10 shows a flow diagram for processing sweet corn. Sweet corn is harvested at the proper maturity stage and is then transported to the processing plant. Once the harvested corn enters the processing plant, it is handled mechanically. Initial processing includes husking and silking. There are several husking machine designs, but all have one thing in common—a pair of rapidly revolving rubbing or milled steel rolls, which catch the husks and remove them. Automatic husking and silking machines have capacities of approximately 7 ton/h. These machines have a live bottom hopper, vibrating pans, and a husking bed of spaced horizontal rolls inclined downward. The angle can be adjusted to accommodate varying degrees of looseness of the corn husk, i.e., be more or less steep. Other adjustments can be made according to the variety and condition of the corn.

The corn is first wilted in a steaming tunnel to loosen the husk, then it is conveyed to the hopper. The feed rate is controlled by a variable-speed motor on the live bottom belt feeder. Ears move from the hopper to a flat vibratory bed conveyor that spaces and aligns the ears. This arrangement minimizes "piggybacking" and maintains end-to-end alignment with the husking rollers, where rubber rollers remove the husks. Vibrating carts above the rolls are adjustable to provide the proper degree of rotational and vibratory action; roll speeds also may be adjusted to aid rotation.

Cooling is an important step. Postharvest changes in the kernels due to rise in the temperature were discussed earlier. Research shows that sweet corn held at refrigerated conditions for 24 hours undergoes minimal changes in the specific gravity and total solids content.

Sorting and washing is followed by blanching and cooling. Two sortings are done in typical canning operations. One is to remove improperly husked ears and return them to the husking department where the culls and trash are discarded. Another is to sort husked corn into Grades A and B. Grade A is used for fancy and choice canned products and Grade B is used for cream-style corn. Washing is done by two methods. One is by means of a silker-brusher washer, and another is by means of a cylinder or conical reel. In general, corn is passed through a heavy spray of water in a rotary washer. The ears are rubbed against each other vigorously so that dirt is loosened and removed.

Blanching of vegetables in steam or hot water during preparation for canning or freezing is done frequently by the vegetable-processing industries. Blanching removes occluded gases

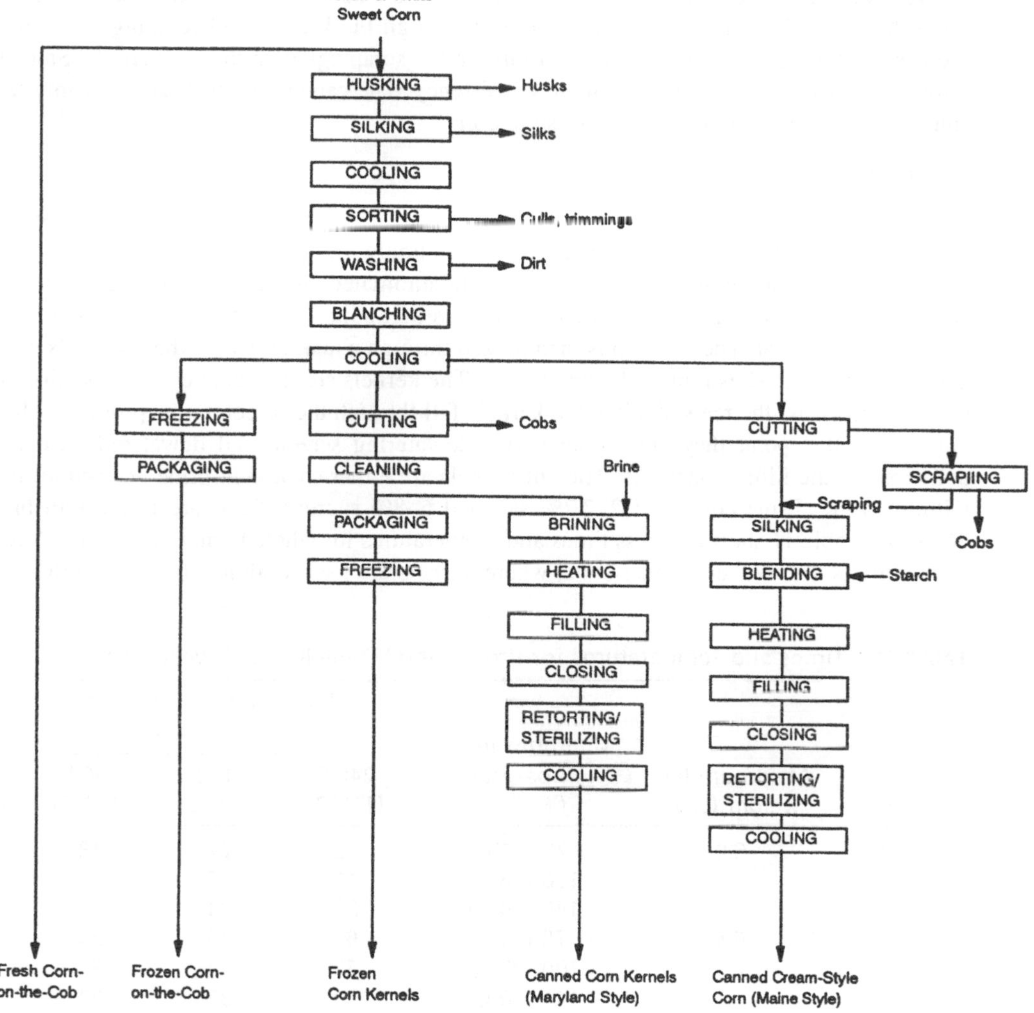

Figure 10 Flow sheet for processing of sweet corn. (From Ref. 79.)

from the tissue, thus reducing the amount of trapped air in the can. Blanching also results in less strain on the cans during thermal processing and produces a much higher vacuum in the can after processing. Blanching of cut corn is done to inhibit peroxidase and lipoxygenase activities after processing. Blanching is achieved by spreading the products in a thin layer on a mesh belt, which passes through a chamber of live steam at atmospheric pressure. Another method is to use a tabular blancher, in which cut corn is conveyed through a tube or pipe heat exchanger until it is heated to the desired temperature.

1. Canning

Sweet corn canning is a major vegetable-processing industry. In the United States approximately 41% of total harvested sweet corn is canned. Corn is packed Maryland style or Maine style.

Maryland style is whole kernel corn consisting of whole kernels cut from the cob and canned in brine. Maine-style corn is obtained by cutting through the kernels and scraping the remaining protions of the kernel from the cobs and mixing the scrapings with the cut kernels. Salt, sugar, starch, and water are added to maintain consistency. The canned product has a creamy texture and is commonly referred to as cream-style corn.

Whole Kernel Corn

Whole kernel corn is defined as a whole kernel or a whole grain or cut kernels consisting of whole or substantially whole cut kernels packed in liquid. In making whole kernel corn, ears are blanched before cutting to coagulate the juice. In automated modern corn-cutting machines, the ears are fed small end first through curved knives that adjust themselves by means of a spring to the size of the ear. The cut corn is then passed through a tank of water where kernels will sink and the floating trash is removed autmatically. The kernels are then conveyed to the silking and washing reels. As the reels revolve, the kernels fall through the screen to the product-collecting pan. From this point they are pumped to a dewatering screen, and dewatered corn is then delivered to the filling machines. The cut kernels are either canned in brine or vacuum-packed without brine. Brine consist of 2–2.7% salt and 6–8% sucrose. Cans are filled with brine at 46–49°C. Table 11 shows process times and temperatures for whole kernel sweet corn. Vacuum-packaging is the another popular style, where kernels are packed almost free of liquid.

TABLE 11 Times and Temperatures for Processing Whole Kernel Sweet Corn

Can size	Maximum fill weight (oz.)	Minimum initial temperature (°F) (°C)	Minimum time (min) at retort temperature		
			240°F (116°C)	245°F (118°C)	250°F (121°C)
211 × 304	6.0	70 (21°C)	42	27	18
		100 (38°C)	41	26	17
		140 (60°C)	38	24	15
211 × 304	6.4	70 (21°C)	60	44	34
		100 (38°C)	57	41	31
		140 (60°C)	53	37	27
303 × 406	12.0	70 (21°C)	52	36	26
		100 (38°C)	50	34	25
		140 (60°C)	46	34	22
303 × 406	13.0	70 (21°C)	57	41	31
		100 (38°C)	55	39	28
		140 (60°C)	50	35	25
404 × 700	35.0	70 (21°C)	73	53	40
		100 (38°C)	69	50	37
		140 (60°C)	63	44	32
603 × 700	76.0	70 (21°C)	95	68	50
		100 (38°C)	87	61	45
		120 (49°C)	82	57	41
		140 (60°C)	76	52	37
		160 (71°C)	70	47	33

These processes apply *only* if the corn is well cleaned. They also apply to succotash prepared from whole kernel corn and beans.
Source: Ref. 80.

Cream-Style Corn

Cream-style corn consists of whole or partly whole cut kernels packed in creamy components from the corn kernels plus other liquids and other ingredients to form a product of creamy consistency. Knives of the corn cutter are adjusted in such a way that kernels are cut in half. Then the ears are passed through a scraper that scrapes the remaining portions of the kernels from the cobs. The scrapings and cut kernels are mixed to form a creamy mixture. In batch-type mixers, the cut, desilked corn is mixed with water, sugar, salt, and starch in a stainless steel tank equipped with rotating paddles. The amount of water added to the batch varies with corn maturity. Under average conditions, 3.8 liters of water are added to 15 liters of corn. Starch is added to give the desired creamy consistency. Table 12 shows process times and temperatures for canning cream-style corn. Starch is premixed with water to avoid lumps. McDermott (81) replaced some of the starch with a mixture of sodium alginate and calcium-binding salt. This reduced the viscosity and improved the heat transfer. The mixture is steam-heated to 71–82°C in a batch mixture. Then it is transferred to another tank where the temperature is increased to 88–93°C, long enough to gelatinize the added starch. The mixture is filled in cans, retorted, and cooled.

2. Frozen Corn

Frozen corn operations are similar to canning except that the ears are carefully sized by ear cutters into the exact length for frozen corn. Blanching is required for all frozen sweet corn to inhibit peroxidase and lipoxygenase enzymes.

TABLE 12 Process Times and Temperatures of Cream-Style Corn or Succotash

Can size	Minimum initial temperature (°F) (°C)	Minimum time (min) at retort temperature		
		240°F (116°C)	245°F (118°C)	250°F (121°C)
211 × 304	140 (60°C)	70	59	52
	160 (71°C)	67	56	49
	180 (82°C)	62	52	45
211 × 400	140 (60°C)	74	63	55
	160 (71°C)	71	59	52
	180 (82°C)	66	55	48
303 × 406	140 (60°C)	95	81	72
	160 (71°C)	88	76	67
	180 (82°C)	82	70	62
307 × 409	140 (60°C)	105	90	81
	160 (71°C)	100	84	74
	180 (82°C)	90	77	69
603 × 700	140 (60°C)	260	235	215
	160 (71°C)	240	215	200
	180 (82°C)	215	195	180

These processes also apply to succotash prepared from cream-style corn and beans. For succotash prepared from whole kernel corn and beans, use processes suggested for whole kernel corn in brine.
Source: Ref. 80.

TABLE 13 Summary of Scoring System

Attributes	Grade A (Fancy)	Grade B (Ex. Std.)	Grade C (Standard)	Substandard
Color	9–10	8	6–7[a]	0–5
Defects	36–40	32–35[b]	28–31[b]	0–27[a]
Tenderness and maturity	45–50	40–44[b]	36–39[b]	0–35[b]
Total score—minimum	90	80	70	—

[a]Partial limiting rule: sample units with score of 6 points may not grade higher than C.
[b]Limiting rule: sample units with these scores may not be of higher grade.
Source: Ref. 4.

Whole Kernel Pack

Whole kernel pack is prepared from fresh, clean, sound, succulent kernels of sweet corn. Corn is husked, silked and sorted, trimmed, washed, and blanched. The cleaned kernels are steam-blanched for 2–3 minutes on a continuous, stainless steel blanching belt. It is necessary to quickly cool the blanched kernels with a spray of water. A summary of the USDA grading is presented in Table 13. U.S. Grade A is made from similar varietal characteristics, good flavor, good color with a minimum score of 90 for color, defects, and tenderness. U.S. grade B or Extra Standard includes kernels with similar varietal characteristic with a minimum score of 80. Kernels should have good color and score no less than 7 points, be reasonably free from defects, and be reasonably tender (Table 13).

Corn on the Cob

Frozen corn on the cob is prepared from sound, properly matured, fresh sweet corn ears by removing husk and silk, sorting, trimming, and washing to assure clean and wholesome product. The cobs are carefully sorted to fit appropriate packaging material, then the cobs are blanched. The blanching time depends on the maturity and the size of the cob. Usually it runs between 6 and 12 minutes of steam-blanching. There are three grades of frozen corn on the cob: U.S. Grade A, U.S. Grade B, and Standard. U.S. Grade A or U.S. Fancy includes ears with similar varietal characteristics that have good flavor and aroma. Cobs are uniform in size and well developed; kernel color is uniform and well developed. U.S. Grade A is partricularly free from all defects, with a product score of not less than 90 points (Table 14). U.S. Grade B or U.S. Extra Standard includes ears of similar varietal characteristic with at least reasonably good flavor, aroma, and color. The ears may lack uniformity of size and development. The product score should not be less than 80 points (Table 14). Standard grade includes all ears that fail to meet the requirements of Grade B.

3. Baby Foods

Creamed corn baby food is available in supermarkets. The general process for producing creamed corn baby food begins with taking frozen corn, breaking it apart, and mixing it with hot water. Ingredients such as rice flour and nonfat dry milk powder are mixed with water to form a slurry. The slurry is strained and transported to the batch tank. The corn slurry is heated by steam injection and strained to remove excess fibers. The particle size is adjusted by grinding the slurry. It is then transported to the batch tank and blended with other ingredient slurry and

TABLE 14 Score Chart for Frozen Corn on the Cob

Attributes	Maximum points	Grade A	Grade B	Substandard
Color	20	18–20	16–17[a]	0–15
Uniformity of size	10	9–10	8	0–7[b]
Development	10	9–10	8	0–7[b]
Defects	30	27–30	24–26[a]	0–23[a]
Tenderness and maturity	30	27–30	24–26[a]	0–23[a]
Minimum score	—	90	80	—

[a]Limiting rule: Products falling into this classification cannot earn a higher grade regardless of total score.
[b]Limiting rule: Products falling into this classification cannot earn a higher grade than Grade B regardless of total score.
Source: Ref. 4.

heated to a prescribed temperature. Consistency is controlled by adding water. The finished batch is then sent through an in-line strainer and metal detector to a filler bowl. The glass jars are filled with product to the appropriate headspace and weight, capped, and then thermally processed in a closed retorn system to a lethality of no less than $F_0 = 6.0$.

ACKNOWLEDGMENTS

We acknowledge the help of Ken Anderson, NC+ Hybrids, Lincoln, Nebraska; David Wysong and J. A. Kalsich, University of Nebraska–Lincoln; and Randy Klein and Ken Wurdeman, Nebraska Corn Board; for providing photographs for use in this chapter. Also, the senior author thanks Gerald Biby, Industrial Agricultural Products Center, and Dr. D. K. Salunkhe for their constant encouragement and support during the preparation of this mansucript.

REFERENCES

1. Mangelsdorf, P. C., *Corn: Its Origin, Evaluation, and Improvement*, Bulknap Press, Cambridge, MA, 1974, p. 2.
2. Huelsen, W. A., *Sweet Corn*, Interscience Publisher, Inc., New York, 1954, p. 409.
3. Galinat, W. C., *The Evolution of Sweet Corn*, University of Massachusetts Research Bulletin, Amherst, 1971, p. 952.
4. *The Almanac of Canning, Freezing, Preserving Industries*, Vol. 2, 71st ed., Edward Ejudge and Sons, Westminister, MD, 1993, p. 714.
5. Winn, J., *Canned Sweet Corn*, annual trade report, USDA/FIS 1994.
6. Wiley, R. C., F. D. Schales, and K. A. Corey, Sweet corn, in *Quality and Preservation of Vegtables* (N. A. M. Eskin, ed.,), CRC Press Inc., Boca Raton, FL, 1989.
7. Wiley, R. C., Sweet corn aroma: Studies of its chemical components and influence on flavor, in *Evaluation of Quality of Fruits and Vegetables* (H. E. Pattee, ed.), AVI Publishing Co., New York, 1985, p. 349.
8. Bonucci, P. A., Production of sweet corn, *US Patent 3971161*, July 27, 1976 (1986).
9. Thompson, H. C., and W. C. Kelley, *Vegetable Crops*, 5th ed., McGraw-Hill Book Company, New York, 1957.

10. Rodrigues, G. R., B. L. Raina, E. B. Pantastico, and M. B. Bhatti, Quality of raw materials for processing-postharvest physiology: Harvest indices, in *Postharvest Physiology, Handling and Utilization of Tropical and Sub-tropical Fruits and Vegetables* (E. B. Pantastico, ed.), The AVI Publishing Co. Inc., Westport, CT, 1975, p. 407.
11. Dale, J. L., J. MaFerran, E. V. Wann, and R. L. Bona, Field evaluation of sweet corn hybrides and inbreds with emphasis on virus disease resistance, *Rep. Ser. of the Agric. Expt. Sta. Univ. of Arkansas* No. 258, 1981.
12. Yumaguchi, M., *World Vegetables: Principles, Production, and Nutritive Values*, The AVI Publishing Co., Westport, CT, 1983, p. 415.
13. Metcalf, C. L., W. P. Flint, and R. L. Metcalf, *Destructive and Useful Insects: Their Habits and Control*, 5th ed., McGraw-Hill Book Company, New York, 1993.
14. Watts, R. L., and G. S. Watts, *The Vegetable Growing Business*, The Orange Judd Publishing Co. Inc., New York, 1954.
15. Nyvall, R. F., Diseases of corn, in *Field Crop Disease Handbook*, Van Nostrand Reinhold, New York, 1989.
16. Culpepper, C. W., and C. A. Magoon, A study of the factors determining quality in sweet corn, *J. Agric. Res. 34*:413 (1927).
17. Ware, G. W., and J. P. McCollum, *Raising Vegetables*, Interstate Printers and Publishers Inc., Danville, IL, 1959.
18. Shoemaker, J. S., *Vegetable Growing*, John Wiley and Sons, Inc., New York, 1949.
19. Pantastico, E. B., H. Subramanyam, M. B. Bhati, N. Ali, and E. K. Akamine, Post harvest physiology: Harvest indices, in *Postharvest Physiology, Handling and Utilization of Tropical and Sub-tropical Fruits and Vegetables* (E. B. Pantastico, ed.), The AVI Publishing Co., Westport, CT, 1975, p. 56.
20. Work, P., and J. Carew, *Vegetable Production and Marketing*, John Wiley and Sons, Inc., New York, 1955.
21. Huelsen, W. A., and W. H. Michaels, The yield complex of sweet corn, *Illinois Agric. Exp. Sta. Bull.* 432 (1937).
22. Geise, C. E., P. G. Homeyerm, and R. G. Tischer, A comparison of three methods for determination of moisture in sweet corn, *Food Technol. 5*:250 1951.
23. Becwar, M. R., N. S. Mansour, and G. W. Varsevald, Microwave drying: A rapid method for determining sweet corn moisture, *HortScience 12*:562 (1977).
24. Drake, S. R., and J. W. Nelson, A comparison of three methods of maturity determination in sweet corn, *HortScience 14*:156 (1979).
25. Scott, G. C., R. O. Belkengren, and E. C. Rithcell, Refractive index as a measure of sweet corn maturity, *Food Ind. 17*:1030 (1945).
26. Garwood, D. L., F. J. McArdle, S. F. Vanderslice, and J. Shannon, Postharvest carbohydrate transformations and processed quality of high sugar maize genotypes, *J. Am. Soc. Hortic. Sci. 101*:400 (1976).
27. Henry, C. H., E. B. Wilcox, D. K. Salunkhe, and F. E. Lindquist, Evaluation of certain methods to determine maturity in relation to yield and quality of yellow sweet corn grown for processing, *Food Technol. 10*:374 (1956).
28. Darbyshire, B., W. A. Muirhead, and R. J. Henry, Water soluble polysaccharide determination as a technique for evaluation of sweet corn maturity, *Aust. J. Agric. Res. 29*:517 (1978).
29. Twigg, B. A., A. Kramer, H. N. Falen, and F. L. Southerland, Objective evaluation of the maturity factor in processed sweet corn, *Food Technol. 10*:171 (1976).
30. Kramer, A., A trimetric test for sweet corn quality, *Proc. Am. Soc. Hortic. Sci. 59*:23 (1952).
31. Showalter, R. K., and H. A. Schomer, Temperature studies of commercial broccoli and sweet corn prepackaging at the shipping point, *Proc. Am. Soc. Hortic. Sci. 54*:325 (1949).
32. Alban, E. K., and R. C. Scott, Postharvest handling and marketing of garden fresh sweet corn, *Res. Circ. Ohio Agric. Exp. Sta.* No. 23 (1954).
33. Ryall, A. L., and W. J. Lipton, *Handling, Transportation and Storage of Fruits and Vegetables*. Vol. 1, *Vegetables and Melons*, The AVI Publishing Co. Inc., Westport, CT, 1972.

34. Flora, L. F., and R. C. Wiley, Sweet corn aroma chemical components and relative importance in the overall flavor response, *J. Food Sci. 39*:770 (1974).
35. Wolf, M. J., C. L. Buzan, M. M. McMasters, and C. E. Rist, Structure of mature corn kernel III. Microscopic structure of the endosperm of dent corn, *Cereal Chem 29*:349 (1952).
36. Wolf, M. J., C. L. Buzan, M. M. McMasters, and C. E. Rist, Structure of mature corn kernel IV. Microscopic structure of germ of dent corn, *Cereal Chem. 29*:362 (1952).
37. Wolf, M. J., C. L. Buzan, M. M. McMasters, and C. E. Rist, Structure of mature corn kernel I. Gross anatomy and structural relationship, *Cereal Chem. 29*:321 (1952).
38. Wolf, M. J., C. L. Buzan, M. M. McMasters, and C. E. Rist, Structure of mature corn kernel II. Microscopic structure of pericarp, seed coat and hilar layer of dent corn, *Cereal Chem. 29*:334 (1952d).
39. Khalil, T., and A. Kramer, Histological and biochemical studies of sweet corn (*Zea mays* L.) pericarp as influenced by maturity and processing, *J. Food Sci. 36*:1064 (1971).
40. Bailey, D. M., The relation of the pericarp to tenderness of the sweet corn, *Proc. Am. Soc. Hortic. Sci. 36*:555 (1938).
41. Puangnak, W., Effect of hybrid, maturity and kernel structure on lipid content, composition and aroma development in sweet corn, M.S. thesis, University of Maryland, College Park, 1976.
42. Munchen, C., *Food Composition and Nutrition Tables 1981/82*, Wissenschaftliche Verlagsgesellschft GmbH, Stuttgart, 1981.
43. *USDA Handbook 8-11*, United States Department of Agriculture, Washington, DC, 1984.
44. Appleman, C. O., Forecasting the date and duration of best canning stage for sweet corn, *Maryland Agric. Exp. Sta. Bull.* No. 254 (1923).
45. Appleman, C. O., and J. M. Arthur, Carbohydrate metabolism in green corn during storage at different temperature, *J. Agric. Res. 17*:135 (1919).
46. Rumpf, G., J. Mawson, and H. Hansen, Gas chromatic analysis of the soluble substances of sweet corn kernels as a method indicating the degree of maturity attained and change in quality during storage, *J. Sci. Food. Agric. 23*:193 (1972).
47. Wall, J. S., and J. W. Paulis, Corn and sorghum grain proteins, in *Advances in Cereal Science and Technology* (Pomeranz, Y., ed.) American Association of Cereal Chemists, St. Paul, MN, 1978, p. 135.
48. Pukrushpan, T., Protein changes in sweet corn cultivars with stages of maturity, *3333 Field Crop Abstract 31*:5989 (1977).
49. Pascual, L., and Wiley, R. C., Fatty acid composition in the major lipid fractions of maturing sweet corn (*Zea mays* L.), *J. Am. Soc. Hortic. Sci. 99*:573 (1974).
50. Flora, L. F., and R. C. Wiley, Effects of various endosperm mutants on oil content of fatty acids composition of whole kernel corn (*Zea mays* L.), *J. Am. Soc. Hortic. Sci. 97*:604 (1972).
51. Hardenburg, R. E., A. E. Watada, and C. Y. Wang, *The Commercial Storage of Fruits, Vegetables, and Florist and Nursery Stocks*, U.S. Dept. Agric. Handbook 66, 1986, p. 58.
52. Farrow, R. P., F. C. Lamb, E. R. Elkins, N. Low, J. Humphrey, and K. Kemper, Nutritive content of canned tomato juice and whole kernel corn, *J. Food Sci. 38*:595 (1973).
53. Lee, C. Y., P. E. McCoon, and J. M. LeBowitz, Vitamin A value of sweet corn, *J. Agric. Food Chem. 29*:1294 (1981).
54. Petzoid, E. N., and F. W. Quackenbush, Zeinoxanthin, a crystalline carotenol from corn gluten, *Arch. Biochem. Biophys. 86*:163 (1960).
55. Carter, E. G. A., and K. J. Carpenter, The available niacin values of foods for rats and relation to analytical values, *J. Nutr. 112*:2091 (1982).
56. Wall, J. S., M. R. Young, and K. J. Carpenter, Transformation of niacin-containing compounds in corn during grain development: Relationship to niacin nutritional availability, *J. Agric. Food Chem. 35*:752 (1987).
57. Wolnik, K. A., F. L. Fricke, S. G. Capar, G. L. Braude, M. W. Meyer, R. D. Satzger, and W. Kuennen, Elements in major raw agricultural crops in the United States. II. Other elements in lettuce, peanuts, potatoes, soybean, sweet corn, and wheat, *J. Agric. Food Chem. 31*:1244 (1983).

58. Smittle, D. A., R. E. Thornton, and B. B. Dean, Sweet corn quality deterioration: A survey of industry practices and possible method of control, *Wash. Agric. Expt. Stat. Circ. 555* (1972).
59. Wagenknecht, A. C., Lipoxidase activity and off-flavor in underblanched frozen corn-on-the-cob, *Food Res. 24*:539 (1959).
60. Lee, Y. C., Lipoxygenase and off-flavor development in some frozen foods, *Korean J. Food Sci. Technol. 13*:53 (1981).
61. Valasco, P. J., M. H. Lim, R. M. Pangborn, and J. R. Whitaker, Enzymes responsible for off-flavor and off-aroma in blanched and frozen stored vegetables, *Appl. Biochem Biotechnol. 11*:118 (1989).
62. Burnette, F. S., Peroxidase and its relationship to food flavor and quality: A review, *J. Food Sci. 42*:1 (1977).
63. Lim, M. H., P. J. Valasco, R. M. Pangborn, and J. R. Whitaker, Enzyme involved in off-aroma formation in broccoli, in *Quality Factors of Fruits and Vegetables* (J. J. Jen, ed.), American Chemical Society, Washington, DC, 1989, p. 72.
64. Theerakulkait, C., and D. M. Barrett, Lipoxygenase in sweet corn germ: isolation and physicochemical properties, *J. Food Sci. 60*:1029 (1995).
65. Theerakulkait, C., D. M. Barrett, and M. R. McDanial, Sweet corn germ enzymes affect odor formation, *J. Food Sci. 60*:1034 (1995).
66. Rodriguez-Saona, L. E., D. M. Barrett, and D. P. Selivonchick, Peroxidase and lipoxygenase influence on stability of polyunsaturated fatty acids in sweet corn (*Zea mays*L.) during frozen storage, *J. Food Sci. 60*:1041 (1995).
67. Wade, J. H., Characterization of aroma components of corn, Ph.D. thesis, University of Georgia, Athens, 1981.
68. Bills, D. D., and T. W. Keenan, Dimethyl sulfide and its precursor in sweet corn, *J. Agric. Food Chem. 16*:643 (1968).
69. Dignan, D. M., and R. C. Wiley, DMS levels int he aroma of cooked frozen sweet corn as affected by cultivar, maturity, blanching, and packaging, *J. Food Sci. 41*:346 (1976).
70. Evans, R. H., Isolation and characterization of high boiling flavor compounds of cooked corn, M.S. thesis, University of Georgia, Athens, 1977.
71. Lee, C. Y., Pyrazine compounds in canned sweet corn flavor, *Food Chem. 3*:319 (1978).
72. Yoshikawa, K., L. M. Libbey, W. Y. Cobb, and E. A. Day, 1-Pyrroline: The odor component of strecker-degraded proline and ornithine, *J. Food Sci. 30*:991 (1965).
73. Sims, W. L., R. F. Kashmire, and O. A. Lorenz, Quality sweet corn production, *Calif. Agr. Expt. Sta. Ext. Serv. Cir. 557*:22 (1971).
74. Stewart, J. K., and W. R. Barger, Effects of cooling method and top icing on the quality of peas and sweet corn, *Proc. Am. Soc. Hortic. Sci. 75*:470 (1960).
75. Pantastico, E. B., T. K. Chattopadhyay, and H. Subramanyam, Storage and commercial operations-postharvest physiology: Harvest indices, in *Postharvest Physiology, Handling and Utilization of Tropical and Sub-tropical Fruits and Vegetables* (E. B. Pantastico, ed.), The AVI Publishing Co., Westport, CT, 1975, p. 314.
76. Wills, R. H., T. H. Lee, D. Graham, W. B. McGlasson, and E. G. Hall, *Postharvest: An Introduction to the Physiology and Handling of Fruits and Vegetables*, Granada Publishing Ltd., London, 1981.
77. Spalding, D. H., D. L. Davis, and W. F. Reeder, Quality of sweet corn stored in controlled atmospheres or under low pressure, *J. Am. Soc. Hort. Sci. 103*:592 (1978).
78. Salunkhe, D. K., Gamma irradiation effects on fruits and vegetables, *Econ. Bot. 15*:28 (1961).
79. Johnson, L. A., Corn: Production, processing, and utilization, in *Handbook of Cereal Science and Technology* (Lorenz, K. J., and Kulp, K., eds.), Marcel Dekker, Inc., New York, 1991.
80. *National Food Processors Bulletin 26L*, National Food Processors Association, Washington, DC, 1982.
81. McDermott, F. X., Creamed corn product, *Food Proc. Rev. 26*:167 (1972).

28

Mushrooms

D. M. SAWANT AND K. M. KATE
Mahatma Phule Agricultural University, Rahuri, India

V. M. DHAMANE
College of Agriculture, Kolhapur, India

I. INTRODUCTION

Mushrooms are a popular food product in the Western world (1). The cultivation of edible mushrooms is now one of the most intensive fermentation processes practiced throughout the world to produce food for human consumption. Mushrooms are highly palatable and add flavor and variety to an otherwise monotonous diet. Among the cultivated mushrooms, *Agaricus bisporus* is the most important (Table 1). The world production of mushrooms in 1993 was 1.1 million metric tons. The United States, France, England, and The Netherlands are the major producers of mushrooms.

II. BOTANY

A. Taxonomy

Agaricus is the most important genus, which includes the most important edible species and only mildly poisonous mushroom species (2–4). Other cultivated mushrooms of the same genus include *A. bisporus* cultivar *albidus* (weakly pigmented) and cultivar *bisporus* (strongly pigmented) and species such as *A. bitorquis*, *A. suberonatus*, and *A. arvensis.*

Volvariella volvacea is another important species, which is commonly known as the straw mushroom, paddy straw mushroom, or Chinese mushroom (5), and belongs to the family Plauteaceae of the Basidiomycetes (6). It is an edible mushroom of the tropics and subtropics. It is also referred to as a "warm mushroom" as it can grow at relatively high temperatures, i.e., vegetative growth at about 32–34°C. It is fast-growing mushroom; the time required from spawning to harvesting is only about 5–10 days. Besides *V. volvaceae*, two other species—*V. diplasa* and *V. esculenta*—are also grown in India (7).

Pleurotus species represent a circumscribed group of higher fungi characterized by fruit bodies with an off-center stalk attached to the pileus, which opens up like an oyster shell during morphogenesis (8). These are popularly called oyster mushrooms (8). Several species of

TABLE 1 Commercial World Production of Edible Mushrooms

Species	Common name	Production (1000 MT)
Agaricus bisporus	Champignon	750
Lentinula edodes	Shiitake mushroom	180
Volvariella volvacea	Chinese mushroom	65
Flammulina velutipes	Winter mushroom	65
Pleurotus spp.	Oyster mushroom	40
Pholiota nemeko	Winter mushroom	20
Auricularia spp.	Ear mushroom	12
Tremella spp.	Jelly mushroom	3
Other species		9

Source: Ref. 1.

Pleurotus have the ability to decompose natural lignino-cellulosic wastes (9–13). *Pleurotus* mushrooms have the advantage of growing over a wide range of temperatures from 15 to 31°C. Strains of *P. ostreatus* known to fruit around 15°C are distributed in many part of Europe; strains of the same species from Florida are capable of fruiting at 25°C. Species like *P. cornucopiae*, *P. cystidiosus*, *P.*, *eryngii*, *P. flabellatus*, *P. fossulatus*, *P. sapidus*, and *P. ulmarius* are distributed in subtropical parts of the world. *Agaricus* and *Volvariella* are considered to be temperate and tropical strains, respectively.

Besides species of *Agaricus*, *Volvariella*, and *Pleurotus*, several species of many genera from the classes Ascomycetes and Basidiomycetes (Table 2 and Table 3) are considered to be edible mushrooms (5). The taxonomic relationships of the 16 genera of cultivated mushrooms listed in Table 2 indicate that the members of the Basidiomycetes belong to 10 families, five orders, and two subclasses. Only *Tuber* is a member of the class Ascomycetes.

B. Morphology

Agaricus mushroom is preferred in the immature button stage with the cap (pileus) unopened, the gills (lamellae) not visible, and the stem (stipe) plump rather than elongated (Fig. 1). In *Agaricus* buttons, the lamellae already are formed but are covered by a membranous layer, the veil, which extends from the cap margin to the stipe (14). Following harvest, however, the mushrooms continue to grow and therefore have a short shelf life. Quality deterioation takes place due to physiological and morphological changes, which lead to breaking of the veil, expansion of the cap, elongation of the stem, and shriveling and darkening of the gills, which makes the mushroom unattractive to consumers and reduces its economic value (14).

Button and egg stages of *Volvariella* species are sold in the market at a premium price. Both stages are ovoid in shape. In the button stage, the whole structure is wrapped in a coat called the universal veil. Inside the universal veil is the closed pileus. The stipe can only be seen in the longitudinal section of the whole structure. In the egg stage, the pileus is pushed out of the veil, which remains to form the volva. The pileus in these two stages is similar to that noticed in the mature and elongation stages. Under a microscope, the lamellae of the egg stage do not bear basidiospores, though there may be basidia during sterigmata formation. In the button stage, only the cystidia and paraphyses are seen (5).

Fruit bodies of *Pleurotus* are characterized by an off-center stalk, which may be small or long or even absent; annulus and volva are lacking. The fruit bodies appear like petals of a flower

TABLE 2 Taxonomic Relationships of Genera of Cultivated Mushrooms

Genus	Family	Order	Subclass	Class
Agaricus	Agaricaceae	Agaricales	Holobasidiomycetidae	Basidiomycetes
Auricularia	Auriculariaceae	Auriculariales	Phragmobasidiomycetidae	Basidiomycetes
Coprinus	Coprinaceae	Agaricales	Holobasidiomycetidae	Basidiomycetes
Dictyophora	Phallaceae	Phallales	Holobasidiomycetidae	Basidiomycetes
Flammulina	Tricholomataceae	Agaricales	Holobasidiomycetidae	Basidiomycetes
Hericium	Hericiaceae	Aphyllophorales	Holobasidiomycetidae	Basidiomycetes
Hypholoma	Hypolomataceae	Agaricales	Holobasidiomycetidae	Basidiomycetes
Kuehneromyces	Strophariaceae	Agaricales	Holobasidiomycetidae	Basidiomycetes
Lentinus	Tricholomatacae	Agaricales	Holobasidiomycetidae	Basidiomycetes
Pholiota	Strophariaceae	Agaricales	Holobasidiomycetidae	Basidiomycetes
Pleurotus	Tricholomataceae	Agaricales	Holobasidiomycetidae	Basidiomycetes
Stropharia	Strophariaceae	Agaricales	Holobasidiomycetidae	Basidiomycetes
Tremella	Tremellaceae	Tremellales	Phragmobasidiomycetidae	Basidiomycetes
Tricholoma	Tricholomataceae	Agaricales	Holobasidiomycetidae	Basidiomycetes
Tuber	Tuberaceae	Tuberales	Hymenoascomycetidae	Ascomycetes
Volvariella	Pluteaceae	Agaricales	Holobasidiomycetidae	Basidiomycetes

Source: Ref. 5.

Table 3 Taxonomic Relationships of Genera That Have Not Yet Been Cultivated

Genus	Family	Order	Subclass	Class
Amanita	Amanitaceae	Agaricales	Holobasidiomycetidae	Basidiomycetes
Armillaria	Trichomataceae	Agaricales	Holobasidiomycetidae	Basidiomycetes
Boletus	Boletaceae	Agaricales	Holobasidiomycetidae	Basidiomycetes
Cantharellus	Cantharellaceae	Aphyllophorales	Holobasidiomycetidae	Basidiomycetes
Clitopilus	Entomolataceae	Agaricales	Holobasidiomycetidae	Basidiomycetes
Hydnum	Hydnaceae	Aphyllophorales	Holobasidiomycetidae	Basidiomycetes
Lactarius	Russulaceae	Agaricales	Holobasidiomycetidae	Basidiomycetes
Lepiota	Lepiotaceae	Agaricales	Holobasidiomycetidae	Basidiomycetes
Marasmius	Tricholomataceae	Agaricales	Holobasidiomycetidae	Basidiomycetes
Morchella	Morchellaceae	Pezizales	Hymenoascomycetaceae	Ascomycetes
Peziza	Pezizaceae	Pezizales	Hymenoascomycetaceae	Ascomycetes
Pholiota (= *Agrocybe*)	Strophariaceae	Agaricales	Holobasidiomycetidae	Basidiomycetes
Psalliota	Agaricaceae	Agaricales	Holobasidiomycetidae	Basidiomycetes
Rhodopaxillus (= *Tricholoma*)	Tricholomataceae	Agaricales	Holobasidiomycetidae	Basidiomycetes
Russula	Russulaceae	Agaricales	Holobasidiomycetidae	Basidiomycetes
Termitomyces	Amnitaceae	Agaricales	Holobasidiomycetidae	Basidiomycetes

Source: Ref. 5.

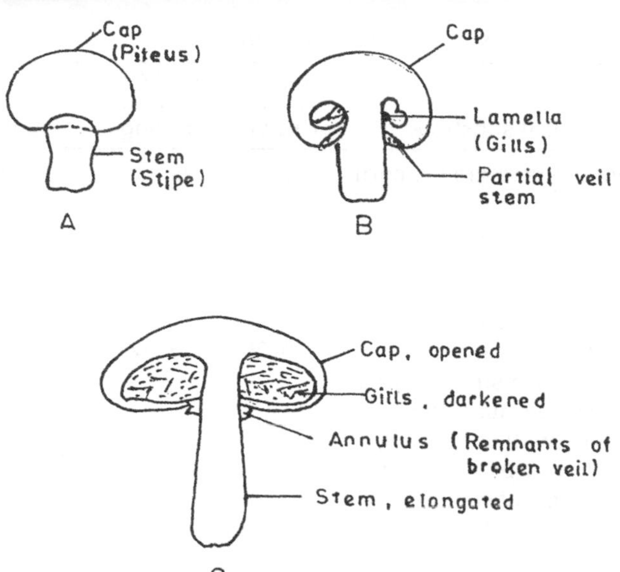

Figure 1 (A) The immature button stage of cultivated commercial mushroom *Agaricus* sp. (B) Vertical section through the button stage of *Agaricus* sp. showing lamellae and partial veil. (C) Vertical section through mature stage of *Agaricus* sp. showing elongated stem and expanded and opened cap with darkened gills on the underside.

in a cluster or individually (8). They open up like an oyster shell with the widest margin away from the stalk. Size varies from species to species and within the same species when cultivated under different climatic and nutritional conditions. Generally, the fruit bodies measure a few to several centimeters in width; the minimum size is about 2–3 cm and maximum size around 15–20 cm. The margin may be smooth, broken, or slightly serrated or dented, depending upon the species (8).

III. PRODUCTION

Spawn is a pure culture of mushroom used as the seeding material for cultivation. It is prepared by inoculating pure culture into sterilized wet natural lignino-cellulosic wastes or onto a starchy material like wheat grains under aseptic conditions. The cultivation aspects of *Agaricus* (15,16), *Pleurotus* (17–24), and *Volvariella* (25–30) are described in detail.

A. Cultural Methods

1. Solid-State Culture

The growing of mushrooms requires a support material or substrate to provide good anchorage, to ensure good aeration, and to provide the nutrients necessary for growth (31). The substrate, temperature, and light requirements of commonly grown mushrooms are presented in Table 4.

The first step in mushroom production is to obtain a pure starter culture of the variety to be grown. The starter is then used to prepare spawn by growing it on a pasteurized substrate. The substrate for species including *Volvariella* and *Pleurotus* are usually based on agricultural

TABLE 4 Growth Conditions for Four Commonly Cultivated Mushrooms

Variety	Typical substrate	Temperature (°C) Growth	Fruiting	Light
Button (or field) mushroom (*Agaricus* species)	Spent brewers grains, corn cobs, hay Rice straw, fertilizer, calcium carbonate Horse manure, brewers grains, gypsum	30	25	No
Straw mushroom (*Volvariella* species)	Used tea leaves, cotton waste, rice straw, coffee hulls, coir dust, sawdust mixed with corn meal; each may be composted or uncomposted and several substrates may be mixed	20–25	28–32	No
Oyster mushroom	Cotton waste, rice straw, coir dust, sawdust mixed with corn meal; each may be composted or uncomposted	25–32	25	Yes
Oak (or Shiitake)	Logs of *Leucaena* species or other trees, sawdust compacted into polypropylene bags	22–27	15–20	For maturation only

Source: Ref. 20.

residues such as chopped straw, sawdust, bagasse, or corn cobs (32–44). Rice bran and 1.0% lime may be added to adjust the acidity of the substrate (3). A building with humidity and temperature-control facilities is also used for mushroom cultivation. Steam boilers are used to pasteurize the beds. Once the beds are prepared, steam is introduced into the growing house for about two hours until the air temperature has risen to 60–62°C. This is maintained for another 2 hours and then lowered to 52°C and held at this temperature for a further 8 hours. Finally, the temperature is allowed to fall to about 35°C over next 12–16 hours and then the bed is ready for adding the spawn. This is added at 0.4% by weight of the bed (2).

Unpasteurized substrate made from straw can be prepared by tying it into bundles, soaking it overnight, and piling it into heaps of three or four layers with spawn broadcast between the layers. The heap is then compacted until 30–60 cm thick. In another method, straw, cotton, and agricultural waste and other substrate are soaked separately and then compacted in layers into wooden frames 30 cm high × 30 cm wide × 100 cm long with spawn between the layers.

In the case of sawdust, fresh sawdust is composted by soaking and mixing it with 1% urea and 1% lime and then storing it in a covered heap for 30–40 days, turning it every 7 days. In a shorter method, 78% sawdust is mixed with 20% rice bran, 1% sugar, and 1% lime and composted for 7 days. The compost is then compressed into blocks or into polypropylene bags. The substrate is steamed at either 100°C for 2–3 hours or 60–70°C for 6–8 hours, cooled, and then spawn is added. The spawn is allowed to grow into full mycelium. This requires 10–15 days for *Volvariella* species, 2–4 weeks for *Pleurotus* and *Agaricus* species, and 2–3 months for *Lentinula* species, depending on the climatic conditions. After full development of mycelium when the mold has fully penetrated the substrate, the mushrooms begin to develop first as buttons and later as umbrellas or flowers. *Agaricus* and *Volvariella* are picked as buttons, the others as flowers.

2. Submerged Culture

The technology of mushroom fermentation in liquid culture was mainly adopted from the antibiotic industry, where fungal pellets are grown in aerated agitated fermenters under controlled conditions of pH and temperature. It is reported that pellets of submerged mycelium develop more flavor than filamentous mycelium due to the improved rheological properties of the culture medium and the autolysis that occurs in the center of the pellets (1).

Several media have been used for submerged mushroom fermentation (1). Agitation and aeration rate, special additives, and the time and method of harvest are also important for growth, pellet formation, and flavor generation. The choice of appropriate inoculum concentration and propagation procedure require special attention, since mushroom mycelium is multicellular and does not produce spores in submerged culture. High biomass yield is also essential for commercial multiplication of mushroom culture (1).

B. Disorders, Contaminants, and Diseases

Mushroom cultivation is a very carefully controlled biological system, which aims to produce a maximum crop of good-quality mushrooms. The success of the system is dependent upon many interacting factors. Disorders, contaminants, and diseases affect yield and quality of mushroom.

1. Disorders

Ryall and Lipton (40) stated that the principal postharvest disorders of cultivated mushrooms result from too slow handling of the product and high temperatures. Excessively high relative humidity causes unattractive elongation of the stem and a slimy surface unless the temperature is low. Slow handling of mushrooms at a high temperature results in brown discoloration of the cap and wilting of the entire structure.

Inadequate oxygen supply and increased atmospheric CO_2 (1–2% v/v) can lead to proliferation of the fruit bodies (41). Similarly, inadequate aeration coupled with lack of suitable illumination is known to cause the developent of long slender stalks in the fruit bodies of *Pleurotus flabellatus* (41). Zadrazil and Schneidereit (42) have observed typical abnormal fruit bodies of *P. ostreatus* grown under extreme climatic conditions.

There are different types of disorders, the causes of some of which are not known. In general, biotic disorders do not result in significant loss in mushroom yield.

Distortion

Many forms of distortion can occur, which are mostly induced by fluctuating or unsatisfactory environmental conditions. These include hollow stem, split stipes (Fig. 2), swollen stems, hard gills, and misshapen mushrooms.

Waterlogging

Symptoms of waterlogging appear as clear water-soaked areas, particularly in the stem. There is exudation of water from waterlogged mushrooms if squeezed, and in severe cases, a spontaneous

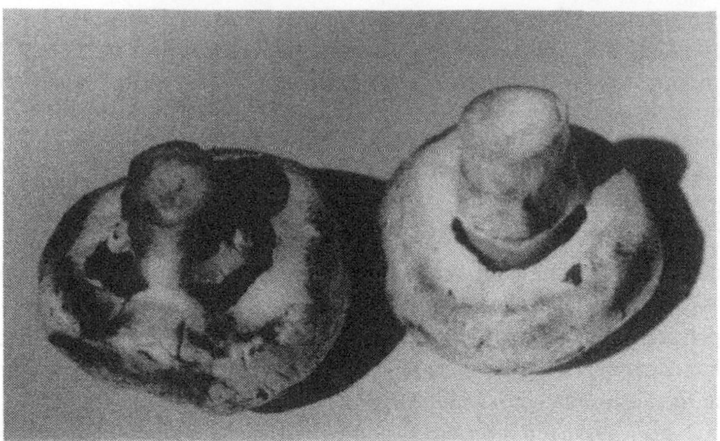

FIGURE 2 Mushroom viewed from below showing splitting stipes.

release of large quantities of clear or colored liquid from the mature mushroom, which subsequently collapses. These symptoms also appear in mushrooms affected by viral and bacterial diseases.

Pinning Disorders

There is a wide range of pinning disorders with more or less identifiable causes. These disorders are more common and often more devastating than other disorders, as they affect both the management of harvest and the total yield. These disorders include mass pinning, clumping, stroma pin death, and rosecomb.

2. Fungal Contaminants

Growth of mushrooms depends on the suitability of the prevailing ecological conditions for the growth of mushroom in preference to other fungi that compete for the same substrate. *Trichoderma* and *Plicaria* are weed molds found in the compost during the cultivation of *Agaricus brunnescens* (45). Fletcher (44) has described the loss of mushroom yield due to the occurrence of *Papulaspora byssina*, *Conidiobolus coronatus*, *Arthrobotrys superba*, *Paecilomyces variotii*, and *Peziza ostrocoderma* as weed molds in mushroom compost. Rajarathnam and Bano (41) observed a number of fungal contaminants on rice straw beds during the growth of *Pleurotus flabellatus*, including *Sclerotinum rolfsii*, *Penicillium digitatum*, and *Mucor javanicus*. Under conditions of poor mushroom mycelium growth, species of *Coprinum* and *Papulaspora* were also recorded (41).

The most common contaminants affecting mushroom production are *Trichoderma viride*, *Chrysosporium luteum* and *C. sulfurum*, *Populaspora byssina*, *Atrobotrys superba*, *Peziza ostracoderma*, *Mucor javanicus*, *Coprinus plateus*, *Stepulariopsis firuicola*, *Chaetomium olivaceum*, *Neuspora crassa*, *Sporendonema purpurescens*, and *Sepedonium* spp.

Trichoderma viride *(Green Mold)*

This mould occurs on the spawned and cased trays. A dense pure white growth of mycelium may appear on the casing surface, developing a green center as spores are produced (Figs. 3 and 4).

FIGURE 3 Contamination of bed with *Trichoderma viride* showing green-colored growth of the pathogen.

Mushrooms grown in or near this mycelium are brown and may crack and distort. It affects the spawn run and checks the pin formation of the mushroom. This fungus grows on decomposed organic matter and dead mushroom tissues. Improper pasteurization of compost and high humidity are also responsible for the spread of this fungus. The spores of this fungus are carried by air, water, and careless handling. It can be avoided by spraying with 0.5% benomyl.

Chrysosporium luteum *and* C. sulfurum *(Mat and Confetti)*

Both of these species are mostly yellow colored and cause considerable yield reduction when they occur in large quantities. Spores of both species become airborne easily and cause trouble if they contaminate casing on compost. Soil is also a common source of both species. Control of this disease is best achieved by paying strict attention to hygiene.

Populaspora byssina *(Brown Plaster Mold)*

This mold occurs on spawned and cropping trays. It produces characteristically large dense roughly circular patches of mycelium on the surface of the casing, initially whitish, later turning brown and powdery. It is also found to colonize the compost. The presence of this fungus is associated with wet compost, which can result from both undercomposting and overcomposting. This fungus spreads very quickly and causes great reduction in yield. Very wet compost, high temperature (20–32°C) during spawn run, and cropping at more than 18°C encourages infection. Properly prepared compost, proper watering, maintaining the required temperature during the spawn run and cropping, and 2% formalin spray can control the fungus.

FIGURE 4 Well-developed colony of *Trichoderma viride* causing blackening and rotting of straw in the bed.

Arthrobotrys superba *(Brown mold)*

This fungus also produces a brown colony on the casing surface at all stages of cropping, tending to be more common at the end of the crop. It is parasitic on nematodes and its presence indicates a high nematode population in the compost and casing. A preparation of this fungus is sold in France to add to compost or casing as a means of improving yield, possibly by attacking nematodes.

Peziza ostracoderma *(Cinnamon Mold)*

This mold generally occurs on the surface of the casing before the first flush. The colonies are circular and white-brown at the center. It causes a reduction in yield by depleting the nutrient status of the compost and also results in a small delay in cropping.

Coprinus plateus *(Inky Cap)*

This disease can be detected by the appearance of a long slender stalk with a small thin cap, which dissolves into black inky liquid. The appearance of this fungus indicates the presence of ammonia in the compost.

Scopulariopsis fimicola *(White Plaster Mold)*

This organism closely resembles brown plaster mold. This fungus produces dense white patches of mycelium and spores on the casing surface and in the compost. Other mold fungi initially produce white mycelium and later change color on aging, but *Scopulariopsis fimicola* remains white. The mycelial growth in this fungus is sometimes so dense that it covers the complete bed surface and reduces yield by competing with mushroom mycelium.

Chaetomium olivaccum *(Olive Green Mold)*

This fungus appears in the compost or spawned bed before casing. Initially it is white, but later it changes to an olive-green color (Fig. 5). It is a fairly common mold and reduces yield in proportion to the extent of colonization of the compost. The fungus is identified by the olive-green fruiting structures, which are about the size of a pinhead and with a rough spiny appearance. These are produced in large numbers on the straw, often well distributed and clearly visible to the naked eye (Fig. 6). Improper pasteurization of the compost, inadequate ventilation, and too wet compost are the reasons for the development of this fungus. Sufficient aeration should be introduced with increasing the temperature above 60°C during

FIGURE 5 Contamination of bed with *Chaetomium olivaceum* showing grey-white mycelial growth of pathogen.

pasteurization. Spraying of trays with 0.2% Thiram and Captan or 0.05% Benlate can check the spread of this mold.

Neospora crassa *(Fire Mold)*

This fungus generally occurs after cook-out, particularly when the crop is left in situ for some time after treatment. The mycelium is first a creamy white but rapidly turns orange. Large numbers of spores are produced, and it is difficult to eliminate once established as it repeatedly reinfects the cropping cycle.

Sporendonema purpurescens *(Red Geotrichum or Lipstick Mold)*

This fungus produces a fine white mycelial growth on the surface of the casing or in the compost, which turns bright pink and finally buff with a powdery appearance as the spores are produced. The spores are airborne. This fungus inhibits the growth of mushroom mycelium and causes crop loss.

Sependonium *spp. (*Sependonium *Yellow Mold)*

This fungus develops in the compost. It is initially white, turning yellow to tan with age. It produces numerous spores, which becomes airborne and can contaminate compost during

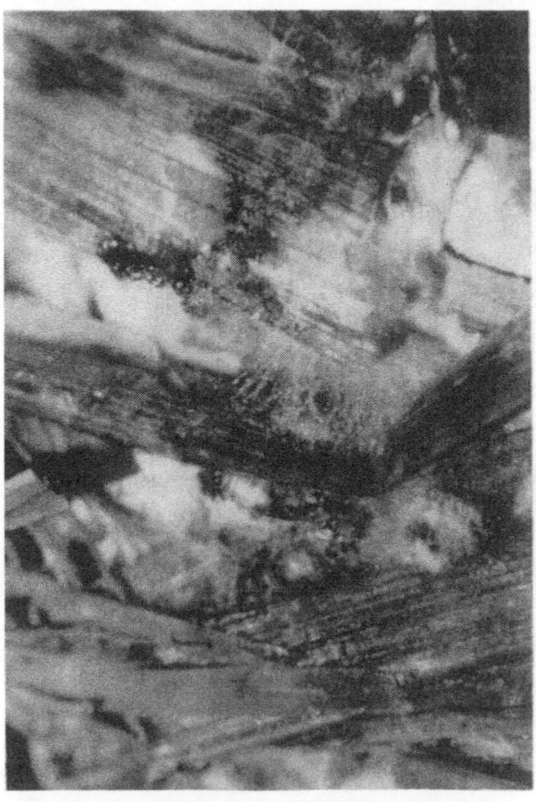

FIGURE 6 Contamination of bed with *Chaetomium olivaceum* causing blackening and rotting of straw in the bed.

preparation. The large spherical spores of this fungus are observed to be heat tolerant. It affects mushroom growth and reduces crop yields.

3. Diseases

Fungi, bacteria, and viruses are the important groups of pathogens inciting several diseases of mushroom crops and causing reduction in yields. Predominant among them are *Mycogene perniciosa* (wet bubble), *Verticillium fungicola* var. *fungicola* (dry bubble), *Hypomyces rosellus* (cobweb), *Diehliomyces microsporus* (false truffle), and *Mortierella bainieri* (shaggy stipe) of fungal origin, *Pseudomonas tolaassi* (bacterial blotch or brown blotch and also mummy disease), and viruses.

Mycogone perniciosa *(Wet Bubble)*

When this fungus infects mushroom, it covers the mushroom with a dense white mat of mycelium, causing the development of distorted mushrooms that are initially white and fluffy but become brown as they age and decay. These are known as sclerodermoid masses. They usually result from the infection of a developing mushroom 10–14 days before symptoms are noticed. Small amber to dark brown drops of liquid develop on the surface of undifferentiated tissue when moisture or humidity is high and results in wet decay. When mature mushrooms are infected, the base of the stalk only may be affected, causing a brown discoloration and forming a white fluffy mycelial outgrowth.

The casing soil may be the source of infection, but other agencies such as unpasteurized compost, high humidity, and high moisture are also responsible for infection and development of the disease. Temperatures above 17°C are favorable for development of the disease. Control of this disease is best achieved by sterilizing diseased mushrooms on the beds with 2% formalin and promptly removing them from the beds. The infected area can be sprayed with 0.2% mancozeb and 0.05% benomyl. Benomyl may be mixed with the casing soil.

Verticillium fungicola *var.* fungicola *(Dry Bubble)*

This fungus causes light brown spots on the cap, which coalesce and cover most of the area of the cap resulting in irregular patches. It produces dark brown spots (Figs. 7 and 8) with diffused edges, which later induce rotting of the affected portion and cap deformation. In advanced stages of the disease, the infection progresses into the stipes, causing development of brown strips in the stipes (Fig. 9).

The primary source of infection is the casing soil or airborne spores. For control of the disease, temperature control during cropping and proper ventilation are advisable. Three sprays of Zineb (0.2%) at the time of casing, pinhead formation, and after two flushes of crops were found most effective (45).

Hypomyces rosellus *(Cobweb)*

This fungus covers affected mushrooms by the growth of mycelium. The affected mushrooms turn brown and rot. The casing soil is the primary source of this fungus. This disease is easily controlled by use of fungicides as soon as it occurs. The affected mushrooms should be carefully removed and affected spots should be treated with a benzimidazole fungicide (0.05%). The transfer of contaminated casing soil to heatlhy crop should be avoided.

Diehliomyces microsporus *(False Truffle)*

This fungus attacks mushroom mycelium and causes mycelial death. The first symptom is the patchy appearance of the mushroom crop with areas of the affected bed failing to produce

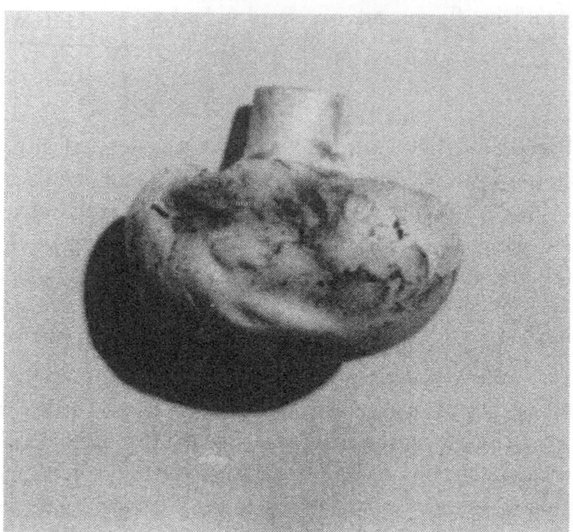

FIGURE 7 Cap spotting caused by *Verticillium fungicola* var. *fungicola*. The spots are dark brown with a diffuse edge.

mushrooms. Severe crop loss results when the crop is affected before, at, or during spawn running, with yield reduction as high as 75%. Control of this fungal disease can be achieved by preventing soil contamination of compost, removing all debris from the boxes and treating them with chemicals between every crop and not allowing the temperature of the compost during spawn run to exceed 21–24°C.

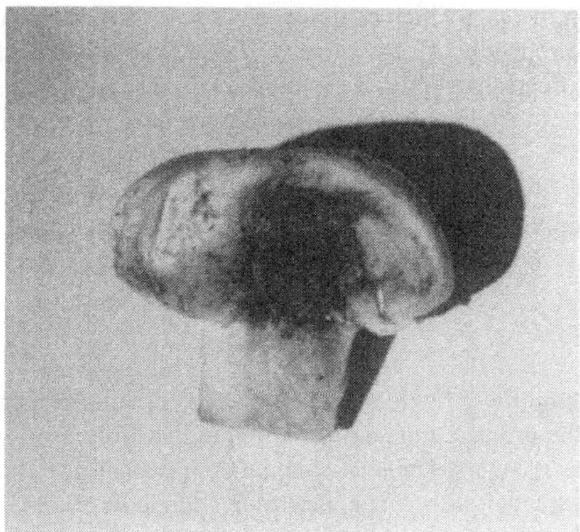

FIGURE 8 Crop spotting caused by *Verticillium fungicola* var. *fungicola* inducing rotting of the affected portion and deformation of cap.

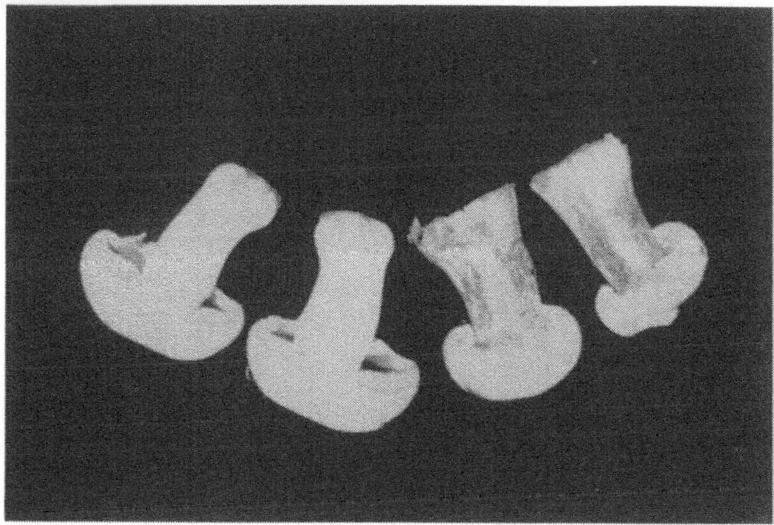

FIGURE 9 Cross section of affected cap and stipes showing infection of *V. fungicola* var. *fungicola* in stripelike pattern in the stipes.

Mortierella bainieri *(Shaggy Stipe)*

The coarse grey-white mycelium of this pathogen grow on the stalk of the mushroom and also over the surrounding casing. The affected stalk is peeled, giving a shaggy appearance. The stalk and cap are discolored, becoming dark brown as the disease progresses. The cap may also develop a brown blotch surrounded by a yellow ring. For control of the disease, all diseased mushrooms should be carefully removed and the affected bed areas treated with Zineb.

Pseudomonas tolaassi *(Bacterial Blotch and Mummy Disease)*

The occurrence and nature of loss in mushroom yield due to bacterial diseases of *Pleurotus* species have been documented (41). The pathogen has been identified as *Pseudomonas tolaassi*, similar to the one affecting *Agaricus* crops. This bacterium causes brown, slightly sunken blotches on the surface of the cap. These spots are irregular, yellow to dark brown (Fig. 10), and coalesce in the later stages. Severely affected mushrooms may be distorted, and the caps may split at the blotchy areas. *Pseudomonas tolaassi* also causes the mummy disease. The most characteristic feature of this disease is its fast rate of spread, which is reported to be 10–25 cm bed length per day. When affected stipes are observed in cross section, they show small dark brown spots. The caps remain small (Fig. 11), and growth ceases after reaching button size. The affected mushrooms later turn grey or brown and become dry, leathery, and mummified (Fig. 12). The internal tissue of the drying mushrooms shows discoloration with dark brown streaks.

Pseudomonas agarici *(Drippy Gill)*

This bacterium infects the gills even before the veil of the mushroom is broken. Affected gills remain underdeveloped showing small brown decaying areas with creamy-white bacterial ooze

Figure 10 Small light brown spots on the cap surface, typical symptoms of bacterial blotch caused by *Puseomonas tolaassi* (From Ref. 45.)

on them, hence the name drippy gill. The disease may originate from infected mycelium, and flies, pickers, and water flash spread the bacterium within the crop.

Viral Diseases

Several viral diseases cause damage to mushrooms, including La France (46), brown disease and watery stipe (47), X-disease (48), and dieback disease (49). Of the six viruses isolated from diseased mushrooms, five are polyhedral, having particles 19, 25, 29, 35, and 50 nm in diameter, and one is bacilliform, with particles measuring 19 × 50 nm (50–52). Various symptoms may be attributed to viral diseases, although none of them are entirely diagnostic. Distortion of the sporophores have been reported including elongation of the stalks, tilting of the caps, and very small caps on normal-sized stalks. Affected crops may show a patchy appearance. In other crops, there may be an overall deterioration and reduction in cropping without very distinct symptoms. Early maturity of sporophores or slight discoloration of mushrooms is also observed. Crop loss has been considerable in severe cases. Strict hygiene is essential if viral diseases are to be controlled.

C. Pests

Mushrooms are also attacked by insect pests. Sciarids, phorids, cecids, mites, nematodes, and springtails infest the fruiting bodies and cause rotting. Some pests cause damage to the spawn and hence lead to reduction in yield. They lay eggs and the larvae feed on the compost, feed on the mycelium, and burrow into the mushroom stalk.

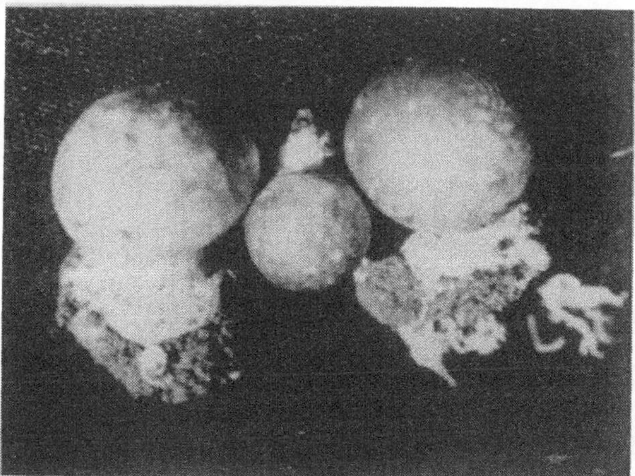

FIGURE 11 Small-sized cap infected with mummy disease.

1. Sciarids (Mushroom Flies)

Several species of sciarids have been reporded as infesting mushroom crops. The most common species are *Lycoriella soluni*, *L. auripila*, *Neosciara pauciseta*, *Sciara carpophilla*, *S. multiseta*, and *S. agaria*. These flies are dark-colored and have slender bodies with long antennae, which are held characteristically erect (Fig. 13). Flies themselves cause little harm to the mushroom bed, but their larvae are more harmful. The most obvious larval damage is tunneling of the stipes. The most serious injury is the attack on developing pinheads and buttons; the mycelium attachment can be severely attacked, causing pinheads to become brown and leathery, or the pinheads can be hollowed out producing a spongelike mass or even be consumed entirely. Control of mushroom flies is best achieved by using sticky traps in the spawn-running room,

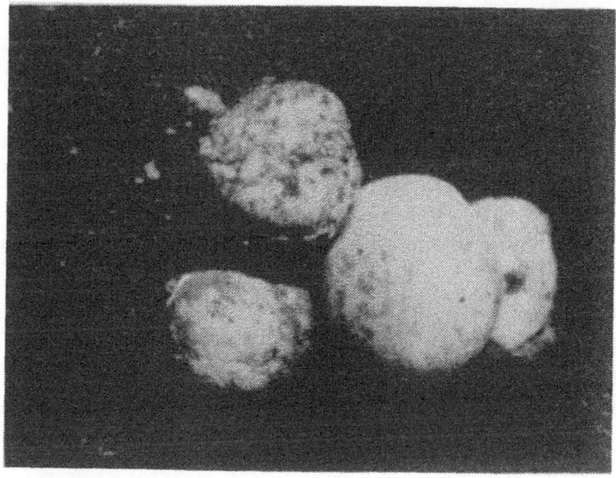

FIGURE 12 Severely infected mushrooms with *Pseudomonas tolaassi* showing mummied mushrooms.

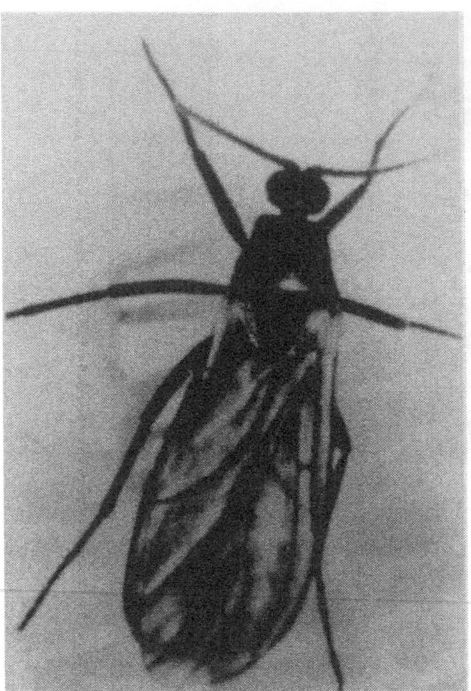

FIGURE 13 Adult mushroom sciarid, *Lycoriella auripila*. (From Ref. 45.)

using knock-down sprays during the first week of spawn-run, and the first week after casing, using pyrethrins during cropping, and removing all spent compost from the farm after mushrooms are completely harvested.

2. Phorids

The common species of phorids attacking mushroom are *Megaselia halterata*, *M. nigra*, *M. agarici*, *M. bovistra*, and *M. flavinervis*. These have short antennae and wing venation (Fig. 14). They are brown-black in color and are generally stouter than sciarids. The larvae are creamy-white legless maggots with pointed heads. The larvae feed solely on mushroom mycelium and do not burrow into mushrooms. The flies are very active near lights and therefore can be a considerable nuisance to pickers. They are also vectors of *V. fungicola* var. *fungicola*, and a relatively low density of flies (75/m^2) is capable of spreading the pathogen and initiating the dry bubble disease.

Recommendations for the control of phorids include incorporation of diazinon into the compost at spawning, use of pyrethrins during cropping to kill flies, use of sticky traps to monitor the number of flies in the spawn-running room, use of effective screens (40 mesh/inch) for ventilation ducts in the spawn-running room, protection of the crop for 3 weeks after spawning with applications of dichlorvos, and removal of the spent compost from the farm after harvesting of mushrooms (4).

3. Cecids

Six species of cecid have been recorded on mushrooms but only four are common: *Heteropeza pygmaea*, *Mycophila speyeri*, *M. barnesi*, and *Lestremia cinerea.* These are tiny orange-black

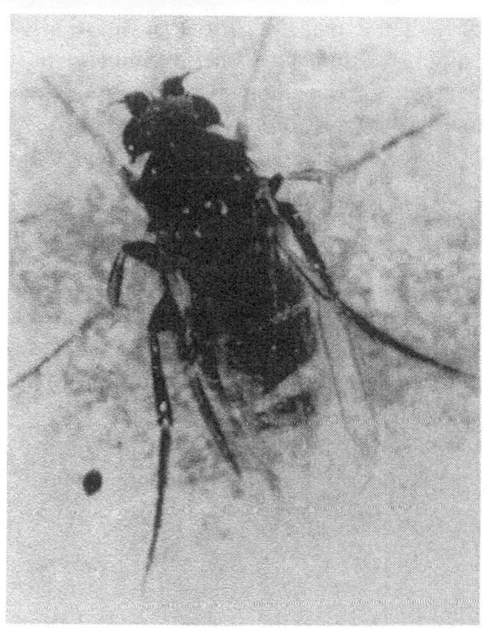

FIGURE 14 Adult mushroom phorid, *Megisella halterata.* (Source: Ref. 45.)

flies, which are seldom seen. They eat the mycelium but mostly feed on the outside of the stipes or at the junction of stipes and gills and cause losses in marketable yield. The bacterial pathogens of the mushroom are also carried on their skin, and these pathogens cause brown, discolored strips on mushroom stipe and gills. The delicate gill tissue can then break down to produce tiny pustules of black fluid. Control of cecids can be achieved by observing strict hygiene on the farm, isolating infested houses with Sudol foot dips, incorporating diazinon in the compost at spawning, dipping wooden trays in 2% sodium pentochlorophorate, and removing all the spent compost from the farm after complete harvesting of mushrooms.

4. Mites

Many species of mites have been recorded on mushrooms. The most common of these are *Tansonemus myceliophagus*, *T. floricolus*, *Pygmephorus americanus*, *T. putrescentiae*, *T. longior*, *Histiostoma gracilipes*, *Parasitus fimetorum*, *Digamasellus fallax*, *Arctoseius cetratus*, *Linopodes antennaepes*, and *Caloglyphus* spp. These mites (*T. myceliphagus* and *T. floricolus*) cause a reddish-brown discoloration and rounding of the base of affected mushroom stipes, sometimes severe on the basal attachments of the mushroom. The mites of *P. americanus* often swarm in large numbers on the surface of the casing and mushrooms, giving them a reddish-brown color, and therefore they have been called red pepper mites. Other species of mites (*Tyrophagus*, *Calogyphus*, *Histiostoma*) do not cause damage to the mushroom but are carriers of the inoculum of pathogens of disease. Species of *Parasitus, Digamasellus* and *Arctoseius* feed on various stages of mushroom pests and can be regarded as beneficial, but they can be a source of irritation to pickers.

5. Nematodes

Three species of mycophagous nematodes, *Ditylenchus mycelophagus*, *Aphelenchoides compostícola*, and *A. oesterocaudatus*, are primary pests of mushroom since they feed extensively on fungi and destroy mushroom mycelial growth. These parastiic nematodes have a spikelike structure called a stylet. The nematodes project the stylet, puncture the mycelium, and suck its components. They puncture the mycelium at several points and thus the mycelium start to lose their growing ability. Nematodes also feed on the mushroom, turning them brown, watery, and stunted. For control of nematodes, compost should be properly prepared and casing soil should be free of nematodes.

6. Springtails

These are quite tiny and cannot be seen with naked eye. They have stout antennae. They can crawl with speed, but they move by springing several inches into air. When they are in a mass, they look like gunpowder on beds. They can feed on both mycelium and sporophores, resulting in minute open pits in the stem and cap. From these pits, dry branched tunnels can be formed. Common species are *Achorutes armatus*, *Lepodocyrtus cyaneus*, *L. lanuginosus*, *Isotoma simplex*, *Prerstoma minuta*, and *Xenylla* sp. The springtails can be controlled by efficient pasteurization of the surface layers of manure during composting and keeping the cropping hosue floor free from all organic debris.

D. Harvesting

Mushrooms are harvested based on consumer preference in the markets. It is customary to harvest mushrooms of *Agaricus bisporus* as soon as caps attain a diameter of 2.5 cm, just before they begin to open. This type is most suitable for canning. The individual mushrooms are plucked up by twisting gently clockwise and then counterclockwise, after which they are pulled up very softly. The lower portion of the stipe is cut with a sharp knife and is kept in a trash box, while the cleaned mushrooms are collected in another box. If there are many pinheads around the mushroom to be picked, then it is advisible to cut that mushroom with a sharp-edged knife so that the nearby pins are not disturbed. Otherwise those pinheads will not grow into buttons and will turn yellow. When all mushrooms of the desired size have been harvested, the next stage is to fill up the holes with sterilized casing soil. The surface of the beds should be kept quite level and new casing should be made firm by giving a gentle pat.

Because the time period between the button and open stage of *Agaricus bitorquis* is quite short, it is advisable to harvest mushrooms of this species at the proper time. As this mushroom requires a high concentration of carbon dioxide and high temperature for its growth, harvesting of this mushroom can be uncomfortable for the picker. Hence, ventilation and circulation of clean air can be temporarily increased before and during picking.

Paddy straw mushrooms (*Volvariella* spp.) are usually not left to grow to their maximum size but are harvested for selling before the volva breaks (i.e., at the button stage) or just after rupture (i.e., at the egg stage). Since these mushrooms grow at rather high temperature and humidity conditions, they are fast growing under normal conditions, and the first crop of mushrooms is usually harvested 10–15 days after spawning. The first crop normally constitutes three successive days of harvesting during which 70–90% of the expected yield is obtained. Five to seven days after the first flush, the second crop is ready for harvesting, and this crop may also require 2–3 days for harvesting, but gives a lower yield of 10–30% of the total crop. In order to harvest mushrooms of good quality, the straw mushrooms are harvested twice a day, in the morning and in the afternoon, and sometimes an additional crop is picked at noon. When the fruiting bodies reach harvesting size, they are carefully separated from the bed/substrate base by lifting, while shaking slightly to the left and right, and then twisting them off. Mushrooms should not be cut off with knives or scissors from the base of the stalk, because the stalks left behind might cause rot or be attacked by pests and contaminated by molds which could, in turn, destroy the mushroom bed.

The harvesting standards for oyster mushrooms (*Pleurotus* spp.) are different for different purposes. They are harvested before the mushrooms show slightly curled edges. For some uses, small caps measuring 20–25 mm are required. The fresh mushrooms are usually sold after packing in plastic cases.

E. Grading and Packaging

The grading and packing of mushrooms are done in the packing house. While some growers prefer to trim the mushroom stalks prior to packing, others grade their crop as they pick, carrying them in groups of three to four chips each to receive a particular grade. Grading may also be done by going around several times, first picking one grade, then another, and then the remaining. The Harding Committee (4) recommended the following grades.

1. *Buttons*—Membranes securely intact, which are likely to remain unbroken for 24 hours.
2. *Cups*—Membrane breaking or broken, i.e., the mushroom in an open stage of maturity, the cap retaining a pronounced downward curve.
3. *Opens*—Those advanced beyond the cup stage.

In all grades, the cap diameter should be 2.5–6.5 ± 1 cm and the stem length no more than 2.5 cm with a 15–cm stem limit in the case of buttons. Other grades, such as small buttons (< 2.5 cm) and large opens (> 6.5 cm), are also used. Top-grade mushrooms are harvested before the veil is broken at room temperature; the white cap and stem elongate while they continue to grow and the veil breaks, exposing the darkcolored gills. The white cap and stem change to a brownish color, making mushrooms unattractive to the consumer.

IV. CHEMICAL COMPOSITION

The nutritional composition of different species of mushroom is presented in Table 5. They contain high proportions of proteins and minerals (53). The relative amounts of calcium and phosphorus are remarkably high. The carbohydrate content ranges from 46.6 to 81.8% of dry

TABLE 5 Proximate Composition of Mushroom

Species	Moisture	Crude protein[a] (N × 4.38)	Fat	Carbohydrates[a]	Fiber[a]	Ash[a]	Energy[a]
Pleurotus flabellatus	91	21.6	1.8	57.4	11.9	10.7	271
Pleurotus ostreatus	73	10.5	1.6	81.8	7.5	6.1	367
Agaricus campestris	89	26.3	1.8	59.8	10.4	12.0	328
Volvariella diplasia	90	28.5	2.6	57.4	17.4	11.5	304
Lentinus edodes	90	17.5	8.0	67.5	8.0	7.0	387

[a]Values on per 100g dry weight basis.
Source: Refs. 55–58.

TABLE 6 Essential Amino Acid[a] Composition of Mushroom

Amino acids	P. flabellatus	A. bisporus	V. diplasia	L. edodes	Egg prot
Leucine	6.2	7.5	5.0	7.9	8.8
Isoleucine	8.3	4.5	7.8	4.9	6.6
Valine	6.6	2.5	9.7	3.7	7.3
Tryptophan	1.3	2.0	1.5	—	1.6
Lysine	7.5	9.1	6.1	4.3	6.4
Threonine	5.9	6.1	8.4	5.9	5.1
Phenylalanine	2.8	4.2	7.0	5.9	5.8
Tyrosine	2.8	3.8	2.2	3.9	4.2
Cystine	1.1	1.0	3.2	—	2.4
Methionine	1.7	0.9	1.2	1.9	3.1
Arginine	9.5	12.1	9.3	7.9	6.5
Histidine	3.0	2.7	4.2	1.9	2.4
Total EAA excluding arginine and histidine	44.2	41.6	50.1	38.4	51.3

[a]g amino acid/100 g of corrected crude protein.
Source: Ref. 63.

weight (53,54). Mushrooms contain pentosans, hexosans (53), mannitol, and α-α-trehalose along with traces of glucose (55–61). Mannitol helps to maintain the osmotic concentration in the fruit body cell, which is necessary to maintain a water content as high as 90% (62).

Mushrooms are a good source of proteins. The digestibility of mushroom proteins is higher than those from spinach and other leafy vegetables. The essential amino acid compositoin of different species of mushrooms is presented in Table 6 (63).

The fat content in various species of mushrooms ranges from 1.08 to 9.4% on a dry weight basis (63). The mineral and vitamin contents in mushroom are presented in Table 7 and 8. *Pleurotus* mushroom is reported to display some interesting biological properties. The production of an antibiotic substance termed pleurotin (Fig. 15) in the culture filtrate of *Pleurotus*

TABLE 7 Mineral Content of Different Species of Mushroom[a]

Mineral	P. flabellatus	A. campestris	V. diplasia	L. edodes
Calcium (mg/100 g)	24	23	58	118
Phosphorus (mg/100 g)	1550	1429	1042	650
Potassium (mg/100 g)	3760	4762	3333	1246
Iron (ppm)	124	186	177	30
Zinc (ppm)	58.6	—	—	—
Copper (ppm)	21.9	12.8	—	—

[a]Data on dry weight basis.
Source: Refs. 63–65.

TABLE 8 Vitamin Content[a] of Different Species of Mushroom

Vitamin (mg/100 g dry wt)	*P. flabellatus*	*A. bisporus*	*Lentinus edodes*	*Volvariella diplasia*
Ascorbic acid	144	82	n.a.	n.a.
Thiamine	1.46	1.44	0.40	0.32
Niacin	73.3	36.19	11.90	59.5
Riboflavin	7.1	4.95	0.90	2.73
Pantothenic acid	33	22.8	n.a.	n.a.
Folic acid[b]	1222	933	n.a.	n.a.

n.a. = Not available.

[a]Mean value.

[b]μg/100 g.

Source: Ref. 66.

griseus has been demonstrated (67). Pleurotin was found to be active against gram-positive bacteria (68).

Several reports are available (69–79) on flavor compounds in mushrooms. About 50 different volatile compounds have been identified in various mushroom species representing a wide variety of chemical structures including aliphatic alcohols, aldehydes, ketones, esters, lactone, mono- and sesquiterpenes, and aromatic compound such as cinamyl acetate. It is generally thought that a series of C_8 compounds are the primary volatiles contributing to the characteristic flavor of edible mushrooms, and that less volatile C_9 and C_{10} compounds are occasionally present in edible mushrooms. Various compounds, including 3-methylbutanal, butanol, 3-methylbutanol, pentanol, hexanol, furfural, phenylacetaldehyde, and terpineol, have been identified as minor volatile components of mushrooms. The flavor characteristics of several mushroom volatiles are given in Table 9. Trans-2-ocetenal has the lowest

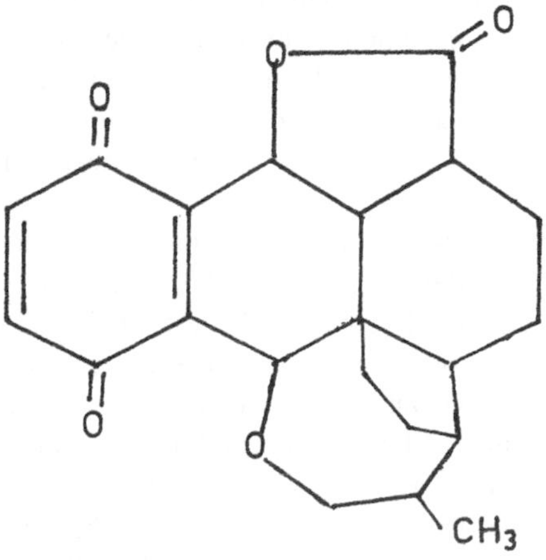

FIGURE 15 Structure of pleurotin.

TABLE 9 Flavor Volatiles Commonly Found in Edible Mushrooms and Their Organoleptic Characteristics

Compound	Threshhold level (μg/ml)	Flavor Description
1-Octanol	0.480	Sweet, detergent of soap
3-Octanol	0.018	Like cod liver oil, weakly nutty
1-Octen-3-ol	0.010	Mushroomlike, raw mushroom, general mushroom, butterlike fungal resinous
Trans-2-octen-1-ol	0.040	Medical, oily, sweet
Trans-2-octenal	0.003	Sweet, phenolic
1-Octen-3-one	0.004	Like boiled mushroom, metallic, wild mushroom
3-Octanone	0.050	Sweet, esterlike fruity, musty, floral
1-Octen-3-yl acetate	0.090	Mushroomlike, soapy
1-Octen-3-yl propionate	0.022	Herbaceous, sweet fruity mushroomlike

Source: Ref. 3.

threshold value, but does not possess a typical mushroom odor, whereas 1-octen-3-ol, 1-octen-3-yl-acetate, and 1-octen-3-yl propionate have low threshold that produce characteristic mushroomlike flavor.

Picardi and Issenberg (72) isolated aroma concentrates characteristic of both raw and cooked flavors of *Agaricus bisporus*. They determined the major volatile constituents of raw mushrooms as 3-octanone, 3-octanol, 1-octen-3-ol, benzaldehyde, octanol, and 2-octen-1-ol. The major volatile constituents of cooked mushrooms were these six compounds plus 1-octen-3-one. During cooking, 1-octen-3-one was formed in detectable amounts only after the mushrooms had been boiled for approximately 15 minutes, and its greatest production occurred after half an hour. A compound isolated from mushroom sporophores of *Armillaria matsutake*, called matsutake alcohol (oct-1-en-3-ol), has been claimed to possess characteristic mushroomlike aroma (80).

The characteristic odor of Shiitake (*Lentinus edodes*), a species of mushroom that has long been a popular flavor food in Japan and China, was identified to be 1,2,3,5,6-pentathiepane (lenthionine) by Wada et al. (81). Threshold values of this compound were in the range of 0.27–0.53 and 12.5–25.0 ppm in water and vegetable oil, respectively. Yasumoto et al. (82) reported the presence of methanethiol, dimethyl sulfide, and dimethyl disulfide in Shiitake. These sulfur compounds also appeared to contribute to the mushroom flavor. Synthesis of lenthionine from formadehyde, methylene chloride, and sodium polysulfide was reported by Morita and Kobayashi (83). A precursor of lenthionine was isolated from the fruiting bodies of the mushroom. It is a sulfur-containing peptide, lentinic acid, that undergoes enzymic reaction to form the odorous substance, lenthionine (84,85).

TABLE 10 Volatile Components Characteristic to Aroma of Mushrooms

Species	Raw	Cooked
Agaricus bisporus	3-Octanone 3-Octanol 1-Octen-3-ol Benzaldehyde Octanol 2-Octen-1-ol	1-Octen-3-one
Armillaria matusutake	1-Octen-3-ol	
Lentinus edodes (Shiitake)	1,2,3,5,6-Pentathiepane (Lenthionine)	

Source: Refs. 55, 56, 80, 81, 83.

Thomas (86) analyzed the flavor of fried mushroom, *Boletus edulis*, and reported nine pyrazines and seven 2-formylphyroles among the 70 constituents (Table 10).

V. STORAGE

Fresh mushrooms (fruit bodies) have a high rate of metabolic activity, which quickly declines and leads to deterioration (53). Several reports (87–92) are available on extending the shelf life of *Agaricus* fruit bodies. Gormley and MacCanna (87) reported that degree of growth (stem elongation and cap opening), discoloration, and changes in texture are the improtant factors in acceptance by the consumers. In the case of *Pleurotus* mushroom, the degree of whiteness, the texture, and the nature of fruit body margin have been considered to be the major characteristics (Table 11) that define the quality of mushroom (52). Bano and Rajarathnam (53) reported that fruit bodies (200 g) of *P. flabellatus* packed in 20-µm-thick polythylene bags (16 × 25 cm) with one pinhole on either side were stored up to 24 hours at ambient temperature (22–28°C). The optimum package density for a 16 × 25 cm polyethylene bag was 200 g. With lower package densities of 50 and 100 g, there was a lower concentration of CO_2 and more discoloration; with higher package density (250 g/bag), there was less browning and a greater accumulation of CO_2. However, packaging of 250 g density caused tearing at the margins of the packed fruit bodies. These results indicated a relationship between the accumulated concentration of CO_2 and the degree of discoloration. It is known that the phenoloxidases responsible for enzymatic browning require oxygen for their activity.

A. Low-Temperature Storage

Fresh mushrooms do not keep well in storage and are, therefore, stored only for short periods (23,24). Deterioration is marked by brown discoloration of the surface, elongation of stalks, and opening of the veils. Freshly picked mushrooms keep in good condition at 0°C for 5 days, at 4–5°C for 2 days, and at 10°C for only 1 day (93). Hence, mushrooms should be cooled as soon as possible (within 5 hours) to 0°C and held at this temperature throughout marketing. Ryall and Lipton (40) recommended hydrocooling to both clean and cool the mushroom simultaneously. Vacuum-cooling can also be adopted, although it causes a slightly higher weight loss (3%) than room-cooling (2.5%). Dehydration is correlated with black stems and open veils. Moisture-retentive film overwraps or film caps usually are beneficial in reducing moisture loss. Deterioration of mushroom can be retarded in overwrapped prepacks (94). Treatment with a solution of salt and sodium bisulfite has been found effective in reducing discoloration (95,96). Pantastico et al. (97) recommended a temperature of 0°C with 95% RH to extend the marketable life of mushrooms for about 10 days.

B. Controlled-Atmosphere Storage

Controlled atmosphere with 6–10% CO_2 has been shown to have beneficial effects on prolonging the shelf life of mushrooms (98,99). Hatton et al. (100) noted the beneficial effect of controlled

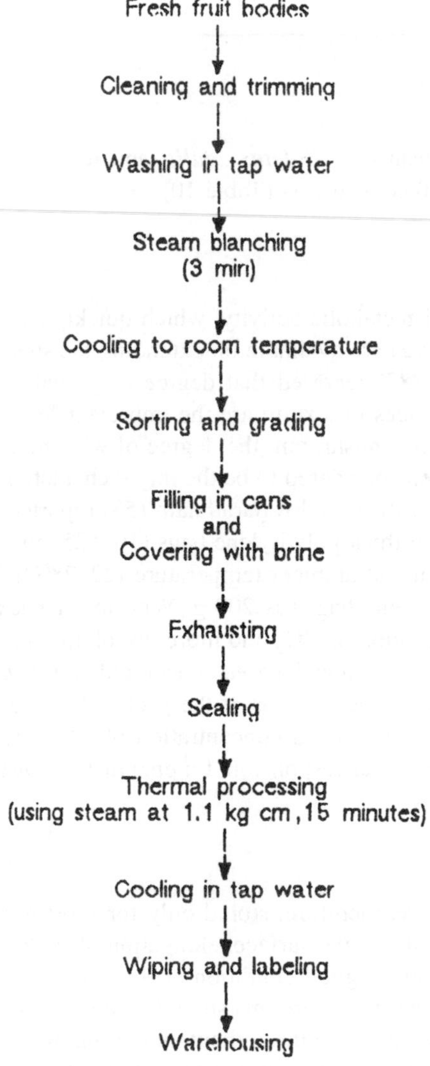

FIGURE 16 Flowchart for canning of fruit bodies. (From Ref. 53.)

atmosphere in preventing mold growth and retarding cap opening at CO_2 levels as high as 20% in the atmosphere, but with less than 10% CO_2 was found to be injurious.

C. Hyperbaric Storage

Robitaille and Badenhop (101) found that neither pressurization nor gradual depressurization over 6 hours injured mushrooms. The high pressure did not effect respiration but significantly reduced moisture loss during storage. The mushrooms depleted O_2 to very low levels and showed high tolerance to CO_2. At 0.035 atm partial pressure, CO_2 reduced browning of mushrooms but not respiration. The same authors (101) developed a completely autonomous storage system with CO_2 removal and automatic O_2 replenishment.

D. Irradiation

One of the major challenges in handling of mushrooms is to control desiccation and opening of caps in fresh mushrooms during storage even at 0–4°C. The opening of caps was shown to be delayed for 10–14 days when mushrooms were irradiated at 10–100 krad (102). It was also demonstrated that mushrooms irradiated at 175–350 krad had no harmful effect on animals. Irradiation up to a maximum dose of 250 krad has been accepted as a commercial tool to treat mushrooms in The Netherlands since 1969 (103). Rodriguez et al. (104) concluded that gamma irradiation at 50 krad or more inhibited browning, respiration, opening of caps, and lengthening of stalks in mushrooms stored at 15°C.

Keresztes et al. (105) have investigated the gills of *Pleurotus ostreatus* that were gamma-irradiated (5.0 kGy dose) to extend the storage life by employing scanning and transmission electron microscopy. Inhibition of spore production in irradiated specimens was found to be caused by the destruction of basidia rather than by the retardation of normal spore development. In *P. ostreatus*, the hymenium appeared to be more sensitive to irradiation than it was in *A. bisporus*. Thomas (14) has reported changes in chemical and textural properties of mushrooms preserved by irradiation.

VI. PROCESSING

Mushrooms can be frozen, canned, air-dried, or freeze-dried (54). Being a highly perishable product, mushrooms should be processed as soon as they are harvested—within a period of few hours. The desirable processing qualities of mushrooms include tender flesh, little tendency to discolor, and an agreeable flavor (42).

A. Canning

Mushrooms for canning are soaked in water for 10–15 minutes and washed with a water spray. They are blanched in water at 80–82°C for 8–10 minutes in order to shrink the product, which allows proper filling of the container and prevents excessive darkening. The blanched and cooled mushrooms are canned and covered with a solution of 1.5% salt and 0.2% citric acid. The acid is used to prevent excessive darkening. The open cans are sealed after heating at 65.5°C. After sealing, the product is heat-processed to prevent the growth of organisms causing botulism.

The quality of canned *P. sajor-caju* mushrooms was studied by Bano (106). A flowchart for canning of the fruit bodies is presented in Figure 16. The product was stored at different

TABLE 11 Cut-Out Examination of Canned *P. sajor-caju* Stored at Different Temperatures

Storage temperature/ period (days)	pH of brine	Internal appearance of can	External appearance of fruit bodies	Drained weight (%)
Stored at 37°C				
2	6.2	L.F.	Light yellow	62.6
4	6.0	L.F.	Light yellow	62.5
6	5.9	M.F.	Yellowish-brown	62.2
8	5.8	H.F.	Brown	59.3
12	N.D.	N.D.	N.D.	N.D.
Stored at 21–30°C				
2	6.2	Normal	Normal	57.9
4	6.0	V.L.F.	Slightly yellow	57.9
6	5.8	L.F.	Light yellow	58.0
8	5.8	M.F.	Yellowish-brown	59.4
12	5.8	H.F.	Brown	61.3
Stored at 2°C				
2	6.1	Normal	Normal	58.0
4	6.0	Normal	Normal	58.0
6	6.0	Normal	Normal	58.0
8	5.8	Normal	Normal	57.8
12	6.0	Normal	Normal	57.8

L.F. = Light feathery; M.F. = medium feathery; N.D. = not determined; V.L.F. = very light feathery; H.F. = heavy feathery.
Source: Ref. 53.

temperatures. It was shown that the product could be stored up to 4, 8, and 12 months, respectively, at 37°C, room temperature (21–30°C), and 2°C (Table 11). Canning of *P. florida* has been attempted with considerable success (107). Stems and caps were separately blanched in steam for 3 minutes and poured while hot with a 2% NaCl brine into cans. Blanched caps were also canned in butter sauce (1 part flour, 3 parts butter, and 9 parts water) and cream sauce (1 part flour, 2 parts butter and 15 parts whole milk). All the cans were processed for 12

TABLE 12 Sensory Evaluation of Canned *P. florida*

Canning medium	Flavor	Texture	Appearance	Overall
Brine	4.0[a]	6.6	5.0	4.6[a]
Butter sauce	7.4	6.6	5.8	7.0
Cream sauce	7.0	6.8	7.2[b]	7.2

[a]Significantly different at 1% level.
[b]Significantly different at 5% level.
Source: Ref. 105.

TABLE 13 Dehydration of *P. flabellatus* by Different Methods

Dryer	Time for drying (h)	Dehydration ratio	Final moisture content (%)	Reconstitution ratio
Cross-flow dryer[a]	5.5	1:7.5	4.5–5.0	1:5.0
Through flow dryer[a,b]	4.5	1:8.3	4.0–4.5	1:5.5
Vacuum shelf dryer[a]	5.0	1:7.0	3.5–4.0	1:6.0
Freeze-drier	7.0	1:9.0	2.3–3.0	1:8.0

[a]Dry temperature: first 2 h, 60–65°C; next 2–3 h, 55–60°C.
[b]Air velocity through the dryer 3–4 m s^{-1}.
[c]Air velocity through the dryer 0.3 m s^{-1}.
Source: Ref. 53.

minutes at 121°C. Brine-canned caps were rated poor, and canning in butter or cream sauces improved acceptability (108) (Table 12). Flavor and overall quality of brine-canned product were rated significantly lower, while the appearance of cream sauce product was rated significantly higher.

B. Dehydration

Mushrooms are not air-dried extensively mainly due to the low quality of the dried product. Canned mushrooms offer a more economical and better-quality alternative. Mushrooms are, however, freeze-dried for use in dehydrated soups and specialty products. Mushrooms can be dehydrated using various treatments before drying to avoid browning and to achieve better hydration. In order to check the browning, a series of treatments such as SO_2 treatment in solution (KMS + citric acid) and steam-blanching are given. Even though water-blanching was found very effective in prevention of browning, there was a loss in weight and flavor was also drastically lost. Hence, steam-blanching might be preferred. Bano and Rajarathnam (53) tried different methods for dehydrating *P. flabellatus*. Drying in vacuum shelf driers at low temperature (35°C) was found to be better (Table 13). The best dehydrated product was obtained by freeze-drying (109–112).

C. Freezing and Freeze-Drying

Harvested mushrooms are precooled by growers to 2–4°C. They are sorted, washed, and the base of the stem is cut off. High levels of chlorine, up to 50 ppm, in wash water have been found useful in reducing microbial load. The mushrooms are graded according to size and frozen. Mushrooms contain high levels of polyphenol oxidase, which causes the undesirable discoloration that appears in a bruised or cut mushroom. Prevention of enzymatic discoloration is one of the most important problems faced in the freezing of mushrooms. Blanching results in appreciable shrinkage and produces an undesirable grey color. Mushrooms are given a short blanch in hot water primarily to aid in reducing microbial contamination rather than for enzyme inactivation. Rapid freezing maintains the white color. IQF freezing is recommended to produce a high-quality frozen product. Vacuum-packaging inhibits oxidative discoloration. Fang et al.

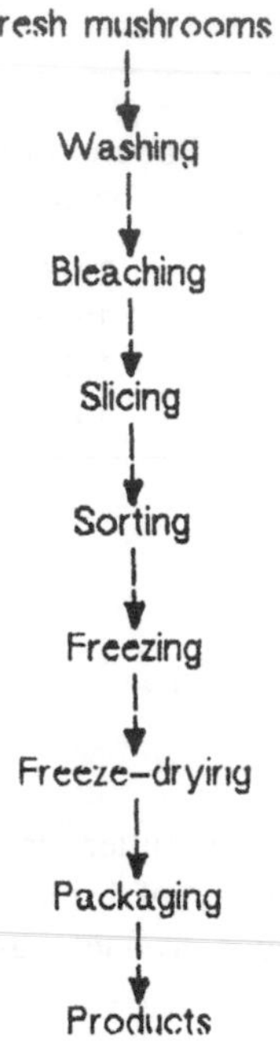

FIGURE 17 Commercial freeze-drying process for mushroom. (From Ref. 113.)

(110) reported that polyphenoloxidase activity in sliced mushrooms was inhibited by dipping in a solution of metabisulfite containing 200 ppm SO_2, dipping in 2% NaCl solution, blanching in boiling water for 2 minutes, followed by evaporative cooling. The product was frozen with Freon-12 at –30°C for 60 minutes and dried to 3% moisture. The product had a better texture than one made by a slow-freezing process. However, the blanching process caused considerable loss of water, soluble solids, and ascorbic acid. This also resulted in a lighter color of freeze-dried products after rehydration. A procedure adopted by commercial freeze-drying plants for mushrooms (113) is presented in Figure 17. Luh and Eidels (114) reported the storage stability and chemical changes in freeze-dried mushrooms slices packed in plastic-laminate and aluminum-film combinations. It was observed that the product packed in aluminum foil combination pouches under nitrogen and stored at 20°C or lower had better quality retention. The freeze-dried product contains about 3% moisture (114).

D. Pickling

Singh and Bano (115) have studied the suitability of *Pleurotus* for picking. Since mushroom is a nonacidic food, vinegar is added, both to increase the taste and protect from microbial attack. The flow diagram of this process is given in Figure 18. The recipe contained 1 kg of chopped fruit bodies, 12 g of cumin seeds, 12 g of fenugreek seeds, 12 g of coriander seeds, 7 g turmeric powder, 10 g mustard seeds, 100 g green chilies, 50 g of table salt, and 500 ml of sesame oil. The stored product was found to contain 0.9% acidity as acetic acid. There was no rancidity or off-flavor development.

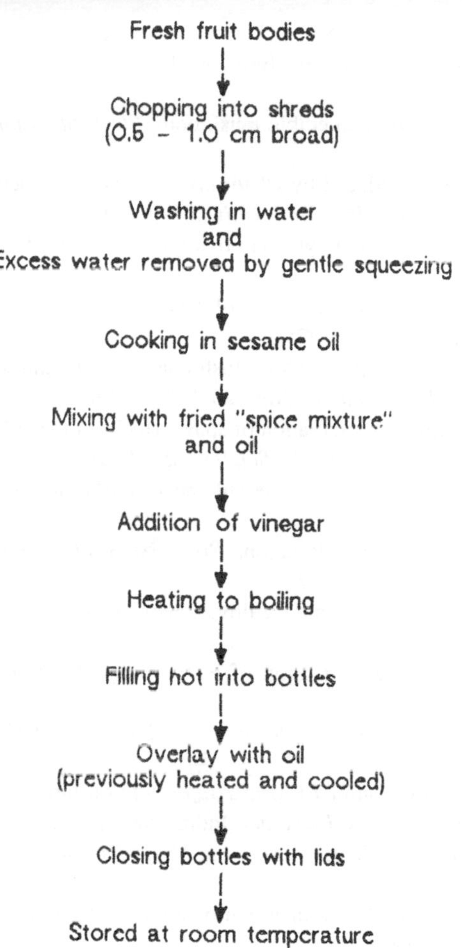

FIGURE 18 Flowchart for preparation of pickle from fruit bodies. (From Ref. 53.)

REFERENCES

1. Hadar, Y., and C. G. Dosoretz, Mushroom mycelium as a potential source of food flavour, *Trends Food Sci. Technol. 2*(9):214 (1991).
2. Anderson, E. E., and C. R. Feller, The food value of mushrooms (*A. campestris*), *Proc. Am. Soc. Hort. Sci. 41*:301 (1942).
3. Fellow, P., A mouldy old business spawn some money, *Food Chain-Intermed. Technol.* (March 14):3 (1995).
4. Atkins, F. C., *Mushroom Growing Today*, MacMillan, New York, 1966.
5. Chang, S. T., and P. G. Miles. *Edible Mushrooms and Their Cultivation*, CBS Publ. and Distributors, Delhi, India, 1993, p. 345.
6. Singer, R., *Mushroom and Truffles: Botany Cultivation and Utilization*, Leonard Hill, London, 1961.
7. Bahal, N., *Handbook on Mushroom*, Oxford and IBH Publishing Co., New Delhi, India, 1984, p. 123.
8. Rajarathnam, S., and Z. Bano, *Pleurotus* mushrooms, Part I, Morphology, life cycle, taxonomy, breeding and cultivation, *CRC Crit. Rev. Food Sci. Nutr. 26*(2):157 (1987).

9. Rajarathnam, S., D. B. Wankhede, and M. V. Patwardhan, Some chemical and biochemical changes in straw constituents during growth of *Pleurotus flabellatus* (Berk, K. Br.) Sacc, *Eur. J. Appl. Microbiol. 8*:125 (1979).
10. Madan, M., and R. Bisaria, Cellulolytic enzymes from an edible mushroom, *Pleurotus sajor-caju*, *Biotechnol. Lett. 5*:601 (1983).
11. Hong, J. S., Studies on hemicellulolytic enzymes produced by *Pleurotus ostreatus* I. Properties of crude hemicellulolytic enzymes, *Bull. Agric. Coll. Jeonbuq. Natl. Univ. 7*:89 (1976).
12. Toyama, N., and K. Ogawa, Comparative studies on cellulolytic and oxidizing enzyme activities of edible and inedible wood rotters, *Mush. Sci. 9*:745 (1976).
13. Ulezlo, I. V., T. M. Uporova, and R. V. Feniksova, Oxidative enzymes of the lignin-degrading fungus, *Pleurotus ostreatus*, *Prikil. Biokhim Mikrobiol. 11*(4):535 (1975).
14. Thomas, P., Radiation preservation of foods of plant origin Part VI. Mushrooms, tomato, minor fruits and vegetables dried fruits and nuts, *CRC Crit. Rev. Food Sci. Nutr. 26*(4):313 (1988).
15. San Antonio, J. P., and S. W. Hwang, Liquid nitrogen preservation of spawn stocks of the cultivated mushroom, *Agaricus bisporus* (Lange) Sing, *J. Am. Soc. Hortic. Sci. 95*:565 (1970).
16. San Antonio, J. P., Origin and improvement of the spawn of the cultivated mushroom, *Agaricus brunnescens* Peck., *Hort. Rev. 6*:45 (1984).
17. Jandaik, C. L., Problems and prospects of *Pleurotus* cultivation, *Proc. 1st Symp. Survey and Cutlivation of Edible Mushrooms in India*, Vol. II, 1976, p. 67.
18. Bano, Z., S. Rajarathnam, and N. Nagaraja, Some aspects on the cultivation of *Pleurotus flabellatus* in India, *Mushroom Sci. 10*:597 (1979).
19. Kurtzman, R. H., Jr., A vertical tray system for the cultivation of edible fungi, *Mushroom Sci. 10*(2):429 (1979).
20. Quimio, T. H., Indoor cultivation of *Pleurotus* mushroom, *Tech. Bull. U.P. Los Banos*, College of Agriculture, Laguna, Philippines, 1978.
21. Kandaswamy, T. K., and K. Sivaprakasam, Cultivation of *Pleurotus sajor-caju* on farm wastes in Tamil Nadu, *RRAI Symp. Punjab Agricultural University, Ludhiana*, India, 1980, p. 363.
22. Singh, R. P., Cultivation of *Pleurotus sajor-caju* (Fr.) Sing, Mushroom, *Mushroom Newslett. Trop. 3*(3):8 (1983).
23. Bano, Z., N. Nagaraja, S. Rajarathnam, and M. V. Patwardhan, Cultivation of *Pleurotus* spp. in village model hut, *Indian Food Packer 33*(6):19 (1979).
24. Mantel, E. F. K., The development of mushroom production and its importance as food in India, in *Sub-tropical and Tropical Horticulture: Botany, Cultivation and Utilization*, InterScience, New York, 1961.
25. Bahal, N., Polybag method—a new technique for cultivation of paddy straw mushroom (*Volvariella vulvacea*), *Curr. Sci. 50*:378 (1982).
26. Chang, S. T., *Volvariella vulvacea, The Biology and Cultivation of Edible Mushroom*, Academic Press, New York, 1978.
27. Garcha, H. S., *Mushroom Growing*, Punjab Agricultural University, Ludhiana, 1980.
28. Ramaswamy, K., and T. K. Kandaswamy, Spawn composition and spawning methods on yield of straw mushroom, *Indian Mushroom Sci. 1*:277 (1978).
29. Ramakrishnan, K., D. Lilithakumar, N. Shanmughen, and C. S. Krishnamurthy. A simple technique for increasing the yield of straw mushroom, *Volvariella displasia*, *Madras Agric. J. 55*:144 (1968).
30. Thomas, K. H., T. S. Ramkrishnan, and I. L. Narasimhan, Paddy straw mushroom, *Madras Agric. J. 31*:57 (1943).
31. Fritsche, G., The development of promising new *Agaricus bisporus* varieties, *Groenten Fruit 37*(1):59 (1981).
32. Bahukhandi, D., and R. L. Munjal, Cultivation of *Pleurotus* sp. on different agricultural residues, *Indian Phytopathol. 42*(4):492 (1989).

33. Bhandari, T. P. S., R. N. Singh, and B. L. Verma, Cultivation of oyster mushroom on different substrate, *Indian Phytopathol. 44*(4):555 (1991).
34. Chauhan, S., and D. C. Pant, Effect of spawn substrates and storage conditions on sporophore production in *Pleurotus sajor-caju, Indian J. Mycol. Plant Pathol. 18*:321 (1988).
35. Ganju, V., L. N. Nair, and K. G. Mukherji, Cultivation of *Pleurotus sajor-caju* (Fr.), Singer on different substrate, *J. Rec. Adv. App. Sci. 3*(2):478 (1988).
36. Pal, J., and Y. S. Paul, Cultivation of *P. sajor-caju* on *P. sajor-caju* on legume wastes, *Indian J. Mycol. Plant Pathol. 15*:76 (1985).
37. Kumar, D. S. M., and M. K. Biswas, Oyster mushrooms, *Kisan World* (March):45 (1992).
38. Qin, S. X., H. L. Zhang, L. K. Ren, and X. J. Yn, Effect of different cultivation materials on nutritive composition of *Pleurotus* fruiting bodies, *Edible Fungi China 3*:12 (1989).
39. Upadhyay, R. C., and H. S. Sohi, Apple oomace a good substrate for cultivation of edible mushroom, *Current Sci. 57*:1189 (1988).
40. Ryall, A. L., and W. J. Lipton, *Handling, Transportation and Storage of Fruits and Vegetables*, Vol. I, *Vegetables and Melons*, AVI Pub., Westport, CT, 1972.
41. Rajarathnan, S., and Z. Bano, *Pleurotus* mushrooms, Part IB Pathology, in vitro and in vivo growth requirements and world status, *CRC Crit. Rev. Food Sci. Nutr. 26*:243 (1987).
42. Zadrazil, P., and M. Schneidereit, Die Grundlagen für die Inkulturnahme einer bisher nicht kultivierten *Pleurotus-Art, Champignon 12*:25 (1972).
43. Gandy, D. G., Weed moulds, *Mush. J. 23*:428 (1974).
44. Fletcher, J. T., Mushroom, moulds, and management, *Mush. J. 56*:52 (1977).
45. Fletcher, J. T., P. I. White, and R. H. Gaze, *Mushrooms Pest and Disease Control*, Intercept Ltd., England, 1986, p. 156.
46. Sinden, J. W., and E. Hausar, Report on two new mushroom diseases, *Mush. Sci. 1*:52 (1960).
47. Gandy, D. G., Watery stipe of cultivated mushrooms, *Nature Lond. 185*:482 (1960).
48. Kneebone, L. K., J. D. Lockard, and R. A. Hagar, Infectivity studies with X-disease, *Mush. Sci. 5*:461 (1962).
49. Gandy, D. G., and M. Hollings, Dieback of mushrooms: A disease associated with a virus, *Ann. Rep. Glasshouse Crops Res. Inst. 1961*:103 (1969).
50. Hollings, M., and O. M. Stone. Viruses in fungi, *Sci. Prog.* (Oxford) *57*:371 (1969).
51. Hollings, M., and O. M. Stone, Viruses that infect fungi, *Ann. Rev. Phytopathol. 9*:93 (1971).
52. Nair, N. G., Observations on virus disease of cultivated mushroom, *Agaricus bisporus* in Australia, *Mush. Sci. 8*:155 (1972).
53. Bano, Z., and S. Rajarathnam, *Pleurotus* mushrooms, Part II. Chemical composition, nutritional value, post-harvest physiology, preservation and role as human food, *CRC Crit. Rev. Food Sci. Nutr. 27*:87 (1988).
54. Salunkhe, D. K., and B. B. Desai, Mushroom in *Post-harvest Biotechnology of Vegetables*, Vol. II, CRC Press, Boca Raton, FL, 1984, p. 147.
55. Mendel, J. B., The chemical composition and nutritive value of some edible American fungi, *Am. J. Physiol. 1*:225 (1918).
56. Anderson, E. E., and C. R. Fellers, The food value of mushroom (*A. campestris*), *Proc. Am. Soc. Hortic. Sci. 41*:301 (1942).
57. Bano, Z., K. S. Srinivasan, and N. S. Singh. Essential amino acid composition of proteins of mushroom, *V. volvacea, J. Food Sci. Technol. 8*:180 (1971).
58. Sugimori, T., Y. Oyama, and T. Omichi, Studies on basidiomycetes I. Production of mycelium and fruiting body from noncarbohydrate organic substances, *J. Ferment. Technol. 49*:435 (1971).
59. Yoshida, H., T. Sugahara, and J. Hayoshi, Free sugars and free sugar alcohols of mushroom, *Nippon Shokuhin Kogya Gakkaishi 31*:765 (1984).
60. Pfyffer, G. E., and D. M. Rast, the polyol pattern of some fungi not hitherto investigated for sugar alcohol, *Exp. Mycol. 4*:160 (1980).

61. Hammond, J. B. W., The composition of fresh and stored oyster mushroom (*Pleurotus ostreatus*), *Phytochemistry 19*:2565 (1980).
62. Rajarathnam, S., Studies on the pathological and biochemical aspects during cultivation and storage of the white oyster mushroom *Pleurotus flabellatus*, Ph.D. thesis, University of Mysore, 1981.
63. Bano, Z., and S. Rajarathnam, *Pleurotus* mushroom as a nutritious food, in *Tropical Mushrooms: Biological Nature and Cultivation Methods* (S. T. Chang and T. H. Quimio, eds.), The Chinese University Press, Hong Kong, 1982, p. 363.
64. Bano, Z., K. V. Nagaraja, S. Vibhakar, and O. P. Kapur, Mineral and heavy metal contents in the sporophores (*Pleurotus* sp.) *Mush. Newslett. Trop. 2*(2):3 (1981).
65. Andriano, F. T., and R. A. Cruz, The chemical composition of Philippines mushroom, *Phil. J. Agric. 4*:1 (1933).
66. Bano, Z., and S. Rajarathnam, Vitamin values of *Pleurotus* mushrooms, *Qual. Plant. Plant Foods Hum. Nutr. 36*:11 (1986).
67. Grandjean, J., and R. Hule, Structure of pleurotin, benzoquinone extracted from *Pleurotus* griseus, *Tetrahydron Lett.* 1893 (1974).
68. Robbins, W. J., F. K. Kawanagh, and A. Hervey, Antibiotic substances from basidomycetes I. *Pleurotus griseus*, *Proc. Natl. Acad. Sci. 33*:171 (1947).
69. Maga, J. A., Mushroom flavour, *J. Agric. Food. Chem. 16*:517 (1976).
70. Dijkstra, F. Y., Studies on mushroom flavours, II. Some flavor compounds in fresh, canned and dried edible mushroom, *Z. Lebensm. Unters. Forsch. 160*:401 (1976).
71. Cronin, D., and M Ward, The characterization of some mushroom volatiles, *J. Sci. Food Agric.* 22:477 (1971).
72. Picardi, S. M., and P. Issenberg, Investigations of some volatile constituents of mushroom (*Agaricus bisporus*) changes which occur during heating, *J. Agric. Food Chem. 21*:959 (1973).
73. Pyyasalo, H., Identification of volatile compounds in seven edible fresh mushrooms, *Acta. Chem. Scand, Ser. B. 30*:235 (1976).
74. Pyyasalo, H., and M. Suikko, Odour characterization and treshold values of some volatile compounds in fresh mushrooms, *Lebensm. Wiss. Technol. 9*:371 (1976).
75. Dijkstra, F. Y., and T. O. Wiken, Studies on mushroom flavours I. Organoleptic significance of constituents of the cultivated mushroom (*A. bisporus*), *Z. Lebensm. Unters. Forsch. 160*:255 (1976).
76. Sulkowska, J., and E. Kaminski, Effects of different drying methods on quality and content of aromatic volatiles in dried mushrooms (*A. bisporus*), *Acta Aliment. Pol. 3*:409 (1977).
77. Card, A., and C. Avisse, Comparative study of the aroma of raw and cooked mushroom (*A. bisporus*), *Ann. Technol. Agric. 126*:287 (1977).
78. Dressl,. R., D. Bahri, and K. H. Engel, Formation of eight carbon compounds in mushroom (*A. campestric*), *J. Agric. Food Chem. 30*:89 (1982).
79. Drawert, F. R. G. Berger, and K. Neuhauser, Biosynthesis of flavor compounds by micro-organisms IV Characterization of the major principles of the odor of *Pleurotus eousmus*, *Eur. J. Appl. Microbiol. 18*:124 (1983).
80. Murahashi, S., Odorous principles of (Armillia) Matsutake. I. *Sci. Pap. Inst. Phys. Chem. Res. Jpn. 30*:263 (1936).
81. Wada, S., H. Nakatani, and K. Morita. A new aroma-bearing substance from Shitake, an edible mushroom, *J. Food Sci. 32*:559 (1967).
82. Yasumoto, K., K. Iwami, Y. Baba, and H. Mitsuda, Analysis of highly volatile compounds and determination of formaldehyde in Shitake mushroom, *J. Jpn. Soc. Food Nutr. 24*:463 (1971).
83. Morita, K., and S. Kobayashi, Isolation and synthesis of lenthionine an odorous substance of Shitake, an edible mushroom, *Tetrahedron Lett. 6*:573 (1966).
84. Yasumoto, K., K. Iwami, and H. Hitsuda, A new sulfur-containing peptide from *Lentinus edodes* acting as a precursor for lenthionine, *Agric. Biol. Chem. 35*:2059 (1971).

85. Yasumoto, K., K. Iwami, and H. Mitsuda, Enzyme catalyzed evolution of lenthionine from lentinic acid, *Agric. Biol. Chem. 35*:2070 (1971).
86. Thomas, A. F., An analysis of the flavor of the dried mushroom *Boletus edulis, J. Agric. Food Chem. 21*:955 (1973).
87. Gormley, T. R., and C. MacCanna, Prepackaging and shelf life of mushroom, *Ir. J. Agric. Res. 6*:255 (1967).
88. Nicholas, R., and J. B. W. Hammond, Storage of mushroom in prepack the effect of changes in carbon dioxide and oxygen on quality, *J. Sci. Food Agric. 24*:1371 (1974).
89. Nicholas, R., and J. B. W. Hammond, Investigation on storage of pre-packed mushrooms, *Mushroom J. 24*:473 (1974).
90. Nicholas, R., and J. B. W. Hammond, Storage of *Agaricus edulis* sporophores in prepacks, *Mushroom J. 45*:285 (1976).
91. Gormley, T. R., Chill storage of mushrooms, *J. Sci. Food Agric. 26*:401 (1976).
92. Rajarathnam, S., Z. Bano, and M. V. Patwardhan, Postharvest Physiology and storage of the white oyster mushroom (*P. flabellatus*), *J. Food Technol. 18*:153 (1983).
93. Hardenburg, R. E., A. E. Watada, and C. Y. Wang, *The Commercial Storage of Fruits, Vegetables and Florist and Nursery Stocks*, U.S. Department of Agriculture, Agricultural Research Service Handbook No. 68, 1986, p. 62.
94. Nichols, R., and J. B. W. Hammond, The relationship between respiration, atmosphere and quality in intact and perforated mushroom prepacks, *J. Food Technol. 10*:427 (1975).
95. Hughes, D. H., Mushroom discoloration research at University of Delaware, *Mush. Sci. 4*:447 (1959).
96. Goodman, R. N., Chemical (antioxidant) controls for discoloration of mushrooms, *Mo. Agr. Expt. Sta. Bull.* 688 1957, p. 4.
97. Pantastico, E. B., T. K. Chattopadhyay, and H. Subramanyam, Storage and storage operations, in *Post-harvest Physiology, Handling and Utilization of Tropical and Sub-tropical Fruits and Vegetables* (E. B. Pantastico, ed.), AVI Pub. Co., Westport, CT, 1975, p. 314.
98. Bramlage, W. J., and W. J. Lipton, *Gamma Radiation of Vegetables to Extend Market Life*, U.S. Dept. Agr. Market Res. Rep. 703, 1965, p. 16.
99. Dilley, D. R., The hypobaric concept for controlled atmosphere storage, *Proc. 2nd Natl. Controlled Atmos. Res. Conf.* (D. H. Dewey, ed.) Mich. State Univ. Hort. Rept. 28, 1977, p. 29.
100. Hatton, T. S., Jr., E. B. Pantastico, and E. K. Akamine, Controlled atmosphere storage, Part-III. Individual commodity requirements, in *Post-harvest Physiology, Handling and Utilization of Tropical and Sub-tropical Fruits and Vegetables* (E. B. Pantastico, ed.), AVI Pub. Co., Westport, CT, 1975, p. 201.
101. Robitaille, H. A., and A. F. Badenhop, Mushroom response to postharvest hyperbaric storage, *J. Food Sci. 46*(1):249 (1981).
102. Training manual on food irradiation technology and techniques, *IAEA Tech. Rep. Ser. No. 114* (1970).
103. Pablo, I. S., E. K. Akamine, and K. Chachin, Irradiation, in *Postharvest Physiology, Handling and Utilization of Tropical and Sub-tropical Fruits and Vegetables* (E. B. Pantastico, ed.), AVI Pub. Co., Westport, CT, 1975, p. 219.
104. Rodriguez, R., B. L. Raina, E. B. Pantastico, and M. B. Bhatti, Quality of raw-materials for processing, in *Postharvest Physiology, Handling and Utilization of Tropical and Sub-tropical Fruits and Vegetables* (E. B. Pantastico, ed.), AVI Pub. Co., Westport, CT, 1975, p. 467.
105. Keresztes, A., J. Kovacs, and E. Kovacs, Effect of ionizing irradiation and storage on mushroom ultra-structure I. The gills of *Agaricus bisporus* and *P. ostreatus*, *Food Microstruct. 4*(2):349 (1985).
106. Bano, Z., Unpublished data, Central Food Technological Research Institute, Mysore, India, 1986.
107. Oddson, L., and P. Jelen, Food processing potential of the oyster mushroom *P. florida*, *Can. Inst. Food Sci. Technol. J. 14*(1):36 (1981).
108. Oddson, L., An Evaluation of the oyster mushrom (*P. florida*) for food processing potential, M.Sc. thesis, Department of Food Science, University of Alberta, Edmonton, Canada, 1980.

109. Luh, B. S., B. Feinberg, and J. J. Meehan, Freezing preservation of vegetables, in *Commercial Vegetables in Processing* (B. S. Luh and J. G. Woodroof, eds.), The AVI Pub. Co. Inc., Westport, CT, 1982, p. 317.
110. Fang, T. T., P. Footrakul, and B. S. Luh, Effects of blanching chemical treatments and freezing methods on quality of freeze-dried mushrooms, *J. Food Sci. 36*:1044 (1971).
111. Kompany, E., and F. Rene, Effect of freezing conditions on aroma retention in frozen and freeze-dried mushrooms (*Agaricus bisporus*), *J. Food Sci. Technol. 32*(4):278 (1995).
112. Suguna, S., M. Usha, V. V. Sreenarayanan, R. Raghupathy, and L. Gothandapani, Dehydration of mushroom by sun-drying thin-layer drying, fluidized bed drying and solar cabinet drying, *J. Food Sci. Technol. 32*(4):284 (1995).
113. Li, C. F., Freeze drying, in *Commercial Vegetable Processing* (B. S. Luh and J. G. Woodroof, eds.), The AVI Pub. Co. Inc., Westport, CT, 1982, p. 416.
114. Luh, B. S., and L. Eidels, Chemical changes in freeze dried mushrooms (*Agaricus bisporus*), *Fruchtsaft-Ind. 14*:58 (1969).
115. Singh, N. S., and Z. Bano, Standardization of mushroom (*Pleurotus* species) pickle in oil, *Indian Food Packer 31*(5):18 (1977).

29

Minor Vegetables

P. M. Kotecha and S. S. Kadam
Mahatma Phule Agricultural University, Rahuri, India

I. BREADFRUIT

Breadfruit is a widely grown nutritious tree fruit. It is more important as a subsistence crop than as a commercial crop in most areas of the world, especially in the Pacific Islands, where it is an important staple crop (1). The fruit is typically roasted or boiled, and is occasionally fried as chips.

A. Botany

Breadfruit is a member of the genus *Artocarpus* (family Moraceae), which contains about 50 species of trees that grow in the hot, moist regions of the Southeast Asian tropics and in the Pacific Islands. *A. altilis* (synonym *A. communis*) and *A. marianensis* as well as possible hybrids grow wild and are limited in distribution to Micronesia (1).

The fruit of breadfruit is a highly specialized structure, a syncarp composed of 1500–2000 flowers attached to the fruit axis or core. The core contains numerous latex tubes and larger vascular bundles, which discolor rapidly upon cutting due to oxidative enzyme activity. The bulk of the fruit is formed from the persistent perianthus of each flower. The perianthus fuse together except at the base, forming a cavity contining the true fruit and its enclosed ovary and seed. As the fruit develops, this area grows vigorously and become fleshy at maturity, forming the edible portion of the fruit. The rind is usually stained with latex exudations at maturity. Breadfruit seeds are thin walled, obovoid, 1–2 cm thick, and embedded in the pulp (1).

The creamy white or pale yellow flesh is soft and pulpy when mature, surrounding a spongy core. Since many cultivars of breadfruit are seedless, it has been inferred that fruit development is due to parthenocarpy. Seedless cultivars often have numerous, tiny aborted seeds surrounding the core. A few seeded forms also exist, which usually contain from one to several normal or aborted seeds.

B. Production

Breadfruit can be successfully grown in warm areas having a well-distributed annual rainfall of 200–250 cm, temperature range of 16–38°, and 70–80% humidity. Lateritic red loams are best suited for breadfruit. The plant requires a high humus content in the soil. In shallow soils, it may grow satisfactorily, but sooner or later a decline sets in, resulting in the death of the tree. Good drainage of the soil is essential for breadfruit, as impeded drainage stunts the trees and water stagnation around the roots causes premature dropping of fruits (2).

Breadfruit is propagated through seeds in the case of the seeded variety. The seeds lose viability soon after removal from the fruits. They should therefore be sown immediately after extraction from the ripe fruits. Propagation of the seedless variety is effected through several vegetative methods, the most common being planting of root cuttings and air-layering of root suckers. Formation of root suckers is stimulated by deliberate injury to the roots. Sprouted root portions are separated from the mother plant and planted during the rainy season. Root cuttings taken between October and March have shown success ranging from 29 to 90% (3). Root cuttings about 2.5 cm in diameter and 22.5 cm long when planted horizontally give 90% success as compared to 40% success when planted vertically (3). Breadfruit can be propagated by grafting tender branches on seedlings of *A. hirsutus* as stock. Budding is also successful. Active buds of breadfruit trees can be easily grafted to seedlings of wild jackfruit plant. The trees are planted 15–20 m apart in circular pits of one cubic meter. Young plants should be properly watered and shaded but they cannot stand stagnation. In coastal regions, although the plants do not require much irrigation, young plants may be watered at frequent intervals during hot spells. After the plants have been established, farmyard manure (FYM) plus wood ash or a mixture of ammonium sulfate, muriate of potash, and superphosphate may be applied during the monsoon (4). In addition to this basal manure, 2 kg of superphosphate should be applied to each bearing tree annually to enhance the number and size of the fruits.

No serious diseases or pests have so far been noticed on breadfruit. Soft rot caused by *Rhizopus artocarpi* Racib is the common fungal disease affecting the tree (4). This disease causes rotting and dropping of fruits at all stages of development. Prophylactic spraying with copper fungicides is effective against fungal diseases (4). Shedding of immature fruits due to fungal diseases can be controlled by spraying the tender fruit with 1% Bordeaux mixture twice at intervals of a month after setting of fruit (4).

A full-grown and a well-bearing tree produces about 150–200 fruits a year, although the average annual yield is reported to be 50–100 fruits weighing 23–46 kg. The fruit is green in the initial stages and turns yellowish-green on maturity. Harvesting is best done by lowering the fruits within letting them fall from the tree.

C. Chemical Composition

Breadfruit is an important source of energy, being high in carbohydrates but low in fats and protein. When enough breadfruit is eaten daily to provide most of the caloric needs, it is a good source of phosphorous, iron, calcium, ascorbic acid, and the B vitamins—thiamine, nicotonic acid, and riboflavin. Breadfruit has little yellow pigment and is not a good source of vitamin A (3). It is a better source of calcium, riboflavin, nicotinic acid, phosphorus, and ascorbic acid. The seeds are a good source of protein (8%) and are low in fat (3–5%) compared to nuts such as peanuts or almonds. They are a poor source of ascorbic acid. The nutrient composition of flesh and seeds, is presented in Table 1.

D. Storage

Fruits are typically picked when in a mature but firm starchy stage. Sweet ripe fruits are rarely eaten, although a few traditional dishes utilize fruit at this stage. Fruits soften within 1–3 days, although a few cultivars may keep for as long as 10 days (3). Fruits may be submerged in water

TABLE 1 Proximate Composition of Flesh and Seeds of Breadfruit

Constituent	Flesh	Seeds
Water (%)	69.1	61.9
Food energy (kcal)	121.0	156.0
Protein (g)	1.3	7.94
Carbohydrate (g)	28.1	38.2
Fat (g)	0.4	4.68
Calcium (mg)	23.2	48.3
Phosphorus (mg)	47.2	89.4
Iron (mg)	0.63	0.13
Thiamine (mg)	0.09	0.08
Riboflavin (mg)	0.06	1.84
Nicotinic acid (mg)	1.28	1.9
Vitamin C (mg)	8.7	1.9

Source: Ref. 3.

to retard softening, but the outer surface splits and softens, reducing the amount available for consumption. Preliminary studies to extend the shelf life of breadfruit demonstrated that the fruit can be stored in polyethylene bags at low temperature, however, this tropical fruit is prone to chilling injury at temperatures below 13°C (3). Cooked breadfruit can be frozen, and this storage method deserves greater attention as it may provide a simple, effective means to better utilize this crop (4).

E. Processing

1. Drying

Drying is a common method used to store breadfruit. Mature or ripe fruits are boiled until soft, then sliced into pieces which are dried in the sun for 3–4 days. The dried breadfruit is eaten soon after drying or prepared for long-term storage by first roasting whole fruits in a fire. After the fruits are peeled and cut into bite-sized pieces, the pieces are dried on racks for 8–24 hours over a fire. Breadfruit dried in this manner can be stored for up to a year in leaf-lined baskets and indefinitely in airtight plastic or glass containers. Dried breadfruit is usually eaten without additional preparation, but it may be ground into a flour and mixed with water or coconut milk to make a porridge.

2. Fermentation

The fermented, preserved breadfruit is called *ma*, *masi*, or *bwiru* (4). Pit storage is a semi-anaerobic fermentation process involving intense acidification, which reduces fruit to a sour paste. Fermented breadfruit can be stored for 1–2 years in the pit, and the leaves are replaced as needed during the storage period. A cleaner, more uniform product can be prepared by placing the softened fruits in airtight plastic containers (4).

II. JACKFRUIT

Jackfruit is widely cultivated through the tropical lowlands in both the hemispheres. It is commonly grown in Burma and Malaysia and to a considerable extent in Brazil (5). Jackfruit as a commercial crop is not popular due to a wide variation in fruit quality, the long gestation period of plants raised from seeds, absence of a commercial method of vegetative propagation, and the widespread belief that excessive consumption leads to certain digestive ailments. It is a comparatively cheap fruit and favored by poor people when the price of staple food is high. Jackfruit is not grown in a regular orchard. Hence, reliable statistics of area and production are not available.

A. Botany

Jackfruit (*Artocarpus heterophyllus* Lam syn. *A. integrifolia*) belongs to the family Moraceae. The chromosome number is $2n = 56$. The tree is 8–11 m tall with a straight stem, evergreen, with glossy leaves. On young plants, the leaves are lobed, but they are fully expanded on the mature tree. The inflorescence is always solitary and peduncles have unisexual flowers. The male spikes develop from the main or secondary branches, while female spikes arise from "foot stalks" on the trunks and also on primary and secondary branches. Jackfruit is formed by a large number of individual flowers, and the fruit, therefore, is called a "multiple fruit."

Jackfruit is a cross-pollinated plant. Hence, it has innumerable types or forms depending on fruit characteristics. Cultivated types are classified into two groups: soft- and firm-fleshed. Some distinct types are capable of maintaining their identity even after propagation by seeds (6), e.g., *Rudraleshi* and Singapore or Ceylon Jack. A variety named Muttam Vakira produces fruits of 7.0 kg average weight, 46 cm in length, and 23 cm in width (5).

B. Production

Propagation of jackfruit by seed is most common. This leads to immense variation, with the result that it is difficult to find jackfruits of standard quality or performance in any plantation or region. Moreover, seedlings take many years to come to fruiting. A number of vegetative propagation methods, including cutting and air-layering, budding, inarch grafting (7–9) and in vitro propagation by tissue culture (10–12), have been tried with certain limitations.

Commonly, the square system is followed for planting. A hexagonal system is also followed in less fertile soil. A spacing of 12×12 m is followed in fertile soil. However, high-density planting is practiced in heavy soil (5). Jackfruit is an evergreen tree bearing on the mature wood of trunk and branches, hence no regular training or pruning is practiced except for removing dead and diseased parts.

Jackfruit is mostly grown as a rainfed crop. Hence regular irrigation is seldom needed. If grafts are planted, they are hand-watered during winter and summer for the first 2 years. For quick growth of trees, manures and fertilizers may be added twice a year before and after the rainy season. It is advisable to apply 80 kg of FYM per tree per year. According to the nutrient status of the soil and tree growth, chemical fertilizers are applied. These are spread and thoroughly mixed with the soil.

Jackfruit is not grown in an orchard, hence no regular intercultural operations are practiced. However, it is necessary to weed out the areas under the canopy. If planted in regular layouts, it would be advisable to grow an intercrop such as *Stylosanthus hamata* or fodder shrubs like *Sesbania grandiflora* and *Subabul* (Luecaena).

No serious pests have been observed on jackfruit. However, diseases like soft rot of fruits and pink disease have been noticed. Soft rot is caused by the fungus *Rhizopus artocarpi*, which affects the male spikes and young fruits resulting in blackening, rotting, and shedding of premature fruits. This disease can be controlled by spraying the young fruits with Zineb and Elatox (13). In pink disease, pinkish as well as whitish growth is seen on the branches and shoots.

The incrustation gradually extends and ultimately surrounds the stem and branches. In order to control this disease, the affected parts are cut off and Bordeaux paste or coal tar is applied at the cut ends. A prophylactic spray of 1% Bordeaux mixture before the onset of monsoon reduces the incidence of this disease. Although number of pests are known, shoot and trunk borer, brown weevil (*Ochyromera artocarpi*), mealybug, and jack scales are important pests of jackfruit.

Jackfruit seedlings start fruiting after 8–10 years, whereas the grafts may fruit starting at 3–5 years. Fully mature but unripe fruits are harvested, and the fruit maturity is judged by dull appearance and dull sound. The season extends to August. Yield varies from tree to tree and according to the age of the tree. On average, 50–100 fruits of medium size (6–10 kg) are borne on adult trees. Jackfruit trees live for more than 80 years. No systematic grading is practiced. The fruits are transported to nearby markets in cartloads. For distant markets, the fruits are transported by truck. Generally, no packing is used. Jackfruit prices depend upon the size, quality, type of fruit, and season (14).

C. Composition

Ripe jackfruit contains 14.5% sugars, of which α-glucose, β-glucose, fructose, and sucrose constitute 3.63, 2.33, 1.74, and 6.90%, respectively (15). The composition of jackfruit is presented in Table 2. The seeds are rich in carbohydrates and are also a rich source of vitamins B_1 and B_2. The volatile components of jackfruit are listed by Berry and Kalra (15).

D. Storage and Processing

The initial quality and stage of maturity influence the storage life of jackfruit. Pink disease caused by *Botrybasidium salmonicolor* (Berk. and Br.) Venkatraman affects jackfruit. Fruit rot is caused by *Phytophthora palmivora*. *Rhizopus* rot of jackfruit caused by *R. artocarpi* Racib has been reported in India (9). *Bactocera rufomaculala* and larvae of the moth *Perina nuda* have been reported to infest jackfruit.

The ripe fruit is consumed as a dessert fruit. Jackfruit chips are prepared by frying ripe or semi-ripe slices in margarine. Jackfruit leather is also prepared from the firm-flesh type fruits.

TABLE 2 Nutrient Composition of Jackfruit (per 100 g edible portion)

Constituent	Content
Moisture (g)	84.0
Protein (g)	2.6
Fat (g)	0.3
Carbohydrate (g)	9.4
Thiamine (mg)	0.05
Riboflavin (mg)	0.04
Niacin (mg)	0.2
Ascorbic acid (mg)	14
Calcium (mg)	30
Phosphorus (mg)	40
Iron (mg)	1.7
Energy (kcal)	51

Source: Ref. 5.

Jackfruit bulbs can be preserved successfuly by canning them in syrup (16). A palatable beverage concentrate can be made from jackfruit pulp by adding sugar, citric acid, and water.

III. PLANTAIN

The banana was probably among the first domestically grown plants. This occurred in southeast Asia, probably Malaya. At that time the vegetative parts were used as food, since the fruits were inedible. Africa is the largest producer of plantain in the world. The world production of plantains is given in Table 3 (17).

A. Botany

Plantain is a member of the genus *Musa* (family *Musaceae*). Plants are treelike huge perennial monocotyledenous herbs with a basal corm. They grow 2–6 m in height. The pseudostem, which is trunklike, is composed of leaf sheaths. The underground stem is a corm with very short internodes. Leaves are formed in spiral succession with the inflorescence, the final organ, emerging through the base of the pseudostem. Three to four buds, called suckers, develop from the corm. The fruit is a berry; the edible fruit of commerce is parthenocarpic and therefore seedless.

B. Production

Plantain can be cultivated in a wide variety of climates varying from tropical to dry subtropical, mostly between 30°N and 30°S of the equator. Winter mean temperatures should not fall below

TABLE 3 World Production of Plantain

Continents/Country	Production (1000 MT)
World	27902
Africa	20539
North and Central America	1552
South America	5037
Asia	770
Oceania	5
Uganda	8488
Rwanda	2700
Colombia	2682
Zaire	2291
Nigeria	1400
Ghana	1322
Equador	950
Paraguay	710
Dominican Republic	566
Sri Lanka	495
Guinea	429
Haiti	270
Panama	100

Source: Ref. 17.

16°C, and about 1250 mm of rainfall per year are required. Strong winds will damage the leaves and reduce crop yields.

Plantains require soil with good drainage and aeration. The soil need not be deep, as most of the adventitious fibrous roots are found in the top 30 cm. Soils with high organic matter, low salinity, and pH range of slightly acid to slightly alkaline are preferred; those containing 0.05% or more NaCl are toxic. Plantain can be propagated exclusively by vegetative means, either by means of suckers or by large corms. Spacing varies from 3 to 6 m between plants. Plants should be supplied with small amounts of N during growth and with P and K twice a year. If rainfall is less than 100 mm per month, supplemental irrigation is required, especially during the reproductive phase of growth.

Bunches are usually harvested when fruits are firm and green, about 80–90 days after the female flowers open. The yield of plantain varies from 7 to 16 MT/ha; this varies with spacing. Bunches are cut by hands in the field and placed into padded boxes to minimize physical damage during transport. Plantain usually requires no ethylene treatment as a firm-textured and starchy pulp is desired. To delay ripening during transport, plantain can be packed in boxes lined with 40 μm polyethylene, which is sealed to create a modified atmosphere. A C_2H_4 absorber may also be enclosed.

C. Chemical Composition

Plantain is an important source of energy, being high in carbohydrates but low in fats and protein. It is a good source of phosphorus, iron, calcium, and ascorbic acid. The nutrient composition of plantain is given in Table 4.

D. Storage

As a chilling-sensitive crop, the fruits should be held at or above 12–13°C, the recommended storage temperature. To prevent ripening, ethylene should be removed from the atmosphere during transport and storage. In controlled-atmosphere (CA) storage, the O_2 concentration should be from 1 to 3%, CO_2 from 5 to 10%, and with little C_2H_4 as possible. Reduction of pressure

TABLE 4 Nutrient Composition of Plantain (per 100 g edible portion)

Constituent	Content
Moisture (g)	83.2
Protein (g)	1.4
Fat (g)	14.0
Carbohydrate (g)	99
Thiamine (mg)	0.05
Riboflavin (mg)	0.02
Niacin (mg)	0.3
Ascorbic acid (mg)	24
Calcium (mg)	10
Phosphorus (mg)	29
Iron (mg)	0.6
Energy (kcal)	64

Source: Ref. 18.

(hypobaric) to ½ atmosphere and 16°C in the storage room will double the shelf life of plantains (18). At the same temperature but at ⅓ atmosphere pressure, the storage life is doubled again. These effects of hypobaric conditions are due to reduced O_2 tensions and accelerated C_2H_4 removal.

E. Food Uses

Unlike the ordinary banana, which is eaten when ripe, the plantain even when ripe is starchy and unpalatable unless cooked. Boiling, steaming, or baking removes the astringency caused by tannins. The fruits are also used in the making of beer, wine, and distilled products. The hearts of the stem (pseudostem) are used in India as vegetables. The male bud of many varieties, after removal of the fibrous bracts, are boiled and eaten as vegetables in Southeast Asia.

IV. OTHER VEGETABLES

A. Bamboo Shoot

Bamboo is a perennial belonging to the grass (Graminae) family. The emerging shoots of the genera *Phyllostachys*, *Denrocalamus*, *Bambusa*, and *Sasa* are harvested and used as vegetables. The important edible bamboos are *P. edulus*, *P. bambusoides*, and *P. duleis*.

Bamboo is commercially produced in the Orient and Japan using traditional and forced cultivation practices. Both methods of cultivation require that emerging shoots be covered with soil or compost to prevent the photoactivated synthesis of bitter and potentially toxic cyanogenic glycosides.

The tender bamboo shoot is surrounded by a fibrous sheath; the internal tissue structure resembles that of other monocotyledonous plants, being comprised primarily of parenchymal tissue and fibrovascular bundles. The chemical and nutrient composition of bamboo shoot is presented in Table 5 (19).

Fresh bamboo shoots are prepared for consumption by removing the sheaths, cutting them

TABLE 5 Nutrient Composition of Bamboo Shoot (per 100 g edible portion)

Constituent	Content
Water (g)	91.00
Protein (g)	2.60
Fat (g)	0.30
Carbohydrate (g)	5.20
Vitamin A (I.U.)	20
Vitamin C (mg)	4.0
Calcium (mg)	13
Iron (mg)	0.50
Magnesium (mg)	3
Phosphorus (mg)	59
Sodium (mg)	4
Potassium (mg)	533
Energy (kcal)	113

Source: Ref. 19.

lengthwise if they are large, and boiling them for 30 minutes. This extended boiling time serves to tenderize the shoots and to leach out bitter flavors caused by the presence of cyanogenic glycosides. Precooked bamboo may be canned for long-term storage and retail in domestic and foreign markets.

B. Cardoon

Three major varieties of cardoon (*Cynara cardunculus*) commonly grown for consumption are Large Smooth, Large Smooth Spanish, and Ivory White Smooth. Cardoon, like globe artichoke, is a thistlelike, herbaceous perennial having large, woolly, deep-lobed, prickly leaves. It is a member of the composite (Compositae) family. The edible portion consists of the leaf stalks. The chemical and nutrient composition of this cardoon is given in Table 6.

Cardoon cannot be stored for extended periods in the fresh state owing to its high respiration rate; tissue temperature must be reduced to approximately 0°C soon after harvest. Since cardoon is also subject to rapid moisture loss, it should be maintained in a high-humidity environment (90–95% RH) throughout storage and distribution. Cardoon leaf stalks are prepared and eaten in a variety of ways, e.g., raw, steamed, boiled, or deep-fat-fried. In addition, cardoon can be frozen, dried, or canned.

C. Drumstick

This is a small genus of quick-growing trees distributed in India, Arabia, Asia Minor, and Africa. Two species are recorded from India, of which one, *Moringa oleifera* Lam., is widely cultivated in the tropics for its edible fruits.

The tree is indigenous to northwest India and is plentiful on recent alluvial land in or near sandy beds of rivers and streams. It is often cultivated in hedges and home gardens. It grows in all types of soils, except stiff clays, and thrives best in the tropical climate of south India. The tree can be propagated by seeds or from cuttings—the latter are preferred. Plants raised from seeds produce fruits of inferior quality.

TABLE 6 Nutrient Composition of Cardoon (per 100 g edible portion)

Constituent	Content
Water (g)	94.00
Protein (g)	0.70
Fat (g)	0.10
Carbohydrate (g)	4.89
Vitamin A (I.U.)	120
Vitamin C (mg)	2.0
Calcium (mg)	70
Iron (mg)	0.70
Magnesium (mg)	42
Phosphorus (mg)	23
Sodium (mg)	170
Potassium (mg)	400
Energy (kJ)	84

Source: Ref. 19.

TABLE 7 Nutrient Composition of Drumstick (per 100 g edible portion)

Content	Pods	Leaves
Moisture (g)	86.9	75.9
Protein (g)	2.5	6.7
Fat (g)	0.1	1.7
Carbohydrate (g)	3.7	12.5
Vitamin A (I.U.)	363	22374
Thiamine (mg)	0.50	0.06
Riboflavin (mg)	0.70	0.05
Niacin (mg)	0.2	0.8
Ascorbic acid (mg)	120	220
Calcium (mg)	30	440
Phosphorus (mg)	110	70
Iron (mg)	5.3	70
Energy (kcal)	26	92

Source: Ref. 18.

There are not many named varieties of this tree in India. A type named *Jaffna*, grown in parts of south India, produces fruits 60–90 cm in length. *Chavakacheri murunga*, also Jaffna type, bears fruits as long as 90–120 cm; *Chem murunga* is a type yielding pods with red tips.

The tree is not affected by any serious disease in India. A foot rot caused by *Diplodia* sp. has been observed in Madras. Two caterpillars and a stem borer are known to affect the tree. Of these, the hairy caterpillar, which causes defoliation, can be controlled by spraying the tree with fish oil rosin soap or by burning with a lighted torch (4).

The drumstick tree is valued mainly for its tender pods, which are esteemed as vegetables. The nutrient composition of drumstick pods (18) and leaves is given in Table 7. They are chopped or sliced and used in culinary preparation; they are also pickled. A small quantity of drumstick pods are canned in India and marketed in Europe and the United States. Fermented pods are also used in some parts of India. The flowers and tender leaves are used as pot herbs. The seeds are consumed after frying; they are said to taste like peanuts. Tree roots are used as a condiment or garnish in the same way as horseradish.

All parts of the tree are considered medicinal and are used in the treatment of ascites, rheumatism, venomous bites, and as cardiac and circulatory stimulants (18). The root of the young tree and the root bark are rubefacient and vesicant. The leaves are rich in vitamins A and C and are considered useful in treating scurvy and catarrhal infections (18). A paste of the leaves is used as an external application for wounds. Flowers are used as tonic, diuretic, and cholagogue agents. The seeds are considered antipyretic, acrid, and bitter. The seed oil is applied in rheumatism and gout (18).

REFERENCES

1. Ragone, D., Breadfruit, in *Encyclopaedia of Food Science, Food Technology and Nutrition* (R. MaCrae, R. K. Robinson, and M. J. Sadler, eds.), Academic Press, London, 1993, p. 2180.
2. Murai, M. F., Pen, and C. D. Miller, *Some Tropical South Pacific Island Foods*, Research Extension Series, Honolulu, University of Hawaii Press, 1958.

3. Wootom, M., and F. Tumaalii, Bread fruit production, utilization and composition: a review, *Food Technol.* (Australia) *36*:464 (1984).
4. Anonymous, Breadfruit, in *Wealth of India*, Council of Scientific and Industrial Research Publication, 1980.
5. Samaddar, H. N., Jackfruit, in *Fruits of India. Tropical and Subtropical* (T. K. Bose, ed.), Naya Prokash, Calcutta, 1985, p. 487.
6. Sen, P. K., and T. C. Bose, Effects of growth substances on rooting of jackfruit (*Artocarpus integrifolis* Linn.F.) layerings, *Indian J. Agric. 3*:43 (1959).
7. Gunjate, R. T., D. T. Kokkai, and V. P. Limaye, Epicotyl grafting in jackfruit, *Current Sci. 49*:667 (1980).
8. Madhav Rao, V. N., The jackfruit in India, *Farm Bull. 34*:18 (1965).
9. Singh, K., *Farm Information Bulletin No. 71*, Directorate of Extension Ministry of Agriculture, New Delhi, 1972.
10. Salunkhe, D. K., and B. B. Desai, Jackfruit, in *Postharvest Technology of Fruits*, Vol. II, CRC Press, Boca Raton, FL, 1985, p. 127.
11. Siddappa, G. S., Utilization of jackfruit and orange, *Indian Coffee 15*:130 (1971).
12. Rajmohan, K., and N. Mohanakumaran, Influence of explant source on the *in vitro* propagation of jack (*Artocarpus heterophyllus* Lam.), *Agril. Res. J. Kerala 26*:169 (1988).
13. Singh, N. I., and K. U. Singh, Efficacy of certain fungicides against *Rhizopus* rot of jackfruit, *Indian Phytopath. 42*:465 (1989).
14. Bajpai, P. N., Studies on flowering and fruit development in jackfruit, *Hort. Adv. 7*:38 (1968).
15. Berry, S. K., and G. L. Kalra, Chemistry and technology of jack fruit—a review, *Indian Food Packer 42*(3):62 (1988).
16. Chin, A. H. G., and Z. Nashirwan, Jack fruit, for canning in syrup, *MARDI Res. J. 17*(2):266 (1989).
17. FAO, *Food and Agriculture Organization Production Year Book*, 1993, p. 166.
18. Susanta, K. R., and A. K. Chakrabarti, Vegetables of tropical climate, in *Encyclopaedia of Food Science, Food Technology and Nutrition* (R. MaCrae, R. K. Robinson, and M. J. Sadler, eds.). Academic Press, London, 1993, p. 4743.
19. Smith, J. L., R. L. Jackman, and R. Y. Yada, Stem and other vegetagles, in *Encyclopaedia of Food Science, Food Technology and Nutrition* (R. MaCrae, R. K. Robinson, and M. J. Sadler, eds.), Academic Press, London, 1993, p. 4737.

30

Vegetables in Human Nutrition

S. S. Kadam
Mahatma Phule Agricultural University, Rahuri, India
D. K. Salunkhe
Utah State University, Logan, Utah

I. INTRODUCTION

With the continued increase in the population and pressure on land use, the adequate supply of nutritious food to all the people has been a matter of serious concern in many developing countries (1). It is estimated that more than 460 million people in the world are severely malnourished. The reasons put forth for this situation include high population density, poor socioeconomic status, inadequate sanitary and health facilities, and nonavailability of enough good-quality food (2). According to Borgstrom (3), the present human population is already much greater than the earth can maintain as free human beings. While the world population will double within 25 years, food production can only increase arithmetically because of scarce resources of energy, water, and fertilizers (2). There is a growing recognition that an increase in production alone will not solve the food problem. The food that is already produced must be conserved in an edible and nutritionally adequate form until it is distributed and consumed by those who need it (4). Many foods required for good nutrition, such as fruits and vegetables, are highly perishable and therefore suffer heavy losses after their harvest, particularly during postharvest handling (Table 1). The problem of food preservation is especially critical in hot and humid parts of the world, including developing countries facing problems of malnutrition and undernutrition. The world food supply can be increased substantially if the postharvest losses of food are prevented (6). The nutrition of low-income residents of third world countries can be improved markedly if perishable foods can be preserved utilizing available postharvest technology (4).

II. NUTRIENT SOURCES

Vegetables are the major source of dietary fiber, minerals, and vitamins. In addition, they contribute proteins, fats, and carbohydrates to the human nutrition (4). Vegetables are rich in

TABLE 1 Postharvest Losses of Vegetables During Commercial Handling in Two Major Cities of Brazil

	Sao Paolo[a]		Recife[b]	
Vegetable	Wholesale	Retail	Wholesale	Retail
Cabbage	19.0	14.0	23	27.5
Carrot	15.0	8.0	15.6	15.5
Beet	12.0	7.0	26.1	15.0
Green bean	19.0	5.0	17.1	17.3
Okra	23.0	8.0	17.7	17.2
Pepper	22.0	10.0	25.4	11.5
Tomato	24.0	13.0	7.6	30.0

[a]Average annual temperature 23°C.
[b]Average annual temperature 31°C.
Source: Ref. 5.

calcium, phosphorus, iron, and magnesium. Certain leafy vegetables such as amaranth, portulaca, and basella contain appreciable quantities of oxalic acid, which converts calcium from a soluble to an insoluble form and reduces their bioavailability (7).

Vegetables are good sources of vitamin C, vitamin A, thiamine, riboflavin, niacin, pantothenic acid, and folic acid (4). Carrot, sweet potato, spinach, and tomato are rich in vitamin A, and most green vegetables contain about 40–80 μg of folic acid per 100 g (4).

Potato, parsnip, sweet potato, and other root vegetables are good sources of starch (7). Potatoes contain about 2% protein. The quality of proteins in potato is considered to be excellent (8). In addition, mushrooms (9) and beans (10) are excellent sources of proteins. The protein content of mushrooms and beans ranges from 16 to 25% on a dry weight basis (9). The proteins in potatoes and mushrooms are highly digestible, whereas some vegetables like peas, beans, and spinach are reported to have less digestible proteins (9).

Leafy vegetables and roots provide the dietary fiber essential for bowel movement and possibily for the prevention of diseases such as appendicitis, colon cancer, diabetes, diverticulosis, gallstones, and obesity (10). Dietary fiber, currently redesignated as nonstarch polysaccharides, refers to the indigestible residue of plant foods (11). It is a single entity but consists of a wide range of complex polysaccharides like cellulose, hemicellulose, lignin, gums, petcins, and mucins with different chemical, physiochemical, and physiological properties. They possess the capacity to imbibe water and swell, thus contributing bulk to the diet. Total dietary fiber can be further classified as soluble and insoluble. Insoluble fiber includes plant cell wall material, cellulose, lignin, and hemicellulose. Soluble fiber includes noncellulosic polysaccharides, pectins, gums, and mucilages (11).

Foods containing dietery fiber require chewing, hence it limits food intake and acts as a natural appetite supressant (12). Soluble dietary fiber causes distention of the stomach, which gives a feeling of satiety. It has a high water-binding capacity and produces viscosity. Pectins and gums account for slower gastric emptying (12). Enzyme activities are affected by the type and amount of fiber eaten. Fiber increases the activity by protecting the enzyme from degradation, it delays absorption of glucose and fat after a meal, and it increases fecal loss of bile acids by binding. Among different fractions of fiber, lignin has strongest binding capacity. Soluble dietary fiber provides a carbohydrate substrate, which stimulates growth of bacteria in the large bowel. It accounts for an increase in fecal bulk. Reduced large intestine transit time is also known to reduce the risk of cancer (11,13).

Vegetables contain some flavor compounds, including sugars, amino acids, organic acids, volatiles such as aromatics, hydrocarbons, aldehydes, acetals, ketones, alcohols, esters, and sulfur compounds. When used along with other food items, these flavoring compounds make food more palatable (10). Onions and garlic are used as flavoring agents in a variety of soups, sausage, and curries. The pungency of garlic is due to the volatile sulfur compound diallyl disulfide, which is produced by the action of allinase enzyme on the amino acid alliin present in garlic (10). Several volatile sulfur compounds are responsible for the characteristic flavor of cole crops. Dimethy trisulfide has been indicated as a major aroma component in cooked cole vegetables (7).

Vegetables contain a variety of pigments. Tomato is rich in lycopene. Carrot and sweet potato are rich sources of carotene, and red cultivars of carrot contain anthocyanin pigment. The red color of garden beet is due to betacyanin. It also contains the yellow pigment betaxanthin (10).

III. TOXIC CONSTITUENTS AND ANTINUTRIENTS

Potato tubers, when exposed to unfavorable atmospheric conditions during storage, produce toxic constituents, namely solanin and chaconine (14), which are bitter in taste and have toxic effects (Table 2) when consumed in a significant quantity. Some leafy vegetables such as amaranthus, portulaca, and basella contain appreciable quantities of oxalic acid (7). This binds with calcium and forms calcium oxalate, thus reducing the bioavailability of calcium. Calcium oxalate present in the form of fine crystals in the leaves of taro causes itching of the skin and a pricking sensation of the tongue and throat (10). The anthra-quinenones of rhubarb are mainly in the root, but human poisoning generally occurs from eating rhubarb leaves. The leaf poison is commonly thought to be oxalate, but other factors—possibly the quinones—are also involved (15).

TABLE 2 Toxicity of α-Solanine in Various Species of Laboratory Animals After Different Routes of Exposure

	α-Solanine dose		
Experiment	Administration	Amount	Effects
Sheep	Oral	225 mg/kg	Toxic
	Oral	500 mg/kg	Lethal
	Intravenous	17 mg/kg	Toxic
	Intravenous	50 mg/kg	Lethal
Rat	Gastric intubation	590 mg/kg	50% death within 24 hr
	Intraperitoneal	75 mg/kg	50% death in few hours
Mice	Oral	100 mg/kg	Nontoxic
	Intraperitoneal	42 mg/kg	50% death in 7 days
	Intraperitoneal	10 mg/kg	Toxic
Rhesus monkey	Intraperitoneal	20 mg/kg/d	Death 2 hr after 2nd injection
	Intraperitoneal	40 mg TGA/kg	Death 48 hr after treatment
Man	Oral	2.8 mg/kg	Toxic
	Oral	20–25 mg/kg	Toxic

Source: Ref. 14.

TABLE 3 Salicylic Acid Content of Fresh Vegetables

Vegetable	Salicylic acid (mg/kg)
Sweet corn (whole kernels)	0.10
French stringless beans	0.08
Garlic	0.08
Cauliflower	0.07
Red king potatoes	0.06
Tomato	0.08
Capsicum (red peppers)	0.04
Sweet potato	0.04
Cabbage	0.01
Rhubarb	0.03

Source: Ref. 17.

In terms of human lives lost from phenols originating in plants, salicylic acid is probably the most dangerous. According to Smith and Smith (16) accidental and suicidal consumption of salicylates produces death of the order of four per million per year. Robertson and Kermode (17) determined the salicylic acid content in a wide range of fresh and canned vegetables. The concentration ranged from 0.01 mg/kg in cabbage to 0.10 mg/kg in whole kernel sweet corn (Table 3). In canned products, the salycilic acid levels ranged from 0.02 mg/kg in peas to 0.82 mg/kg in cream-style sweet corn (Table 4). Feingold (18) reported that low molecular weight chemicals including salicylates are associated with hyperactivity in humans.

Cassava roots contain a toxic hydrocyanic glucoside, which on enzymatic hydrolysis liberates hydrocyanic acid (18). However, proper processing helps to reduce the levels of this toxic compound to safe level. Taro and yam contain an acrid substance, calcium oxylate crystals, which cause irritation to the tongue and mouth (10). This compound can be removed by cooking

TABLE 4 Salicylic Acid Content of Commercially Canned Foods

Food	Salicylic acid (mg/kg)
Cream-style sweet corn	0.82
Corn only	0.73
Brine only	0.46
Tomato soup	0.08
Tomato puree	0.07
Sliced beet root	0.07
Tomato sauce	0.05
Garden peas	0.02

Source: Ref. 17.

the roots in boiling water and discarding the water. Some species of *Dioscorea* contain the toxic alkaloid dioscorine, which can be destroyed by roasting or boiling (10).

Peas and beans are known to contain certain antinutritional factors such as trypsin inhibitors, lectin, phytate, and polyphenols. The protease inhibitors are known to reduce digestibility of dietary proteins. Phytate can combine with calcium and iron and reduce their bioavailability (19).

IV. THERAPEUTIC PROPERTIES

Vegetables are a well-known source of different types of flavonoids, which include flavonones, flavones, flavonols, anthocyanins, catechins, and biflavans (20). Apart from fat-soluble tocopherols, flavonoids are the most common and most active antioxidant compounds in human food, being active in hydrophilic as well as in lipophilic systems (12). This has considerable practical impact on nutrition. It protects the flavonoid-containing foodstuff from oxidative deterioration, prolongs shelf life and keeping quality, and improves taste, acceptability, and wholesomeness of mixed dishes by inhibiting the oxidation of accompanying lipids (20). Flavanoids inhibit the activity of copper-containing enzymes and is useful in vascular disorders. They increase the stability of structural proteins of connective tissue and reduce fragility of fibrous membranes (21).

Flavonoids increase or stabilize the biological activity of ascorbic acid. All vegetable foods rich in flavonoids are marked by the unusual stability of their vitamin C. This effect is said to be mediated by three mechanisms—a simple oxidation-inhibiting effect, the binding of copper and hence inhibition of enzyme-catalyzed oxidation, and binding of ascorbic acid with a flavonoid, which is very stable (22,23). The ascorbic acid–stabilizing effect of flavonoids results in an increased production of ascorbic acid-2-sulfate, which improves membrane stability by synthesis of sulfated mucopolysaccharides and enhances cholesterol excretion as cholesteryl sulfate. Such a mechanism would help to explain the stabilizing effect of flavonoids on mesenchymal structures and their capacity to reduce hypercholesterolemia (24).

Flavonoids also have twofold anticarcinogenic activity. They have cytostatic properties and can reduce tumor growth. They also provide biochemical protection to cells against damage from carcinogenic substances (20).

Vegetables are also a very important source of carotenoids as a whole—β-carotene in particular—and vitamin C. Recent dietary studies of human populations have consistently shown protective effects of carotenoids against lung cancer. Approximately 10% of carotenoids are converted into vitamin A by enzymatic cleavage in the body. Because vitamin A is essential for maintenance of normal epithelial cellular differentiation, the carotenoids are important in cancer prevention because of their provitamin activity and most importantly because of their antioxidant function. Similarly, vitamin C is the most abundant water-soluble antioxidant in the body. Carotenoids and vitamin C are also known to reduce cancer risk by enhancing tumor surveillance by the immune system (13).

Some vegetables are known to contain a phenolic constituent, ellagic acid. This has been shown to be effective as an antimutagen and anticarcinogen and potential inhibitor of common cancer inducers (25).

Onion and garlic are known to reduce low-density lipoproteins and induce the formation of high-density-lipoproteins (10). The increase in high-density lipoprotein is more evident with consumption of white and yellow onions and is reduced with cooking. The allicin that is

TABLE 5 Retention of Ascorbic Acid, Thiamine, Riboflavin, Niacin, and Carotene in Vegetable Canning

Product	After blanching				After processing			
	No. of observation	Max. (%)	Min. (%)	Mean (%)	No. of observation	Max. (%)	Min. (%)	Mean (%)
Asparagus								
Ascorbic acid	26	100	74	95	32	100	80	92
Thiamine	12	100	79	92	31	85	60	67
Riboflavin	12	100	72	90	26	100	65	88
Niacin	8	100	77	94	22	100	77	96
Green beans								
Ascorbic acid	38	90	50	74	41	75	40	55
Thiamine	34	100	82	91	41	90	55	71
Riboflavin	29	100	70	95	30	100	85	96
Niacin	29	100	60	93	30	100	80	92
Carotene	—	—	—	—	9	96	81	87
Lima beans								
Ascorbic acid	12	83	54	72	10	100	60	76
Thiamine	12	77	36	58	15	67	32	47
Riboflavin	8	100	59	76	12	100	50	87
Niacin	8	98	68	81	11	100	77	85

Source: Ref. 23.

produced in garlic when the tissue is injured by cutting or crushing has been reported to have hypocholesterolaemic action (10). It has been indicated that allicin, allistalin, garlicin, diallyl disulfide, diallyl trisulfide, and essential oils are active against certain bacteria and fungi.

Cole vegetables contain a group of compounds known as indoles which have been recently linked with the prevention of cancers of the colon, rectum, and breast (7). Some species of *Cucumis*, *Luffs*, *Coccinia*, and *Momordica* exhibit varying degrees of bitterness in fruits, leaves, and twigs. This is caused by terpenes called momordicins. Owing to its bitter principle, the bitter gourd is used in ayurvedic medicine. A hypoglycemic ingredient, Cheratin, isolated from bitter gourd, lowers blood sugars (7). A sapogenic compound called diosgenin present in some species of yam is used in the production of cortisone and contraceptive drugs (10).

TABLE 6 Retention of Ascorbic Acid, Thiamine, Riboflavin, Niacin, and Carotene in Canned Tomato Juice

Vitamin	No. of observation	Retention after processing		
		Max.(%)	Min.(%)	Mean(%)
Ascorbic acid	90	90	35	67
Thiamine	18	100	73	89
Riboflavin	17	100	86	97
Niacin	17	100	83	98
Carotene	7	74	60	67

Source: Ref. 23.

TABLE 7 Retention (%) of Vitamins in Canned Foods During Storage

Product	Storage conditions		Ascorbic acid	Carotene	Niacin	Riboflavin	Thiamine
	0°F	Months					
Yellow corn	50	12	98	85	89	80	90
	65	12	94	87	89	80	86
	80	12	89	84	91	78	74
	50	24	92	69	91	71	89
	65	24	89	72	90	68	76
	80	24	81	87	96	61	60
Peas, Alaska	50	12	91	95	82	91	91
	65	12	89	91	77	84	86
	80	12	84	95	82	82	75
	50	24	90	93	99	80	89
	65	24	88	89	87	73	85
	80	24	81	98	85	68	68
Spinach	50	12	93	91	100	92	96
	65	12	91	90	103	89	89
	80	12	86	84	99	85	76
	56	24	90	80	96	82	90
	65	24	88	80	100	80	82
	80	24	81	81	101	69	71
Tomato	50	12	95	94	91	94	94
	65	12	94	98	93	95	93
	80	12	82	95	93	91	82
	50	24	89	75	88	96	91
	65	24	87	75	88	98	87
	80	24	70	74	85	97	70

Source: Ref. 24.

V. NUTRIENT STABILITY IN PROCESSED PRODUCTS

Vegetables are processed into dehydrated, canned, and fermented products by employing different methods of food preservation (6). Vegetables can be preserved by canning, which involves heat treatment. Signfiicant losses (23,24) of nutrients, particularly vitamins, occur during canning (Tables 5 and 6), and these losses increase during storage (Table 7). Carbohydrates in vegetables of low pH are partially hydrolyzed to simple sugars during the retorting operation and subsequent storage. Thermal processing can either increase or decrease the digestibility of proteins (6). Heat-induced protein denaturation and/or destruction of antinutrients increase the ease with which proteins are hydrolyzed by intestinal proteases, but heat can also reduce protein quality if the temperature is too high and of prolonged duration. The losses in protein availability as well as nutritional quality are accelerated by access to oxygen and water vapor (6).

Vegetables are dehydrated using various methods. Some vitamins are lost during dehydration (6). The extent of loss depends on the care exercised in preparing the material before dehydration, the dehydration process used, and the storage conditions of the dehydrated products (6). Vitamins A and C are sensitive to oxidative degradation. Thamine can be destroyed by heat and sulfuring and riboflavin by light (6). Processes and products that enhance nutritional quality as well as sensory appeal have been developed. Packaging is an important factor in the retention of quality (25). The advances in food packaging have aided in maintaining the nutritional and sensory properties of processed vegetables during storage.

REFERENCES

1. Chavan, J. K., and S. S. Kadam, Nutritional improvement of cereals by sprouting, *CRC Crit. Rev. Food Sci. Nutr. 28*(5):40 (1989).
2. Salunkhe, D. K., and B. B. Desai. *Postharvest Biotechnology of Vegetables*, Vol. II, CRC Press, Boca Raton, FL, 1984, p. 161.
3. Borgstrom, G., The food and people dilemma, in *The Man-Environment System in the Late Twentieth Century* (W. L. Thomas, ed.), Duxbury Press, MA, 1973, p. 140.
4. Salunkhe, D. K., H. R. Bolin, and N. R. Reddy, *Storage Processing and Nutritional Quality of Fruits and Vegetables*, Vol. I—*Fresh Fruits and Vegetables*, CRC Press, Boca Raton, FL, 1991, p. 2.
5. Gorgatti-Netto, A., Postharvest losses: Extent of the problem, *Food Nutr. Bull. 1*(3):34 (1979).
6. Salunkhe, D. K., H. R Bolin, and N. R. Reddy, *Storage, Processing and Nutritional Quality of Fruits and Vegetables*, 2nd ed., Vol. II, *Processed Fruits and Vegetables*, CRC Press, Boca Raton, 1991, p. 162.
7. Roy, S. K., and A. K. Chakrabarti, Vegetables of temperate climate—commercial and dietary importance, in *Encyclopaedia of Food Sci., Food Technology and Nutrition* (R. MaCrae, R. K. Robinson, and M. J. Sadler, eds.), Academic Press, London, 1993, p. 4715.
8. Salunkhe, D. K., and S. S. Kadam, Introduction, in *Potato: Production, Processing and Products* (D. K. Salunkhe, S. S. Kadam, and S. J. Jadhav, eds.), CRC Press, Boca Raton, FL, 1991, p. 1.
9. Bano Z., and S. Rajarathnam, *Pleurotus* mushrooms, Part II. Chemical composition, nutritional value, postharvest physiology preservation and role as human food, *CRC Crit. Rev. Food Sci. Nutr. 27*(2):87 (1988).
10. Roy, S. K., and A. K. Chakrabarti, Vegetables of tropical climate—commercial and dietary importance, in *Encyclopaedia of Food Science, Food Technology and Nutrition* (R. MaCrae, R. K. Robinson, and M. J. Sadler, eds.), Academic Press, London, 1993, p. 4743.
11. Dreher, M. L., *Handbook of Dietary Fiber*, Marcel Dekker, New York, 1987, p. 285.
12. Prakash, J., Fruits: The lesser known health benefits, Proc. National Seminar on Postharvest Technology of Fruits, held at Bangalore, August 7-9, 1995.
13. Dickerson, J. W. T., Cancer, in *Encyclopaedia of Food Science, Food Technology and Nutrition* (R. MaCrae, R, K. Robinson, and M. J. Sadler, eds.), Academic Press, London, 1993, p. 607.
14. Jadhav, S. J., A. Kumar, and J. K. Chavan, Glycoalkaloids, in *Potato, Production, Processing and Products*, CRC Press, Boca Raton, FL, 1991, p. 203.
15. Singleton, V. L., and F. H. Kratzer, Plant phenolics, in *Toxicants Occurring Naturally in Foods*, 2nd ed., Report of the Committee on Food Protection, Food and Nutrition Board, National Research Council, National Academy of Sciences, Washington, DC, 1973, p. 309.
16. Smith, M. J. H., and P. K. Smith, *The Salicylates: A Critical Bibliographic Review*, John Wiley and Sons, New York, 1966.
17. Robertson, G. L., and W. J. Kermode, Salicylic acid in fresh and canned fruits and vegetables, *J. Sci. Food Agric. 32*:833 (1981).
18. Feingold, B. F., Food additives and child development, *Hosp. Pract. 8*:10 (1972).
19. Bokanga, M., Biotechnology and cassava processing in Africa, *Food Technol. 49*:86 (1995).
20. Sathe, S. K., asnd D. K. Salunkhe, Technology of removal of unwanted components of dry legumes, in *Handbook of World Food Legumes: Nutritional Chemistry, Processing Technology and Utilization*, Vol. III (D. K. Salunkhe and S. S. Kadam, eds.), CRC Press, Boca Raton, FL, 1989, p. 249.
21. Kuhnau, J, The flavonoids: A class of semi-essential food components and their role in human nutrition, *World Rev. Nutr., Diet. 24*:117 (1976).

22. Gabor, M., The anti-inflammatory action of flavonoids, 3rd Hungarian Bioflavonoid Symp., Hungarian Academy of Sciences, Budapest, Hungary, 1971.
23. Harper, K. A., A. D. Morton, and E. J. Rolfe, Phenolic compounds of black currant juice and their protective effect on ascorbic acid III. Mechanism of ascorbic acid oxidation and its inhibition by flavonoids, *J. Food Technol. 4*:255 (1969).
24. Lewis, E. J., and B. M. Watts, Antioxidant and copper binding properties of onions, *Food Res. 23*:274 (1958).
25. Mumma, R. O., and A. J. Verlangieri, *In vivo* sulfatation of cholesterol by ascorbic acid 3-sulfate as a possible explanation for the hypocholesteremic effects of ascorbic acid, *Fed. Proc. 30*:370 (1971).
26. Anonymous, *Retention of Nutrients During Canning*, Research Laboratories, National Canners Association, Washington, DC, 1955.
27. Cameron, E. J., R. W. Pilcher, and L. E. Clijcorn, Nutrient retention during canned food production, *Am. J. Pub. Health 39*:759 (1949).
28. Kadam, S. S., and D. K. Salunkhe, Fruits in human nutrition, in *Handbook of Fruit Science and Technology* (D. K. Salunkhe and S. S. Kadam, eds.), Marcel Dekker, New York, 1995, p. 614.

Index

ISBN 0-8247-0105-4
EAN
9 780824 701055

90000>

Printed and bound by CPI Group (UK) Ltd, Croydon, CR0 4YY
23/10/2024
01778259-0009